FUNDAMENTALS
OF OPTICS

FRANCIS A. JENKINS
Late Professor of Physics
University of California, Berkeley

HARVEY E. WHITE
Professor of Physics, Emeritus
Director of the Lawrence Hall
of Science, Emeritus
University of California, Berkeley

Fundamentals of Optics

FOURTH EDITION

McGRAW-HILL BOOK COMPANY

Auckland Bogotá Guatemala Hamburg Lisbon
London Madrid Mexico New Delhi Panama Paris San Juan
São Paulo Singapore Sydney Tokyo

FUNDAMENTALS OF OPTICS
INTERNATIONAL EDITION 1981

4 5 6 7 8 9 20 ANL 20 9

Library of Congress Cataloging in Publication Data

Jenkins, Francis Arthur, dates.
Fundamentals of optics.

First ed. published in 1937 under title: Fundamentals of physical optics.
Includes index.
1. Optics. I. White, Harvey Elliott, date joint author. II. Title.
QC355.2.J46 1976 535 75-26989
ISBN 0-07-032330-5

When ordering this title use ISBN 0-07-085346-0

Printed in Singapore

CONTENTS

PREFACE TO THE THIRD EDITION

The chief objectives in preparing this new edition have been simplification and modernization. Experience on the part of the authors and of the many other users of the book over the last two decades has shown that many passages and mathematical derivations were overly cumbersome, thereby losing the emphasis they should have had. As an example of the steps taken to rectify this defect, the chapter on reflection has been entirely rewritten in simpler form and placed ahead of the more difficult aspects of polarized light. Furthermore, by expressing frequency and wavelength in circular measure, and by introducing the complex notation in a few places, it has been possible to abbreviate the derivations in wave theory to make room for new material.

In any branch of physics fashions change as they are influenced by the development of the field as a whole. Thus, in optics the notions of wave packet, line width, and coherence length are given more prominence because of their importance in quantum mechanics. For the same reason, our students now usually learn to deal with complex quantities at an earlier stage, and we have felt justified in giving some examples of how helpful these can be. Because of the increasing use of concentric optics, as well as graphical methods of ray tracing, these subjects have been introduced in the chapters on geometrical optics. The elegant relationships between geometrical optics and particle mechanics, as in the electron microscope and quadrupole lenses,

In Part Two, Wave Optics, Chapter 11 has been modified to give a better approach to the subject of wave motion. In Chapter 16 a section has been added on the correlation interferometer. Some of the major features of recent developments have been added at the end of Chapter 28: modern wave optics, spatial filtering, the phase-contrast microscope, and schlieren optics.

In Part Three, Quantum Optics, three new chapters have been added as important new developments: Chapter 29, Light Quanta and Their Origin; Chapter 30, Lasers; and Chapter 31, Holography.

I wish to take this opportunity of thanking Dr. Donald H. White for his assistance in gathering much of the new material used in this the fourth edition.

HARVEY E. WHITE

PREFACE TO THE FOURTH EDITION

This fourth edition is written primarily to be used as a textbook by college students majoring in one of the physical sciences. The first, second, and third editions were written by Francis A. Jenkins and Harvey E. White while teaching optics in the physics department at the University of California, Berkeley. With the passing of Professor Jenkins in 1960 this fourth edition has been revised by Harvey E. White.

A considerable number of innovative ideas and new concepts have been developed in the field of optics since the third edition was published in 1957, thereby requiring a sizable amount of new material. Three new chapters, a number of new sections on modern optics, a number of new references, and all new problems at the ends of all chapters have been added to bring the fourth edition up to date.

Fizeau's experiments on the speed of light in air and Foucault's experiments on the speed of light in stationary matter have been moved to Chapter 1. This serves as a better introduction to the important concept of refractive index and leaves the rest of Chapter 19 relatively unchanged.

In Part One, Geometrical Optics, the long and tedious calculations of ray tracing, using logarithms, has been replaced by direct calculations using the relatively new electronic calculators, thereby permitting lens design engineers to program larger computers.

could not be developed because of lack of space; the instructor may wish to supplement the text in this direction. The same may be true of the rather too brief treatments of some subjects where old principles have recently come into prominence, as in Čerenkov radiation, the echelle grating, and multilayer films.

A difficulty that must present itself to the authors of all textbooks at this level is that of avoiding the impression that the subject is a definitive, closed body of knowledge. If the student can be persuaded to read the original literature to any extent, this impression soon fades. To encourage such reading, we have inserted many references, to original papers as well as to books, throughout the text. An entirely new set of problems, representing a rather greater spread of difficulty than heretofore, is included.

It is not possible to mention all those who have assisted us by suggestions for improvement. Specific errors or omissions have been pointed out by L. W. Alvarez, W. A. Bowers, J. E. Mack, W. C. Price, R. S. Shankland, and J. M. Stone, while H. S. Coleman, J. W. Ellis, F. S. Harris, Jr., R. Kingslake, C. F. J. Overhage, and R. E. Worley have each contributed several valuable ideas. We wish to express our gratitude to all of these, as well as to T. L. Jenkins, who suggested the simplification of certain derivations and checked the answers to many of the problems.

FRANCIS A. JENKINS

HARVEY E. WHITE

FUNDAMENTALS
OF OPTICS

PART ONE

Geometrical Optics

1
PROPERTIES OF LIGHT

All the known properties of light are described in terms of the experiments by which they were discovered and the many and varied demonstrations by which they are frequently illustrated. Numerous though these properties are, their demonstrations can be grouped together and classified under one of three heads: geometrical optics, wave optics, and quantum optics, each of which may be subdivided as follows:

Geometrical optics
- Rectilinear propagation
- Finite speed
- Reflection
- Refraction
- Dispersion

Wave optics
- Interference
- Diffraction
- Electromagnetic character
- Polarization
- Double refraction

Quantum optics
- Atomic orbits
- Probability densities
- Energy levels
- Quanta
- Lasers

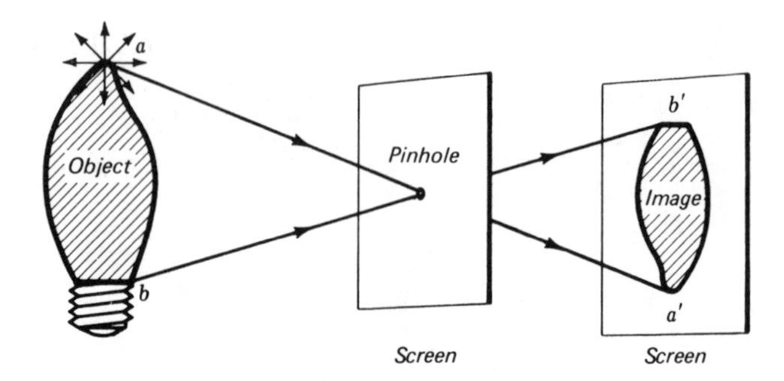

FIGURE 1A
A demonstration experiment illustrating the principle that light rays travel in straight lines. The rectilinear propagation of light.

The first group of phenomena classified as *geometrical optics* are treated in the first 10 chapters of this text and are most easily described in terms of straight lines and plane geometry. The second group, *wave optics*, deals with the wave nature of light, and is treated in Chaps. 11 to 28. The third group, *quantum optics*, deals with light as made up of tiny bundles of energy called *quanta*, and is treated from the optical standpoint in Chaps. 29 to 33.

1.1 THE RECTILINEAR PROPAGATION OF LIGHT

The rectilinear propagation of light is the technical terminology applied to the principle that "light travels in straight lines." The fact that objects can be made to cast fairly sharp shadows may be considered a good demonstration of this principle. Another illustration is found in the pinhole camera. In this simple and inexpensive device the image of a stationary object is formed on a photographic film or plate by light passing through a small opening, as diagramed in Fig. 1A. In this figure the object is an ornamental light bulb emitting white light. To see how an image is formed, consider the rays of light emanating from a single point a near the top of the bulb. Of the many rays of light radiating in many directions the ray that travels in the exact direction of the hole passes through to the point a' near the bottom of the image screen. Similarly, a ray leaving b near the bottom of the bulb and passing through the hole will arrive at b', near the top of the image screen. Thus it can be seen how an inverted image of the entire bulb is formed.

If the image screen is moved closer to the pinhole screen, the image will be proportionately smaller, whereas if it is moved farther away, the image will be proportionately larger. Excellent sharp photographs of stationary objects can be made with this arrangement. By making a pinhole in one end of a small box and placing a photographic film or plate at the other end, taking several time exposures as trial runs, good pictures are attainable. For good, sharp photographs the hole

FIGURE 1B
Photograph of the University of California Hospital, San Francisco, taken with a pinhole camera. Plate distance 9.5 cm; Panchromatic film; exposure 3.0 min; square hole = 0.33 mm.

must be very small, because its size determines the amount of blurring in the image. A small square hole is quite satisfactory. A piece of household aluminum foil is folded twice and the corner fold cut off with a razor blade, leaving good clean edges. After several such trials, and examination with a magnifying glass, a good square hole can be selected. The photograph reproduced in Fig. 1B was taken with such a pinhole camera. Note the undistorted perspective lines as well as the depth of focus in the picture.

1.2 THE SPEED OF LIGHT

The ancient astronomers believed that light traveled with an infinite speed. Any major event that occurred among the distant stars was believed to be observable instantly at all other points in the universe.

It is said that around 1600 Galileo tried to measure the speed of light but was not successful. He stationed himself on a hilltop with a lamp and his assistant on a distant hilltop with another lamp. The plan was for Galileo to uncover his lamp at an agreed signal, thereby sending a flash of light toward his assistant. Upon seeing the light the assistant was to uncover his lamp, sending a flash of light back to Galileo, who observed the total elapsed time. Many repetitions of this experiment, performed at greater and greater distances between the two observers, convinced Galileo that light must travel at an infinite speed.

We now know that the speed of light is *finite* and that it has an approximate value of

$$v = 300{,}000 \text{ km/s} = 186{,}400 \text{ mi/s}$$

In 1849 the French physicist Fizeau* became the first man to measure the speed of light here on earth. His apparatus is believed to have looked like Fig. 1C. His account of this experiment is quite detailed, but no diagram of his apparatus is given in his notes.

An intense beam of light from a source S is first reflected from a half-silvered mirror G and then brought to a focus at the point O by means of lens L_1. The diverging beam from O is made into a parallel beam by lens L_2. After traveling a distance of 8.67 km to a distant lens L_3 and mirror M, the light is reflected back toward the source. This returning beam retraces its path through L_2, O, and L_1, half of it passing through G and entering the observer's eye at E.

The function of the toothed wheel is to cut the light beam into short pulses and to measure the time required for these pulses to travel to the distant mirror and back. When the wheel is at rest, light is permitted to pass through one of the openings at O.

* Armand H. L. Fizeau (1819–1896), French physicist, was born of a wealthy French family that enabled him to be financially independent. Instead of shunning work, however, he devoted his life to diligent scientific experiment. His most important achievement was the measurement of the speed of light in 1849, carried on in Paris between Montmartre and Suresnes. He also gave the correct explanation of the Doppler principle as applied to light coming from the stars and showed how the effect could be used to measure stellar velocities. He carried out his experiments on the velocity of light in a moving medium in 1851 and showed that light is dragged along by a moving stream of water.

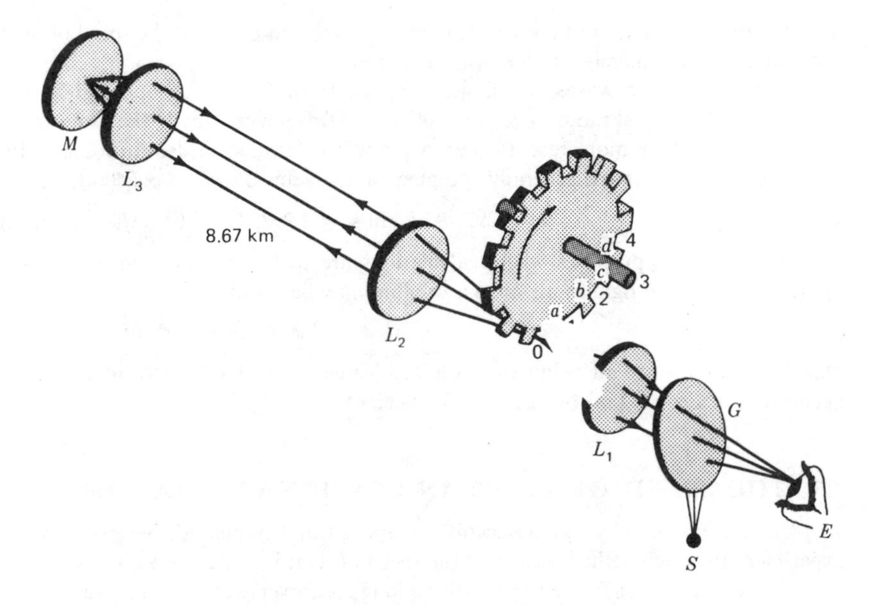

FIGURE 1C
Experimental arrangement described by the French physicist Fizeau, with which he determined the speed of light in air in 1849.

In this position all lenses and the distant mirror are aligned so that an image of the light source S can be seen by the observer at E.

The wheel is then set rotating with slowly increasing speed. At some point the light passing through O will return just in time to be stopped by tooth a. At this same speed light passing through opening 1 will return in time to be stopped by the next tooth b. Under these circumstances the light S is completely eclipsed from the observer. At twice this speed the light will reappear and reach a maximum intensity. This condition occurs when the light pulses getting through openings 1, 2, 3, 4, . . . return just in time to get through openings 2, 3, 4, 5, . . . , respectively.

Since the wheel contained 720 teeth, Fizeau found the maximum intensity to occur when its speed was 25 rev/s. The time required for each light pulse to travel over and back could be calculated by $(\frac{1}{720})(\frac{1}{25}) = 1/18{,}000$ s. From the measured distance over and back of 17.34 km, this gave a speed of

$$v = \frac{d}{t} = \frac{17.34 \text{ km}}{1/18{,}000 \text{ s}} = 312{,}000 \text{ km/s}$$

In the years that followed Fizeau's first experiments on the speed of light, a number of experimenters improved on his apparatus and obtained more and more accurate values for this universal constant. About three-quarters of a century passed, however, before A. A. Michelson, and others following him, applied new and improved methods to visible light, radio waves, and microwaves and obtained the speed of light accurate to approximately six significant figures.

Electromagnetic waves of all wavelengths, from X rays at one end of the spectrum to the longest radio waves, are believed to travel with exactly the same speed in a vacuum. These more recent experiments will be treated in detail in Chap. 19, but we give here the most generally accepted value of this universal constant,

$$c = 299{,}792.5 \text{ km/s} = 2.997925 + 10^8 \text{ m/s} \qquad (1a)$$

For practical purposes where calculations are to be made to four significant figures, the speed of light in air or in a vacuum may be taken to be

$$c = 3.0 + 10^8 \text{ m/s} \qquad (1b)$$

One is often justified in using this rounded value since it differs from the more accurate value in Eq. (1a) by less than 0.1 percent.

1.3 THE SPEED OF LIGHT IN STATIONARY MATTER

In 1850, the French physicist Foucault* completed and published the results of an experiment in which he had measured the speed of light in water. Foucault's experiment was of great importance for it settled a long controversy over the nature of light. Newton and his followers in England and on the Continent believed light to be made up of small particles emitted by every light source. The Dutch physicist Huygens, on the other hand, believed light to be composed of waves, similar to water or sound waves.

According to Newton's corpuscular theory, light should travel faster in an optically dense medium like water than in a less dense medium like air. Huygens' wave theory required light to travel slower in the more optically dense medium. Upon sending a beam of light back and forth through a long tube containing water, Foucault found the speed of light to be less than in air. This result was considered by many to be a strong confirmation of the wave theory.

Foucault's apparatus for this experiment is shown in Fig. 1D. Light coming through a slit S is reflected from a plane rotating mirror R to the equidistant concave mirrors M_1 and M_2. When R is in the position 1, the light travels to M_1, back along the same path to R, through the lens L, and by reflection to the eye at E. When R is in position 2, the light travels the lower path through an auxiliary lens L' and tube

* Jean Bernard Leon Foucault (1819–1868), French physicist. After studying medicine he became interested in experimental physics and with A. H. L. Fizeau carried out experiments on the speed of light. After working together for some time, they quarreled over the best method to use for "chopping" up a light beam, and thereafter went their respective ways. Fizeau (using a toothed wheel) and Foucault (using a rotating mirror) did admirable work, each supplementing the work of the other. With a rotating mirror Foucault in 1850 was able to measure the speed of light in a number of different media. In 1851 he demonstrated the earth's rotation by the rotation of the plane of oscillation of a long, freely suspended, heavy pendulum. For the development of this device, known today as a *Foucault pendulum*, and his invention of the gyroscope, he received the Copley medal of the Royal Society of London, in 1855. He also discovered the eddy currents induced in a copper disk moving in a strong magnetic field and invented the optical polarizer which bears his name.

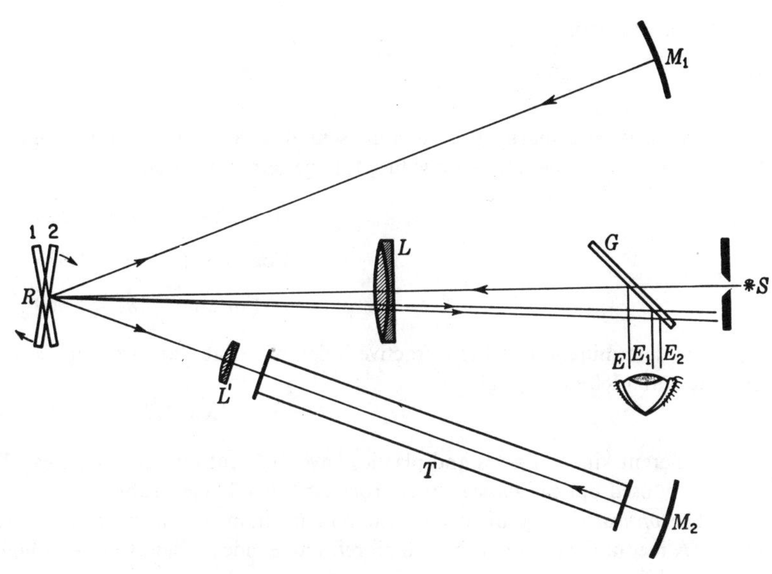

FIGURE 1D
Foucault's apparatus for determining the speed of light in water.

T to M_2, back to R, through L to G, and then to the eye E. If now the tube T is filled with water and the mirror is set into rotation, there will be a displacement of the images from E to E_1 and E_2. Foucault observed that the light ray through the tube was more displaced than the other. This means that it takes the light longer to travel the lower path through water than it does the upper path through the air.

The image observed was due to a fine wire parallel to, and stretched across, the slit. Since sharp images were desired at E_1 and E_2, the auxiliary lens L' was necessary to avoid bending the light rays at the ends of the tube T.

Over 40 years later the American physicist Michelson (first American Nobel laureate 1907) measured the speed of light in air and water. For water he found the value of 225,000 km/s, which is just three-fourths the speed in a vacuum. In ordinary optical glass, the speed was still lower, about two-thirds the speed in a vacuum.

The speed of light in air at normal temperature and pressure is about 87 km/s less than in a vacuum, or $v = 299{,}706$ km/s. For many practical purposes this difference may be neglected and the speed of light in air taken to be the same as in a vacuum, $v = 3.0 \times 10^8$ m/s.

1.4 THE REFRACTIVE INDEX

The index of refraction, or refractive index, of any optical medium is defined as the ratio between the speed of light in a vacuum and the speed of light in the medium:

$$\text{Refractive index} = \frac{\text{speed in vacuum}}{\text{speed in medium}} \tag{1c}$$

In algebraic symbols

$$n = \frac{c}{v} \tag{1d}$$

The letter n is customarily used to represent this ratio. Using the speeds given in Sec. 1.3, we obtain the following values for the refractive indices:

$$\text{For glass:} \qquad n = 1.520 \tag{1e}$$

$$\text{For water:} \qquad n = 1.333 \tag{1f}$$

$$\text{For air:} \qquad n = 1.000 \tag{1g}$$

Accurate determination of the refractive index of air at standard temperature (0°C) and pressure (760 mmHg) give

$$n = 1.000292 \qquad \text{for air} \tag{1h}$$

Different kinds of glass and plastics have different refractive indices. The most commonly used optical glasses range from 1.52 to 1.72 (see Table 1A).

The *optical density* of any transparent medium is a measure of its refractive index. A medium with a relatively high refractive index is said to have a high optical density, while one with a low index is said to have a low optical density.

1.5 OPTICAL PATH

To derive one of the most fundamental principles in geometric optics, it is appropriate to define a quantity called the *optical path*. The path d of a ray of light in any medium is given by the product *velocity* times *time*:

$$d = vt$$

Since by definition $n = c/v$, which gives $v = c/n$, we can write

$$d = \frac{c}{n}t \qquad \text{or} \qquad nd = ct$$

The product nd is called the *optical path* Δ:

$$\Delta = nd$$

The optical path represents the distance light travels in a vacuum in the same time it travels a distance d in the medium. If a light ray travels through a series of optical media of thickness $d, d', d'', \ldots$ and refractive indices $n, n', n'', \ldots$, the total optical path is just the sum of the separate values:

$$\Delta = nd + n'd' + n''d'' + \cdots \tag{1i}$$

A diagram illustrating the meaning of optical path is shown in Fig. 1E. Three media of length d, d', and d'', with refractive indices n, n', and n'', respectively, are shown touching each other. Line AB shows the length of the actual light path through these media, while the line CD shows the distance Δ, the distance light would travel in a vacuum in the same amount of time t.

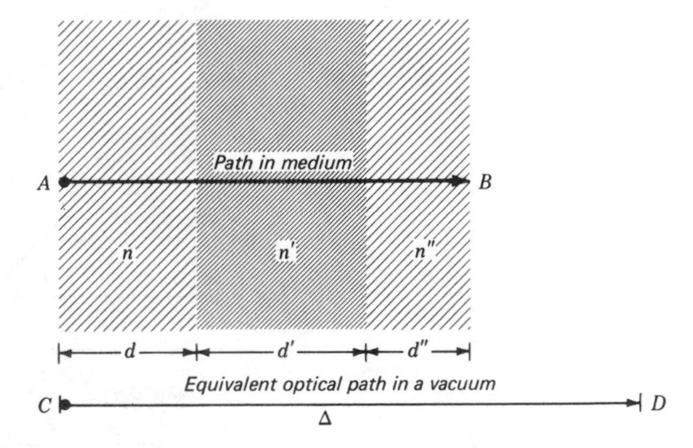

FIGURE 1E
The optical path through a series of optical media.

1.6 LAWS OF REFLECTION AND REFRACTION

Whenever a ray of light is incident on the boundary separating two different media, part of the ray is reflected back into the first medium and the remainder is refracted (bent in its path) as it enters the second medium (see Fig. 1F). The directions taken by these rays can best be described by two well-established laws of nature.

According to the simplest of these laws, the angle at which the incident ray strikes the interface MM' is exactly equal to the angle the reflected ray makes with the same interface. Instead of measuring the angle of incidence and the angle of reflection from the interface MM', it is customary to measure both from a common line perpendicular to this surface. This line NN' in the diagram is called the *normal.* As the angle of incidence ϕ increases, the angle of reflection also increases by exactly the same amount, so that for all angles of incidence

● $$\textit{angle of incidence} = \textit{angle of reflection} \tag{1j}$$

A second and equally important part of this law stipulates that the reflected ray lies in the plane of incidence and on the opposite side of the normal, the plane of incidence being defined as the plane containing the incident ray and the normal. In other words, the incident ray, the normal, and the reflected ray all lie in the same plane, which is perpendicular to the interface separating the two media.

The second law is concerned with the incident and refracted rays of light, and states that the sine of the angle of incidence and the sine of the angle of refraction bear a constant ratio one to the other, for all angles of incidence:

$$\frac{\sin \phi}{\sin \phi'} = \text{const} \tag{1k}$$

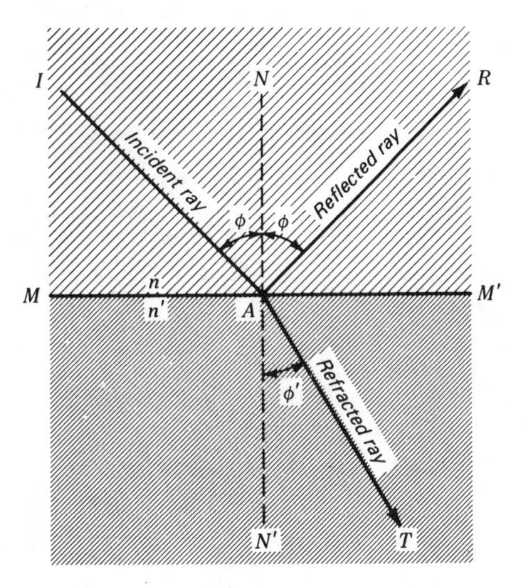

FIGURE 1F
Reflection and refraction at the boundary separating two media with refractive indices n and n', respectively.

Furthermore, the refracted ray also lies in the plane of incidence and on the opposite side of the normal. This relationship, experimentally established by Snell,* is known as *Snell's law*. In addition the constant is found to have exactly the ratio of the refractive indices of the two media n and n'. Hence we can write

$$\frac{\sin \phi}{\sin \phi'} = \frac{n'}{n} \tag{1l}$$

which can be written in the symmetrical form

● $$n \sin \phi = n' \sin \phi' \tag{1m}$$

By Eqs. (1c) and (1d) the refractive indices of different optical media are defined as

$$n = \frac{c}{v} \quad \text{and} \quad n' = \frac{c}{v'} \tag{1n}$$

where c is the speed of light in a vacuum ($c = 2.997925 + 10^8$ m/s) and v and v' are the speeds of light in the two media.

* Willebrord Snell (1591–1626), Dutch astronomer and mathematician, was born at Leyden. At twenty-one he succeeded his father as professor of mathematics at the University of Leyden. In 1617, he determined the size of the earth from measurements of its curvature between Alkmaar and Bergen-op-Zoom. He announced what is essentially the law of refraction in an unpublished paper in 1621. His geometrical construction requires that the ratios of the cosecants of ϕ and ϕ' be constant. Descartes was the first to use the ratio of the sines, and the law is known as Descartes' law in France.

By the substitution of Eqs. (1c) in Eq. (1l), we obtain,

$$\frac{\sin \phi}{\sin \phi'} = \frac{v}{v'} \tag{1o}$$

If one or both indices are different from unity, the ratio n'/n is often called the *relative index* n' and Snell's law can be written

$$\frac{\sin \phi}{\sin \phi'} = n' \tag{1p}$$

If the first medium is a vacuum, $n = 1.0$, the relative index has just the value of the second index and Eq. (1p) is again valid. If the first index is air at normal temperature and pressure ($n = 1.000292$), and if three-figure accuracy is satisfactory, Eq. (1p) is again used.

Wherever practical, we shall use unprimed symbols to refer to the first medium, primed symbols for the second medium, double primed symbols for the third medium, etc. When the angles of incidence and refraction are very small, a good approximation is obtained by setting the sines of angles equal to the angles themselves, obtaining

$$\frac{\phi}{\phi'} = \frac{n'}{n} \tag{1q}$$

1.7 GRAPHICAL CONSTRUCTION FOR REFRACTION

A simple method for tracing a ray of light across a boundary separating two optically transparent media is shown in Fig. 1G. Because the principles involved in this construction are readily extended to complicated optical systems, the method is useful in the preliminary design of many different kinds of optical instruments.

After the line GH is drawn, representing the boundary separating the two media of index n and n', the angle of incidence ϕ of the incident ray JA is selected and the construction proceeds as follows. At one side of the drawing, and as reasonably close as possible, a line OR is drawn parallel to JA. With a point of origin O, two circular arcs are drawn with their radii proportional to the two indices n and n', respectively. Through the point of intersection R a line is drawn parallel to the boundary normal NN', intersecting the arc n' at P. The line OP is next drawn in; parallel to it, through A, the refracted ray AB is drawn. The angle β between the incident and refracted ray, called the *angle of deviation*, is given by

$$\beta = \phi - \phi' \tag{1r}$$

To prove that this construction follows Snell's law exactly, we apply the law of sines to the triangle ORP:

$$\frac{OR}{\sin \phi'} = \frac{OP}{\sin (\pi - \phi)}$$

Since $\sin (\pi - \phi) = \sin \phi$, $OR = n$, and $OP = n'$, substitution gives directly

$$\frac{n}{\sin \phi'} = \frac{n'}{\sin \phi} \tag{1s}$$

which is Snell's law [Eq. (1l)].

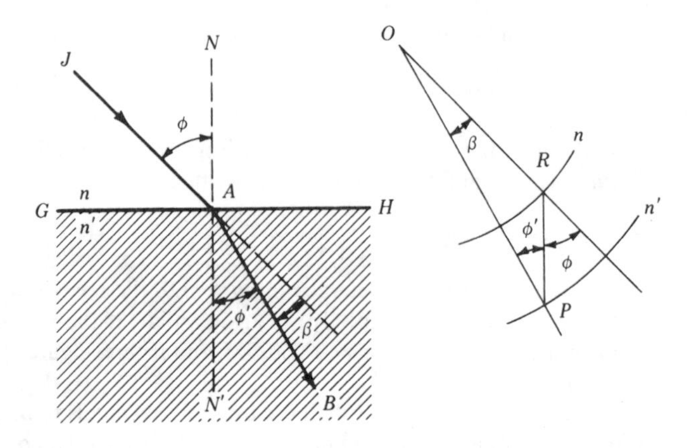

FIGURE 1G
Graphical construction for refraction at a smooth surface separating two media of index n and n'.

1.8 THE PRINCIPLE OF REVERSIBILITY

The symmetry of Eqs. (1j) and (1m) with respect to the symbols used shows at once that if a reflected or refracted ray is reversed in direction, it will retrace its original path. For any given pair of media with indices n and n' any one value of ϕ is correlated with a corresponding value of n'. This will be equally true when the ray is reversed and ϕ' becomes the angle of incidence in the medium of n'; the angle of refraction will then be ϕ. Since reversibility holds at each reflecting and refracting surface, it holds also for even the most complicated light paths. This useful principle has more than a purely geometrical foundation, and later it will be shown to follow from the application of wave motion to a principle in mechanics.

1.9 FERMAT'S PRINCIPLE

The term *optical path* was introduced in Sec. 1.5, where it was defined as the distance a light ray would travel in a vacuum in the same time it travels from one point to another, a specified distance, through one or more optical media. The real path of a ray of light through a prism, with media of different refractive index on either side, is shown in Fig. 1H. The optical path from the point Q in medium n, through medium n', and to the point Q'' in medium n'' is given by

$$\Delta = nd + n'd' + n''d'' \tag{1t}$$

One can also define an optical path in a medium of continuously varying refractive index by replacing the summation by an integral. The paths of the rays are then curved, and Snell's law of refraction loses its meaning.

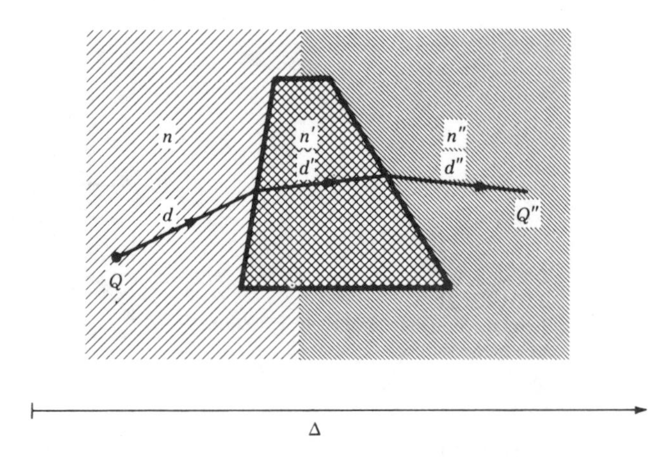

FIGURE 1H
The refraction of light by a prism and the meaning of optical path Δ.

We shall now consider *Fermat's* principle*, which is applicable to any type of variation of n and hence contains within it the laws of reflection and refraction as well:

> *The path taken by a light ray in going from one point to another through any set of media is such as to render its optical path equal, in the first approximation, to other paths closely adjacent to the actual one.*

The other paths must be possible ones in the sense that they may undergo deviations only where there are reflecting or refracting surfaces. Fermat's principle will hold for a ray whose optical path is a minimum with respect to adjacent hypothetical paths. Fermat himself stated that the time required by the light to traverse the path is a minimum and the optical path is a measure of this time. But there are plenty of cases in which the optical path is a maximum or neither a maximum nor a minimum but merely stationary (at a point of inflection) at the position of the true ray.

Consider a ray of light that must pass through a point Q and then, after reflection from a plane surface, pass through a second point Q'' (see Fig. 1I). To find the real path, we first drop a perpendicular to GH and extend it an equal distance on the other side to Q'. The straight line $Q'Q''$ is drawn in, and from its intersection B the line QB

* Pierre de Fermat (1601–1665), French mathematician, born at Beaumont-de-Lomagne. In his youth, with Pascal, he made discoveries about the properties of numbers, on which he later built his method of calculating probabilities. His brilliant researches in the theory of numbers rank him as the founder of modern theory. He also studied the reflection of light and enunciated his principle of least time. His justification for this principle was that nature is economical, but he was unaware of circumstances where exactly the opposite is true. Fermat was a counselor for the parliament of Toulouse, distinguished for both legal knowledge and for strict integrity of conduct. He was also an accomplished general scholar and linguist.

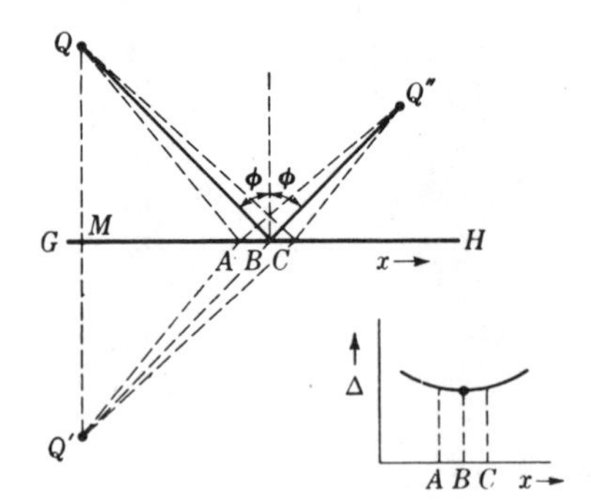

FIGURE 1I
Fermat's principle applied to reflection at a plane surface.

is drawn. The real light path is therefore QBQ'', and, as can be seen from the symmetry relations in the diagram, it obeys the law of reflection.

Consider now adjacent paths to points like A and C on the mirror surface close to B. Since a straight line is the shortest path between two points, both the paths $Q'AQ''$ and $Q'CQ''$ are greater than $Q'BQ''$. By the above construction and equivalent triangles, $QA = Q'A$, and $QC = Q'C$, so that $QAQ'' > QBQ''$ and $QCQ'' > QBQ''$. Therefore the real path QBQ'' is a minimum. A graph of hypothetical paths close to the real path QBQ'', as shown in the lower right of the diagram, indicates the meaning of a minimum, and the flatness of the curve between A and C illustrates that to a first approximation adjacent paths are equal to the real optical path.

Consider finally the optical properties of an ellipsoidal reflector, as shown in Fig. 1J. All rays emanating from a point source Q at one focus are reflected according to the law of reflection and come together at the other focus Q'. Furthermore all paths are equal in length. It will be remembered that an ellipse can be drawn with a string of fixed length with its ends fastened at the foci. Because all optical paths are equal, this is a stationary case, as mentioned above. On the graph in Fig. 1K(b) equal path lengths are represented by a straight horizontal line.

Some attention will be devoted here to other reflecting surfaces like a and c shown dotted in Fig. 1J. If these surfaces are tangent to the ellipsoid at the point B,

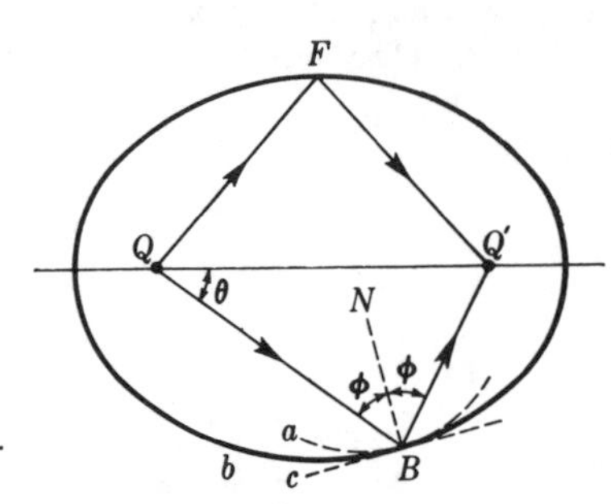

FIGURE 1J
Fermat's principle applied to an elliptical reflector.

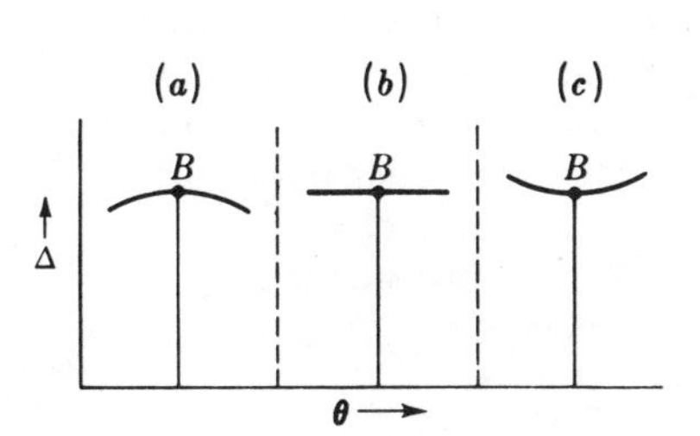

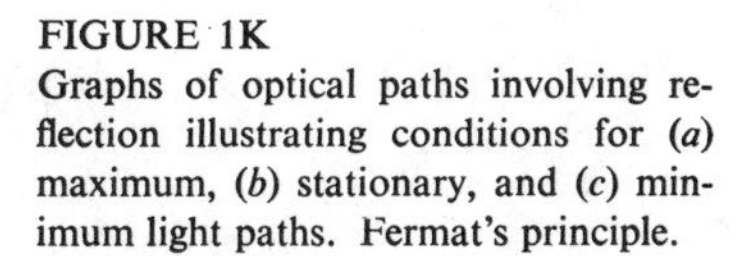
FIGURE 1K
Graphs of optical paths involving reflection illustrating conditions for (a) maximum, (b) stationary, and (c) minimum light paths. Fermat's principle.

the line NB is normal to all three surfaces and QBQ' is a real path for all three. Adjacent paths from Q to points along these mirrors, however, will give a minimum condition for the real path to and from reflector c and a maximum condition for the real path to and from reflector a (see Fig. 1K).

It is readily shown mathematically that both the laws of reflection and refraction follow from Fermat's principle. Figure 1L, which represents the refraction of a ray at a plane surface, can be used to prove the law of refraction [Eq. (1m)]. The length of the optical path between the point Q in the upper medium of index n and another point Q' in the lower medium of index n' passing through any point A on the surface is

$$\Delta = nd + n'd' \tag{1u}$$

where d and d' represent the distances QA and AQ', respectively.

Now if we let h and h' represent perpendicular distances to the surface and p the total length of the x axis intercepted by these perpendiculars, we can invoke the pythagorean theorem concerning right triangles and write

$$d^2 = h^2 + (p - x)^2 \qquad d'^2 = h'^2 + x^2$$

When these values of d and d' are substituted in Eq. (1i), we obtain

$$\Delta = n[h^2 + (p - x)^2]^{1/2} + n'(h'^2 + x^2)^{1/2} \tag{1v}$$

According to Fermat's principle, Δ must be a minimum or a maximum (or in general stationary) for the actual path. One method of finding a minimum or maximum for the optical path is to plot a graph of Δ against x and find at what value of x

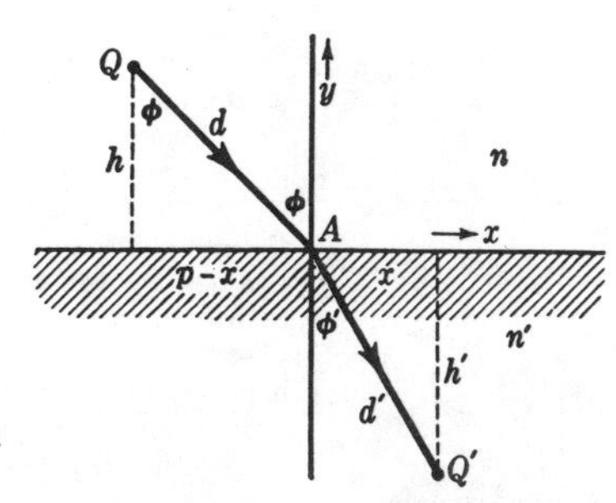

FIGURE 1L
Geometry of a refracted ray used in illustrating Fermat's principle.

a tangent to the curve is parallel to the x axis (see Fig. 1K). The mathematical means for doing the same thing is, first, to differentiate Eq. (1v) with respect to the variable x, thus obtaining an equation for the slope of the graph, and, second, to set this resultant equation equal to zero, thus finding the value of x for which the slope of the curve is zero.

By differentiating Eq. (1v) with respect to x and setting the result equal to zero, we obtain

$$\frac{d\Delta}{dx} = \frac{\frac{1}{2}n}{[h^2 + (p - x)^2]^{1/2}}(-2p + 2x) + \frac{\frac{1}{2}n'}{(h'^2 + x^2)^{1/2}}2x = 0$$

which gives

$$n\frac{p - x}{[h^2 + (p - x)^2]^{1/2}} = n'\frac{x}{(h'^2 + x^2)^{1/2}}$$

or simply

$$n\frac{p - x}{d} = n'\frac{x}{d'}$$

By reference to Fig. 1L it will be seen that the multipliers of n and n' are just the sines of the corresponding angles, so that we have now proved Eq. (1m), namely

$$n \sin \phi = n' \sin \phi' \qquad \text{(1w)}$$

A diagram for reflected light, similar to Fig. 1L, can be drawn and the same mathematics applied to prove the law of reflection.

1.10 COLOR DISPERSION

It is well known to those who have studied elementary physics that refraction causes a separation of white light into its component colors. Thus, as is shown in Fig. 1M, the incident ray of white light gives rise to refracted rays of different colors (really a continuous *spectrum*) each of which has a different value of ϕ'. By Eq. (1m) the value of n' must therefore vary with color. It is customary in the exact specification of indices of refraction to use the particular colors corresponding to certain dark lines in the spectrum of the sun. These *Fraunhofer* lines*, which are designated by the letters A, B, C, . . . , starting at the extreme red end, are given in Table 1A. The ones most commonly used are those in Fig. 1M.

The angular divergence of rays F and C is a measure of the *dispersion* produced, and has been greatly exaggerated in the figure relative to the average *deviation* of the

* Joseph von Fraunhofer (1787–1826) was the son of a Bavarian glazier. He learned glass grinding from his father and entered the field of optics from the practical side. Fraunhofer gained great skill in the manufacture of achromatic lenses and optical instruments. While measuring the refractive index of different kinds of glass and its variation with color or wavelength, he noticed and made use of the yellow D lines of the sodium spectrum. He was one of the first to produce diffraction gratings, and his rare skill with these devices enabled him to produce better spectra than his predecessors. Although the dark lines of the solar spectrum were first observed by W. H. Wollaston, they were carefully observed by Fraunhofer, under high dispersion and resolution, and the wavelengths of the most prominent lines were measured with precision. He mapped 576 of these lines, the principal ones, denoted by the letters A through K, being known by his name.

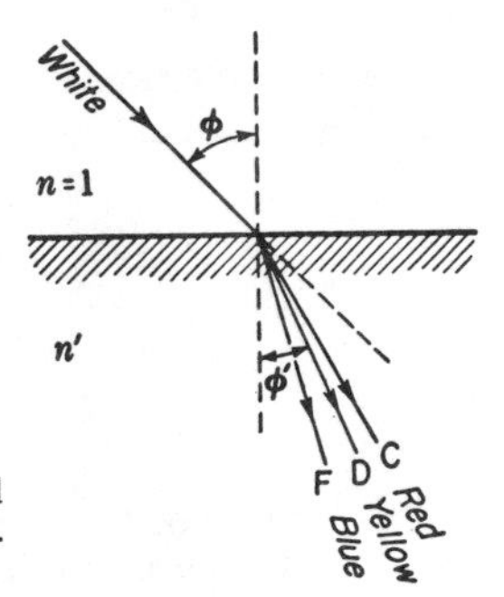

FIGURE 1M
Upon refraction, white light is spread out into a spectrum. This is called dispersion.

spectrum, which is measured by the angle through which ray D is bent. To take a typical case of crown glass, the refractive indices as given in Table 1A are

$$n_F = 1.52933 \qquad n_D = 1.52300 \qquad n_C = 1.52042$$

Now it is readily shown from Eq. (1q) that for a given small angle ϕ the dispersion of the F and C rays ($\phi'_F - \phi'_C$) is proportional to

$$n_F - n_C = 0.00891$$

while the deviation of the D ray ($\phi - \phi'_D$) depends on $n_D - 1$ which is equal to 0.52300. Thus it is nearly 60 times as great. The ratio of these two quantities varies greatly for different kinds of glass and is an important characteristic of any optical substance. It is called the *dispersive power* and is defined by the equation

$$V = \frac{n_F - n_C}{n_D - 1} \tag{1x}$$

The reciprocal of the dispersive power is called the *dispersive index* ν:

$$\nu = \frac{n_D - 1}{n_F - n_C} \tag{1y}$$

For most optical glasses ν lies between 20 and 60 (see Table 1B and Appendix III).

Table 1A FRAUNHOFER'S DESIGNATIONS, ELEMENT SOURCE, WAVELENGTH, AND REFRACTIVE INDEX FOR FOUR OPTICAL GLASSES*

Designation	Chemical element	Wavelength, Å†	Spectacle crown	Light flint	Dense flint	Extra dense flint
C	H	6563	1.52042	1.57208	1.66650	1.71303
D	Na	5892	1.52300	1.57600	1.67050	1.72000
F	H	4861	1.52933	1.58606	1.68059	1.73780
G′	H	4340	1.53435	1.59441	1.68882	1.75324

* For other glasses and crystals see Appendices III and IV.

† To change wavelengths in angstroms (Å) to nanometers (nm), move decimal point one place to the left (see Appendix VI).

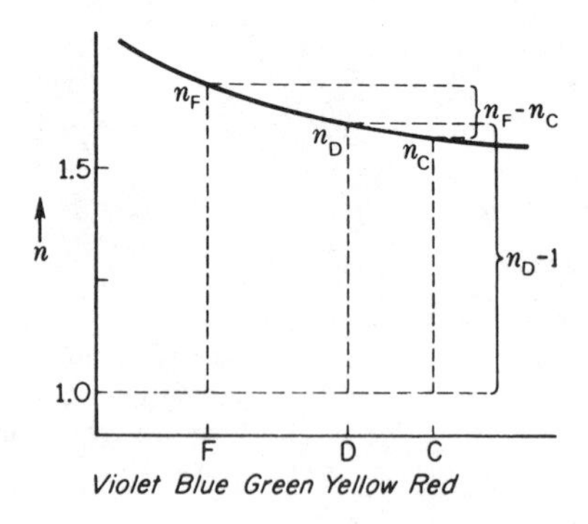

FIGURE 1N
The variation of refractive index with color.

Figure 1N illustrates schematically the type of variation of n with color that is usually encountered for optical materials. The denominator of Eq. (1y), which is a measure of the dispersion, is determined by the difference in the index at two points near the ends of the spectrum. The numerator, which measures the average deviation, represents the magnitude in excess of unity of an intermediate index of refraction.

It is customary in most treatments of geometrical optics to neglect chromatic effects and assume, as we have in the next seven chapters, that the refractive index of each specific element of an optical instrument is that determined for yellow sodium D light.

Table 1B DISPERSION INDEX FOR FOUR OPTICAL GLASSES*

Glass	Spectacle crown	Light flint	Dense flint	Extra dense flint
ν	58.7	41.2	47.6	29.08

* See Table 1A.

PROBLEMS*

1.1 A boy makes a pinhole camera out of a cardboard box with the dimensions 10.0 cm × 10.0 cm × 16.0 cm. A pinhole is located in one end, and a film 8.0 cm × 8.0 cm is placed in the other end. How far away from a tree 25.0 m high should the boy place his camera if the image of the tree is to be 6.0 cm high on the film? *Ans.* 66.7 m

1.2 A physics student wishes to repeat Fizeau's experiment for measuring the speed of light. If he uses a toothed wheel containing 1440 teeth and his distant mirror is located in a laboratory window across the college campus 412.60 m away, how fast must his wheel be rotated if the returning light pulses show the first maximum intensity?

1.3 If the mirror R in Foucault's experiment were to rotate at 12,000 rev/min, find (*a*) the rotational speed of the mirror R in revolutions per second and (*b*) the rotational speed of the sweeping beam RM_1 in radians per second. Find the time it takes the light to traverse the path (*c*) RM_1R and (*d*) RM_2R. What is the observed slit deflection (*e*) EE_1, and (*f*) EE_2? Assume the distances $RM_1 = RM_2 = 6.0$ m, $RS = RE = 6.0$ m, the

* Before solving any problems in this text, read Appendix VI.

length of the water tube $T = 5.0$ m, the refractive index of water is 1.3330, and the speed of light in air is 3.0×10^8 m/s.

1.4 If the refractive index for a piece of optical glass is 1.5250, calculate the speed of light in the glass. *Ans.* 1.9659×10^8 m/s

1.5 Calculate the difference between the speed of light in kilometers per second in a vacuum and the speed of light in air if the refractive index of air is 1.0002340. Use velocity values to seven significant figures.

1.6 If the moon's distance from the earth is 3.840×10^5 km, how long will it take microwaves to travel from the earth to the moon and back again?

1.7 How long does it take light from the sun to reach the earth? Assume the earth's distance from the sun to be 1.50×10^8 km. *Ans.* 500 s, or 8 min 20 s

1.8 A beam of light passes through a block of glass 10.0 cm thick, then through water for a distance of 30.5 cm, and finally through another block of glass 5.0 cm thick. If the refractive index of both pieces of glass is 1.5250 and of water is 1.3330, find the total optical path.

1.9 A water tank is 62.0 cm long inside and has glass ends which are each 2.50 cm thick. If the refractive index of water is 1.3330 and of glass is 1.6240, find the overall optical path.

1.10 A beam of light passes through 285.60 cm of water of index 1.3330, then through 15.40 cm of glass of index 1.6360, and finally through 174.20 cm of oil of index 1.3870. Find to three significant figures (*a*) each of the separate optical paths and (*b*) the total optical path. *Ans.* (*a*) 380.7, 25.19, and 241.6 cm, (*b*) 647 cm

1.11 A ray of light in air is incident on the polished surface of a block of glass at an angle of 10°. (*a*) If the refractive index of the glass is 1.5258, find the angle of refraction to four significant figures. (*b*) Assuming the sines of the angles in Snell's law can be replaced by the angles themselves, what would be the angle of refraction? (*c*) Find the percentage error.

1.12 Find the answers to Prob. 1.11, if the angle of incidence is 45.0° and the refractive index is 1.4265.

1.13 A ray of light in air is incident at an angle of 54.0° on the smooth surface of a piece of glass. (*a*) If the refractive index is 1.5152, find the angle of refraction to four significant figures. (*b*) Find the angle of refraction graphically. (See Fig. Pl.13).
Ans. (*a*) 32.272°, (*b*) 32.3°

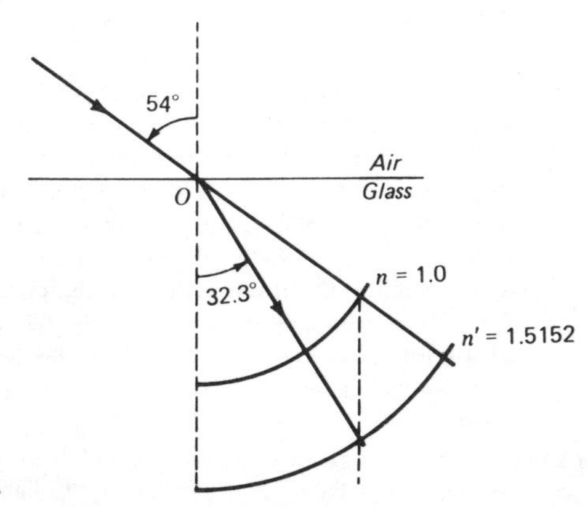

FIGURE P1.13
Graph for part (*b*) of Prob. 1.13.

1.14 A straight hollow pipe exactly 1.250 m long, with glass plates 8.50 mm thick to close the two ends, is thoroughly evacuated. (*a*) If the glass plates have a refractive index of 1.5250, find the overall optical path between the two outer glass surfaces. (*b*) By how much is the optical path increased if the pipe is filled with water of refractive index 1.33300. Give answers to five significant figures.

1.15 Referring to Fig. 1L, the distance $x = 6.0$ cm, $h = 12.0$ cm, $h' = 15.0$ cm, $n = 1.3330$, and $n' = 1.5250$. Find ϕ', ϕ, d, d', p, and Δ, to three significant figures.

Ans. $\phi' = 21.80°$, $\phi = 25.14°$, $d = 13.26$ cm, $d' = 16.16$ cm, $p = 11.63$ cm, $\Delta = 42.3$ cm

1.16 Solve Prob. 1.15 graphically.

1.17 In studying the refraction of light Kepler arrived at a refraction formula

$$\phi = \frac{\phi'}{1 - k \sec \phi'} \qquad \text{where} \qquad k = \frac{n' - 1}{n'}$$

n' being the relative index of refraction. Calculate the angle of incidence ϕ for a piece of glass for which $n' = 1.7320$ and the angle of refraction $\phi' = 32.0°$ according to (*a*) Kepler's formula and (*b*) Snell's law. Note that $\sec \phi' = 1/(\cos \phi')$.

1.18 White light is incident at an angle of 55.0° on the polished surface of a piece of glass. If the refractive indices for red C light and blue F light are $n_C = 1.53828$, and $n_F = 1.54735$, respectively, what is the angular dispersion between these two colors? (*a*) Find the two angles to five significant figures and (*b*) the dispersion to three significant figures. *Ans.* (*a*) $\phi'_C = 32.1753°$, $\phi'_F = 31.9643°$, (*b*) 0.2110°

1.19 A piece of dense flint glass is to be made into a prism. If the refractive indices for red, yellow, and blue light are specified as $n_C = 1.64357$, $n_D = 1.64900$, and $n_F = 1.66270$, find (*a*) the dispersive power and (*b*) the dispersion constant for this glass.

1.20 A block of spectacle crown glass is to be made into a lens. The refractive indices furnished by the glass manufacturer are specified as $n_C = 1.52042$, $n_D = 1.52300$, and $n_F = 1.52933$. Determine the value of (*a*) the dispersion constant and (*b*) the dispersive power.

1.21 A piece of extra dense flint glass is to be made into a prism. The refractive indices furnished by the glass manufacturer are those given in Table 1A. Find the value of (*a*) the dispersive power and (*b*) the dispersion constant.

Ans. (*a*) 0.034403, (*b*) 29.067

1.22 Two plane mirrors are inclined to each other at an angle α. Applying the law of reflection show that any ray whose plane of incidence is perpendicular to the line of intersection of the two mirrors is deviated by two reflections by an angle δ which is independent of the angle of incidence. Express this deviation in terms of α.

1.23 An ellipsoidal mirror has a major axis of 10.0 cm, a minor axis of 8.0 cm, and foci 6.0 cm apart. If there is a point source of light at one focus Q, there are only two rays of light that pass through the point C, midway between B and Q', as shown in the accompanying figure. Draw such an ellipse and graphically determine whether these two paths QBC and QDC are maxima, minima, or stationary.

1.24 A ray of light in air enters the center of one face of a prism at an angle making 55.0° with the normal. Traveling through the glass, the ray is again refracted into the air beyond. Assume the angle between the two prism faces to be 60.0° and the glass to have a refractive index of 1.650. Find the deviation of the ray (*a*) at the first surface and (*b*) the second surface. Find the total deviation (*c*) by calculation and (*d*) graphically.

1.25 One end of a glass rod is ground and polished to the shape of a hemisphere with a diameter of 10.0 cm. Five parallel rays of light 2.0 cm apart and in the same plane are incident on this curved end, with one ray traversing the center of the hemisphere parallel to the rod axis. If the refractive index is 1.5360, calculate the distances from the front surface to the point where the refracted rays cross the axis.

1.26 Crystals of clear strontium titanate are made into semiprecious gems. The refractive indices for different colors of light are as follows:

	Red	Yellow	Blue	Violet
λ, Å	6563	5892	4861	4340
n	2.37287	2.41208	2.49242	2.57168

Calculate the value of (*a*) the dispersion constant and (*b*) the dispersive power. Plot a graph of the wavelength λ against the refractive index n. Use the blue, yellow, and red indices.

2

PLANE SURFACES AND PRISMS

The behavior of a beam of light upon reflection or refraction at a plane surface is of basic importance in geometrical optics. Its study will reveal several of the features that will later have to be considered in the more difficult case of a curved surface. Plane surfaces often occur in nature, e.g., as the cleavage surfaces of crystals or as the surfaces of liquids. Artificial plane surfaces are used in optical instruments to bring about deviations or lateral displacements of rays as well as to break light into its colors. The most important devices of this type are prisms, but before taking up this case of two surfaces inclined to each other, we must examine rather thoroughly what happens at a single plane surface.

2.1 PARALLEL BEAM

In a beam or pencil of parallel light, each ray meets the surface traveling in the same direction. Therefore any one ray may be taken as representative of all the others. The parallel beam remains parallel after reflection or refraction at a plane surface, as shown in Fig. 2A(*a*). Refraction causes a change in width of the beam which is easily seen to be in the ratio $(\cos \phi')/(\cos \phi)$, whereas the reflected beam remains of the same

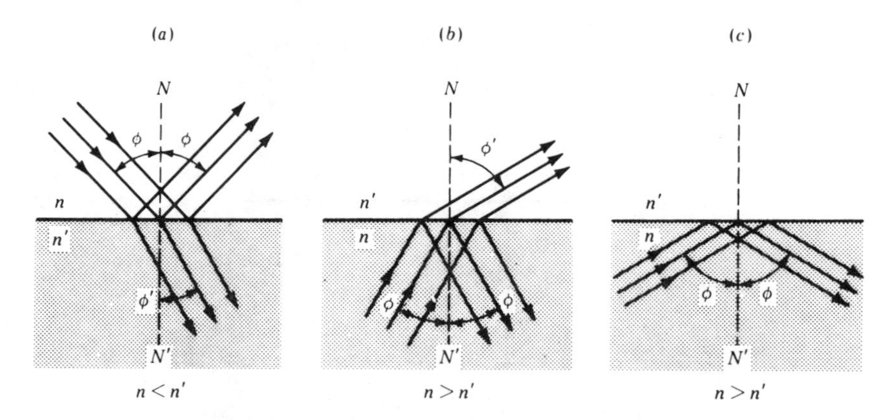

FIGURE 2A
Reflection and refraction of a parallel beam: (*a*) external reflection; (*b*) internal reflection at an angle smaller than the critical angle; (*c*) total reflection at or greater than the critical angle.

width. There is also chromatic dispersion of the refracted beam but not of the reflected one.

Reflection at a surface where n increases, as in Fig. 2A(*a*), is called *external reflection*. It is also frequently termed *rare-to-dense* reflection because the relative magnitudes of n correspond roughly (though not exactly) to those of the actual densities of materials. In Fig. 2A(*b*) is shown a case of *internal reflection* or *dense-to-rare* reflection. In this particular case the refracted beam is narrow because ϕ' is close to 90°.

2.2 THE CRITICAL ANGLE AND TOTAL REFLECTION

We have already seen in Fig. 2A(*a*) that as light passes from one medium like air into another medium like glass or water the angle of refraction is always less than the angle of incidence. While a decrease in angle occurs for all angles of incidence, there exists a range of refracted angles for which no refracted light is possible. A diagram illustrating this principle is shown in Fig. 2B, where for several angles of incidence, from 0 to 90°, the corresponding angles of refraction are shown from 0° to ϕ_c, respectively.

It will be seen that in the limiting case, where the incident rays approach an angle of 90° with the normal, the refracted rays approach a fixed angle ϕ_c beyond which no refracted light is possible. This particular angle ϕ_c, for which $\phi = 90°$, is called the *critical angle*. A formula for calculating the critical angle is obtained by substituting $\phi = 90°$, or $\sin \phi = 1$, in Snell's law [Eq. (1m)],

$$n \times 1 = n' \sin \phi_c$$

• so that

$$\sin \phi_c = \frac{n}{n'} \tag{2a}$$

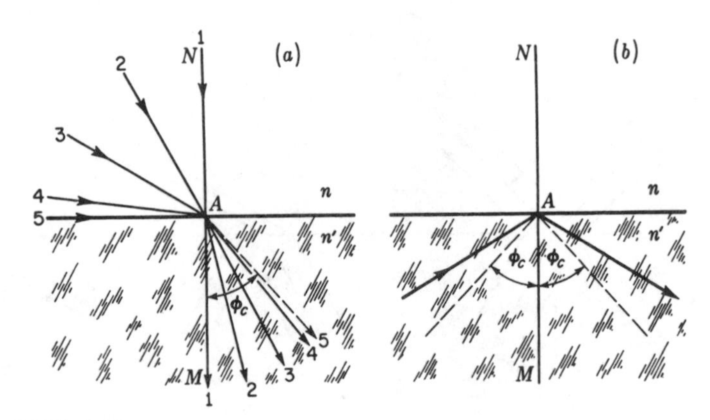

FIGURE 2B
Refraction and total reflection: (*a*) the critical angle is the limiting angle of refraction; (*b*) total reflection beyond the critical angle.

a quantity which is always less than unity. For a common crown glass of index 1.520 surrounded by air $\sin \phi_c = 0.6579$, and $\phi_c = 41°8'$.

If we apply the principle of reversibility of light rays to Fig. 2B(*a*), all incident rays will lie within a cone subtending an angle of $2\phi_c$, while the corresponding refracted rays will lie within a cone of 180°. For angles of incidence greater than ϕ_c there can be no refracted light and every ray undergoes total reflection as shown in Fig. 2B(*b*).

> *The critical angle for the boundary separating two optical media is defined as the smallest angle of incidence, in the medium of greater index, for which light is totally reflected.*

Total reflection is really total in the sense that no energy is lost upon reflection. In any device intended to utilize this property there will, however, be small losses due to absorption in the medium and to reflections at the surfaces where the light enters and leaves the medium. The commonest devices of this kind are called *total reflection prisms*, which are glass prisms with two angles of 45° and one of 90°. As shown in Fig. 2C(*a*), the light usually enters perpendicular to one of the shorter faces, is totally reflected from the hypotenuse, and leaves at right angles to the other short face. This deviates the rays through a right angle. Such a prism may also be used in two other ways which are illustrated in (*b*) and (*c*) of the figure. The Dove prism (*c*) interchanges the two rays, and if the prism is rotated about the direction of the light, they rotate around each other with twice the angular velocity of the prism.

Many other forms of prisms which use total reflection have been devised for special purposes. Two common ones are illustrated in Fig. 2C(*d*) and (*e*). The roof prism accomplishes the same purpose as the total reflection prism (*a*) except that it introduces an extra inversion. The triple mirror (*e*) is made by cutting off the corner of a cube by a plane which makes equal angles with the three faces intersecting at that

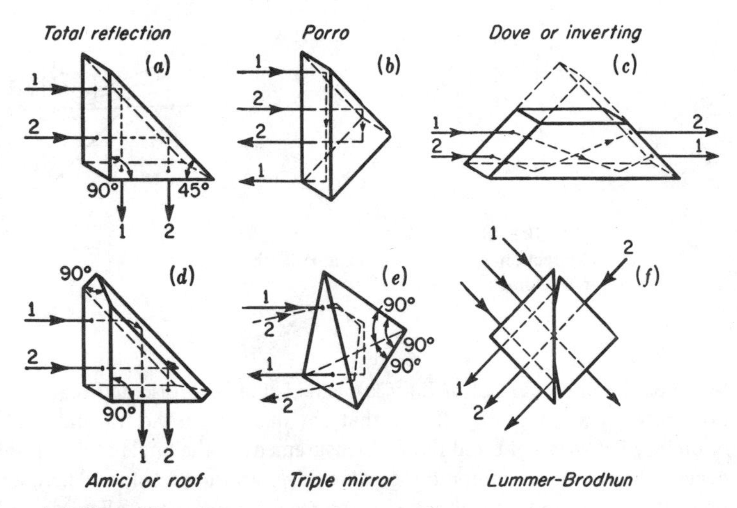

FIGURE 2C
Reflecting prisms utilizing the principle of total reflection.

corner.* It has the useful property that any ray striking it will, after being internally reflected at each of the three faces, be sent back parallel to its original direction.

The Lummer-Brodhun "cube" shown in (*f*) is used in photometry to compare the illumination of two surfaces, one of which is viewed by rays (2) coming directly through the circular region where the prisms are in contact, the other by rays (1) which are totally reflected in the area around this region.

Since, in the examples shown, the angles of incidence can be as small as 45°, it is essential that this exceed the critical angle in order that the reflection be total. Supposing the second medium to be air ($n' = 1$), this requirement sets a lower limit on the value of the index n of the prism. By Eq. (2a) we must have

$$\frac{n'}{n} = \frac{1}{n} \geq \sin 45^\circ$$

so that $n \geq \sqrt{2} = 1.414$. This condition always holds for glass and is even fulfilled for optical materials having low refractive indices such as Lucite ($n = 1.49$) and fused quartz ($n = 1.46$).

The principle of most accurate *refractometers* (instruments for the determination of refractive index) is based on the measurement of the critical angle ϕ_c. In both the Pulfrich and Abbe types a convergent beam strikes the surface between the unknown sample, of index n, and a prism of known index n'. Now n' is greater than n, so the

* A 46-cm array of 100 of these prisms is located on the moon's surface, 3.84×10^8 m from the earth. This retrodirector, placed there during the Apollo 11 moon flight, is used to return light from a laser beam from the earth to a point on the earth close to the source. Such a marker can be used to accurately determine the distance to the moon at different times. See J. E. Foller and E. J. Wampler, The Lunar Reflector, *Sci. Am.*, March 1970, p. 38. For more details see Sec. 30.13.

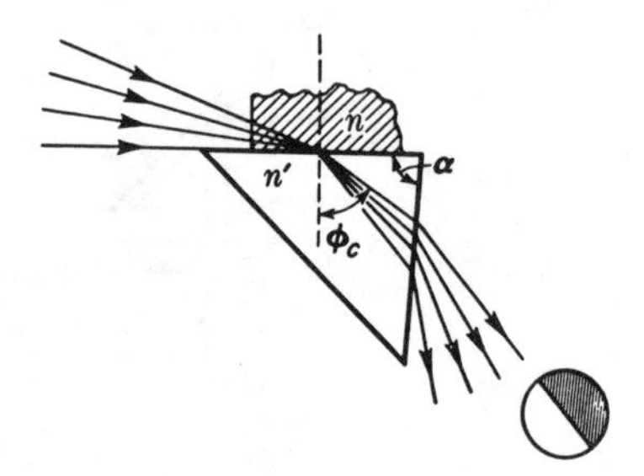

FIGURE 2D
Refraction by the prism in a Pulfrich refractometer.

two must be interchanged in Eq. (2a). The beam is so oriented that some of its rays just graze the surface (Fig. 2D) so that one observes in the transmitted light a sharp boundary between light and dark. Measurement of the angle at which this boundary occurs allows one to compute the value of ϕ_c and hence of n. There are important precautions that must be observed if the results are to be at all accurate.*

2.3 PLANE-PARALLEL PLATE

When a single ray traverses a glass plate with plane surfaces that are parallel to each other, it emerges parallel to its original direction but with a lateral displacement d which increases with the angle of incidence ϕ. Using the notation shown in Fig. 2E, we may apply the law of refraction and some simple trigonometry to find the displacement d. Starting with the right triangle ABE, we can write

$$d = l \sin (\phi - \phi') \qquad \text{(2b)}$$

which, by the trigonometric relation for *the sine of the difference between two angles*, can be written

$$d = l(\sin \phi \cos \phi' - \sin \phi' \cos \phi) \qquad \text{(2c)}$$

From the right triangle ABC we can write

$$l = \frac{t}{\cos \phi'}$$

which, substituted in Eq. (2c), gives

$$d = t\left(\frac{\sin \phi \cos \phi'}{\cos \phi'} - \frac{\sin \phi' \cos \phi}{\cos \phi'}\right) \qquad \text{(2d)}$$

From Snell's law [Eq. (1m)] we obtain

$$\sin \phi' = \frac{n}{n'} \sin \phi$$

* For a valuable description of this and other methods of determining indices of refraction see A. C. Hardy and F. H. Perrin, "Principles of Optics," pp. 359–364, McGraw-Hill Book Company, New York, 1932.

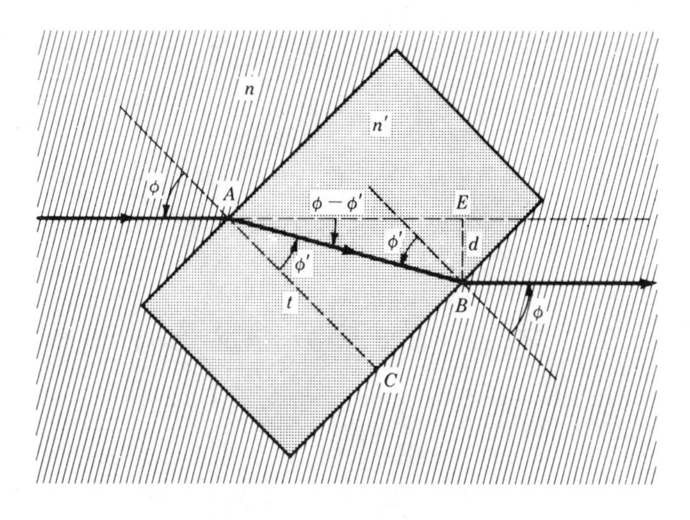

FIGURE 2E
Refraction by a plane-parallel plate.

which upon substitution in Eq. (2d), gives

$$d = t\left(\sin\phi - \frac{\cos\phi}{\cos\phi'}\frac{n}{n'}\sin\phi\right)$$

$$d = t\sin\phi\left(1 - \frac{n}{n'}\frac{\cos\phi}{\cos\phi'}\right) \tag{2e}$$

From 0° up to appreciably large angles, d is nearly proportional to ϕ, for as the ratio of the cosines becomes appreciably less than 1, causing the right-hand factor to increase, the sine factor drops below the angle itself in almost the same proportion.*

2.4 REFRACTION BY A PRISM

In a prism the two surfaces are inclined at some angle α so that the deviation produced by the first surface is not annulled by the second but is further increased. The chromatic dispersion (Sec. 1.10) is also increased, and this is usually the main function of a prism. First let us consider, however, the geometrical optics of the prism for light of a single color, i.e., for *monochromatic light* such as is obtained from a sodium arc.

* This principle is made use of in most of the home moving-picture film-editor devices in common use today. Instead of starting and stopping intermittently, as it does in the normal film projector, the film moves smoothly and continuously through the film-editor gate. A small eight-sided prism, immediately behind the film, produces a stationary image of each picture on the viewing screen of the editor. See Prob. 2.2 at the end of this chapter.

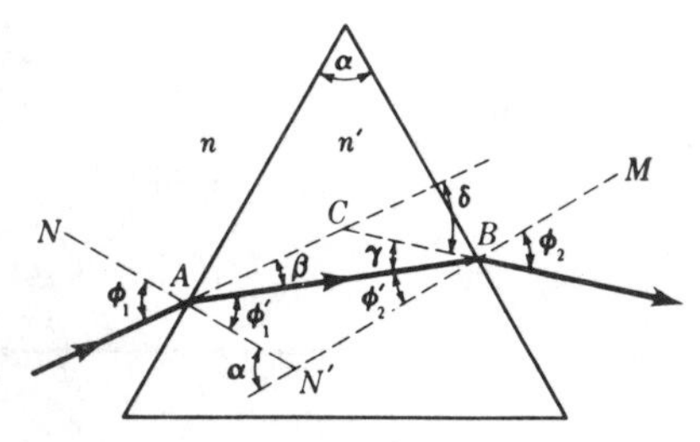

FIGURE 2F
The geometry associated with refraction by a prism.

The solid ray in Fig. 2F shows the path of a ray incident on the first surface at the angle ϕ_1.

Its refraction at the second surface, as well as at the first surface, obeys Snell's law, so that in terms of the angles shown

$$\frac{\sin \phi_1}{\sin \phi_1'} = \frac{n'}{n} = \frac{\sin \phi_2}{\sin \phi_2'} \tag{2f}$$

The angle of deviation produced by the first surface is $\beta = \phi_1 - \phi_1'$, and that produced by the second surface is $\gamma = \phi_2 - \phi_2'$. The total angle of deviation δ between the incident and emergent rays is given by

$$\delta = \beta + \gamma \tag{2g}$$

Since NN' and MN' are perpendicular to the two prism faces, α is also the angle at N'. From triangle ABN' and the exterior angle α, we obtain

$$\alpha = \phi_1' + \phi_2' \tag{2h}$$

Combining the above equations, we obtain

$$\delta = \beta + \gamma = \phi_1 - \phi_1' + \phi_2 - \phi_2' = \phi_1 + \phi_2 - (\phi_1' + \phi_2')$$

or

$$\delta = \phi_1 + \phi_2 - \alpha \tag{2i}$$

2.5 MINIMUM DEVIATION

When the total *angle of deviation* δ for any given prism is calculated by the use of the above equations, it is found to vary considerably with the angle of incidence. The angles thus calculated are in exact agreement with the experimental measurements. If during the time a ray of light is refracted by a prism the prism is rotated continuously in one direction about an axis (A in Fig. 2F) parallel to the refracting edge, the angle of deviation δ will be observed to decrease, reach a minimum, and then increase again, as shown in Fig. 2G.

The smallest deviation angle, called the *angle of minimum deviation* δ_m, occurs at that particular angle of incidence where the refracted ray inside the prism makes equal angles with the two prism faces (see Fig. 2H). In this special case

$$\phi_1 = \phi_2 \qquad \phi_1' = \phi_2' \qquad \beta = \gamma \tag{2j}$$

To prove these angles equal, assume ϕ_1 does not equal ϕ_2 when minimum deviation occurs. By the principle of the reversibility of light rays (see Sec. 1.8),

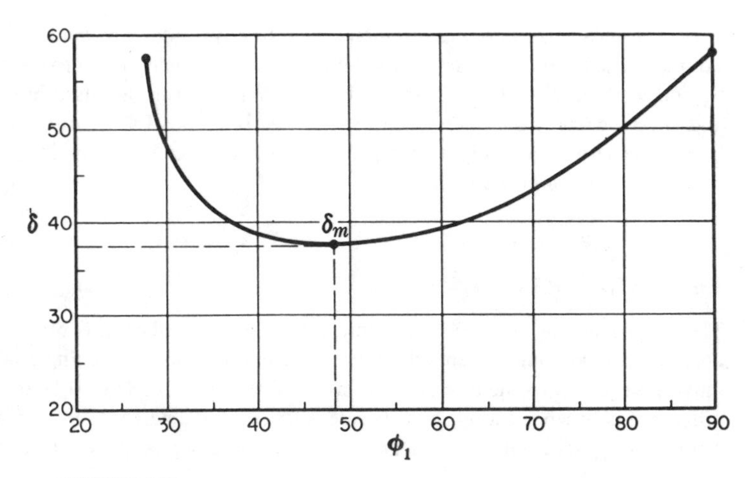

FIGURE 2G
A graph of the deviation produced by a 60° glass prism of index $n' = 1.50$. At minimum deviation $\delta_m = 37.2°$, $\phi_1 = 48.6°$, and $\phi_1{}' = 30.0°$.

there would be two different angles of incidence capable of giving minimum deviation. Since experimentally we find only one, there must be symmetry and the above equalities must hold.

In the triangle ABC in Fig. 2H the exterior angle δ_m equals the sum of the opposite interior angles $\beta + \gamma$. Similarly, for the triangle ABN', the exterior angle α equals the sum $\phi_1' + \phi_2'$. Consequently

$$\alpha = 2\phi_1' \qquad \delta_m = 2\beta \qquad \phi_1 = \phi_1' + \beta$$

Solving these three equations for ϕ_1' and ϕ_1 gives

$$\phi_1' = \tfrac{1}{2}\alpha \qquad \phi_1 = \tfrac{1}{2}(\alpha + \delta_m)$$

Since by Snell's law $n'/n = (\sin \phi_1)/(\sin \phi_1')$,

$$\frac{n'}{n} = \frac{\sin \frac{1}{2}(\alpha + \delta_m)}{\sin \frac{1}{2}\alpha} \tag{2k}$$

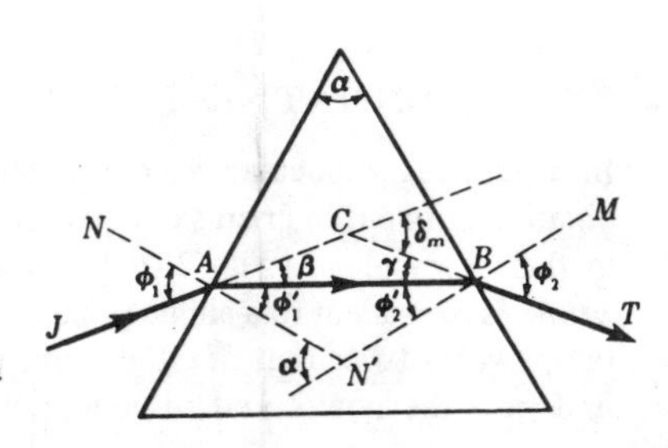

FIGURE 2H
The geometry of a light ray traversing a prism at minimum deviation.

The most accurate measurements of refractive index are made by placing the sample in the form of a prism on the table of a spectrometer and measuring the angles δ_m and α, the former for each color desired. When prisms are used in spectroscopes and spectrographs, they are always set as nearly as possible at minimum deviation because otherwise any slight divergence or convergence of the incident light would cause astigmatism in the image.

2.6 THIN PRISMS

The equations for the prism become much simpler when the refracting angle α becomes small enough to ensure that its sine and the sine of the angle of deviation δ may be set equal to the angles themselves. Even at an angle of 0.1 rad, or 5.7°, the difference between the angle and its sine is less than 0.2 percent. For prisms having a refracting angle of only a few degrees, we can therefore simplify Eq. (2k) by writing

$$\frac{n'}{n} = \frac{\sin \frac{1}{2}(\delta_m + \alpha)}{\sin \frac{1}{2}\alpha} = \frac{\delta_m + \alpha}{\alpha}$$

• and

$$\delta = (n' - 1)\alpha \qquad (2l)$$

Thin prism in air

The subscript on δ has been dropped because such prisms are always used at or near minimum deviation, and n has been dropped because it will be assumed that the surrounding medium is air, $n = 1$.

It is customary to measure the *power* of a prism by the deflection of the ray in centimeters at a distance of 1 m, in which case the unit of power is called the *prism diopter* (D). A prism having a power of 1 prism diopter therefore displaces the ray on a screen 1 m away by 1 cm. In Fig. 2I(*a*) the deflection on the screen is x cm and is numerically equal to the power of the prism. For small values of δ it will be seen that the power in prism diopters is essentially the angle of deviation δ measured in units of 0.01 rad, or 0.573°.

For the dense flint glass of Table 1A, $n'_D = 1.67050$, and Eq. (2l) shows that the refracting angle of a 1-D prism should be

$$\alpha = \frac{0.57300}{0.67050} = 0.85459°$$

2.7 COMBINATIONS OF THIN PRISMS

In measuring binocular accommodation, ophthalmologists make use of a combination of two thin prisms of equal power which can be rotated in opposite directions in their own plane [Fig. 2I(*b*)]. Such a device, known as the *Risley* or *Herschel prism*, is equivalent to a single prism of variable power. When the prisms are parallel, the power is twice that of either one; when they are opposed, the power is zero. To find how the power and direction of deviation depend on the angle between the

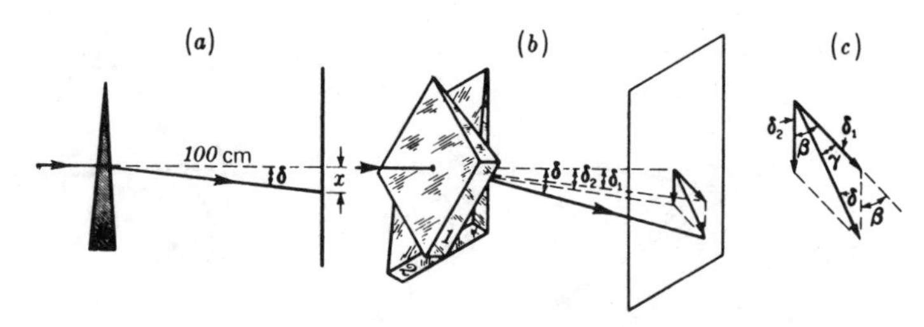

FIGURE 2I
Thin prisms: (*a*) the displacement x in centimeters at a distance of 1 m gives the power of the prism in diopters; (*b*) Risley prism of variable power; (*c*) vector addition of prism deviations.

components, we use the fact that the deviations add vectorially. In Fig. 2I(*c*) it will be seen that the resultant deviation δ will in general be, from the law of cosines,

$$\delta = \sqrt{\delta_1{}^2 + \delta_2{}^2 + 2\delta_1\delta_2 \cos\beta} \qquad \text{(2m)}$$

where β is the angle between the two prisms. To find the angle γ between the resultant deviation and that due to prism 1 alone (or, we may say, between the "equivalent" prism and prism 1) we have the relation

$$\tan\gamma = \frac{\delta_2 \sin\beta}{\delta_1 + \delta_2 \cos\beta} \qquad \text{(2n)}$$

Since almost always $\delta_1 = \delta_2$, we may call the deviation by either component δ_i, and the equations simplify to

$$\delta = \sqrt{2\delta_i{}^2(1 + \cos\beta)} = \sqrt{4\delta_i{}^2 \cos^2\frac{\beta}{2}} = 2\delta_i \cos\frac{\beta}{2} \qquad \text{(2o)}$$

and

$$\tan\gamma = \frac{\sin\beta}{1 + \cos\beta} = \tan\frac{\beta}{2}$$

so that

$$\gamma = \frac{\beta}{2} \qquad \text{(2p)}$$

2.8 GRAPHICAL METHOD OF RAY TRACING

It is often desirable in the process of designing optical instruments to be able to trace rays of light through the system quickly. For prism instruments the principles presented below are extremely useful. Consider first a 60° prism of index $n' = 1.50$ surrounded by air of index $n = 1.00$. After the prism has been drawn to scale, as in Fig. 2J, and the angle of incidence ϕ_1 has been selected, the construction begins as in Fig. 1G.

Line OR is drawn parallel to JA, and, with an origin at O, the two circular arcs are drawn with radii proportional to n and n'. Line RP is drawn parallel to NN',

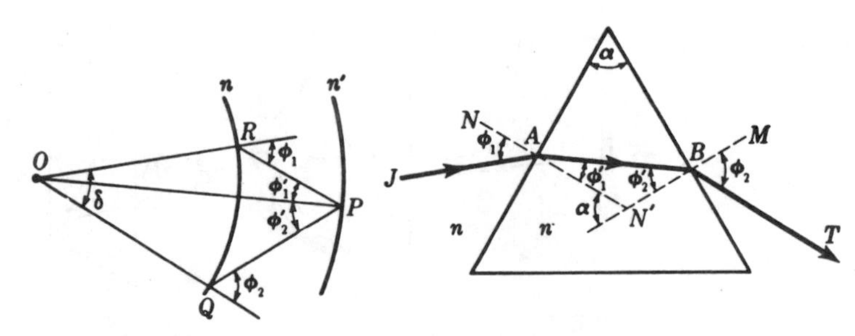

FIGURE 2J
A graphical method for ray tracing through a prism.

and OP is drawn to give the direction of the refracted ray AB. Carrying on from the point P, a line is drawn parallel to MN' to intersect the arc n at Q. The line OQ then gives the correct direction of the final refracted ray BT. In the construction diagram at the left the angle RPQ is equal to the prism angle α, and the angle ROQ is equal to the total angle of deviation δ.

2.9 DIRECT-VISION PRISMS

As an illustration of ray tracing through several prisms, consider the design of an important optical device known as a *direct-vision prism*. The primary function of such an instrument is to produce a visible spectrum the central color of which emerges from the prism parallel to the incident light. The simplest type of such a combination usually consists of a crown-glass prism of index n' and angle α' opposed to a flint-glass prism of index n'' and angle α'', as shown in Fig. 2K.

The indices n' and n'' chosen for the prisms are those for the central color of the spectrum, namely, for the sodium yellow D lines. Let us assume that the angle α'' of the flint prism is selected and the construction proceeds with the light emerging perpendicular to the last surface and the angle α' of the crown prism as the unknown.

The flint prism is first drawn with its second face vertical. The horizontal line OP is next drawn, and, with a center at O, three arcs are drawn with radii proportional to n, n', and n''. Through the intersection at P a line is drawn perpendicular to AC intersecting n' at Q. The line RQ is next drawn, and normal to it the side AB of the crown prism. All directions and angles are now known.

OR gives the direction of the incident ray, OQ the direction of the refracted ray inside the crown prism, OP the direction of the refracted ray inside the flint prism, and finally OP the direction of the emergent ray on the right. The angle α' of the crown prism is the supplement of angle RQP.

If more accurate determinations of angles are required, the construction diagram will be found useful in keeping track of the trigonometric calculations. If the dispersion of white light by the prism combination is desired, the indices n' and n'' for the red and violet light can be drawn in and new ray diagrams constructed proceeding now

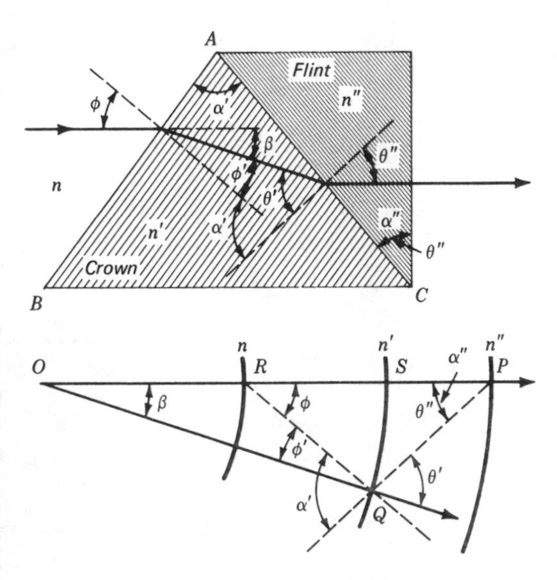

FIGURE 2K
Graphical ray tracing applied to the design of a direct-vision prism.

from left to right in Fig. 2K(*b*). These rays, however, will not emerge perpendicular to the last prism face.

The principles just outlined are readily extended to additional prism combinations like those shown in Fig. 2L. It should be noted that the upper direct-vision prism in Fig. 2L is in principle two prisms of the type shown in Fig. 2K placed back to back.

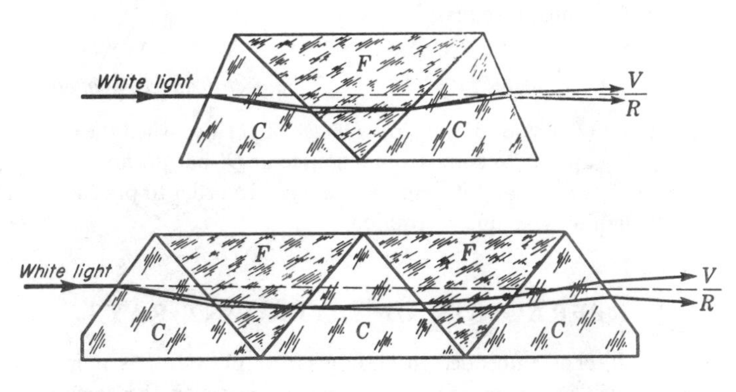

FIGURE 2L
Direct-vision prisms used for producing a spectrum with its central color in line with the incident white light.

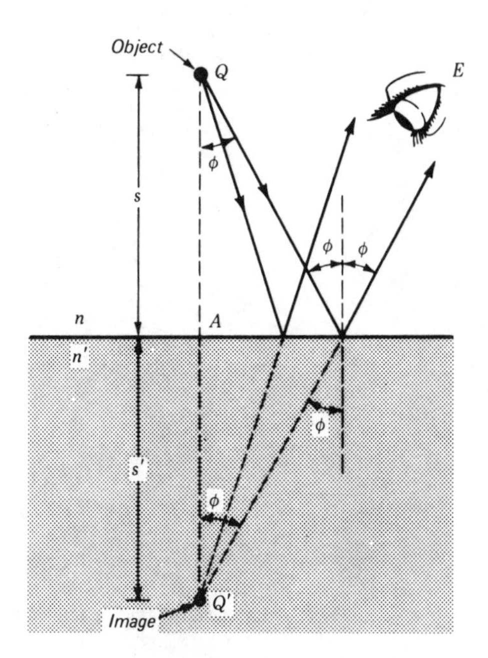

FIGURE 2M
The reflection of divergent rays of light from a plane surface.

2.10 REFLECTION OF DIVERGENT RAYS

When a divergent pencil of light is reflected at a plane surface, it remains divergent. All rays originating from a point Q (Fig. 2M) will after reflection appear to come from another point Q' symmetrically placed behind the mirror. The proof of this proposition follows at once from the application of the law of reflection [Eq. (1j)], according to which all the angles labeled ϕ in the figure must be equal. Under these conditions the distances QA and AQ' along the line QAQ' drawn perpendicular to the surface must be equal; i.e.,

$$s = s'$$

$$\textit{object distance} = \textit{image distance}$$

The point Q' is said to be a *virtual image* of Q since when the eye receives the reflected rays, they appear to come from a source at Q' but do not actually pass through Q', as would be the case if it were a *real image*. In order to produce a real image a surface other than a plane one is required.

2.11 REFRACTION OF DIVERGENT RAYS

If an object is embedded in clear glass or plastic or is immersed in a transparent liquid such as water, the image appears closer to the surface. Fig. 2N has been drawn accurately to scale for an object Q located in water of index 1.3330 at a depth

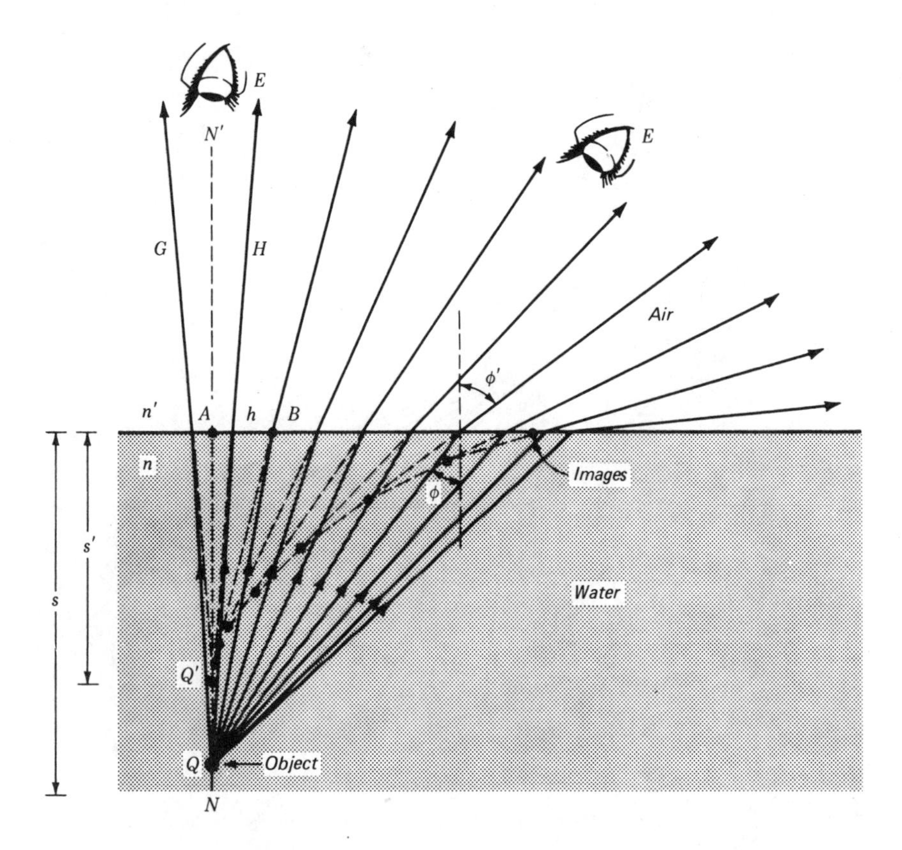

FIGURE 2N
Image positions of an object under water as seen by an observer above; $n > n'$.

s below the surface. Light rays diverging from this object arrive at the surface at angles ϕ. There they are refracted at larger angles ϕ', only to diverge more rapidly as shown. Extending these emergent rays backward, we locate their intersections in pairs. These are image points, or *virtual images*. As the observer changes his position, the virtual image moves closer to the surface and along the curve formed by the successive images.

If the object is located in the less dense medium and is observed from the medium of higher index, we obtain an entirely different view (see Fig. 2O). An object Q in air is observed by an underwater swimmer or fish. Rays of light diverging from any point of this object are refracted according to Snell's law. Extended backward to their intersections, their virtual images are located. Note how far away these images are for large angles of ϕ and ϕ'.

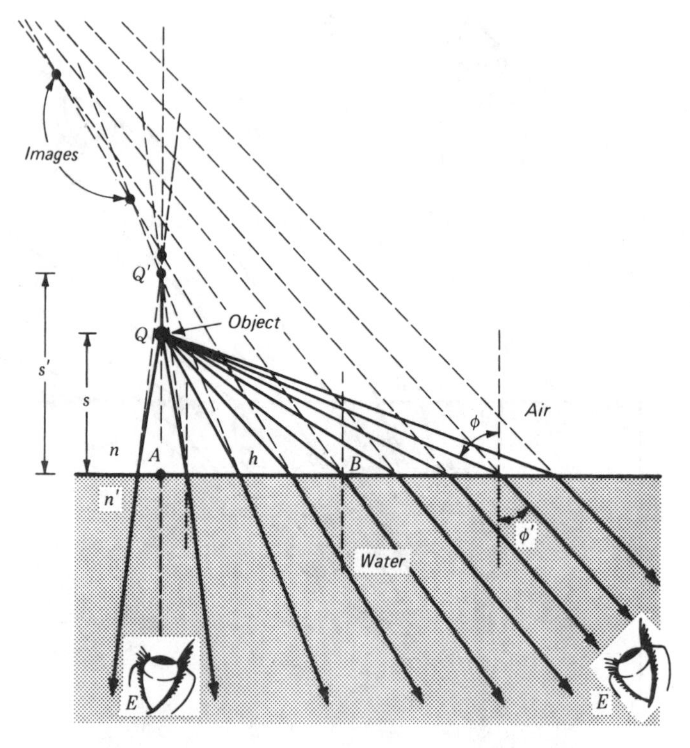

FIGURE 2O
Image positions of an object in air as seen by an observer under water; $n < n'$.

2.12 IMAGES FORMED BY PARAXIAL RAYS

Of particular interest to many observers are the object and image distances s and s' for rays making small angles ϕ and ϕ'.

> *Rays for which angles are small enough to permit setting the cosines equal to unity and the sines and tangents equal to the angles are called paraxial rays.*

Consider the right triangles QAB and $Q'AB$ in Fig. 2N, redrawn in Fig. 2P. Since there is a common side $AB = h$, we can write

$$h = s \tan \phi = s' \tan \phi'$$

From this we find

$$s' = s \frac{\tan \phi}{\tan \phi'} = s \frac{\sin \phi \cos \phi'}{\cos \phi \sin \phi'} \tag{2q}$$

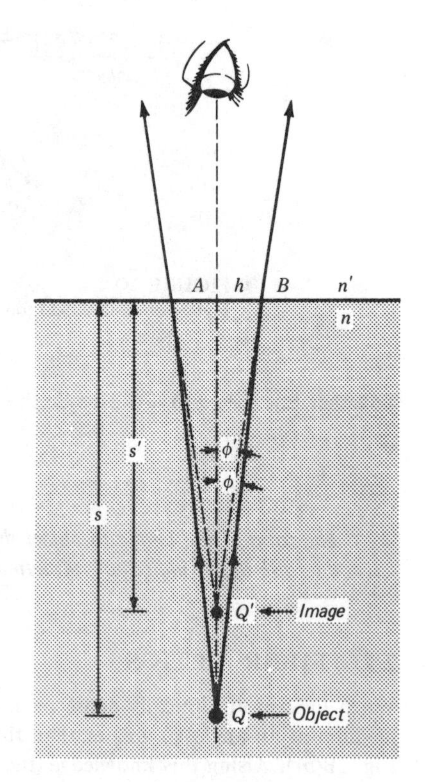

FIGURE 2P
Paraxial rays for an object in water and observed from the air above.

Applying Snell's law,

$$\frac{\sin \phi}{\sin \phi'} = \frac{n'}{n}$$

we obtain on substitution in Eq. (2q)

$$s' = s \frac{n'}{n} \frac{\cos \phi'}{\cos \phi} \tag{2r}$$

For paraxial rays like the ones shown in the diagram, angles ϕ and ϕ' are very small; Eq. (2q) can be written

$$s' = s \frac{\phi}{\phi'} \qquad \text{or} \qquad \frac{s'}{s} = \frac{\phi}{\phi'} \tag{2s}$$

and Eq. (2r) written

$$\frac{\phi}{\phi'} = \frac{n'}{n} \tag{2t}$$

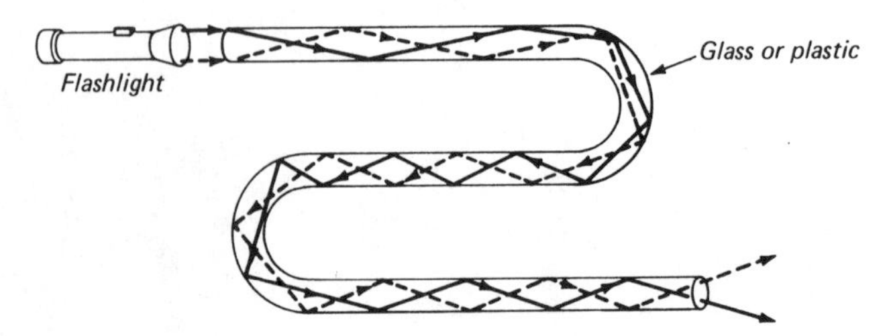

FIGURE 2Q
Light from a flashlight follows a bent transparent rod by total reflection.

Together Eqs. (2s) and (2t) provide the simple relation

$$\frac{s'}{s} = \frac{n'}{n} \qquad \text{(2u)}$$

Paraxial rays

The ratio of the image to object distance for paraxial rays is just equal to the ratio of the indices of refraction.

2.13 FIBER OPTICS

When light in an optically dense medium approaches the boundary of a less dense medium at an angle ϕ, greater than the critical angle ϕ_c, it is totally reflected [see Fig. 2B(*b*)]. Using this knowledge, the British physicist John Tyndall demonstrated that light rays in a tank of water shining through a hole in the side follow the stream of water emerging from the orifice. This effect is commonly observed today in fountains illuminated by lights from under the water. The transmission of light from a flashlight through a glass or plastic rod is shown in Fig. 2Q.

Bundles of tiny rods or fibers of clear glass or plastic provide the basis for the sizable industry of *fiber optics*. Tests on individual fibers over 50 m long show that there are essentially no losses due to reflection on the sides. All attenuation of an incident beam is attributable to reflection from the two ends and absorption by the fiber material.

An ordered array or bundle of tiny transparent fibers can be used to transmit light images around corners and over long distances. A bundle of hundreds and even thousands of fibers is frequently made to follow a path with many turns and ends up at a distant or nearby point (see Fig. 2R). If the individual fibers in a bundle are not arranged in an orderly array as in the figure but are randomly interwoven, the emerging image will be scrambled and meaningless.

Fibers are usually coated with a thin transparent layer of glass or other material of lower refractive index. Total reflection will still take place between the two. This separates the fibers of a bundle from one another, thereby preventing light leakage between touching fibers and at the same time protecting the fire-polished reflecting surfaces.

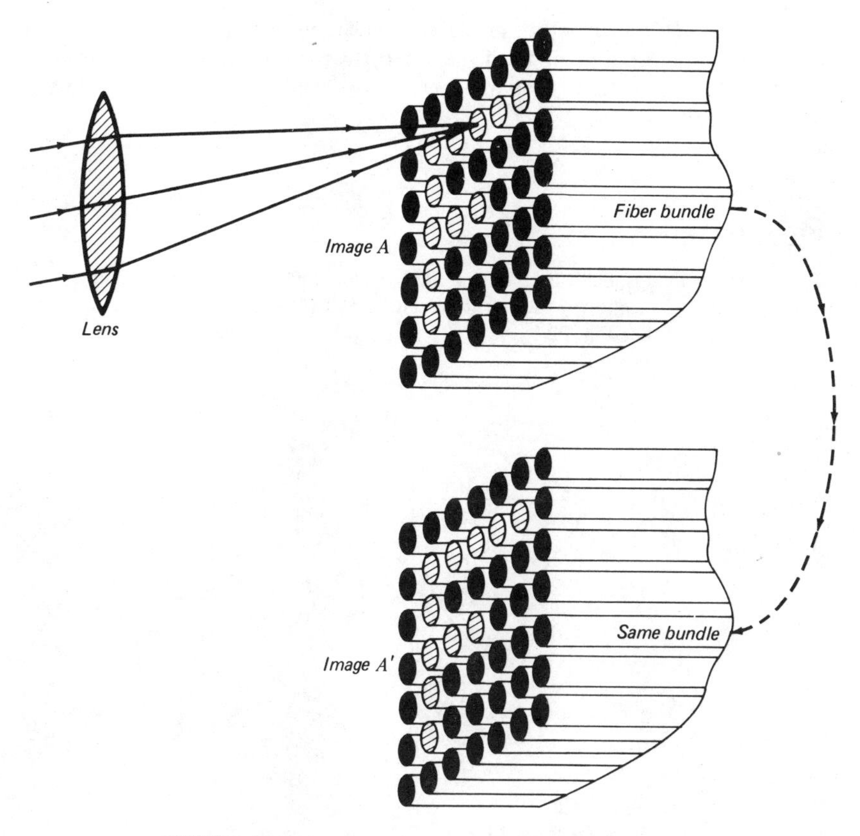

FIGURE 2R
An ordered array of fine glass fibers can be used to transmit images from one end A to the other A' along any curved path.

One method for producing coated fibers is to insert a thick, high-refractive-index glass rod in tubing of lower index. In a special furnace the two are then drawn down to $\frac{1}{1000}$ in. diameter, and the thickness is controlled within narrow limits. A bundle of these fibers can then be fused together to form a solid mass and drawn down a second time so that individual fibers are about 2 μm in diameter. This is about two wavelengths of visible light. Such bundles can resolve approximately 250 lines per millimeter.

If fibers are drawn down until their diameters are close to the wavelength of light, they cease to act like pipes and behave more like waveguides used in conducting microwaves.* Two wavelengths of light is an approximate limit for image

* For an introductory treatment of microwaves and waveguides, see Harvey E. White, "Modern College Physics," 5th ed., pp. 547–551, D. Van Nostrand, Princeton, N.J., 1966. For further details on fiber optics, see Narinder S. Kapany, Fiber Optics, *Sci. Am.*, November, 1960, pp. 71–80.

transmission. Of the numerous practical applications of fiber optics, one of the most important is in the field of medicine. A cystoscope, or catheter-type instrument, enables the surgeon to observe and operate by remote control on tiny areas deep within the body.

PROBLEMS

2.1 A ray of light is incident on a piece of glass at an angle of 45.0°. If the angle of refraction is 25.37°, find (*a*) the refractive index and (*b*) the critical angle. (*c*) Solve (*b*) graphically (See Fig. P2.1). *Ans.* (*a*) 1.6504, (*b*) 37.30°, (*c*) 1.650 and 37.3°

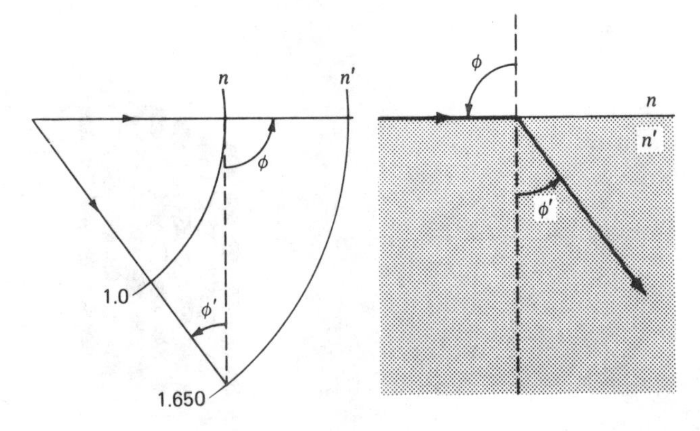

FIGURE P2.1
Graph for Prob. 2.1.

2.2 Calculate the lateral displacements of rays of light incident on a block of glass with parallel sides at the following angles: (*a*) 5.0°, (*b*) 10.0°, (*c*) 15.0°, (*d*) 20.0°, (*e*) 30.0°, and (*f*) 40.0°. (*g*) Plot a graph of d versus ϕ. Assume the glass thickness to be 5.0 cm.

2.3 A rectangular aquarium is to be filled with water. The sides are made of glass plates 8.0 mm thick. Inside, the walls are 35.0 cm apart, and the refractive index of the glass is 1.5250. If a ray of light is incident on one side at an angle of 50.0°, find the lateral displacement produced when the tank is (*a*) empty and (*b*) filled with water.

2.4 A Pulfrich refractometer is used to measure the refractive index of a clear transparent oil. The glass prism has a refractive index of 1.52518 and a refracting angle α of 80.0°. If the boundary between light and dark field makes an angle of 29.36° with the normal to the second face, find the refractive index. *Ans.* 1.3371

2.5 A 55.0° prism made of dense flint glass is used at an angle of incidence of $\phi_1 = 60.0°$. Using the refractive index for D light given in Table 1A, find (*a*) the angle of deviation β at the first surface, (*b*) the angle of deviation γ at the second surface, and (*c*) the total deviation by the prism.

2.6 A 50.0° crown-glass prism has a refractive index $n_D = 1.52300$ for sodium yellow light. If a ray of this yellow light is incident on one surface at an angle of 45.0°, find (*a*) the angle of deviation β at the first surface, (*b*) the angle of deviation γ at the second surface, and (*c*) the total deviation by the prism.

2.7 A 45.0° flint glass prism has a refractive index of 1.6705 for sodium yellow light, and it is adjusted for minimum deviation. Find (*a*) the angle of minimum deviation and (*b*) the angle of incidence. (*c*) Solve graphically.

2.8 A 60.0° prism produces an angle of minimum deviation of 43.60° for blue light. Find (*a*) the refractive index, (*b*) the angle of refraction, and (*c*) the angle of incidence.

Ans. (*a*) 1.572, (*b*) 30.0°, (*c*) 51.81°

2.9 A 55.0° prism has a refractive index of 1.68059 for blue light. (*a*) Graphically determine the angle of deviation for each of the following angles of incidence: 40.0, 45.0, 50.0, 55.0, 60.0, and 65.0°. (*b*) Plot a graph of δ against ϕ (see Fig. 2G).

2.10 Two thin prisms have powers of 6.0 D each. At what angles should their axes be superimposed to produce powers of 2.0, 4.0, 6.0, 8.0, 10.0, and 12.0 D?

Ans. 160.8, 141.1, 120.0, 96.4, 67.1, and 0°

2.11 Two thin prisms of 5.0 and 7.0 D, respectively, are superimposed so their axes make an angle of 75.0° with each other. Find (*a*) the resultant deviation they produce in degrees, (*b*) the power of the resultant deviation in diopters, and (*c*) the angle the resultant makes with the stronger of the two prisms.

2.12 A direct-vision prism is to be made of two elements like the one shown in Fig. 2K. The flint-glass prism of index 1.720 has an angle $\alpha'' = 55.0°$. Find the angle α' for the crown-glass prism if its refractive index is 1.520. Solve by (*a*) graphical methods and (*b*) calculation.

2.13 A coin lies on the bottom of a bathtub. If the water is 36.0 cm deep and the refractive index of water is 1.3330, find the image depth of the coin as seen from straight above. Assume the sines of angles can be replaced by the angles themselves.

3
SPHERICAL SURFACES

Many common optical devices contain not only mirrors and prisms having flat polished surfaces but lenses having spherical surfaces with a wide range of curvatures. Such spherical surfaces, in contrast with plane surfaces treated in the last chapter, are capable of forming real images.

Cross-sectional diagrams of several standard forms of lenses are shown in Fig. 3A. The three *converging*, or *positive*, *lenses*, which are thicker at the center than at the edges, are shown as (*a*) *equiconvex*, (*b*) *plano-convex*, and (*c*) *positive meniscus*. The three *diverging*, or *negative*, *lenses*, which are thinner at the center, are (*d*) *equiconcave*, (*e*) *plano-concave*, and (*f*) *negative meniscus*. Such lenses are usually made of optical glass as free as possible from inhomogeneities, but occasionally other transparent materials like quartz, fluorite, rock salt, and plastics are used. Although we shall see that the spherical form for the surfaces may not be the ideal one in a particular instance, it gives reasonably good images and is much the easiest to grind and polish.

This chapter treats the behavior of refraction at a single spherical surface separating two media of different refractive indices, and the following chapters show how the treatment can be extended to two or more surfaces in succession. These combinations form the basis for the treatment of thin lenses in Chap. 4, thick lenses in Chap. 5, and spherical mirrors in Chap. 6.

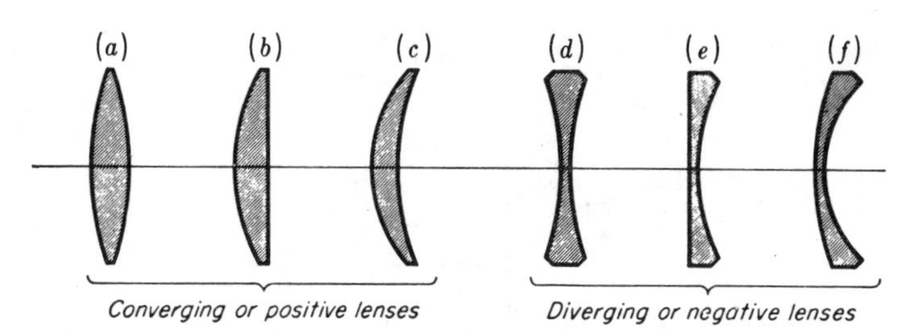

FIGURE 3A
Cross sections of common types of thin lenses.

3.1 FOCAL POINTS AND FOCAL LENGTHS

Characteristic diagrams showing the refraction of light by convex and concave spherical surfaces are given in Fig. 3B. Each ray in being refracted obeys Snell's law as given by Eq. (1m). The principal axis in each diagram is a straight line through the center of curvature C. The point A where the axis crosses the surface is called the *vertex*. In diagram (*a*) rays are shown diverging from a point source F on the axis in the first medium and refracted into a beam everywhere parallel to the axis in the second medium. Diagram (*b*) shows a beam converging in the first medium toward the point F and then refracted into a parallel beam in the second medium. F in each of these two cases is called the *primary focal point*, and the distance f is called the *primary focal length*.

In diagram (*c*) a parallel incident beam is refracted and brought to a focus at the point F', and in diagram (*d*) a parallel incident beam is refracted to diverge as if it came from the point F'. F' in each case is called the *secondary focal point*, and the distance f' is called the *secondary focal length*.

Returning to diagrams (*a*) and (*b*) for reference, we now state that *the primary focal point F is an axial point having the property that any ray coming from it or proceeding toward it travels parallel to the axis after refraction.* Referring to diagrams (*c*) and (*d*), we make the similar statement that *the secondary focal point F′ is an axial point having the property that any incident ray traveling parallel to the axis will, after refraction, proceed toward, or appear to come from, F′.*

A plane perpendicular to the axis and passing through either focal point is called a *focal plane*. The significance of a focal plane is illustrated for a convex surface in Fig. 3.C Parallel incident rays making an angle θ with the axis are brought to a focus in the focal plane at a point Q'. Note that Q' is in line with the undeviated ray through the center of curvature C and that this is the only ray that crosses the boundary at normal incidence.

It is important to note in Fig. 3B that the primary focal length f for the convex surface [diagram (*a*)] is not equal to the secondary focal length f' of the same surface [diagram (*c*)]. It will be shown in Sec. 3.4 that the ratio of the focal lengths f'/f is equal to the ratio n'/n of the corresponding refractive indices [see Eq. (3e)]:

$$\frac{f'}{f} = \frac{n'}{n} \tag{3a}$$

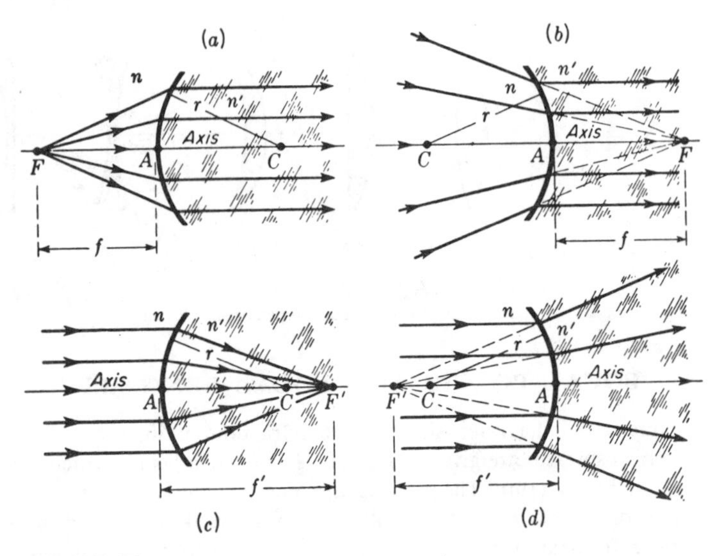

FIGURE 3B
The focal points F and F' and focal lengths f and f' associated with a single spherical refracting surface of radius r separating two media of index n and n'.

In optical diagrams it is common practice to show incident light rays traveling from left to right. A convex surface therefore is one in which the center of curvature C lies to the right of the vertex, while a concave surface is one in which C lies to the left of the vertex.

If we apply the principle of the reversibility of light rays to the diagrams in Fig. 3B, we should turn each diagram end-for-end. Diagram (*a*), for example, would then become a concave surface with converging properties, while diagram (*b*) would become a convex surface with diverging properties. Note that we would then have the incident rays in the denser medium, i.e., the medium of greater refractive index.

3.2 IMAGE FORMATION

A diagram illustrating image formation by a single refracting surface is given in Fig. 3D. It has been drawn for the case in which the first medium is air with an index $n = 1$ and the second medium is glass with an index $n' = 1.60$. The focal lengths f and f' therefore have the ratio 1:1.60 [see Eq. (3a)]. Experimentally it is observed that if the object is moved closer to the primary focal plane, the image will be formed farther to the right away from F' and will be larger, i.e., magnified. If the object is moved to the left, farther away from F, the image will be found closer to F' and will be smaller in size.

All rays coming from the object point Q are shown brought to a focus at Q'.

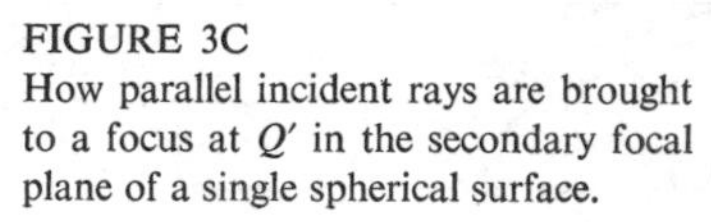

FIGURE 3C
How parallel incident rays are brought to a focus at Q' in the secondary focal plane of a single spherical surface.

Rays from any other object point like M will also be brought to a focus at a corresponding image point like M'. This ideal condition never holds exactly for any actual case. Departures from it give rise to slight defects of the image known as *aberrations.* The elimination of aberrations is the major problem of geometrical optics and will be treated in detail in Chap. 9.

If the rays considered are restricted to *paraxial rays*, a good image is formed with *monochromatic light. Paraxial rays* are defined as *those rays which make very small angles with the axis and lie close to the axis throughout the distance from object to image* (see Sec. 2.12). The formulas given in this chapter are to be taken as applying to images formed only by paraxial rays.

3.3 VIRTUAL IMAGES

The image $M'Q'$ in Fig. 3D is a real image in the sense that if a flat screen is located there, a sharply defined image of the object MQ will be formed on the screen. Not all images, however, can be formed on a screen, as is illustrated in Fig. 3E. Light rays from an object point Q are shown refracted by a concave spherical surface separating the two media of index $n = 1.0$ and $n' = 1.50$, respectively. The focal lengths have the ratio 1:1.50.

Since the refracted rays are diverging, they will not come to a focus at any point. To an observer's eye located at the right, however, such rays will appear to be coming from the common point Q'. In other words, Q' is the image point corresponding to the object point Q. Similarly M' is the image point corresponding to the object point M. Since the refracted rays do not come from Q' but only appear to do so, no image can be formed on a screen placed at M'. For this reason such an image is said to be *virtual.*

3.4 CONJUGATE POINTS AND PLANES

The principle of the reversibility of light rays has the consequence that if $Q'M'$ in Fig. 3D were an object, an image would be formed at QM. Hence, if any object is placed at the position previously occupied by its image, it will be imaged at the position previously occupied by the object. The object and image are thus interchangeable, or conjugate. Any pair of object and image points such as M and M'

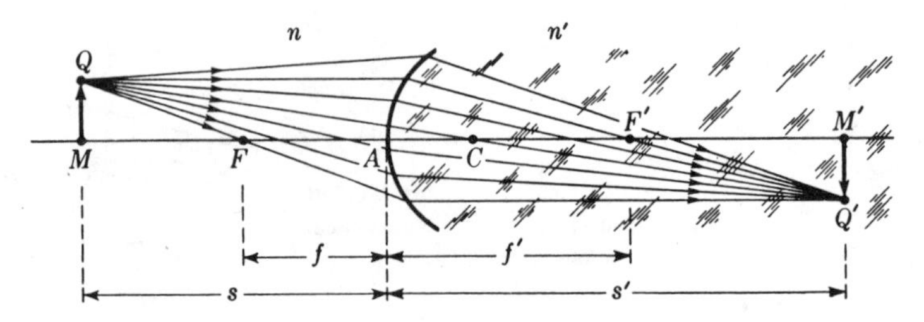

FIGURE 3D
All rays leaving the object point Q and passing through the refracting surface are brought to a focus at the image point Q'.

in Fig. 3D are called *conjugate points*, and planes through these points perpendicular to the axis are called *conjugate planes*.

If one is given the radius of curvature r of a spherical surface separating two media of index n and n', respectively, as well as the position of an object, there are three general methods that may be employed to determine the position and size of the image: (1) graphical methods, (2) experiment, and (3) calculation using the formula

$$\frac{n}{s} + \frac{n'}{s'} = \frac{n' - n}{r} \tag{3b}$$

In this equation s is the object distance and s' is the image distance. This equation, called the *gaussian formula for a single spherical surface*, is derived in Sec. 3.10.

EXAMPLE 1 The end of a solid glass rod of index 1.50 is ground and polished to a hemispherical surface of radius 1 cm. A small object is placed in air on the axis 4 cm to the left of the vertex. Find the position of the image. Assume $n = 1.00$ for air.

SOLUTION The given quantities are $n = 1.0$, $n' = 1.50$, $r = +1.0$ cm, and $s = +4.0$ cm. The unknown quantity is s'. By direct substitution of the given quantities in Eq. (3b) we obtain

$$\frac{1}{4} + \frac{1.50}{s'} = \frac{1.50 - 1.00}{1} \qquad \frac{1.50}{s'} = \frac{0.50}{1} - \frac{1}{4}$$

from which $s' = 6.0$ cm. One concludes, therefore, that a real image is formed in the glass rod 6 cm to the right of the vertex.

As an object M is brought closer to the primary focal point, Eq. (3b) shows that the distance AM' of the image from the vertex becomes steadily greater and that in the limit when the object reaches F the refracted rays are parallel and the image is formed at infinity. Then we have $s' = \infty$, and Eq. (3b) becomes

$$\frac{n}{s} + \frac{n'}{\infty} = \frac{n' - n}{r}$$

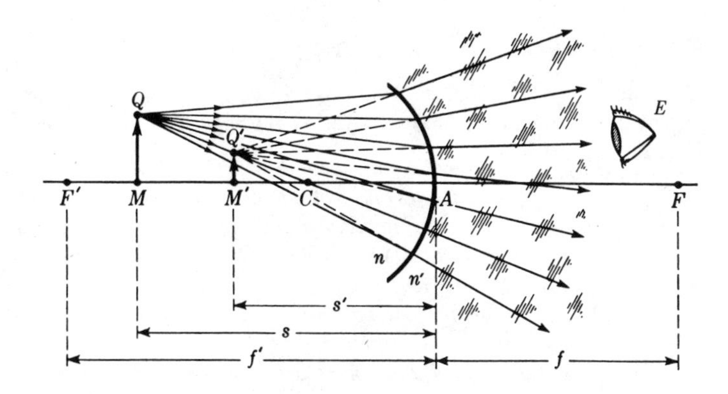

FIGURE 3E
All rays leaving the object point Q, and passing through the refracting surface appear to be coming from the virtual image point Q'.

Since this particular object distance is called the primary focal length f, we may write

$$\frac{n}{f} = \frac{n' - n}{r} \tag{3c}$$

Similarly, if the object distance is made larger and eventually approaches infinity, the image distance diminishes and becomes equal to f' in the limit, $s = \infty$. Then

$$\frac{n}{\infty} + \frac{n'}{s'} = \frac{n' - n}{r}$$

or, since this value of s' represents the secondary focal length f',

$$\frac{n'}{f'} = \frac{n' - n}{r} \tag{3d}$$

Equating the left-hand members of Eqs. (3c) and (3d), we obtain

$$\frac{n}{f} = \frac{n'}{f'} \qquad \text{or} \qquad \frac{n'}{n} = \frac{f'}{f} \tag{3e}$$

When $(n' - n)/r$ in Eq. (3b) is replaced by n/f or by n'/f' according to Eqs. (3c) and (3d), there results

$$\frac{n}{s} + \frac{n'}{s'} = \frac{n}{f} \qquad \text{or} \qquad \frac{n}{s} + \frac{n'}{s'} = \frac{n'}{f'} \tag{3f}$$

Both these equations give the conjugate distances for a single spherical surface.

3.5 CONVENTION OF SIGNS

The following set of sign conventions will be adhered to throughout the following chapters on geometrical optics, and it would be well to have them firmly in mind:

1 All figures are drawn with the light traveling from left to right.
2 All object distances (s) are considered positive when they are measured to the left of the vertex and negative when they are measured to the right.
3 All image distances (s') are positive when they are measured to the right of the vertex and negative when to the left.
4 Both focal lengths are positive for a converging system and negative for a diverging system.
5 Object and image dimensions are positive when measured upward from the axis and negative when measured downward.
6 All convex surfaces are taken as having a positive radius, and all concave surfaces are taken as having a negative radius.

EXAMPLE 2 A concave surface with a radius of 4 cm separates two media of refractive index $n = 1.00$ and $n' = 1.50$. An object is located in the first medium at a distance of 10 cm from the vertex. Find (*a*) the primary focal length, (*b*) the secondary focal length, and (*c*) the image distance.

SOLUTION The given quantities are $n = 1.0$, $n' = 1.50$, $r = -4.0$ cm, and $s = +10.0$ cm. The unknown quantities are f, f', and s'. (*a*) We use Eq. (3c) directly to obtain

$$\frac{1.0}{f} = \frac{1.5 - 1.0}{-4} \qquad \text{or} \qquad f = \frac{-4.0}{0.5} = -8.0 \text{ cm}$$

(*b*) We use Eq. (3d) directly and obtain

$$\frac{1.5}{f'} = \frac{1.5 - 1.0}{-4} \qquad \text{or} \qquad f' = \frac{-6.0}{0.5} = -12.0 \text{ cm}$$

Note that in this problem both focal lengths are negative and that the ratio f/f' is 1/1.5 as required by Eq. (3a). The minus signs indicate a diverging system similar to Fig. 3E.

(*c*) We use Eq. (3f) and obtain, by direct substitution,

$$\frac{1.0}{10} + \frac{1.5}{s'} = \frac{1.0}{-8.0} \qquad \text{giving} \qquad s' = -6.66 \text{ cm}$$

The image is located 6.66 cm from the vertex A, and the minus sign shows it is to the left of A and therefore virtual, as shown in Fig. 3E.

3.6 GRAPHICAL CONSTRUCTIONS. THE PARALLEL-RAY METHOD

It would be well to point out here that although the above formulas hold for all possible object and image distances, they apply only to images formed by paraxial rays. For such rays the refraction occurs at or very near the vertex of the spherical

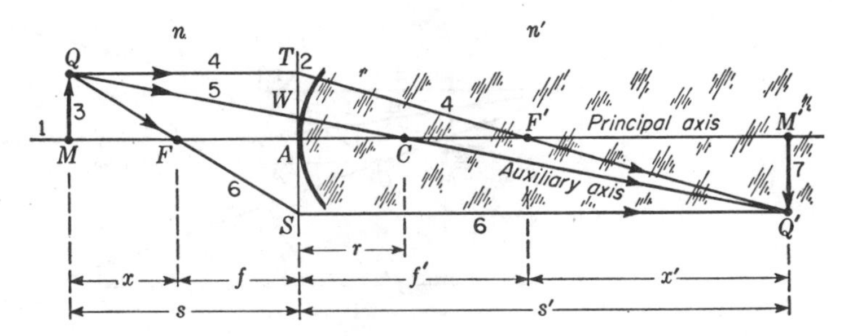

FIGURE 3F
Parallel-ray method for graphically locating the image formed by a single spherical surface.

surface, so that the correct geometrical relations are obtained in graphical solutions by drawing all rays as though they were refracted at the plane through the vertex A and normal to the axis.

The parallel-ray method of construction is illustrated in Figs. 3F and 3G for convex and concave surfaces, respectively. Consider the light emitted from the highest point Q of the object in Fig. 3F. Of the rays emanating from this point in different directions the one (QT) traveling parallel to the axis will by definition of the focal point be refracted to pass through F'. The ray QC passing through the center of curvature is undeviated because it crosses the boundary perpendicular to the surface.

These two rays are sufficient to locate the tip of the image at Q', and the rest of the image lies in the conjugate plane through this point. All other paraxial rays from Q, refracted by the surface, will also be brought to a focus Q'. As a check we note that the ray QS, which passes through the point F, will (by definition of the primary focal point) be refracted parallel to the axis and will cross the others at Q' as shown in the figure.

This method is called the *parallel-ray method.* The numbers 1, 2, 3, . . . indicate the order in which the lines are customarily drawn.

When the method just described is applied to a diverging system, as shown in Fig. 3G, similar procedures are carried out. Ray QT, drawn parallel to the axis, is refracted as if it came from F'. Ray QS, directed toward F, is refracted parallel to the axis. Finally ray QW, passing through C, goes on undeviated. Extending all these refracted rays back to the left finds them intersecting at the common point Q'. $Q'M'$ is therefore the image of the object QM. Note that $Q'M'$ is not a real image since it cannot be formed on a screen.

In both these figures the medium to the right of the spherical surface has the greater index; i.e., we have made $n' > n$. If in Fig. 3F the medium on the left were to have the greater index, so that $n' < n$, the surface would have a diverging effect and each of the focal points F and F' would lie on the opposite side of the vertex from that shown, just as they do in Fig. 3G. Similarly, if we made $n' < n$ in Fig. 3G, the surface would have a converging effect and the focal points would lie as they do in Fig. 3F.

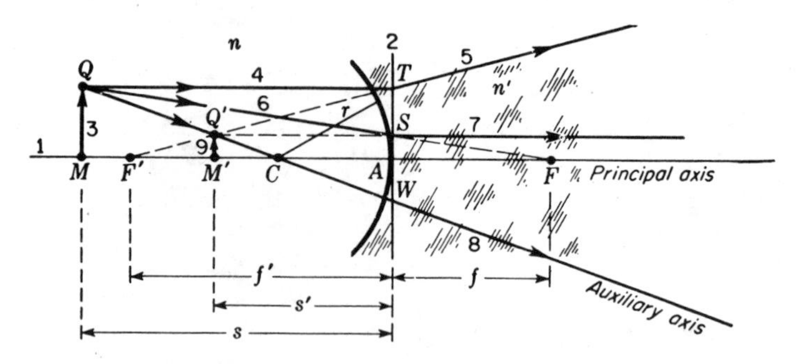

FIGURE 3G
The parallel-ray method applied to a concave spherical surface having diverging properties.

Since any ray through the center of curvature is undeviated and has all the properties of the principal axis, it may be called an *auxiliary axis*.

3.7 OBLIQUE-RAY METHODS

Method 1 In more complicated optical systems that are treated in the following chapters it is convenient to be able graphically to trace a ray across a spherical boundary for any given angle of incidence. The oblique-ray methods permit this to be done with considerable ease. In these constructions one is free to choose any two rays coming from a common object point and, after tracing them through the system, find where they finally intersect. This intersection is then the image point.

Let *MT* in Fig. 3H represent any ray incident on the surface from the left. Through the center of curvature *C* a dashed line *RC* is drawn, parallel to *MT*, and extended to the point where it crosses the secondary focal plane. The line *TX* is then drawn as the refracted ray and extended to the point where it crosses the axis at *M'*. Since the axis may here be considered as a second ray of light, *M* represents an axial object point and *M'* its conjugate image point.

The principle involved in this construction is the following. If *MT* and *RA* were parallel incident rays of light, they would (after refraction and by the definition of focal planes) intersect the secondary focal plane *WF'* at *X*. Since *RA* is directed toward *C*, the refracted ray *ACX* remains undeviated from its original direction.

Method 2 This method is shown in Fig. 3I. After drawing the axis *MM'* and the arc representing the spherical surface with a center *C*, any line such as 1 is drawn to represent any oblique ray of light. Next, an auxiliary diagram is started by drawing *XZ* parallel to the axis. With an origin at *O*, line intervals *OK* and *OL* are laid off proportional to *n* and *n'*, respectively, and perpendiculars are drawn through *K*, *L*, and *A*. From here the construction proceeds in the order of the numbers 1, 2, 3, 4, 5, and 6. Line 2 is drawn through *O* parallel to line 1, line 4 is drawn through *J* parallel to line 3, and line 6 is drawn through *T* parallel to line 5.

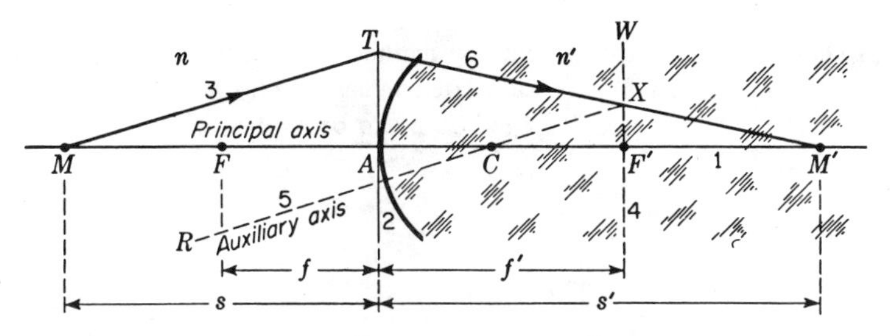

FIGURE 3H
The oblique-ray method for graphically locating images formed by a single spherical surface.

A proof for this construction is readily obtained by writing down proportionalities from three pairs of similar triangles in the two diagrams. These proportionalities are

$$\frac{h}{s} = \frac{i}{n} \qquad \frac{h}{s'} = \frac{j}{n'} \qquad \frac{h}{r} = \frac{i + j}{n' - n}$$

We now transpose n and n' to the left in all three equations.

$$\frac{hn}{s} = i \qquad \frac{hn'}{s'} = j \qquad \frac{h(n' - n)}{r} = i + j$$

We finally add the first two equations and for the right-hand side substitute the third equality:

$$\frac{hn}{s} + \frac{hn'}{s'} = i + j \qquad \text{and} \qquad \frac{n}{s} + \frac{n'}{s'} = \frac{n' - n}{r}$$

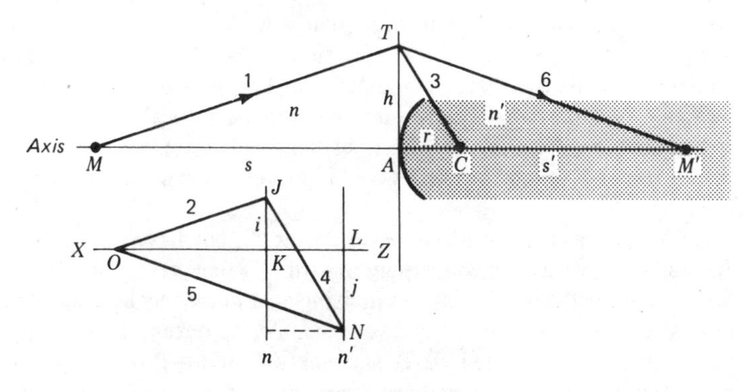

FIGURE 3I
The auxiliary-diagram method for graphically locating images formed by paraxial rays.

It should be noted that to employ method 1 the secondary focal length f' must be known, or it must first be calculated from the known radius of curvature and the refractive indices n and n'. Method 2 can be applied without knowing either of the focal lengths.

3.8 MAGNIFICATION

In any optical system the ratio between the transverse dimension of the final image and the corresponding dimension of the original object is called the *lateral magnification*. To determine the relative size of the image formed by a single spherical surface, reference is made to the geometry of Fig. 3F. Here the undeviated ray 5 forms two similar right triangles QMC and $Q'M'C$.

The theorem of the proportionality of corresponding sides requires that

$$\frac{M'Q'}{MQ} = \frac{CM'}{CM} \qquad \text{or} \qquad \frac{-y'}{y} = \frac{s' - r}{s + r}$$

We now define y'/y as the lateral magnification m and obtain

$$m = \frac{y'}{y} = -\frac{s' - r}{s + r} \tag{3g}$$

If m is positive, the image will be virtual and erect, while if it is negative, the image is real and inverted.

3.9 REDUCED VERGENCE

In the formulas for a single spherical refracting surface, Eqs. (3b) to (3f), the distances s, s', r, f, and f' appear in the denominators. The reciprocals $1/s$, $1/s'$, $1/r$, $1/f$, and $1/f'$ actually represent curvatures of which s, s', r, f, and f' are the radii.

Reference to Fig. 3J will show that if we think of M in the left-hand diagram as a point source of waves, their refraction by the spherical boundary causes them to converge toward the image point M'. In the right-hand diagram plane waves are refracted so as to converge toward the secondary focal point F'. Note that these curved lines representing the crests of light waves are everywhere perpendicular to the corresponding light rays that could have been drawn from object point to image point.

As the waves from M strike the vertex A, they have a radius s and a curvature $1/s$, and as they leave A, converging toward M', they have a radius s' and a curvature $1/s'$. Similarly the incident waves arriving at A in the second diagram have an infinite radius ∞ and a curvature of $1/\infty$, or zero. At the vertex where they leave the surface, the radius of the refracted waves is equal to f' and their curvature is equal to $1/f'$.

The gaussian formulas may therefore be considered as involving the addition and subtraction of quantities proportional to the curvatures of spherical surfaces. When these curvatures rather than radii are used, the formulas become simpler in

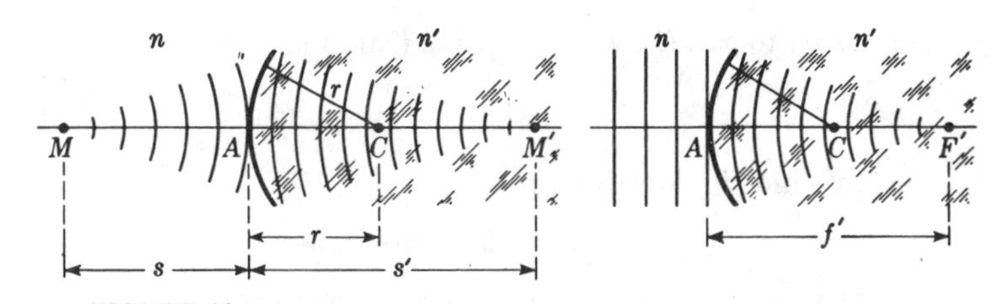

FIGURE 3J
The refraction of light waves at a single spherical surface.

form and for some purposes more convenient. We therefore introduce at this point the quantities

$$V = \frac{n}{s} \qquad V' = \frac{n'}{s'} \qquad K = \frac{1}{r} \qquad P = \frac{n}{f} \qquad P = \frac{n'}{f'} \tag{3h}$$

The first two, V and V', are called *reduced vergences* because they are direct measures of the convergence and divergence of the object and image wave fronts, respectively. For a divergent wave from the object s is positive, and so is the vergence V. For a convergent wave, on the other hand, s is negative, and so is its vergence. For a converging wave front toward the image, V' is positive, and for a diverging wave front, V' is negative. Note that in each case the refractive index involved is that of the medium in which the wave front is located.

The third quantity K is the curvature of the refracting surface (reciprocal of its radius), while the fourth and fifth quantities are, according to Eq. (3e), equal and define the refracting *power*. *When all distances are measured in meters, the reduced vergences V and V', the curvature K, and the power P are in units called diopters.* We may think of V as the power of the object wave front just as it touches the refracting surface and V' as the power of the corresponding image wave front which is tangent to the refracting surface. In these new terms, Eq. (3b) becomes

$$V + V' = P \tag{3i}$$

where

$$P = \frac{n' - n}{r} \qquad \text{or} \qquad P = (n' - n)K \tag{3j}$$

EXAMPLE 3 One end of a glass rod of refractive index 1.50 is ground and polished with a convex spherical surface of radius 10 cm. An object is placed in the air on the axis 40 cm to the left of the vertex. Find (*a*) the power of the surface and (*b*) the position of the image.

SOLUTION The given quantities are $n = 1.0$, $n' = 1.50$, $r = +10.0$ cm, and $s = +40.0$ cm. The unknown quantities are P and s'. To find the solution to (*a*), we make use of Eq. (3j), substitute the given distance in meters, and obtain

$$P = \frac{1.50 - 1.00}{0.10} = +5.0 \text{ D}$$

For the answer to part (*b*), we first use Eq. (3h) to find the vergence V.

$$V = \frac{1.00}{0.40} = +2.5 \text{ D}$$

Direct substitution in Eq. (3i) gives

$$2.5 + V' = 5 \qquad \text{from which} \qquad V' = +2.5 \text{ D}$$

To find the image distance, we have $V' = n'/s'$, so that

$$s' = \frac{n'}{V'} = \frac{1.50}{2.5} = +0.60 \text{ m} = +60 \text{ cm}$$

This answer should be verified by the student, using one of the graphical methods of construction drawn to a convenient scale.

3.10 DERIVATION OF THE GAUSSIAN FORMULA

The basic equation (3b) is of sufficient importance to warrant its derivation in some detail. While there are many ways of performing a derivation, a method involving oblique rays will be given here. In Fig. 3K an oblique ray from an axial object point M is shown incident on the surface at an angle ϕ and refracted at an angle ϕ'. The refracted ray crosses the axis at the image point M'. If the incident and refracted rays MT and TM' are paraxial, the angles ϕ and ϕ' will be small enough to permit putting the sines of the two angles equal to the angles themselves; for Snell's law we write

$$\frac{\phi}{\phi'} = \frac{n'}{n} \tag{3k}$$

Since ϕ is an exterior angle of the triangle MTC and equals the sum of the opposite interior angles,

$$\phi = \alpha + \beta \tag{3l}$$

Similarly β is an exterior angle of the triangle TCM', so that $\beta = \phi' + \gamma$ and

$$\phi' = \beta - \gamma \tag{3m}$$

Substituting these values of ϕ and ϕ' in Eq. (3k) and multiplying out, we obtain

$$n'\beta - n'\gamma = n\alpha + n\beta \qquad \text{or} \qquad n\alpha + n'\gamma = (n' - n)\beta$$

For paraxial rays α, β, and γ are very small angles, and we may set $\alpha = h/s$, $\beta = h/r$, and $\gamma = h/s'$. Substituting these values in the last equation, we obtain

$$n\frac{h}{s} + n'\frac{h}{s'} = (n' - n)\frac{h}{r}$$

By canceling h throughout we obtain the desired equation,

$$\frac{n}{s} + \frac{n'}{s'} = \frac{n' - n}{r} \tag{3n}$$

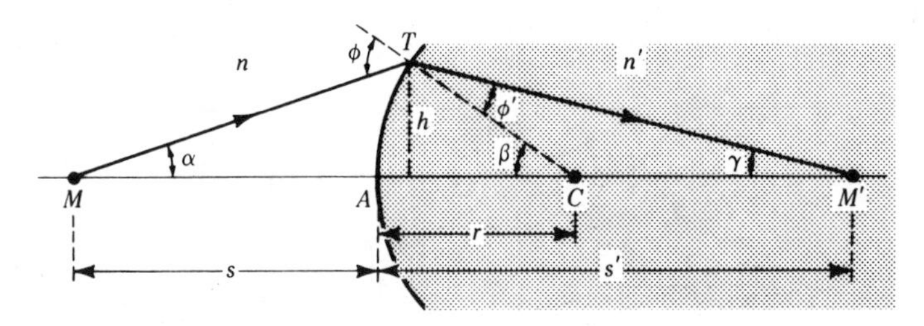

FIGURE 3K
Geometry for the derivation of the paraxial formula used in locating images.

3.11 NOMOGRAPHY

The term *nomograph* is derived from the greek words *nomos*, meaning *law*, and *graphein*, meaning *to write*. In physics the term applies to certain graphical representations of physical laws, which are intended to simplify or speed up calculations. Figure 3L is a nomograph relating *object* and *image distances* as given by Eq. (3f), namely,

$$\frac{n}{s} + \frac{n'}{s'} = \frac{n}{f} \tag{3o}$$

Its simplicity and usefulness become apparent when it is seen that any straight line drawn across the figure will intersect the three lines at values related by the above equation.

EXAMPLE 4 One end of a plastic rod of index 1.5 is ground and polished to a radius of +2.0 cm. If an object in air is located on the axis 12.0 cm from the vertex, what is the image distance?

SOLUTION The given quantities are $n = 1.0$, $n' = 1.50$, $r = +2.0$ cm, and $s = +12.0$ cm. The unknown is s'. By direct substitution and the use of Eq. (3o) we obtain

$$\frac{s}{n} = \frac{12}{1} = +12.0 \qquad \text{and} \qquad \frac{f}{n} = \frac{r}{n' - n} = \frac{2}{1.5 - 1} = +4.0$$

If the straight edge of a ruler is now placed on $s/n = +12.0$ and $f/n = +4.0$, it will intersect the third line at $s'/n' = +6.0$. Since $n' = 1.5$, s' is equal to 6×1.5, or +9.0 cm.

A little study of this nomograph will show that it applies to all object and image distances, real or virtual, and to all surfaces with positive or negative radii of curvature. Furthermore we shall find in Chap. 4 that it can be applied to all thin lenses by setting n and n' equal to unity. For thin lenses the three axes represent s, s', and f directly, and no calculations are necessary.

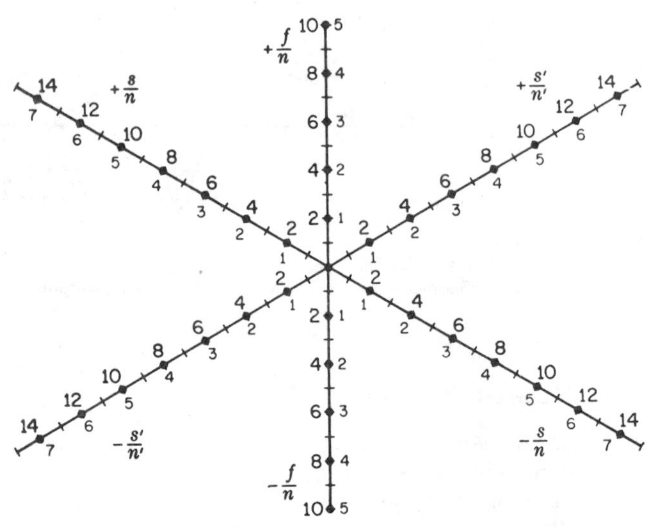

FIGURE 3L
Nomograph for determining the object or image distance for a single spherical surface or for a thin lens.

PROBLEMS

3.1 The left end of a long glass rod of index 1.6350 is ground and polished to a convex spherical surface of radius 2.50 cm. A small object is located in the air and on the axis 9.0 cm from the vertex. Find (*a*) the primary and secondary focal lengths, (*b*) the power of the surface, (*c*) the image distance, and (*d*) the lateral magnification.

Ans. (*a*) $+3.937$ and $+6.43$ cm, (*b*) $+25.40$ D, (*c*) $+11.44$ cm, (*d*) -0.777

3.2 Solve Prob. 3.1 graphically. (*a*) Find the image distance by the oblique-ray method 1. (*b*) Find the relative size of the image by the parallel-ray method.

3.3 The left end of a long plastic rod of index 1.530 is ground and polished to a convex spherical surface of radius 2.650 cm. An object 2.50 cm high is located in the air and on the axis 16.0 cm from the vertex. Find (*a*) the primary and secondary focal lengths, (*b*) the power of the surface, (*c*) the image distance, and (*d*) the size of the image.

3.4 Solve Prob. 3.3 graphically. (*a*) Find the image distance by the oblique-ray method 1. (*b*) Find the size of the image by the parallel-ray method.

3.5 The left end of a water trough has a transparent surface of radius -2.0 cm. A small object 2.5 cm high is located in the air and on the axis 10.0 cm from the vertex. Find (*a*) the primary and secondary focal lengths, (*b*) the power of the surface, (*c*) the image distance, and (*d*) the size of the image. Assume water to have an index 1.3330.

Ans. (*a*) -6.01 and -8.01 cm, (*b*) -16.65 D, (*c*) -5.0 cm, (*d*) $+0.938$ cm

3.6 Solve Prob. 3.5 graphically. (*a*) Find the image distance by the oblique-ray method 1. (*b*) Find the size of the image by the parallel-ray method.

3.7 The left end of a long plastic rod of index 1.480 is ground and polished to a spherical surface of radius -2.60 cm. An object 2.50 cm high is located in the air and on the axis 12.0 cm′ from the vertex. Find (*a*) the primary and secondary focal lengths, (*b*) the power of the surface, (*c*) the image distance, and (*d*) the size of the image.

3.8 Solve Prob. 3.7 graphically. (*a*) Find the image distance by the oblique-ray method 1. (*b*) Find the size of the image by the parallel-ray method.

3.9 The left end of a long glass rod of index 1.620 is polished to a convex surface of radius +1.20 cm and then submerged in water of index 1.3330. A small object 2.50 cm high is located in the water 10.0 cm in front of the vertex. Calculate (*a*) the primary and secondary focal lengths, (*b*) the power of the surface, (*c*) the image distance, and (*d*) the size of the image.

Ans. (*a*) +5.57 and +6.77 cm, (*b*) 23.91 D, (*c*) +15.31 cm, (*d*) −3.150 cm

3.10 Solve Prob. 3.9 graphically. (*a*) Find the image distance by the oblique-ray method 2. (*b*) Find the size of the image by the parallel-ray method.

3.11 A glass rod 2.50 cm long and of index 1.70 has both ends polished to spherical surfaces with radii $r_1 = +2.80$ cm and $r_2 = -2.80$ cm. An object 2.0 cm high is located on the axis 8.0 cm from the first vertex. Find (*a*) the primary and secondary focal lengths for each of the surfaces, (*b*) the image distance for the first surface, (*c*) the object distance for the second surface, and (*d*) the final image distance from the second vertex.

3.12 Solve Prob. 3.11 graphically after calculating the answer to part (*a*).

3.13 A parallel beam of light enters a clear plastic bead 2.50 cm in diameter and index 1.440. At what point beyond the bead are these rays brought to a focus?

Ans. 0.795 cm

3.14 Solve Prob. 3.13 graphically by the method illustrated in Fig. 3I.

3.15 A clear crystal bead of index 1.720, and radius 1.50 cm is submerged in a clear liquid of index 1.360. If a parallel beam of light in the liquid is allowed to enter the bead, at what point beyond the other side will the light be brought to a focus?

3.16 Solve Prob. 3.15 graphically by the method illustrated in Fig. 3I.

3.17 A hollow glass cell is made of thin glass in the form of an equiconcave lens. The radii of the two surfaces are 1.650 cm, and the distance between the two vertices is 1.850 cm. When sealed airtight, this cell is submerged in water of index 1.3330. Calculate (*a*) the focal lengths of each surface and (*b*) the power of each surface.

Ans. (*a*) $f_1 = +6.60$ cm, $f_1' = +4.95$ cm, $f_2 = +4.95$ cm, and $f_2' = +6.60$ cm, (*b*) $P_1 = +20.18$ D, and $P_2 = +20.18$ D

3.18 A spherical surface with a radius of +2.650 cm is polished on the end of a glass rod of index 1.560. Find its power when placed in (*a*) air, (*b*) water of index 1.3330, (*c*) oil of index 1.480, and (*d*) an organic liquid of index 1.780.

4

THIN LENSES

Diagrams of several standard forms of thin lenses were shown in Fig. 3A as illustrations of the fact that most lenses have surfaces that are spherical in form. Some surfaces are convex, others are concave, and still others are plane. When light passes through any lens, refraction at each of its surfaces contributes to its image-forming properties, according to the principles put forward in Chap. 3. Not only does each individual surface have its own primary and secondary focal points and planes, but the lens as a whole has its own pair of focal points and focal planes.

A thin lens may be defined as one whose thickness is considered small in comparison with the distances generally associated with its optical properties. Such distances are, for example, radii of curvature of the two spherical surfaces, primary and secondary focal lengths, and object and image distances.

4.1 FOCAL POINTS AND FOCAL LENGTHS

Diagrams showing the refraction of light by an equiconvex lens and by an equiconcave lens are given in Fig. 4A. The axis in each case is a straight line through the geometrical center of the lens and perpendicular to the two faces at the points of intersection. For spherical lenses this line joins the centers of curvature of the two

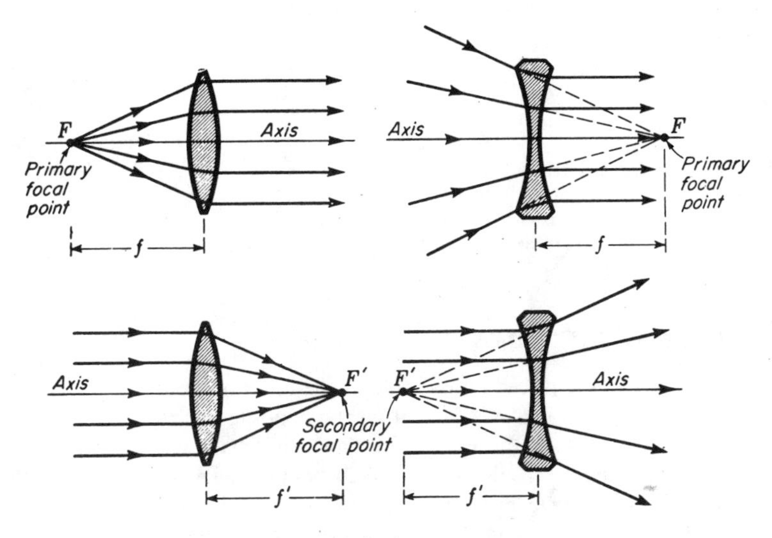

FIGURE 4A
Ray diagrams illustrating the primary and secondary focal points F and F' and the corresponding focal lengths f and f' of thin lenses.

surfaces. *The primary focal point F is an axial point having the property that any ray coming from it or proceeding toward it travels parallel to the axis after refraction.*

Every thin lens in air has two focal points, one on each side of the lens and equidistant from the center. This can be seen by symmetry in the cases of equiconvex and equiconcave lenses, but it is also true for other forms provided the lenses may be regarded as thin. *The secondary focal point F' is an axial point having the property that any incident ray traveling parallel to the axis will, after refraction, proceed toward, or appear to come from, F'.* The two lower diagrams in Fig. 4A are given for the purpose of illustrating this definition. In analogy to the case of a single spherical surface (see Chap. 3), a plane perpendicular to the axis and passing through a focal point is called a *focal plane*. The significance of the focal plane is illustrated for a converging lens in Fig. 4B. Parallel incident rays making an angle θ with the axis are brought to a focus at a point Q' in line with the chief ray. The chief ray in this case is defined as the ray which passes through the center of the lens.

The distance between the center of a lens and either of its focal points is its focal length. These distances, designated f and f', usually measured in centimeters or inches, have a positive sign for converging lenses and a negative sign for diverging lenses. It should be noted in Fig. 4A that the primary focal point F for a converging lens lies to the left of the lens, whereas for a diverging lens it lies to the right. For a lens with the same medium on both sides, we have, by the reversibility of light rays,

$$f = f'$$

Note carefully the difference between a thin lens in air, where the focal lengths are equal, and a single spherical surface, where the two focal lengths have the ratio of the two refractive indices [see Eq. (3a)].

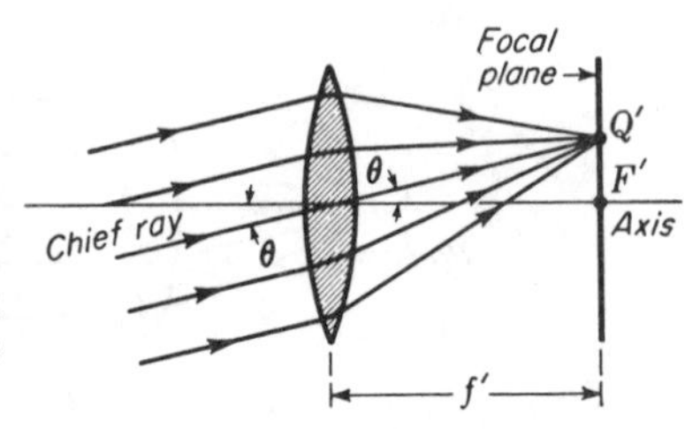

FIGURE 4B
How parallel incident rays are brought to a focus at the secondary focal plane of a thin lens.

4.2 IMAGE FORMATION

When an object is placed on one side or the other of a converging lens and beyond the focal plane, an image is formed on the opposite side (see Fig. 4C). If the object is moved closer to the primary focal plane, the image will be formed farther away from the secondary focal plane and will be larger, i.e., magnified. If the object is moved farther away from F, the image will be formed closer to F' and will be smaller.

In Fig. 4C all the rays coming from an object point Q are shown as brought to a focus Q', and the rays from another point M are brought to a focus at M'. Such ideal conditions and the formulas given in this chapter hold only for paraxial rays, i.e., rays close to lens axis and making small angles with it.

4.3 CONJUGATE POINTS AND PLANES

If the principle of the reversibility of light rays is applied to Fig. 4C, we observe that $Q'M'$ becomes the object and QM becomes its image. The object and image are therefore *conjugate*, just as they are for a single spherical surface (see Sec. 3.4). Any pair of object and image points such as M and M' in Fig. 4C are called *conjugate points*, and planes through these points perpendicular to the axis are called *conjugate planes*.

If we know the focal length of a thin lens and the position of an object, there are three methods of determining the position of the image: (1) graphical construction, (2) experiment, and (3) use of the lens formula

$$\bullet \qquad \frac{1}{s} + \frac{1}{s'} = \frac{1}{f} \qquad (4a)$$

Here s is the object distance, s' is the image distance, and f is the focal length, all measured to or from the center of the lens. This lens equation will be derived in Sec. 4.14. We now consider the graphical methods.

4.4 THE PARALLEL-RAY METHOD

The parallel-ray method is illustrated in Fig. 4D. Consider the light emitted from the extreme point Q on the object. Of the rays emanating from this point in different directions the one (QT) traveling parallel to the axis will by definition of the focal point

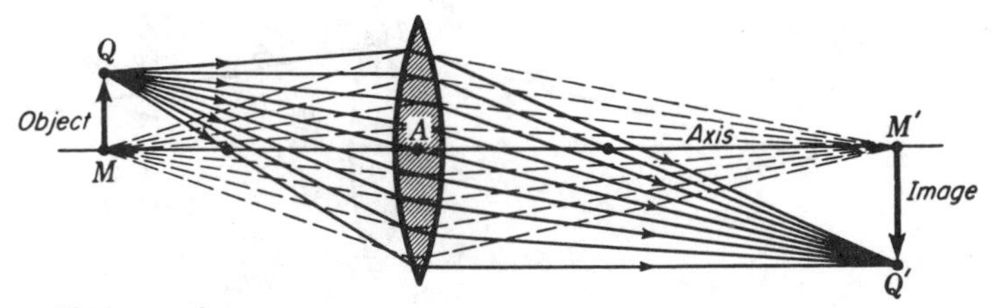

FIGURE 4C
Image formation by an ideal thin lens. All rays from an object point Q which pass through the lens are refracted to pass through the image point Q'.

be refracted to pass through F'. The ray QA, which goes through the lens center where the faces are parallel, is undeviated and meets the other ray at some point Q'. These two rays are sufficient to locate the tip of the image at Q', and the rest of the image lies in the conjugate plane through this point. All other rays from Q will also be brought to a focus at Q'. As a check, we note that the ray QF which passes through the primary focal point will by definition of F be refracted parallel to the axis and will cross the others at Q' as shown in the figure. The numbers 1, 2, 3, etc., in Fig. 4D indicate the order in which the lines are customarily drawn.

4.5 THE OBLIQUE-RAY METHOD

Let MT in Fig. 4E represent any ray incident on the lens from the left. It is refracted in the direction TX and crosses the axis at M'. The point X is located at the intersection between the secondary focal plane $F'W$ and the dashed line RR' drawn through the center of the lens parallel to MT.

The order in which each step of the construction is made is again indicated by the numbers 1, 2, 3, ... The principle involved in this method may be understood by reference to Fig. 4B. Parallel rays incident on the lens are always brought to a

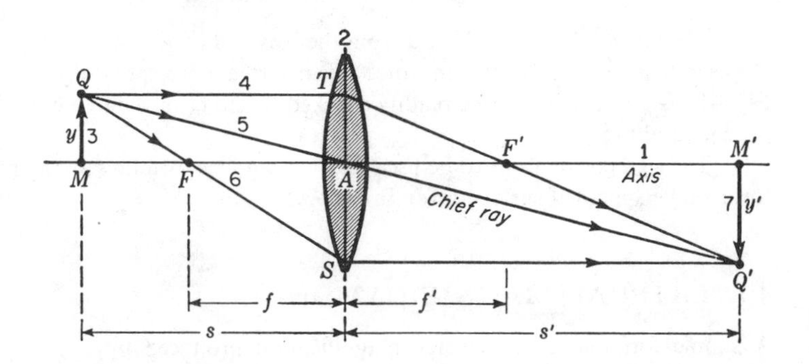

FIGURE 4D
The parallel-ray method for graphically locating the image formed by a thin lens.

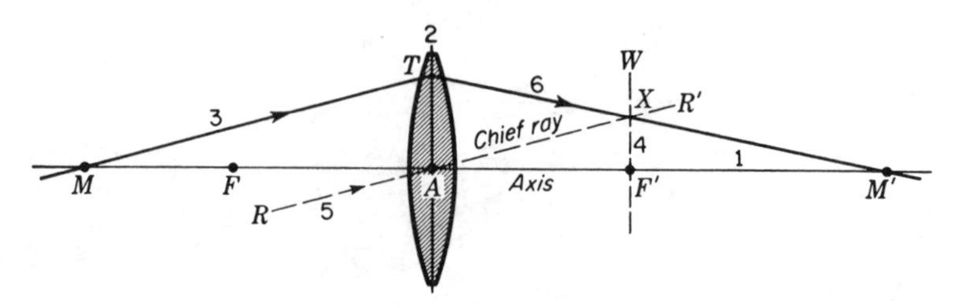

FIGURE 4E
The oblique-ray method for graphically locating the image formed by a thin lens.

focus at the focal plane, the ray through the center being the only one undeviated. Therefore, if we actually have rays diverging from M, as in Fig. 4E, we can find the direction of any one of them after it passes through the lens by making it intersect the parallel line RR' through A in the focal plane. This construction locates X and the position of the image M'. Note that RR' is not an actual ray in this case and is treated as such only as a means of locating the point X.

4.6 USE OF THE LENS FORMULA

To illustrate the application of Eq. (4a) to find the image position, we select an example in which all quantities occurring in the equation have a positive sign. Let an object be located 6.0 cm in front of a positive lens of focal length +4.0 cm. The given quantities are $s = +6.0$ cm and $f = +4.0$ cm, and the unknown is s'. As a first step we transpose Eq. (4a) by solving for s':

$$s' = \frac{s \times f}{s - f} \tag{4b}$$

From direct substitution of the given quantities in this equation we have

$$s' = \frac{(+6) \times (+4)}{(+6) - (+4)} = +12.0 \text{ cm}$$

The image is formed 12.0 cm from the lens and is *real*, as it will always be when s' has a positive sign. In this instance it is inverted, corresponding to the diagram in Fig. 4C. These results can be readily checked by either of the two graphical methods presented above.

The sign conventions to be used for the thin-lens formulas are identical to those for a single spherical surface given in Sec. 3.5.

4.7 LATERAL MAGNIFICATION

A simple formula for the image magnification produced by a single lens can be derived from the geometry of Fig. 4D. By construction it is seen that the right triangles QMA and $Q'M'A$ are similar. Corresponding sides are therefore proportional to each other, so that

$$\frac{M'Q'}{MQ} = \frac{AM'}{AM}$$

where AM' is the image distance s' and AM is the object distance s. Taking upward directions as positive, $y = MQ$, and $y' = -M'Q'$; so we have by direct substitution $y'/y = -s'/s$. The lateral magnification is therefore

$$m = \frac{y'}{y} = -\frac{s'}{s} \tag{4c}$$

When s and s' are both positive, as in Fig. 4D, the negative sign of the magnification signifies an inverted image.

4.8 VIRTUAL IMAGES

The images formed by the converging lenses in Figs. 4C and 4D are real in that they can be made visible on a screen. They are characterized by the fact that rays of light are actually brought to a focus in the plane of the image. A virtual image cannot be formed on a screen (see Sec. 3.3). The rays from a given point on the object do not actually come together at the corresponding point in the image; instead they must be projected backward to find this point. Virtual images are produced with converging lenses when the object is placed between the focal point and the lens and with diverging lenses when the object is in any position. Examples are shown in Figs. 4F and 4G.

Figure 4F shows the parallel-ray construction for a positive lens used as a magnifier, or reading glass. Rays emanating from Q are refracted by the lens but are not sufficiently deviated to come to a real focus. To the observer's eye at E these rays appear to be coming from a point Q' on the far side of the lens. This point represents a virtual image, because the rays do not actually pass through Q'; they only appear to come from there. Here the image is right side up and magnified. In the construction of this figure, ray QT parallel to the axis is refracted through F', while ray QA through the center of the lens is undeviated. These two rays when extended backward intersect at Q'. The third ray QS, traveling outward as though it came from F, actually misses the lens, but if the latter were larger, the ray would be refracted parallel to the axis, as shown. When projected backward, it also intersects the other projections at Q'.

EXAMPLE 1 If an object is located 6.0 cm in front of a lens of focal length +10.0 cm, where will the image be formed?

SOLUTION The given quantities are $s = +6.0$ cm, and $f = +10.0$ cm, while the unknown quantities are s' and m. By making direct substitutions in Eq. (4b) we obtain

$$s' = \frac{(+6) \times (+10)}{(+6) - (+10)} = \frac{+60}{-4} = -15.0 \text{ cm}$$

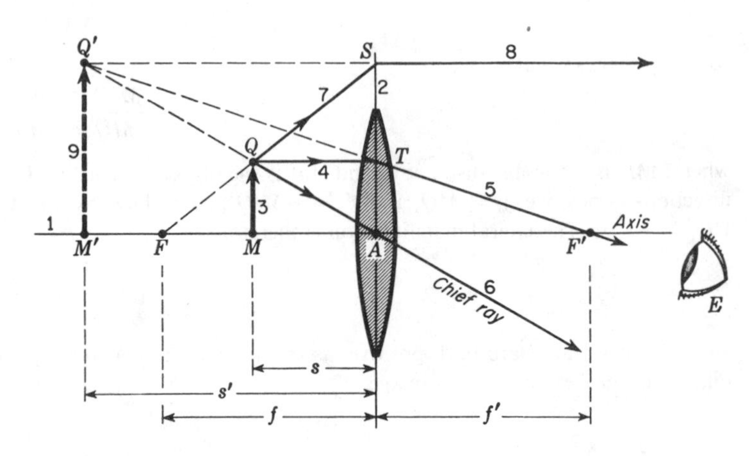

FIGURE 4F
The parallel-ray method for graphically locating the virtual image formed by a positive lens when the object is between the primary focal point and the lens.

The minus sign indicates that the image lies to the left of the lens. Such an image is always virtual. The magnification is obtained by the use of Eq. (4c),

$$m = -\frac{s'}{s} = -\frac{-15}{+6} = +2.50\times$$

The positive sign means that the image is erect.

With the negative lens shown in Fig. 4G the image is virtual for all positions of the object, is always smaller than the object, and lies closer to the lens than the object. As is seen from the diagram, rays diverging from the object point Q are made more divergent by the lens. To the observer's eye at E these rays appear to be coming from the point Q' on the far side of but close to the lens. In applying the lens formula to a diverging lens it must be remembered that the focal length f is negative.

EXAMPLE 2 An object is placed 12.0 cm in front of a diverging lens of focal length 6.0 cm. Find the image.

SOLUTION The given quantities are $s = +12.0$ cm and $f = -6.0$ cm, while the unknown quantities are s' and m. We substitute directly in Eq. (4b), to obtain

$$s' = \frac{(+12) \times (-6)}{(+12) - (-6)} = \frac{-72}{+18}$$

from which $s' = -4.0$ cm. For the image size Eq. (4c) gives

$$m = -\frac{s'}{s} = -\frac{-4}{12} = +\tfrac{1}{3}\times$$

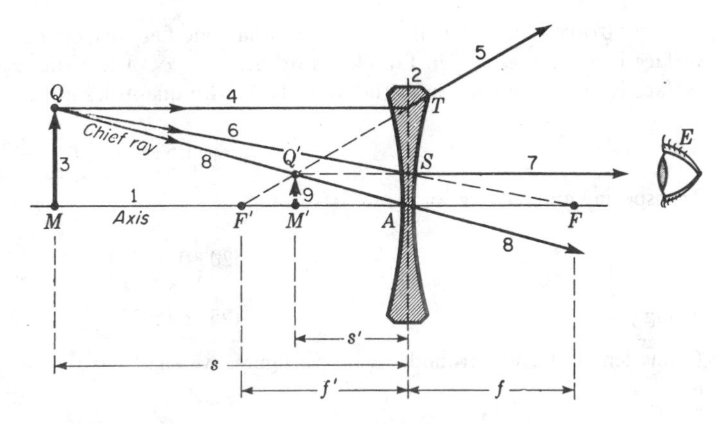

FIGURE 4G
The parallel-ray method for graphically locating the virtual image formed by a negative lens.

The image is therefore to the left of the lens, virtual, erect, and one-third the size of the object.

4.9 LENS MAKERS' FORMULA

If a lens is to be ground to some specified focal length, the refractive index of the glass must be known. It is customary for manufacturers of optical glass to specify the refractive index for yellow sodium light, the D line. Supposing the index to be known, the radii of curvature must be so chosen as to satisfy the equation

$$\frac{1}{f} = (n - 1)\left(\frac{1}{r_1} - \frac{1}{r_2}\right) \tag{4d}$$

As the rays travel from left to right through a lens, *all convex surfaces are taken as having a positive radius and all concave surfaces a negative radius.* For an equiconvex lens like the one in Fig. 3A(*a*), r_1 for the first surface is positive and r_2 for the second surface negative. Substituting the value of $1/f$ from Eq. (4a), we write

$$\frac{1}{s} + \frac{1}{s'} = (n - 1)\left(\frac{1}{r_1} - \frac{1}{r_2}\right) \tag{4e}$$

EXAMPLE 3 A plano-convex lens having a focal length of 25.0 cm [Fig. 3A(*b*)] is to be made of glass of refractive index $n = 1.520$. Calculate the radius of curvature of the grinding and polishing tools that must be used to make this lens.

SOLUTION Since a plano-convex lens has one flat surface, the radius for that surface is infinite, and r_1 in Eq. (4d) is replaced by ∞. The radius r_2 of the second surface is the unknown. Substitution of the known quantities in Eq. (4d) gives

$$\frac{1}{25} = (1.520 - 1)\left(\frac{1}{\infty} - \frac{1}{r_2}\right)$$

Transposing and solving for r_2, we have

$$\frac{1}{25} = 0.520\left(0 - \frac{1}{r_2}\right) = -\frac{0.520}{r_2}$$

giving
$$r_2 = -(25 \times 0.520) = -13.0 \text{ cm}$$

If this lens is turned around, as in the figure, we shall have $r_1 = +13.0$ cm and $r_2 = \infty$.

4.10 THIN-LENS COMBINATIONS

The principles of image formation presented in the preceding sections of this chapter are readily extended to optical systems involving two or more thin lenses. Consider, for example, two converging lenses spaced some distance apart as shown in Fig. 4H. Here an object Q_1M_1 is located at a given distance s_1 in front of the first lens, and an image $Q_2'M_2'$ is formed some unknown distance s_2' from the second lens. We first apply the graphical methods to find this image distance and then show how to calculate it by the use of the thin-lens formula.

The first step in applying the graphical method is to disregard the presence of the second lens and find the image produced by the first lens alone. In the diagram the parallel-ray method, as applied to the object point Q_1, locates a real and inverted image at Q_1'. Any two of the three incident rays 3, 5, and 6 are sufficient for this purpose. Once Q_1' is located, we know that all the rays leaving Q_1 will, upon refraction by the first lens, be directed toward Q_1'. Making use of this fact, we construct a fourth ray by drawing line 9 back from Q_1' through A_2 to W. Line 10 is then drawn in connecting W and Q_1.

The second step is to imagine the second lens in place and to make the following changes. Since ray 9 is seen to pass through the center of lens 2, it will emerge without deviation from its previous direction. Since ray 7 between the lenses is parallel to the axis, it will upon refraction by the second lens pass through its secondary focal point F_2'. The intersection of rays 9 and 11 locates the final image point Q_2'. Q_1 and Q_1' are conjugate points for the first lens, Q_2 and Q_2' are conjugate points for the second lens, and Q_1 and Q_2' are conjugate for the combination of lenses. When the image $Q_2'M_2'$ is drawn in, corresponding pairs of conjugate points on the axis are M_1 and M_1', M_2 and M_2', and M_1 and M_2'.

The oblique-ray method given in Fig. 4E is applied to the same two lenses in Fig. 4I. A single ray is traced from the object point M to the final image point M_2'. The lines are drawn in the order indicated. The dotted line 6 is drawn through A_1 parallel to ray 4 to locate the point R_1'. The dotted line 9 is drawn through A_2 parallel to ray 7 to locate R_2'. This construction gives the same conjugate points along the

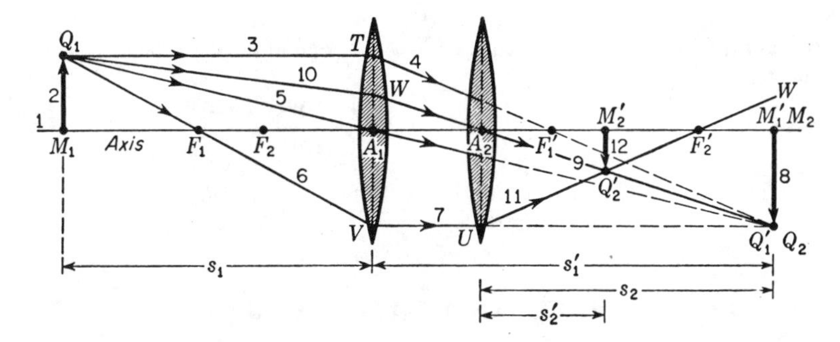

FIGURE 4H
The parallel-ray method for graphically locating the final image formed by two thin lenses.

axis. Note that the axis itself is considered as the second light ray in locating the image point M_2'.

By way of comparison and as a check on the graphical solutions, we can assign specific values to the focal lengths of the lenses and apply the thin-lens formula to find the image. Assume that the two lenses have focal lengths of $+3$ and $+4$ cm, respectively, that they are placed 2 cm apart, and that the object is located 4 cm in front of the first lens.

We begin the solution by applying Eq. (4b) to the first lens alone. The given quantities to be substituted are $s_1 = +4$ cm and $f_1 = +3$ cm:

$$s_1' = \frac{s_1 \times f_1}{s_1 - f_1} = \frac{(+4) \times (+3)}{(+4) - (+3)} = +12 \text{ cm}$$

The image formed by the first lens alone is, therefore, *real* and 12.0 cm to the right of A_1. The image becomes the object for the second lens, and since it is only 10.0 cm from A_2, the object distance s_2 becomes -10.0 cm. The minus sign is necessary and results from the fact that the object distance is measured to the right of the lens. We say that the image produced by the first lens becomes the object for the second lens. Since the rays are converging toward the image of the first lens, the object for the second lens is virtual and its distance therefore has a negative value.

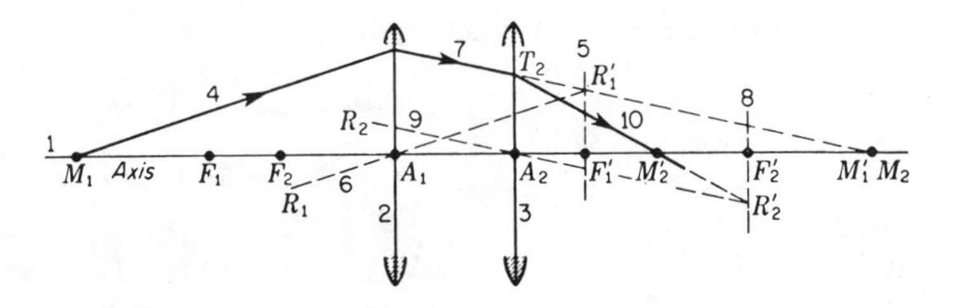

FIGURE 4I
The oblique-ray method for graphically locating the final image formed by two thin lenses.

Applying the lens formula [Eq. (4b)] to the second lens, we have $s_2 = -10.0$ cm and $f_2 = +4.0$ cm:

$$s_2' = \frac{(-10) \times (+4)}{(-10) - (+4)} = +2.86 \text{ cm}$$

The final image is 2.86 cm to the right of lens 2 and is *real*.

4.11 OBJECT SPACE AND IMAGE SPACE

For every position of the object there is a corresponding position for the image. Since the image may be either real or virtual and may lie on either side of the lens, the *image space* extends from infinity in one direction to infinity in the other. But object and image points are conjugate; so the same argument holds for the *object space*. In view of their complete overlapping, one might wonder how it is that the distinction between object and image space is made. This is done by defining everything that pertains to the rays before they have passed through the refracting system as belonging to the object space and everything that pertains to them afterward as belonging to the image space. Referring to Fig. 4H, the object Q_1 and the rays Q_1T, Q_1A_1, and Q_1V are all in the object space for the first lens. Once these rays leave that lens, they are in the image space of the first lens, as is also the image Q_1'. This space is also the object space for the second lens. Once the rays leave the second lens, they and the image point Q_2' are in the image space øf the second lens.

4.12 THE POWER OF A THIN LENS

The concept and measurement of lens power correspond to those used in the treatment of reduced vergence and the power of a single surface as given in Sec. 3.9. The power of a thin lens in diopters is given by the reciprocal of the focal length in meters:

• $$P = \frac{1}{f} \quad \text{diopters} = \frac{1}{\text{focal length, m}} \tag{4f}$$

For example, a lens with a focal length of +50.0 cm has a power of 1/0.50 m = +2 D ($P = +2.0$ D), whereas one of −20.0 cm focal length has a power of 1/0.20 m = −5 D ($P = -5.0$ D), etc. Converging lenses have a positive power, while diverging lenses have a negative power.

By making use of the lens makers' formula [Eq. (4d)] we can write

$$P = (n - 1)\left(\frac{1}{r_1} - \frac{1}{r_2}\right) \tag{4g}$$

where r_1 and r_2 are the two radii, measured in meters, and n is the refractive index of the glass.

EXAMPLE 4 The radii of both surfaces of an equiconvex lens of index 1.60 are equal to 8.0 cm. Find its power.

SOLUTION The given quantities to be used in Eq. (4g) are $n = 1.60$, $r_1 = 0.080$ m, and $r_2 = -0.080$ m (see Fig. 3A for the shape of an equiconvex lens).

$$P = (n - 1)\left(\frac{1}{r_1} - \frac{1}{r_2}\right) = (1.60 - 1)\left(\frac{1}{0.080} - \frac{1}{-0.080}\right) = 0.60\,\frac{2}{0.080} = +15.0\text{ D}$$

Spectacle lenses are made to the nearest quarter of a diopter, thereby reducing the number of grinding and polishing tools required in the optical shops. Furthermore, the sides next to the eyes are always concave to permit free movement of the eyelashes and yet to keep as close and as normal to the axis of the eye as possible. *Note:* It is important to insert a plus or minus sign in front of the number specifying lens power; thus, $P = +3.0$ D, $P = -4.5$ D, etc.

4.13 THIN LENSES IN CONTACT

When two thin lenses are placed in contact, as shown in Fig. 4J, the combination will act as a single lens with two focal points symmetrically located at F and F' on opposite sides. Parallel incoming rays are shown refracted by the first lens toward its secondary focal point F_1'. Further refraction by the second lens brings the rays together at F'. This latter is defined as the secondary focal point of the combination, and its distance from the center is defined as the combination's secondary focal length f'.

If we now apply the simple lens formula (4a) to the rays as they enter and leave the second lens L_2, we note that for the second lens alone f_1' is the object distance (taken with a negative sign), f' is the image distance, and f_2' is the focal length. When Eq. (4a) is applied, these substitutions for s, s', and f, respectively, give

$$\frac{1}{-f_1'} + \frac{1}{f'} = \frac{1}{f_2'} \qquad \text{or} \qquad \frac{1}{f'} = \frac{1}{f_1'} + \frac{1}{f_2'}$$

Since we have assumed that the lenses are in air, the primary focal lengths are all equal to their respective secondary focal lengths and we can drop all primes and write

$$\frac{1}{f} = \frac{1}{f_1} + \frac{1}{f_2} \tag{4h}$$

In words, *the reciprocal of the focal length of a thin-lens combination is equal to the sum of the reciprocals of the focal lengths of the individual lenses.* Since by Eq. (4f) we can write $P_1 = 1/f_1$, $P_2 = 1/f_2$, and $P = 1/f$, we obtain for the power of the combination

● $$P = P_1 + P_2 \tag{4i}$$

In general, *when thin lenses are placed in contact, the power of the combination is given by the sum of the powers of the individual lenses.*

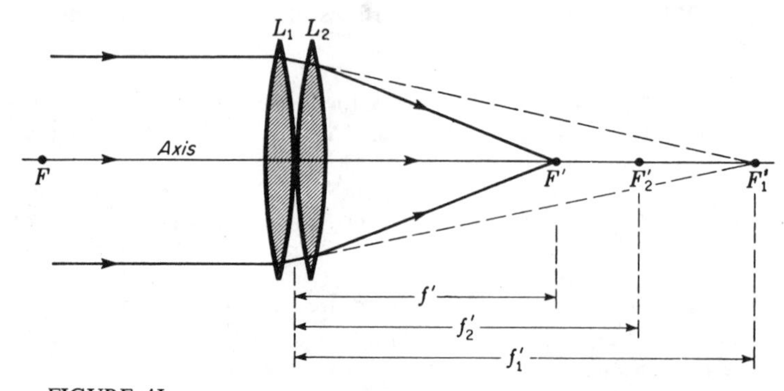

FIGURE 4J
The power of a combination of thin lenses in contact is equal to the sum of the powers of the individual lenses.

4.14 DERIVATION OF THE LENS FORMULA

A derivation of Eq. (4a), the *lens formula*, is readily obtained from the geometry of Fig. 4D. The necessary features of the diagram are repeated in Fig. 4K, which shows only two rays leading from the object of height y to the image of height y'. Let s and s' represent the object and image distances from the lens center and x and x' their respective distances from the focal points F and F'.

From similar triangles $Q'TS$ and $F'TA$ the proportionality between corresponding sides gives

$$\frac{y - y'}{s'} = \frac{y}{f'}$$

Note that $y - y'$ is written instead of $y + y'$ because y', by the convention of signs, is a negative quantity. From the similar triangles QTS and FAS,

$$\frac{y - y'}{s} = \frac{-y'}{f}$$

The sum of these two equations is

$$\frac{y - y'}{s} + \frac{y - y'}{s'} = \frac{y}{f'} - \frac{y'}{f}$$

Since $f = f'$, the two terms on the right can be combined and $y - y'$ canceled out, yielding the desired equation,

$$\frac{1}{s} + \frac{1}{s'} = \frac{1}{f}$$

This is the lens formula in the *gaussian* form*.

* Karl Friedrich Gauss (1777–1855), German astronomer and physicist, was chiefly known for his contributions in the mathematical theory of magnetism. Coming from a poor family, he received support for his education because of his obvious mathematical ability. In 1841 he published the first general treatment of the first-order theory of lenses in his now famous papers, "Dioptrische Untersuchungen."

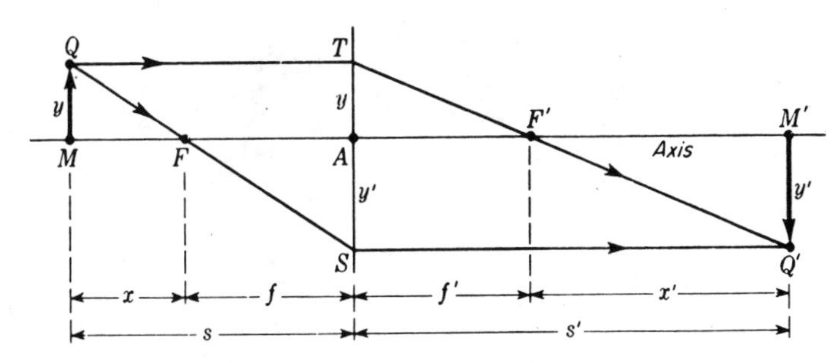

FIGURE 4K
The geometry used for the derivation of thin-lens formulas.

Another form of the lens formula, the *newtonian form*, is obtained in an analogous way from two other sets of similar triangles, QMF and FAS on the one hand and TAF' and $F'M'Q'$ on the other. We find

$$\frac{y}{x} = \frac{-y'}{f} \quad \text{and} \quad \frac{-y'}{x'} = \frac{y}{f} \tag{4j}$$

Multiplication of one equation by the other gives

$$xx' = f^2$$

In the gaussian formula the object distances are measured from the lens, while in the newtonian formula they are measured from the focal points. Object distances (s or x) are positive if the object lies to the left of its reference point (A or F, respectively), while image distances (s' or x') are positive if the image lies to the right of its reference point (A or F', respectively).

The lateral magnification as given by Eq. (4c) corresponds to the gaussian form. When distances are measured from focal points, one should use the newtonian form, which can be obtained directly from Eq. (4j):

$$m = \frac{y'}{y} = -\frac{f}{x} = -\frac{x'}{f} \tag{4k}$$

In the more general case where the medium on the two sides of the lens is different, it will be shown in the next section that the primary and secondary focal distances f and f' are different, being in the same ratio as the two refractive indices. The newtonian lens formula then takes the symmetrical form

$$xx' = ff'$$

4.15 DERIVATION OF THE LENS MAKERS' FORMULA

The geometry required for this derivation is shown in Fig. 4L. Let n, n', and n'' represent the refractive indices of the three media as shown, f_1 and f_1' the focal lengths for the first surface alone, and f_2' and f_2'' the focal lengths for the second surface alone.

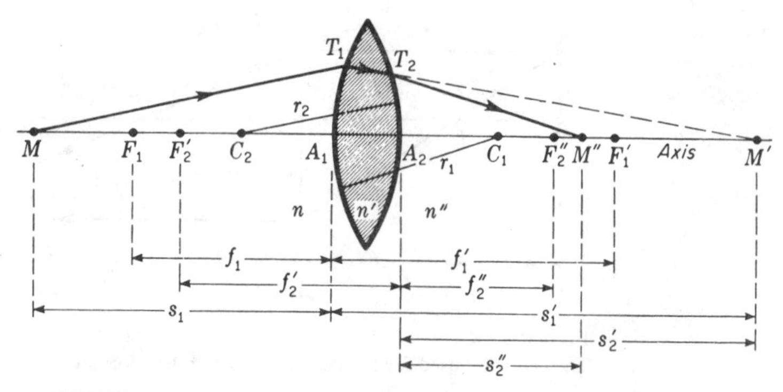

FIGURE 4L
Each surface of a thin lens has its own focal points and focal lengths, as well as separate object and image distances.

The oblique ray MT_1 is incident on the first surface as though it came from an axial object point M at a distance s_1 from the vertex A_1. At T_1 the ray is refracted according to Eq. (3b) and is directed toward the conjugate point M':

$$\frac{n}{s_1} + \frac{n'}{s_1'} = \frac{n' - n}{r_1} \tag{4l}$$

Arriving at T_2, the same ray is refracted in the new direction T_2M''. For this second surface the object ray T_1T_2 has for its object distance s_2', and the refracted ray gives an image distance of s_2''. When Eq. (3b) is applied to this second refracting surface,

$$\frac{n'}{s_2'} + \frac{n''}{s_2''} = \frac{n'' - n'}{r_2} \tag{4m}$$

If we now assume the lens thickness to be negligibly small compared with the object and image distances, we note the image distance s_1' for the first surface becomes equal in magnitude to the object distance s_2' for the second surface. Since M' is a virtual object for the second surface, the sign of the object distance for this surface is negative. As a consequence we can set $s_1' = -s_2'$ and write

$$\frac{n'}{s_1'} = -\frac{n'}{s_2'}$$

If we now add Eqs. (4l) and (4m) and substitute this equality, we obtain

$$\frac{n}{s_1} + \frac{n''}{s_2''} = \frac{n' - n}{r_1} + \frac{n'' - n'}{r_2} \tag{4n}$$

If we now call s_1 the object distance and designate it s as in Fig. 4M and call s_2'' the image distance and designate it s'', we can write Eq. (4n) as

$$\frac{n}{s} + \frac{n''}{s''} = \frac{n' - n}{r_1} + \frac{n'' - n'}{r_2} \tag{4o}$$

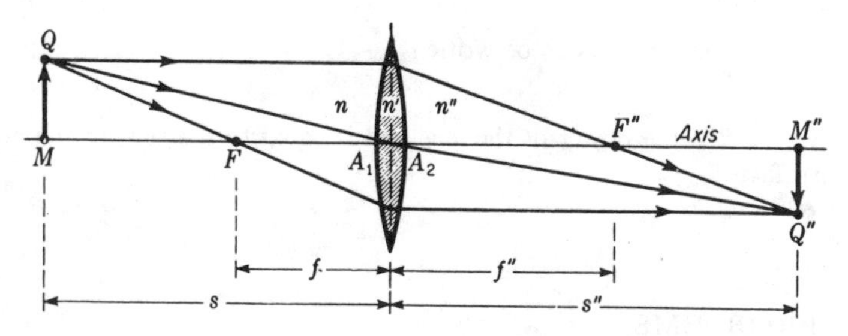

FIGURE 4M
When the media on the two sides of a thin lens have different indices, the primary and secondary focal lengths are not equal and the ray through the lens center is deviated.

This is the general formula for a thin lens having different media on the two sides. For such cases we can follow the procedure given in Sec. 3.4 and define primary and secondary focal points F and F'', and the corresponding focal lengths f and f'', by setting s or s'' equal to infinity. When this is done, we obtain

$$\frac{n}{f} = \frac{n' - n}{r_1} + \frac{n'' - n'}{r_2} = \frac{n''}{f''} \tag{4p}$$

In words, the focal lengths have the ratio of the refractive indices of the two media n and n'' (see Fig. 4M)

$$\frac{f}{f''} = \frac{n}{n''} \tag{4q}$$

If the medium on both sides of the lens is the same, $n = n''$, Eq. (4o) reduces to

$$\frac{n}{s} + \frac{n''}{s''} = (n' - n)\left(\frac{1}{r_1} - \frac{1}{r_2}\right) \tag{4r}$$

Note: The minus sign in the last factor arises when n'' and n' are reversed for the removal of like terms in the last factor of Eq. (4o).

Finally, if the surrounding medium is air ($n = 1$), we obtain the lens makers' formula

$$\frac{1}{s} + \frac{1}{s''} = (n' - 1)\left(\frac{1}{r_1} - \frac{1}{r_2}\right) \tag{4s}$$

In the power notation of Eq. (3i), the general formula [Eq. (4o)] can be written

$$V + V'' = P_1 + P_2 \tag{4t}$$

where $$V = \frac{n}{s} \qquad V'' = \frac{n''}{s''} \qquad P_1 = \frac{n' - n}{r_1} \qquad P_2 = \frac{n'' - n'}{r_2} \tag{4u}$$

Equation (4t) can be written

$$V + V'' = P \qquad (4v)$$

where P is the power of the lens and is equal to the sum of the powers of the two surfaces:

$$P = P_1 + P_2 \qquad (4w)$$

PROBLEMS

4.1 An object located 12.0 cm in front of a thin lens has its image formed on the opposite side 42.0 cm from the lens. Calculate (*a*) the focal length of the lens and (*b*) the lens power. *Ans.* (*a*) +9.33 cm, (*b*) +10.72 D

4.2 An object 2.50 cm high is placed 12.0 cm in front of a thin lens of focal length 3.0 cm. Calculate (*a*) the image distance, (*b*) the magnification, and (*c*) the nature of the image. (*d*) Check your answers by a graph.

4.3 The two faces of a thin lens have radii $r_1 = +10.0$ cm and $r_2 = -25.0$ cm, respectively. The lens is made of glass of index 1.740. Calculate (*a*) the focal length and (*b*) the power of the lens.

4.4 An object 3.50 cm high is located 10.0 cm in front of a lens whose focal length $f = -6.0$ cm. Calculate (*a*) the power of the lens, (*b*) the image distance, and (*c*) the lateral magnification. Graphically locate the image by (*d*) the parallel-ray method and (*e*) the oblique-ray method.

4.5 An equiconcave lens is to be made of flint glass of index 1.750. Calculate the radii of curvature if it is to have a power of −3.0 D. *Ans.* Both 50.0-cm radius

4.6 A plano-convex lens is to be made of light flint glass of index 1.680. Calculate the radius of curvature necessary to give the lens a power of 4.5 D.

4.7 Two lenses with focal lengths $f_1 = +5.0$ cm and $f_2 = +10.0$ cm are located 5.0 cm apart. If an object 2.50 cm high is located 15.0 cm in front of the first lens, find (*a*) the position and (*b*) the size of the final image.
Ans. (*a*) +2.00 cm from second lens, (*b*) −1.0 cm

4.8 A converging lens is used to focus a sharp image of a candle flame on a screen. Without moving the candle flame a second lens with radii $r_1 = +10.0$ cm and $r_2 = -20.0$ cm and index 1.650 is placed in the converging beam 30.0 cm from the screen. (*a*) Calculate the power of the second lens. (*b*) How close to the second lens should the screen now be placed to obtain a sharp image of the flame? (*c*) Make a graph of this experiment.

4.9 A double-convex lens is to be made of glass having a refractive index of 1.580. If one surface is to have twice the radius of the other and the focal length is to be +6.0 cm, find the radii.

4.10 Two lenses having focal lengths $f_1 = +9.0$ cm and $f_2 = -18.0$ cm are placed 3.0 cm apart. If an object 2.50 cm high is located 20.0 cm in front of the first lens, calculate (*a*) the position and (*b*) the size of the final image. (*c*) Check your answer graphically.

4.11 A lantern slide 8.0 cm high is located 3.50 m from a projection screen. What is the focal length of the lens that will be required to project an image 1.0 m high?

4.12 An object is located 1.60 m from a white screen. A lens of what focal length will be required to form a real and inverted image on the screen with a magnification of −6.0?
Ans. 19.59 cm

4.13 Three thin lenses have powers +1.50, −2.80, and 3.40 D, respectively. What are all the possible powers that can be obtained with these three lenses using one, two, or three at a time in contact?

4.14 Two thin lenses having the following radii of curvature and index are placed in contact. For the first lens $r_1 = +12.0$ cm, $r_2 = -18.0$ cm, and $n = 1.560$, and for the second lens $r_1 = -30.0$ cm, $r_2 = +20.0$ cm, and $n = 1.650$. Find their (*a*) individual powers, (*b*) combined power, (*c*) individual focal lengths, and (*d*) combined focal length.

4.15 An object 2.50 cm high is located 15.0 cm in front of a lens of +5.0 cm focal length. A lens with a focal length of −12.0 cm is placed 2.50 cm beyond this converging lens. Find (*a*) the position and (*b*) the size of the final image.

Ans. (*a*) +8.57 cm, (*b*) −2.143 cm

4.16 An object 2.50 cm high is located 8.0 cm in front of a lens of −2.40 cm focal length. A lens of +5.0 cm focal length is placed 1.50 cm behind the first lens. Find (*a*) the position and (*b*) the size of the final image. (*c*) Draw a graph.

4.17 Three lenses with focal lengths of +8.40, −4.60, and +6.20 cm, respectively, are located one behind each other in this order and 2.0 cm apart. (*a*) If parallel light is incident on the first lens, how far behind the third lens will the light be brought to a focus? (*b*) Draw a scale diagram.

4.18 An object 3.50 cm high is located 8.0 cm in front of a lens of −7.0 cm focal length. A lens of +4.50 cm focal length is placed 3.5 cm behind the first lens. Find (*a*) the position and (*b*) the size of the image. (*c*) Make a diagram to scale.

5
THICK LENSES

When the thickness of a lens cannot be considered small compared with its focal length, some of the thin-lens formulas of Chap. 4 are no longer applicable. The lens must be treated as a *thick lens*. This term is used not only for a single homogeneous lens with two spherical surfaces separated by an appreciable distance but also for any system of coaxial surfaces which is treated as a unit. The thick lens may therefore include several component lenses, which may or may not be in contact. We have already investigated one case which comes under this category, namely, the combination of a pair of thin lenses spaced some distance apart, as shown in Fig. 4H.

5.1 TWO SPHERICAL SURFACES

A simple form of thick lens comprises two spherical surfaces as shown in Fig. 5A. A treatment of the image-forming capabilities of such a system follows directly from procedures outlined in Chaps. 3 and 4. Each surface, acting as an image-forming component, contributes to the final image formed by the system as a whole.

Let n, n', and n'' represent the refractive indices of three media separated by two

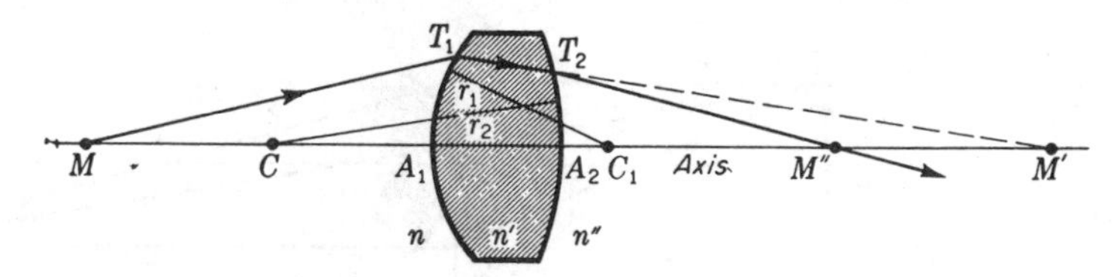

FIGURE 5A
Details of the refraction of a ray at both surfaces of a lens.

spherical surfaces of radius r_1 and r_2. A light ray from an axial object point M is shown refracted by the first surface in a direction T_1M' and then further refracted by the second surface in a direction T_2M''. Since the lens axis may be considered as a second ray of light originating at M and passing through the system, M'' is the final image of the object point M. Hence M and M'' are conjugate points for the thick lens as a whole, and all rays from M should come to a focus at M''.

We shall first consider the parallel-ray method for graphically locating an image formed by a thick lens and then apply the general formulas already given for calculating image distances. The formulas to be used are (see Sec. 3.4)

$$\frac{n}{s_1} + \frac{n'}{s_1'} = \frac{n' - n}{r_1} \qquad \frac{n'}{s_2'} + \frac{n''}{s_2''} = \frac{n'' - n'}{r_2} \tag{5a}$$

For first surface   *For second surface*

5.2 THE PARALLEL-RAY METHOD

The parallel-ray method of graphical construction, applied to a thick lens of two surfaces, is shown in Fig. 5B. Although the diagram is usually drawn as one, it has been separated into two parts here to simplify its explanation. The points F_1 and F_1' represent the primary and secondary focal points of the first surface, and F_2' and F_2'' represent the primary and secondary focal points of the second surface, respectively.

Diagram (*a*) is constructed by applying the method of Fig. 3F to the first surface alone and extending the refracted rays as far as is necessary to locate the image $M'Q'$. This real image, $M'Q'$, then becomes the object for the second surface, as shown in diagram (*b*). The procedure is similar to that given for two thin lenses in Fig. 4H. Ray 5 in diagram (*b*), refracted parallel to the axis by the first surface, is refracted as ray 7 through the secondary focal point F_2'' of the second surface.

Rays 8 and 9 are obtained by drawing a line from Q' back through C_2 and then, through the intersection B, drawing the line BQ. The intersection of rays 7 and 8 locates the final image point Q'' and the final image $M''Q''$.

EXAMPLE 1 An equiconvex lens 2 cm thick and having radii of curvature of 2 cm is mounted in the end of a water tank. An object in air is placed on the axis of the

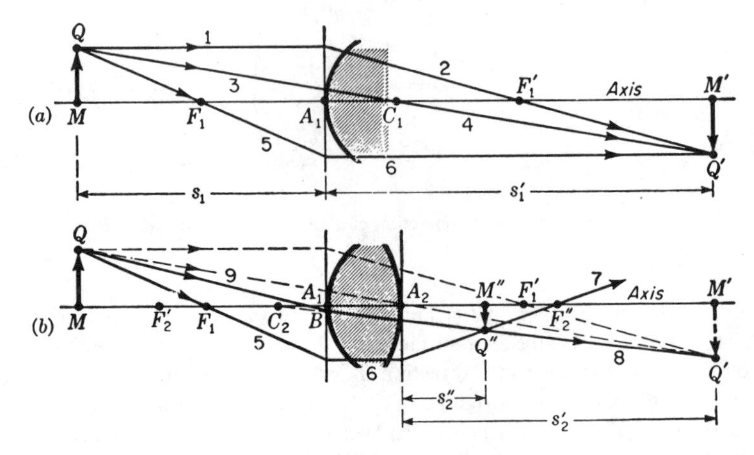

FIGURE 5B
The parallel-ray method for graphically locating the image formed by a thick lens.

lens 5 cm from its vertex. Find the position of the final image. Assume refractive indices of 1.00, 1.50, and 1.33 for air, glass, and water, respectively.

SOLUTION The relative dimensions in this problem are approximately those shown in Fig. 5B(*b*). If we apply Eq. (5a) to the first surface alone, we find the image distance to be

$$\frac{1.00}{5} + \frac{1.50}{s_1'} = \frac{1.50 - 1.00}{2} \qquad \text{or} \qquad s_1' = +30 \text{ cm}$$

When the same equation is applied to the second surface, we note that the object distance is s_1' minus the lens thickness, or 28 cm, and that since it pertains to a virtual object it has a negative sign. The substitutions to be made are, therefore, $s_2 = -28$ cm, $n' = 1.50$, $n'' = 1.33$, and $r_2 = -2.0$ cm.

$$\frac{1.50}{-28} + \frac{1.33}{s_2''} = \frac{1.33 - 1.50}{-2} \qquad \text{or} \qquad s_2'' = +9.6 \text{ cm}$$

Particular attention should be paid to the signs of the various quantities in this second step. Because the second surface is concave toward the incident light, r_2 must have a negative sign. The incident rays in the glass belong to an object point M', which is virtual, and thus s_2', being to the right of the vertex A_2, must also be negative. The final image is formed in the water ($n'' = 1.33$) at a distance $+9.6$ cm from the second vertex. The positive sign of the resultant signifies that the image is *real*.

It should be noted that Eqs. (5a) hold for paraxial rays only. The diagrams in Fig. 5B, showing all refraction as taking place at vertical lines through the vertices A_1 and A_2, are likewise restricted to paraxial rays.

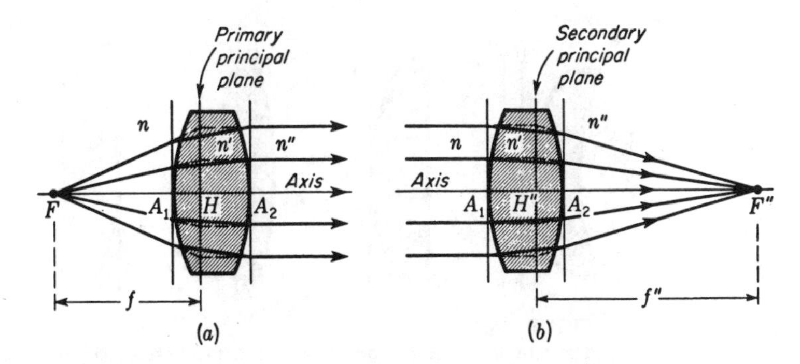

FIGURE 5C
Ray diagrams showing the primary and secondary principal planes of a thick lens.

5.3 FOCAL POINTS AND PRINCIPAL POINTS

Diagrams showing the characteristics of the two focal points of a thick lens are given in Fig. 5C. In the first diagram diverging rays from the primary focal point F emerge parallel to the axis, while in the second diagram parallel incident rays are brought to a focus at the secondary focal point F''. In each case the incident and refracted rays have been extended to their point of intersection between the surfaces. Transverse planes through these intersections constitute *primary* and *secondary principal planes.* These planes cross the axis at points H and H'', called the *principal points.* It will be noticed that there is a point-for-point correspondence between the two principal planes, so that each is an erect image of the other and both are the same size. For this reason they have sometimes been called *unit planes.* They are best defined by saying that *the principal planes are two planes having unit positive lateral magnification.*

The focal lengths, as shown in the figure, are measured from the focal points F and F'' to their respective principal points H and H'' and not to their respective vertices A_1 and A_2. If the medium is the same on both sides of the lens, $n'' = n$, the primary focal length f is exactly equal to the secondary focal length f''.

If the media on the two sides of the lens are different so that n'' is not equal to n, the two focal lengths are different and have the ratio of their corresponding refractive indices:

$$\frac{n''}{n} = \frac{f''}{f} \tag{5b}$$

In general the focal points and principal points are not symmetrically located with respect to the lens but are at different distances from the vertices. This is true even when the media on both sides are the same and the focal lengths are equal. As a lens with a given material and focal length is "bent" (see Fig. 5D), departing in either direction from the symmetrical shape of an equiconvex lens, the principal points are shifted. For meniscus lenses of considerable thickness and curvature, H and H'' may be completely outside the lens.

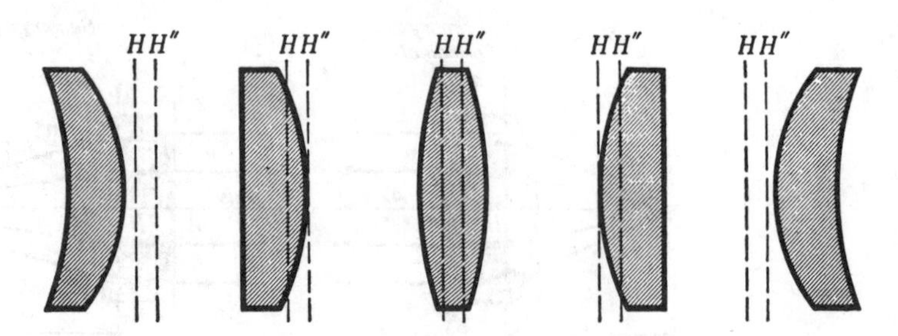

FIGURE 5D
The variation of the positions of the primary and secondary principal planes as a thick lens of fixed focal length is subjected to "bending."

5.4 CONJUGATE RELATIONS

In order to trace any ray through a thick lens, the positions of the focal points and principal points must first be determined. Once this has been done, either graphically or by computation, the parallel-ray construction can be used to locate the image as shown in Fig. 5E. The construction procedure follows that given in Fig. 4M for a thin lens, except that here all rays in the region between the two principal planes are drawn parallel to the axis.

By a comparison of the two figures and from the derivations of Eqs. (4n) and (4o), it will be found that, provided the object and image distances are measured to or from the principal points, we can apply the gaussian lens formula

$$\frac{n}{s} + \frac{n''}{s''} = \frac{n}{f} = \frac{n''}{f''} \tag{5c}$$

or by Eq. (3h)

$$V + V'' = P$$

In the special case where the media on the two sides of the lens are the same, so that $n'' = n$, we find $f'' = f$ and Eq. (5c) becomes

$$\frac{1}{s} + \frac{1}{s''} = \frac{1}{f} = \frac{1}{f''} \tag{5d}$$

Figure 5F shows that for the purposes of graphical construction the lens may be regarded as replaced by its two principal planes. Often the image distance is the unknown, and Eq. (5c) can be written in the more useful form

$$s'' = \frac{n''}{n} \frac{s \times f}{s - f} \tag{5e}$$

5.5 THE OBLIQUE-RAY METHOD

The oblique-ray method of construction may be used to find graphically the focal points of a thick lens. As an illustration, consider a glass lens of index 1.50, thickness 2.0 cm, and radii $r_1 = +3.0$ cm, $r_2 = -5.0$ cm, surrounded by air of index $n = 1.00$.

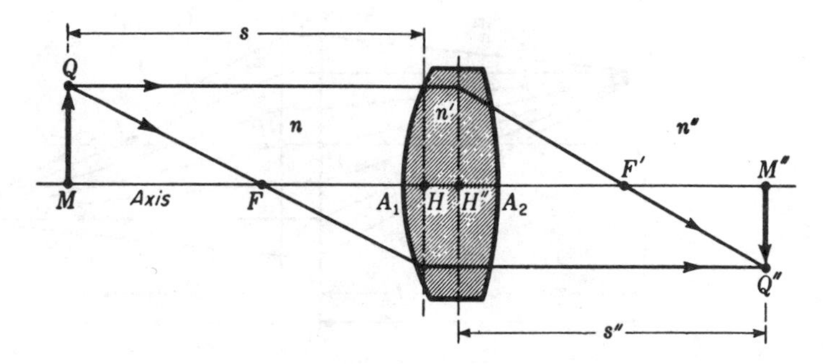

FIGURE 5E
The parallel-ray method of construction for graphically locating an image formed by a thick lens.

The first step is to calculate the primary and secondary focal lengths of each surface separately by the use of the formulas for a single spherical surface [Eqs. (3c) and (3d)]. Using the present notation, these are

$$\frac{n}{f_1} = \frac{n'}{f_1'} = \frac{n' - n}{r_1} \quad \text{and} \quad \frac{n'}{f_2'} = \frac{n''}{f_2''} = \frac{n'' - n'}{r_2} \tag{5f}$$

The given quantities are

$$r_1 = +3.0 \text{ cm} \qquad r_2 = -5.0 \text{ cm} \qquad d = 2.0 \text{ cm} \qquad n' = 1.50 \qquad n'' = n = 1.00$$

By substituting these values in Eqs. (5f) we obtain

$$f_1 = +6.0 \text{ cm} \qquad f_1' = +9.0 \text{ cm} \qquad f_2' = +15.0 \text{ cm} \qquad f_2'' = +10.0 \text{ cm}$$

With these focal lengths known, the lens axis can be drawn as in Fig. 5G and the known points measured off to some suitable scale. After drawing lines 2 and 3 through the lens vertices, a parallel incident ray 4 is selected. Upon refraction at the first surface the ray takes the new direction 5 toward F_1', the secondary focal point of that surface. After line 6 is drawn through F_2'', line 7 is drawn through C_2 parallel to ray 5. The point B where line 6 crosses line 7 determines the direction of the final refracted ray 8. The intersection of ray 8 with the axis locates the secondary focal point F'' of the lens, while its intersection N'' with the incident ray locates the corresponding secondary principal plane H''.

By turning the lens around and repeating this procedure, the position of the primary focal point F and the position of the primary principal point H can be determined. The student will find it well worthwhile to carry out this construction and to check the results by measuring the focal lengths to verify the fact that they are equal. It is to be noted that, in accordance with the assumption of paraxial rays, all refraction is assumed to occur at the plane tangent to the boundary at its vertex.

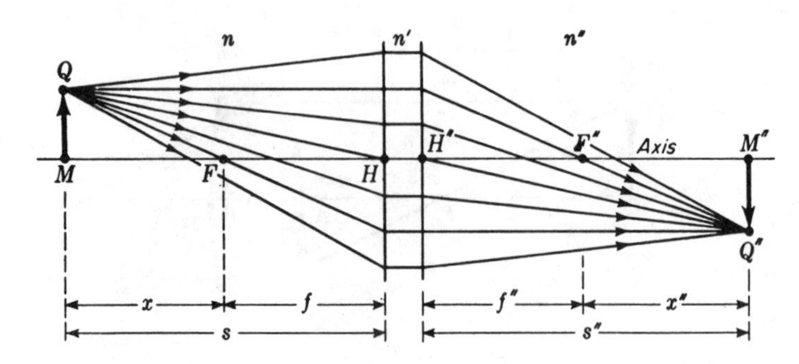

FIGURE 5F
Principal planes and antiprincipal planes are planes of unit magnification.

5.6 GENERAL THICK-LENS FORMULAS

A set of formulas that can be used for the calculation of important constants generally associated with a thick lens is presented below in the form of two equivalent sets.

Gaussian formulas	Power formulas	
$\frac{n}{f} = \frac{n'}{f_1'} + \frac{n''}{f_2''} - \frac{dn''}{f_1'f_2''} = \frac{n''}{f''}$	$P = P_1 + P_2 - \frac{d}{n'} P_1 P_2$	(5g)
$A_1F = -f\left(1 - \frac{d}{f_2'}\right)$	$A_1F = -\frac{n}{P}\left(1 - \frac{d}{n'} P_2\right)$	(5h)
$A_1H = +f\frac{d}{f_2'}$	$A_1H = +\frac{n}{P}\frac{d}{n'} P_2$	(5i)
$A_2F'' = +f''\left(1 - \frac{d}{f_1'}\right)$	$A_2F'' = +\frac{n''}{P}\left(1 - \frac{d}{n'} P_1\right)$	(5j)
$A_2H'' = -f''\frac{d}{f_1'}$	$A_2H'' = -\frac{n''}{P}\frac{d}{n'} P_1$	(5k)

These equations are derived from geometrical relations that can be obtained from a diagram like Fig. 5G. As an illustration, the gaussian equation (5k) is derived as follows. From the two similar right triangles $T_1A_1F_1'$ and $T_2A_2F_1'$, we can write corresponding sides as proportions

$$\frac{A_1F_1'}{A_1T_1} = \frac{A_2F_1'}{A_2T_2} \quad \text{or} \quad \frac{f_1'}{h} = \frac{f_1' - d}{j}$$

and, from the two similar right triangles $N''H''F''$ and T_2A_2F'', we can write the proportions

$$\frac{H''F''}{H''N''} = \frac{A_2F''}{A_2T_2} \quad \text{or} \quad \frac{f''}{h} = \frac{f'' - H''A_2}{j}$$

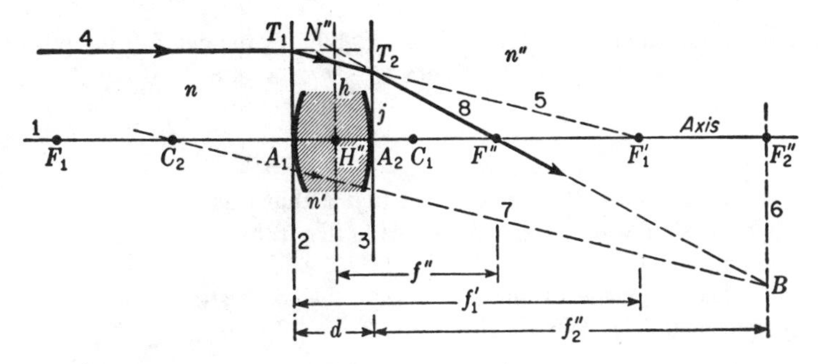

FIGURE 5G
The oblique-ray method for graphically tracing paraxial rays through a thick lens.

If we solve each equation for j/h and then equate the right-hand sides of the resultant equations, we obtain

$$\frac{f_1' - d}{f_1'} = \frac{f'' - H''A_2}{f''} \qquad \text{or} \qquad H''A_2 = f''\frac{d}{f_1'}$$

If we now reverse the segment $H''A_2$ to A_2H'' by changing the sign from + to −, we obtain

$$A_2H'' = -f''\frac{d}{f_1'}$$

In terms of surface power and lens power,

$$P_1 = \frac{n}{f_1} = \frac{n'}{f_1'} \qquad P_2 = \frac{n'}{f_2'} = \frac{n''}{f_2''} \qquad P = \frac{n}{f} = \frac{n''}{f''} \tag{5l}$$

the same equation can be written

$$A_2H'' = -\frac{n''}{P}\frac{d}{n'}P_1$$

In the design of certain optical systems it is convenient to know the *vertex power* of a lens. This power, sometimes called *effective power*, is given as

$$P_v = \frac{P}{1 - dP_1/n'} \tag{5m}$$

and is defined as the reciprocal of the distance from the back surface of the lens to the secondary focal point. This distance is commonly called the *back focal length*. Since $P_v = 1/A_2F''$, the above equation for vertex power is obtained by inverting Eq. (5j). In the inversion the lens is assumed to be in air so that $n'' = 1$.

In a similar way the distance from the primary focal point to the front surface of the lens is called the *front focal length*, and the reciprocal of this distance is called the *neutralizing power*. $P_n = 1/A_1F$. Calling P_n the neutralizing power, we can take the reciprocal of Eq. (5h) to obtain

$$P_n = \frac{P}{1 - dP_2/n'} \tag{5n}$$

The name is derived from the fact that a thin lens of this specified power and of opposite sign will, upon contact with the front surface, give zero power to the combination.

The following example will serve as an illustration of the use of thick-lens formulas applied to two surfaces.

EXAMPLE 2 A lens has the following specifications: $r_1 = +1.5$ cm, $r_2 = +1.5$ cm, $d = 2.0$ cm, $n = 1.00$, $n' = 1.60$, and $n'' = 1.30$. Find (*a*) the primary and secondary focal lengths of the separate surfaces, (*b*) the primary and secondary focal lengths of the system, and (*c*) the primary and secondary principal points.

SOLUTION (*a*) To apply the gaussian formulas, we first calculate the individual focal lengths of the surfaces by means of Eq. (5f).

$$\frac{n}{f_1} = \frac{n' - n}{r_1} = \frac{1.60 - 1.00}{1.5} \qquad f_1 = \frac{1.00}{0.40} = +2.50 \text{ cm}$$

$$= 0.400 \qquad f_1' = \frac{1.60}{0.40} = +4.00 \text{ cm}$$

$$\frac{n'}{f_2'} = \frac{n'' - n'}{r_2} = \frac{1.30 - 1.60}{1.5} \qquad f_2' = \frac{1.60}{-0.20} = -8.00 \text{ cm}$$

$$= -0.200 \qquad f_2'' = \frac{1.30}{-0.20} = -6.50 \text{ cm}$$

(*b*) The focal lengths of the system are calculated from Eq. (5g).

$$\frac{n}{f} = \frac{n'}{f_1'} + \frac{n''}{f_2''} - \frac{d}{f_1'}\frac{n''}{f_2''} = \frac{1.60}{4.00} + \frac{1.30}{-6.50} - \frac{2.00}{4.00}\frac{1.30}{-6.50}$$

$$\frac{n}{f} = 0.40 - 0.20 + 0.10 = 0.30$$

$$\text{or } f = \frac{1.00}{0.30} = +3.333 \text{ cm} \qquad \text{and} \qquad f'' = \frac{n''}{0.30} = \frac{1.30}{0.30} = +4.333 \text{ cm}$$

The focal points of the system are given by Eqs. (5h) and (5j).

$$A_1F = -f\left(1 - \frac{d}{f_2'}\right) = -3.333\left(1 - \frac{2.0}{-8.0}\right) = -4.166 \text{ cm}$$

$$A_2F'' = +f''\left(1 - \frac{d}{f_1'}\right) = +4.33\left(1 - \frac{2.0}{4.0}\right) = +2.167 \text{ cm}$$

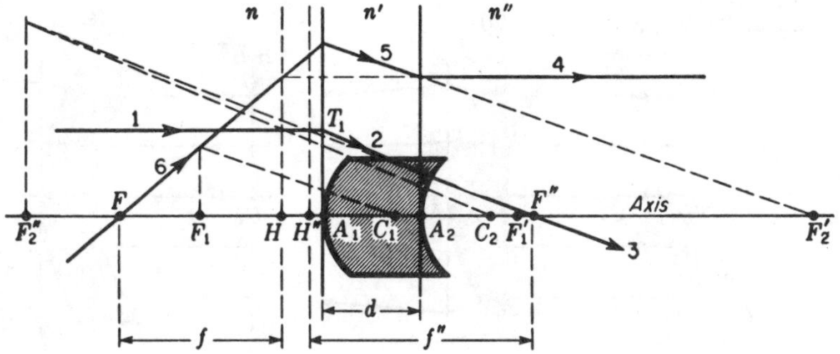

FIGURE 5H
A graphical construction for locating the focal points and principal points of a thick lens.

(*c*) The principal points are given by Eqs. (5i) and (5k).

$$A_1H = +f\frac{d}{f_2'} = +3.33\,\frac{2.0}{-8.0} = -0.833\text{ cm}$$

$$A_2H'' = -f''\frac{d}{f_1'} = -4.33\,\frac{2.0}{4.0} = -2.167\text{ cm}$$

Positive signs represent distances measured to the right of the reference vertex and negative signs those measured to the left.

By subtracting the magnitudes of the two intervals A_1F and A_1H, the primary focal length $FH = 4.166 - 0.833 = 3.333$ cm is obtained and serves as a check on the calculations in part (*b*). Similarly the addition of the two intervals A_2F'' and A_2H'' gives the secondary focal length

$$H''F'' = 2.167 + 2.167 = 4.334\text{ cm}$$

The graphical solution of this same problem is shown in Fig. 5H. After the axis is drawn and the lens vertices A_1 and A_2 and the centers C_1 and C_2 are located, the individual focal points F_1, F_1', F_2', and F_2'' are laid off according to the results in part (*a*). The parallel ray 1 is refracted at the first surface toward F_1'. The oblique-ray method is applied to this ray 2 at the second surface, and the final ray 3 is obtained. The point where ray 3 crosses the axis locates the secondary focal point F'', and the point where its backward extension intersects ray 1 locates the secondary principal plane H''. Ray 4 is constructed backward by drawing it parallel to the axis and from right to left. The first refraction gives ray 5 up and to the left as if it came from F_2'. The oblique-ray method applied to ray 5 at the left-hand surface yields ray 6. The point where ray 6 crosses the axis locates F, and the point where it crosses the extension of ray 4 locates H. Hence parts (*b*) and (*c*) of the problem are solved graphically, and they check with the calculated values.

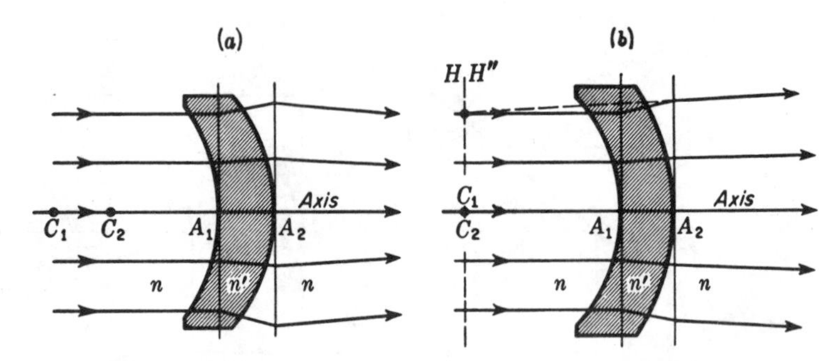

FIGURE 5I
Special thick lenses: (*a*) a positive lens with equal radii of curvature; (*b*) a negative lens with concentric surfaces.

5.7 SPECIAL THICK LENSES

Two special lenses of some interest as well as practical importance are presented here. The first, as shown in Fig. 5I, is a lens with spherical surfaces of equal radii, $r_1 = r_2$. A lens of this description, surrounded by a medium of lower index, $n' > n$, has a small but positive power. Its principal planes are located some distance from and to the right of the lens, and their spacing HH'' is equal to the lens thickness d. If the surrounding medium has a greater index, as in the case of an air space between the surfaces of two lenses of equal index, $n' < n$, the power is again positive but the principal planes lie some distance to the left of the lens and a distance d apart.

The second special case is that of a *concentric lens*, both surfaces having the same center of curvature. Where such a lens is surrounded by a medium of lower index, $n' > n$, the system has a negative power with a long focal length and the principal points coincide with the common center of curvature of the two surfaces. In other words, it acts like a thin lens located at C_1C_2.

5.8 NODAL POINTS AND OPTICAL CENTER

Of all the rays that pass through a lens from an off-axis object point to its corresponding image point, there will always be one for which the direction of the ray in the image space is the same as that in the object space; i.e., the segments of the ray before reaching the lens, and after leaving it, are parallel. The two points at which these segments, if projected, intersect the axis are called the *nodal points*, and the transverse planes through them are called the *nodal planes*. This third pair of points and their associated planes are shown in Fig. 5J, which also shows the optical center of the lens at C. It is readily shown that if the medium on both sides is the same, the nodal points N and N'' coincide with the principal points H and H'' but if the two media have different indices, the principal points and the nodal points will be separate. Since the incident and emergent rays make equal angles with the axis, the nodal points are called conjugate points of *unit positive angular magnification*.

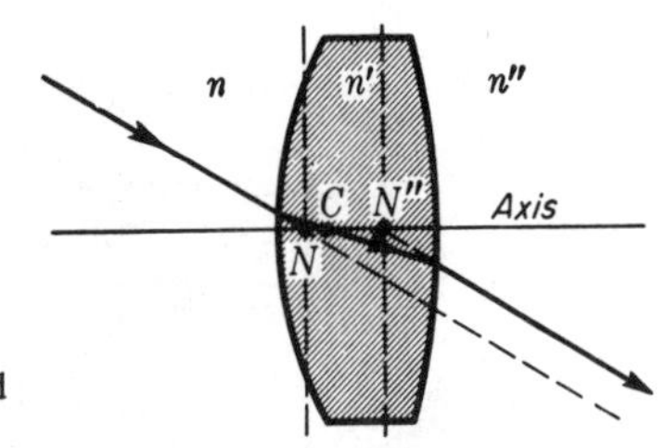

FIGURE 5J
The significance of the nodal points and nodal planes of a thick lens.

If the ray is to emerge parallel to its original direction, the two surface elements of the lens where it enters and leaves must be parallel to each other so that the effect is like that of a plane-parallel plate. A line between these two points crosses the axis at the *optical center C*. It is therefore through the optical center that the undeviated ray must be drawn in all cases. It has the interesting property that its position, depending as it does only on the radii of curvature and thickness of the lens, does not vary with color of the light. All the six cardinal points (Sec. 5.9) will in general have a slightly different position for each color.

Figure 5K will help to clarify the different significance of the nodal points and the principal points. This figure is drawn for $n'' \neq n$, so that the two sets of points are separate. Ray 11 through the secondary nodal point is parallel to ray 10, the latter being incident in the direction of the primary nodal point N. On the other hand both these segments intersect the principal planes at the same distance above the principal points H and H''. From the small parallelogram at the center of the diagram, it is observed that the distance between nodal planes is exactly equal to the distance between principal planes. In general, therefore,

● $$NN'' = HH'' \tag{5o}$$

Furthermore in this case, where the initial and final values of the refractive index differ, the focal lengths, which are measured from the principal points, are no longer equal. The primary focal length FH is equal to the distance $N''F''$, while the secondary focal length $H''F''$ is equal to FN:

● $$f = FH = N''F'' \qquad \text{and} \qquad f'' = H''F'' = FN \tag{5p}$$

Nodal points can be determined graphically, as shown in Fig. 5K, by measuring off the distance $ZQ = HH'' = Z'Q''$ and drawing straight lines through QZ' and ZQ''. From the geometry of this diagram, the lateral magnification y'/y is given by

$$m = \frac{y''}{y} = -\frac{s'' - HN}{s + HN} \tag{5q}$$

● where $$HN = f''\frac{n'' - n}{n''} \tag{5r}$$

When the object and image distances s and s'' are, as usual, measured from their corresponding principal points H and H'', Eq. (5c) is valid for paraxial rays.

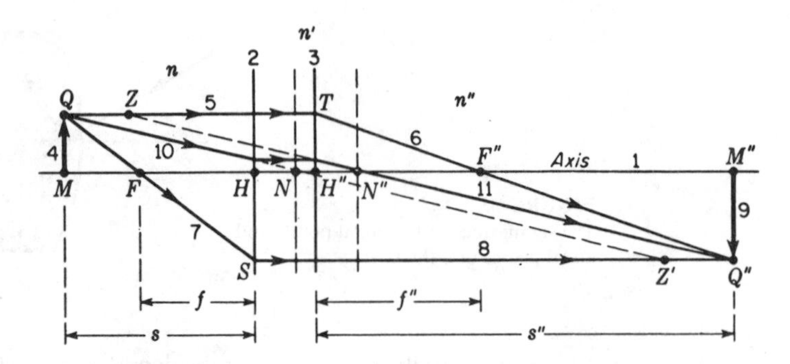

FIGURE 5K
The parallel-ray method of graphically locating the nodal points and planes of a thick lens.

The distance from the first vertex to the primary nodal point is given by

$$A_1N = f\left(\frac{d}{f_2'} + \frac{n'' - n}{n}\right) \tag{5s}$$

EXAMPLE 3 Find the nodal points of the thick lens given in Example 2.

SOLUTION To locate the primary nodal point N, we may use Eq. (5r) and substitute the given values of $n = 1.00$ and $n'' = 1.30$ and the already calculated value of $f'' = +4.333$ cm,

$$HN = 4.333\,\frac{1.30 - 1.00}{1.30} = +1.00 \text{ cm}$$

Hence the nodal points N and N'' are 1.00 cm to the right of their respective principal points H and H''.

5.9 OTHER CARDINAL POINTS

In thick-lens problems a knowledge of the six cardinal points, comprising the focal points, principal points, and nodal points, is always adequate to obtain solutions. Other points of lesser importance but still of some interest are (1) negative principal points and (2) negative nodal points. *Negative principal points* are conjugate points for which the lateral magnification is unity and negative. For a lens in air they lie at twice the focal length and on opposite sides of the lens. *Negative nodal points* lie as far from the focal points as the ordinary cardinal nodal points but on opposite sides. Their position is such that the angular magnification is unity and negative. Although a knowledge of these two pairs of cardinal points is not essential to the solution of optical problems, in certain cases considerable simplification is achieved by using them.

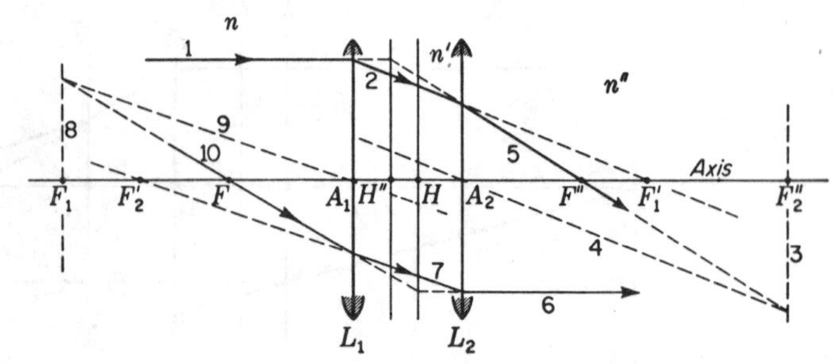

FIGURE 5L
Focal points and principal points of a system involving two thin lenses.

5.10 THIN-LENS COMBINATION AS A THICK LENS

A combination of two or more thin lenses may also be referred to as a thick lens because the optical properties of a set of coaxially mounted lenses can be conveniently treated in terms of only two focal points and two principal points. If the object space and image space have the same refractive index (and this is nearly always the case), the nodal points and planes coincide with the principal points and planes.

A combination of two thin lenses with focal lengths of 8.0 and 9.0 cm, respectively, is shown in Fig. 5L. By the oblique-ray method the focal points F and F'' and the principal points H and H'' have been determined graphically. In doing so the refraction at each lens was considered in the same way as the refraction at the individual surfaces of the thick lens of Fig. 5G. There is a strong resemblance between these two diagrams; i.e., for a thin lens we assume that all the deviation occurs at one plane, just as for a single surface. This assumption is justified only when the separation of the principal planes of the lens can be neglected. The definition of a thin lens is just a statement of this fact: *a thin lens is one in which the two principal planes and the optical center coincide at the geometrical center of the lens.* The locations of the centers of the two lenses in this example are labeled A_1 and A_2 in Fig. 5L.

A diagram for a combination of a positive and a negative lens is given in Fig. 5M. The construction lines are not shown, but the graphical procedure used in determining the paths of the two rays is the same as that shown in Fig. 5L. Note here that the final principal points H and H'' lie outside the interlens space but that the focal lengths f and f'' measured from these points are as usual equal. The lower ray, although shown traveling from left to right, is graphically constructed by drawing it from right to left.

The positions of the cardinal points of a combination of two thin lenses in air can be calculated by means of the thick-lens formulas given in Sec. 5.6. As used for thin lenses in place of individual refracting surfaces, A_1 and A_2 become the two lens centers, while f_1, f_2 and P_1, P_2 become their separate focal lengths and powers, respectively. The latter are given by

$$P_1 = \frac{n_1 - n}{r_1} + \frac{n' - n_1}{r_1'} = \frac{n}{f_1} \qquad P_2 = \frac{n_2 - n'}{r_2} + \frac{n'' - n_2}{r_2'} = \frac{n'}{f_2'} \tag{5t}$$

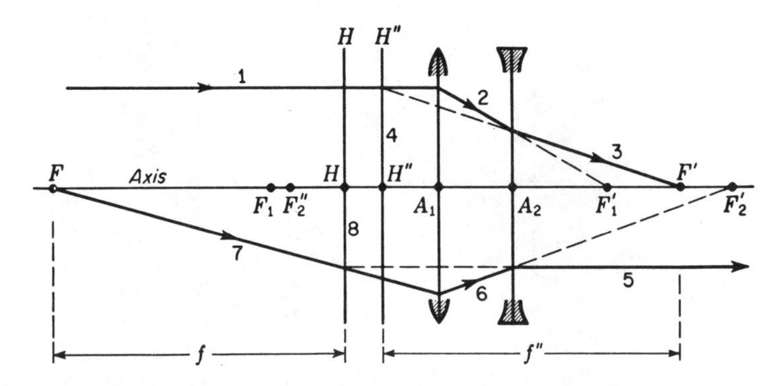

FIGURE 5M
The oblique-ray method applied to positive and negative thin lenses in combination.

where r_1 and r_1' are the radii of the first lens of index n_1 and r_2 and r_2' are the radii of the second lens of index n_2. The surrounding media have indices n, n', and n'' (see Fig. 5L). The other formulas, Eqs. (5g) to (5k), remain unchanged.

To illustrate the use of these formulas, let us consider the following problem on a lens combination similar to that shown in Fig. 5M.

EXAMPLE 4 An equiconvex lens with radii of 4 cm and index $n_1 = 1.50$ is located 2.0 cm in front of an equiconcave lens with radii of 6.0 cm and index $n_2 = 1.60$. The lenses are to be considered as thin. The surrounding media have indices $n = 1.00$, $n' = 1.33$, and $n'' = 1.00$. Find (*a*) the power, (*b*) the focal lengths, (*c*) the focal points, and (*d*) the principal points of the system.

SOLUTION (*a*) In this instance we shall solve the problem by the use of the power formulas. By Eqs. (5t) the powers of the two lenses in their surrounding media are

$$P_1 = \frac{1.50 - 1.00}{0.04} + \frac{1.33 - 1.50}{-0.04} = 12.50 + 4.17 = +16.67 \text{ D}$$

$$P_2 = \frac{1.60 - 1.33}{-0.06} + \frac{1.00 - 1.60}{0.06} = -4.45 - 10.0 = -14.45 \text{ D}$$

By Eq. (5g), we obtain

$$P = 16.67 - 14.45 + 0.015 \times 16.67 \times 14.45$$

or

$$P = +5.84 \text{ D}$$

(*b*) Using Eq. (5l), we find

$$f = \frac{n}{P} = \frac{1.00}{5.84} = 0.171 \text{ m} = 17.1 \text{ cm}$$

$$f'' = \frac{n''}{P} = \frac{1.00}{5.84} = 0.171 \text{ m} = 17.1 \text{ cm}$$

(*c*) By Eqs. (5h) to (5k) we obtain

$$A_1F = -\frac{1.00}{5.84}(1 + 0.015 \times 14.45) = -0.208 \text{ m} = -20.8 \text{ cm}$$

$$A_1H = +\frac{1.00}{5.84}0.015\,(-14.45) = -0.037 \text{ m} = -3.7 \text{ cm}$$

$$A_2F'' = +\frac{1.00}{5.84}(1 - 0.015 \times 16.67) = +0.128 \text{ m} = +12.8 \text{ cm}$$

(*d*) The principal points are

$$A_2H'' = -\frac{1.00}{5.84}0.015 \times 16.67 = -0.043 \text{ m} = -4.3 \text{ cm}$$

As a check on these results we find that the difference between the first two intervals A_1F and A_1H gives the primary focal length $FH = 17.1$ cm. Similarly the sum of the second two intervals A_2F'' and A_2H'' gives the secondary focal length $H''F'' = 17.1$ cm.

5.11 THICK-LENS COMBINATIONS

The problem of calculating the positions of the cardinal points of a thick lens consisting of a combination of several component lenses of appreciable thickness is one of considerable difficulty, but one which can be solved by use of the principles already given. In a combination of two lenses such as that in Fig. 5L, if the individual lenses cannot be considered as thin, each must be represented by a pair of principal planes. There are thus two pairs of principal points, H_1 and H_1' for the first lens and H_2' and H_2'' for the second, and the problem is to combine these to find a single pair H and H'' for the combination and to determine the focal lengths. By carrying out a construction similar to Fig. 5G for each lens separately, it is possible to locate the principal points and focal points of each. Then the construction of Fig. 5L can be accomplished, taking account of the unit magnification between principal planes.

Formulas can be given for the analytical solution of this problem, but because of their complexity they will not be given here.* Instead, we shall describe a method of determining the positions of the cardinal points of any thick lens by direct experiment.

5.12 NODAL SLIDE

The nodal points of a single lens or of a combination of lenses can be located experimentally by mounting the system on a nodal slide. This is merely a horizontal support which permits rotation of the lens about any desired point on its axis. As is shown in

* These equations are given, for example, in G. S. Monk, "Light, Principles and Experiments," Dover Publications, Inc., New York, 1963.

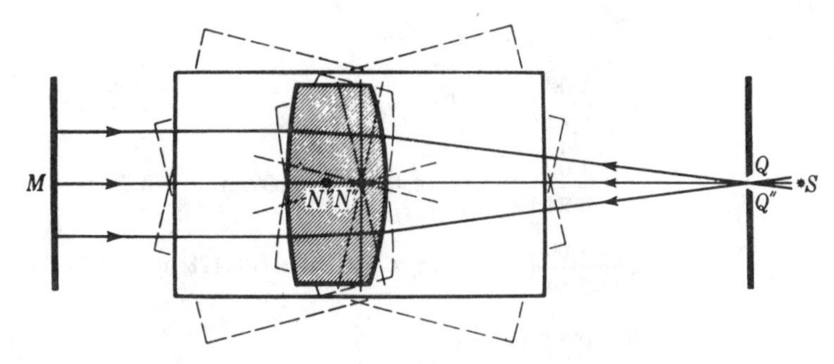

FIGURE 5N
The use of the nodal slide in locating nodal points.

Fig. 5N, light from a source S is sent through a slit Q, adjusted to lie at the secondary focal point of the lens. Emerging as a parallel beam, this light is reflected back on itself by a fixed plane mirror M, passing again through the lens system and being brought to a focus at Q''. This image of the slit is formed slightly to one side of the slit itself on the white face of one of the slit jaws. The nodal slide carrying the lens system is now rotated back and forth and the lens repeatedly shifted, until the rotation produces no motion of the image Q''. When this condition is reached, the axis of rotation N'' locates one nodal point. By turning the nodal slide end-for-end and repeating the process, the other nodal point N is found. When performed in air, this experiment of course locates the principal points as well, and the distance $N''Q''$ is an accurate measure of the focal length.

The principle of this method of rotation about a nodal point is illustrated in Fig. 5O. In the first diagram ray 4 along the axis passes through N and N'' to the focus at Q''. In the second diagram the lens system has been rotated about N'' and the same bundle of rays passes through the lens, coming to a focus at the same point Q''. Ray 3 is now directed toward N and ray 4 toward N''. When projected across from the plane of N to that of N'', the rays still converge toward Q'' even though F'' is now shifted to one side. Note that ray 3 approaches N in exactly the same direction that it leaves N'', corresponding to the defining condition for the nodal points.

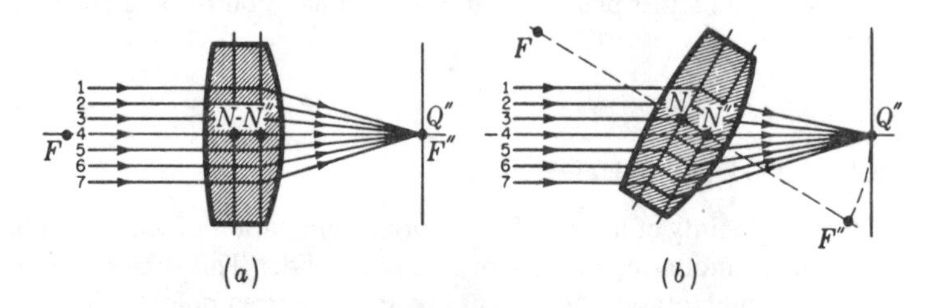

FIGURE 5O
Rotation of a lens about its secondary nodal point shifts the refracted rays but not the image.

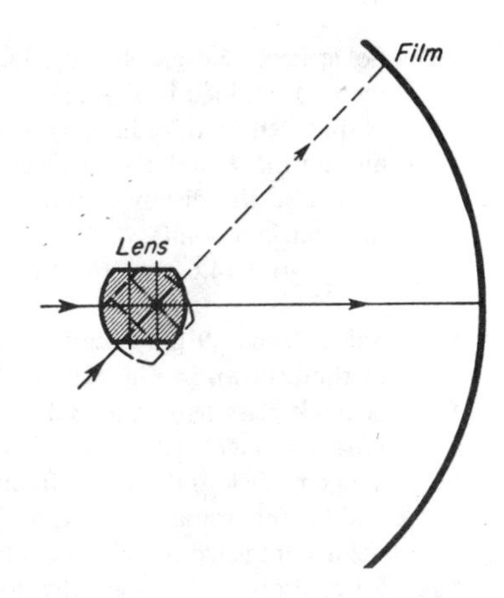

FIGURE 5P
In the panoramic camera the lens rotates about a nodal point as a center.

If a camera lens is pivoted about its secondary nodal point and a long strip of photographic film is curved to a circular arc of radius f'', a continuous picture covering a very wide angle can be taken. Such an instrument, shown schematically in Fig. 5P, is known as a *panoramic camera*. The shutter usually consists of a vertical slit just in front of the film, which moves with the rotation so that it always remains centered on the lens axis.

PROBLEMS

In Probs. 1 to 23 if the primary and secondary focal lengths of each of the two elements of the optical system are not already given, they must first be calculated.

5.1 An equiconvex lens located in air has radii of 5.20 cm, an index of 1.680, and a thickness of 3.50 cm. Calculate (*a*) the focal length and (*b*) the power of the lens. Find (*c*) the distances from the vertices to the focal points and (*d*) the principal points.
Ans. (*a*) +4.43 cm, (*b*) +22.59 D, (*c*) $A_1F = -3.222$ cm, and $A_2F'' = +3.222$ cm, (*d*) $A_1H = +1.206$ cm, and $A_2H'' = -1.206$ cm

5.2 Solve Prob. 5.1 graphically, locating the focal points and principal points.

5.3 A plano-convex lens 2.80 cm thick is made of glass of index 1.530. If the second surface has a radius of 3.50 cm, find (*a*) the focal length of the lens and (*b*) the power of the lens. Find the distances from the vertices to (*c*) the focal points and (*d*) the principal points.

5.4 Solve Prob. 5.3 graphically, locating the focal points and principal points.

5.5 A glass lens with radii $r_1 = +2.50$ cm and $r_2 = +4.50$ cm has a thickness of 2.90 cm and an index of 1.630. Calculate (*a*) the focal length and (*b*) the power of the lens. Find the distances from the vertices to (*c*) the focal points and (*d*) the principal points.
Ans. (*a*) +5.73 cm, (*b*) +17.46 D, (*c*) $A_1F = -7.163$ cm, and $A_2F'' = +3.162$ cm, (*d*) $A_1H = -1.433$ cm, and $A_2H'' = -2.568$ cm

5.6 Solve Prob. 5.5 graphically, locating the focal points and principal points.

5.7 A glass lens with radii $r_1 = +6.50$ cm and $r_2 = +3.20$ cm has a thickness of 2.80 cm and an index of 1.560. Calculate (*a*) the focal length and (*b*) the power of this lens in air. Find the distances from the vertices to (*c*) the focal points and (*d*) the principal points.

5.8 Solve Prob. 5.8 graphically, locating the focal points and principal points. Use the method outlined in Fig. 5H.

5.9 A thick lens with radii $r_1 = -4.50$ cm and $r_2 = -3.60$ cm has a thickness of 3.0 cm and an index of 1.560. Calculate (*a*) the focal length and (*b*) the power of the lens. Find also the distances from the vertices to the corresponding (*c*) focal points and (*d*) principal points.

Ans. (*a*) +14.64 cm, (*b*) +6.83 D, (*c*) $A_1F = -10.26$ cm, and $A_2F'' = +18.14$ cm, (*d*) $A_1H = +4.38$ cm, and $A_2H'' = +3.502$ cm

5.10 Solve Prob. 5.9 graphically, locating the focal points and principal points. Use the method shown in Fig. 5H.

5.11 A thick glass lens is placed in the end of a tank containing a transparent liquid of refractive index 1.420. The lens, with radii $r_1 = +3.80$ cm and $r_2 = -1.90$ cm, is 4.60 cm thick and has a refractive index of 1.620. If r_2 is in contact with the liquid, find (*a*) the primary and secondary focal lengths and (*b*) the power of the lens. Find the distances from the vertices to (*c*) the focal points and (*d*) the principal points.

5.12 Solve Prob. 5.11 graphically, locating the focal points and principal points. Use the method shown in Fig. 5H.

5.13 A glass lens 3.20 cm thick has radii $r_1 = +4.50$ cm and $r_2 = -2.20$ cm and an index of 1.630. If r_1 is in contact with the air and r_2 is in contact with a transparent oil of index 1.350, find (*a*) the primary and secondary focal lengths and (*b*) the power of the system. Find the distances from the vertices to the (*c*) focal points, (*d*) principal points, and (*e*) nodal points.

5.14 Solve Prob. 5.13 graphically, locating the six cardinal points of the optical system. Use the methods of Fig. 5H.

5.15 A glass lens with radii $r_1 = +3.0$ cm and $r_2 = +3.0$ cm has an index of 1.60, and a thickness of 3.0 cm. It is placed in the end of a tank so that air is in contact with face r_1, and a transparent oil of index 1.30 is in contact with face r_2. Find (*a*) the primary and secondary focal lengths and (*b*) the power of the system as a lens. Calculate the positions of (*c*) focal points, (*d*) principal points, and (*e*) nodal points.

Ans. (*a*) +7.27 and +9.46 cm, (*b*) $P = +13.75$ D,
(*c*) $A_1F = -8.64$ cm, $A_2F'' = -3.546$ cm,
(*d*) $A_1H = -1.364$ cm and $A_2H'' = +5.91$ cm,
(*e*) $HN = +2.182 = H''N''$

5.16 Solve Prob. 5.15 graphically, locating the six cardinal points of the optical system.

5.17 A glass lens 4.50 cm thick, and index 1.70, has radii of $r_1 = +3.0$ cm and $r_2 = +3.50$ cm. If a liquid of index 1.320 is in contact with r_1 and a very dense, transparent, oil of index 2.20 is in contact with r_2, find (*a*) the primary and secondary focal lengths and (*b*) the power of this optical system. Also find the distances from the two vertices to (*c*) the principal points, (*d*) the focal points, and (*e*) the nodal points. (*f*) If an object is located in the liquid of index 1.320 and 13.50 cm from r_1, find the image position.

5.18 Solve Prob. 5.17 graphically, locating the six cardinal points of the lens system and the image distance.

5.19 Two thin lenses in air, with focal lengths of +8.0 and +10.0 cm, respectively, are placed 3.0 cm apart. For this optical combination, find (*a*) the focal lengths, (*b*) the power, and distances from the lens centers to (*c*) the focal points and (*d*) the principal points.

Ans. (*a*) $f_1 = f_2 = +5.33$ cm, (*b*) +18.75 D,
(*c*) $A_1F = -3.733$ cm, and $A_2F'' = +3.333$ cm,
(*d*) $A_1H = +1.60$ cm, and $A_2H'' = -2.0$ cm

5.20 Solve Prob. 5.19 graphically, locating the focal points and principal points. Use the method of Fig. 5L.

5.21 Two thin lenses with focal lengths $f_1 = +24.0$ cm and $f_2 = -6.0$ cm, respectively, are mounted in a holder so their centers are 4.0 cm apart. If air surrounds both lenses, find the (*a*) focal length, (*b*) the power, and (*c*) the distances from the lens centers to the focal points and principal points.

5.22 Solve Prob. 5.21 graphically, locating the focal points and principal points. Use the method of Fig. 5M.

5.23 A lens with equal radii of curvature, $r_1 = r_2 = +4.0$ cm, is 3.50 cm thick and has an index of 1.650. If the lens is surrounded by air, find (*a*) the power and (*b*) the focal length of this thick lens. Calculate the positions of (*c*) the focal points and (*d*) the principal points.

Ans. (*a*) $+6.03$ D, (*b*) $f = f'' = +16.60$ cm,
(*c*) $A_1F = -22.48$ cm, and $A_2F'' = +10.72$ cm,
(*d*) $A_1H = -5.88$ cm, and $A_2H'' = -5.88$ cm

5.24 Solve Prob. 5.23 graphically, locating the focal points and principal points. Use the method shown in Fig. 5H.

5.25 With Fig. 5G as a guide, make a diagram locating the secondary focal point. From similar triangles in your diagram derive Eq. (5j).

5.26 With Fig. 5J as a guide, make a diagram locating the primary focal point. From similar triangles in your diagram derive Eq. (5h).

6

SPHERICAL MIRRORS

A spherical reflecting surface has image-forming properties similar to those of a thin lens or of a single refracting surface. The image from a spherical mirror is in some respects superior to that from a lens, notably in the absence of chromatic effects due to dispersion that always accompany the refraction of white light. Therefore mirrors are occasionally used in place of lenses in optical instruments, but their applications are not so broad as those of lenses because they do not offer the same possibilities for correction of the other aberrations of the image (see Chap. 9).

Because of the simplicity of the law of reflection compared with the law of refraction, the quantitative study of image formation by mirrors is easier than for lenses. Many features are the same, and these we shall pass over rapidly, putting the chief emphasis upon those characteristics which are different. To begin with, we restrict the discussion to images formed by paraxial rays.

6.1 FOCAL POINT AND FOCAL LENGTH

Diagrams showing the reflection of a parallel beam of light by a concave mirror and by a convex one are given in Fig. 6A. A ray striking the mirror at some point such as T obeys the law of reflection $\phi'' = \phi$. All rays are shown as brought to a common

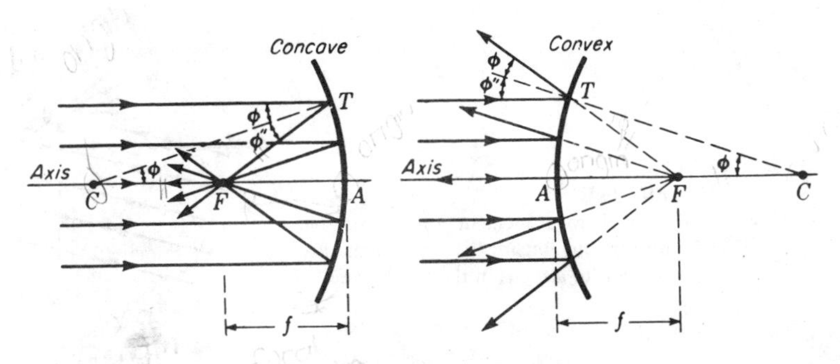

FIGURE 6A
The primary and secondary focal points of spherical mirrors coincide.

focus at F, although this will be strictly true only for paraxial rays. The point F is called the *focal point* and the distance FA the *focal length*. In the second diagram the reflected rays diverge as though they came from a common point F. Since the angle TCA also equals ϕ, the triangle TCF is isosceles, and in general $CF = FT$. But for very small angles ϕ (paraxial rays), FT approaches equality with FA. Hence

$$FA = \tfrac{1}{2}(CA)$$

$$f = -\tfrac{1}{2}r \qquad (6a)$$

and the focal length equals one-half the radius of curvature [see also Eq. (6d)].

The negative sign is introduced in Eq. (6a) so that the focal length of a concave mirror, which behaves like a positive or converging lens, will also be positive. According to the sign convention of Sec. 3.5, the radius of curvature is negative in this case. The focal length of a convex mirror, which has a positive radius, will then come out to be negative. This sign convention is chosen as being consistent with that used for lenses; it gives converging properties to a mirror with positive f and diverging properties to a mirror with negative f. By the principle of reversibility it can be seen from Fig. 6A that the primary and secondary focal points of a mirror coincide. In other words, it has but one focal point.

As before, a transverse plane through the focal point is called the focal plane. Its properties, as shown in Fig. 6B, are similar to those of either focal plane of a lens; e.g., parallel rays incident at any angle with the optic axis are brought to a focus at some point in the focal plane. The image Q' of a distant off-axis point object occurs at the intersection with the focal plane of that ray which goes through the center of curvature C.

6.2 GRAPHICAL CONSTRUCTIONS

Figure 6C, which illustrates the formation of a real image by a concave mirror, is self-explanatory. When the object MQ is moved toward the center of curvature C, the image also approaches C and increases in size until when it reaches C, it is the

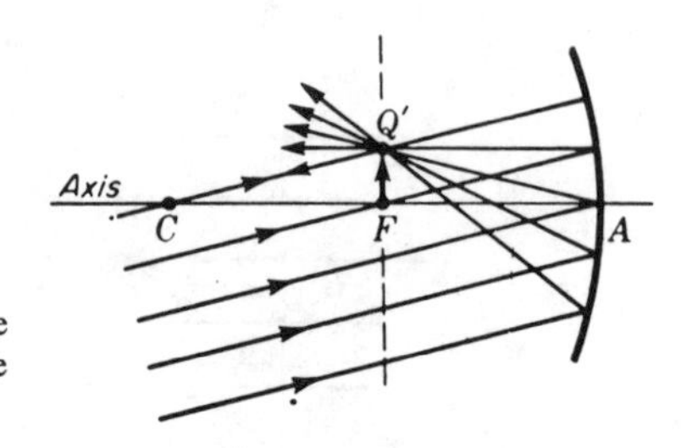

FIGURE 6B
Parallel rays incident on a concave mirror but inclined to the axis are brought to a focus in the focal plane.

same size as the object. The conditions when the object is between C and F can be deduced from the interchangeability of object and image as applied to this diagram. When the object is inside the focal point, the image is virtual, as with a converging lens. The methods of graphically constructing the image follow the same principles as those used for lenses, including the fact that paraxial rays must be represented as being deflected at the tangent plane instead of at the actual surface.

An interesting experiment can be performed with a large concave mirror set up under the condition of unit magnification, as shown in Fig. 6D. A bouquet of flowers is suspended upside down in a box and illuminated by a shaded lamp S. The large mirror is placed with its center of curvature C at the top surface of the stand, on which a real vase is placed. The observer's eye at E sees a perfect reproduction of the bouquet, not merely as a picture but as a faithful three-dimensional replica, which creates a strong illusion that it is a real object. As shown in the diagram, the rays diverge from points on the image just as they would if the real object were in the same position.

The parallel-ray method of construction is given for a concave mirror in Fig. 6E. Three rays leaving Q are, after reflection, brought to the conjugate point Q'. The image is real, inverted, and smaller than the object. Ray 4 drawn parallel to the axis is, by definition of the focal point, reflected through F. Ray 6 drawn through F is reflected parallel to the axis, and ray 8 through the center of curvature strikes the

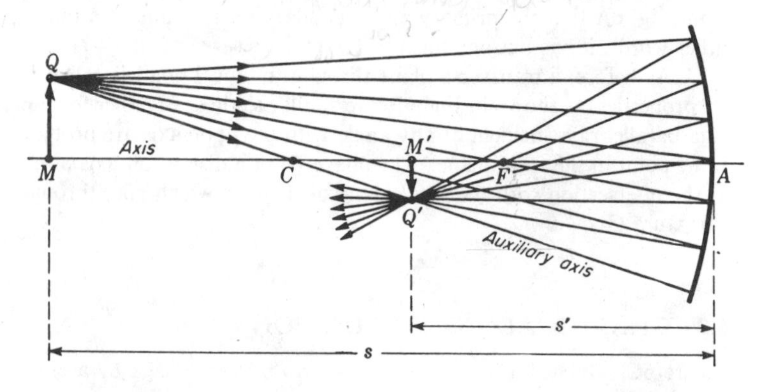

FIGURE 6C
Real image due to a concave mirror.

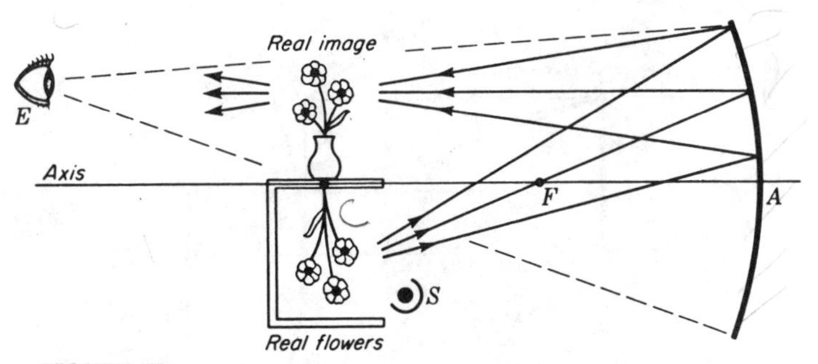

FIGURE 6D
Experimental arrangement for an optical illusion produced by a real image of unit magnification. The three-dimensional image shows parallax, as the real flowers do, and the image is so real that the eye cannot detect the difference between the real image and the real object.

mirror normally and is reflected back on itself. The crossing point of any two of these rays is sufficient to locate the image.

A similar procedure is applied to a convex mirror in Fig. 6F. The rays from the object point Q, after reflection, diverge from the conjugate point Q'. Ray 4, starting parallel to the axis, is reflected as if it came from F. Ray 6 toward the center of curvature C is reflected back on itself, while ray 7 going toward F is reflected parallel to the axis. Since the rays never pass through Q', the image $Q'M'$ in this case is virtual.

The oblique-ray method can also be used for mirrors, as illustrated in Fig. 6G for a concave mirror. After drawing the axis 1 and the mirror 2, we lay out the points C and F and draw a ray 3 making any arbitrary angle with the axis. Through F, the broken line 4 is then drawn parallel to 3. Where this line intersects the mirror at S, a parallel ray 6 is drawn backward to intersect the focal plane at P. Ray 7 is then drawn through TP and intersects the axis at M'. By this construction M and M' are conjugate points, and 3 and 7 are the parts of the ray in object and image spaces. The principle involved in this construction is obvious from the fact that if 3 and 4

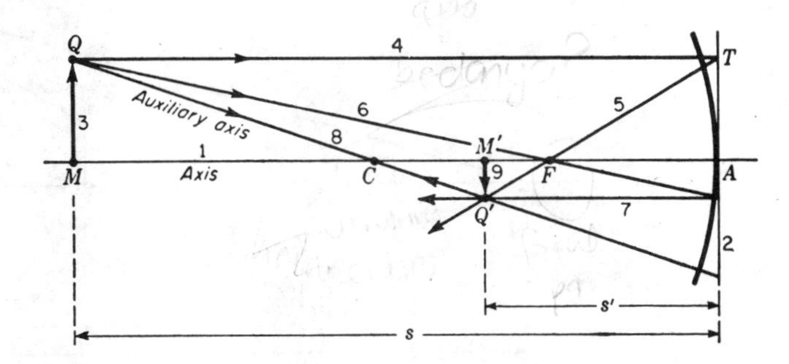

FIGURE 6E
Parallel-ray method for graphically locating the image formed by a concave mirror.

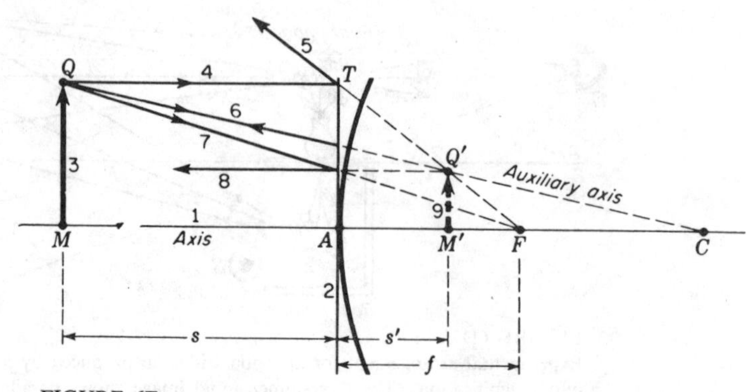

FIGURE 6F
Parallel-ray method for graphically locating the image formed by a convex mirror.

were parallel incident rays, they would come to a focus at P in the focal plane. If in place of ray 4 another ray were drawn through C and parallel to ray 3, it too would cross the focal plane at P. A ray through the center of curvature would be reflected directly back upon itself.

6.3 MIRROR FORMULAS

In order to be able to apply the standard lens formulas of the preceding chapters to spherical mirrors with as little change as possible, we must adhere to the following sign conventions:

1 Distances measured from left to right are positive while those measured from right to left are negative.

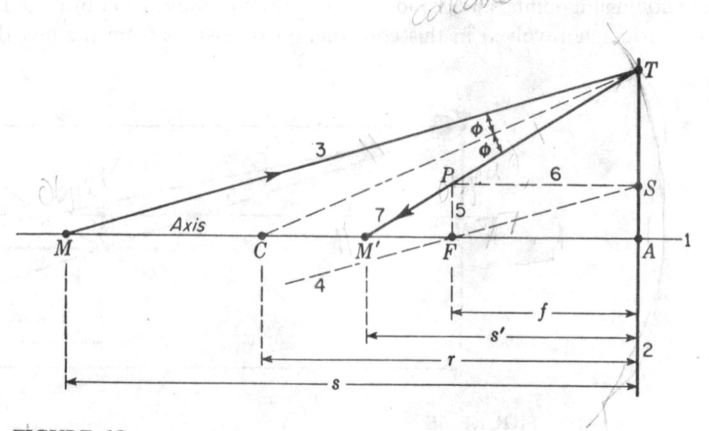

FIGURE 6G
Oblique-ray method for locating the image formed by a concave mirror.

2 *Incident rays travel from left to right and reflected rays from right to left.*
3 *The focal length is measured from the focal point to the vertex.* This gives f a positive sign for concave mirrors and a negative sign for convex mirrors.
4 *The radius is measured from the vertex to the center of curvature.* This makes r negative for concave mirrors and positive for convex mirrors.
5 *Object distances s and image distances s' are measured from the object and from the image respectively to the vertex.* This makes both s and s' positive and the object and image real when they lie to the left of the vertex; they are negative and virtual when they lie to the right.

The last of these sign conventions implies that for mirrors the object space and the image space coincide completely, the actual rays of light always lying in the space to the left of the mirror. Since the refractive index of the image space is the same as that of the object space, the n' of the previous equations becomes numerically equal to n.

The following is a simple derivation of the formula giving the conjugate relations for a mirror. In Fig. 6G it is observed that by the law of reflection the radius CT bisects the angle MTM'. Using a well-known geometrical theorem, we can then write the proportion

$$\frac{MC}{MT} = \frac{CM'}{M'T}$$

Now, for paraxial rays, $MT \approx MA = s$ and $M'T \approx M'A = s'$, where the symbol $\approx$ means "is approximately equal to." Also, from the diagram,

$$MC = MA - CA = s + r$$

and $$CM' = CA - M'A = -r - s' = -(s' + r)$$

Substituting in the above proportion gives

$$\frac{s + r}{s} = -\frac{s' + r}{s'}$$

which can easily be put in the form

$$\frac{1}{s} + \frac{1}{s'} = -\frac{2}{r} \qquad \text{(6b)}$$

Mirror formula

The primary focal point is defined as that axial object point for which the image is formed at infinity, so substituting $s = f$ and $s' = \infty$ in Eq. (6b), we have

$$\frac{1}{f} + \frac{1}{\infty} = -\frac{2}{r}$$

from which $$\frac{1}{f} = -\frac{2}{r} \qquad \text{or} \qquad f = -\frac{r}{2} \qquad \text{(6c)}$$

The secondary focal point is defined as the image point of an infinitely distant object point. This is $s' = f'$ and $s = \infty$, so that

$$\frac{1}{\infty} + \frac{1}{f'} = -\frac{2}{r}$$

from which

$$\frac{1}{f'} = -\frac{2}{r} \qquad \text{or} \qquad f' = -\frac{r}{2} \tag{6d}$$

Therefore the primary and secondary focal points fall together, and the magnitude of the focal length is one-half the radius of curvature. When $-2/r$ is replaced by $1/f$, Eq. (6b) becomes

$$\frac{1}{s} + \frac{1}{s'} = \frac{1}{f} \tag{6e}$$

just as for lenses.

The lateral magnification of the image from a mirror can be evaluated from the geometry of Fig. 6C. From the proportionality of sides in the similar triangles $Q'AM'$ and QAM we find that $-y'/y = s'/s$, giving

$$m = \frac{y'}{y} = -\frac{s'}{s} \tag{6f}$$

EXAMPLE 1 An object 2.0 cm high is situated 10.0 cm in front of a concave mirror of radius 16.0 cm. Find (*a*) the focal length of the mirror, (*b*) the position of the image, and (*c*) the lateral magnification.

SOLUTION The given quantities are $y = +2.0$ cm, $s = +10.0$ cm, and $r = -16.0$ cm. The unknown quantities are f, s', and m. (*a*) By Eq. (6c),

$$f = -\frac{-16}{2} = +8.0 \text{ cm}$$

(*b*) By Eq. (6e),

$$\frac{1}{10} + \frac{1}{s'} = \frac{1}{8} \qquad \text{or} \qquad \frac{1}{s'} = \frac{1}{8} - \frac{1}{10} = \frac{1}{40}$$

giving

$$s' = +40.0 \text{ cm}$$

(*c*) By Eq. (6f),

$$m = -\tfrac{40}{10} = -4$$

The image occurs 40.0 cm to the left of the mirror, is 4 times the size of the object, and is real and inverted.

6.4 POWER OF MIRRORS

The power notation that was used in Sec. 4.12 to describe the image-forming properties of lenses can readily be extended to spherical mirrors as follows. As definitions, we let

$$P = \frac{1}{f} \qquad V = \frac{1}{s} \qquad V' = \frac{1}{s'} \qquad K = \frac{1}{r} \tag{6g}$$

Equations (6b), (6e), (6c), and (6f) then take the forms

$$V + V' = -2K \tag{6h}$$

$$V + V' = P \tag{6i}$$

$$P = -2K \tag{6j}$$

$$m = \frac{y'}{y} = -\frac{V}{V'} \tag{6k}$$

EXAMPLE 2 An object is located 20.0 cm in front of a convex mirror of radius 50.0 cm. Calculate (*a*) the power of the mirror, (*b*) the position of the image, and (*c*) its magnification.

SOLUTION Expressing all distances in meters, we have

$$K = \frac{1}{0.50} = +2\text{ D} \qquad \text{and} \qquad V = \frac{1}{0.20} = +5\text{ D}$$

(*a*) By Eq. (6j),

$$P = -2K = -4\text{ D}$$

(*b*) By Eq. (6i),

$$5 + V' = -4 \qquad \text{or} \qquad V' = -9\text{ D}$$

or

$$s' = \frac{1}{V'} = -\frac{1}{9} = -0.111\text{ m} = -11.1\text{ cm}$$

(*c*) By Eq. (6k),

$$m = -\frac{5}{-9} = +0.555$$

The power $P = -4$ D, and the image is virtual and erect. It is located 11.1 cm to the right of the mirror and has a magnification of 0.555×.

6.5 THICK MIRRORS

The term *thick mirror* is applied to a lens system in which one of the spherical surfaces is a reflector. Under these circumstances the light passing through the system is reflected by the mirror back through the lens system, from which it emerges finally into the space from which it entered the lens. Three common forms of optical systems that may be classified as thick mirrors are shown in Fig. 6H. In each case the surface farthest to the right has been drawn with a heavier line than the others, designating the reflecting surface. A parallel incident ray is also traced through each system to where it crosses the axis, thus locating the focal point.

In addition to a focal point and focal plane every thick mirror has a principal point and a principal plane. Two graphical methods by which principal points and planes can be located are given below. The *oblique-ray method* is applied to (*a*) the thin lens and mirror combination in Fig. 6I, while the *auxiliary-diagram method* is applied to (*b*) the thick lens and mirror combination in Fig. 6J.

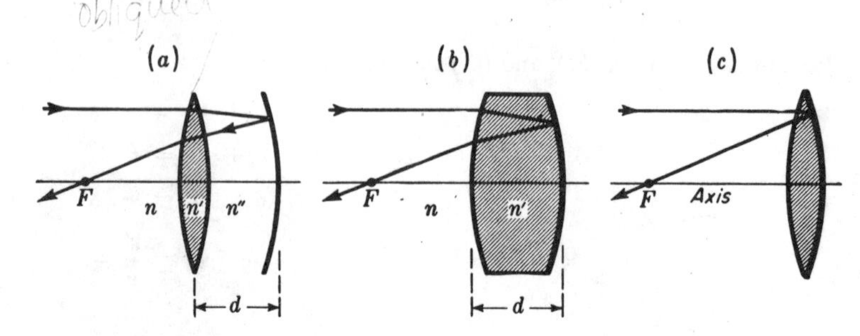

FIGURE 6H
Diagrams of several types of thick mirrors, showing the location of their respective focal points.

In the first illustration the lens is considered *thin* so that its own principal points may be assumed to coincide at H_1, its center. An incident ray parallel to the axis is refracted by the lens, reflected by the mirror, and again refracted by the lens before it crosses the axis of the system at F. The point T where the incident and final rays, when extended, cross each other locates the principal plane, and H represents the principal point. If we follow the sign conventions for a single mirror (Sec. 6.3), the focal length f of this particular combination is positive and is given by the interval FH.

In the second illustration (Fig. 6J) the incident ray is refracted by the first surface, reflected by the second, and finally refracted a second time by the first surface to a point F where it crosses the axis. The point T where the incident and final rays intersect locates the principal plane and principal point H.

The graphical ray-tracing construction for this case, shown in the auxiliary diagram in Fig. 6J, is started by drawing XZ parallel to the axis. With the origin O near the center, intervals proportional to n and n' are measured off in both directions

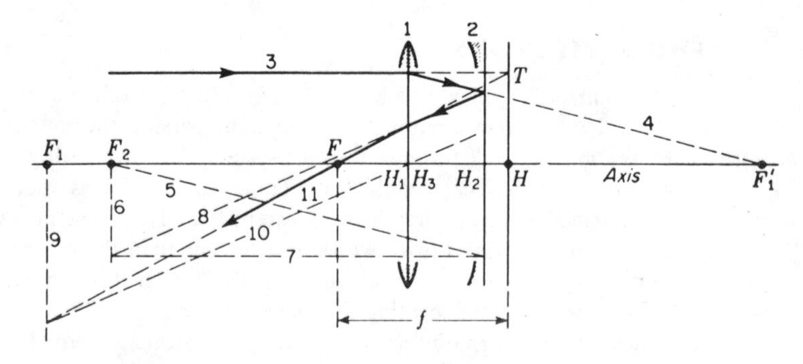

FIGURE 6I
Oblique-ray construction for locating the focal point and principal point of a thick mirror.

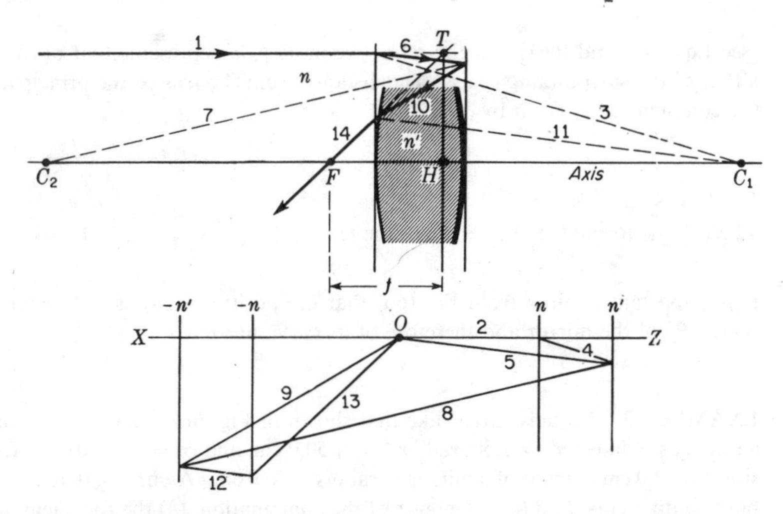

FIGURE 6J
Auxiliary-diagram method of graphically locating the focal point and principal point of a thick mirror.

along XZ. After the vertical lines representing n and n' are drawn, the remaining lines are drawn in the order of the numbers 1, 2, 3, ... Each even-numbered line is drawn parallel to its preceding odd-numbered line. The proof that this construction is exact for paraxial rays is similar to that given for Fig. 3I.

6.6 THICK-MIRROR FORMULAS

These formulas will be given in the power notation for case (*a*) shown in Fig. 6H. When r_1, r_2, and r_3 are the radii of the three surfaces consecutively from left to right, the power of the combination can be shown* to be given by

$$P = (1 - cP_1)(2P_1 + P_2 - cP_1P_2) \tag{6l}$$

where, for the case in diagram (*a*) only and $n'' = n$,

$$P_1 = (n' - n)(K_1 - K_2) \tag{6m}$$

$$P_2 = -2nK_3 \tag{6n}$$

and

$$K_1 = \frac{1}{r_1} \qquad K_2 = \frac{1}{r_2} \qquad K_3 = \frac{1}{r_3}$$

* For a derivation of these equations, see J. P. C. Southall, "Mirrors, Prisms, and Lenses," 3d ed., p. 379, The Macmillan Company, New York, 1936.

[see Eqs. (4p) and (6d)]. Of the refractive indices, n' represents that of the lens and n that of the surrounding space. The distance from the lens to the principal point of the combination is given by

$$H_1H = \frac{c}{1 - cP_1} \tag{6o}$$

where H_1 is located at the center of the lens and

$$c = \frac{d}{n} \tag{6p}$$

It is important to note from Eq. (6o) that the position of H is independent of the power P_2 of the mirror and therefore of its curvature K_3.

EXAMPLE 3 A thick mirror like that shown in Fig. 6H(*a*) has as one component a thin lens of index $n' = 1.50$, radii $r_1 = +50.0$ cm, and $r_2 = -50.0$ cm. This lens is situated 10.0 cm in front of a mirror of radius -50.0 cm. Assuming that air surrounds both components, find (*a*) the power of the combination, (*b*) the focal length, and (*c*) the principal point.

SOLUTION (*a*) By Eq. (6m), the power of the lens is

$$P_1 = (1.50 - 1)\left(\frac{1}{0.50} - \frac{1}{-0.50}\right) = +2 \text{ D}$$

Equation (6n) gives for the power of the mirror

$$P_2 = -2\,\frac{1}{-0.50} = +4 \text{ D}$$

From Eq. (6p),

$$c = \frac{d}{n} = \frac{0.10}{1} = 0.10 \text{ m}$$

Finally the power of the combination is given by Eq. (6l) as

$$P = (1 - 0.10 \times 2)(2 \times 2 + 4 - 0.10 \times 2 \times 4)$$
$$= 0.8(4 + 4 - 0.8) = +5.76 \text{ D}$$

(*b*) A power of $+5.76$ D corresponds to a focal length

$$f = \frac{1}{P} = \frac{1}{5.76} = 0.173 \text{ m} = +17.3 \text{ cm}$$

(*c*) The position of the principal point H is determined from Eq. (6o) through the distance

$$H_1H = \frac{0.10}{1 - 0.10 \times 2} = \frac{0.10}{0.80} = 0.125 \text{ m} = +12.5 \text{ cm}$$

It is therefore 12.5 cm to the right of the lens, or 2.5 cm in back of the mirror.

6.7 OTHER THICK MIRRORS

As a second illustration of a thick mirror, consider the thick lens silvered on the back, as shown in Fig. 6H(*b*). A comparison of this system with the one in diagram (*a*) shows that Eqs. (6l) to (6p) will apply if the powers P_1 and P_2 are properly defined. For diagram (*b*), P_1 refers to the power of the first surface alone, and P_2 refers to the power of the second surface as a mirror of radius r_2 in a medium of index n'. In other words,

$$P_1 = \frac{n' - n}{r_1} \qquad P_2 = -\frac{2n'}{r_2} \qquad \text{and} \qquad c = \frac{d}{n'} \tag{6q}$$

With these definitions the power of thick mirror (*b*) is given by Eq. (6l) and the principal point by Eq. (6o).

The third illustration of a thick mirror consists of a thin lens silvered on the back surface as shown in Fig. 6H(*c*). This system may be looked upon (1) as a special case of diagram (*a*), where the mirror has the same radius as the back surface of the thin lens and the spacing *d* is reduced to zero, or (2) as a special case of diagram (*b*), where the thickness is reduced to practically zero. In either case Eq. (6l) reduces to

● $$P = 2P_1 + P_2 \tag{6r}$$

and the principal point H coincides with H_1 at the common center of the lens and mirror. P_1 represents the power of the thin lens in air and P_2 the power of the mirror in air, or P_1 represents the power of the first surface of radius r_1 and P_2 represents the power of the second surface as a mirror of radius r_2 in a medium of index n' [see Eq. (6q)].

6.8 SPHERICAL ABERRATION

The discussion of a single spherical mirror in the preceding sections has been confined to paraxial rays. Within this rather narrow limitation, sharp images of objects at any distance may be formed on a screen, since bundles of parallel rays close to the axis and making only small angles with it are brought to a sharp focus in the focal plane. If, however, the light is not confined to the paraxial region, all rays from one object point do not come to a focus at a common point and we have an undesirable effect known as *spherical aberration*. The phenomenon is illustrated in Fig. 6K, where parallel incident rays at increasing distances h cross the axis closer to the mirror. The envelope of all rays forms what is known as a *caustic surface*. If a small screen is placed at the paraxial focal plane F and then moved toward the mirror, a point is reached where the size of the circular image spot is a minimum. This disklike spot is indicated in the diagram and is called the *circle of least confusion*.

The proof that rays from an outer zone of a concave mirror cross the axis inside the paraxial focal point can be simply given by reference to Fig. 6L. According to the law of reflection applied to the ray incident at T, the angle of reflection ϕ'' is equal to the angle of incidence ϕ. This in turn is equal to the angle TCA. Having

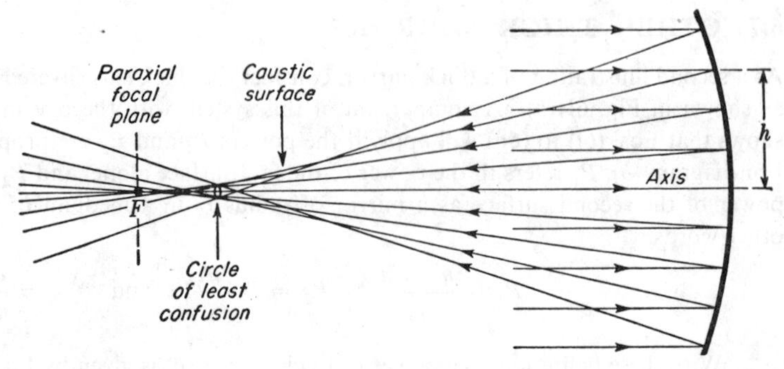

FIGURE 6K
Spherical aberration of a concave spherical mirror.

two equal angles, triangle CTX is isosceles, and hence $CX = XT$. Since a straight line is the shortest path between two points,

$$CT < CX + XT$$

Now CT is the radius of the mirror and equals CA, so that

$$CA < 2CX$$

Therefore

$$\tfrac{1}{2}CA < CX$$

The geometry of the figure shows that as T is moved toward A, the point X approaches F, and in the limit $CX = XA = FA = \frac{1}{2}CA$.

Over the past years numerous methods of reducing spherical aberration have been devised. If instead of a spherical surface the mirror form is that of a paraboloid of revolution, rays parallel to the axis are all brought to a focus at the same point as in Fig. 6M(a). Another method is the one shown later in Fig. 10Q of inserting a *corrector plate* in front of a spherical mirror, thereby deviating the rays by the proper amount prior to reflection. With the plate located at the center of curvature of the mirror, a very useful optical arrangement known as the *Schmidt system* is obtained. Still a third system, known as a *Mangin mirror*, is shown in Fig. 6M(b). Here a

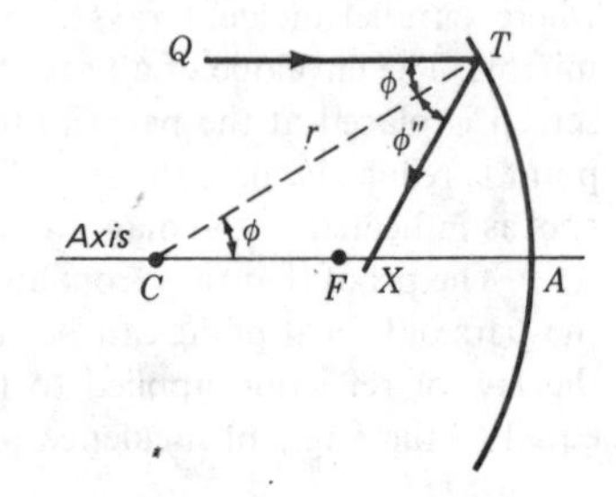

FIGURE 6L
Geometry showing how marginal rays parallel to the axis of a spherical mirror cross the axis inside the focal point.

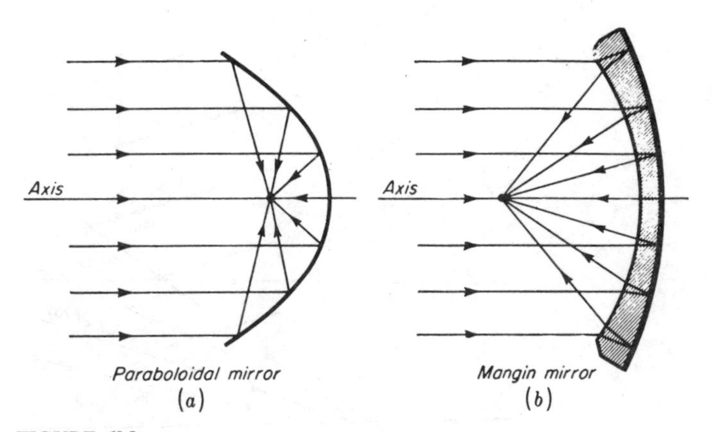

FIGURE 6M
(*a*) Concave parabolic mirror and (*b*) concave spherical mirror, corrected for spherical aberration.

meniscus lens is employed in which both surfaces are spherical. When the back surface is silvered to form the concave mirror, all parallel rays are brought to a reasonably good focus.

6.9 ASTIGMATISM

This defect of the image occurs when an object point lies some distance from the axis of a concave or convex mirror. The incident rays, whether parallel or not, make an appreciable angle ϕ with the mirror axis. The result is that, instead of a point image, two mutually perpendicular line images are formed. This effect is known as astigmatism and is illustrated by a perspective diagram in Fig. 6N. Here the incoming rays are parallel while the reflected rays are converging toward two lines S and T. The reflected rays in the vertical or *tangential* plane $RASE$ are seen to cross or to focus at T, while the fan of rays in the horizontal or *sagittal* plane $JAKE$ cross or focus at S. If a screen is placed at E and moved toward the mirror, the image will become a vertical line at S, a circular disk at L, and a horizontal line at T.

If the positions of the T and S images of distant object points are determined for a wide variety of angles, their loci will form a paraboloidal and a plane surface respectively, as shown in Fig. 6O. As the obliquity of the rays decreases and they approach the axis, the line images not only come closer together as they approach the paraxial focal plane but they shorten in length. The amount of astigmatism for any pencil of rays is given by the distance between the T and S surfaces measured along the chief ray.

Equations giving the two astigmatic image positions are*

$$\frac{1}{s} + \frac{1}{s'_T} = -\frac{2}{r \cos \phi} \qquad \frac{1}{s} + \frac{1}{s'_S} = -\frac{2 \cos \phi}{r}$$

* For a derivation of these equations, see G. S. Monk, "Light, Principles and Experiments," Dover Publications, Inc., New York, 1963.

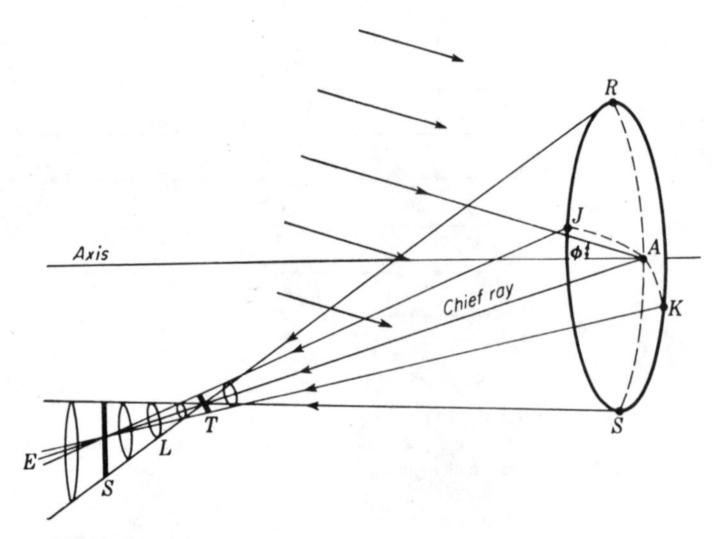

FIGURE 6N
Astigmatic images of an off-axis object point at infinity, as formed by a concave spherical mirror. The lines T and S are perpendicular to each other.

In both equations s and s' are measured along the chief ray. The angle ϕ is the angle of obliquity of the chief ray, and r is the radius of curvature of the mirror.

The Schmidt optical system, which will be discussed later (Fig. 10Q), and the Mangin mirror shown in Fig. 6M(b) constitute instruments in which the astigmatism of a spherical mirror is reduced to a minimum. While the two focal surfaces T and S exist for these devices, they lie very close together and the loci of their mean position (such as L in Fig. 6N) form a nearly spherical surface. The center of this spherical surface is located at the center of curvature of the mirror, as shown in Fig. 10Q.

A paraboloidal mirror is free from spherical aberration even for large apertures but shows unusually large astigmatic $S - T$ differences off the axis. For this reason paraboloidal reflectors are limited in their use to devices that require a small angular spread, such as astronomical telescopes and searchlights.

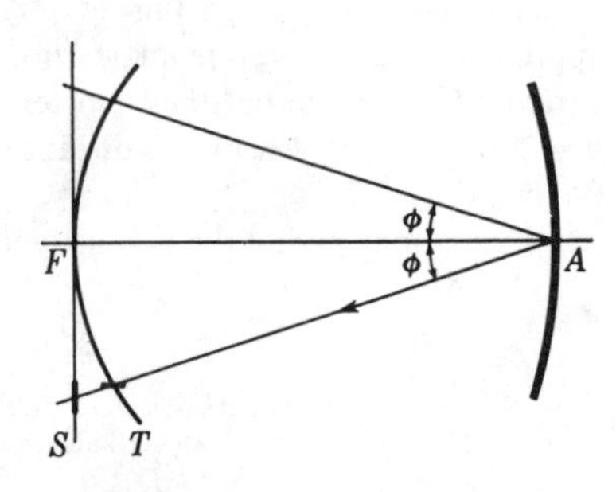

FIGURE 6O
Astigmatic surfaces for a concave spherical mirror.

PROBLEMS

6.1 A spherical mirror has a radius of -24.0 cm. An object 3.0 cm high is located in front of the mirror at a distance of (*a*) 48.0 cm, (*b*) 36.0 cm, (*c*) 24.0 cm, (*d*) 12.0 cm, and (*e*) 6.0 cm. Find the image distance for each of these object distances.

Ans. (*a*) $+16.0$ cm, (*b*) $+18.0$ cm, (*c*) $+24.0$ cm, (*d*) $+12.0$ cm, (*e*) -12.0 cm

6.2 Solve Prob. 6.1 graphically.

6.3 A spherical mirror has a radius of -15.0 cm. An object 2.50 cm high is located in front of the mirror at a distance of (*a*) 45.0 cm, (*b*) 30.0 cm, (*c*) 15.0 cm, (*d*) 10.0 cm, and (*e*) 5.0 cm. Find the image distance for each of these object distances.

6.4 Solve Prob. 6.3 graphically.

6.5 The radius of a spherical mirror is $+18.0$ cm. An object 4.0 cm high is located in front of the mirror at a distance of (*a*) 36.0 cm, (*b*) 24.0 cm, and (*c*) 12.0 cm. Find the image distance and image size for each of these object distances.

Ans. (*a*) -7.20 cm from vertex and $+0.80$ cm high, (*b*) -6.55 cm from vertex and $+1.092$ cm high, (*c*) -5.40 cm from vertex and $+1.712$ cm high

6.6 Solve Prob. 6.5 graphically.

6.7 The radius of a spherical mirror is $+8.0$ cm. An object 3.50 cm high is located in front of the mirror at a distance of (*a*) 16.0 cm, (*b*) 8.0 cm, (*c*) 4.0 cm, and (*d*) 2.0 cm. Find the image distance and image size for each of these object distances.

6.8 Solve Prob. 6.7 graphically.

6.9 A concave mirror is to be used to focus the image of a tree on a photographic film 8.50 m away from the tree. If a lateral magnification of $-\frac{1}{20}$ is desired, what should be the radius of curvature of the mirror? *Ans.* -85.2 cm

6.10 A thin equiconvex lens of index 1.530 and radii of 16.0 cm is silvered on one side. Find the (*a*) focal length and (*b*) power of this system if light enters the unsilvered side.

6.11 A thin lens of index 1.650 has radii $r_1 = +5.0$ cm and $r_2 = -15.0$ cm. If the second surface is silvered, what is (*a*) the focal length and (*b*) the power of the system?

6.12 A thin lens of index 1.720 located in air has radii $r_1 = -6.0$ cm and $r_2 = -12.0$ cm. If the second surface is silvered, what is the power of the system? Use the special-case formulas (6q) and (6r).

6.13 A thin lens with a focal length of $+10.0$ cm is located 2.00 cm in front of a spherical mirror with a radius of -18.0 cm. Find (*a*) the power, (*b*) the focal length, (*c*) the principal point, and (*d*) the focal point of this thick-mirror optical system.

Ans. (*a*) $+23.11$ D, (*b*) $+4.33$ cm, (*c*) $H_1H = +2.50$ cm, (*d*) -1.83 cm

6.14 Solve Prob. 6.13 graphically. Use the method shown in Fig. 6I.

6.15 A thin lens with a focal length of -12.30 cm is placed 2.50 cm in front of a spherical mirror of radius -9.20 cm. Find the power of (*a*) the first lens and (*b*) the second lens. Calculate (*c*) the power of the system and (*d*) its focal length. Locate (*e*) the principal point and (*f*) the focal point.

6.16 Solve Prob. 6.15 graphically. Use the method of Fig. 6I.

6.17 A thick lens of index 1.560 has radii $r_1 = +15.0$ cm and $r_2 = -30.0$ cm. If the second surface is silvered and the lens is 5.0 cm thick, find (*a*) the power, (*b*) the focal length, (*c*) the principal point, and (*d*) the focal point.

Ans. (*a*) $+14.67$ D, (*b*) $+6.82$ cm, (*c*) $H_1H = +3.640$ cm, (*d*) $H_1F = +3.180$ cm

6.18 Solve Prob. 6.17 graphically.

6.19 A lens 4.50 cm thick has an index of 1.720 and radii $r_1 = -6.0$ cm and $r_2 = -12.0$ cm. If the second surface is silvered, find (*a*) the power, (*b*) the focal length, (*c*) the position of the principal point, and (*d*) the position of the focal point.

6.20 Solve Prob. 6.19 graphically.

6.21 The curved surface of a plano-convex lens has a radius of 20.0 cm. The refractive index of the glass is 1.650, and the thickness is 2.750 cm. If the curved surface is silvered, find (*a*) the power, (*b*) the focal length, (*c*) the principal point, and (*d*) the focal point. *Ans.* (*a*) +16.50 D, (*b*) +6.06 cm, (*c*) +1.667 cm, (*d*) +4.394 cm

6.22 Solve Prob. 6.21 graphically. Use the method shown in Fig. 6J.

6.23 If the plane surface of the lens given in Prob. 6.21 is silvered in place of the curved surface, what are the answers to parts (*a*) to (*d*)?

6.24 Solve Prob. 6.23 graphically. Use the method shown in Fig. 6I.

6.25 An object is located 20.0 cm in front of a mirror of radius −16.0 cm. Plot a graph of the two astigmatic surfaces for (*a*) $\phi = 0°$, (*b*) $\phi = 10.0°$, (*c*) $\phi = 20.0°$, and (*d*) $\phi = 30.0°$.

6.26 Plot a graph of the two astigmatic surfaces for a spherical mirror having a radius of −20.0 cm. Assume parallel incident light, and show the curves for (*a*) $\phi = 0°$, (*b*) $\phi = 10.0°$, (*c*) $\phi = 20.0°$, and (*d*) $\phi = 30.0°$.

7

THE EFFECTS OF STOPS

One subject in geometrical optics, though very important from a practical standpoint, is frequently neglected because it does not directly concern the size, position, and sharpness of the image. This is the question of the *field of view*, which determines how much of the surface of a broad object can be seen through an optical system. In treating the field of view it is of primary importance to understand how and where the bundle of rays traversing the system is limited. The effect of stops or diaphragms, which will always exist (even if only as the rims of lenses or mirrors), must be investigated.

7.1 FIELD STOP AND APERTURE STOP

In Fig. 7A a single lens with two stops is shown forming the image of a distinct object. Three bundles of parallel rays from three different points on the object are shown brought to a focus in the focal plane of the lens. It can be seen from these bundles that the stop close to the lens limits the size of each bundle of rays, while the stop just in front of the focal plane limits the angle at which the incident bundles can get through

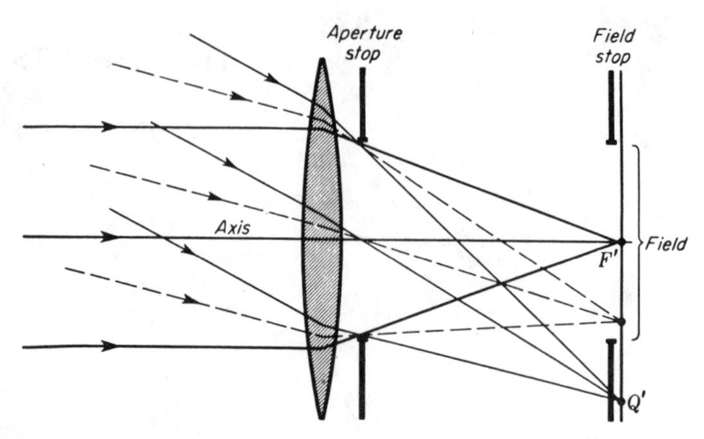

FIGURE 7A
The difference between a field stop and an aperture stop.

to this plane. The first is called an *aperture stop*. It obviously determines the amount of light reaching any given point in the image and therefore controls the brightness of the latter. The second, or *field stop*, determines the extent of the object, or the field, that will be represented in the image.

7.2 ENTRANCE AND EXIT PUPILS

A stop $P'E'L'$ placed behind the lens as in Fig. 7B is in the image space and limits the image rays. By a graphical construction or by the lens formula, the image of this real stop, as formed by the lens, is found to lie at the position PEL shown by the broken lines. Since $P'E'L'$ is inside the focal plane, its image PEL lies in the object space and is virtual and erect. It is called the *entrance pupil*, while the real aperture $P'E'L'$ is, as we have seen, called the aperture stop. When it lies in the image space, as it does here, it becomes the *exit pupil*. (For a treatment of object and image spaces see Sec. 4.11.)

It should be emphasized that P and P', E and E', and L and L' are pairs of conjugate points. Any ray in the object space directed through one of these points will after refraction pass through its conjugate point in the image space. Ray IT directed toward P is refracted through P', ray KR directed toward E is refracted through E', and ray NU directed toward L is refracted through L'. The image point Q' is located graphically by the broken line JQ', parallel to the others and passing undeviated through the optical center A. The aperture stop $P'E'L'$ in the position shown also functions to some extent as a field stop, but the edges of the field will not be sharply limited. The diaphragm which acts as a field stop is usually made to coincide with a real or virtual image, so that the edges will appear sharp.

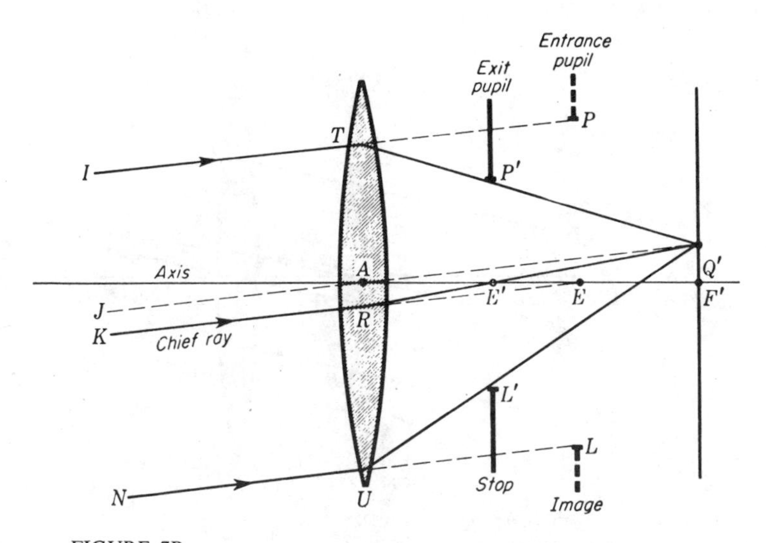

FIGURE 7B
How an aperture stop and its image can become the exit and entrance pupils, respectively, of a system.

7.3 CHIEF RAY

Any ray in the object space that passes through the center of the entrance pupil is called a *chief ray*. Such a ray after refraction also passes through the center of the exit pupil. In any actual optical instrument the chief ray rarely passes through the center of any lens itself. The points E and E' at which the chief ray crosses the axis are known as the *entrance-pupil point* and the *exit-pupil point*. The former, as we shall see, is particularly important in determining the field of view.

7.4 FRONT STOP

In certain types of photographic lenses a stop is placed close to the lens, either before it (*front stop*) or behind it (*rear stop*). One of the functions of such a stop, as will be seen in Chap. 9, is to improve the quality of the image formed on the photographic film. With a front stop as shown in Fig. 7C its small size and its location in the object space make it the entrance pupil. Its image $P'E'L'$ formed by the lens is in the image space and constitutes the exit pupil. Parallel rays IT, JW, and NU have been drawn through the two edges of the entrance pupil and through its center. The lens causes these rays to converge toward the screen as though they had come from the conjugate points P', E', and L' in the exit pupil. Their intersection at the image point Q' occurs where the undeviated ray KA crosses the secondary focal plane. Note that the chief ray is directed through the center of the entrance pupil in the object space and

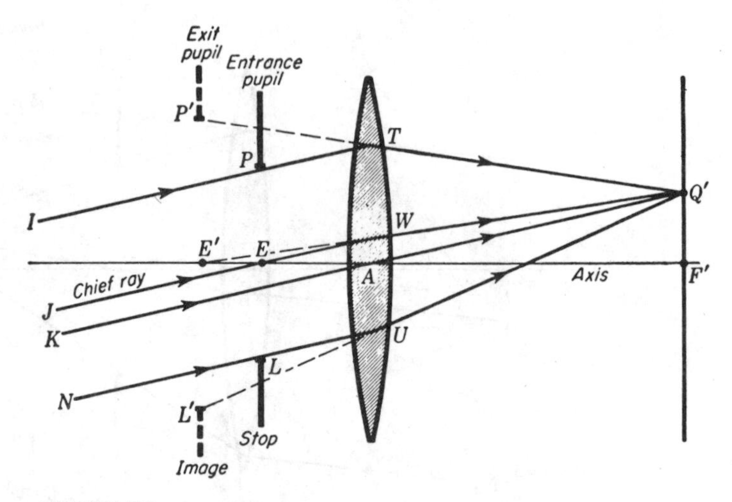

FIGURE 7C
A front stop and its image can become the respective entrance and exit pupils of a system.

emerges from the lens as though it had come from the center of the exit pupil in the image space.

While a certain stop of an optical system may limit the rays getting through the system from one object point, it may not be the aperture stop for other object points at different distances away along the axis. For example, in Fig. 7D a lens with a front stop is shown with an object point at M. For this point the periphery of the lens itself becomes the aperture stop, and since it limits the object rays, it is the entrance pupil. Its image, which is again the lens periphery, is also the exit pupil. The lens margin is therefore the aperture stop, the entrance pupil, and the exit pupil for the point M. If this object point were to lie to the left of Z, PEL would become the entrance pupil and the aperture stop and its image $P'E'L'$ the exit pupil.

In the preliminary design of an optical instrument it may not be known which element of the system will constitute the aperture stop. As a result the marginal rays for each element must be investigated one after the other to determine which one actually does the limiting. Regardless of the number of elements the system has, it will usually contain but one limiting aperture stop. Once this stop is located, *the entrance pupil of the entire system is the image of the aperture stop formed by all lenses preceding it and the exit pupil is the image formed by all lenses following it.* Figures 7B and 7C, where there is only a single lens either before or behind the stop, should be studied in connection with this statement.

7.5 STOP BETWEEN TWO LENSES

A common arrangement in photographic lenses is to have two separate lens elements with a variable stop, or iris diaphragm, between them. Figure 7E is a diagram representing such a combination, and in it the elements 1 and 2 are thin lenses while

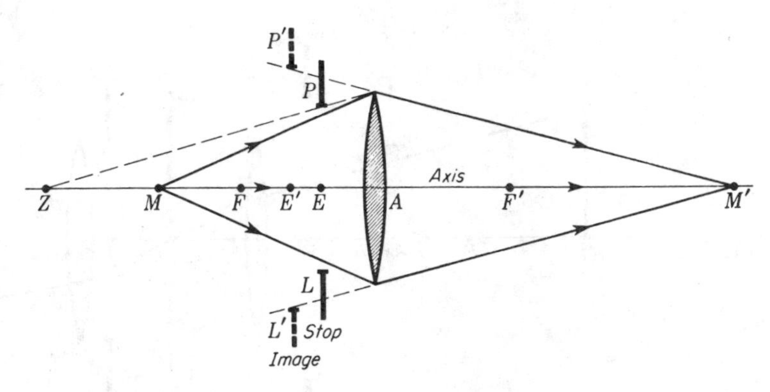

FIGURE 7D
The entrance and exit pupils are not the same for all object and image points.

$P_0E_0L_0$ is the stop. By definition the entrance pupil of this system is the image of the stop formed by lens 1. This image is virtual, erect, and located at PEL. Similarly by definition the exit pupil of the entire system is the image of the stop formed by lens 2. This image, located at $P'E'L'$, is also virtual and erect. The entrance pupil PEL lies in the object space of lens 1, the stop $P_0E_0L_0$ lies in the image space of lens 1 as well as in the object space of lens 2, and the exit pupil $P'E'L'$ lies in the image space of lens 2. Points P_0 and P, E_0 and E, and L_0 and L are conjugate pairs of points for the first lens, while P_0 and P', E_0 and E', and L_0 and L' are conjugate pairs for the second lens. This makes points like P and P' conjugate for the whole system. If a point object is located on the axis at M, rays MP and ML limit the bundle that will get through the system. At the first lens these rays are refracted

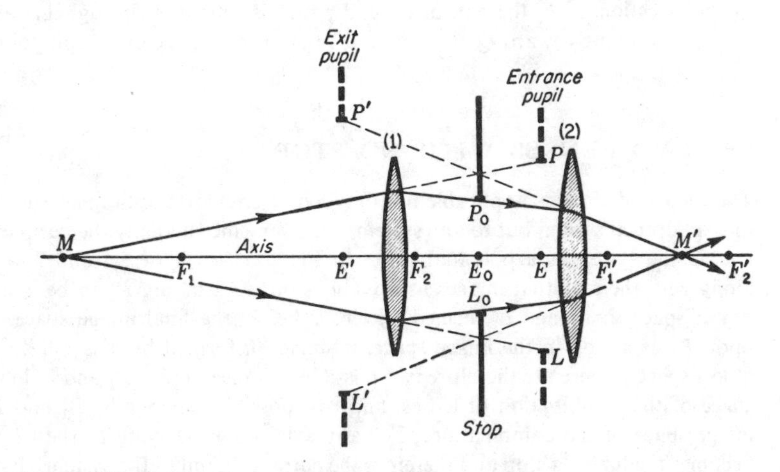

FIGURE 7E
Stop between two lenses. The entrance pupil of a system is in its object space, while the exit pupil is in its image space.

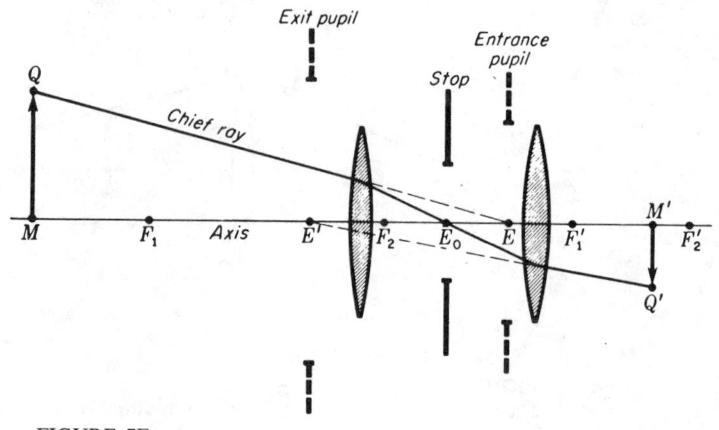

FIGURE 7F
The direction taken by any chief ray is such that it passes through the centers of the entrance pupil, the stop, and the exit pupil.

through P_0 and L_0, and at the second lens they are again refracted in such directions that they appear to come from P' and L' as shown. The purpose of using primed and unprimed symbols to designate exit and entrance pupils respectively should now be clear; one lies in the image space, the other in the object space, and they are conjugate images.

The same optical system is shown again in Fig. 7F for the purpose of illustrating the path of a chief ray. Of the many rays that can start from any specified object point Q and traverse the entire system, a chief ray is one which approaches the lens in the direction of E, the entrance pupil point, is refracted through E_0, and finally emerges traveling toward Q' as though it came from E', the exit pupil point.

7.6 TWO LENSES WITH NO STOP

The theory of stops is applicable not only where circular diaphragms are introduced into an optical system but to any system whatever, since actually the periphery of any lens in the system is a potential stop. In Fig. 7G two lenses, 1 and 2, are shown, along with their mutual images as possible stops. Assuming P_1 to be a stop in the object space, its image P' formed by lens 2 lies in the final image space. Looking upon P_2 as a stop in the image space, its image P formed by lens 1 lies in the first object space. There are therefore two possible entrance pupils, P_1 and P, in the object space of the combination of lenses, and two possible exit pupils, P_2 and P', in the image space of the combination. For any axial point M lying to the left of Z, P_1 becomes the limiting stop and therefore the entrance pupil of the system. Its image P' becomes the exit pupil. If, on the other hand, M lies to the right of Z, P becomes the entrance pupil and P_2 the exit pupil.

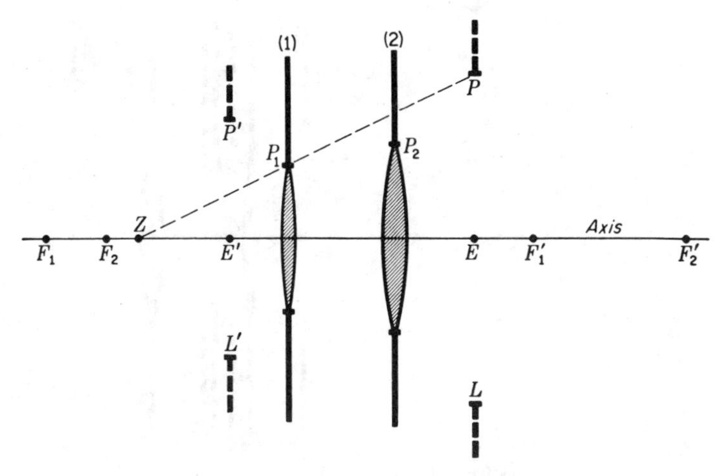

FIGURE 7G
The margin of any lens may be the aperture stop of the system.

7.7 DETERMINATION OF THE APERTURE STOP

In the system of two lenses with a stop between them represented in Figs. 7E and 7F, the lenses were made sufficiently large not to become aperture stops. If, however, they are not large compared with the stop, as may well be the case with a camera lens when the iris diaphragm is wide open, the system of stops and pupils may become similar to those shown in Fig. 7H. This system consists of two lenses and a stop, each one of which, along with its various images, is a potential aperture stop. P_1' is the virtual image of the first lens formed by lens 2, P_0' the virtual image of the stop P formed by lens 2, P_0 the virtual image of P formed by lens 1, and P_2 the virtual image of the second lens formed by lens 1. In other words, when looking through the system from the left one would see the first lens, the stop, and the second lens in the apparent positions P_1, P_0, and P_2. Looking from the right, one would see them at P_1', P_0', and P_2'. Of all these stops P_0, P_1, and P_2 are potential entrance pupils located in the object space of the system.

For all axial object points lying to the left of X, P_1 limits the entering bundle of rays to the smallest angle and hence constitutes the entrance pupil of the system. In general the object of which it is the image will be the aperture stop, which in this case is the aperture P_1 of lens 1 itself. The image of the entrance pupil formed by the entire lens system, namely P_1', constitutes the exit pupil. For object points lying between X and Z, P_0 becomes the entrance pupil, P the aperture stop, and P_0' the exit pupil. Finally, for points to the right of Z, P_2 is the entrance pupil while P_2' is both the aperture stop and the exit pupil. It is apparent from this discussion that the aperture stop of any system may change with a change in the object position. The general rule is that *the aperture stop of the system is determined by that stop or image of a stop which subtends the smallest angle as seen from the object point*. If it is determined by an image, the aperture stop itself is the corresponding object. In most actual

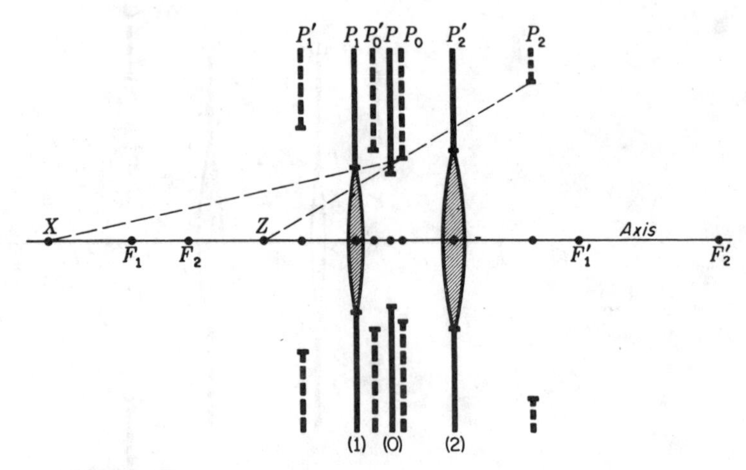

FIGURE 7H
A system composed of several elements has a number of possible stops and pupils.

optical instruments the effective stop does not change over the range of object positions normally covered by the instrument in use.

Having established the methods of determining the positions of the aperture stop and of the entrance and exit pupils, we may now take up the two important properties of an optical system, field of view and brightness. To begin with, let us consider the former property.

7.8 FIELD OF VIEW

When one looks out at a landscape through a window, the field of view outside is limited by the size of the window and by the position of the observer. In Fig. 7I the eye of the observer is shown at E, the window opening at JK, and the observed field at GH. In this simple illustration the window is the field stop (Sec. 7.1). When the eye is moved closer to the window, the angular field α is widened, while when it is moved farther away, the field is narrowed. It is common practice with optical instruments to specify the field of view in terms of the angle α and to express this angle in degrees. The angle θ which the extreme rays entering the system make with the axis is called the *half-field angle* and limits the width of the object that can be seen. This object field includes the angle 2θ, and in this instance is the same as the image field of angular width α.

7.9 FIELD OF A PLANE MIRROR

The field of view afforded by a plane mirror is very similar to that of a simple window. As shown in Fig. 7J, TU represents a plane mirror, and $P'E'L'$ the pupil of the observer's eye, which here constitutes the exit pupil. The entrance pupil PEL is the

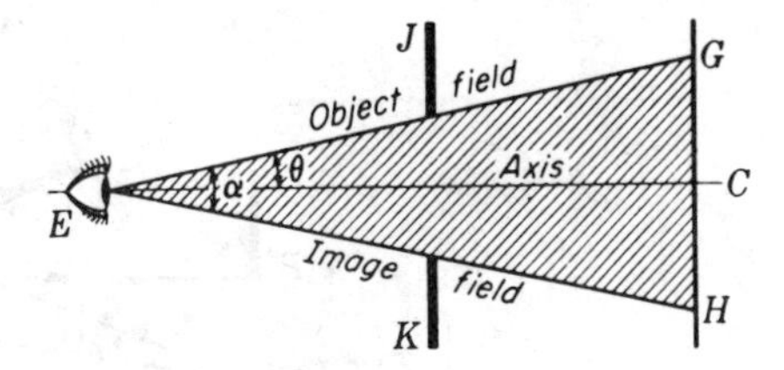

FIGURE 7I
Field of view through a window.

virtual image of the eye pupil formed by the mirror and is located just as far behind the mirror as the actual pupil is in front of it. The chief rays $E'T$ and $E'U$ limit the field of view in image space, while the corresponding incident rays ER and ES define the field of view in object space. The latter show the limits of the field in which an object can be situated and still be visible to the eye. In this case also, although not in general, it subtends the same angle as the image field.

The formation of the image of an object point Q within this field is also illustrated. From this point three rays have been drawn toward the points P, E, and L in the entrance pupil. Where these rays encounter the mirror, the reflected rays are drawn toward the conjugate points P', E', and L' in the exit pupil. The object Q and the entrance pupil PEL are in the object space, while the image Q' and the exit pupil $P'E'L'$ are in the image space. If Q happens to be located close to RT, only part of the bundle of rays defined by the entrance pupil will be intercepted by the mirror and will be reflected into the exit pupil. In defining the field of view it is customary to use the chief ray RTE', although in the present case this distinction is not important because of the relative smallness of the pupil of the eye. Its size is obviously greatly exaggerated in the diagram.

Since the limiting chief ray is directed toward the entrance pupil point E, the

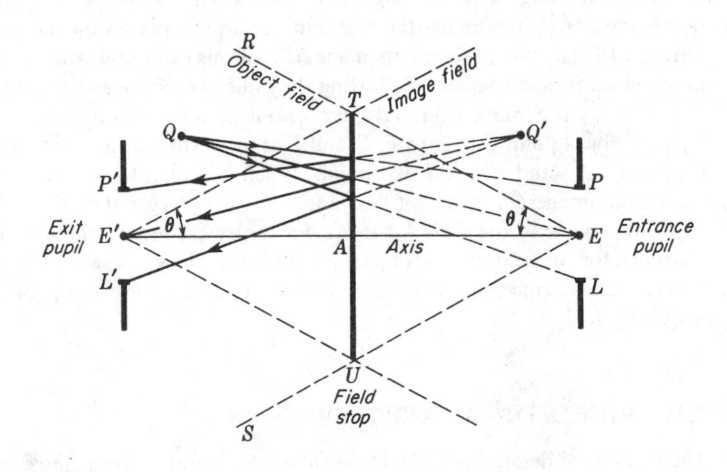

FIGURE 7J
Field of view looking in a plane mirror.

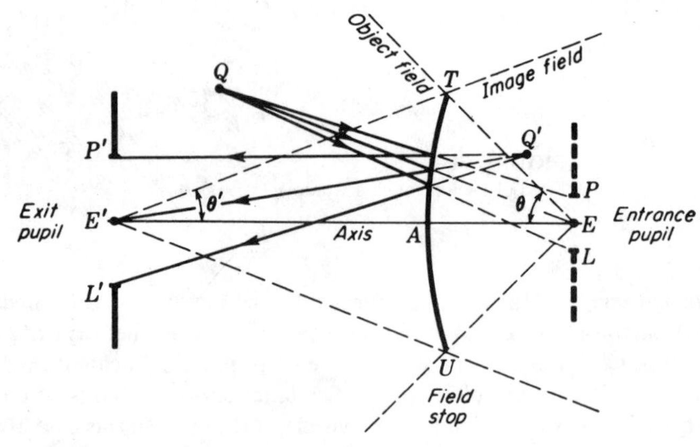

FIGURE 7K
Field of view looking in a convex mirror.

half-field angle θ is in general determined by the smallest angle subtended at E by any stop, or image of a stop, in the object space. *The stop determined in this way is the field stop of the system.* For a single mirror the field stop is the border of the mirror itself.

7.10 FIELD OF A CONVEX MIRROR

When the mirror has a curvature, the situation is little changed except that the object field and the image field no longer subtend the same angle ($\theta \neq \theta'$ in Fig. 7K). In this figure $P'E'L'$ represents the real pupil of an eye placed on the axis of a convex mirror TU. The mirror forms an image PEL of this exit pupil, and this is the entrance pupil which is now smaller. Following the same procedure as for a plane mirror, the lines limiting the image field and the object field have been drawn. Rays emanating from an object point Q toward P, E, and L of the entrance pupil are shown as reflected toward P', E', and L' in the exit pupil. When extended backward, these rays locate the virtual image Q'. The half-field angle θ is here larger than θ', which determines the field of view to the eye. A similar but somewhat more complicated diagram can be drawn for the field of view of a concave mirror. This case will be left as an exercise for the student, since it is very similar to that of a converging lens to be discussed next. See Prob. 7.12.

7.11 FIELD OF A POSITIVE LENS

The method of determining the half-field angles θ and θ' for a single converging lens is shown in Fig. 7L. The pupil of the eye, as an exit pupil, is situated on the right, and its real inverted image appears at the left. The chief rays through the entrance pupil

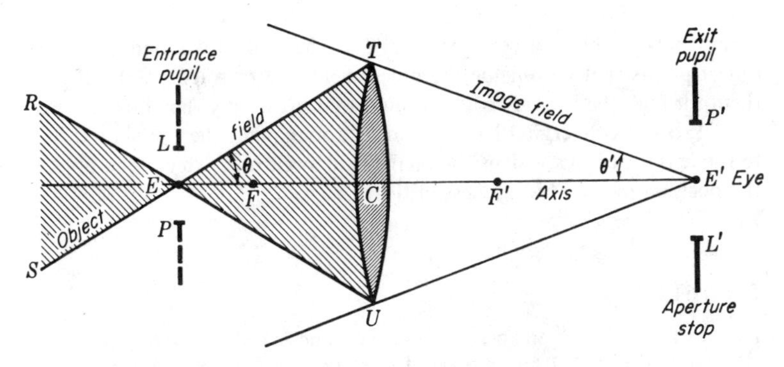

FIGURE 7L
Field of view looking in a converging lens.

point E which are incident at the periphery of the lens are refracted through the conjugate point E'.

The shaded areas, or rather cones, ETU and ERS mark the boundaries within which any object must lie in order to be seen in the image field. The field stop in this case is the lens TU itself, since it determines the half-field angle subtended at the entrance pupil point. If the eye, and therefore the exit pupil, is moved closer to the lens, thereby increasing the image-field angle θ', the inverted entrance pupil moves to the left, causing a lengthening of the object-field cone ETU.

The same lens has been redrawn in Fig. 7M, where an object QM is shown in a position inside the primary focal point. Through each of the three points P, E, and L, rays are drawn from Q to the lens. From there the refracted rays are directed through the corresponding points P', E', and L' on the exit pupil. By extending them backward to their common intersection, the virtual image is located at Q'. The oblique-ray or parallel-ray methods of construction (not shown) may be used to confirm this position of the image. It will be noted that if objects are to be placed near the entrance pupil point E, they must be very small; otherwise only a part of

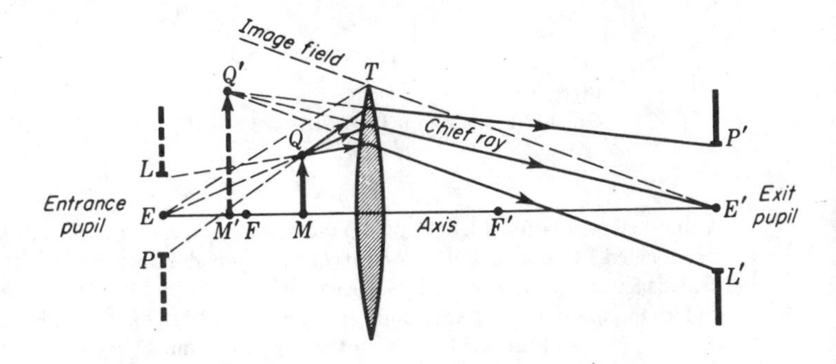

FIGURE 7M
Image formation within the field of a converging-lens system.

them will be visible to an eye placed at E'. The student will find it instructive to select object points that lie outside the object field and to trace graphically the rays from them through the lens. It will be found that invariably they miss the exit pupil.

When a converging lens is used as a magnifier, the eye should be placed close to the lens, since this widens the image-field angle and extends the object field so that the position of the object is less critical.

PROBLEMS

7.1 A thin lens with an aperture of 4.80 cm and a focal length of +3.50 cm has a 3.0-cm stop located 1.50 cm in front of it. An object 1.50 cm high is located with its lower end on the axis 8.0 cm in front of the lens. Locate graphically and by formula (*a*) the position and (*b*) the size of the exit pupil. (*c*) Locate the image of the object graphically by drawing the two marginal rays and the chief ray from the top end of the object.

Ans. (*a*) $s' = -2.625$ cm, (*b*) 5.250 cm, (*c*) See Fig. P7.1; $y' = -1.167$ cm; $s' = +6.222$ cm

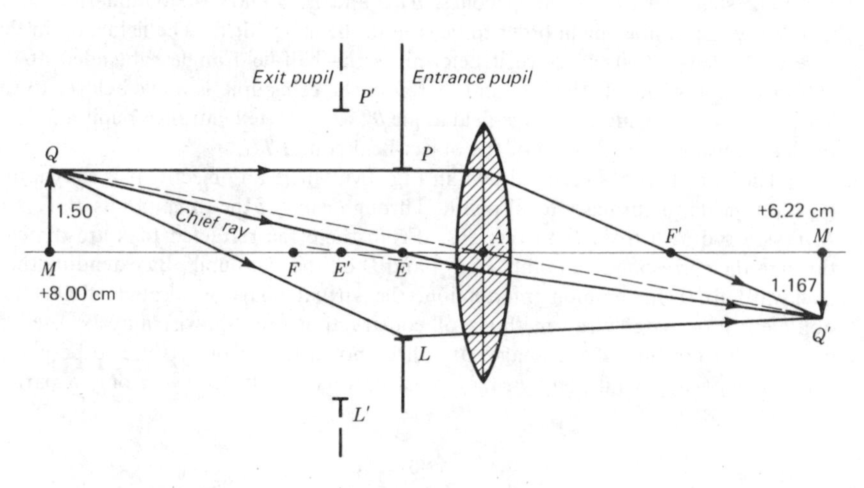

FIGURE P7.1
Graphical solution for Prob. 7.1.

7.2 A thin lens with a focal length of +5.0 cm and an aperture of 6.0 cm has a 3.80-cm stop located 1.60 cm behind it. An object 2.20 cm high is located with its lower end on the axis 8.0 cm in front of the lens. Locate graphically and by formula (*a*) the position and (*b*) the size of the entrance pupil. (*c*) Locate the object graphically by drawing two marginal rays and the chief ray from the top end of the object.

7.3 A thin lens with a focal length of −6.0 cm and an aperture of 7.0 cm has a 3.0-cm stop located 3.0 cm in front of it. An object 2.0 cm high is located with its lower end on the axis 10.0 cm in front of the lens. Find graphically and by formula (*a*) the position and (*b*) the size of the exit pupil. (*c*) Graphically locate the image by drawing the two marginal rays and the chief ray from the top of the object.

7.4 A thin lens with a focal length of +6.0 cm has an aperture of 6.0 cm. A 6.0-cm stop is located 2.0 cm in front of the lens, and a 4.0-cm stop is located 2.0 cm behind the lens. An object 4.0 cm high is centrally located on the axis 12.0 cm in front of the lens. Find the images of the two stops, and determine (*a*) the stop of the system, (*b*) its size, and (*c*) its position with respect to the lens. (*d*) Locate the image and determine its size by drawing the two marginal rays and the chief ray from the top end of the object. (*e*) Solve graphically (Fig. P7.4).

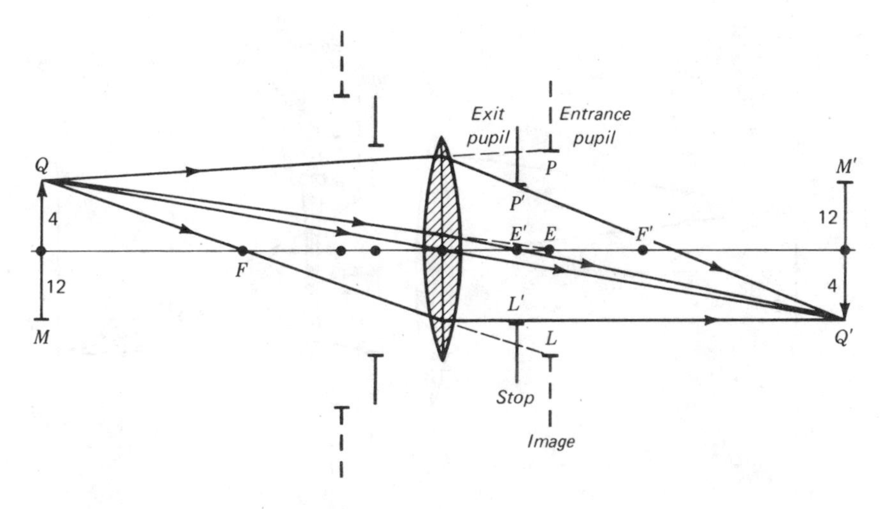

FIGURE P7.4
Graphical solution for Prob. 7.4.

Ans. (*a*) the 4.0-cm stop is the stop of the system, and its image, which is in object space, is the entrance pupil; (*b*) 6.0 cm; (*c*) 3.0 cm behind the lens; (*d*) $y' = -4.0$ cm, $s' = +12.0$ cm, real and inverted

7.5 Two thin lenses with focal lengths of +5.0 and +7.0 cm and apertures of 8.0 and 9.0 cm, respectively, are located 3.50 cm apart. A stop 5.0 cm in diameter is located between the two lenses and 2.0 cm from the first lens. An object 4.0 cm high is located with its center 10.0 cm in front of the first lens. Find graphically and by formula (*a*) the position and (*b*) the size of the entrance pupil. Find (*c*) the position and (*d*) the size of the exit pupil. Find (*e*) the position and (*f*) the size of the final image. Draw the two marginal rays and the chief ray from the top end of the object to the image.

7.6 Two thin lenses with focal lengths of +7.0 and +6.0 cm and apertures of 9.0 and 8.0 cm, respectively, are located 5.0 cm apart. A stop of 6.0 cm diameter is located between the two lenses 2.0 cm from A_1. An object 6.0 cm high is located with its center 9.0 cm in front of the first lens. Find graphically and by formula (*a*) the position and (*b*) the size of the entrance pupil. Find (*c*) the position and (*d*) the size of the exit pupil. Find (*e*) the position and (*f*) the size of the final image. Draw two marginal rays and the chief ray from the top end of the object to the image.

7.7 A thin lens with an aperture of 6.0 cm and focal length of -10.0 cm is located 4.0 cm behind another thin lens with an aperture of 8.0 cm and a focal length of $+5.0$ cm. An object 4.0 cm high is located with its center on the axis $+12.0$ cm in front of the first lens, and a stop 5.0 cm in diameter is located midway between the lenses. Calculate and graphically find (*a*) the size and position of the entrance pupil, (*b*) the size and position of the exit pupil, and (*c*) the size and position of the final image. See Fig. P7.7 below.

Ans. (*a*) $+8.33$ and -3.333 cm, (*b*) $+4.17$ and -1.667 cm, (*c*) $+5.26$ and $+8.42$ cm

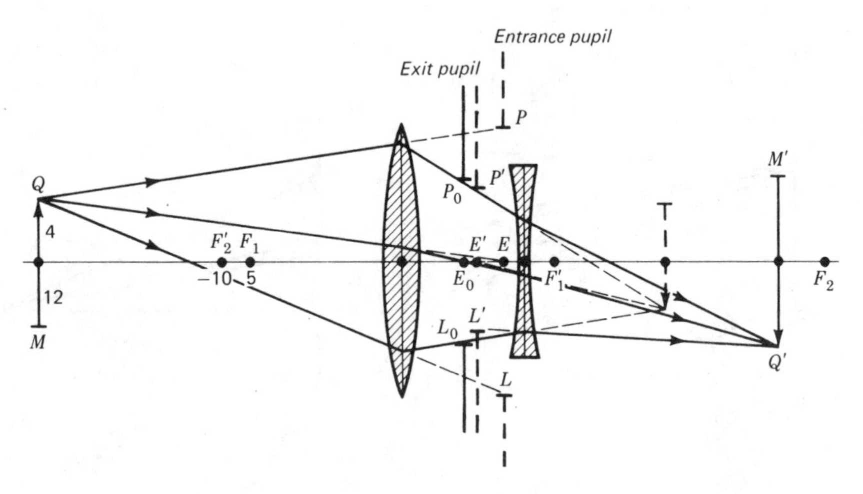

FIGURE P7.7
Graphical solution for Prob. 7.7.

7.8 A thin lens with a focal length of 9.0 cm and an aperture of 6.0 cm is located 4.50 cm in front of a diverging lens with a focal length of -8.0 cm and an aperture of 6.0 cm. For light incident on the first lens parallel to the axis calculate the position and size of (*a*) the entrance pupil and (*b*) the exit pupil. (*c*) Solve graphically. Find (*d*) the focal point of the system, (*e*) the principal point to which it is measured, and (*f*) the focal length.

7.9 A clear glass marble is ground into a Coddington magnifier lens (see Fig. 10J). The sphere diameter is 2.40 cm, the refractive index is 1.52, and the cylinder diameter is 1.80 cm. The central groove is ground 0.30 cm deep. Find (*a*) the position and (*b*) the size of the entrance pupil; (*c*) the position and (*d*) the size of the exit pupil; (*e*) the focal length of the magnifier; (*f*) the position of the focal point; and (*g*) the position of the principal point.

7.10 An exit pupil with a 5.0-cm aperture is located 10.0 cm in front of a spherical mirror with a radius of curvature of $+16.0$ cm. An object 3.0 cm high is centrally located on the axis 7.0 cm in front of the mirror. Find graphically (*a*) the entrance pupil, (*b*) the image of the object, and (*c*) the minimum aperture for the mirror required to see the entire object from all points of the exit pupil (see Fig. P7.10).

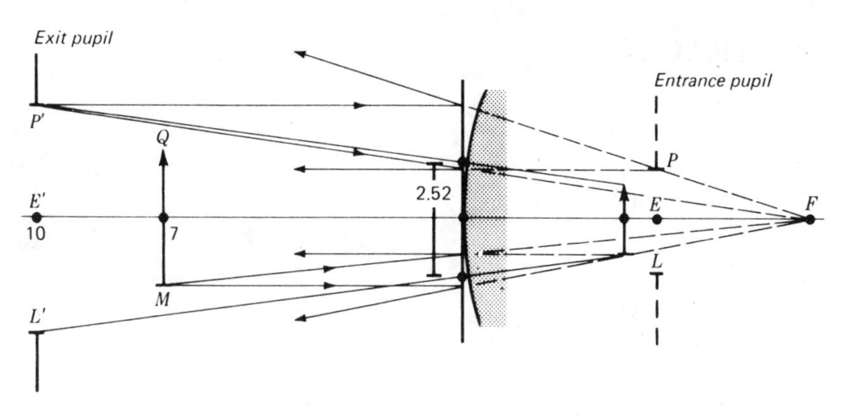

FIGURE P7.10
Graphical solution for Prob. 7.10.

Ans. (*a*) $AE = -4.44$ cm, $PL = 2.22$ cm, (*b*) -3.73 cm, $QM = +1.60$ cm, (*c*) 2.52 cm

7.11 An exit pupil 4.0 cm in diameter is located 8.0 cm in front of a spherical mirror of radius +14.0 cm. An object 3.0 cm high is centrally located on the axis +5.0 cm in front of the mirror. Graphically determine (*a*) the size and (*b*) the position of the entrance pupil. Find (*c*) the position and (*d*) the size of the image by drawing the two marginal rays and chief ray from the bottom of the object.

7.12 An exit pupil with a 10.0-cm aperture is located 48.0 cm in front of a concave spherical mirror with a radius of −30.0 cm. An object 5.0 cm high is centrally placed on the axis 36.0 cm in front of the mirror. Graphically find (*a*) the position and (*b*) the size of the entrance pupil. Find also (*c*) the position and (*d*) the size of the image by drawing the two marginal rays and the chief ray from the top of the object.

7.13 A lens with an aperture of 2.0 cm and a focal length of +3.0 cm is used as a magnifier. An object 1.60 cm high is centrally located on the axis 2.0 cm to the left of the lens, and an exit pupil 1.0 cm high is centrally located on the right 1.50 cm from the lens. Graphically locate (*a*) the position and (*b*) the size of the entrance pupil. Find also from your graph (*c*) the image position and (*d*) the image size. (*e*) Calculate the magnification.

8

RAY TRACING

Up to this point the discussion of image formation by a system of one or more spherical surfaces has been confined to the consideration of paraxial rays. With this limitation it has been possible to derive relatively simple methods of calculating and constructing the position and size of the image. In practice the apertures of most lenses are so large that paraxial rays constitute only a very small fraction of all the effective rays. It is therefore important to consider what happens to rays that are not paraxial. The straightforward method of attacking this problem is to trace the paths of the rays through the system, accurately applying Snell's law to the refraction at each surface.

8.1 OBLIQUE RAYS

All rays which lie in a plane through the principal axis and are not paraxial are called *oblique rays*. When the law of refraction is accurately applied to a number of rays through one or more coaxial surfaces, the position of the image point is found to vary with the obliquity of the rays. This leads to a blurring of the image known as lens aberrations, and the study of these aberrations will be the subject of the following chapter. Experience shows that it is possible, by properly choosing the radii and positions of spherical refracting surfaces, to reduce the aberrations greatly. Only in this way have optical instruments been designed and constructed having large usable apertures and at the same time good image-forming qualities.

Lens designers follow three general lines of approach to the problem of finding the optimum conditions. The first is to use graphical methods to find the approximate radii and spacing of the surfaces that should be used for the particular problem at hand. The second is to use well-known aberration formulas to calculate the approximate shapes and spacings. If the results of these methods of approach do not produce image-forming systems of sufficiently high quality and better definition is required, the third method, known as *ray tracing*, is applied. It consists in finding the exact paths of several representative rays through the system selected. Some of these rays will be paraxial and some oblique, and each is traced from the object to the image.

If the results are not satisfactory, the surfaces are moved, the radii are changed, and the process is repeated until an apparent minimum of aberration is obtained. Until recent years this was a long and tedious trial process, requiring in some cases hundreds of hours of work. Five-, six-, or seven-place logarithms were required, and certain standard tabular forms were printed by the different designers for recording the calculations and results. Recent researches in electronics have led to the development of high-speed calculators capable of ray tracing through complicated systems in a very short time. Such calculators undoubtedly lead today to the design and production of new and better high-quality optical systems.

In this chapter we shall first consider the method of *graphical ray tracing* and then the method of *calculation ray tracing*. Lens aberrations and the approximate methods using aberration formulas will be treated in Chap. 9.

8.2 GRAPHICAL METHOD FOR RAY TRACING

The graphical method for ray tracing to be presented here is an extension of the procedure given in Sec. 1.10 and shown for refraction at plane surfaces in Figs. 1G and 2J. It is important to note that while the principles used follow Snell's law exactly, the accuracy of the results obtained depends upon the precision with which the operator makes his drawing. A good drawing board, with T square and triangles, or a drafting machine is therefore essential; as large a drawing board as is feasible is to be preferred. The use of a sharp pencil is a necessity.

The diagrams in Fig. 8A illustrate the construction for refraction at a single spherical surface separating two media of index n and n'. After the axis and the surface with a center at C are drawn, any incident ray like 1 is selected for tracing. An auxiliary diagram is now constructed below, comparable in size, and with its axis parallel to that of the main diagram. With the point O as a center two circular arcs are drawn with radii proportional to the refractive indices. Succeeding steps of the construction are carried out in the following order: Line 2 is drawn through O parallel to ray 1. Line 3 is drawn through points T and C. Line 4 is drawn through N parallel to line 3 and extended to where it intersects the arc n' at Q. Line 5 connects O and Q, and line 6 is drawn through T parallel to line 5.

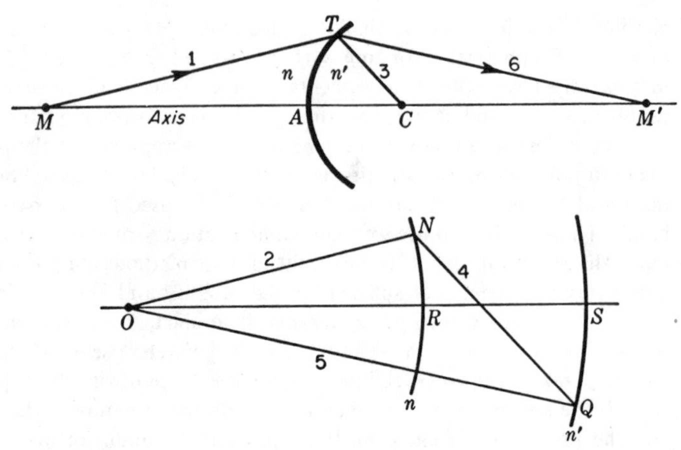

FIGURE 8A
A graphical method for ray tracing through a single spherical surface. The method is exact and obeys Snell's law for all rays.

In this diagram the radial line TC is normal to the surface at the point T and corresponds to the normal NN' in Fig. 1G. The proof that such construction follows Snell's law exactly is given in Sec. 1.10.

The graphical method applied to a system involving a series of coaxial spherical surfaces is shown in Fig. 8B. Two thick lenses having indices n' and n'', respectively, are surrounded by air of index $n = 1.00$. In the auxiliary diagram below arcs are

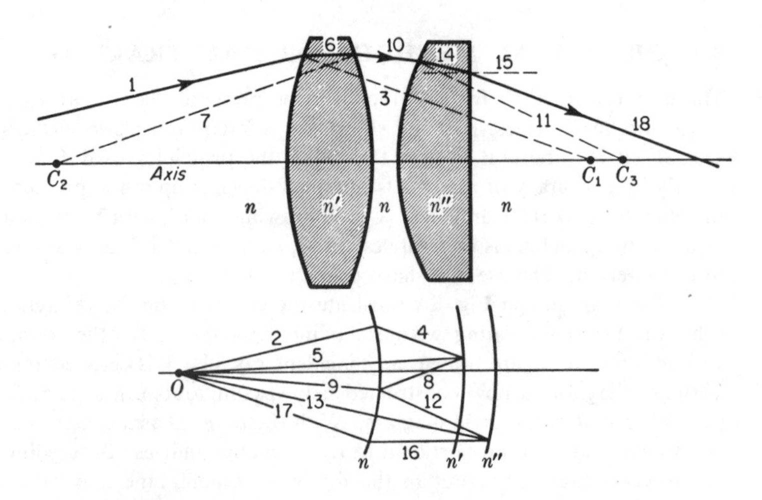

FIGURE 8B
Exact graphical method for ray tracing through a centered system of spherical refracting surfaces.

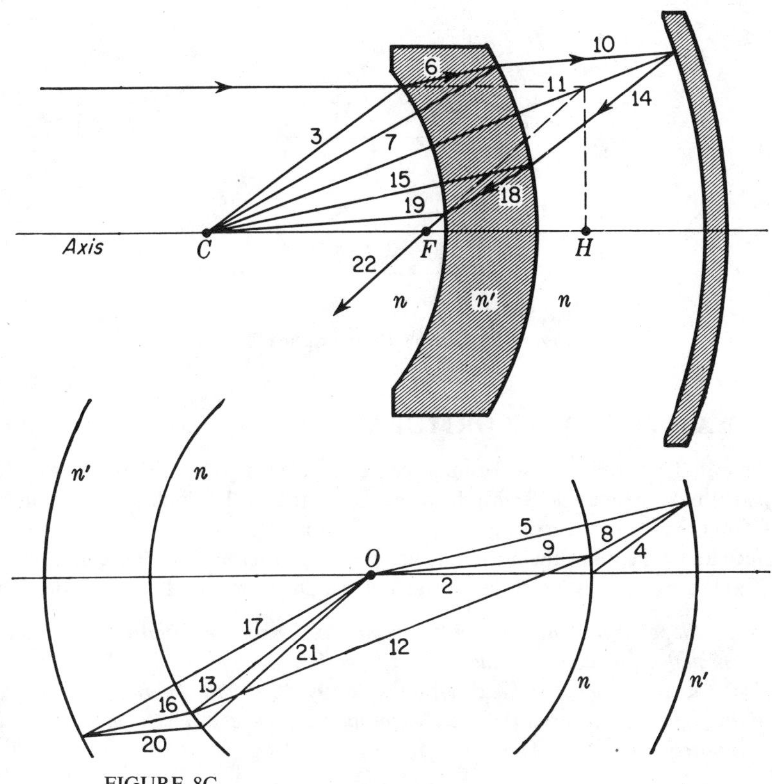

FIGURE 8C
Ray tracing through a thick mirror.

drawn for the three indices n, n', and n''. All lines are drawn in parallel pairs as before and in consecutive order starting with the incident light ray 1. Each even-numbered line is drawn parallel to the odd-numbered line just preceding it, ending up with the final ray 18. Note that the radius of the fourth surface is infinite and line 15 drawn toward its center at infinity is parallel to the axis. The latter is in keeping with the procedures in Figs. 1G, 2J, and 2K.

When the graphical method of ray tracing is applied to a thick mirror, the arcs representing the various known indices are drawn on both sides of the origin, as shown in Fig. 8C. Again in this case the lines are drawn in parallel pairs with each even-numbered line parallel to its preceding odd-numbered line. Where the ray is reflected by the concave mirror, the rays 10 and 14 must make equal angles with the normal. Note that in the auxiliary diagram the corresponding lines 9, 12, and 13 form an isosceles triangle. The particular optical arrangement shown here is known as a concentric optical system. The fact that all surfaces have a common center of curvature gives rise to some very interesting and useful optical properties (see Sec. 10.21).

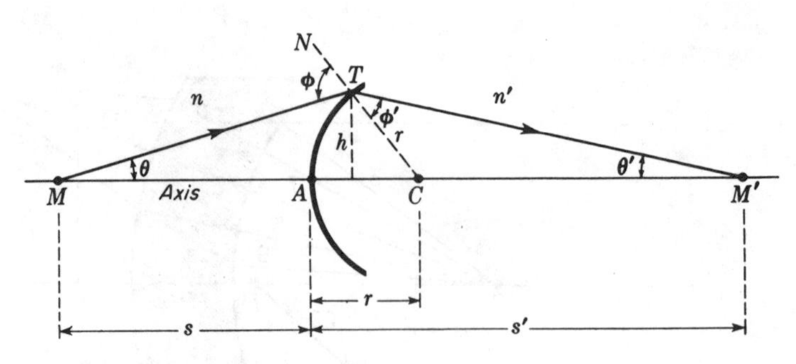

FIGURE 8D
Geometry used in deriving the ray-tracing formulas.

8.3 RAY-TRACING FORMULAS

A diagram from which these formulas can be derived is given in Fig. 8D. An oblique ray MT making an angle θ with the axis is refracted by the single spherical surface at T so that it crosses the axis again at M'. The line TC is the radius of the refracting surface and constitutes the normal from which the angles of incidence and refraction at T are measured. As regards the signs of the angles involved, we consider that

> *1 Slope angles are positive when the axis must be rotated counterclockwise through an angle of less than $\pi/2$ to bring it into coincidence with the ray.*
> *2 Angles of incidence and refraction are positive when the radius of the surface must be rotated counterclockwise through an angle of less than $\pi/2$ to bring it into coincidence with the ray.*

Accordingly, angles θ, ϕ, and ϕ' in Fig. 8D are positive, while angle θ' is negative.

Applying the law of sines to the triangle MTC, one obtains

$$\frac{\sin(\pi - \phi)}{r + s} = \frac{\sin\theta}{r}$$

Since the sine of the supplement of an angle equals the sine of the angle itself,

$$\frac{\sin\phi}{r + s} = \frac{\sin\theta}{r}$$

Solving for $\sin\phi$, we find

● $$\sin\phi = \frac{r + s}{r}\sin\theta \qquad (8a)$$

Now by Snell's law the angle of refraction ϕ' in terms of the angle of incidence ϕ is given by

● $$\sin\phi' = \frac{n}{n'}\sin\phi \qquad (8b)$$

In the triangle MTM' the sum of all interior angles must equal π. Therefore

$$\theta + (\pi - \phi) + \phi' + (-\theta') = \pi$$

which, upon solving for θ', gives

$$\theta' = \phi' + \theta - \phi \tag{8c}$$

This equation allows us to calculate the slope angle of the refracted ray. To find where the ray crosses the axis and the image distance s', the law of sines may be applied to the triangle TCM', giving

$$\frac{-\sin\theta'}{r} = \frac{\sin\phi'}{s' - r}$$

The image distance is therefore

$$s' = r - r\frac{\sin\phi'}{\sin\theta'} \tag{8d}$$

An important special case is that in which the incident ray is parallel to the axis. Under this simplifying condition it may be seen from Fig. 8E that

$$\sin\phi = \frac{h}{r} \tag{8e}$$

where h is the height of the incident ray PT above the axis. For the triangle TCM', the sum of the two interior angles ϕ' and θ' equals the exterior angle at C. When the angles are assigned their proper signs, this gives

$$\theta' = \phi' - \phi \tag{8f}$$

The six equations above which are numbered form an important set by which any oblique ray lying in a *meridian plane* can be traced through a number of coaxial spherical surfaces. A meridian plane is defined as any plane containing the axis of the system. While most of the rays emanating from an extraaxial object point do not lie in a meridian plane, the image-forming properties of an optical system can usually be determined from properly chosen meridian rays. *Skew rays*, or rays that are not confined to a meridian plane, do not intersect the axis and are difficult to trace.

8.4 SAMPLE RAY-TRACING CALCULATIONS

For a single spherical refracting surface, either concave or convex, Eqs. (8a), (8b), (8c), and (8d), respectively, are used to find the image distance s'. If the incident light is parallel to the axis, Eqs. (8e), (8b), (8f), and (8d) are used, in that order. This second set of equations will be used for the sample calculations in the example below.

A desk calculator is by far the least time-consuming way of solving ray-tracing problems, and if a computer capable of being programmed is available, the time can be even shorter. Seven-place logarithms can be used, but the process is long, tedious, and subject to frequent errors. If logarithm tables are used, the subtraction of one logarithm from another to find a quotient can be avoided by employing the cologarithms of all quantities occurring in the denominator. Thus the operations are reduced to those of addition.

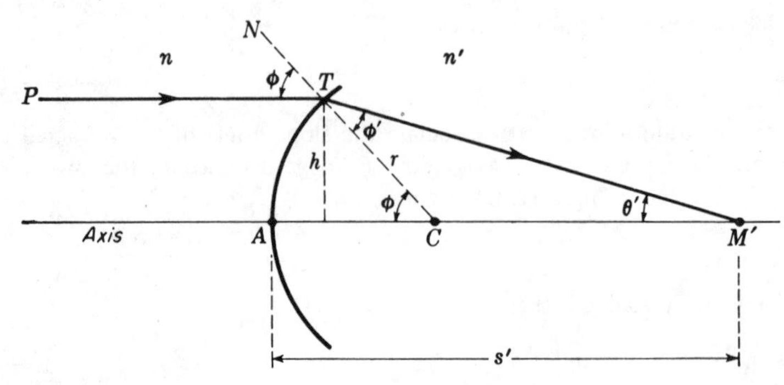

FIGURE 8E
Geometry for ray tracing with parallel incident light.

EXAMPLE 1 A convex spherical surface of radius $r = +5.0$ cm is ground and polished on the end of a large cylindrical glass rod of index 1.67200. Assume incident light parallel to the axis by using rays at heights of (*a*) 3.0 cm, (*b*) 2.0 cm, (*c*) 1.0 cm, and (*d*) 0.

SOLUTION It is convenient to set these given quantities up in tabular form, as shown in Table 8A.

Table 8A RAY-TRACING CALCULATIONS FOR A SINGLE CONVEX SPHERICAL SURFACE*
$r = +5.0$ cm $n = 1.0$ cm $n' = 1.67200$

Eq.	Unknown	Relationship	$h = 3.0$	$h = 2.0$	$h = 1.0$	$h = 0$
(8e)	$\sin\phi$	$\frac{h}{r}$	+0.6000000	0.4000000	0.2000000	0.6000000
(8b)	$\sin\phi'$	$\frac{n}{n'}\sin\phi$	+0.3588517	0.2392344	0.1196172	0.3588517
		ϕ	+36.869898°	23.578178°	11.536959°	
		ϕ'	+21.029692°	13.841356°	6.8700110°	
(8f)	θ'	$\phi' - \phi$	−15.840206°	9.7368220°	4.6669480°	
		$\sin\theta'$	−0.2729554	0.1691228	0.0813636	0.2411483
(8d)	$r - s'$	$r\frac{\sin\phi'}{\sin\theta'}$	−6.5734494	7.0728015	7.3507809	7.4404775
		s'	+11.573449	12.072802	12.350781	12.440478

* Although the refractive index of air at normal temperature and pressure is 1.000292, it is customary in ray tracing to use the value 1.000000.

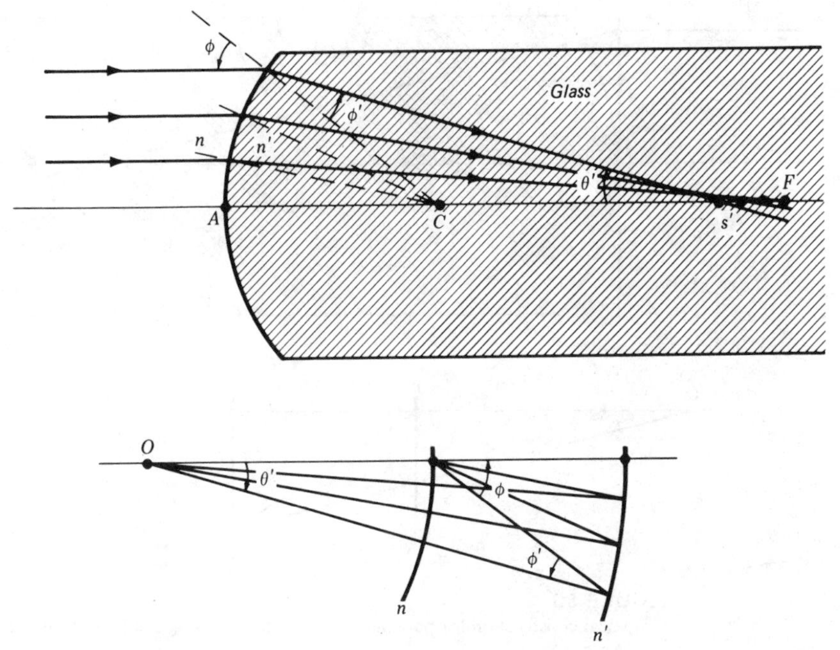

FIGURE 8F
Graphical construction for parallel rays refracted at a single spherical surface.

The equation numbers in the first column and the *unknowns* and *known quantities* in the next two columns clearly show what is being calculated and how it is used in the following lines. A graphical solution of this example is shown in Fig. 8F.

At or near $h = 0$ we are dealing with paraxial rays, where all angles are extremely small. Here the *sines of angles* and the *angles* themselves are interchangeable. Hence Eq. (8f) can be written

$$\sin\theta' = \sin\phi' - \sin\phi \tag{8g}$$

For $h = 0$, therefore, the following procedure should be used. First select the number that corresponds to one of the values of $\sin\phi$ in another column. For example, in the column headed $h = 3.0$ cm, we find $\sin\phi = 0.6000000$, and under it $\sin\phi' = 0.3588517$. The difference between these two values is, by Eq. (8g), entered for $\sin\theta'$ as 0.2411483. To find the value 7.4404775 in row (8d), multiply 0.3588517 by 5.0 and divide by 0.2411483. Adding $r = 5.0$ cm, we obtain 12.440478, the paraxial value of s' given in the last row. The first three figures of s' are found graphically in Fig. 8G.

Let us now see how the above equations and procedures are employed to calculate the image distances for a thick lens with two surfaces (see Fig. 8H).

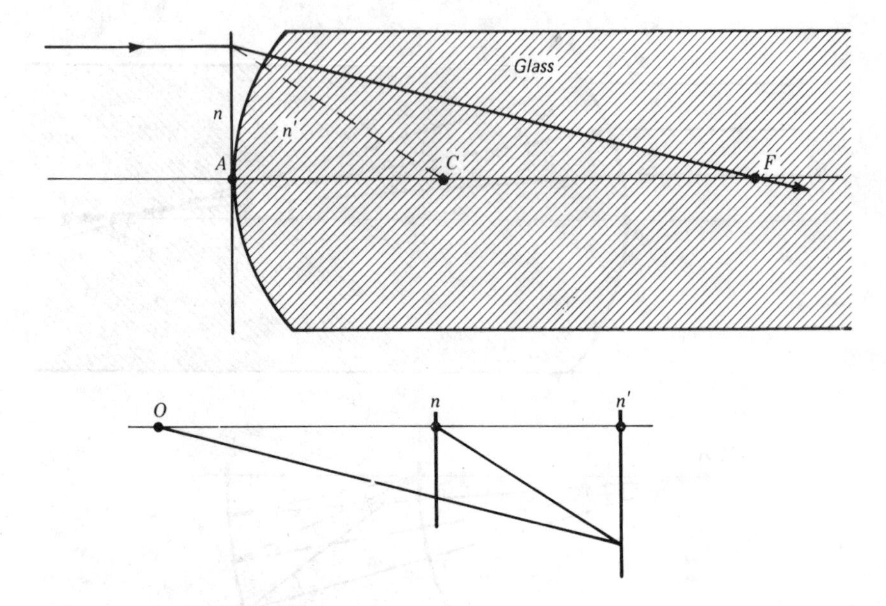

FIGURE 8G
Graphical construction for paraxial rays refracted at a single spherical surface.

EXAMPLE 2 A double convex lens 3.0 cm thick and with radii $r_1 = +15.0$ cm and $r_2 = -15.0$ cm has a refractive index of 1.62300. If rays of light parallel to the axis are incident on the first surface at heights of 6.0, 4.0, 2.0, and 0 cm, find the image distances s_2' (*a*) by calculation and (*b*) graphically.

SOLUTION (*a*) For the first surface, with light incident parallel to the axis, we use the same four equations as in the preceding example. With subscripts of 1 on r, ϕ, ϕ' and s', these become

$$\sin \phi_1 = \frac{h}{r_1} \tag{8h}$$

$$\sin \phi_1' = \frac{n}{n'} \sin \phi_1 \tag{8i}$$

$$\theta' = \phi_1' - \phi_1 \tag{8j}$$

$$r_1 - s_1' = r_1 \frac{\sin \phi_1}{\sin \theta'} \tag{8k}$$

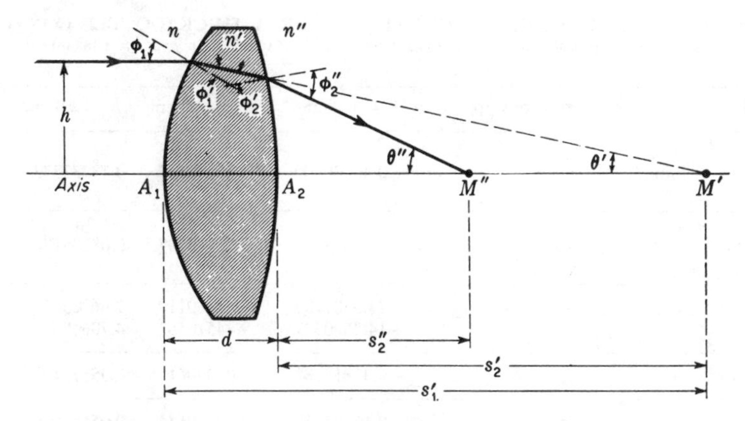

FIGURE 8H
Geometry involved in the use of ray-tracing formulas for a thick lens.

Since the images for the first surface become the objects for the second surface, the lens thickness is subtracted, the sign is changed, and we obtain

$$s_2' = d - s_1' \tag{8l}$$

For refraction at the second surface we use Eqs. (8a), (8b), (8c), and (8d) and with subscripts of 2 obtain

$$\sin \phi_2' = \frac{r_2 + s_2'}{r_2} \sin \theta' \tag{8m}$$

$$\sin \phi_2'' = \frac{n'}{n''} \sin \phi_2' \tag{8n}$$

$$\theta'' = \phi_2'' + \theta' - \phi_2' \tag{8o}$$

$$r_2 - s_2'' = r_2 \frac{\sin \phi_2''}{\sin \theta'} \tag{8p}$$

At or near $h = 0$ we are again dealing with paraxial rays, and all angles are extremely small. Since the sines of angles and the angles themselves are interchangeable, Eqs. (8j) and (8o) can be written

$$\sin \theta' = \sin \phi_1 - \sin \phi_1' \tag{8q}$$

$$\sin \theta'' = \sin \phi_2'' + \sin \theta' - \sin \phi_2' \tag{8r}$$

For paraxial rays we use Eqs. (8h) through (8p), with Eqs. (8q) and (8r) substituted for Eqs. (8j) and (8o), and follow the same procedure as in Example 1. First

Table 8B RAY-TRACING CALCULATIONS FOR A THICK DOUBLE CONVEX LENS*

$r_1 = +15.0$ cm $\quad r_2 = -15.0$ cm $\quad d = 3.0$ cm $\quad n = n'' = 1.00000 \quad n' = 1.62500$

Eq.	Unknown	Relationship	$h = 6.0$ cm	$h = 4.0$ cm	$h = 2.0$ cm	$h = 0$
(8h)	$\sin\phi_1$	$\dfrac{h}{r_1}$	+0.40000000	0.26666667	0.13333333	0.40000000
(8i)	$\sin\phi_1'$	$\dfrac{n}{n'}\sin\phi$	+0.24615385	0.16410257	0.08205128	0.24615385
		ϕ_1	+23.5781785°	15.4660119°	7.6622555°	
		ϕ_1'	+14.2500327°	9.4451058°	4.7064843°	
(8j)	θ'	$\phi_1' - \phi_1$	−9.3281458°	6.0209061°	2.9557712°	
		$\sin\theta'$	−0.16208858	0.10489134	0.05156506	0.15384615
(8k)	$r_1 - s_1'$	$r_1\dfrac{\sin\phi_1'}{\sin\theta'}$	-22.7795601	23.4675230	23.8682656	24.0000010
		s_1'	+37.7795601	38.4675230	38.8682656	39.0000010
(8l)	s_2'	$d - s_1'$	−34.7795601	35.4675230	35.8682565	36.0000010
		$r_2 + s_2'$	−49.7795601	50.4675230	50.8682656	51.0000010
		$\dfrac{r_2 + s_2'}{r_2}$	+3.3186373	3.3645015	3.3912177	3.4000007
(8m)	$\sin\phi_2'$	$\dfrac{r_2 + s_2'}{r_2}\sin\theta'$	−0.5379132	0.35290707	0.17486834	0.5230770
(8n)	$\sin\phi_2''$	$\dfrac{n'}{n''}\sin\phi_2'$	−0.8741091	0.5737371	0.28416105	0.8500002
		ϕ''	−60.9397126°	35.0112384°	16.5087070°	
		θ'	−9.3281458°	6.0209061°	2.9557712°	
		ϕ_2'	+32.5416940°	20.6652279°	10.0709964°	
(8o)	θ''	$\phi_2'' + \theta' - \phi_2'$	−37.7261644°	20.3669166°	9.3934818°	
		$\sin\theta''$	−0.6118882	0.34803079	0.16321370	0.4807694
(8p)	$r_2 - s_2''$	$r_2\dfrac{\sin\phi_2''}{\sin\theta''}$	−21.4281571	24.7278596	26.1155513	26.519997
		s_2''	+6.4281571	9.7278596	11.1155513	11.519997
		$\delta s_2''$	5.0918399	1.7921374	0.4044457	0

* Although the refractive index for air is 1.000292, the value for a vacuum is used here.

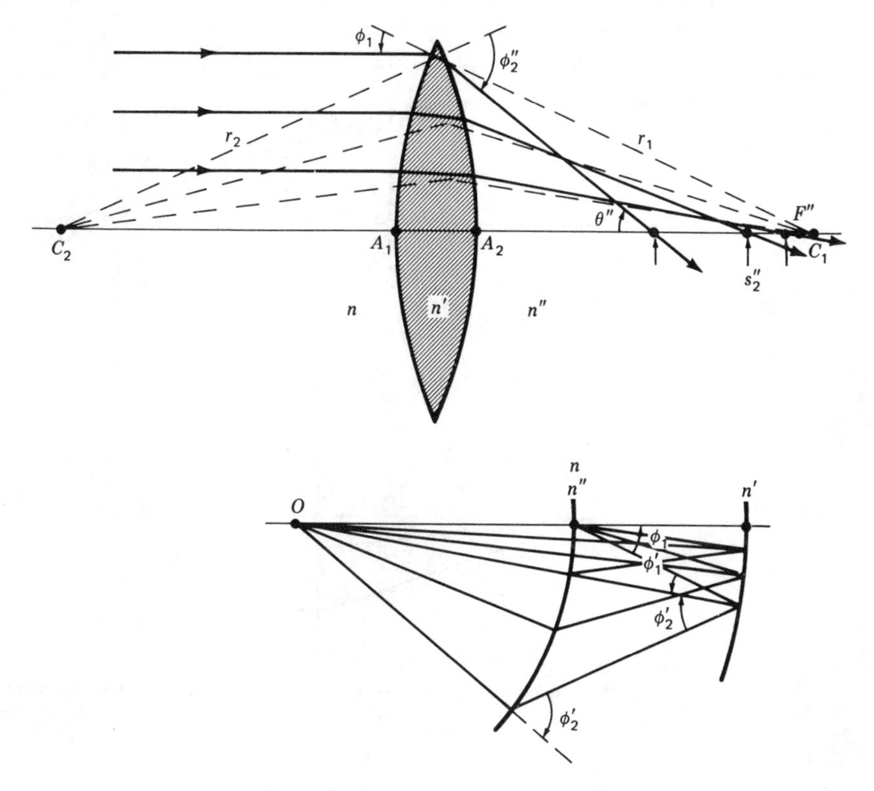

FIGURE 8I
Graphical solution for ray tracing through a thick lens; see Example 2.

find the number that corresponds to one of the values of sin ϕ_1 in another column. For example, in the column headed $h = 6.0$ cm we find sin $\phi_1 = +0.4000000$ and under it sin $\phi_1' = +0.24615385$. The difference between these two numbers is, by Eq. (8q), entered for sin θ' as -0.15384615. To find the value 24.0000010 in row (8k) we multiply 0.24615385 by 15.0 and divide by 0.15384615. Adding $r_1 = 15.0$ cm, we obtain 39.0000010 cm. From here on through Eq. (8n) we use only values in the last column, obtaining sin $\phi_2'' = -0.8500002$. For Eq. (8o) we now use Eq. (8r) and with values of sin $\phi_2'' = -0.8500002$, sin $\theta' = -0.15384615$, and sin $\phi_2' = -0.5230770$, find sin $\theta'' = -0.4807694$.

The final figures show that when parallel rays are incident on the lens at heights of 6.0, 4.0, 2.0, and 0 cm, the axial intercepts are, to seven significant figures, at $s_2'' = +6.428157$, $+9.727860$, $+11.115551$ and 11.519997 cm, respectively.

(*b*) Graphical solutions to this problem are given in Figs. 8I and 8J. It will be

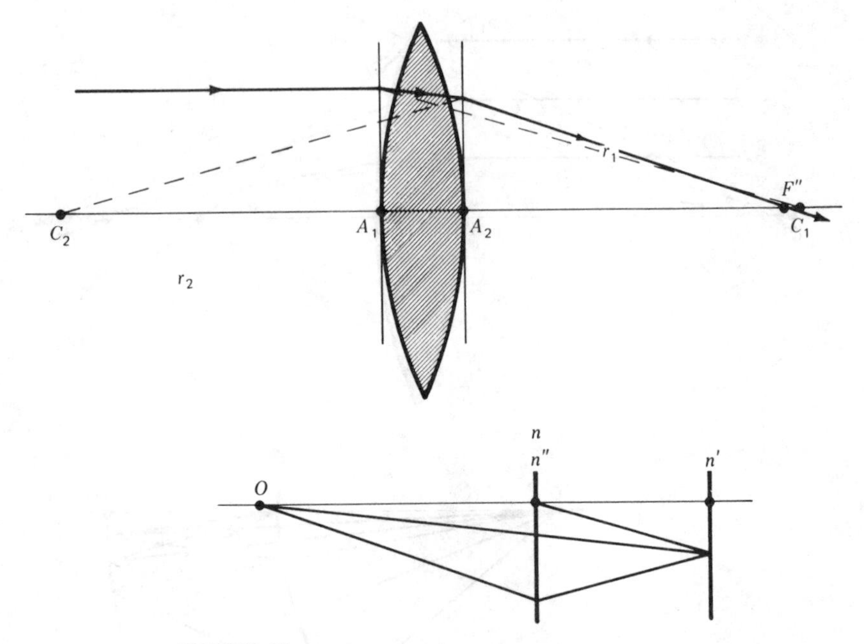

FIGURE 8J
Graphical solution for paraxial rays passing through a thick lens; see Example 2.

seen that the distance from the lens vertex to the focal point is not a constant but varies slightly for different zones of the lens (see Fig. 8K). This defect in the image-forming properties of all lenses with spherical surfaces is called *spherical aberration* and will be treated in detail in the next chapter. The focal distances s_1' and s_2'' for $h = 0$ and for $\theta = 0$ in Table 8B are identical with the values obtained with the paraxial-ray formulas given in Sec. 5.1.

Whenever a plane surface is encountered, refraction is traced exactly by means of Eq. (1l). If, for example, the second surface of a lens is plane, Snell's law becomes

$$\sin \theta'' = \frac{n'}{n''} \sin \theta'$$

and Eq. (2q) becomes

$$s_2'' = s_2' \frac{\tan \theta'}{\tan \theta''}$$

where $\theta'' = \phi_2''$ and $\theta' = \phi_2'$. The calculations are carried out by tabulating the proper values as in Table 8B.

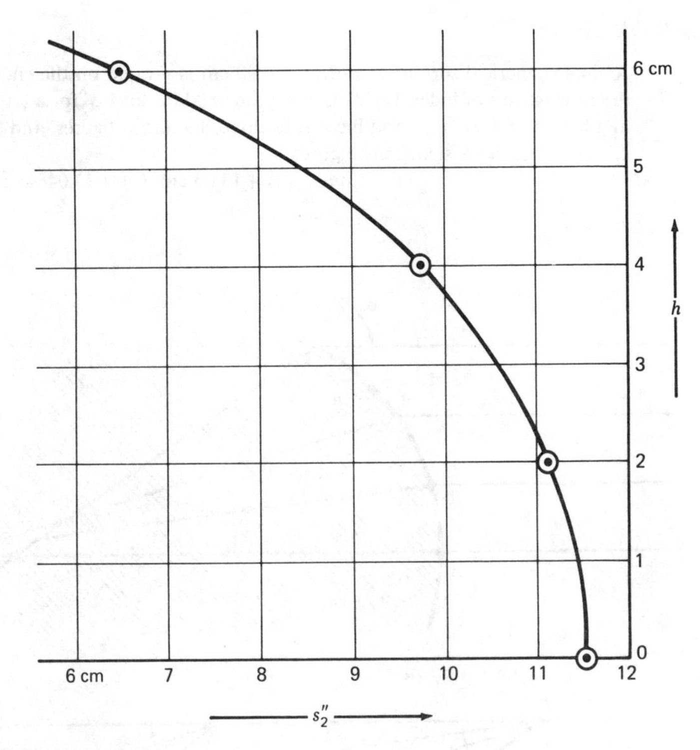

FIGURE 8K
Change in focal length for parallel rays entering a double convex glass lens in air; see Example 2.

During the early nineteen-thirties, T. Smith developed a useful set of equations for handling ray tracing in complex thick-lens systems. The simple form of the ray-tracing equations, Eqs. (8a) to (8f), and the respective way in which they are applied to surface after surface suggested the use of matrices. The successive refractions and transfers can then be carried out mathematically with matrix operators.

Although these preliminary developments went unnoticed by lens designers for almost 30 years, the matrix approach began to be used in the nineteen-sixties. Although the matrix treatment is beyond the scope of this book, some students may find it useful to look into.*

* For a detailed development of the matrix method of ray tracing, see K. Hallbach, Matrix Representation of Gaussian Optics, *Am. J. Phys.*, **32**:90 (1964); W. Brouwer, "Matrix Methods in Optical Instrument Design"; E. L. O'Neill, "Introduction to Statistical Optics," and A. Nussbaum, "Geometrical Optics."

PROBLEMS

8.1 A single spherical surface of radius +6.50 cm is ground on the end of a large cylindrical glass rod of index 1.65820. Find the axial distance s' for a parallel incident ray at a height of 6.0 cm (*a*) graphically, to three significant figures, and (*b*) by ray-tracing calculations, to six significant figures.

Ans. (*a*) +13.05 cm, (*b*) +13.04646 cm (see Fig. P8.1)

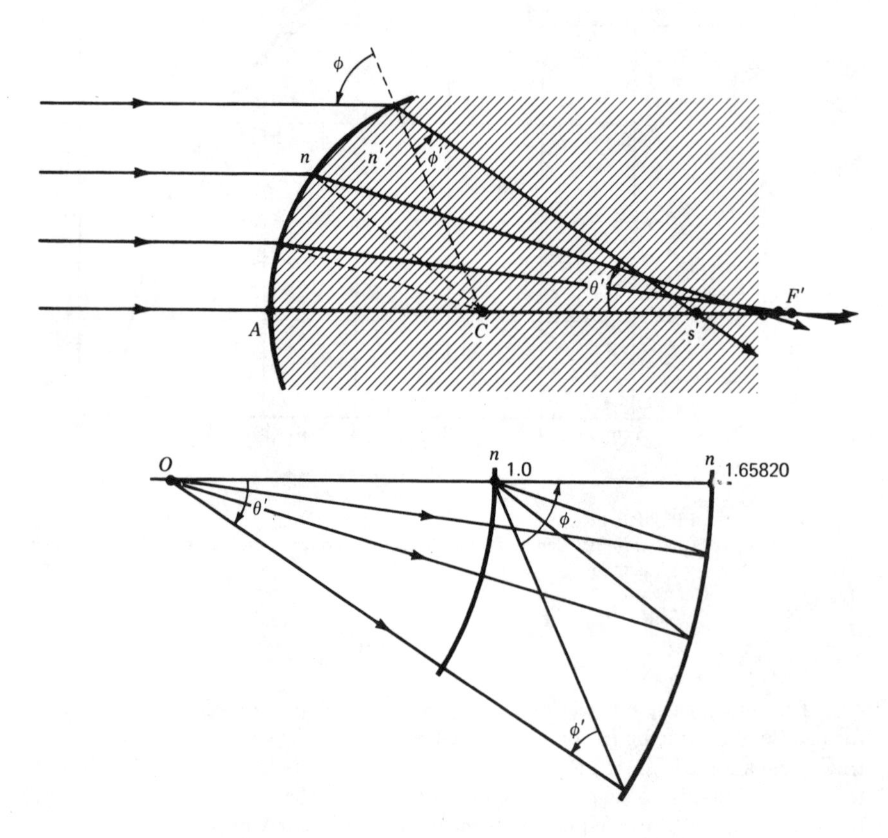

FIGURE P8.1
Graphical solution for Probs. 8.1 to 8.3.

8.2 Solve Prob. 8.1 for a ray at a height of 4.0 cm.

Ans. (*a*) +15.15 cm, (*b*) +15.14873 cm (see Fig. P8.1)

8.3 Solve Prob. 8.1 for a ray at a height of 2.0 cm.

Ans. (*a*) +16.09 cm, (*b*) +16.08820 cm (see Fig. P8.1)

8.4 Solve Prob. 8.1 for a bundle of paraxial rays ($h = 0$).

8.5 A single concave spherical surface of radius −7.0 cm is ground on one end of a large cylindrical glass rod of index 1.68500. Find the axial distance s' for a parallel incident ray at a height of 6.0 cm (*a*) graphically, to three significant figures, and (*b*) by ray-tracing calculations, to six significant figures.

8.6 Solve Prob. 8.5 for a ray at a height of 4.0 cm.

8.7 Solve Prob. 8.5 for a ray at a height of 2.0 cm.

8.8 Solve Prob. 8.5 for a bundle of paraxial rays, $h = 0$.

8.9 A single spherical surface is ground and polished on the end of a large cylindrical glass rod of index 1.82500. The radius of curvature $r = +8.0$ cm. The rod is immersed in a thin oil of index 1.35600. Find the axial distances s' for parallel incident rays at a height of (*a*) 6.0 cm, (*b*) 4.0 cm, (*c*) 2.0 cm, and (*d*) 0 cm. Solve graphically and by calculation.

Ans. (*a*) +25.54043 cm, (*b*) +28.85935 cm, (*c*) +30.58603 cm, (*d*) +31.13007 (see Fig. P8.9)

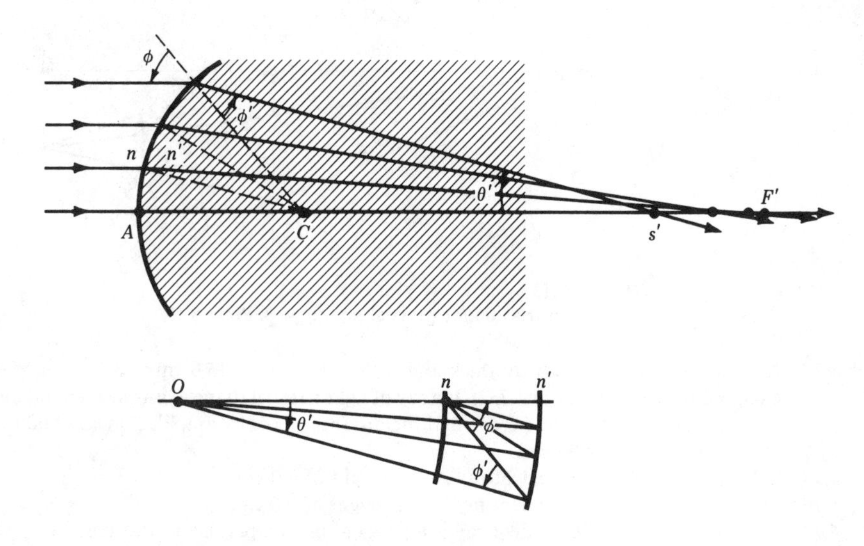

FIGURE P8.9
Graphical solution for Prob. 8.9.

8.10 A double convex lens 6.0 cm thick and with radii $r_1 = +16.0$ cm and $r_2 = -20.0$ cm has a refractive index of 1.750. If a ray of light parallel to the axis is incident on the first surface at a height of 6.0 cm, find (*a*) by graphical means and (*b*) by calculation to six figures, the distance s_2''.

8.11 Solve Prob. 8.10 if the incident ray is at a height of 4.0 cm.

8.12 Solve Prob. 8.10 if the incident ray is at a height of 2.0 cm.

8.13 Solve Prob. 8.10 for paraxial rays, $h = 0$.

Ans. (*a*) +10.71 cm, (*b*) +10.71225 cm (see Fig. P8.13)

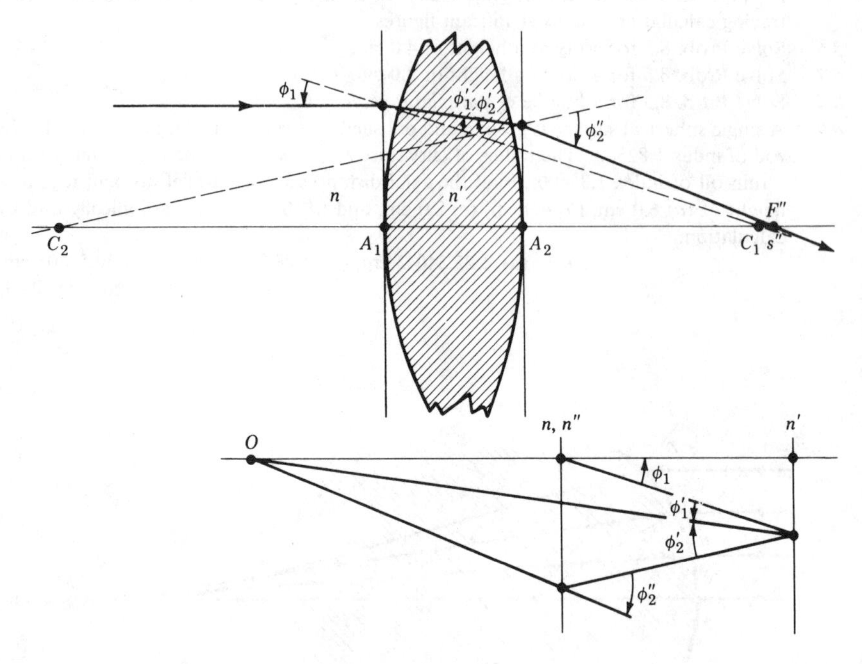

FIGURE P8.13
Graphical construction for paraxial rays, $h = 0$, Prob. 8.13.

8.14 A double concave lens 1.0 cm thick and with radii $r_1 = -15.0$ cm and $r_2 = +15.0$ cm has a refractive index of 1.732. If a ray of light parallel to the axis is incident on the first surface at a height of 5.0 cm, find the distance s_2'' (*a*) by graphical means and (*b*) by calculation, to six figures.

8.15 Solve Prob. 8.14 if the incident ray is at a height of 4.0 cm.

8.16 Solve Prob. 8.14 if the incident ray is at a height of 2.0 cm.

8.17 Solve Prob. 8.14 if the incident ray is a paraxial bundle parallel to the axis.

8.18 A double convex lens of index 1.63700 has radii $r_1 = +13.50$ cm and $r_2 = -13.50$ cm and forms one end of a tank containing oil of index 1.42500. If face r_2 is in contact with the oil and air is in contact with r_1, find the axial crossover points s_2'' for parallel incoming light at heights of (*a*) 6.0 cm, (*b*) 4.50 cm, (*c*) 3.0 cm, and (*d*) 1.50 cm, and (*e*) 0 cm. Solve graphically, and by calculations using ray-tracing methods.

Ans. (*a*) +17.4514 cm, (*b*) +19.06432 cm, (*c*) +19.9898 cm, (*d*) 20.4842 cm, (*e*) +20.6408 cm (see Fig. P8.18)

8.19 A plano-convex lens 3.0 cm thick is silvered on the flat side to form a thick mirror. If $r_1 = +15.0$ cm and $r_2 = \infty$ and the index of the glass is 1.50000, find (*a*) by graphical means and (*b*) by ray-tracing calculations the distance s_2'' for a ray parallel to the axis and at a height of 6.0 cm.

8.20 Solve Prob. 8.19 for a paraxial bundle of rays close to the axis, $h = 0$.

Ans. (*a*) +13.93 cm, (*b*) +13.92857 cm (see Fig. P8.20)

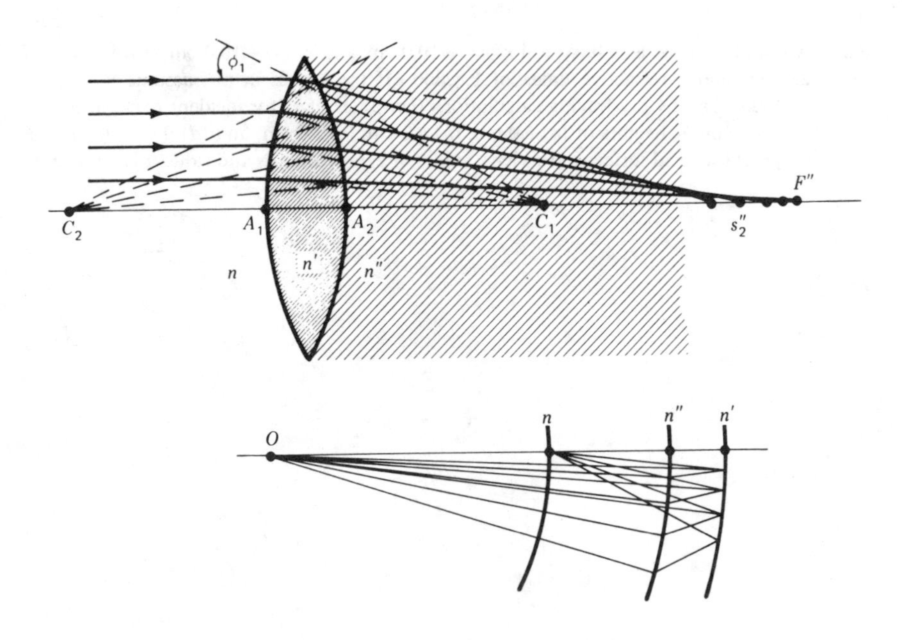

FIGURE P8.18
Graphical solution for Prob. 8.18.

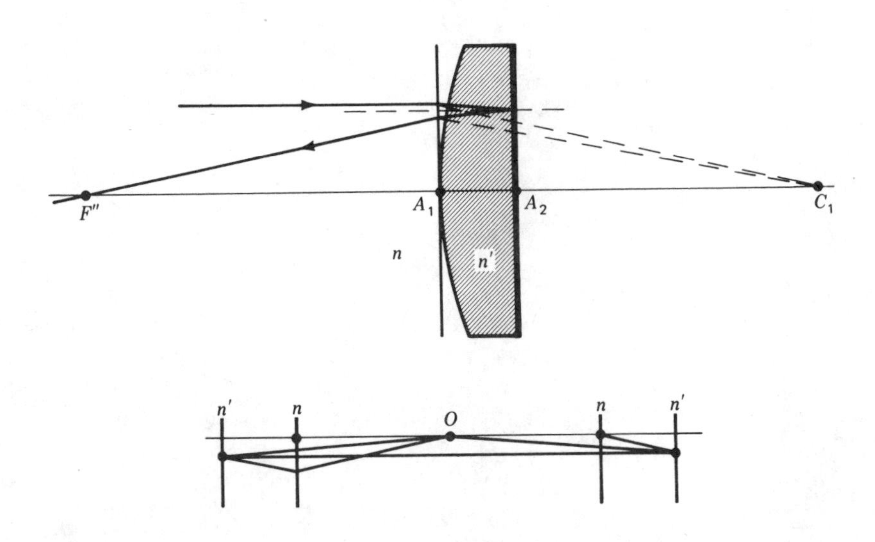

FIGURE P8.20
Paraxial ray tracing to find the focal point F''.

8.21 A double convex lens has radii $r_1 = +10.0$ cm and $r_2 = -10.0$ cm, a thickness of 2.0 cm, and a refractive index of 1.52300. If the lens is in air, assume $n = n'' = 1.00000$. Find by ray-tracing calculations the values of s_2'' for incident rays parallel to the axis and at heights of (*a*) 6.0 cm, (*b*) 4.0 cm, (*c*) 2.0 cm, and (*d*) 0 cm. (*e*) Plot a graph of the focal surface of this lens, values of s_2'' horizontally and values of h vertically.

9

LENS ABERRATIONS

The processes of ray tracing presented in the last chapter serve to emphasize the inability of the paraxial-ray formulas of the Gauss theory to give an accurate account of image detail. A wide beam of rays incident on a lens parallel to the axis, for example, is not brought to a focus at a unique point. The resulting image defect is known as *spherical aberration.* The gaussian formulas developed and used in the preceding chapters give, therefore, only an idealized account of the images produced with lenses of wide aperture.

When ray tracing is applied to object points located farther and farther off the axis, the observed image defects become more and more pronounced. The methods of reducing these aberrations to a minimum, and thereby permitting the formation of reasonably satisfactory images, are one of the chief problems of geometrical optics. It would be impossible within the scope of this book to give all the details of the extensive mathematical theory involved in this problem.* Instead we shall attempt to

* For a more thorough account of lens aberrations the reader is referred to A. E. Conrady, "Applied Optics and Optical Design," vol. 1, Oxford University Press, New York, 1929; reprinted (paperback) vols. 1 and 2, Dover Publications, Inc., New York, 1960.

show how most of the aberrations manifest themselves and at the same time discuss some of the known formulas to see how they may be used in the design of high-quality optical systems.

9.1 EXPANSION OF THE SINE. FIRST-ORDER THEORY

In order to formulate a satisfactory theory of lens aberrations, many theoreticians have found it convenient to start with the correct and precise ray-tracing formulas, as given in Eqs. (8a) through (8f), and to expand the sines of each angle into a power series. An expansion of the sine of an angle by Maclaurin's theorem gives

$$\sin\theta = \theta - \frac{\theta^3}{3!} + \frac{\theta^5}{5!} - \frac{\theta^7}{7!} + \frac{\theta^9}{9!} \cdots \tag{9a}$$

For small angles this is a rapidly converging series. Each member is small compared with the preceding member. It shows that for paraxial rays where the slope angles are very small we may, to a first approximation, neglect all terms beyond the first and write

$$\sin\theta = \theta$$

When θ is small, the other angles ϕ, ϕ', and θ' are also small, provided the ray lies close to the axis. By substituting θ for $\sin\theta$, ϕ for $\sin\phi$, and θ' for $\sin\theta'$, in Eqs. (8a), (8b), and (8d), we obtain

$$\phi = \frac{r + s}{r}\,\theta \qquad \phi' = \frac{n}{n'}\,\phi$$

$$\theta' = \phi' + \theta - \phi \qquad s' = r - r\,\frac{\phi'}{\theta'}$$

By the algebraic substitution of the first equation in the second, the resultant equation in the third, and this resultant in the fourth, all angles can be eliminated. The final equation obtained by these substitutions is none other than the gaussian formula,

$$\frac{n}{s} + \frac{n'}{s'} = \frac{n' - n}{r}$$

This equation and others developed from it form the basis of what is usually called *first-order theory*.

The justification for writing $\sin\theta = \theta$, etc., for all small angles is illustrated in Fig. 9A and in Table 9A. For an angle of 10°, for example, the arc length θ is only 0.5 percent greater than sin 10°, while for 40° it is about 10 percent greater. These differences are measures of spherical aberration and, therefore, of image defects.

Table 9A VALUES OF sin θ AND ITS FIRST THREE EXPANSION TERMS

	sin θ	θ	$\frac{\theta^3}{3!}$	$\frac{\theta^5}{5!}$
10°	0.1736482	0.1745329	0.0008861	0.0000135
20°	0.3420201	0.3490658	0.0070888	0.0000432
30°	0.5000000	0.5235988	0.0239246	0.0003280
40°	0.6427876	0.6981316	0.0567088	0.0013829

9.2 THIRD-ORDER THEORY OF ABERRATIONS

If all the sines of angles in the ray-tracing formulas [Eqs. (8a) to (8f)] are replaced by the first two terms of the series in Eq. (9a), the resultant equations, in whatever form they are given, represent the results of *third-order theory*. Thus sin θ is replaced by $\theta - \theta^3/3!$, sin ϕ is replaced by $\phi - \phi^3/3!$, etc. The resulting equations give a reasonably accurate account of the principal aberrations.

In this theory the aberration of any ray, i.e., its deviation from the path prescribed by the gaussian formulas, is expressed in terms of five sums, S_1 to S_5, called the *Seidel sums*. If a lens were to be free of all defects in its ability to form images, all five of these sums would have to equal zero. No optical system can be made to satisfy all these conditions at once. Therefore it is customary to treat each sum separately, and the vanishing of certain ones corresponds to the absence of certain aberrations. Thus, if for a given axial object point the Seidel sum $S_1 = 0$, there is no *spherical aberration* at the corresponding image point. If both $S_1 = 0$ and $S_2 = 0$, the system will also be free of *coma*. If, in addition to $S_1 = 0$ and $S_2 = 0$, the sums $S_3 = 0$ and $S_4 = 0$ as well, the images will be free of *astigmatism* and *curvature of field*. If finally S_5 could be made to vanish, there would be no *distortion* of the image. These aberrations are also known as the *five monochromatic aberrations* because they exist for any specified color and refractive index. Additional image defects occur when the

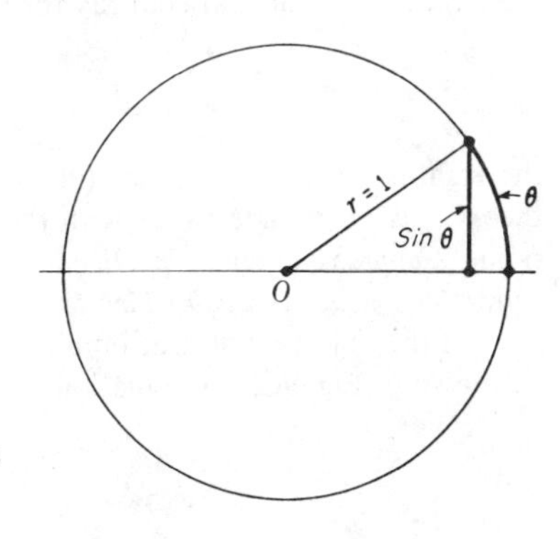

FIGURE 9A
The arc of an angle θ in relation to its sine.

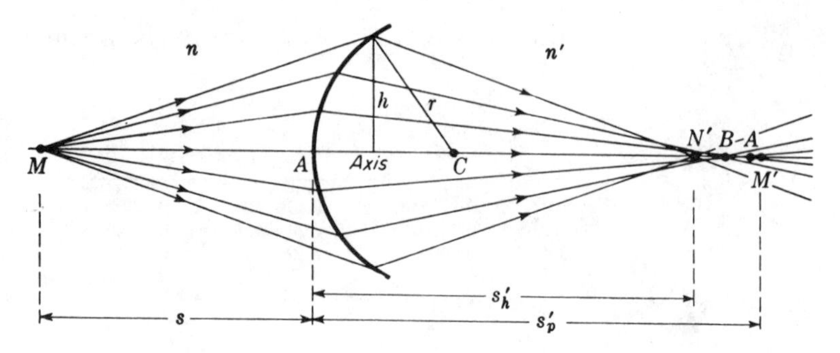

FIGURE 9B
The spherical aberration in the image of an axial object point as formed by a single spherical refracting surface.

light contains various colors. We shall first discuss each of the monochromatic aberrations and then take up the chromatic effects.

9.3 SPHERICAL ABERRATION OF A SINGLE SURFACE

This is a term introduced in Sec. 6.8, and shown in Fig. 6K, to describe the blurring of the image formed when parallel light is incident on a spherical mirror. A similar blurring of the image that occurs upon refraction by spherical surfaces will now be discussed. In Fig. 9B M is an object point on the axis of a single spherical refracting surface, and M' is its paraxial image point. Oblique rays incident on the surface in a zone of radius h are brought to a focus closer to, and at a distance of s'_h from, the vertex A.

The distance $N'M'$, as shown in the diagram, is a measure of the *longitudinal spherical aberration*, and its magnitude is found from the third-order formula

$$\frac{n}{s} + \frac{n'}{s'_h} = \frac{n' - n}{r} + \left[\frac{h^2 n^2 r}{2f'n'}\left(\frac{1}{s} + \frac{1}{r}\right)^2 \left(\frac{1}{r} + \frac{n' - n}{ns}\right)\right] \tag{9b}$$

Since from the paraxial-ray formula, Eq. (3b), we have

$$\frac{n}{s} + \frac{n'}{s'_p} = \frac{n' - n}{r}$$

the right-hand bracket in Eq. (9b) is a measure of the deviations from first-order theory. Its magnitude varies with the position of the object point and for any fixed point is *approximately proportional to h^2, the square of the radius of the zone on the refracting surface through which the rays pass.*

If the object point is at infinity so that the incident rays are parallel to the axis, as shown in Fig. 9C, this equation reduces to

$$\frac{n'}{s'_h} = \frac{n'}{f'} + \frac{h^2 n^2}{2f'r^2 n'} \tag{9c}$$

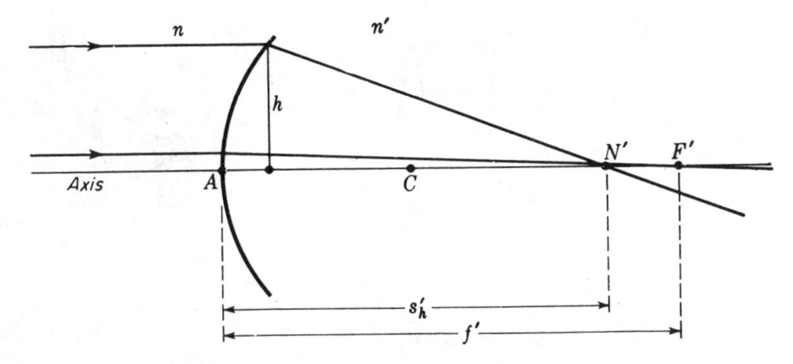

FIGURE 9C
Longitudinal spherical aberration for parallel light incident on a single spherical refracting surface.

Again the magnitude of the aberration is proportional to h^2, the square of the height of the ray above the axis.

9.4 SPHERICAL ABERRATION OF A THIN LENS

The existence of spherical aberration for a single spherical surface indicates that it may also occur in combinations of such surfaces, e.g., in a thin lens. Since many of the lenses in optical instruments are used to focus parallel incident or emergent rays, it is usual for comparison purposes to determine the spherical aberration for parallel incident light. Figure 9D(*a*) illustrates this special case and shows the position of the paraxial focal point F' as well as the focal points A, B, and C for zones of increasing diameter. Diagram (*b*) in Fig. 9D illustrates the difference between *longitudinal spherical aberration*, abbreviated Long. SA, and *lateral spherical aberration*, abbreviated Lat. SA.

As a measure of the actual magnitudes involved in longitudinal spherical aberration, we can use the figures calculated by ray-tracing methods for some of the lenses in the preceding chapter. For example, the focal lengths for three zones of a double convex lens can be taken from Table 8B. The results are $+11.52000$ cm for paraxial rays, $+11.11555$ cm for rays from zone $h = 2.0$ cm, $+9.72786$ cm from zone $h = 4.0$ cm, and $+6.42816$ cm from zone $h = 6.0$ cm. These give a longitudinal spherical aberration of $+1.79214$ cm for the 4.0-cm zone, or about 15.6 percent of the paraxial focal length. A graph showing this variation of focal length with zone radius is given in Fig. 9E(*a*). For small h the curve approximates a parabola, and since the marginal rays intersect the axis to the left of the paraxial focal point, the spherical aberration is said to be *positive*. A similar curve, drawn for a typical double concave lens of nearly the same dimensions, is shown in Fig. 9E(*b*). Bending to the right, this lens is said to have *negative spherical aberration*.

A series of positive lenses of the same diameter and paraxial focal length but of

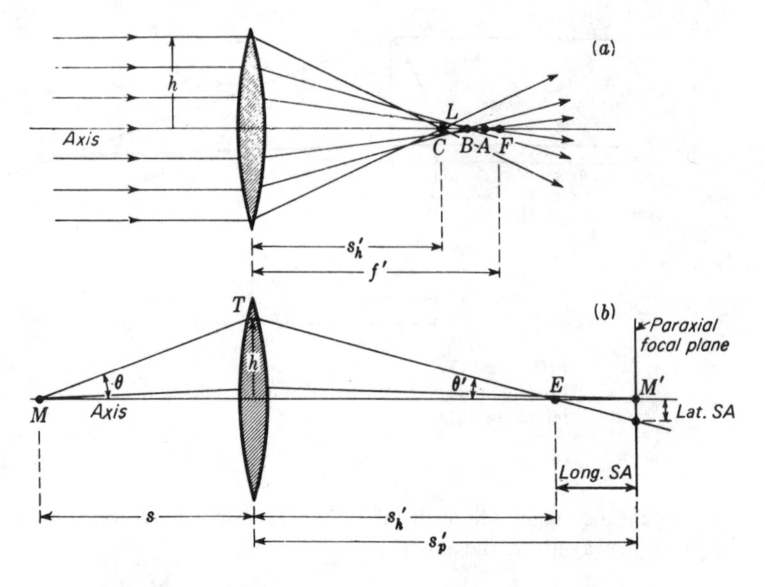

FIGURE 9D
Illustrations of lateral and longitudinal spherical aberration of a lens.

different shape is presented in Fig. 9F(*a*). The alteration of shape represented in this series is known as *bending* the lens. Each lens is labeled by a number q called its *shape factor*, defined by the formula

$$q = \frac{r_2 + r_1}{r_2 - r_1} \tag{9d}$$

As an example, if the two radii of a converging meniscus lens are $r_1 = -15.0$ cm and $r_2 = -5.0$ cm, it has a shape factor

$$q = \frac{-5 - 15}{-5 + 15} = -2$$

The usual reason for considering the bending of a lens is to find that shape for which the spherical aberration is a minimum. That such a minimum exists is shown by the graphs of Fig. 9F(*b*). These curves are drawn for the same lenses as shown in (*a*), and the values were taken from Table 9B. They were calculated by the ray-tracing methods of Table 8B. It will be noted that lens 5, for which the shape factor q is $+0.5$, has the least spherical aberration. The amount of this aberration for the ray having $h = 1.0$ cm is shown for the same series of lenses by the curves of Fig. 9G. Over the range of shape factors from about $q = +0.4$ to $q = +1.0$ the spherical aberration varies only slightly, since it is close to a minimum. At no point, however, does it go to zero. We see therefore that by choosing the proper radii for the two

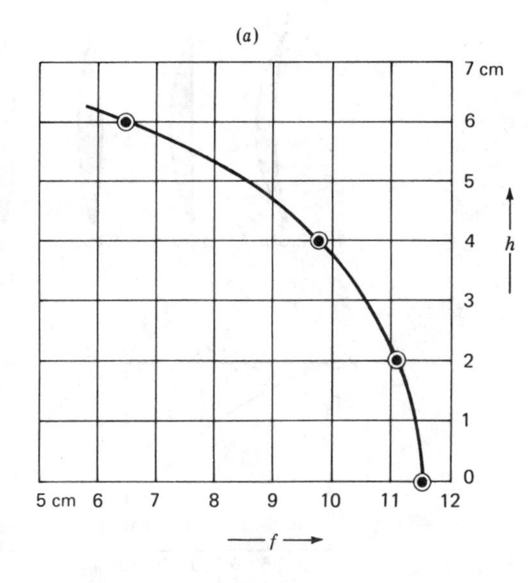

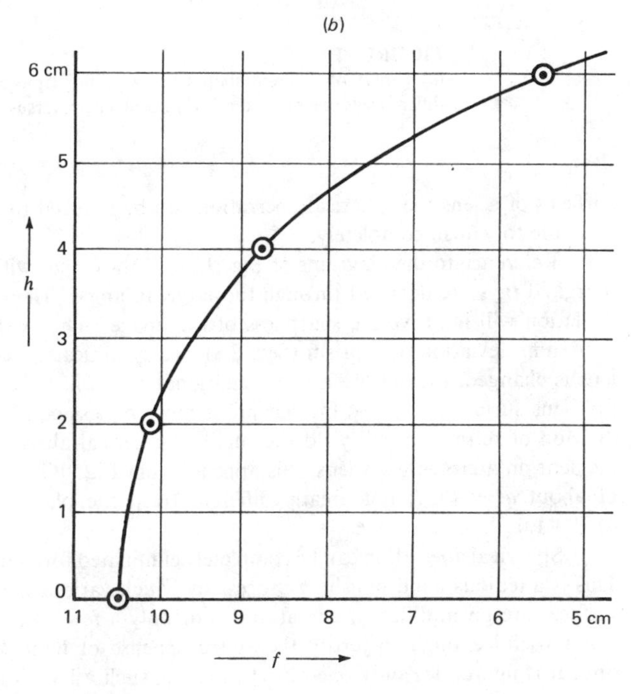

FIGURE 9E
Change in focal length of two glass lenses in air: (*a*) double convex and (*b*) double concave.

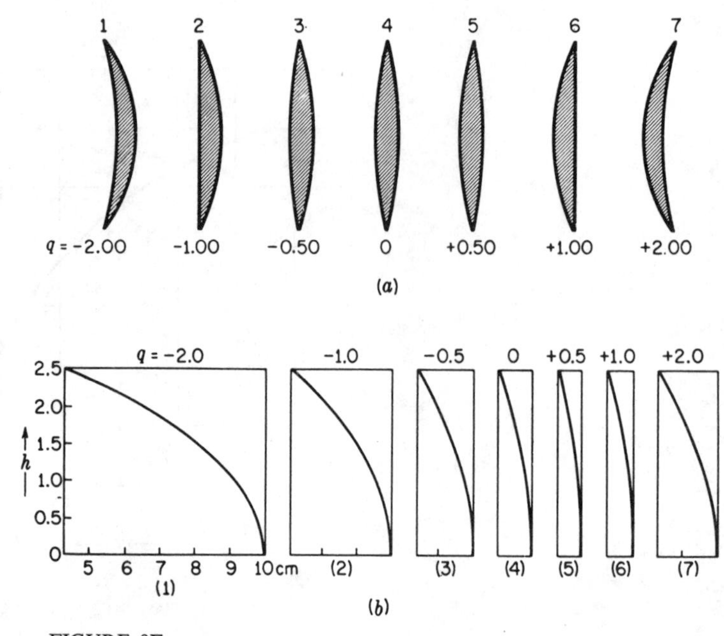

FIGURE 9F
(*a*) Lenses of different shapes but with the same power or focal length. The difference is one of bending. (*b*) Focal length versus ray height h for these lenses.

surfaces of a lens the spherical aberration can be reduced to a minimum but cannot be made to vanish completely.

Reference to the diagrams of Fig. 9D will show that with spherical surfaces the marginal rays are deviated through too large an angle. Hence any reduction of this deviation will improve the sharpness of the image. The existence of a condition of minimum deviation in a prism (Sec. 2.8) clearly indicates that when the shape of a lens is changed, the deviation of the marginal rays will be least when they enter the first lens surface and leave the second at more or less equal angles. Such an equal division of refraction will yield the smallest spherical aberration. For parallel light incident on a crown-glass lens, this appears from Fig. 9G to occur at a shape factor of about $q = +0.7$, not greatly different from the plano-convex lens, for which $q = +1.0$.

Spherical aberration can be completely eliminated for a single lens by *asphering*. This is a tedious hand-polishing process by which various zones of one or both lens surfaces are given different curvatures. For only a few special instruments are such lenses useful enough to justify the added expense of hand figuring. Furthermore, since it is figured for only one object distance, such a lens is not free from spherical aberration for other distances. The most common practice in lens design is to adhere to the simple spherical surfaces and to reduce the spherical aberration by a proper choice of radii.

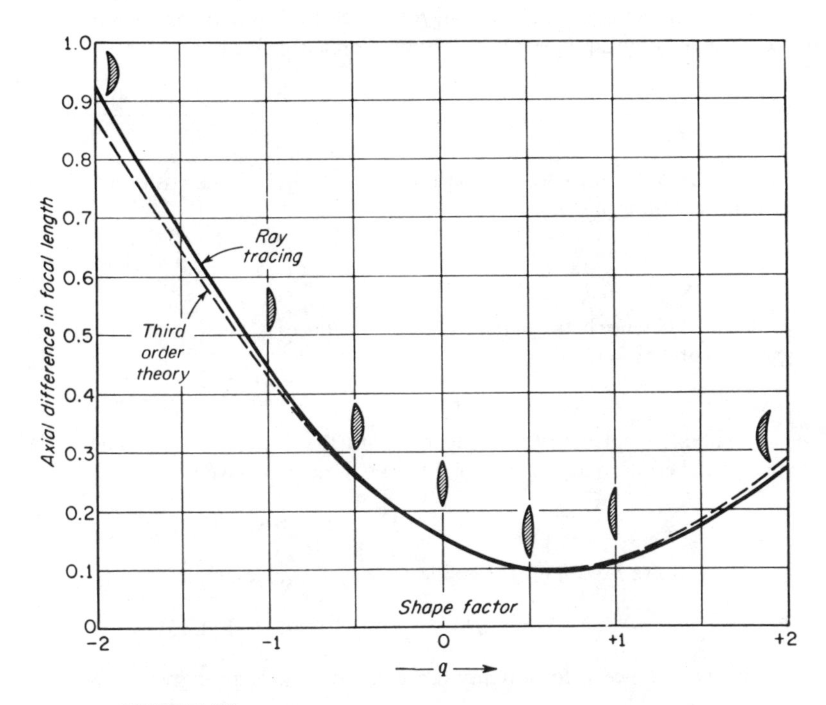

FIGURE 9G
A graph of the spherical aberration for lenses of different shape but the same focal length. For the lenses shown $h = 1$ cm, $f = +10$ cm, $d = 2$ cm, and $n' = 1.51700$.

9.5 RESULTS OF THIRD-ORDER THEORY

Although the derivation of an equation for spherical aberration from third-order theory is too lengthy to be given here, some of the resulting equations are of interest. For a thin lens we have the reasonably simple formula

$$L_s = \frac{h^2}{8f^3}\frac{1}{n(n-1)}$$

$$\left[\frac{n+2}{n-1}q^2 + 4(n+1)pq + (3n+2)(n-1)p^2 + \frac{n^3}{n-1}\right] \qquad \text{(9e)}$$

where $L_s = \dfrac{1}{s_h'} - \dfrac{1}{s_p'}$

As shown in Fig. 9D(*b*), s_h' is the image distance for an oblique ray traversing the lens at a distance h from the axis, s_p' is the image distance for paraxial rays, and f the paraxial focal length. The constant p is called the *position factor*, and q is the shape factor defined by Eq. (9d). The position factor is defined as

$$p = \frac{s' - s}{s' + s} \tag{9f}$$

Making use of the first-order equation $1/f = 1/s + 1/s'$, the position factor may also be expressed in terms of f as

$$p = \frac{2f}{s} - 1 = 1 - \frac{2f}{s'} \tag{9g}$$

The difference between the two image distances, $s'_p - s'_h$, is called the longitudinal spherical aberration:

$$\text{Long. SA} = s'_p - s'_h$$

The intercept of the oblique ray with the paraxial focal plane is the lateral spherical aberration and from Fig. 9D(*b*) is seen to be given by

$$\text{Lat. SA} = (s'_p - s'_h) \tan \theta'$$

If we solve Eq. (9e) for the difference $s'_p - s'_h$, we obtain

$$\text{Long. SA} = s'_p s'_h L_s \qquad \text{and} \qquad \text{Lat. SA} = s'_p h L_s \tag{9h}$$

The image distance s'_h for any ray through any zone is given by

$$s'_h = \frac{s'_p}{1 + s'_p L_s}$$

A comparison of the third-order theory with the exact results of ray tracing is included in Fig. 9G. When the shape factor is not far from that corresponding to the minimum, the agreement is remarkably good. The numerical results of third-order theory for the seven lenses of Fig. 9F are presented in the last column of Table 9B.

Equations useful in lens design are obtained by finding the shape factor that

Table 9B SPHERICAL ABERRATION OF LENSES HAVING THE SAME FOCAL LENGTH BUT DIFFERENT SHAPES q

Lens thickness = 1 cm $\quad f = 10$ cm $\quad n = 1.5000 \quad h = 1$ cm

Shape of lens	r_1	r_2	q	Ray tracing	Third-order theory
Concavo-convex	−10.000	− 3.333	−2.00	0.92	0.88
Plano-convex	∞	− 5.000	−1.00	0.45	0.43
Double convex	20.000	− 6.666	−0.50	0.26	0.26
Equiconvex	10.000	−10.000	0	0.15	0.15
Double convex	6.666	−20.000	+0.50	0.10	0.10
Plano-convex	5.000	∞	+1.00	0.11	0.11
Concavo-convex	3.333	10.000	+2.00	0.27	0.29

will make Eq. (9e) a minimum. This may be done by differentiating with respect to the shape factor and equating to zero:

$$\frac{dL_s}{dq} = \frac{h^2}{8f^3} \frac{2(n+2)q + 4(n-1)(n+1)p}{n(n-1)^2}$$

Equating to zero and solving for q, we obtain

$$q = -\frac{2(n^2 - 1)p}{n + 2} \tag{9i}$$

as the required relation between shape and position factors to produce minimum spherical aberration. As a rule a lens is designed for some particular pair of object and image distances so that p can be calculated from Eq. (9f). For a lens of a given n the shape factor that will produce a minimum lateral spherical aberration can be obtained at once from Eq. (9i). In order to determine the radii that will correspond to such a calculated shape factor and still yield the proper focal length, one can then use the lens maker's formula

$$\frac{1}{s} + \frac{1}{s'} = (n - 1)\left(\frac{1}{r_1} - \frac{1}{r_2}\right) = \frac{1}{f}$$

Substitution of values of s, s' and r_1, r_2 from Eqs. (9g) and (9d) gives the following useful set of equations, due to Coddington:

$$\begin{aligned} s &= \frac{2f}{1+p} & s' &= \frac{2f}{1-p} \\ r_1 &= \frac{2f(n-1)}{q+1} & r_2 &= \frac{2f(n-1)}{q-1} \end{aligned} \tag{9j}$$

The last two relations give the radii in terms of q and f. Division of one of these by the other gives

$$\frac{r_1}{r_2} = \frac{q-1}{q+1} \tag{9k}$$

As a problem let us suppose that a single lens is to be made with a focal length of 10.0 cm and that we wish to find the radii of the surfaces which will give the minimum spherical aberration for parallel incident light. For simplicity we shall assume that the glass has an index $n = 1.50$. In using Eq. (9i) the position factor p and the shape factor q must first be determined. Substitution of $s = \infty$ and $s' =$ 10.0 cm in Eq. (9f) gives

$$p = \frac{10 - \infty}{10 + \infty} = -1$$

It can be seen that if s is not infinite but is allowed to approach infinity, the ratio $(s' + s)/(s' - s)$ will approach the value -1 and will in the limit be equal to this. Substituting this position factor in Eq. (9i), we obtain

$$q = -\frac{2(2.25 - 1)(-1)}{1.5 + 2} = \frac{2.5}{3.5} = 0.714$$

This value falls at the minimum of the curve of Fig. 9G. The ratio of the two radii is given by Eq. (9k) as

$$\frac{r_1}{r_2} = \frac{0.714 - 1}{0.714 + 1} = \frac{-0.286}{1.714} = -0.167$$

The negative sign means that the surfaces curve in opposite directions, and the numerical value indicates a ratio of the radii of about 6:1. Their individual values are found from Eq. (9j) to be

$$r_1 = \frac{10}{1.714} = 5.83 \text{ cm} \qquad \text{and} \qquad r_2 = \frac{10}{0.286} = -35.0 \text{ cm}$$

Such a lens lies between lenses 5 and 6 in Fig. 9F and has essentially the same amount of spherical aberration as either one. For this reason plano-convex lenses are often employed in optical instruments with the convex side facing the parallel incident rays. Should such a lens be turned around so that the flat side is toward the incident light, its shape factor becomes $q = -1.0$ and the spherical aberration increases about fourfold.

Although spherical aberration cannot be entirely eliminated for a single spherical lens, it is possible to do so for a combination of two or more lenses of opposite sign. The amount of spherical aberration introduced by one lens of such a combination must be equal and opposite to that introduced by the other. If, for example, the doublet is to have a positive power and no spherical aberration, the positive lens should have the greater power and its shape should be at or near that for minimum spherical aberration, while the negative lens should have a smaller power and its shape should not be near that for the minimum. Neutralization by such an arrangement is possible because spherical aberration varies as the cube of the focal length, and therefore changes sign with the sign of f [see Eq. (9e)]. In a cemented lens of two elements, the two interfaces should have the same radius. The other two may then be varied and used to correct for spherical aberration. With four radii to manipulate, other aberrations like chromatic aberration can be reduced at the same time. This subject will be considered in Sec. 9.13.

9.6 FIFTH-ORDER SPHERICAL ABERRATION

The two curves that were given in Fig. 9G show that, for a lens having a shape factor anywhere near the optimum, the agreement between the exact results of ray tracing and the approximate results of third-order theory is remarkably good. For larger values of h, however, and for shapes further removed from the optimum appreciable differences occur. This indicates the necessity of including the fifth-order terms in the theory. The third-order equation (9e) shows that spherical aberration should be proportional to h^2, so that the curves in Fig. 9F(b) should be parabolas. Nevertheless accurate measurements show that for larger h departures from proportionality to h^2 do occur and that spherical aberration is more closely represented by an equation of the form

$$\text{Long. SA} = ah^2 + bh^4 \qquad (9l)$$

where a and b are constants. The term ah^2 represents the third-order effect and bh^4 the fifth-order effect. Some numerical results for a single lens, indicating the necessity for the inclusion of the latter term, are shown in Table 9C. The *boldface* values in the fourth row are the true values for longitudinal spherical aberration, obtained by ray-tracing methods, while those in the last row correspond to a parabola that has been fitted at $h = 1.0$ cm to the equation

$$\text{Long. SA} = a'h^2$$

with $a' = 0.11530 \text{ cm}^{-1}$.

The first row gives the third-order corrections ah^2 and the second row the fifth-order corrections bh^4. The third row contains the values calculated from Eq. (9l) by fitting the curve at the two points $h = 1.0$ cm and $h = 2.0$ cm. Assuming the values 0.11530 and 0.48208 at these points, the constants become

$$a = 0.11356 \qquad \text{and} \qquad b = 0.00174$$

A comparison of the totals in the third row with the correct values in the fourth row reveals the excellent agreement of the latter with Eq. (9l). Graphs of the values in rows 2 and 3 are given in Fig. 9H and show the negligible contribution of the fifth-order correction at small values of h. If only the third-order aberration were present in a lens, it would be possible to combine a positive and a negative lens having equal aberrations to obtain a combination corrected for all zones. Because they actually would have different amounts of fifth-order aberration, however, such a combination can be corrected for one zone only.

A graph illustrating the spherical aberration of a cemented doublet which is corrected for the marginal zone is shown in Fig. 9H(c). It will be seen that the curve comes to zero only at the origin and at the margin. The combination becomes badly overcorrected if the aperture is further increased. The plane of best focus lies a little to the left of the paraxial and marginal focal points, and its position (the vertical broken line) corresponds to that of the circle of least confusion.

Let a and b in Eq. (9l) represent the constants for a thin-lens doublet. If the combination is to be corrected at the margin, i.e., for a ray at the height h_m, we must have

$$\text{Long. SA} = ah_m{}^2 + bh_m{}^4 = 0$$

or

$$a = -bh_m{}^2$$

Table 9C FIFTH-ORDER CORRECTION TO SPHERICAL ABERRATION
$f = 10.0$ cm $\quad r_1 = +5.0$ cm $\quad r_2 = \infty \quad n = 1.500 \quad d = 1.0$ cm

Row h, cm	0.5	1.0	1.5	2.0	2.5	3.0
1 ah^2	0.02839	0.11356	0.25551	0.45424	0.70975	1.02204
2 bh^4	0.00011	0.00174	0.00881	0.02784	0.06797	0.14094
3 $ah^2 + bh^4$	0.02850	**0.11530**	0.26432	**0.48208**	0.77772	1.16928
4 Ray tracing	**0.02897**	**0.11530**	**0.26515**	**0.48208**	**0.77973**	**1.16781**
5 Parabola	0.02882	**0.11530**	0.25942	0.46120	0.71812	1.03770

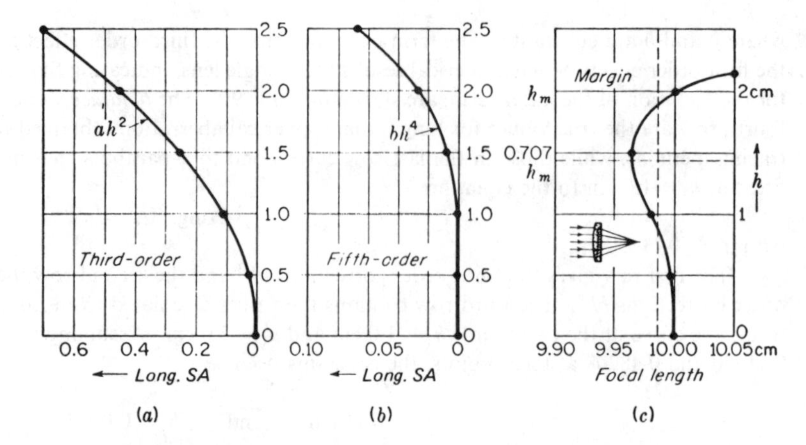

FIGURE 9H
(*a*) Third-order and (*b*) fifth-order contributions to longitudinal spherical aberration. (*c*) Longitudinal spherical aberration of a corrected doublet as used in telescopes.

Substitution in Eq. (9l) yields

$$\text{Long. SA} = -bh_m{}^2h^2 + bh^4$$

where h_m is fixed and h may take any value between 0 and h_m. To find where this expression has a maximum value, we differentiate with respect to h and equate to zero, as follows:

$$\frac{d(\text{Long. SA})}{dh} = -2bh_m{}^2h + 4bh^3 = 0$$

Dividing by $-2bh$, we obtain

$$h = h_m\sqrt{\tfrac{1}{2}} = 0.707h_m$$

as the radius of the zone at which the aberration reaches a maximum [see Fig. 9H(*c*)]. In lens design spherical aberration is always investigated by tracing a ray through the combination for the zone of radius $0.707h_m$.

9.7 COMA

The second of the monochromatic aberrations of third-order theory is called *coma*. It derives its name from the cometlike appearance of the image of an object point located just off the lens axis. Although the lens may be corrected for spherical aberration and may bring all rays to a good focus on the axis, the quality of the images of points just off the axis will not be sharp unless the lens is also corrected for coma. Figure 9I illustrates this lens defect for a single object point infinitely distant and off the axis. Of the fan of rays in the meridian plane that is shown, only those through the center of the lens form an image at A'. Two rays through the margin

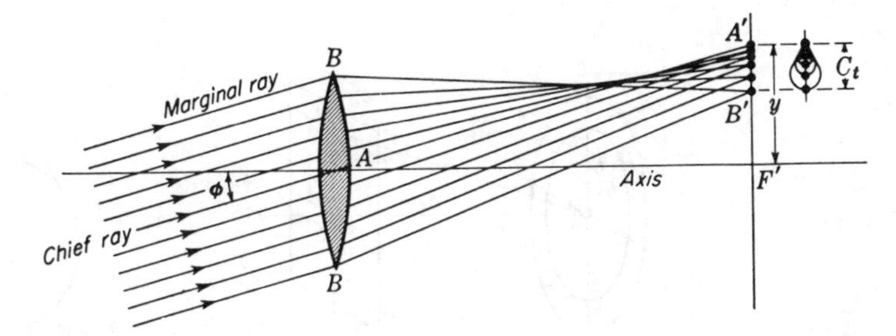

FIGURE 9I
Coma, the second of the five monochromatic aberrations of a lens. Only the tangential fan of rays is shown.

come together at B'. Thus it appears that the magnification is different for different parts of the lens. If the magnification for the outer rays through a lens is greater than that for the central rays, the coma is said to be *positive*, while if the reverse is true as in the diagram, the coma is said to be *negative*.

The shape of the image of an off-axis object point is shown at the upper right in Fig. 9I. Each of the circles represents an image from a different zone of the lens. Details of the formation of the *comatic circle* by the light from one zone of the lens are shown in Fig. 9J. Rays 1, which correspond to the *tangential* rays B in Fig. 9I, cross at 1 on the comatic circle, while rays 3, called the *sagittal rays*, cross at the top of that circle. In general all points on a comatic circle are formed by the crossing of pairs of rays passing through two diametrically opposite points of the same zone. Third-order theory shows that the radius of a comatic circle is given by

$$C_s = \frac{jh^2}{f^3}(Gp + Wq) \qquad \text{(9m)}$$

where j, h, and f are the distances indicated in Fig. 9K(a) and p and q are the Coddington position and shape factors given by Eqs. (9f) and (9d). The other two constants are defined as

$$G = \frac{3(2n + 1)}{4n} \qquad \text{and} \qquad W = \frac{3(n + 1)}{4n(n - 1)}$$

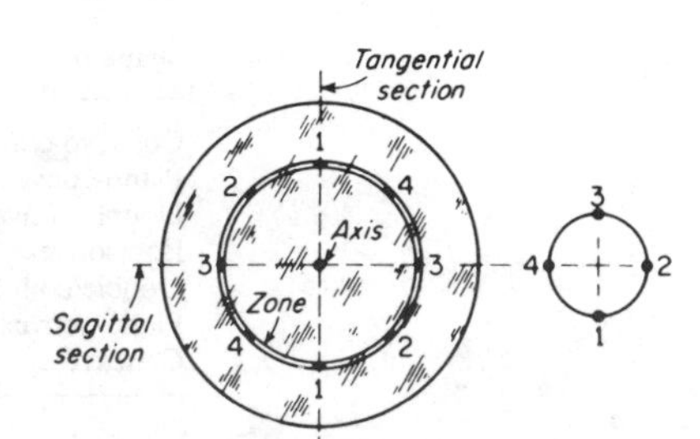

FIGURE 9J
Each zone of a lens forms a ring-shaped image called a comatic circle.

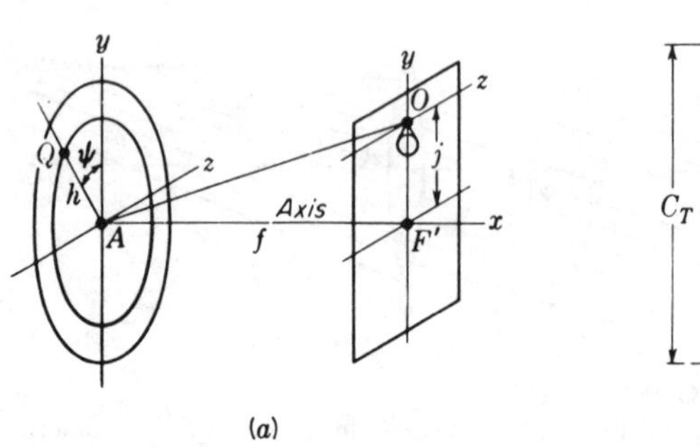

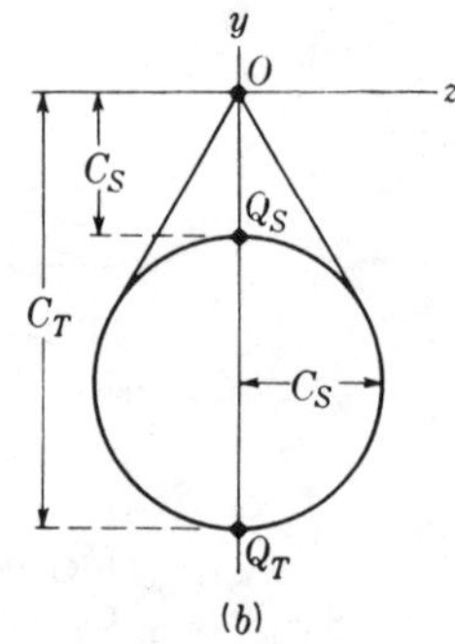

FIGURE 9K
Geometry of coma, showing the relative magnitudes of sagittal and tangential magnifications.

The shape of the comatic figure is given by

$$y = C_S(2 + \cos 2\psi) \qquad z = C_S \sin 2\psi$$

which shows that the tangential coma C_T is three times the sagittal coma C_S [see Fig. 9K(b)]. Thus

$$C_T = 3C_S$$

To see how coma is affected by changing the shape of a lens a graph of the height of the comatic figure C_T is plotted against the shape factor q in Fig. 9L. The numerical values plotted in this graph are calculated from Eq. (9m) and listed in Table 9D.

A parallel beam of light is assumed to be incident on the lens at an angle of 11° with the axis. The values of the longitudinal spherical aberration, given for comparison purposes, are also calculated from third-order theory, Eq. (9e), and assume

Table 9D COMPARISON OF COMA AND SPHERICAL ABERRATION FOR LENSES OF THE SAME FOCAL LENGTH BUT DIFFERENT SHAPE FACTOR
$h = 1.0$ cm $\quad f = +10.0$ cm $\quad y = 2.0$ cm
$n = 1.5000$

Shape of lens	Shape factor	Coma, cm	Spherical aberration, cm
Concavo-convex	−2.0	−0.0420	+0.88
Plano-convex	−1.0	−0.0270	+0.43
Double convex	−0.5	−0.0195	+0.26
Equiconvex	0	−0.0120	+0.15
Double convex	+0.5	−0.0045	+0.10
Plano-convex	+1.0	+0.0030	+0.11
Concavo-convex	+2.0	+0.0180	+0.29

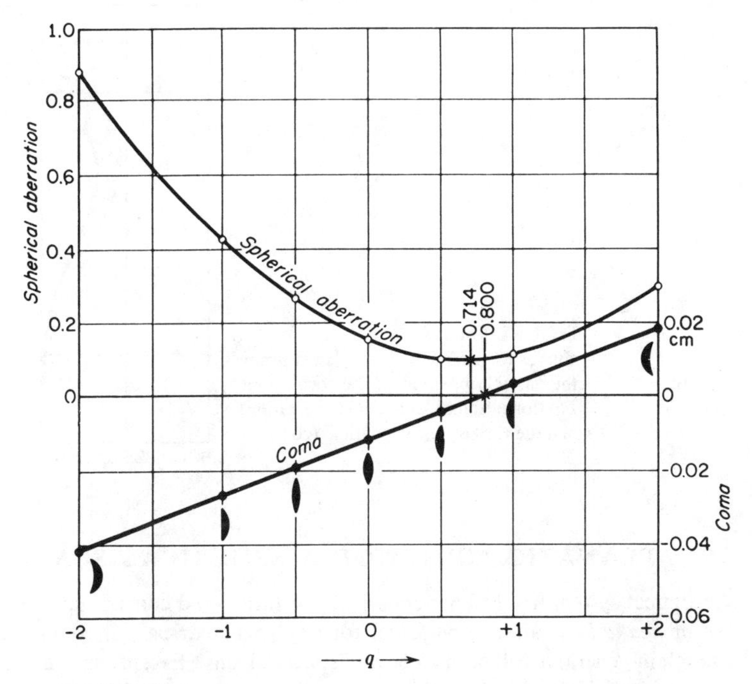

FIGURE 9L
Graphs comparing coma with longitudinal spherical aberration for a series of lenses having different shapes.

parallel light incident on the lens parallel to the axis and passing through the same zone.

The fact that the line representing coma crosses the zero axis indicates that a single lens can be made that is entirely free of this aberration. It is important to note, for the lenses shown, that the shape factor $q = 0.800$ for no coma is so near the shape factor $q = 0.714$ for minimum spherical aberration that a single lens designed for $C_T = 0$ will have practically the minimum amount of spherical aberration.

In order to calculate the value of q that will make Eq. (9m) vanish, C_S is set equal to zero. There results

$$q = -\frac{G}{W}p \tag{9n}$$

If the shape and position factors of a single lens obey this relation, the lens is coma-free. A doublet designed to correct for spherical aberration can at the same time be corrected for coma. A graph showing the residual spherical aberration and coma for a telescope objective is given in Fig. 9M.

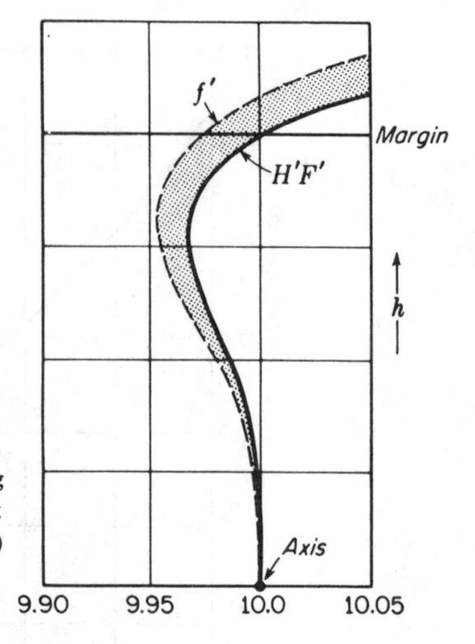

FIGURE 9M
Curves for a cemented doublet, showing the variable position of the focal point F' (longitudinal spherical aberration) and the variable focal length f' (coma = $H'F' - f'$).

9.8 APLANATIC POINTS OF A SPHERICAL SURFACE

An optical system free of both spherical aberration and coma is said to be *aplanatic*. An *aplanatic lens* can also be found for any particular pair of conjugate points, although in general it will need to be an aspherical lens. Except for a few special cases, no lens combination with spherical surfaces is completely free of both these aberrations.

One special case which is of considerable importance in microscopy is that of a single spherical refracting surface. To demonstrate the existence of aplanatic points for a single surface, a useful construction, originally discovered by Huygens, will first be described. In Fig. 9N(*a*) the ray RT represents any ray in the first medium, of index n, incident on the surface at T and making an angle ϕ with the normal NC. Around C as a center and with radii

$$\rho = r\frac{n'}{n} \qquad \text{and} \qquad \rho' = r\frac{n}{n'} \tag{9o}$$

the broken circular arcs are drawn as shown. Where RT, when produced, intersects the larger circle, a line JC is drawn, and this intersects the smaller circle at K. Then TK gives the direction of the refracted ray in accordance with the law of refraction.* Furthermore any ray whatever directed toward J will be refracted through K.

The aplanatic points of a single surface are located where the two construction circles cross the axis [see Fig. 9N(*b*)]. All rays initially traveling toward M will pass through M', and similarly all rays diverging from M' will after refraction appear to originate at M. The application of this principle to a microscope is illustrated in Fig. 9O. A drop of oil having the same index as the hemispherical lens is placed on the microscope slide and the lens lowered into contact as shown. All rays from an object

* For a proof of this proposition, see J. P. C. Southall, "Mirrors, Prisms, and Lenses," 3d ed., p. 512, The Macmillan Company, New York, 1936.

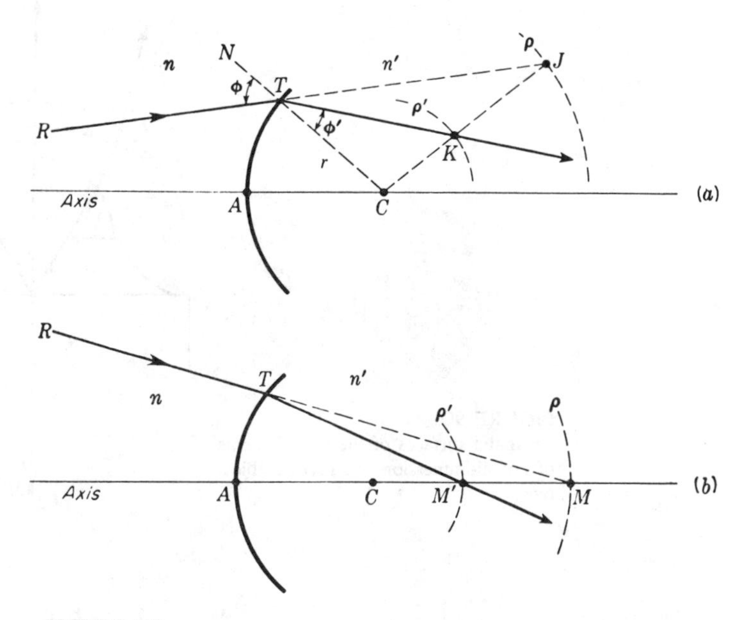

FIGURE 9N
(*a*) A graphical construction for refraction at a single spherical surface; $\rho = rn'/n$ and $\rho' = rn/n'$. (*b*) Location of the aplanatic points of a single spherical surface.

at M leave the hemispherical surface after refraction as though they came from M', and this introduces a lateral magnification of $M'A/MA$. If a second lens is added which has the center of its concave surface at M' (and therefore is normal to all rays), refraction at its upper surface, of radius $n' \times CM'$, will give added magnification without introducing spherical aberration. This property of the upper lens, however, holds strictly only for rays from the single point M, and not for points adjacent to it. There is a limit to this process which is set by chromatic aberration (see Sec. 9.13).

9.9 ASTIGMATISM

If the first two Seidel sums vanish, all rays from points on or very close to the axis of a lens will form point images and there will be no spherical aberration or coma. When the object point lies at some distance away from the axis, however, a point image will be formed only if the third sum S_3 is zero. If the lens fails to satisfy this third condition, it is said to be afflicted with *astigmatism*, and the resulting blurred images are said to be astigmatic. The formation of real astigmatic images from a concave spherical mirror is discussed in Sec. 6.9. To help understand the formation of astigmatic images by a lens, a ray diagram has been drawn in perspective in Fig. 9P(*a*). Considering the rays from a point object Q, all those in the fan contained in the vertical

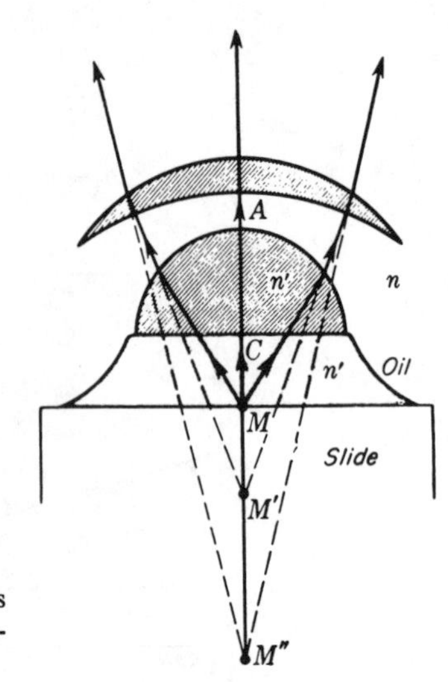

FIGURE 9O
Aplanatic surfaces of the first elements of an oil-immersion microscope objective.

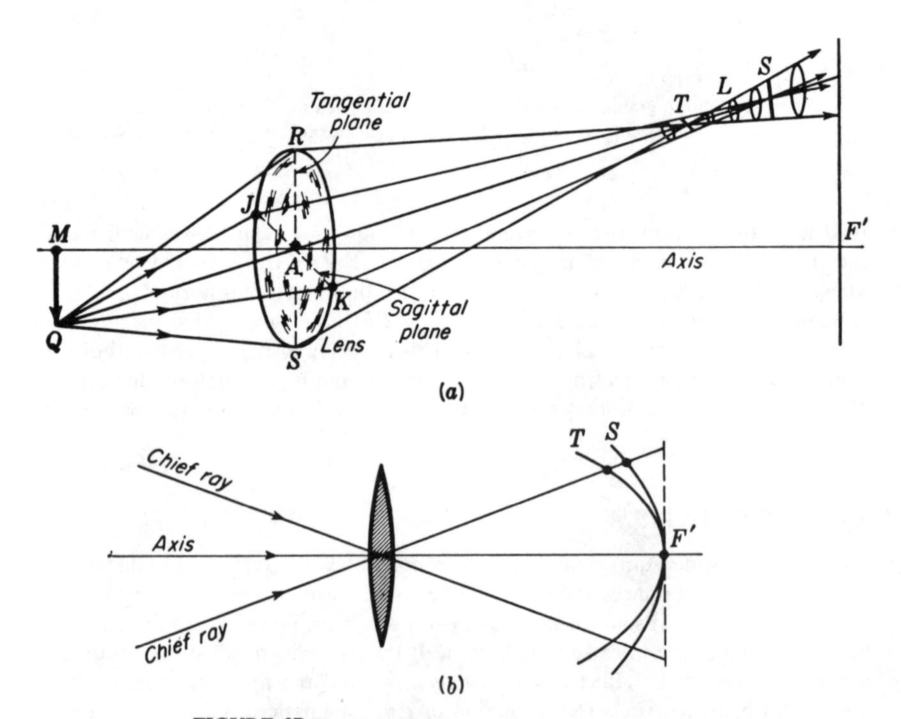

FIGURE 9P
(*a*) Perspective diagram showing the two focal lines which constitute the image of an off-axis object point Q. (*b*) Loci of the tangential and sagittal images. The two surfaces approximate paraboloids of revolution.

or tangential plane cross at T, while the fan of rays in the horizontal or sagittal plane crosses at S. The tangential and sagittal planes intersect the lens in RS and JK, respectively. Rays in these planes are chosen because they locate the two *focal lines* T and S formed by all rays going through the lens. These are perpendicular to their respective tangential and sagittal planes. At L the image is approximately disk-shaped, and constitutes the circle of least confusion for this case.

If the positions of the T and S images are determined for a wide field of distant object points, their loci will form paraboloidal surfaces whose sections are shown in Fig. 9P(*b*). The amount of astigmatism, or *astigmatic difference*, for any pencil of rays is given by the distance between these two surfaces measured along the chief ray. On the axis, where the two surfaces come together, the astigmatic difference is zero; away from the axis it increases approximately as the square of the image height. Astigmatism is said to be positive when the T surface lies to the left of S, as shown in the diagram. It should be noted that for a concave mirror (Fig. 6O), the sagittal surface is a plane coinciding with the paraxial focal plane.

If (as in Fig. 9Q) the object is a spoked wheel in a plane perpendicular to the axis with its center at M, the rim would be found to be in focus on the T surface while the spokes would be in focus on the S surface. It is for this reason that the terms *tangential* and *sagittal* are applied to the planes and images. On the surface T all images will be lines parallel to the rim as shown at the left in Fig. 9Q, while on the surface S all images will be lines parallel to the spokes as shown at the right.

Equations giving the astigmatic image distances for a single refracting surface are*

$$\frac{n\cos^2\phi}{s} + \frac{n'\cos^2\phi'}{s'_T} = \frac{n'\cos\phi' - n\cos\phi}{r}$$
$$\frac{n}{s} + \frac{n'}{s'_S} = \frac{n'\cos\phi' - n\cos\phi}{r} \tag{9p}$$

where ϕ and ϕ' are the angles of incidence and refraction of the chief ray, r the radius of curvature, s the object distance, and s_T and s_S the T and S image distances, the latter being measured along the chief ray. For a spherical mirror these equations reduce to

$$\frac{1}{s} + \frac{1}{s'_T} = \frac{1}{f\cos\phi} \qquad \text{and} \qquad \frac{1}{s} + \frac{1}{s'_S} = \frac{\cos\phi}{f}$$

Coddington has shown that for a thin lens in air with an aperture stop at the lens, the positions of the tangential and sagittal images are given by

$$\frac{1}{s} + \frac{1}{s'_T} = \frac{1}{\cos\phi}\left(\frac{n\cos\phi'}{\cos\phi} - 1\right)\left(\frac{1}{r_1} - \frac{1}{r_2}\right)$$
$$\frac{1}{s} + \frac{1}{s'_S} = \cos\phi\left(\frac{n\cos\phi'}{\cos\phi} - 1\right)\left(\frac{1}{r_1} - \frac{1}{r_2}\right) \tag{9q}$$

* For a derivation of these formulas see G. S. Monk, "Light, Principles and Experiments," Dover Publications, Inc., New York, 1963.

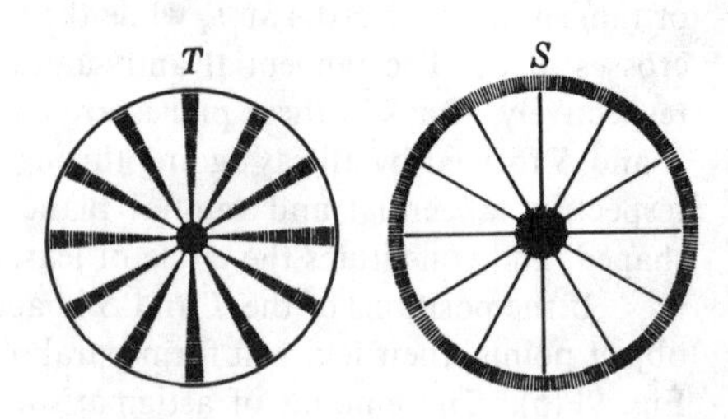

FIGURE 9Q
Astigmatic images of a spoked wheel.

The angle ϕ is the angle of obliquity of the incident chief rays, and ϕ' the angle of this ray within the lens. Therefore $n = \sin \phi / \sin \phi'$. The application of these formulas to thin lenses shows that the astigmatism is approximately proportional to the focal length and is very little improved by changing the shape.

Although a contact doublet composed of one positive and one negative lens shows considerable astigmatism, the introduction of another element consisting of a stop *or* a lens can be made to greatly reduce it. By the proper spacing of the lens elements of any optical system or by the proper location of a stop if one is used, the curvature of the astigmatic image surfaces can be changed considerably. Four important stages in the flattening of the astigmatic surfaces due to these alterations are shown in Fig. 9R. Diagram (*a*) represents the normal shape of the T and S surfaces for a contact doublet or a single lens. In diagram (*b*) the separation of lens elements is such that the two surfaces fall together at P. Further alteration of the lens shapes and their spacing can be made and the T and S curves straightened, as in diagram (*c*), or moved still farther apart until they are bisected by the normal plane through the focal point F', as in diagram (*d*). Of these four arrangements, only the second is free of astigmatism. The single paraboloidal surface P, over which point images are formed, is called the *Petzval surface*.

9.10 CURVATURE OF FIELD

If for an optical system the first three Seidel sums are zero, the system will form point images of point objects on as well as off the axis. Under these circumstances the images fall on the curved Petzval surface where the tangential and sagittal surfaces come together, as in Fig. 9R(*b*). Even though astigmatism is corrected for such a system, the focal surface is curved. If a flat screen is placed in position B, the center of the field will be in sharp focus but the edges will be quite blurred. With a screen at A, the center of the field and the field margins will be blurred, while sharp focus will be obtained about halfway out.

Mathematically a Petzval surface exists for every optical system, and if the powers and refractive indices of the lenses remain fixed, the shape of the Petzval surface cannot be changed by altering the shape factors of the lenses or their spacing. Such alterations, however, will change the shapes of the T and S surfaces, but always in such a way that the ratio of the distances PT and PS is 3 : 1. It will be noted that

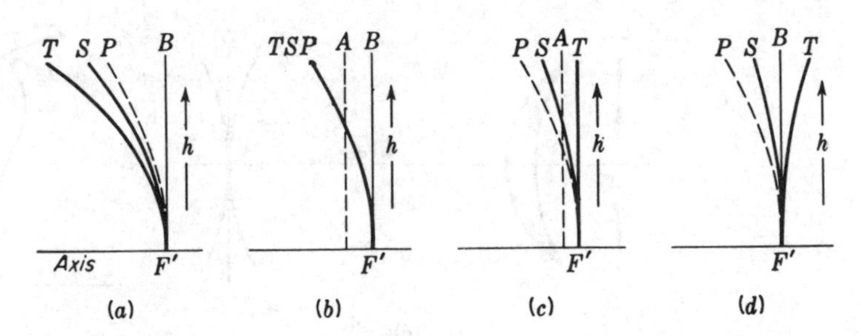

FIGURE 9R
Diagrams showing the astigmatic surfaces T and S in relation to the fixed Petzval surface P as the spacing between lenses (or between lens and stop) is changed.

this ratio is maintained throughout Fig. 9R. If a system is designed to make the T surface flat, as in Fig. 9R(*c*), the 3 : 1 ratio of distances requires the S surface to be curved, but not strongly so. If a screen is placed at a compromise position A, the images over the entire field will be in reasonably good focus. This condition of correction is commonly used for certain types of photographic lenses. If more negative astigmatism is introduced, the condition shown in Fig. 9R(*d*) is reached, in which the T surface is convex and the S surface is concave by an equal amount. In this case a screen placed at the paraxial focus will show considerable blurring at the field edges.

Curvature of field can be corrected for a single lens by means of a stop. Acting as a second element of the system, a stop limits the rays from each object point in such a way that the paths of the chief rays from different points go through different parts of the lens [Fig. 9S(*a*)]. Certain manufacturers of inexpensive box cameras employ a single meniscus lens and a stop and with them obtain reasonably good images. The stop is located in front of the lens, with the light incident on the concave surface. Although the compromise field is flat and sharp focus is obtained at the center, astigmatism gives rise to blurred images at the margins.

In complex lens systems it is possible, because of differences in third- and fifth-order corrections, to control the astigmatism and cause the tangential and sagittal surfaces to come together at an outer zone as well as at the center of the field. Typical curves for the camera objective called an *anastigmat* are shown in Fig. 9S(*b*). Experience has shown that the best state of correction is obtained by making the crossover point, called the *node*, occur at a relatively short distance in front of the focal plane.

9.11 DISTORTION

Even if an optical system were designed so that the first four Seidel sums were zero, it could still be affected by the fifth aberration known as *distortion*. To be free of distortion a system must have uniform lateral magnification over its entire field. A pinhole camera is ideal in this respect, for it shows no distortion; all straight lines

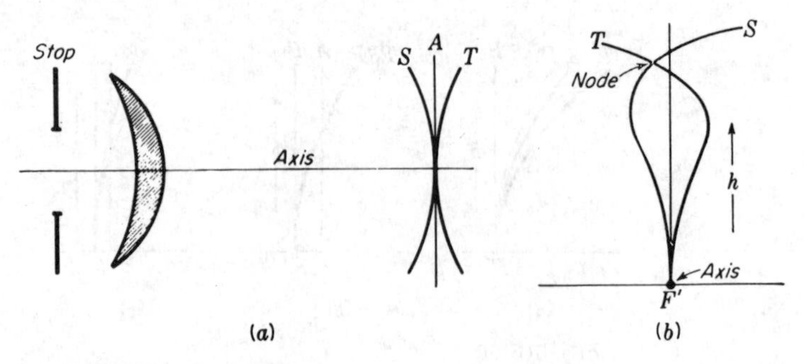

FIGURE 9S
(*a*) A properly located stop may be used to reduce field curvature. (*b*) Astigmatic surfaces for an anastigmat camera lens.

connecting each pair of conjugate points in the object and image planes pass through the opening. Constant magnification for a pinhole camera as well as for a lens implies, as may be seen from Fig. 9T(*a*), that

$$\frac{\tan \phi'}{\tan \phi} = \text{const}$$

The common forms of image distortion produced by lenses are illustrated in the lower part of Fig. 9T. Diagram (*b*) represents the undistorted image of an object consisting of a rectangular wire mesh. The second diagram shows *barrel* distortion, which arises when the magnification decreases toward the edge of the field. The third diagram represents *pincushion* distortion, corresponding to a greater magnification at the borders.

A single thin lens is practically free of distortion for all object distances. It cannot, however, be free of all the other aberrations at the same time. If a stop is placed in front of or behind a thin lens, distortion is invariably introduced; if it is placed at the lens, there is no distortion. Frequently in the design of good camera lenses astigmatism, as well as distortion, is corrected for by a nearly symmetrical arrangement of two lens elements with a stop between them.

To illustrate the principles involved, consider the lens shown in Fig. 9U(*a*), which has a front stop. Rays from object points like M, at or near the axis, go through the central part of the lens, while rays from off-axis object points like Q_2 are refracted only by the upper half. In the latter case the stop decreases the ratio of image to object distances measured along the chief ray, thereby reducing the lateral magnification below that obtaining for object points near the axis. This system therefore suffers from barrel distortion. When the lens and stop are turned around, as in Fig. 9U(*b*), the ratio of image to object distances is seen to increase as the object point lies farther off the axis. The result is increased magnification and pincushion distortion.

By combining two identical lenses with a stop midway between them as in Fig. 9U(*c*), a system is obtained which because of its symmetry is free from distortion for unit magnification. With other magnifications, however, the lenses must be corrected for spherical aberration with respect to the entrance and exit pupils. These two pupils S' and S'' coincide with the principal planes of the combination. Such a corrected

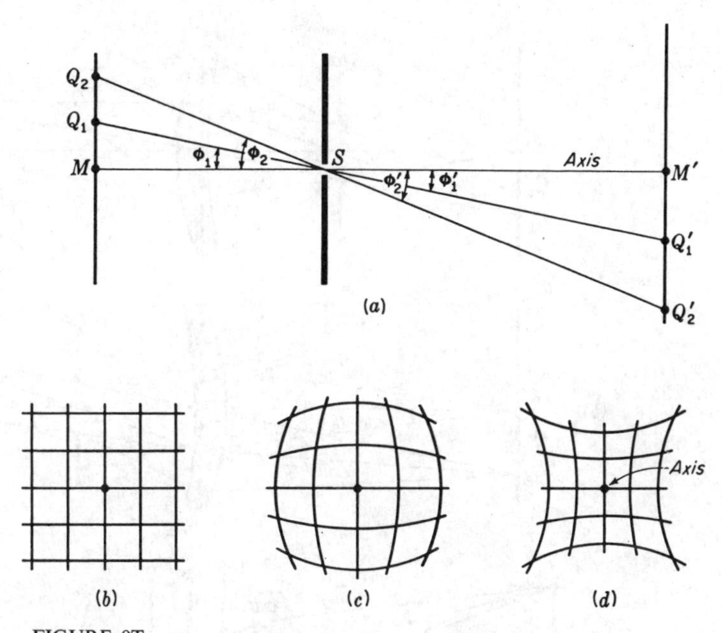

FIGURE 9T
(*a*) A pinhole camera shows no distortion. Images of a rectangular object screen shown with (*b*) no distortion, (*c*) barrel distortion, and (*d*) pincushion distortion.

lens system is called an *orthoscopic doublet*, or rapid rectilinear lens. Because this combination cannot be corrected for spherical aberration for the object and image planes and for the entrance and exit pupils at the same time, the lens suffers from this aberration as well as from astigmatism. Photographic lenses of this type are discussed in Sec. 10.5.

Summarizing very briefly the various methods of correcting for aberrations, spherical aberration and coma can be corrected by using a contact doublet of the proper shape; astigmatism and curvature of field require for their correction the use of several separated components; and distortion can be minimized by the proper placement of a stop.

9.12 THE SINE THEOREM AND ABBE'S SINE CONDITIONS

In Chap. 3 it was found that the lateral magnification produced by a *single spherical surface* is given by the relation

$$m = \frac{y'}{y} = -\frac{s' - r}{s + r}$$

This equation follows from the similarity of triangles MQC and $M'Q'C$ in Fig. 3F.

From Eq. (8a) we obtain the exact relation

$$s + r = r\,\frac{\sin \phi}{\sin \theta}$$

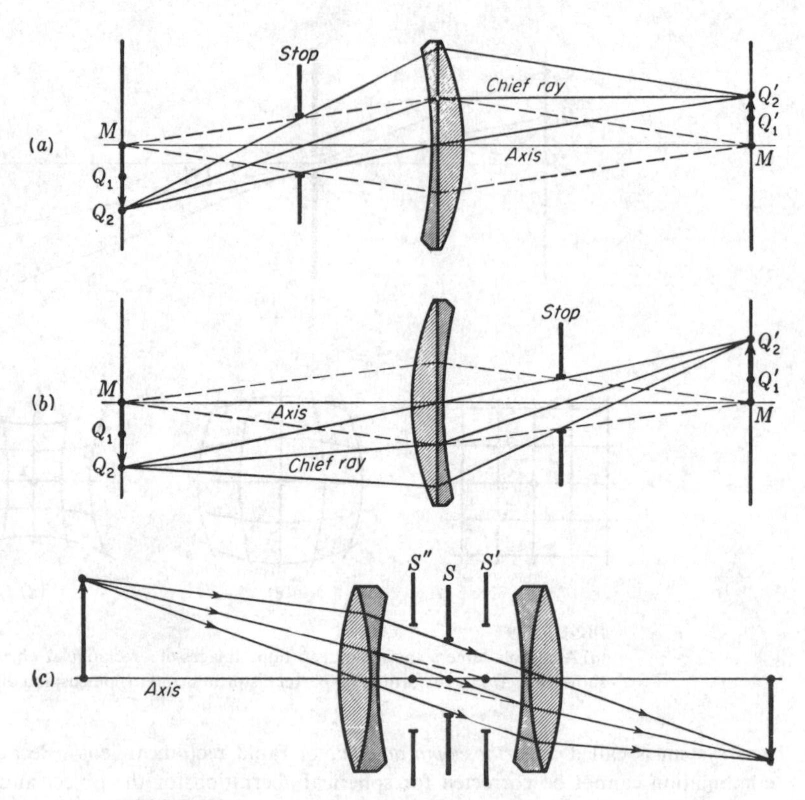

FIGURE 9U
(*a*) A stop in front of a lens giving rise to barrel distortion. (*b*) A stop behind a lens giving rise to pincushion distortion. (*c*) A symmetrical doublet with a stop between is relatively free of distortion.

and from Eq. (8d)

$$s' - r = -r\frac{\sin\phi'}{\sin\theta'}$$

If we substitute these two equations in the first equation, we obtain

$$\frac{y'}{y} = \frac{\sin\phi'\,\sin\theta}{\sin\theta'\,\sin\phi}$$

According to Snell's law

$$\frac{\sin\phi'}{\sin\phi} = \frac{n}{n'}$$

which upon substitution gives

$$\frac{y'}{y} = \frac{n\sin\theta}{n'\sin\theta'}$$

or

$$ny\sin\theta = n'y'\sin\theta' \qquad \textit{Sine theorem}$$

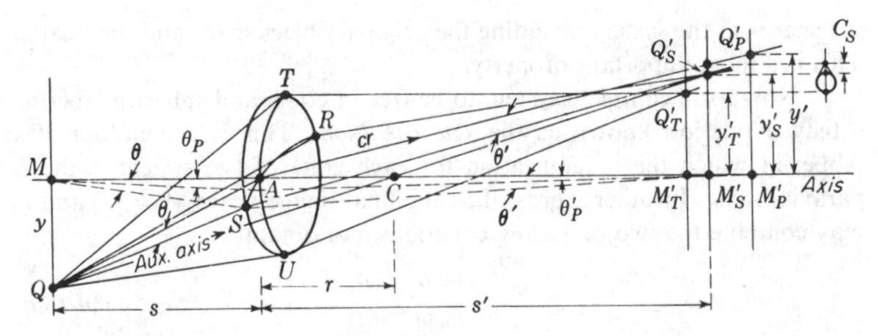

FIGURE 9V
Refraction at a spherical surface illustrating the sine theorem as it applies to coma.

Here y and y' are the object and image heights, n and n' are the indices of the object and image spaces, and θ and θ' are the slope angles of the ray in these two spaces, respectively (see Fig. 9V). This very general theorem applies to all rays, no matter how large the angles θ and θ' may be.

For paraxial rays where θ and θ' are both small, $\sin\theta$ and $\sin\theta'$ can be replaced by θ_P and θ'_P, respectively, to give

$$ny\theta_P = n'y'\theta'_P \qquad \textit{Lagrange theorem}$$

a relation referred to as the Lagrange theorem. In both these theorems all quantities on the left side refer to object space, while those on the right side refer to image space.

Figure 9V shows a pair of sagittal rays QR and QS from the object point Q through one zone of a single refracting surface. These two particular rays, after refraction, come to a focus at a point Q'_S on the auxiliary axis. On the other hand, a pair of tangential rays QT and QU through the same zone come to a focus at Q'_T, while paraxial rays come to a focus at Q'_P. Because of the general spherical aberration and astigmatism of the single surface the paraxial, the sagittal, and the tangential focal planes do not coincide. The conventional comatic figure shown at the right in Fig. 9V arises only in the absence of spherical aberration and astigmatism.

Since coma is confined to lateral displacements in the image in which y and y' are relatively small, we can neglect astigmatism and apply the above theorems to the single surface as follows: Note that θ and θ' for the object point Q, which are the slope angles of the zonal rays QS and Q'_SS relative to the chief ray (cr), are virtually equal to the slope angles of the rays from the axial object point M through the same zone of the surface. We can therefore apply the sine theorem to find the sagittal image magnification for any zone and obtain

$$m_S = \frac{y'_S}{y} = \frac{n\sin\theta}{n'\sin\theta'}$$

where $y'_S = Q'_SM'_S$, in Fig. 9V.

To show that the sine theorem and the Lagrange theorem can be extended to a complete optical system containing two or more lens surfaces, we recognize that in the image space of the first lens surface the two products are $n'_1y'_1\sin\theta'_1$ and $n'_1y'_1\theta'_{P1}$, respectively. These products are identical for the object space of the second surface because $n'_1 = n_2$, $y'_1 = y_2$, and $\theta'_1 = \theta_2$; hence the products are *invariant* for all the spaces in the system including the original object space and the final image space. This is a most important property.

Now for a complete system to be free of coma and spherical aberration it must satisfy a relation known as the *sine condition.* This is a condition discovered by Abbe, in which the magnification for each zone of the system is the same as for paraxial rays. In other words, if in the final image space $y'_s = y'$, and $m_s = m$, we may combine the two preceding equations and obtain

$$\frac{\sin \theta}{\sin \theta'} = \frac{\theta_P}{\theta'_P} = \text{const} \qquad \textit{Sine condition} \qquad (9r)$$

Any optical system is therefore free of coma, if in the absence of spherical aberration $(\sin \theta)/(\sin \theta') = \text{const}$ for all values of θ. In lens design coma is sometimes tested for by plotting the ratio $(\sin \theta)/(\sin \theta')$ against the height of the incident ray. Because most lenses are used with parallel incident or emergent light, it is customary to replace $\sin \theta$ by h, the height of the ray above the axis, and to write the sine condition in the special form

$$\frac{h}{\sin \theta'} = \text{const} \qquad (9s)$$

The ray diagram in Fig. 9W shows that the constant in this equation is the focal distance measured along the image ray, which we here call f'. To prevent coma, f' must be the same for all values of h. Since freedom from spherical aberration requires that all rays cross the axis at F', an accompanying freedom from coma requires that the principal "plane" be a spherical surface (represented by the dotted line in the figure) of radius f'. It is thus seen that whereas spherical aberration is concerned with the crossing of the rays at the focal point, coma is concerned with the shape of the principal surface. It should be noted that the aplanatic points of a single spherical surface (see Sec. 9.8) are unique in that they are entirely free of spherical aberration and coma and satisfy the sine condition exactly.

9.13 CHROMATIC ABERRATION

In the discussion of the third-order theory given in the preceding sections, no account has been taken of the change of refractive index with color. The assumption that n is constant amounts to investigating the behavior of the lens for monochromatic light only. Because the refractive index of all transparent media varies with color, a single lens forms not only one image of an object but a series of images, one for each color of light present in the beam. Such a series of colored images of an infinitely distant object point on the axis of the lens is represented diagramatically in Fig. 9X(*a*). The prismatic action of the lens, which increases toward its edge, is such as to cause dispersion and to bring the violet light to a focus nearest to the lens.

As a consequence of the variation of focal length of a lens with color, the lateral magnification must vary as well. This may be seen by the diagram of Fig. 9X(*c*), which shows only the red and violet image heights of an off-axis object point Q. The horizontal distance between the axial images is called *axial* or *longitudinal chromatic*

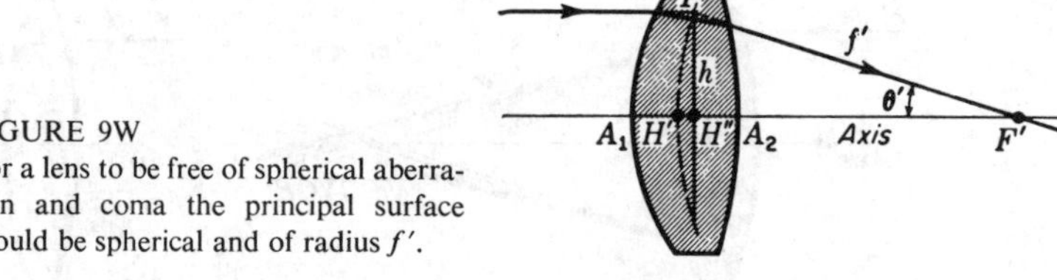

FIGURE 9W
For a lens to be free of spherical aberration and coma the principal surface should be spherical and of radius f'.

aberration, while the vertical difference in height is called *lateral chromatic aberration*. Because these aberrations are often comparable in magnitude with the Seidel aberrations, correction for both lateral and longitudinal color is of considerable importance. As an indication of relative magnitudes, it may be noted that the longitudinal chromatic aberration of an equiconvex lens of spectacle crown glass having a focal length of 10.0 cm and a diameter of 3.0 cm is exactly the same (2.50 mm) as the spherical aberration of marginal rays in the same lens.

While there are several general methods for correcting chromatic aberration, the method of employing two thin lenses in contact, one made of crown glass and the other of flint glass, is the commonest and will be considered first. The usual form of such an *achromatic doublet* is shown in Fig. 9X(*b*). The crown-glass lens, which has a large positive power, has the same dispersion as the flint-glass lens, for which the power is smaller and negative. The combined power is therefore positive, while the dispersion is neutralized, thereby bringing all colors to approximately the same focus. The possibility of achromatizing such a combination rests upon the fact that the dispersions produced by different kinds of glass are not proportional to the deviations they produce (Sec. 1.4). In other words, the dispersive powers $1/v$ differ for different materials.

Typical dispersion curves showing the variation of n with color are plotted for a number of common optical glasses in Fig. 9Y, and the actual values of the index n for the different Fraunhofer lines are presented in Table 9E. The peak of the visual

Table 9E REFRACTIVE INDICES OF TYPICAL OPTICAL MEDIA FOR FOUR COLORS

Medium	Designation	ICT type	v	n_C	n_D	n_F	n_G
Borosilicate crown	BSC	500/664	66.4	1.49776	1.50000	1.50529	1.50937
Borosilicate crown	BSC-2	517/645	64.5	1.51462	1.51700	1.52264	1.52708
Spectacle crown	SPC-1	523/587	58.7	1.52042	1.52300	1.52933	1.53435
Light barium crown	LBC-1	541/599	59.7	1.53828	1.54100	1.54735	1.55249
Telescope flint	TF	530/516	51.6	1.52762	1.53050	1.53790	1.54379
Dense barium flint	DBF	670/475	47.5	1.66650	1.67050	1.68059	1.68882
Light flint	LF	576/412	41.2	1.57208	1.57600	1.58606	1.59441
Dense flint	DF-2	617/366	36.6	1.61216	1.61700	1.62901	1.63923
Dense flint	DF-4	649/338	33.9	1.64357	1.64900	1.66270	1.67456
Extra dense flint	EDF-3	720/291	29.1	1.71303	1.72000	1.73780	1.75324
Fused quartz	SiO_2		67.9		1.4585		
Crystal quartz (*O* ray)	SiO_2		70.0		1.5443		
Fluorite	CaF_2		95.4		1.4338		

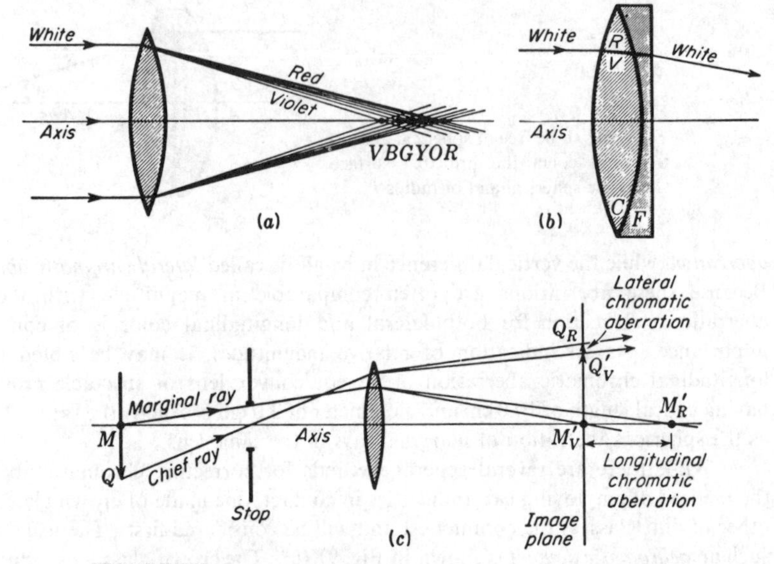

FIGURE 9X
(*a*) Chromatic aberration of a single lens. (*b*) A cemented doublet corrected for chromatic aberration. (*c*) Illustrating the difference between longitudinal chromatic aberration and lateral chromatic aberration.

brightness curve* in Fig. 9Y occurs not far from the yellow D line. It is for this reason that the index n_D has been chosen by optical designers as the basic index for ray tracing and for the specification of focal lengths. Two other indices, one on either side of n_D, are then chosen for purposes of achromatization. As indicated in the table, the ones most often used are n_C for the red end of the spectrum and n_F or n_G for the blue end.

For two thin lenses in contact, the resultant focal length f_D or power P_D of the combination for the D line is given by Eqs. (4h) and (4i):

$$\frac{1}{f_D} = \frac{1}{f'_D} + \frac{1}{f''_D} \qquad \text{or} \qquad P_D = P'_D + P''_D \tag{9t}$$

where the index D indicates that the quantity depends on n_D, f'_D and P'_D refer to the focal length and power of the crown-glass component, and f''_D and P''_D to the focal length and power of the flint-glass component. In terms of indices of refraction and radii of curvature, the power form of the equation becomes

$$P_D = (n'_D - 1)\left(\frac{1}{r'_1} - \frac{1}{r'_2}\right) + (n''_D - 1)\left(\frac{1}{r''_1} - \frac{1}{r''_2}\right) \tag{9u}$$

* Brightness is a sensory magnitude in light just as loudness is a sensory magnitude in sound. Over a considerable range both vary approximately as the logarithm of the energy. The curve shown represents the logarithms of the standard luminosity curve.

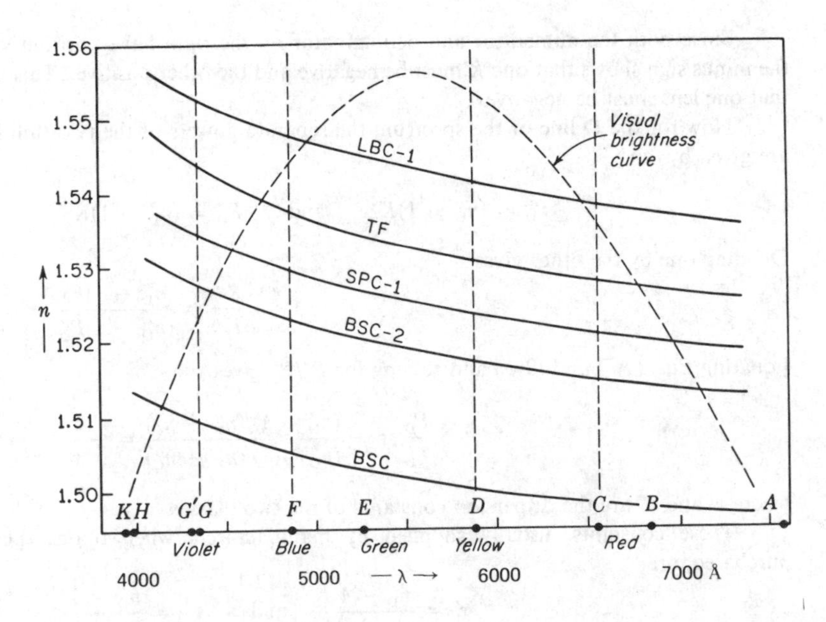

FIGURE 9Y
Graphs of the refractive indices of several kinds of optical glass. These are called dispersion curves.

For convenience let

$$K' = \left(\frac{1}{r_1'} - \frac{1}{r_2'}\right) \quad \text{and} \quad K'' = \left(\frac{1}{r_1''} - \frac{1}{r_2''}\right) \tag{9u'}$$

Then Eq. (9u) can be written more simply as

$$P_D = (n_D' - 1)K' + (n_D'' - 1)K'' \tag{9v}$$

Similarly, for any other colors or wavelengths like the F and C spectrum lines, we can write

$$\begin{aligned} P_F &= (n_F' - 1)K' + (n_F'' - 1)K'' \\ P_C &= (n_C' - 1)K' + (n_C'' - 1)K'' \end{aligned} \tag{9v'}$$

To make the combination achromatic we make the resultant focal length the same for F and C light. This means, making $P_F = P_C$,

$$(n_F' - 1)K' + (n_F'' - 1)K'' = (n_C' - 1)K' + (n_C'' - 1)K''$$

Multiplying out and canceling gives

$$\frac{K'}{K''} = -\frac{n_F'' - n_C''}{n_F' - n_C'} \tag{9v''}$$

Since both the numerator and denominator on the right have position values, the minus sign shows that one K must be negative and the other positive. This means that one lens must be negative.

Now for the D line of the spectrum the separate powers of the two thin lenses are given by

$$P'_D = (n'_D - 1)K' \quad \text{and} \quad P''_D = (n''_D - 1)K'' \tag{9w}$$

Dividing one by the other gives

$$\frac{K'}{K''} = \frac{(n''_D - 1)P'_D}{(n'_D - 1)P''_D} \tag{9w'}$$

Equating Eqs. (9v″) and (9w′) and solving for P''_D/P'_D gives

$$\frac{P''_D}{P'_D} = -\frac{(n''_D - 1)/(n''_F - n''_C)}{(n'_D - 1)/(n'_F - n'_C)} = -\frac{\nu''}{\nu'} \tag{9w''}$$

where ν' and ν'' are the dispersion constants of the two glasses.

These constants, usually supplied by manufacturers when optical glass is purchased, are

$$\nu' = \frac{n'_D - 1}{n'_F - n'_C} \quad \text{and} \quad \nu'' = \frac{n''_D - 1}{n''_F - n''_C} \tag{9x}$$

Values of ν for several common types of glass are given in Table 9E. Since the dispersive powers are all positive, the negative sign in Eq. (9w″) indicates that the powers of the two lenses must be of opposite sign. In other words, if one lens is converging, the other must be diverging. From the extreme members of Eq. (9w″) we obtain

$$\frac{P'_D}{\nu'} + \frac{P''_D}{\nu''} = 0 \quad \text{or} \quad \nu'f' + \nu''f'' = 0 \tag{9x'}$$

Substituting the value of P'_D or that of P''_D from Eq. (9t) in Eq. (9x′), we obtain

$$P'_D = P_D \frac{\nu'}{\nu' - \nu''} \quad \text{and} \quad P''_D = -P_D \frac{\nu''}{\nu' - \nu''} \tag{9x''}$$

The use of the above formulas to calculate the radii for a desired achromatic lens involves the following steps:

1 *A focal length f_D and a power P_D are specified.*
2 *The types of crown and flint glass to be used are selected.*
3 *If they are not already known, the dispersion constants ν' and ν'' are calculated from Eq.* (9x).
4 *P'_D and P''_D are calculated from Eq.* (9x″).
5 *The values of K' and K'' are determined by Eq.* (9w).
6 *The radii are then found from Eq.* (9u′).

Calculation 6 is usually made with other aberrations in mind.

EXAMPLE An achromatic lens having a focal length of 10.0 cm is to be made as a cemented doublet using crown and flint glasses having the following indices:

Glass	n_C	n_D	n_F	n_G
Crown	1.50868	1.51100	1.51673	1.52121
Flint	1.61611	1.62100	1.63327	1.64369

Find the radii of curvature for both lenses if the crown-glass lens is to be equiconvex and the combination is to be corrected for the C and F lines.

SOLUTION The focal length of 10.0 cm is equivalent to a power of +10 D. The dispersion constants ν' and ν'' are, from Eq. (9x),

$$\nu' = \frac{1.51100 - 1.00000}{1.51673 - 1.50868} = 63.4783 \qquad \nu'' = \frac{1.62100 - 1.00000}{1.63327 - 1.61611} = 36.1888$$

Applying Eq. (9x′), we find that the powers of the two lenses must be

$$P'_D = 10 \frac{63.4783}{63.4783 - 36.1888} = +23.2611 \text{ D}$$

$$P''_D = -10 \frac{36.1888}{63.4783 - 36.1888} = -13.2611 \text{ D}$$

The fact that the sum of these two powers $P_D = +10.0000$ D serves as a check on the calculations to this point. Knowing the power required in each lens, we are now free to choose any pair of radii that will give such a power. If two or more surfaces can be made to have the same radius, the necessary number of grinding and polishing tools will be reduced. For this reason the positive element is often made equiconvex, as it is here. Letting $r'_1 = -r'_2$, we apply Eq. (9u′) and then Eq. (9w) to obtain

$$K' = \frac{1}{r'_1} - \frac{1}{r'_2} = \frac{2}{r'_1} = \frac{P'_D}{n'_D - 1} = \frac{23.2611}{0.51100} = 45.5207$$

from which $\quad r'_1 = 0.0439361 \text{ m} = 4.39361 \text{ cm}$

Since the lens is to be cemented, one surface of the negative lens must fit a surface of the positive lens. This leaves the radius of the last surface to be adjusted to give the proper power of −13.2611 D. Therefore we let $r''_1 = -r'_1$ and apply Eqs. (9u′) and (9w) as before, to find

$$K'' = \frac{1}{r''_1} - \frac{1}{r''_2} = -\frac{1}{0.0439361} - \frac{1}{r''_2} = \frac{P''_D}{n''_D - 1} = \frac{-13.2611}{0.62100} = -21.3544$$

This gives

$$\frac{1}{r''_2} = 21.3544 - \frac{1}{0.0439361} = 21.3544 - 22.7603$$

and $\quad \dfrac{1}{r''_2} = -1.4059 \qquad r''_2 = -0.71129 \text{ m} = -71.13 \text{ cm}$

The required radii are therefore

$$r_1' = 4.39 \text{ cm} \qquad r_1'' = -4.39 \text{ cm}$$

$$r_2' = -4.39 \text{ cm} \qquad r_2'' = -71.13 \text{ cm}$$

It will be noted that, with the crown-glass element of this achromat placed toward incident parallel light, the two exposed surfaces are close to what they should be for minimum spherical aberration and coma. This emphasizes the importance of choosing glasses having the proper dispersive powers.

To see how well this lens has been achromatized, we now calculate its focal length for the three colors corresponding to the C, F, and G′ lines. By Eq. (9v′)

$$P_C = (n_C' - 1)K' + (n_C'' - 1)K''$$

$$= 0.50868 \times 45.5207 + 0.61611(-21.3544)$$

$$= 23.1555 - 13.1567$$

giving

$$f_C = 10.0012 \text{ cm}$$

Similarly for the colors corresponding to the F and G′ lines we obtain

$$P_F = +9.9988 \text{ D} \qquad \text{or} \qquad f_F = 10.0012 \text{ cm}$$

$$P_{G'} = +9.9804 \text{ D} \qquad \text{or} \qquad f_{G'} = 10.0196 \text{ cm}$$

The differences between f_C, f_D, and f_F are negligibly small, but $f_{G'}$ is about $\frac{1}{5}$ mm larger than the others. This difference for light outside the region of the C and F lines results in a small circular zone of color about each image point which is called the *secondary spectrum*.

Although the lens in our example would appear to have been corrected for longitudinal chromatic aberration, it has actually been corrected for lateral chromatic aberration. Equal focal lengths for different colors will produce equal magnification, but the different colored images along the axis will coincide only if the principal points also coincide. Practically speaking, the principal points of a thin lens are so close together that both types of chromatic aberration can be assumed to have been corrected by the above arrangement. In a thick lens, however, longitudinal chromatic aberration is absent if the colors corrected for come together at the same axial image point as shown in Fig. 9Z(*a*). Because the principal points for blue and red, H_b' and H_r', do not coincide, the focal lengths are not equal and the magnification is different for different colors. Consequently the images formed in different colors will have different sizes. This is the lateral chromatic aberration or lateral color mentioned at the beginning of this section.

9.14 SEPARATED DOUBLET

Another method of obtaining an achromatic system is to employ two thin lenses made of the same glass and separated by a distance equal to half the sum of their focal lengths. To see why this is true, we begin with the thick-lens formula, Eq. (5g), as

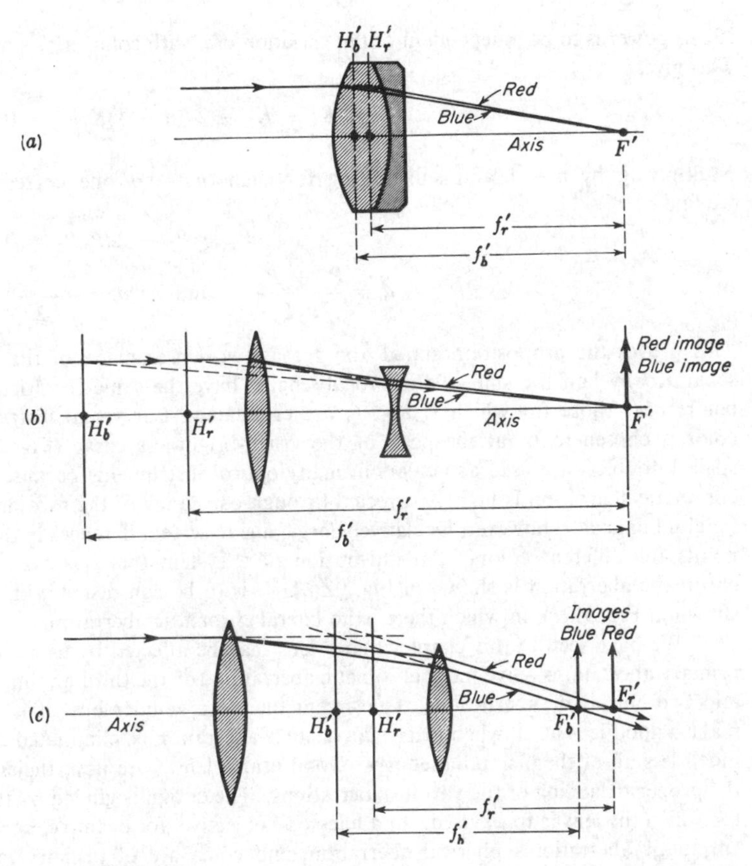

FIGURE 9Z
(*a*) Cemented doublet corrected for longitudinal chromatic aberration. (*b*) Separated doublet corrected for longitudinal chromatic aberration. (*c*) Separated doublet corrected for lateral chromatic aberration.

applied to two thin lenses separated by a distance d:

$$\frac{1}{f} = \frac{1}{f_1} + \frac{1}{f_2} - \frac{d}{f_1 f_2} \qquad \text{or} \qquad P = P_1 + P_2 - dP_1P_2 \tag{9y}$$

which, by analogy with Eq. (9v), may be written

$$P = (n_1 - 1)K_1 + (n_2 - 1)K_2 - d(n_1 - 1)(n_2 - 1)K_1K_2$$

The subscripts 1 and 2 are used here in place of the primes to designate the two lenses, and the K's are given by Eq. (9u'). Since the two lenses are of the same kind of glass, we set $n_1 = n_2$, so that

$$P = (n - 1)(K_1 + K_2) - d(n - 1)^2K_1K_2$$

If this power is to be independent of the variation of n with color, dP/dn must vanish. This gives

$$\frac{dP}{dn} = K_1 + K_2 - 2d(n - 1)K_1K_2 = 0$$

Multiplying by $n - 1$ and substituting for each $(n - 1)K$ the corresponding P, we find

$$P_1 + P_2 - 2dP_1P_2 = 0$$

or

$$d = \frac{P_1 + P_2}{2P_1P_2} \quad \text{and} \quad d = \frac{f_1 + f_2}{2} \tag{9z}$$

This proves the proposition stated above that two lenses made of the same glass separated by half the sum of their focal lengths have the same focal length for all colors near those for which f_1 and f_2 are calculated. For visual instruments this color is chosen to be at the peak of the visual-brightness curve (Fig. 9Y). Such spaced doublets are used as oculars in many optical instruments because the lateral chromatic aberration is highly corrected through constancy of the focal length. The longitudinal color, however, is relatively large, due to wide differences in the principal points for different colors. An illustration of a system that has no longitudinal chromatic aberration is shown in Fig. 9Z(*b*). It is to be contrasted with the system shown in Fig. 9Z(*c*), in which there is no lateral chromatic aberration.

We have seen in this chapter that a lens may be affected by as many as seven primary aberrations—five monochromatic aberrations of the third and higher orders and two chromatic aberrations. One might therefore wonder how it is possible to make a good lens at all when rarely can a single aberration be eliminated completely, much less all of them simultaneously. Good usable lenses are nevertheless made by the proper balancing of the various aberrations. The design is guided by the purpose for which the lens is to be used. In a telescope objective, for example, correction for chromatic aberration, spherical aberration, and coma are of primary importance. On the other hand astigmatism, curvature of field, and distortion are not as serious because the field over which the objective is to be used is relatively small. For a good camera lens of wide aperture and field, the situation is almost exactly reversed.*

PROBLEMS

9.1 A convex spherical surface with a radius of +8.0 cm is ground and polished on the end of a glass rod. If the rod is in air and the refractive index of the glass is 1.620, calculate the (*a*) longitudinal spherical aberration and (*b*) the lateral spherical aberration. Assume the height of the incident ray to be 6.0 cm.

Ans. (*a*) +2.0233 cm, (*b*) −0.6430 cm

* Other treatments of the subject of aberrations will be found in A. C. Hardy and F. H. Perrin, "The Principles of Optics," McGraw-Hill Book Company, New York, 1932; G. S. Monk, "Light, Principles and Experiments," Dover Publications, Inc., New York, 1963; D. H. Jacobs, "Fundamentals of Optical Engineering," McGraw-Hill Book Company, New York, 1943; A. E. Conrady, "Applied Optics and Optical Design," Dover Publications, Inc., New York, 1963; E. Hecht and A. Zajac, "Optics," Addison-Wesley Publishing Company, Inc., Reading, Mass., 1974.

9.2 A single spherical surface with a radius of +20.0 cm is polished on the end of a glass rod. If the rod is in air and the refractive index of the glass is 1.750, find (*a*) the longitudinal spherical aberration and (*b*) the lateral spherical aberration. Assume the height of the incident ray to be 6.0 cm.

9.3 A thin lens with a refractive index of 1.60 has radii $r_1 = +45.0$ cm and $r_2 = -15.0$ cm. If light is incident on this lens parallel to the axis, find (*a*) the focal length for paraxial rays, (*b*) the longitudinal spherical aberration, and (*c*) the lateral spherical aberration for a ray at a height of 2.50 cm.

9.4 A thin glass lens with radii $r_1 = -12.0$ cm and $r_2 = +12.0$ cm has a refractive index of 1.850. If parallel incident light falls on the first surface at a height of 2.50 cm, find (*a*) the paraxial focal length, (*b*) the position factor, (*c*) the shape factor, (*d*) the longitudinal spherical aberration, and (*e*) the lateral spherical aberration.

Ans. (*a*) −7.0588 cm, (*b*) −1.0, (*c*) 0, (*d*) −0.85741 cm, (*e*) −0.345652

9.5 A thin lens of index 1.6250 has radii $r_1 = +8.0$ cm and $r_2 = -8.0$ cm. Find (*a*) the position factor, (*b*) the shape factor, (*c*) the paraxial focal length, (*d*) the longitudinal spherical aberration, and (*e*) the lateral spherical aberration for an axial object point 32.0 cm in front of the lens and for rays through a zone of radius $h = 2.0$ cm.

9.6 A thin lens with an index of 1.7620 has radii $r_1 = +40.0$ cm and $r_2 = -10.0$ cm. If this lens is to be used with parallel incident light, find (*a*) the position factor, (*b*) the shape factor, (*c*) the paraxial focal length, (*d*) the longitudinal spherical aberration, and (*e*) the lateral spherical aberration for a ray at a height of 2.0 cm.

9.7 A thin plano-convex lens has a refractive index of 1.52300. The second surface has a radius of −10.0 cm. If light is incident at a height of 2.0 cm on the flat surface, parallel to the axis, find (*a*) the position factor, (*b*) the shape factor, (*c*) the paraxial focal length, (*d*) the longitudinal spherical aberration, and (*e*) the lateral spherical aberration.

Ans. (*a*) −1.0, (*b*) −1.0, (*c*) +19.12046 cm, (*d*) +0.84766 cm, (*e*) −0.092778 cm

9.8 Find the answers to Prob. 9.7 if the lens is turned around so the incident light is incident on the convex surface.

9.9 A piece of optical glass with a refractive index of 1.5230 is to be made into a lens having a focal length of +24.0 cm. If it is to be used with parallel incident light and is to have a minimum spherical aberration, what would be (*a*) the position factor, (*b*) the shape factor, (*c*) the radius of the first surface, and (*d*) the radius of the second surface?

9.10 A piece of dense flint glass with an index of 1.7930 is to be made into a diverging lens with a focal length of −20.0 cm. If it is to be used with parallel incident light and is to have a minimum spherical aberration, what would be (*a*) the position factor, (*b*) the shape factor, (*c*) the radius of the first surface, and (*d*) the radius of the second surface?

9.11 A thin lens 5.0 cm in diameter has a refractive index of 1.6520 and radii $r_1 = +15.0$ cm and $r_2 = -30.0$ cm. Find (*a*) the position factor, (*b*) the shape factor, (*c*) the G factor, (*d*) the W factor, and (*e*) the height of the comatic figure if the paraxial image point for parallel incident light is 5.0 cm off the principal axis. Give answers to four significant figures. *Ans.* (*a*) $p = -1.0$, (*b*) $q = +0.33333$, (*c*) $G = +1.9540$, (*d*) $W = +1.8466$, (*e*) $C_T = -0.13911$ cm

9.12 A thin lens 6.50 cm in diameter has a refractive index of 1.5230 and radii $r_1 = -15.0$ cm and $r_2 = +30.0$ cm. Find (*a*) the focal length of the lens, (*b*) the position factor, (*c*) the shape factor, (*d*) the G factor, (*e*) the W factor, and (*f*) the height of the comatic figure if the lens is used with parallel incident light and brings the paraxial image point to focus 3.0 cm off the principal axis.

9.13 A thin lens is to be made of a piece of optical crown glass of index 1.6750 and is to have a focal length of 5.0 cm. An object is to be mounted 25.0 cm in front of this lens and a real image is to be formed on a white screen. Calculate (*a*) the image distance and (*b*) the position factor. If this lens is to have a minimum spherical aberration for this object-image distance relation, find (*c*) the shape factor, (*d*) the radius of r_1, and (*e*) the radius of r_2.

9.14 A thin glass lens of index 1.5250 is to be completely free of coma for an object placed 15.0 cm in front of the lens and the image formed 75.0 cm behind the lens. Find (*a*) the focal length of the lens, (*b*) the position factor, (*c*) the shape factor, (*d*) the radius of the first surface, and (*e*) the radius of the second surface.

Ans. (*a*) +12.50 cm, (*b*) +0.6667, (*c*) −0.5614, (*d*) +29.924 cm, (*e*) −8.406 cm

9.15 A thin lens is made of flint glass of index 1.6520. If its focal length is +12.50 cm, and an object is to be placed 50.0 cm in front of the lens, find (*a*) the image distance, (*b*) the position factor, (*c*) the shape factor, (*d*) the radius of the first surface, and (*e*) the radius of the second surface. The image is to show no coma.

9.16 A thin diverging lens is to be made of crown glass of index 1.5230 and have a focal length of −12.0 cm. If an object is to be placed 20.0 cm in front of this lens, and the image is to be free of coma, find (*a*) the image distance, (*b*) the position factor, (*c*) the shape factor, (*d*) the radius of the first surface, and (*e*) the radius of the second surface.

9.17 A meniscus lens 0.750 cm thick and of index 1.520 is to be aplanatic for two points on the concave side of the lens. If the closest of these two points is to be 5.0 cm from the closest vertex, find (*a*) the radii of the two lens surfaces and (*b*) the distance from the closest vertex to the farther aplanatic point.

Ans. (*a*) $r_1 = -3.4682$ cm and $r_2 = -5.0$ cm, (*b*) 7.990 cm

9.18 A meniscus lens 0.650 cm thick, and refractive index 1.585 is to be made of such a shape that it is aplanatic for two points 5.0 cm apart (see Fig. 9O). Find (*a*) the two radii of curvature and (*b*) the distances from the concave surface to the two points.

9.19 Apply Abbe's sine condition to the rays traced through the first lens surface in Table 8B and give the values of $h/(\sin\theta')$ for $h = 1.50$, $h = 1.0$, $h = 0.50$, and $h = 0.00$.

9.20 Apply Abbe's sine condition to the rays traced through the second surface of the lens in Table 8B and give the values of $(\sin\theta')/(\sin\theta'')$ for all four rays.

Ans. 0.26490, 0.30139, 0.31594, and 0.32000

9.21 An achromatic lens with a focal length of 25.0 cm is to be made of crown and flint glasses of the types BSC-2 and DF-2 (see Table 9E for indices). If the crown glass lens is to be equiconvex and the combination is to be cemented, find (*a*) the ν values, (*b*) the two lens powers for sodium light, and (*c*) the radii of the four lens surfaces to correct for the C and F lines.

9.22 An achromatic lens with a focal length of 16.0 cm is to be made of crown and flint glasses of the types BSC and DF-4 (see Table 9E). If the flint glass lens is to have its outer face flat and the combination is to be cemented, find (*a*) the power of the lens, (*b*) the ν values of the two glasses, (*c*) the powers of the two component lenses for sodium yellow light, and (*d*) the radii of the three remaining lens surfaces. The lens is to be corrected for C and F light.

9.23 An achromatic lens is to be made of SPC-1 and DF-4 glasses and is to have a focal length of 12.50 cm (see Table 9E). If the flint glass lens is to have its outer face flat and the two lenses are to be cemented, find (*a*) the power of the lens, (*b*) the ν values of the two glasses, (*c*) the powers of the two lenses, and (*d*) the radii of the three curved surfaces. The lens is to be corrected for C and G′ light.

Ans. (*a*) +8.0 D, (*b*) 37.5449 and 20.9422, (*c*) +18.09104 and −10.09104D, (*d*) $r_1' = +5.25147$ cm, $r_2' = -6.43145$ cm, and $r_1'' = -6.43145$ cm

9.24 An achromatic lens is to be made of two pieces of LBC-1 and EDF-3 optical glass with refractive indices given in Table 9E. If the crown-glass lens is to be equiconvex and the two lenses are to be cemented, find (*a*) the power of the final lens if its focal length is to be 8.0 cm, (*b*) the dispersion constants of the glasses, (*c*) the separate powers of the lenses, (*d*) the radii of all four lens faces, and (*e*) the focal lengths of the final lens for C, D, F, and G′ light. The lens is to be corrected for C and G′ light. Plot a graph of wavelength λ versus focal length *f*. Assume $\lambda_C = 6563$ Å, $\lambda_D = 5892$ Å, $\lambda_F = 4861$ Å, and $\lambda_G = 4307$ Å.

9.25 An achromatic lens is to be made of two pieces of optical glass with refractive indices given for BSC-2 and DF-4 in Table 9E. If its focal length is to be +20.0 cm, the second surface of the flint glass lens is to be flat, and the lens is to be cemented, find (*a*) the power of the completed lens, (*b*) the dispersion constants of the glass, (*c*) the separate powers of the lenses, (*d*) the radii of the four surfaces, and (*e*) the focal lengths of the completed lens for C, D, F, and G′ light. The lens is to be corrected for C and F light. (*f*) Plot a graph of wavelength λ versus focal length *f*. Assume $\lambda_C = 6563$, Å $\lambda_D = 5892$ Å, $\lambda_F = 4861$ Å, and $\lambda_G = 4307$ Å.

10
OPTICAL INSTRUMENTS

The design of efficient optical instruments is the ultimate purpose of geometrical optics. The principles governing the formation of images by a single lens, and occasionally by simple combinations of lenses, have been set forth in the previous chapters. These principles find a wide variety of applications in the many practical combinations of lenses, frequently including also mirrors or prisms, which fall in the category of optical instruments. This subject is one of such large scope and has developed so many ramifications that in a book devoted to the fundamentals of optics it is only possible to describe the principles involved in a few standard types of instruments. In this chapter a description will be given of the more important features of camera lenses, magnifiers, microscopes, telescopes, and oculars. These will serve to illustrate some applications of the basic ideas already discussed and will, it is hoped, be of interest to the student who has used, or expects to use, some of these instruments.

10.1 THE HUMAN EYE

As human beings our sense of vision is one of our most prized possessions. For those of us that enjoy normal vision this marvelous gift of nature is the most useful of all recording instruments, yet in a few instances it should not be relied upon to tell the

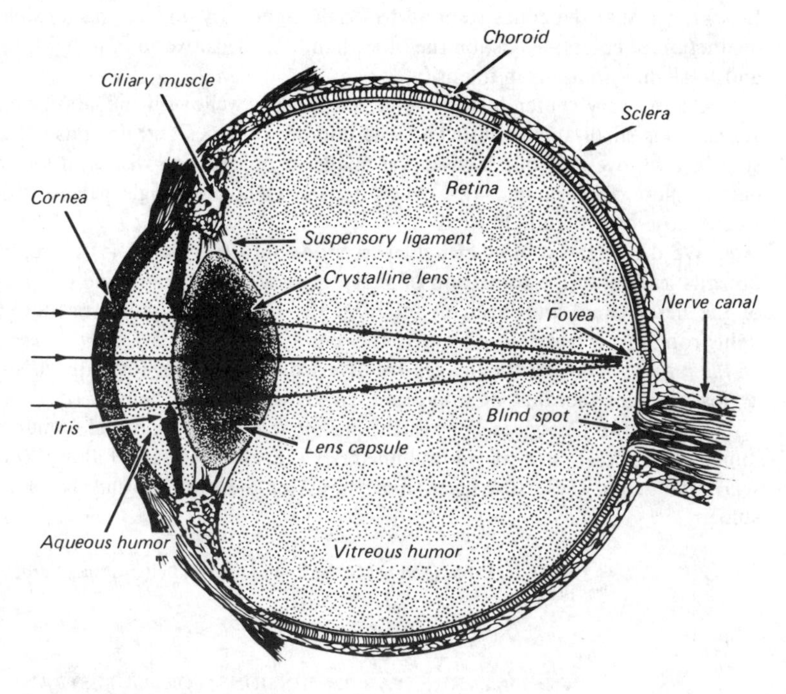

FIGURE 10A
A cross-sectional diagram of a human eye, showing the principal optical components and the retina.

truth. As an illustration of how unreliable vision can be, mention should be made of a whole group of phenomena known as *optical illusions*.*

In spite of these imperfections in our vision, most of us are able to enjoy the beauties of color, form, and motion, all made possible by illumination with visible white light. The eye is like a fine camera, with a *shutter*, *iris*, and *lens system* on one side and a sensitive film called the *retina* on the other (see Fig. 10A). The function of the lens system is to focus on the retina an image of objects to be seen. Like a camera, the iris diaphragm opens wider for faint light and closes down for bright sunlight. The pigment determining the color of the eye is in the iris.

The retina of the eye contains hundreds of *cones* and *rods*, whose function it is to receive light pulses and change them into electric currents. How these electric currents are produced by the cones and rods, and how they are translated by the brain into what we call vision is only partially understood by scientists working in the field.

* See H. E. White, "Modern College Physics," 6th ed., pp. 20–26, D. Van Nostrand, New York, 1972, and N. F. Beeler and F. M. Branley, "Experiments in Optical Illusion," Thomas Y. Crowell Co., New York, 1951.

It is known that the cones respond to bright light only and are responsible for our distinction of colors. Rods, on the other hand, are sensitive to faint light, to motion, and to slight variations in intensity.

At the very center of the retina is a slightly yellowish indentation called the *fovea.* This small area contains a large number of cones and no rods. It is on this spot in each eye that one focuses the image of objects one wishes to see in minute detail. Note, for example, that when one looks at any single word on this page, words close by are quite blurred.

We divide the subject of light perception into two parts: (1) the optical components leading to the formation of sharp images on the retina and (2) the property of the nerve canal and brain to interpret the electrical impulses produced. When light from any object enters the eye, the lens system forms a *real but inverted image on the retina.* While all the images are inverted, as shown in Fig. 10B, it is of course a most amazing fact that we interpret them by the brain as being *erect.*

Figure 10B also gives some of the pertinent facts for a normal human eye. The dimensions shown are all in millimeters, and the diagram follows *Gullstrand's schematic eye.** Table 10A gives dimensions for the eye that may be of use to the student.

* See H. H. Emsley, "Visual Optics," 3d ed., p. 346, Butterworths, Scarborough, Ont., 1955.

Table 10A PRINCIPAL DIMENSIONS FOR GULLSTRAND'S SCHEMATIC EYE
Overall power of eye = 58.64 D

	Refractive index	Axis position, mm	Radius curvature, mm
Cornea, anterior	1.376	0	7.7
and posterior		0.5	6.8
Aqueous humor	1.336		
Vitreous humor	1.336		
Lens:			
Cortex, anterior	1.386	3.6	10.0
and posterior		7.2	−6.0
Core, anterior	1.406	4.15	7.9
and posterior		6.57	5.8
Cardinal points:			
AH		1.348	
AH′		1.602	
AN		7.08	
AN′		7.33	
AF		−15.70	
AF′		24.38	

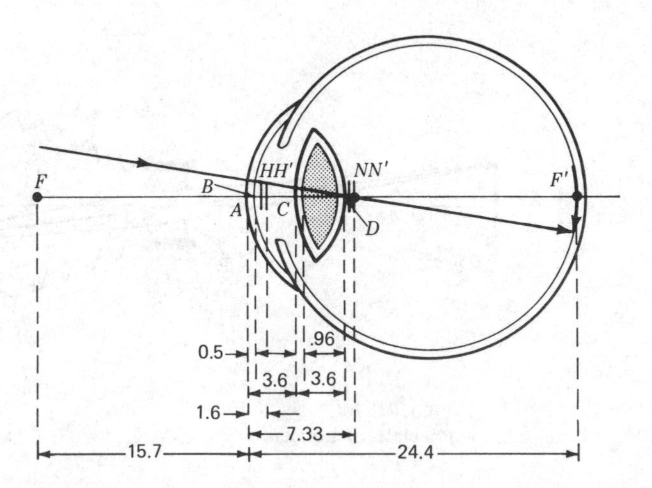

FIGURE 10B
Schematic eye as developed by Gullstrand, showing the real and inverted image on the retina (dimensions are in millimeters).

10.2 CAMERAS AND PHOTOGRAPHIC OBJECTIVES

The fundamental principle of the camera is that of a positive lens forming a real image, as shown in Fig. 10C. Sharp images of distant or nearby objects are formed on a photographic film or plate, which is later developed and printed to obtain the final picture. Where the scene to be taken involves stationary objects, the cheapest camera lens (if it is stopped down almost to a pinhole and a time exposure used) can yield photographs of excellent definition. If, however, the subjects are moving relative to the camera (and this includes the case where the camera is held in the hand), extremely short exposure times are often imperative and lenses of large aperture become a necessity. The most important feature of a good camera, therefore, is that it be equipped with a lens of high relative aperture capable of covering as large an angular field as possible. Because a lens of large aperture is subject to many aberrations, designers of photographic objectives have resorted to the compromises as regards correction that best suit their particular needs. It is the intention here, therefore, to discuss briefly some of these purposes and compromises in connection with a few of the hundreds of well-known makes of photographic objectives.

10.3 SPEED OF LENSES

The amount of light per unit area reflected or emitted by an object being photographed is called its *brightness* or *luminance B*. The amount of light per unit area falling on a photographic film or plate is called the *illuminance E*. The illuminance *E* depends upon

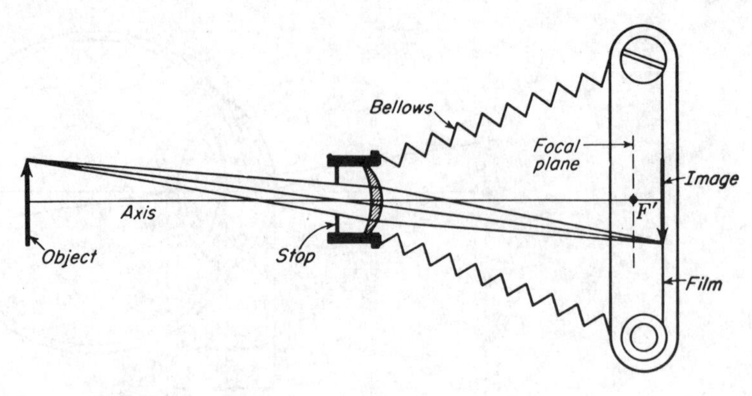

FIGURE 10C
Principles of a camera.

three factors: the brightness B of the object, the area of the *entrance pupil* of the lens, $\pi a^2/4$, and the *focal length* of the lens f (see Fig. 10D).

The light entering a camera is proportional to the brightness of the object, is proportional to the area of the entrance pupil, and is inversely proportional to the square of the focal length. As an equation,

$$E = kB\,\frac{\pi a^2/4}{f^2}$$

where k is a proportionality constant and a is the diameter of the entrance pupil. For any given object to be photographed, we can write

$$E \propto \frac{a^2}{f^2} \tag{10a}$$

It can be seen from Fig. 10C that if f is doubled, the light will be spread over 4 times the area, thereby reducing the illuminance on the film to one-quarter. If the lens diameter is doubled, the lens area is quadrupled, the light falling on the film is quadrupled, while the film area and picture size remain unchanged.

In words, the ratio $(a/f)^2$ is a direct measure of the speed of a camera lens. Instead of specifying this ratio, however, it is customary in the photographic world to specify the *focal ratio*, or *f value*:

● $$f\text{ value} = \frac{f}{a} \tag{10b}$$

Thus a lens which has a focal length of 10.0 cm and a linear aperture of 2.0 cm is said to have an f value of 5, or, as it is usually stated, the lens is an $f/5$ lens.

In order to take pictures of faintly illuminated subjects, or of ones which are in rapid motion and require a very short exposure, a lens of small f value is required. Thus an $f/2$ lens is "faster" than an $f/4.5$ lens (or than an $f/2$ lens stopped down to $f/4.5$) in the ratio $(4.5/2)^2 = 5.06$. A lens of such large relative aperture is difficult to design, as we shall see.

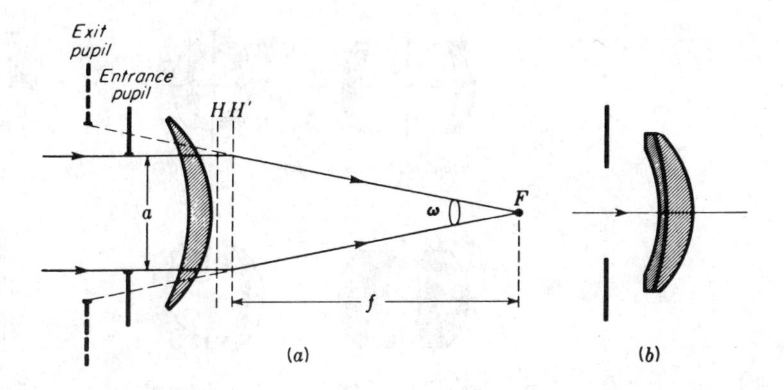

FIGURE 10D
(*a*) Geometry for determining the speed of a lens. (*b*) An achromatic meniscus lens with a front stop.

10.4 MENISCUS LENSES

Many of the cheapest cameras employ a single positive meniscus lens with a fixed stop such as was shown in Fig. 10D(*a*). Developed in about 1812 and called a *landscape lens*, this simple optical device exhibits considerable spherical aberration, thereby limiting its useful aperture to about $f/11$. Off the lens axis, the astigmatism limits the field to about 40°. The proper location of the stop results in a flat field, but with only a single lens there is always considerable chromatic aberration.

By using a cemented doublet as shown in Fig. 10D(*b*), lateral chromatism can be corrected. Instead of correcting for the C and F lines of the spectrum, however, the combination is usually corrected for the yellow D line, near the peak sensitivity of the eye, and the blue G′ line, near the peak sensitivity of many photographic emulsions. Called *DG achromatism*, this type of correction produces the best photographic definition at the sharpest visual focus. In some designs the lens and stop are turned around as in the arrangement of Fig. 9U(*b*).

10.5 SYMMETRICAL LENSES

Symmetrical lenses consist of two identical sets of thick lenses with a stop midway between them; a number of these are illustrated in Fig. 10E. In general, each half of the lens is corrected for lateral chromatic aberration, and by putting them together, curvature of field and distortion are eliminated, as was explained in Sec. 9.11. In the rapid rectilinear lens, flattening of the field was made possible only by the introduction of considerable astigmatism, while spherical aberration limited the aperture to about $f/8$. By introducing three different glasses, as in the Goerz Dagor, each half of the lens could be corrected for lateral color, astigmatism, and spherical aberration. When combined they are corrected for coma, lateral color, curvature, and distortion. Zeiss calls this lens a Triple Protar, while Goerz calls it the DAGor, signifying Double

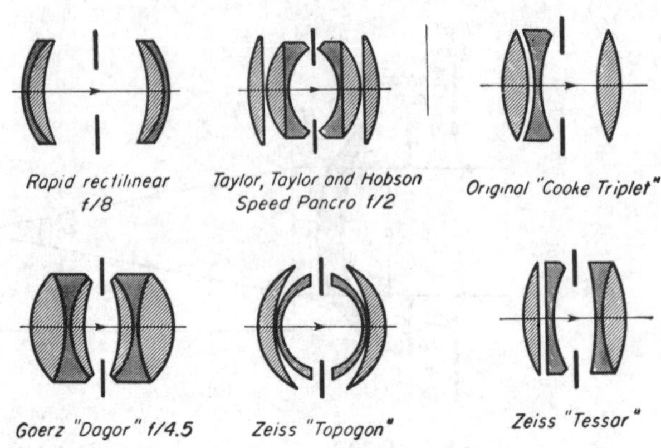

FIGURE 10E
Symmetrical and unsymmetrical camera lenses.

Anastigmat Goerz. The Speed Panchro lens developed by Taylor, Taylor, and Hobson in 1920 is noteworthy because of its fine central definition combined with the high speed of $f/2$ and even $f/1.5$. The Zeiss Topogon lens is but one of a number of special "wide-angle" lenses, particularly useful in aerial photography. Additional characteristics of symmetrical lenses are (1) the large number of lenses employed, and (2) the rather deep curves, which are expensive to produce.

The greater the number of free glass surfaces in a lens, the greater the amount of light lost by reflection. The f value alone, therefore, is not the sole factor in the relative speeds of objectives. The development in recent years of lens coatings that practically eliminate reflection at normal incidence has offered greater freedom in the use of more elements in the design of camera lenses (see Sec. 14.6).

10.6 TRIPLET ANASTIGMATS

A great step forward in photographic lens design was made in 1893 when H. D. Taylor of Cooke and Sons developed the Cooke Triplet (Fig. 10E). The fundamental principles involved in this system follow from the fact that (1) the power which a given lens contributes to a system of lenses is proportional to the height at which marginal rays pass through the lens, whereas (2) the contribution each lens makes to field curvature is proportional to the power of the lens regardless of the distance of the rays from the axis. Hence astigmatism and curvature of field can be eliminated by making the power of the central flint element equal and opposite to the sum of the powers of the crown elements. By spacing the negative lens between the two positive lenses, the marginal rays can be made to pass through the negative lens so close to the axis that the system has an appreciable positive power. A proper selection of dispersions and radii enables additional corrections to be made for color and spherical aberration. The Tessar, one of the best known modern photographic objectives, was developed by Zeiss in 1902. Made in many forms to meet various requirements, the system has a general structure similar to that of a Cooke Triplet in which the rear

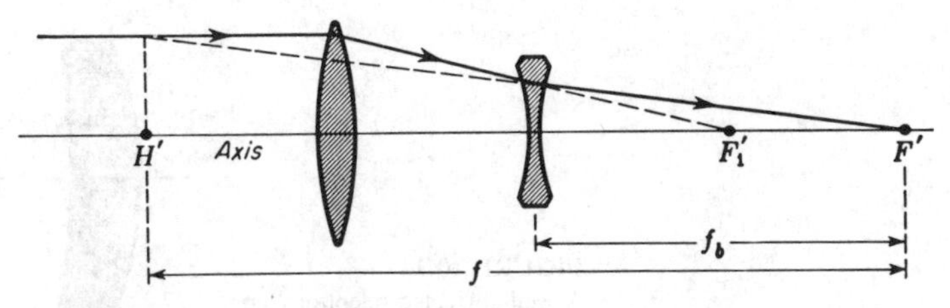

FIGURE 10F
Principles of the telephoto lens.

crown lens is replaced by a doublet. The Leitz Hector, working at $f/2$, is also of the Cooke Triplet type, but each element is replaced by a compound lens. This very fast lens is excellent in a motion-picture camera.

10.7 TELEPHOTO LENSES

Since the image size for a distant object is directly proportional to the focal length of the lens, a telephoto lens which is designed to give a large image is a special type of objective with a longer effective focal length than that normally used with the same camera. Because this would require a greater extension of the bellows than most cameras will permit, the principle of a single highly corrected thick lens is modified as follows. As is shown in Fig. 10F by the refraction of an incident parallel ray, with two such lenses considerably separated the principal point H' can be placed well in front of the first lens, thereby giving a long focal length $H'F'$ with a short lens-to-focal-plane distance (f_b in Fig. 10F). The latter distance, or the *back focal length* as it is usually called, is measured from the real lens to the focal plane, as shown.

Although the focal lengths of older types of telephoto lenses could be varied by changing the distance between the front and rear elements, these lenses are almost always made with a fixed focal length. Flexibility is then obtained by having a set of lenses. This has become necessary through the desire for lenses of greater speed and better correction of the aberrations. A Cooke Telephoto as produced by Taylor, Taylor, and Hobson is shown in Fig. 10G.

10.8 MAGNIFIERS

The magnifier is a positive lens whose function it is to increase the size of the retinal image over and above that which is formed with the unaided eye. The apparent size of any object as seen with the unaided eye depends on the angle subtended by the object (see Fig. 10H). As the object is brought closer to the eye, from A to B to C in the diagram, accommodation permits the eye to change its power and to form a larger and larger retinal image. There is a limit to how close an object may come to the eye if the latter is still to have sufficient accommodation to produce a sharp image.

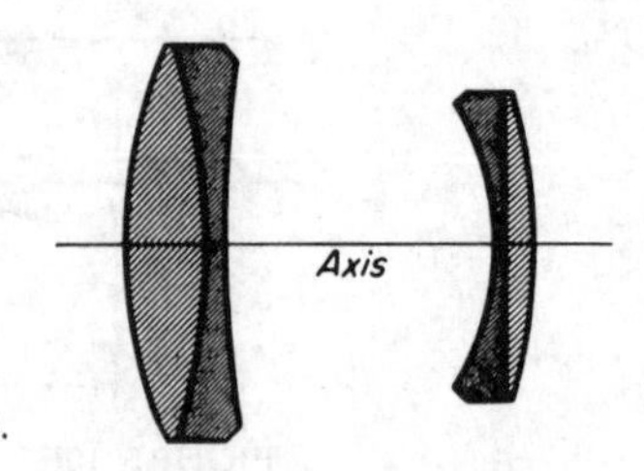

FIGURE 10G
A well-corrected telephoto lens.

Although the nearest point varies widely with various individuals, 25.0 cm is taken to be the standard *near point*, sometimes called the *distance of most distinct vision.* At this distance, indicated in Fig. 10I(*a*), the angle subtended by object or image will be called θ.

If a positive lens is now placed before the eye in the same position, as in diagram (*b*), the object y can be brought much closer to the eye and an image subtending a larger angle θ' will be formed on the retina. What the positive lens has done is to form a virtual image y' of the object y and the eye is able to focus upon this virtual image. Any lens used in this manner is called a *magnifier* or *simple microscope.* If the object y is located at F, the focal point of the magnifier, the virtual image y' will be located at infinity and the eye will be accommodated for distant vision as is illustrated in Fig. 10I(*c*). If the object is properly located a short distance inside of F as in diagram (*b*), the virtual image may be formed at the distance of most distinct vision and a slightly greater magnification obtained, as will now be shown.

The angular magnification M is defined as the ratio of the angle θ' subtended by the image to the angle θ subtended by the object.

$$M = \frac{\theta'}{\theta} \tag{10c}$$

From diagram (*b*), the object distance s is obtained by the regular thin-lens formula as

$$\frac{1}{s} + \frac{1}{-25} = \frac{1}{f} \qquad \text{or} \qquad \frac{1}{s} = \frac{25 + f}{25f}$$

From the right triangles, the angles θ and θ' are given by

$$\tan\theta = \frac{y}{25} \qquad \text{and} \qquad \tan\theta' = \frac{y}{s} = y\,\frac{25 + f}{25f}$$

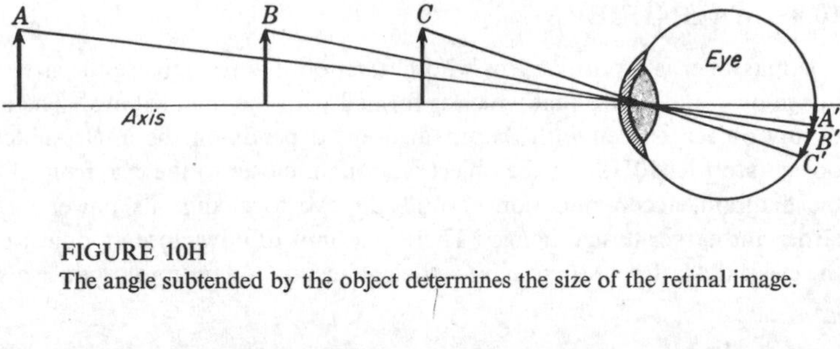

FIGURE 10H
The angle subtended by the object determines the size of the retinal image.

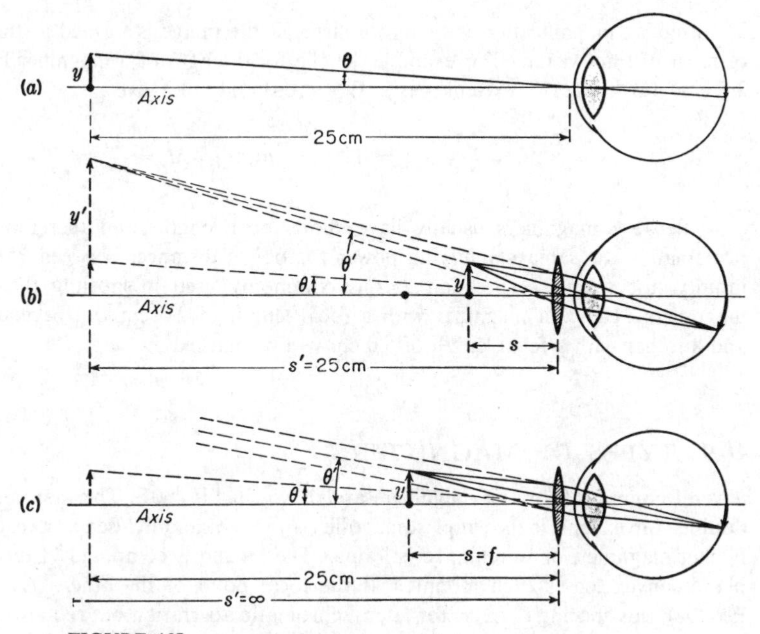

FIGURE 10I
The angle subtended by (*a*) an object at the near point to the naked eye, (*b*) the virtual image of an object inside the focal point, (*c*) the virtual image of an object at the focal point.

For small angles the tangents can be replaced by the angles themselves to give approximate relations

$$\theta = \frac{y}{25} \quad \text{and} \quad \theta' = y\,\frac{25 + f}{25f}$$

giving for the magnification, from Eq. (10c),

$$M = \frac{\theta'}{\theta} = \frac{25}{f} + 1 \tag{10d}$$

In diagram (*c*) the object distance s is equal to the focal length, and the small angles θ and θ' are given by

$$\theta = \frac{y}{25} \quad \text{and} \quad \theta' = \frac{y}{f}$$

giving for the magnification

$$M = \frac{\theta'}{\theta} = \frac{25}{f} \tag{10e}$$

The angular magnification is therefore larger if the image is formed at the distance of most distinct vision. For example, let the focal length of a magnifier be 1 in. or 2.5 cm. For these two extreme cases, Eqs. (10d) and (10e) give

$$M = \frac{25}{2.5} + 1 = 11\times \qquad \text{and} \qquad M = \frac{25}{2.5} = 10\times$$

Because magnifiers usually have short focal lengths and therefore give approximately the same magnifying power for object distances between 25.0 cm and infinity, the simpler expression $25/f$ is commonly used in labeling the power of magnifiers. Hence a magnifier with a focal length of 2.5 cm will be marked 10× and another with a focal length of 5.0 cm will be marked 5×, etc.

10.9 TYPES OF MAGNIFIERS

Several common forms of magnifiers are shown in Fig. 10J. The first, an ordinary double-convex lens, is the simplest magnifier and is commonly used as a reading glass, pocket magnifier, or watchmaker's loupe. The second is composed of two identical plano-convex lenses each mounted at the focal point of the other. As shown by Eq. (9z), this spacing corrects for lateral chromatic aberration but requires the object to be located at one of the lens faces. To overcome this difficulty, color correction is sacrificed to some extent by placing the lenses slightly closer together, but even then the working distance or back focal length [see Eq. (5m)] is extremely short.

The third magnifier, cut from a sphere of solid glass, is commonly credited to Coddington but was originally made by Sir David Brewster. It too has a relatively short working distance, as can be seen by the marginal rays, but the image quality is remarkably good due in part to the central groove acting as a stop. Some of the best magnifiers of today are cemented triplets, like those shown in the last two diagrams. These lenses are symmetrical to permit their use either side up. They have a relatively large working distance and are made with powers up to 20×.

10.10 SPECTACLE LENSES

The ability of the human eye to focus on nearby and distant objects, attributed to the *crystalline lens*, is most prominent in children. Changing the shape of the lens is accomplished by a rather complicated system of ligaments and muscles. Due to tension in the lens capsule, the crystalline lens, if completely free, would tend to become spherical in shape. Surrounding the edge of the lens is an annular ring called the *sciliary muscle*, which on contracting squeezes the lens, causing it to bulge. In effect this reduces the focal length of the lens, bringing nearby objects to a sharp focus on the retina.

If the sciliary muscle relaxes, the *suspensory ligaments* pull outward on the lens periphery, causing it to flatten. This increases the focal length, bringing distant

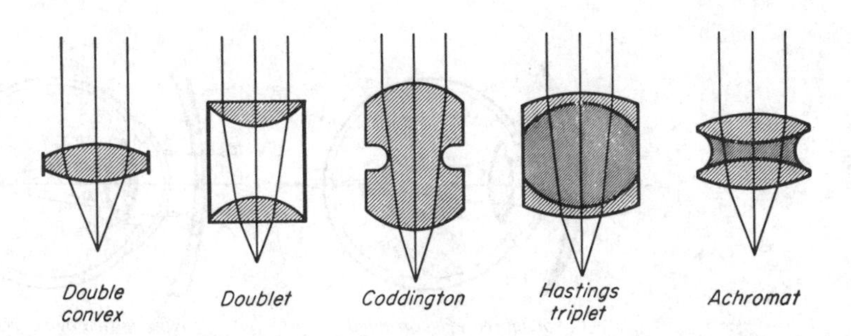

FIGURE 10J
Common types of magnifiers.

objects to focus on the retina. This ability is part of the process of vision called *accommodation*.

As a person grows older, the crystalline lens becomes harder and harder, and the muscles that control its shape grow weaker and weaker, thus making accommodation more and more difficult. This condition is referred to as *presbyopia*. When the length of the eyeball is such that incident parallel light rays converge to a point behind the retina, the person is far-sighted and is said to have *hypermetropia* [see Fig. 10K(*a*)]. When parallel rays come to a focus in front of the retina, as in diagram (*b*), the person is near-sighted and is said to have *myopia*.

In order to correct these defects in one's vision, a converging lens of the appropriate focal length is placed in front of the *hypermetropic eye*, and a diverging lens is placed in front of the *myopic eye*. A positive lens adds some convergence to the rays just before they reach the cornea, thereby enabling the person to see distant objects in sharp focus [see Fig. 10L(*a*)]. A diverging lens in front of the myopic eye can bring distant objects to a sharp focus.

In ophthalmology and optometry it is customary to specify the focal length of spectacle lenses in *diopters*. *The power of any lens in diopters is defined as the reciprocal*

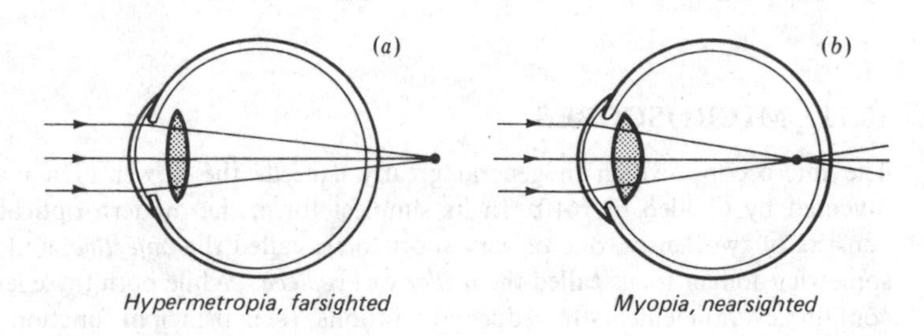

FIGURE 10K
Typical eye defects, largely present in the adult population.

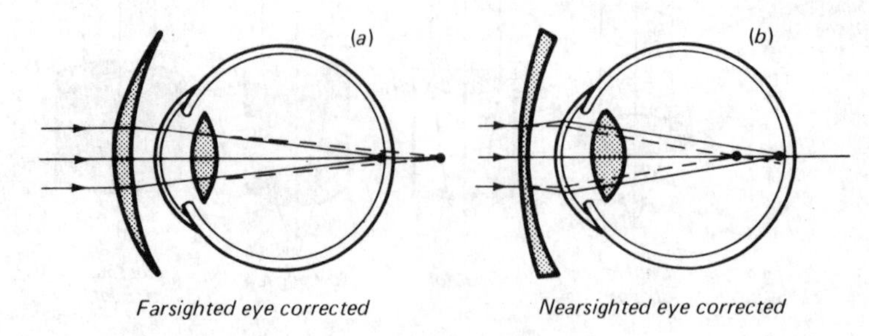

FIGURE 10L
Typical eye defects can be corrected by spectacle lenses.

of the focal length in meters. The symbol for lens power is P, and the unit diopter is abbreviated D. See Sec. 4.12 and Eq. (4f).

$$\text{Diopter} = \frac{1 \text{ m}}{\text{focal length in meters}}$$

$$P = \frac{1}{f} \tag{10f}$$

The lens with by far the greatest power in the eye is the cornea, with 43.0 D. The entire optical system of the eye has a power of 58.6 D. See Table 10A and Fig. 10B.

EXAMPLE A converging lens has a focal length of 27.0 cm. What would be its power in diopters?

SOLUTION Direct substitution of the given quantity, $f = 0.270$ m, in Eq. (10f) gives

$$P = \frac{1}{0.270 \text{ m}} = +3.70 \text{ D}$$

This answer is read, *plus three point seven zero diopters.*

10.11 MICROSCOPES

The microscope, which in general greatly exceeds the power of a magnifier, was invented by Galileo in 1610. In its simplest form, the modern optical microscope consists of two lenses, one of very short focus called the *objective*, and the other of somewhat longer focus called the *ocular* or *eyepiece*. While both these lenses actually contain several elements to reduce aberrations, their principal function is illustrated by single lenses in Fig. 10M. The object (1) is located just outside the focal point of the objective so there is formed a real magnified image at (2). This image becomes the object for the second lens, the eyepiece. Functioning as a magnifier, the eyepiece

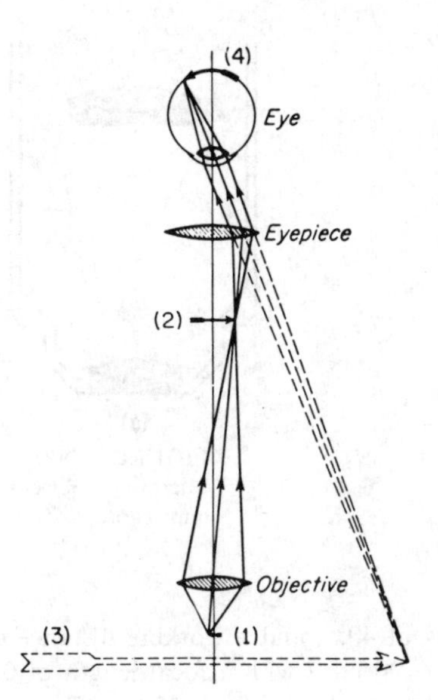

FIGURE 10M
Principles of the microscope, shown with the eyepiece adjusted to produce the image at the distance of most distinct vision.

forms a large virtual image at (3). This image becomes the object for the eye itself, which forms the final real image on the retina at (4).

Since the function of the objective is to form the magnified image that is observed through the eyepiece, the overall magnification of the instrument becomes the product of the linear magnification m_1 of the objective and the angular magnification M_2 of the eyepiece. By Eqs. (4k) and (10e), these are given separately by

$$m_1 = -\frac{x'}{f_1} \quad \text{and} \quad M_2 = \frac{25}{f_2}$$

The overall magnification is therefore

$$M = -\frac{x'}{f_1}\frac{25}{f_2} \tag{10g}$$

It is customary among manufacturers to label objectives and eyepieces according to their separate magnifications m_1 and M_2.

10.12 MICROSCOPE OBJECTIVES

A high-quality microscope is usually equipped with a turret nose carrying three objectives, each of a different magnifying power. By turning the turret, any one of the three objectives can be rotated into proper alignment with the eyepiece. Diagrams of three typical objectives are shown in Fig. 10N. The first, composed of two cemented achromats, is corrected for spherical aberration and coma and has a focal length of 1.6 cm, a magnification of 10×, and a working distance of 0.7 cm. The second is also an achromatic objective with a focal length of 0.4 cm, a magnification

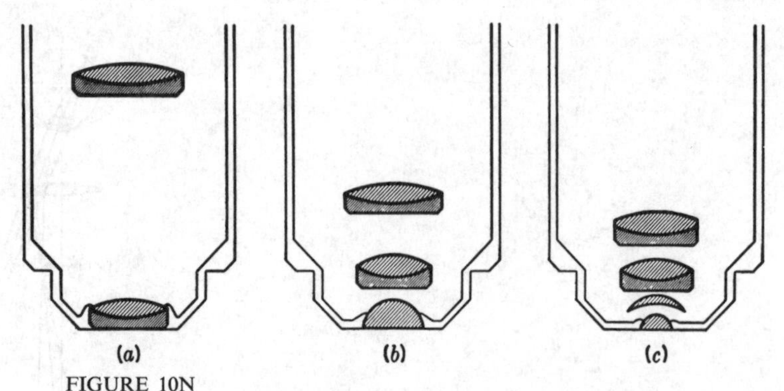

FIGURE 10N
Microscope objectives: (*a*) low-power, (*b*) medium-power, and (*c*) high-power oil immersion.

of 40×, and a working distance of 0.6 cm. The third is an oil-immersion type of objective with a focal length of 0.16 cm, a magnification of 100×, and a working distance of only 0.035 cm. Great care must be exercised in using this last type of lens to prevent scratching of the hemispherical bottom lens. Although oil immersion makes the two lowest lenses aplanatic (see Fig. 9O), lateral chromatic aberration is present. The latter is corrected by the use of a compensating ocular, as will be explained in Sec. 10.18.

10.13 ASTRONOMICAL TELESCOPES

Historically the first telescope was probably constructed in Holland in 1608 by an obscure spectacle-lens grinder, Hans Lippershey. A few months later Galileo, upon hearing that objects at a distance could be made to appear close at hand by means of two lenses, designed and made with his own hands the first authentic telescope. The elements of this telescope are still in existence and may be seen on exhibit in Florence. The principle of the astronomical telescopes of today is the same as that of these early devices. A diagram of an elementary telescope is shown in Fig. 10O. Rays from one point of the distant object are shown entering a long-focus objective lens as a parallel beam. These rays are brought to a focus and form a point image at Q'. Assuming the distant object to be an upright arrow, this image is real and inverted as shown. The eyepiece has the same function in the telescope that it has in a microscope, namely, that of a magnifier. If the eyepiece is moved to a position where this real image lies just inside its primary focal plane F_2, a magnified virtual image at Q'' may be seen by the eye at the near point, 25.0 cm. Normally, however, the real image is made to coincide with the focal points of both lenses, with the result that the image rays leave the eyepiece as a parallel bundle and the virtual image is at infinity. The final image is always the one formed on the retina by rays which appear to have come from Q''. Figure 10P is a diagram of the telescope adjusted in this manner.

In all astronomical telescopes the objective lens is the aperture stop. It is there-

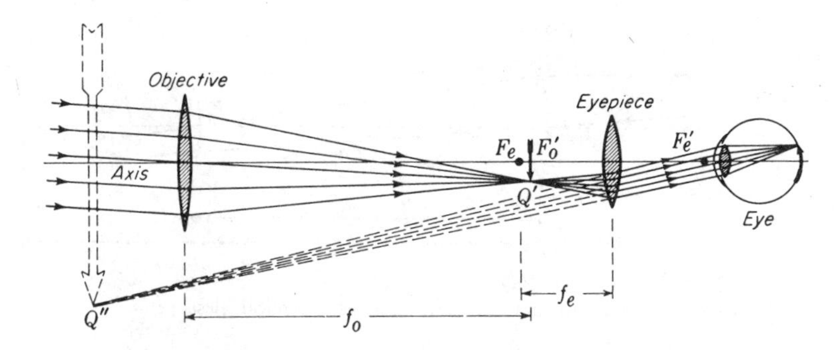

FIGURE 10O
Principles of the astronomical telescope, shown with the eyepiece adjusted to produce the image at the distance of most distinct vision.

fore the entrance pupil, and its image as formed by all the lenses to its right (here, only the eyepiece) is the exit pupil. These elements are shown in Fig. 10Q, which traces the path of one ray incident parallel to the axis and of a chief ray from a distant off-axis object point. The distance from the eye lens, i.e., the last lens of the ocular, to the exit pupil is called the *eye relief* and should normally be about 8.0 mm.

The magnifying power of a telescope is defined as the ratio between the angle subtended at the eye by the final image Q'' and the angle subtended at the eye by the object itself. The object, not shown in Fig. 10Q, subtends an angle θ at the objective and would subtend approximately the same angle to the unaided eye. The angle subtended at the eye by the final image is the θ'. By definition [see Eq. (10e)]

$$M = \frac{\theta'}{\theta}$$

The angle θ is the object-field angle, and θ' is the image-field angle. In other words, θ is the total angular field taken in by the telescope while θ' is the angle that the field

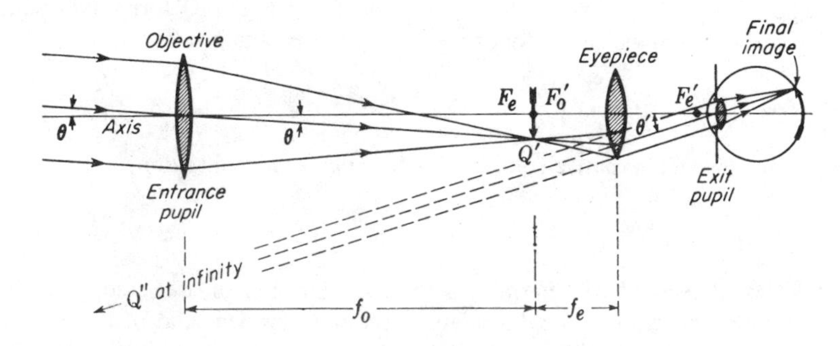

FIGURE 10P
Principles of the astronomical telescope, shown with the eyepiece adjusted to produce the image at infinity.

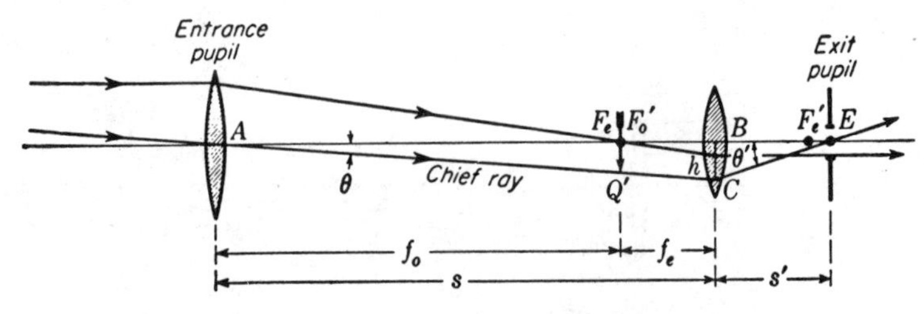

FIGURE 10Q
Entrance and exit pupils of an astronomical telescope.

appears to cover (Sec. 7.11). From the right triangles ABC and EBC, in Fig. 10Q,

$$\tan\theta = \frac{h}{s} \qquad \text{and} \qquad \tan\theta' = -\frac{h}{s'} \tag{10h}$$

Applying the general lens formula $1/s + 1/s' = 1/f$, we have

$$\frac{1}{s'} = \frac{f_O}{f_E(f_O + f_E)} \tag{10i}$$

which, substituted in Eq. (10h), gives

$$\tan\theta = \frac{h}{f_O + f_E} \qquad \text{and} \qquad \tan\theta' = -\frac{hf_O}{f_E(f_O + f_E)}$$

For small angles, $\tan\theta \approx \theta$ and $\tan\theta' \approx \theta'$. Substituting them in Eq. (10g), we obtain

● $$M = \frac{\theta'}{\theta} = -\frac{f_O}{f_E} \tag{10j}$$

Hence the magnifying power of a telescope is just the ratio of the focal lengths of objective and eyepiece respectively, the minus sign signifying an inverted image.

If D and d represent the diameters of the objective and exit pupil respectively, the marginal ray passing through F_O' and F_E in Fig. 10Q forms two similar right triangles, from which the following proportion is obtained

$$-\frac{f_O}{f_E} = \frac{D}{d}$$

giving, as an alternative equation for the angular magnification,

$$M = \frac{D}{d} \tag{10k}$$

A useful method of determining the magnification of a telescope is therefore to measure the ratio of the diameters of the objective lens and of the exit pupil. The latter is readily found by focusing the telescope for infinity and then turning it toward the sky. A thin sheet of white paper held behind the eyepiece and moved back and forth will locate a sharply defined disk of light. This, the exit pupil, is commonly called the *Ramsden circle*. Its size, relative to that of the pupil of the eye, is of great importance in determining the brightness of the image and the resolving power of the instrument (see Sec. 15.9).

Another method of measuring the magnification of a telescope is to sight through the telescope with one eye, observing at the same time the distant object directly with the other eye. With a little practice the image seen in the telescope can be made to overlap the smaller direct image, thereby affording a straightforward comparison of the relative heights of image and object. The object field of the astronomical telescope is determined by the angle subtended at the center of the objective by the eyepiece aperture. In other words, the eyepiece is the field stop of the system. In Fig. 10Q the angle θ is the half-field angle (Sec. 7.8).

10.14 OCULARS AND EYEPIECES

Although a simple magnifier of one of the types shown in Fig. 10J may be used as an eyepiece for a microscope or telescope, it is customary to design special lens combinations for each particular instrument. Such eyepieces are commonly called *oculars*. One of the most important considerations in the design of oculars is correction for lateral chromatic aberration. It is for this reason that the basic structure of most of them involves two lenses of the same glass separated by a distance equal to half the sum of their focal lengths [see Eq. (9z)].

The two most popular oculars based on this principle are known as the *Huygens eyepiece* and the *Ramsden eyepiece* (Fig. 10R). In both these systems the lens nearest the eye is called the *eye lens*, while the lens nearest the objective is called the *field lens*.

10.15 HUYGENS EYEPIECE

In eyepieces of this design the two lenses are usually made of spectacle crown glass with a focal-length ratio f_f/f_e varying from 1.5 to 3.0. As shown in Fig. 10R(*a*), rays from an objective to the left (and not shown) are converging to a real image point Q. The field lens refracts these rays to a real image at Q', from which they diverge again to be refracted by the eye lens into a parallel beam. In most telescopes the objective of the instrument is the entrance pupil of the entire system. The exit pupil or *eyepoint* is, therefore, the image of the objective formed by the eyepiece and is located at the position marked "Exit pupil" in the figure. Here the chief ray crosses the axis of the ocular. A field stop *FS* is often located at Q', the primary focal point of the eye lens, and if cross hairs or a reticle are to be employed, they are mounted in this plane. Although the eyepiece as a whole is corrected for lateral chromatic aberration, the individual lenses are not, so that the image of the cross hairs or reticle formed by the eye lens alone will show considerable distortion and color. Huygens eyepieces with reticles are used in some microscopes, but in this case the reticle is small and is confined to the center of the field. The Huygens eyepiece shows some spherical aberration, astigmatism, and a rather large amount of longitudinal color and pincushion distortion. In general, the eye relief, i.e., the distance between the eye lens of the ocular and the exit pupil, is too small for comfort.

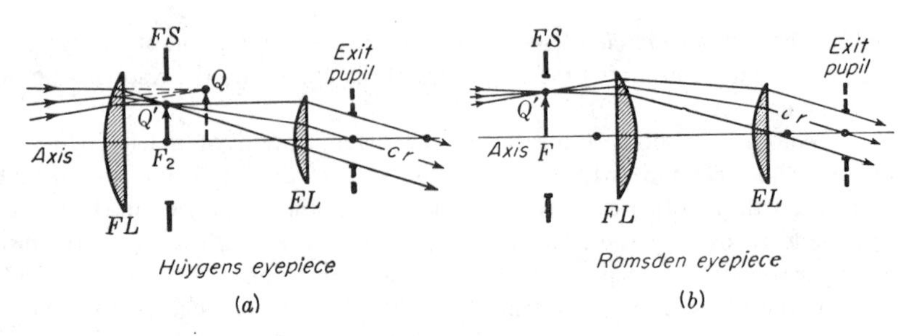

FIGURE 10R
Common eyepieces used in optical instruments.

10.16 RAMSDEN EYEPIECE

In eyepieces of this type as well, the two lenses are usually made of the same kind of glass, but here they have equal focal lengths. To correct for lateral color, their separation should be equal to the focal length. Since the first focal plane of the system coincides with the field lens, a reticle or cross hairs must be located there. Under some conditions this is considered desirable, but the fact that any dust particles on the lens surface would also be seen in sharp focus is an undesirable feature. To overcome this difficulty, the lenses are usually moved a little closer together, thus moving the focal plane forward at some sacrifice of lateral achromatism.

The paths of the rays through a Ramsden eyepiece are shown in Fig. 10R(*b*). The image formed by an objective (not shown) is located at the first focal point F, and it is here that a field stop FS and a reticle or cross hairs are often located. After refraction by both lenses, parallel rays emerge and reach the eye at or near the exit pupil. With regard to aberrations, the Ramsden eyepiece has more lateral color than the Huygens eyepiece, but the longitudinal color is only about half as great. It has about one-fifth the spherical aberration, about half the distortion, and no coma. One important advantage over the Huygens ocular is its 50 percent greater eye relief.

10.17 KELLNER OR ACHROMATIZED RAMSDEN EYEPIECE

Because of the many desirable features of the Ramsden eyepiece, various attempts have been made to improve its chromatic defects. This aberration can be almost eliminated by making the eye lens a cemented doublet (Fig. 10S). Such eyepieces are commonly used in prism binoculars, because the slight amount of lateral color is removed and spherical aberration is reduced through the aberration characteristics of the Porro prisms [See Fig. 2C(*b*)].

10.18 SPECIAL EYEPIECES

The orthoscopic eyepiece shown in the middle diagram of Fig. 10S is characterized by its wide field and high magnification. It is usually employed in high-power telescopes and range finders. Its name is derived from the freedom from distortion

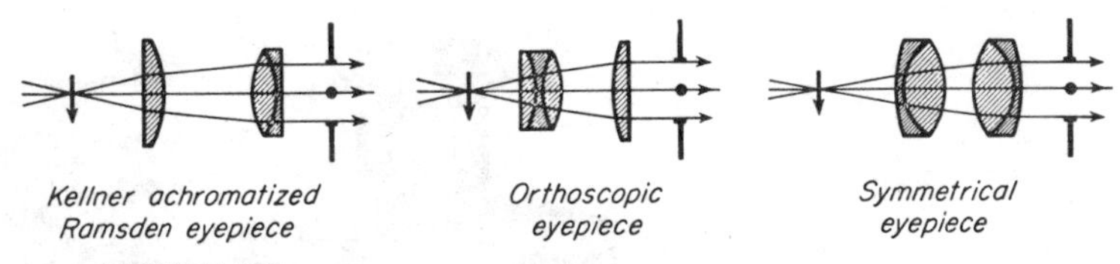

FIGURE 10S
Three types of achromatized eyepieces.

characterizing the system. The symmetrical eyepiece shown at the right in Fig. 10S has a larger aperture than a Kellner of the same focal length. This results in a wider field as well as a long eye relief, hence its frequent use in various types of telescopic gun sights. The danger of having a short eye relief with a recoiling gun should be obvious.

Since lateral chromatic aberration, as well as the other aberrations of an eyepiece, is affected by altering the separation of the two elements, some oculars are provided with means for making this distance adjustable. Some microscopes come equipped with a set of such compensating eyepieces, thereby permitting the undercorrection of lateral color in any objective to be neutralized by an overcorrection of the eyepiece.

10.19 PRISM BINOCULARS

Prism binoculars are in reality a pair of identical telescopes mounted side by side, one for each of the two eyes. Such an instrument is shown in Fig. 10T with part of the case cut away to show the optical parts. The objectives are cemented achromatic pairs, while the oculars are Kellner or achromatized Ramsden eyepieces. The dotted lines show the path of an axial ray through one pair of Porro prisms. The first prism reinverts the image and the second turns it left for right, thereby finally giving an image in the proper position. The doubling back of the light rays has the further advantage of enabling longer focus objectives to be used in short tubes, with consequent higher magnification.

There are four general features that go to make up good binoculars: (1) magnification, (2) field of view, (3) light-gathering power, and (4) size and weight. For hand-held use, binoculars with five-, six-, seven-, or eightfold magnification are most generally used. Glasses with powers above 8 are desirable, but require a rigid mount to hold them steady. For powers less than 4, lens aberrations usually offset the magnification, and the average person can usually see better with the unaided eyes. *The field of view is determined by the eyepiece aperture and should be as large as is practicable.* For seven-power binoculars a 6° object field is considered large, since in the eyepiece the same field is spread over an angle of $7 \times 6°$, or 42°.

The diameter of the objective lenses determines the light-gathering power. Large diameters are important only at night when there is little light available. Binoculars with the specification 6×30 have a magnification of 6 and objective lenses

FIGURE 10T
Diagram of prism binoculars, showing the lenses and totally reflecting Porro prisms.

with an effective diameter of 30.0 mm. The specification 7 × 50 means a magnification of 7 and objectives 50.0 mm in diameter. Although glasses with the latter specifications are excellent for day or night use, they are considerably larger and more cumbersome than the daytime glasses specified as 6 × 30 or 8 × 30. For general civilian use, the latter two are much the most useful.

The diameters of the field and eye lenses of the oculars (*FL* and *EL* in Fig. 10R) *determine the size of field of view.*

10.20 THE KELLNER-SCHMIDT OPTICAL SYSTEM

The Kellner-Schmidt optical system combines a concave spherical mirror with an aspheric lens as shown in Fig. 10U. Kellner devised and patented* this optical system in 1910 as a high-quality source of parallel light. Years later Schmidt introduced the system as a high-speed camera, and it has since become known as a Schmidt camera. While Schmidt was the first to emphasize the importance of placing the corrector plate at the mirror's center of curvature, Kellner shows it in this position in his patent drawing.

The purpose of the lens is to refract incoming parallel rays in such directions that after reflection from the spherical mirror they all come to a focus at the same axial point *F*. This *corrector plate* therefore eliminates the spherical aberration of the mirror. With the lens located at the center of curvature of the mirror, parallel rays

* U.S. Patent 969,785, 1910.

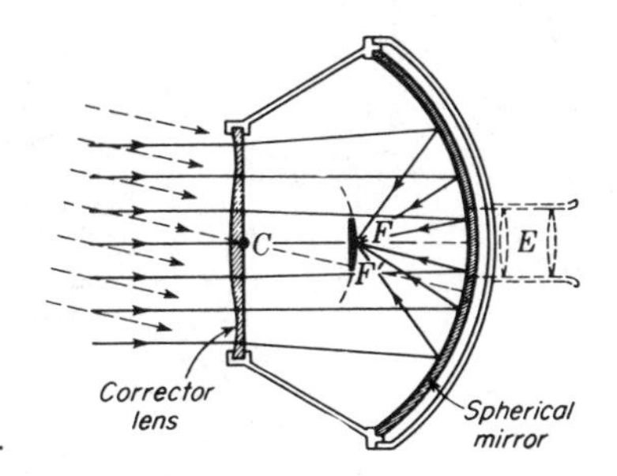

FIGURE 10U
Kellner-Schmidt optical system.

entering the system at large angles with the axis are brought to a relatively good focus at other points like F'. The focal surface of such a system is spherical, with its center of curvature at C.

Such an optical system has several remarkable and useful properties. First as a camera, with a small film at the center or with a larger film curved to fit the focal surface, it has the very high speed of $f/0.5$. Because of this phenomenal speed, Schmidt systems are used by astronomers to obtain photographs of faint stars or comets. They are used for similar reasons in television receivers to project small images from an oscilloscope tube onto a relatively large screen. In this case the convex oscilloscope screen is curved to the focal surface so that the light from the image screen is reflected by the mirror and passes through the corrector lens to the observing screen.

If a convex silvered mirror is located at FF', rays from any distant source entering the system will form a point image on the focal surface and after reflection will again emerge as a parallel bundle in the exact direction of the source. When used in this manner the device is called an *autocollimator*. If the focal surface is coated with fluorescent paint, ultraviolet light from a distant invisible source will form a bright spot at some point on FF' and the visible light emitted from this spot will emerge only in the direction of the source. If a hole is made in the center of the large mirror, an eyepiece may be inserted in the rear to view the fluorescent screen and any ultraviolet source may be seen as a visible source. As such, the device becomes a fast, wide-angled, ultraviolet telescope.

10.21 CONCENTRIC OPTICAL SYSTEMS*

The recent development and use of concentric optical systems warrants at least some mention of their remarkable optical properties. Such systems have the general form of a concave mirror and a concentric lens of the type shown in Fig. 5I. As the title implies, and as is shown in Fig. 10V, all surfaces have a common center of curvature C.

The purpose of the concentric lens is to reduce spherical aberration to a minimum. Off-axis rays traversing the lens are bent away from the axis and (by the

* A. Bouwers, "Achievements in Optics," Elsevier Press, Inc., Houston, Tex., 1950.

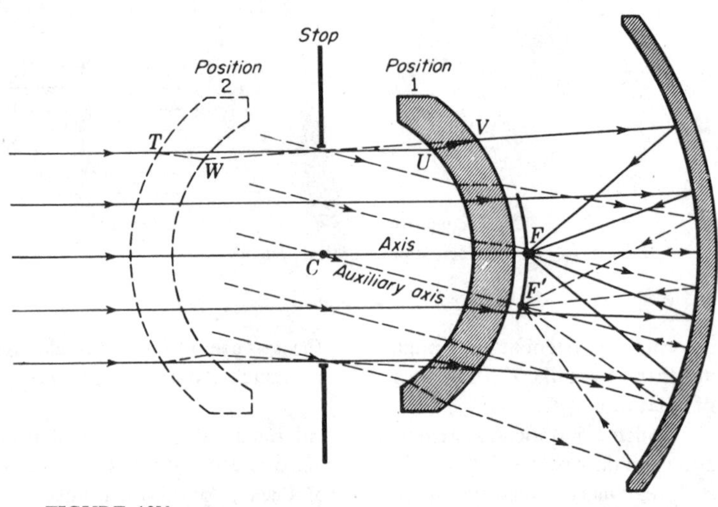

FIGURE 10V
Concentric optical system.

proper choice of radii, refractive index, and lens thickness) can be made to cross the axis at the paraxial focal point *F*. Since any ray through *C* may be considered as an *axis*, the focal surface is also a sphere with *C* as a center of curvature. In some applications the back surface of the lens is made to be the focal surface.

Since the principal planes of the concentric lens both coincide with a plane through *C* perpendicular to the axial ray of any bundle, it is as if the corrector were a thin lens located at *C* and oriented at the proper angle for all incident parallel beams.

Since there are no oblique and no sagittal rays, the system is free of coma and astigmatism. The complete performance of the system is known as soon as the imagery of an axial object point is known. Here lies the essential advantage over the Kellner-Schmidt system. Chromatic aberrations resulting from the lens are small as long as the focal length is long compared with that of the mirror, and this is nearly always the case.

Other important features of the concentric system can be seen from the diagram. There is an unusually small decrease in image brightness with increasing angle of incidence. The corrector lens may be placed in front of *C*, in position 2. In this position the same optical performance is realized. Finally, a concentric convex mirror may be placed about halfway between the lens and the mirror. The reflected light is then brought to a focus through a hole in the center of the large mirror. This latter arrangement, among other things, makes an excellent objective system for a reflecting microscope.

Today many high-precision optical instruments and devices employ the Kellner-Schmidt and concentric optical systems. The Armed Services research laboratories have developed missile-tracking systems and missile-homing-guidance systems employing ultraviolet, visible, and infrared light. Fine telephoto objectives and compact commerical telescopes are also on the market.*

PROBLEMS

10.1 A clear glass bead in the form of a sphere is exactly 2.0 cm in diameter. If the glass has a refractive index of 1.5250, find by calculation (*a*) its focal length, (*b*) its magnifying power, (*c*) its back focal length, and (*d*) the location of its secondary principal point. (*e*) Solve graphically.

10.2 A magnifier is made with two thin plano-convex lenses, each with a focal length of +2.50 cm and spaced 1.50 cm apart with their convex surfaces facing each other [see Fig. 10H(*b*)]. Apply gaussian formulas to find (*a*) the focal length, (*b*) the magnifying power, (*c*) the back focal length, and (*d*) the position of the secondary principal point. (*e*) Solve graphically.

10.3 A Ramsden eyepiece is made of two thin plano-convex lenses, each with a focal length of 3.50 cm and spaced 2.50 cm apart. Applying the thin-lens formulas, find (*a*) its focal length, (*b*) its magnifying power, and (*c*) its back focal length.

10.4 A Ramsden eyepiece is made of two thin lenses, each with a focal length of 36.0 mm and spaced 28.0 mm apart. Applying the thin-lens formulas, find (*a*) its focal length, (*b*) its magnifying power, and (*c*) its back focal length.

Ans. (*a*) +29.46 mm, (*b*) +8.49×, (*c*) 6.55 mm

10.5 A Huygens eyepiece is made of two thin lenses of the same glass, with focal lengths of +2.50 and 1.50 cm, respectively. If the lenses are spaced to correct for chromatic aberration (see Sec. 9.14), find (*a*) the focal length of the eyepiece, (*b*) the magnification, and (*c*) the back focal length of the ocular. (*d*) Make a scale diagram.

10.6 A microscope has an ocular marked +15× and an objective with a focal length of +4.5 mm. What is the total magnification if the objective forms its image 16.0 cm beyond its secondary focal plane?

10.7 A microscope is fitted with an ocular having a focal length of 12.0 mm and an objective with a focal length of 3.20 mm. If the objective forms its image 16.0 cm beyond its secondary focal plane, find the total magnification. *Ans.* −1042×

10.8 The objective and ocular of a microscope are 20.0 cm apart. The focal length of the objective is 7.0 mm, and the focal length of the ocular is 5.0 mm. Treating these as though they were thin lenses, find (*a*) the distance from the objective to the object viewed, (*b*) the linear magnification by the objective, and (*c*) the overall magnification if the final image is formed at infinity.

10.9 The ocular and objective of a microscope have focal lengths of +5.20 and +8.20 mm, respectively, and are located 18.0 cm apart. Treating these as thin lenses, find (*a*) the distance from the objective to the object viewed, (*b*) the linear magnification produced by the objective, and (*c*) the overall magnification if the final image is formed at infinity.

10.10 An objective of an astronomical telescope has a diameter of 12.5 cm and a focal length of 85.0 cm. When it is used with an eyepiece having a focal length of 2.50 cm and a diameter of 1.50 cm, what will be (*a*) the angular magnification, (*b*) the diameter

* See J. J. Villa, Catadioptic Lenses, *Opt. Spectra*, March 1968, p. 57.

of the exit pupil, (*c*) the object-field angle, (*d*) the image-field angle, and (*e*) the eye relief? *Ans.* (*a*) 34.0, (*b*) 0.3676 cm, (*c*) 0.491°, (*d*) 16.70°, (*e*) 2.574 cm

10.11 The objective of a small astronomical telescope has a focal length of +40.0 cm and a diameter of 4.0 cm. When it is used with an eyepiece having a focal length of +12.50 mm and a diameter of 10.0 mm, find (*a*) the angular magnification, (*b*) the diameter of the exit pupil, (*c*) the object-field angle, (*d*) the image-field angle, and (*e*) the eye relief.

10.12 The objective lenses of a pair of binoculars have a focal length 26.50 cm and an aperture of 65.0 mm. The oculars have a focal length of 25.0 mm and an aperture of 12.50 mm. Find (*a*) the angular magnification, (*b*) the diameter of the exit pupils, (*c*) the object-field angle, (*d*) the image-field angle, (*e*) the eye relief, and (*f*) the field at a distance of 1000 m.

PART TWO

Wave Optics

11

VIBRATIONS AND WAVES

The world around us is filled with waves. Some of them we can see or hear, but many more our senses of sight or hearing cannot detect. In the submicroscopic world, atoms and molecules are made up of electrons, protons, neutrons, and mesons that move around as waves within their boundaries. Appropriately stimulated, these same atoms and molecules emit waves we call γ rays, X rays, light waves, heat waves, and radio waves.

In our world of macroscopic bodies, water waves and sound waves are produced by moving masses of considerable size. Earthquakes produce waves as the result of sudden shifts in land masses. Water waves are produced by the wind or ships as they pass by. Sound waves are the result of quick movements of objects in the air.

Any motion that repeats itself in equal intervals of time is called *periodic motion.* The swinging of a clock pendulum, the vibrations of the prongs of a tuning fork, and a mass dancing from the lower end of a coiled spring are but three examples. These particular motions and many others like them that occur in nature are referred to as *simple harmonic motion* (SHM).

11.1 SIMPLE HARMONIC MOTION

Simple harmonic motion is defined as the projection on any diameter of a graph point moving in a circle with uniform speed. The motion is illustrated in Fig. 11A. The *graph point p* moves around the circle of radius a with a uniform speed v. If at every instant of time a normal is drawn to the diameter AB, the intercept P, called the *mass point*, moves with SHM.

Moving back and forth along the line AB, the mass point is continually changing speed v_x. Starting from rest at the end points A or B, the speed increases until it reaches C. From there it slows down again coming to rest at the other end of its path. The return of the mass point is a repetition of this motion in reverse.

The displacement of an object undergoing SHM is defined as the distance from its equilibrium position C to the point P. It will be seen in Fig. 11A that the displacement x varies in magnitude from zero up to its maximum value a, which is the radius of the *circle of reference.*

The maximum displacement a is called the *amplitude*, and the time required to make one complete vibration is called the *period.* If a vibration starts at B, it is completed when the mass point P moves across to A and back again to B. If it starts at C and moves to B and back to C, only half a vibration has been completed. The amplitude a is measured in meters, or a fraction thereof, while the period is measured in seconds.

The frequency of vibration is defined as the number of complete vibrations per second. If a particular vibrating body completes one vibration in $\frac{1}{3}$ s, the period $T = \frac{1}{3}$ s and it will make three complete vibrations in 1 s. If a body makes 10 vibrations in 1 s, its period will be $T = \frac{1}{10}$ s. In other words, the frequency of vibration ν and the period T are reciprocals of each other:

$$\text{frequency} = \frac{1}{\text{period}} \qquad \text{period} = \frac{1}{\text{frequency}}$$

In algebraic symbols,

$$\nu = \frac{1}{T} \qquad T = \frac{1}{\nu} \tag{11a}$$

If the vibration of a body is described in terms of the graph point p, moving in a circle, the frequency is given by the number of *revolutions per second*, or *cycles per second*

$$1 \text{ cycle/second} = 1 \text{ vibration/second} \tag{11b}$$

now called the hertz* (Hz)

$$1 \text{ vib/s} = 1 \text{ Hz} \tag{11c}$$

* Heinrich Rudolf Hertz (1857–1894), German physicist, was born at Hamburg. He studied physics under Helmholtz in Berlin, at whose suggestion he first became interested in Maxwell's electromagnetic theory. His researches with electromagnetic waves which made his name famous were carried out at Karlsruhe Polytechnic between 1885 and 1889. As professor of physics at the University of Bonn, after 1889, he experimented with electrical discharges through gases and narrowly missed the discovery of X rays described by Röntgen a few years later. By his premature death, science lost one of its most promising disciples.

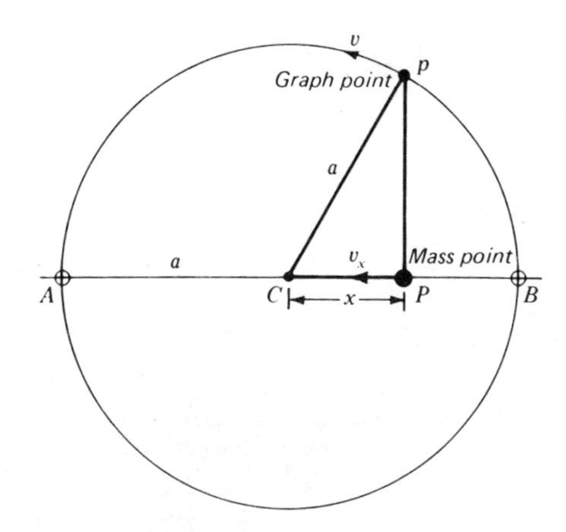

FIGURE 11A
Simple harmonic motion along a straight line AB.

11.2 THE THEORY OF SIMPLE HARMONIC MOTION

At this point we present the theory of SHM and derive an equation for the period of vibrating bodies. In Fig. 11B we see that the displacement x is given by

$$x = a \cos \theta$$

As the graph point p moves with constant speed v, the radius vector a rotates with constant angular speed ω, so that the angle θ changes at a constant rate

$$x = a \cos \omega t \qquad \text{(11d)}$$

The graph point p, moving with a speed v, travels once around the circle of reference, a distance equal to $2\pi a$, in the time of one period T. We now use the relation in mechanics that time equals distance divided by speed, and obtain

$$T = \frac{2\pi a}{v} \qquad \text{(11e)}$$

To obtain the angular speed ω of the graph point in terms of the period, we have

$$T = \frac{2\pi}{\omega} \qquad \text{or} \qquad \omega = \frac{2\pi}{T} \qquad \text{(11f)}$$

An object moving in a circle with uniform speed v has a *centripetal acceleration* toward the center, given by

$$a_c = \frac{v^2}{a} \qquad \text{(11g)}$$

Since this acceleration a_c continually changes the direction of the motion, its component a_x along the diameter, or x axis, changes in magnitude and is given by $a_x = a_c \cos \theta$. Substituting in Eq. (11g), we find

$$a_x = \frac{v^2}{a} \cos \theta$$

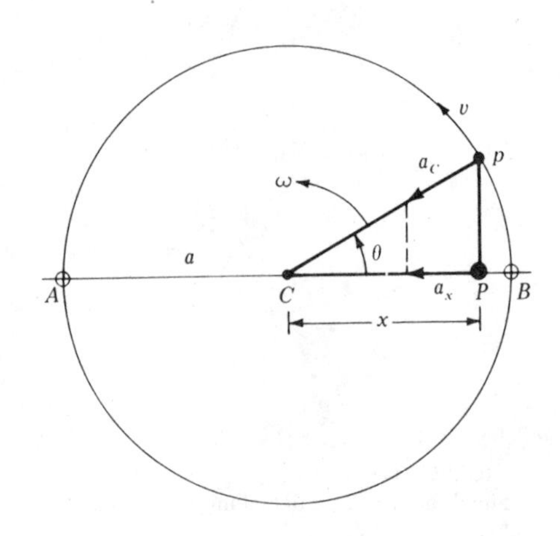

FIGURE 11B
The acceleration a_x of any mass moving with simple harmonic motion is toward a position of equilibrium C.

From the right triangle CPp, $\cos\theta = x/a$, direct substitution gives

$$a_x = \frac{v^2}{a}\frac{x}{a} \quad \text{or} \quad a_x = \frac{v^2}{a^2}x$$

We now multiply both sides of the equation by $a^2/a_x v^2$, take the square root of both sides of the equation, and obtain

$$\frac{a^2}{v^2} = \frac{x}{a_x} \quad \text{and} \quad \frac{a}{v} = \sqrt{\frac{x}{a_x}}$$

For a/v in Eq. (11e) we now substitute x/a_x and obtain for the period of any SHM the relation

$$T = 2\pi\sqrt{\frac{x}{a_x}} \tag{11h}$$

If the displacement is to the right of C, its value is $+x$, and if the acceleration is to the left, its value is $-a_x$. Conversely, when the displacement is to left of C, we have $-x$, and the acceleration is to the right, or $+a_x$. This is the reason for writing

$$T = 2\pi\sqrt{-\frac{x}{a_x}} \tag{11i}$$

11.3 STRETCHING OF A COILED SPRING

As an illustration of the relationships generally applied to vibrating sources, we consider in some detail the stretching of a coiled spring, followed by its vibration with SHM when the stretching force is suddenly released (see Fig. 11C).

As a laboratory experiment, one end of a meterstick is placed at marker Q. A force of 2.0 newtons (N) is applied to the spring, stretching it a distance of 1.25 cm.

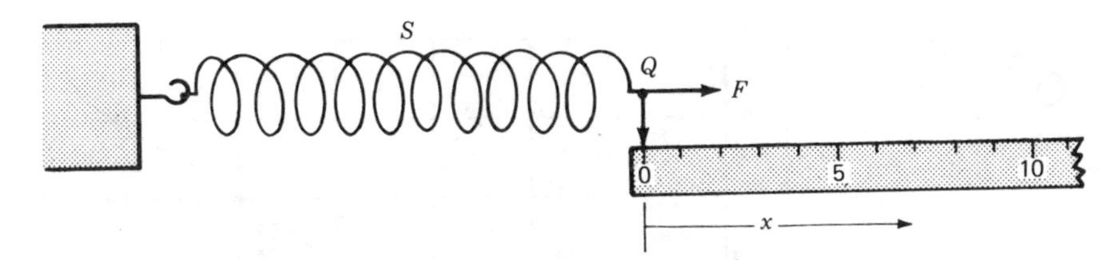

FIGURE 11C
An experiment for measuring the distance x a coiled spring S stretches for different values of the applied force.

When a total force of 4.0 N is applied, the total stretch is 2.50 cm. By applying forces of 6.0, 8.0, and 10.0 N, respectively, the total distances recorded are as shown in Table 11A.

Plotting these data on graph paper produces a straight line, as shown in Fig. 11D. Properly interpreted, this graph means that the applied force F and the displacement of the spring x are directly proportional to each other, and we can write

$$F \propto x \qquad \text{or} \qquad F = kx$$

The proportionality constant k is the slope of the straight line and is a direct measure of the stiffness of the spring. The experimental value of k in this experiment is calculated as follows:

$$k = \frac{F}{x} = \frac{10 \text{ N}}{0.0625 \text{ m}} = 160 \text{ N/m} \qquad (11j)$$

The stiffer the spring, the larger its *stretch constant* k.

Within the limits of this experiment, the spring exerts an equal and opposite force $-F$, as the reaction to the applied force $+F$. For the spring, $-F = kx$, and we can write

$$F = -kx \qquad (11k)$$

Table 11A RECORDED DATA FOR STRETCHING A COILED SPRING

$\frac{F}{\text{N}}$	$\frac{x}{\text{m}}$
0	0
2	0.0125
4	0.0250
6	0.0375
8	0.0500
10	0.0625

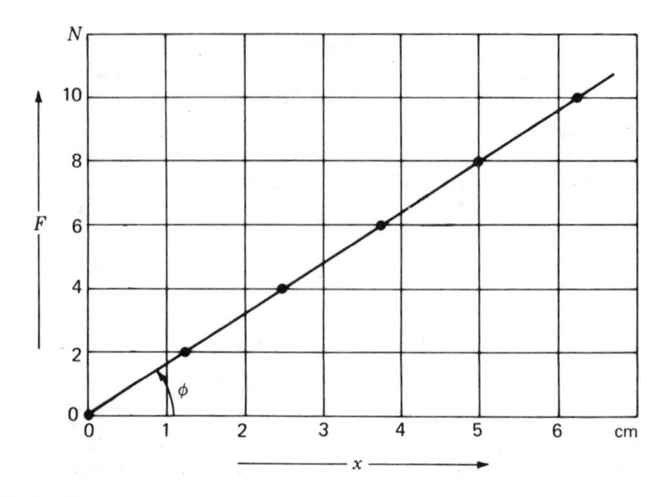

FIGURE 11D
Experimental results on the stretching of a coiled spring as shown in Fig. 11C. This is a demonstration of Hooke's law.

The fact that we obtain a straight line graph in Fig. 11D shows that the stretching of a spring obeys *Hooke's law*.* This is typical of nearly all elastic bodies as long as the body is not permanently deformed, indicating that the forces applied had been carried beyond the *elastic limit*.

Since the *work done* in stretching the spring is given by the *force* multiplied by the *distance* and the force here varies linearly with the distance,

$$\text{Work} = \int F\,dx \tag{11l}$$

As can be seen in Fig. 11E, the *average force* is given by $\frac{1}{2}F$. This, multiplied by the distance x through which it acts, gives the area under the curve, which is the work done†

$$W = \tfrac{1}{2}Fx \tag{11m}$$

If we now replace F by its equivalent value kx from Eq. (11j), we obtain

$$W = \tfrac{1}{2}kx^2 \tag{11n}$$

* Robert Hooke (1635–1703), English experimental physicist, is known principally for his contributions to the wave theory of light, universal gravitation, and atmospheric pressure. He originated many physical ideas but perfected few of them. Hooke's scientific achievements would undoubtedly have received greater acclaim if his efforts had been confined to fewer subjects. He had an irritable temper and made many virulent attacks on Newton and other men of science, claiming that work published by them was due to him.

† In most elementary physics texts it is shown that the area under the curve of a graph, where F is plotted against x, is equal to the total work done.

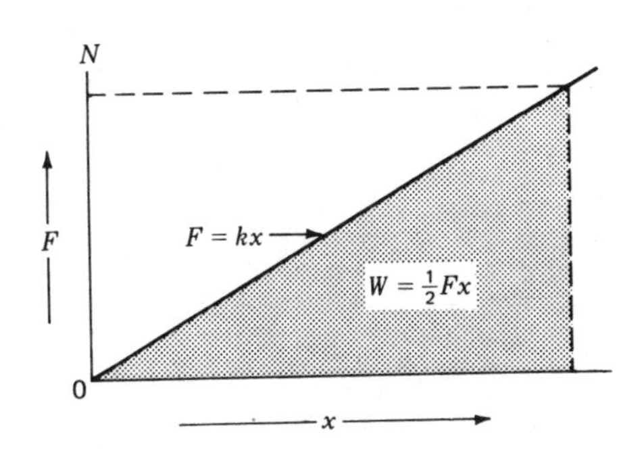

FIGURE 11E
The work done and the energy stored in stretching a spring are given by the area under the graph line $F = kx$.

This relation shows that if the stretch of a spring increases twofold, the energy required, or stored, is increased fourfold, and increasing the displacement threefold increases the energy ninefold.

11.4 VIBRATING SPRING

All bodies in nature are elastic, some more so than others. If a distorting force is applied to change the shape of a body and its shape is not permanently altered, upon release of the force it will be set in vibration.

This property is demonstrated in Fig. 11F by a mass m suspended from the lower end of a spring. In diagram (a) a force F has been applied to stretch the spring a distance a. Upon release, the mass moves up and down with SHM. In diagram (c), m is at its highest point and the spring is shown compressed. The amplitude of the vibration is determined by the distance the spring is stretched from its equilibrium position, and the period of vibration T is given by

● $$T = 2\pi \sqrt{\frac{m}{k}} \qquad (11o)$$

where k is the stiffness of the spring and m is the mass of the vibrating body. This equation shows that if a stiffer spring is used, k being in the denominator, the period is decreased and the vibration frequency is increased. If the mass m is increased, the period is increased and the frequency is decreased.

Since the stretching of the spring obeys Hooke's law, we can apply Eq. (11k). Using the *force equation* from mechanics,

$$F = ma$$

and replacing F in Eq. (11k) by ma, we obtain

$$ma = -kx \qquad \text{or} \qquad \frac{-x}{a} = \frac{m}{k} \qquad (11p)$$

Hence by the replacement of $-x/a$ by m/k in Eq. (11i) we obtain Eq. (11o).

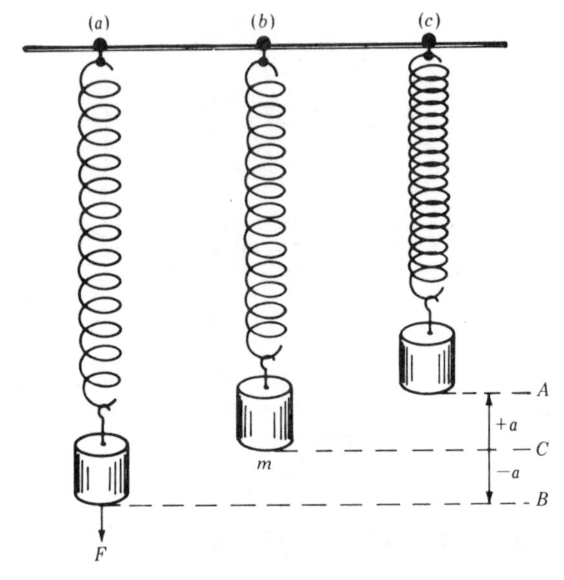

FIGURE 11F
A mass m suspended from a coiled spring is shown in three positions as it vibrates up and down with simple harmonic motion.

EXAMPLE 1 If a 4.0-kg mass is suspended from the lower end of a coiled spring, as shown in Fig. 11F, it stretches a distance of 18.0 cm. If the spring is then extended farther and released, it will be set vibrating up and down with SHM. Find (*a*) the spring constant k, (*b*) the period T, (*c*) the frequency ν, and (*d*) the *total energy* stored in the vibrating system.

SOLUTION The given quantities in the mks system of units are $m = 4.0$ kg, $x = 0.180$ m; the acceleration due to gravity is $g = 9.80$ m/s^2.

(*a*) We can use Eq. (11g), solve for the value of k, and substitute the appropriate values:

$$k = \frac{-F}{x} = \frac{4.0 \times 9.80}{0.180} = 217.8 \text{ N/m}$$

(*b*) We can use Eq. (11o), and upon direct substitution of the known values obtain

$$T = 2\pi\sqrt{\frac{m}{k}} \qquad T = 2\pi\sqrt{\frac{4.0 \text{ kg}}{217.8 \text{ N/m}}}$$

$$T = 0.852 \text{ s}$$

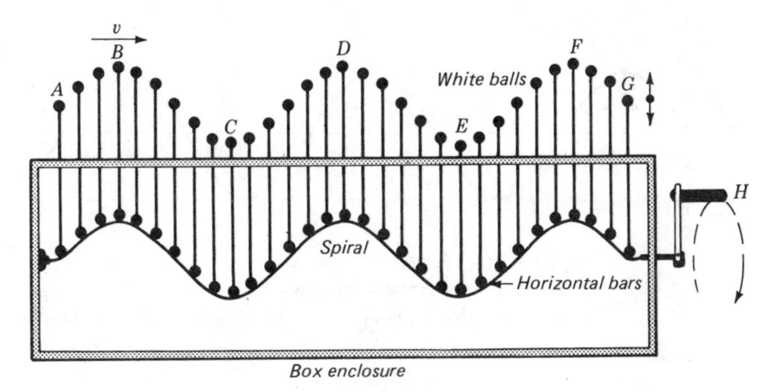

FIGURE 11G
Machine for demonstrating transverse waves.

(*c*) Since the frequency is the reciprocal of the period,

$$v = \frac{1}{T} = \frac{1}{0.852} = 1.174 \text{ Hz}$$

(*d*) The total energy stored in the vibrating system is given by Eq. (11n). By substitution of the given quantities we obtain

$$W = \tfrac{1}{2}kx^2 = \tfrac{1}{2}[(217.8)(0.180)^2] = 3.528 \text{ N m} = 3.528 \text{ J}$$

This answer is read three point five two eight joules.

11.5 TRANSVERSE WAVES

All light waves are classified as *transverse waves*. Transverse waves are those in which each small part of the wave vibrates along a line perpendicular to the direction of propagation and all parts are vibrating in the same plane. A wave machine for demonstrating transverse waves is shown in Fig. 11G. When the handle H is turned clockwise the small white balls at the top of the vertical rods move up and down with SHM. As each ball moves along a vertical line, the wave form *ABCDEFG* moves to the right. When the handle is turned counterclockwise, the wave form moves to the left. In either case each ball performs the exact same motion along its line of vibration, the difference being that each ball is slightly behind or ahead of its neighbor.

When a source vibrates with SHM and sends out transverse waves through a homogeneous medium, they have the general appearance of the waves shown in Fig. 11H. The distance between two similar points of any two consecutive wave forms is called the *wavelength* λ. One wavelength, for example, is equal to the distance between two *wave crests* or two *wave troughs*.

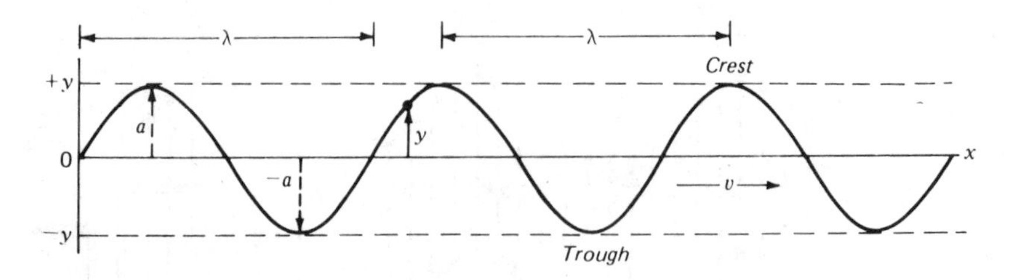

FIGURE 11H
Diagram of a transverse wave, vibrating in the plane of the page, showing the wavelength λ, the amplitude a, the displacement y, and the speed v.

The displacement y of any given point along a wave, at any given instant in time, is given by the vertical distance of that point from its equilibrium position. The value is continually changing from + to − to +, etc. The amplitude of any wave is given by the letter a in Fig. 11H, and is defined as the *maximum value of the displacement y*.

The frequency of a train of waves is given by the number of waves passing by, or arriving at, any given point per second, and is specified in *hertz*, or in vibrations per second. From the definition of frequency ν and the wavelength λ, the speed of the waves v is given by the wave equation

$$v = \nu\lambda \tag{11q}$$

The length of one wave times the number of waves per second equals the distance the waves will travel in 1 s.

11.6 SINE WAVES

The simplest kind of wave train is that for which the motions of all points along the wave have displacements y given by the *sine* or *cosine* of some uniformly increasing function. This in effect describes what we have called SHM.

Consider transverse waves in which the motions of all parts are perpendicular to the direction of propagation. The displacement y of any point on the wave is then given by

$$y = a \sin \frac{2\pi x}{\lambda} \tag{11r}$$

A graph of this equation is shown in Fig. 11I, and the significance of the constants a and λ is clear. To make the wave move to the right with a velocity v, we introduce the time t as follows:

$$y = a \sin \frac{2\pi}{\lambda}(x - vt) \tag{11s}$$

Any particle of the wave, such as P in the diagram, will carry out SHM and will occupy successive positions P, P', P'', P''', etc., as the wave moves.

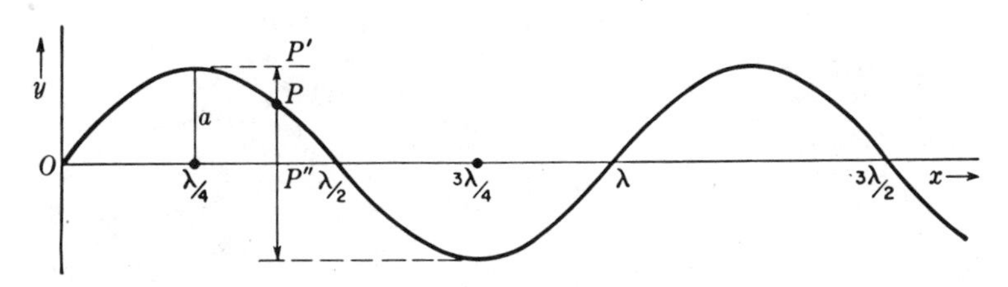

FIGURE 11I
Contour of a sine wave at time $t = 0$.

The time for one complete vibration of any one point is the same as any other point. Furthermore, the period T and its reciprocal the frequency ν are given by the wave equation (11q):

$$v = \nu\lambda = \frac{\lambda}{T} \tag{11t}$$

If we substitute several of these variables in Eq. (11s), we can obtain useful equations for wave motion in general:

$$y = a \sin 2\pi\left(\frac{t}{T} - \frac{x}{\lambda}\right)$$

$$y = a \sin \frac{2\pi}{T}\left(t - \frac{x}{v}\right) \tag{11u}$$

$$y = a \sin 2\pi\nu\left(t - \frac{x}{v}\right)$$

11.7 PHASE ANGLES

In wave motion the instantaneous displacement and direction of propagation are described by specifying the position of the graph point on the circle of reference (see Fig. 11J). The angle θ, measured counterclockwise from the $+x$ axis, specifying the position is called the *phase angle*. As an example consider a point moving up and down along the y axis, as shown in Fig. 11J. The position of the mass point P is given by the projection of the graph point p_1 on the y axis. From the right triangle PpC on the diagram

$$y = a \sin \theta \tag{11v}$$

With the graph point moving at constant speed v, the angular speed ω is constant, and we can write for any angle θ

$$\theta = \omega t$$

Substitution in Eq. (11v) gives

$$y = a \sin \omega t \tag{11w}$$

At time $t = 0$ the graph point is at $+p_0$ and the mass point is at P_0. At some later time t when the mass point is at P, the graph point is at p and we must modify Eq. (11w) by adding the angle α as follows:

$$y = a \sin (\omega t + \alpha) \tag{11x}$$

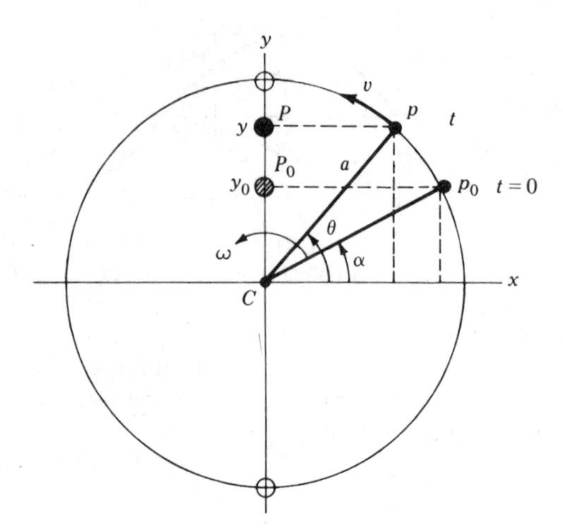

FIGURE 11J
Simple harmonic motion along the y axis, showing the circle of reference, the initial phase angle α, the angular speed ω, and the point P_0 at time $t = 0$.

The angle α is a constant and is called the *initial phase angle*. As the point p moves around the circle, the angle ωt increases at a uniform rate and is always measured from the starting angle α. The total quantity in parentheses is the total angle measured from the $+x$ axis.

It is customary to express all angles in *radian measure* rather than in degrees.

EXAMPLE 2 A given point is vibrating with SHM with a period of 5.0 s and an amplitude of 3.0 cm. If the initial phase angle is $\pi/3$ rad, (60°), find (*a*) the initial displacement and (*b*) the displacement after 12.0 s. (*c*) Make a graph.

SOLUTION (*a*) Since the graph point makes one revolution in 5.0 s, the angular speed ω is 2π rad in 5.0 s, or $2\pi/5$ rad/s [see Eq. (11f)]. At the time $t = 0$, direct substitution in Eq. (11x) gives

$$y = 3 \sin\left(\frac{2\pi}{5} 0 + \frac{\pi}{3}\right)$$

(*b*) After 12.0 s, substitution in Eq. (11x) gives

$$y = 3 \sin\left(\frac{2\pi}{5} 12 + \frac{\pi}{3}\right)$$

$$= 3 \sin\left(4.8\,\pi + \frac{\pi}{3}\right)$$

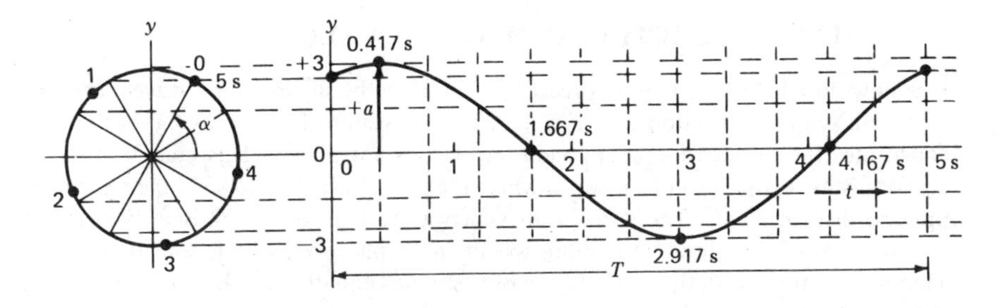

FIGURE 11K
Graph for the example $T = 5.0$ s, $a = 3.0$ cm, and $\alpha = \pi/3$ rad.

The total phase angle of $4.8\pi + \pi/3$ is equivalent to $864° + 60°$, or 924°, and measured from the $+x$ axis places the graph point 24° below the $-x$ axis on the circle of reference. This angle gives

$$\sin 24° = 0.407$$

and

$$y = 3\,(-0.407)$$

or

$$y = -1.220\,\text{cm}$$

A plot of this example is shown in Fig. 11K. The time T is plotted horizontally, and the displacement is plotted vertically for the first complete vibration, or 5.0 s. The up-and-down motion is traced out to show the starting point and initial phase angle and the time when the motion reaches its first maximum and minimum displacement and when the displacement is zero. The amplitude $a = 3.0$ cm is seen near the left side and is equal to the radius of the circle of reference.

A useful and concise way of expressing the equation for simple harmonic waves is in terms of the *angular frequency* $\omega = 2\pi\nu$ and the *propagation number* $k = 2\pi/\lambda$. Equation (11u) then becomes

$$y = a \sin (kx - \omega t) = a \sin (\omega t - kx + \pi)$$

$$= a \cos \left(\omega t - kx + \frac{\pi}{2} \right)$$

The addition of a constant to the quantity in parentheses is of little physical significance, since such a constant can be eliminated by suitably adjusting the zero of the time scale. Thus the equations when written

$$y = a \cos (\omega t - kx) \qquad \text{and} \qquad y = a \sin (\omega t - kx) \tag{11y}$$

will describe the wave of Fig. 11I, if the curve applies at times $t = T/4$ and $T/2$, respectively, instead of at $t = 0$.

11.8 PHASE VELOCITY AND WAVE VELOCITY

It is now possible to state more precisely what actually moves with a wave. The discussion given in connection with Fig. 11K may be summed up by saying that a wave constitutes the progression of a condition of constant phase. This condition might be, for instance, the crest of the wave, where the phase is such as to yield the maximum upward displacement. The speed with which a crest moves along is usually called the wave velocity, although the more specific term phase velocity is sometimes used. That it is identical with the quantity v in our previous equations is shown by evaluating the rate of change of the x coordinate under the condition that the phase remain constant. When the form of the phase in Eq. (11y) is used, the latter requirement becomes

$$\omega t - kx = \text{const}$$

and the wave velocity

$$v = \frac{dx}{dt} = \frac{\omega}{k} \tag{11z}$$

Substitution of $\omega = 2\pi\nu$ and $k = 2\pi/\lambda$ gives agreement with Eq. (11q). For a wave traveling toward $-x$, the constant phase takes the form $\omega t + kx$, and the corresponding $v = -\omega/k$.

The ratio ω/k for a given kind of wave depends on the physical properties of the medium in which the waves travel and also, in general, on the frequency ω itself. For transverse elastic waves involving distortions small enough for the forces to obey Hooke's law, the wave velocity is independent of frequency and is given simply as

$$v = \sqrt{\frac{N}{\rho}} \tag{11za}$$

N being the shear modulus and ρ the density. The proof of this relation is not difficult. From Fig. 11L it will be seen that the sheet of small thickness δx is sheared through the angle α. The shear modulus is the constant ratio of stress to strain. The strain is measured by tan α, so that

$$\text{Strain} = \frac{\delta f}{\delta x}$$

where f is the function giving the shape of the wave at a particular instant. The stress is the tangential force F per unit area acting on the surface of the sheet, and this by Hooke's law must equal the product of the shear modulus and the strain, so that

$$\text{Stress} = F_x = N\frac{\delta f}{\delta x}$$

Because of the curvature of the wave, the stress will vary with x, and the force acting on the left side of the sheet will not be exactly balanced by the force acting on its right side. The net force per unit area is

$$F_x - F_{x+\delta x} = \frac{\partial F}{\partial x}\,\delta x = N\frac{\partial^2 f}{\partial x^2}\,\delta x$$

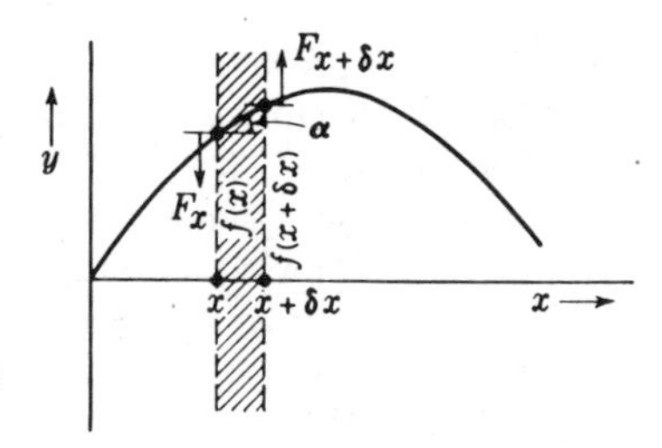

FIGURE 11L
The geometry and mechanics for the shear involved in a transverse wave.

We now apply Newton's second law of motion, equating this force to the product of the mass and the acceleration of unit area of the sheet:

$$N \frac{\partial^2 f}{\partial x^2} \delta x = \rho \, \delta x \frac{\partial^2 f}{\partial t^2}$$

From the fact that they can be polarized (Chap. 24), light waves are known to be transverse waves. Measurements show that their velocity in a vacuum is approximately 3×10^{10} cm/s. If one assumes them to be elastic waves, as was commonly done in the nineteenth century, the question arises: What medium transmits them? In the early *elastic-solid theory*, a medium called the "ether", having the property of a high ratio of rigidity to density, was assumed to occupy all space. Its density was supposed to increase in material substances to account for the lower velocity. There are obvious objections to such a hypothesis. For example, in spite of its resistance to shear, which had to be postulated because light waves are transverse, the ether produces no detectable effects on the motions of astronomical bodies. All the difficulties disappeared when Maxwell developed the present *electromagnetic theory* of light (Chap. 20). Here the mechanical displacement of an element of the medium is replaced by a variation of the electric field (or more generally of the dielectric displacement) at the corresponding point.

The elastic-solid theory was successful in explaining a number of properties of light. There are many parallelisms in the two theories, and much of the mathematics of the earlier theory can be rewritten in electromagnetic terms without difficulty. Consequently, we shall frequently find mechanical analogies useful in understanding the behavior of light. In fact, for the material presented in the next seven chapters, it is immaterial what type of waves are assumed.

11.9 AMPLITUDE AND INTENSITY

Waves transport energy, and the amount of it that flows per second across unit area perpendicular to the direction of travel is called the *intensity* of the wave. If the wave flows continuously with the velocity v, there is a definite energy density, or total energy per unit volume. All the energy contained in a column of the medium of unit cross section and of length v will pass through the unit of area in 1 s. Thus the intensity is given by the product of v and the energy density. Either the energy density or the intensity is proportional to the square of the amplitude and to the square of the frequency. To prove this proposition for sine waves in an elastic medium, it is necessary only to determine the vibrational energy of a single particle executing simple harmonic motion.

Consider for example the particle P in Fig. 11I. At the time for which the figure is drawn, it is moving upward and possesses both kinetic and potential energy. A little later it will have the position P'. Here it is instantaneously at rest, with zero kinetic energy and the maximum potential energy. As it subsequently moves downward, it gains kinetic energy, while the potential energy decreases in such a way that the total energy stays constant. When it reaches the center, at P'', the energy is all kinetic. Hence we may find the total energy either from the maximum potential energy at P' or from the maximum kinetic energy at P''. The latter procedure gives the desired result most easily.

According to Eq. (11y), the displacement of a particular particle varies with time according to the relation

$$y = a \sin(\omega t - \alpha)$$

where α is the value of kx for that particle. The velocity of the particle is

$$\frac{dy}{dt} = \omega a \cos(\omega t - \alpha)$$

When $y = 0$, the sine vanishes and the cosine has its maximum value. Then the velocity becomes $-\omega a$, and the maximum kinetic energy

$$\tfrac{1}{2}m\left[\frac{dy}{dt}\right]^2_{\max} = \tfrac{1}{2}m\omega^2 a^2$$

Since this is also the total energy of the particle and is proportional to the energy per unit volume, it follows that

$$\text{Energy density} \approx \omega^2 a^2 \tag{11zb}$$

The intensity, v times this quantity, will then also be proportional to ω^2 and a^2.

In spherical waves, the intensity decreases as the inverse square of the distance from the source. This law follows directly from the fact that, provided there is no conversion of the energy into other forms, the same amount must pass through any sphere with the source as its center. Since the area of a sphere increases as the square of its radius, the energy per unit area at a distance r from the source, or the intensity, will vary as $1/r^2$. The amplitude must then vary as $1/r$, and one may write the equation of a spherical wave as

$$y = \frac{a}{r}\sin(\omega t - kr) \tag{11zc}$$

Here a means the amplitude at unit distance from the source.

If any of the energy is transformed to heat, i.e., if there is *absorption*, the amplitude and intensity of plane waves will not be constant but will decrease as the wave passes through the medium. Similarly with spherical waves, the loss of intensity will be more rapid than is required by the inverse-square law. For plane waves, the fraction dI/I of the intensity lost in traversing an infinitesimal thickness dx is proportional to dx, so that

$$\frac{dI}{I} = -\alpha \, dx$$

To obtain the decrease in traversing a finite thickness x, the equation is integrated to give

$$\int_0^x \frac{dI}{I} = -\alpha \int_0^x dx$$

Evaluating these definite integrals, we find

● $$I_x = I_0 e^{-\alpha x} \qquad \text{(11zd)}$$

This law, which has been attributed to both Bouguer* and Lambert,† we shall refer to as the *exponential law of absorption.* Figure 11M is a plot of the intensity against thickness according to this law for a medium having $\alpha = 0.4$ per centimeter. The wave equations can be modified to take account of absorption by multiplying the amplitude by the factor $e^{-\alpha x/2}$, since the amplitude varies with the square root of the intensity.

For light, the intensity can be expressed in joules per square meter per second. Full sunlight, for example, has an intensity in these units of about 1.4×10^3. Here it is important to realize that not all this energy flux affects the eye, and not all that does is equally efficient. Hence the intensity as defined above does not necessarily correspond to the sensation of brightness, and it is more usual to find light flux expressed in visual units.‡ The intensity and the amplitude are the purely physical quantities, however, and according to modern theory the latter must be expressed in electrical units. Thus it may be shown that according to equations to be derived in Chap. 20 the amplitude in a beam of sunlight having the above-mentioned value of the intensity represents an electric field strength of 7.3 V/cm and an accompanying magnetic field of 2.4×10^{-7} tesla (T).

The amplitude of light always decreases more or less rapidly with the distance traversed. Only for plane waves traveling in vacuum, such as the light from a star coming through outer space, is it nearly constant. The inverse-square law of intensities may be assumed to hold for a small source in air at distances greater than about 10 times the lateral dimension of the source. Then the finite size of the source causes an error of less than 0.1 percent in computing the intensity, and for laboratory distances the absorption of air may be neglected. In greater thicknesses, however, all "transparent" substances absorb an appreciable fraction of the energy. We shall take up this subject again in some detail in Chap. 22.

* Pierre Bouguer (1698–1758). Royal Professor of Hydrography at Le Havre.

† Johann Lambert (1728–1777). German physicist, astronomer, and mathematician. Worked primarily in the field of heat radiation. Another law, which is always called Lambert's law, refers to the variation with angle of the radiation from a surface.

‡ See, for example, F. W. Sears, "Principles of Physics," vol. 3, "Optics," 3d ed., chap. 13, Addison-Wesley Publishing Company, Inc., Reading, Mass., 1948.

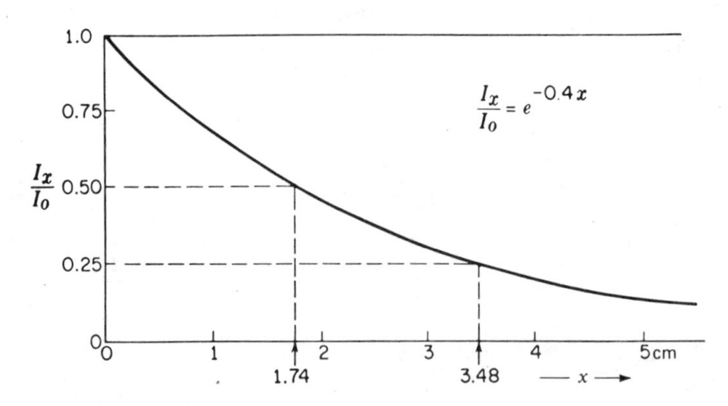

FIGURE 11M
Logarithmic decrease of intensity in an absorbing medium.

11.10 FREQUENCY AND WAVELENGTH

Any wave motion is generated by some sort of vibrating source, and the frequency of the waves is equal to that of the source. The wavelength in a given medium is then determined by the velocity in that medium and by Eq. (11t) is obtained by dividing the velocity by the frequency. Passage from one medium to another causes a change in the wavelength in the same proportion as it does in the velocity, since the frequency is not altered. If we remember that a wave front represents a surface on which the phase of motion is constant, it should be clear that, regardless of any changes of velocity, two different wave fronts are separated by a certain number of waves. That is, the length of any ray between two such surfaces is the same, provided this length is expressed in wavelengths in the appropriate media.

As applied to light, the last statement is equivalent to saying that the optical path is the same along all rays drawn between two wave fronts. For since wavelengths are proportional to velocities, we have

$$\frac{\lambda}{\lambda_m} = \frac{c}{v} = n$$

when the light passes from a vacuum, where it has wavelength λ and velocity c, into a medium where the corresponding quantities are λ_m and v. The optical path corresponding to a distance d in any medium is therefore

$$nd = \frac{\lambda}{\lambda_m} d$$

or the number of wavelengths in that distance multiplied by the wavelength in vacuum. It is customary in optics and spectroscopy to refer to the wavelength of a particular radiation, of a single spectral line, for example, as its wavelength in air under normal conditions. This we shall designate by λ (without subscript), and except in rare circumstances it may be taken as the same as the wavelength in vacuum.

The wavelengths of visible light extend between about 4×10^{-7} m or 400 nm for the extreme violet and 7.2×10^{-7} m or 700 nm for the deep red. Just as the ear becomes insensitive to sound above a certain frequency, so the eye fails to respond to light vibrations of frequencies greater than that of the extreme violet or less than that of the extreme red. The limits, of course, depend somewhat upon the individual, and there is evidence that most persons can see an image with light of wavelength as short as 300 nm, but this is a case of fluorescence in the retina. In this case the light appears to be bluish gray in color and is harmful to the eye. Radiation of wavelength shorter than that of the visible is termed *ultraviolet light* down to a wavelength of about 5 nm and beyond this we are in the region of X rays to 6×10^{-1} nm. Shorter than these, in turn, are the γ rays from radioactive substances. On the long-wavelength side of the visible lies the infrared, which may be said to merge into the radio waves at about 1×10^{6} nm. Figure 11N shows the names which have been given to the various regions of the spectrum of radiation, though we know that no real lines of demarcation exist. It is convenient to use the same units of length throughout such an enormous range. Hence wavelengths are now generally expressed in nanometers (nm) or angstroms (Å) (see Appendix VI).*

It will be seen that visible light covers an almost insignificant fraction of this range. Therefore, although all these radiations are similar in nature, differing only in wavelength, the term *light* is conventionally extended only to the adjacent portions of the spectrum, namely, the ultraviolet and infrared. Many of the results that we shall discuss for light are common to the whole range of radiation, but naturally there are qualitative differences in behavior between the very long and very short waves, which we shall occasionally point out. The divisions between the different types of radiation are purely formal and are roughly fixed by the fact that in the laboratory the different types are generated and detected in different ways. Thus the infrared is emitted copiously by hot bodies and is detected by an energy-measuring instrument such as the thermopile. The shortest radio waves are generated by electric discharges between fine metallic particles immersed in oil and are detected by electrical devices. Nichols and Tear, in 1917, produced infrared waves having wavelengths up to 4.2×10^{5} nm and radio waves down to 2.2×10^{5} nm. The two regions may therefore be said to overlap, keeping in mind, however, that the waves themselves are of the same nature for both. The same holds true for the boundaries of all the other regions of the spectrum.

In sound and other mechanical waves, a change of wavelength occurs when the source has a translational motion. The waves sent out in the direction of motion are shortened, and, in the opposite direction, lengthened. No change is produced in the velocity of the waves themselves; so a stationary observer receives a frequency which is larger or smaller than that of the source. If, on the other hand, the source is at rest and the observer in motion, a change of frequency is also observed, but for a different reason. Here there is no change of wavelength, but the frequency is altered by the change in relative velocity of the waves with respect to the observer. The two cases involve approximately the same change of frequency for the same speed of motion,

* A. J. Ångstrom (1814–1874). Professor of physics at Uppsala, Sweden. Chiefly known for his famous atlas of the solar spectrum, which was used for many years as the standard for wavelength determinations.

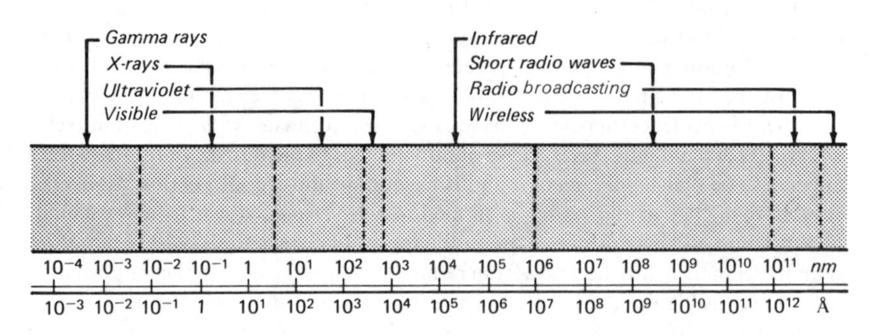

FIGURE 11N
Scale of wavelengths for the range of known electromagnetic waves.

provided this is small compared with the velocity of the waves. These phenomena are known as the *doppler effect** and are most commonly experienced in sound as changes in the acoustic pitch.

Doppler mistakenly attributed the different colors of stars to their motions toward or away from the earth. Because the velocity of light is so large, an appreciable change in color would require that a star have a component of velocity in the line of sight impossibly large compared with the measured velocities at right angles to it. For most stars, the latter usually range between 10 and 30 km/s, with a few as high as 300 km/s. Since light travels at nearly 300,000 km/s, the expected shifts of frequency are small. Furthermore, it makes little difference whether one assumes that the observer or the source is in motion. Suppose that the earth were moving with a velocity u directly toward a fixed star. An observer would then receive u/λ waves in addition to the number $\nu = c/\lambda$ that would reach him if he were at rest. The apparent frequency would be

$$\nu' = \frac{c + u}{\lambda} = \nu\left(1 + \frac{u}{c}\right) \qquad \text{(11ze)}$$

With the velocities mentioned, this would differ from the true frequency by less than 1 part in 1000. A good spectroscope can, however, easily detect and permit the measurement of such a shift as a displacement of the spectrum lines. In fact, this legitimate application of Doppler's principle has become a powerful method of studying the radial velocities of stars. Figure 11O shows an example where the spectrum of μ Cassiopeiae, in the center strip, is compared with the lines of iron from a laboratory source, photographed above and below. All the iron lines also appear in the stellar spectrum as white lines (absorption lines) but are shifted toward the left, i.e., toward

* Christian Johann Doppler (1803–1853). Native of Salzburg, Austria. At the age of thirty-two, unable to secure a position, he was about to emigrate to America. However, at that time he was made professor of mathematics at the Realschule in Prague and later became professor of experimental physics at the University of Vienna.

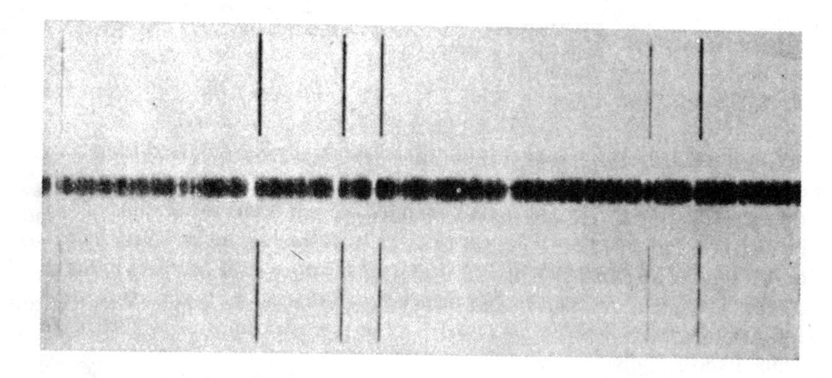

FIGURE 11O
Doppler shift of spectrum lines in a star. Both spectra are negatives. (*Courtesy of McKellar.*)

shorter wavelengths. Measurement shows that the increase of frequency corresponds to a velocity of approach of 115 km/s, which is unusually high for stars in our own galaxy. The spectra of other galaxies (spiral nebulae) all show displacements toward the red, which for the most distant ones amount to several hundred angstroms. Such values would indicate recessional velocities of tens of thousands of kilometers per second, and have been so interpreted. It is rather interesting that here there is enough reddening to change the color of the object, as postulated by Doppler, but in this case it occurs for objects far too faint to be seen by the naked eye.

In the laboratory, there have been found two ways of achieving velocities sufficient to produce detectable doppler shifts. By reflecting light from mirrors mounted on the rim of a wheel rotating at high speed, one can produce speeds of a virtual source as high as 400 m/s. Much larger values are attained by beams of atoms moving in vacuum, as will be discussed later in Sec. 19.15. There, it is also shown that with the abandonment of the material ether necessitated by relativity theory the distinction between the cases of source in motion and of observer in motion disappears. Relativity leads to an equation which is substantially Eq. (11ze) with u representing the relative velocity of approach or recession.

11.11 WAVE PACKETS

No source of waves vibrates indefinitely, as would be required for it to produce a true sine wave. More commonly the vibrations die out because of the dissipation of energy or are interrupted in some way. Then a group of waves of finite length, such as that illustrated in Fig. 11P, is produced. The mathematical representation of a wave packet of this type is rather more complex and will be briefly discussed in the next chapter. Since wave packets are of frequent occurrence, however, some features of their behavior should be mentioned here. In the first place, the wavelength is

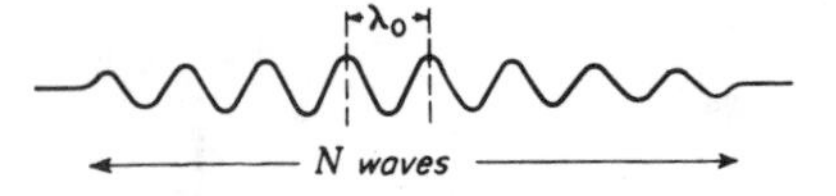

FIGURE 11P
Example of a wave packet.

not well defined. If the packet is sent through any device for measuring wavelengths, e.g., light through a diffraction grating, it will be found to yield a continuous spread over a certain range $\Delta\lambda$. The maximum intensity will occur at the value of λ_0 indicated in Fig. 11P, but energy will appear in other wavelengths, the intensity dying off more or less rapidly on either side of λ_0. The larger the number N of waves in the group, the smaller the spread $\Delta\lambda$, and in fact theory shows that $\Delta\lambda/\lambda_0$ is approximately equal to $1/N$. Hence only when N is very large may we consider the wave to have an accurately defined wavelength.

If the medium through which the packet travels is such that the velocity depends on frequency, two further phenomena will be observed. The individual wave crests will travel with a velocity different from that of the packet as a whole, and the packet will spread out as it progresses. We then have two velocities, the wave (or phase) velocity and the group velocity. The relation between these will be derived in Sec. 12.7.

In light sources, the radiating atoms emit wave trains of finite length. Usually, because of collisions or damping arising from other causes, these packets are very short. According to the theorem mentioned above, the consequence is that the spectrum lines will not be very narrow but will have an appreciable width $\Delta\lambda$. A measurement of this width will yield the effective "lifetime" of the electromagnetic oscillators in the atoms and the average length of the wave packets. A low-pressure discharge through the vapor of mercury containing the single isotope ^{198}Hg yields very sharp spectral lines, of width about 0.005 Å. Taking the wavelength of one of the brightest lines, 5461 Å, we may estimate that there are roughly 10^6 waves in a packet and that the packets themselves are some 50 cm long.

PROBLEMS

11.1 A coil spring hangs from the ceiling as shown in Fig. 11F. When a mass of 50.0 g is fastened to the lower end, the spring is stretched a distance of 15.89 cm. If the mass is now pulled down another 5 cm and released, it will vibrate up and down with SHM. Find (*a*) the spring constant, (*b*) the period of vibration, (*c*) the frequency, (*d*) the angular velocity of a graph point drawn for the vibration, (*e*) the maximum velocity of the mass, and (*f*) its maximum acceleration. (*g*) Plot a graph of the vibration for the time interval $t = 0$ to $t = 3.0$ s if the initial phase angle is 270°. (*h*) Find the time to reach the first maximum and (*i*) the total energy of vibration. (*j*) Write down an equation for the motion.

Ans. (*a*) 30.837 N/m, (*b*) 0.8001 s, (*c*) 1.2499 Hz, (*d*) 5.027 rad/s, (*e*) 0.39265 m/s, (*f*) 0.4754 m/s^2, (*g*) see Fig. P11.1, (*h*) 4.001 s, (*i*) 3.8546 J, (*j*) $y = 0.050 \sin (5.027t + 270°)$ m

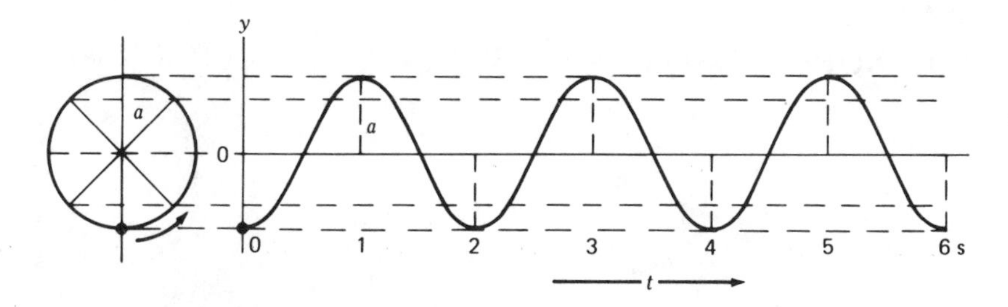

FIGURE P11.1
Graph for part (g) of Prob. 11.1

11.2 A coil spring hangs from the ceiling as shown in Fig. 11F. A mass of 1.60 kg is suspended from the lower end of the spring, stretching it a distance of 12.40 cm. The mass is now pulled down a distance of 4.0 cm more and then released to vibrate in a vertical line. Find (a) the spring constant, (b) the period of vibration, (c) the frequency, (d) the angular velocity of a graph point, (e) the maximum velocity of the mass, and (f) the maximum acceleration. (g) Plot a graph of the vibration for the time interval $t = 0$ to $t = 2.20$ s if the initial phase angle is 225°. (h) Find the time at which the mass first reaches its highest point and (i) the total energy. (j) Write down an equation for the motion.

11.3 A wave is specified by $y = 6 \sin 2\pi(8t - 4x + \frac{3}{4})$. Find ($a$) the amplitude, ($b$) wavelength, ($c$) frequency, ($d$) initial phase angle, and (e) the initial displacement at time $t = 0$ and $x = 0$.

11.4 A wave is specified by $y = 15 \sin 2\pi(4t - 5x + \frac{2}{3})$. Find ($a$) the amplitude, ($b$) the wavelength, ($c$) the frequency, ($d$) the initial phase angle, and (e) the displacement at time $t = 0$ and $x = 0$. *Ans.* (a) 15, (b) $\frac{1}{5}$, (c) 4, (d) 240°, (e) -13.0

12
THE SUPERPOSITION OF WAVES

When two sets of waves are made to cross each other, e.g., the waves created by dropping two stones simultaneously in a quiet pool, interesting and complicated effects are observed. In the region of crossing there are places where the disturbance is practically zero and others where it is greater than that given by either wave alone. A very simple law can be used to explain these effects, which states that the resultant displacement of any point is merely the sum of the displacements due to each wave separately. This is known as the *principle of superposition* and was first clearly stated by Young* in 1802. The truth of this principle is at once evident when we observe that after the waves have passed out of the region of crossing, they appear to have been entirely uninfluenced by the other set of waves. Amplitude, frequency, and all other characteristics are just as if they had crossed an undisturbed space. This could hold only provided the principle of superposition were true. Two different observers can

* Thomas Young (1773–1829). English physician and physicist, usually called the founder of the wave theory of light. An extremely precocious child (he had read the Bible twice through at the age of four), he developed into a brilliant investigator. His work on interference constituted the most important contribution on light since Newton. His early work proved the wave nature of light but was not taken seriously by others until it was corroborated by Fresnel.

see different objects through the same aperture with perfect clearness, whereas the light reaching the two observers had crossed in going through the aperture. The principle is therefore applicable with great precision to light, and we can use it in investigating the disturbance in regions where two or more light waves are superimposed.

12.1 ADDITION OF SIMPLE HARMONIC MOTIONS ALONG THE SAME LINE

Considering first the effect of superimposing two sine waves of the same frequency, the problem resolves itself into finding the resultant motion when a particle executes two simple harmonic motions at the same time. The displacements due to the two waves are here taken to be along the same line, which we shall call the y direction. If the amplitudes of the two waves are a_1 and a_2, these will be the amplitudes of the two periodic motions impressed on the particle, and, according to Eq. (11x) of the last chapter, we can write the separate displacements as follows:

$$\begin{aligned} y_1 &= a_1 \sin(\omega t - \alpha_1) \\ y_2 &= a_2 \sin(\omega t - \alpha_2) \end{aligned} \tag{12a}$$

Note that ω is the same for both waves, since we have assumed them to be of the same frequency. According to the principle of superposition, the resultant displacement y is merely the sum of y_1 and y_2, and we have

$$y = a_1 \sin(\omega t - \alpha_1) + a_2 \sin(\omega t - \alpha_2)$$

When the expression for the sine of the difference of two angles is used, this can be written

$$\begin{aligned} y &= a_1 \sin\omega t \cos\alpha_1 - a_1 \cos\omega t \sin\alpha_1 + a_2 \sin\omega t \cos\alpha_2 - a_2 \cos\omega t \sin\alpha_2 \\ &= (a_1 \cos\alpha_1 + a_2 \cos\alpha_2) \sin\omega t - (a_1 \sin\alpha_1 + a_2 \sin\alpha_2) \cos\omega t \end{aligned} \tag{12b}$$

Now since the a's and α's are constants, we are justified in setting

$$\begin{aligned} a_1 \cos\alpha_1 + a_2 \cos\alpha_2 &= A \cos\theta \\ a_1 \sin\alpha_1 + a_2 \sin\alpha_2 &= A \sin\theta \end{aligned} \tag{12c}$$

provided that constant values of A and θ can be found which satisfy these equations. Squaring and adding Eqs. (12c), we have

$$\begin{aligned} A^2(\cos^2\theta + \sin^2\theta) &= a_1{}^2(\cos^2\alpha_1 + \sin^2\alpha_1) + a_2{}^2(\cos^2\alpha_2 + \sin^2\alpha_2) \\ &\quad + 2a_1a_2(\cos\alpha_1 \cos\alpha_2 + \sin\alpha_1 \sin\alpha_2) \end{aligned}$$

or

$$A^2 = a_1{}^2 + a_2{}^2 + 2a_1a_2 \cos(\alpha_1 - \alpha_2) \tag{12d}$$

Dividing the lower equation (12c) by the upper one, we obtain

$$\tan\theta = \frac{a_1 \sin\alpha_1 + a_2 \sin\alpha_2}{a_1 \cos\alpha_1 + a_2 \cos\alpha_2} \tag{12e}$$

Equation (12d) and (12e) show that values of A and θ exist which satisfy Eqs. (12c), and we can rewrite Eq. (12b), substituting the right-hand members of Eq. (12c). This gives

$$y = A \cos \theta \sin \omega t - A \sin \theta \cos \omega t$$

which has the form of the sine of the difference of two angles and can be expressed as

$$y = A \sin (\omega t - \theta) \qquad (12f)$$

This equation is the same as either of our original equations for the separate simple harmonic motions but contains a new amplitude A and a new phase constant θ. Hence we have the important result that the sum of two simple harmonic motions of the same frequency and along the same line is also a simple harmonic motion of the same frequency. The amplitude and phase constant of the resultant motion can easily be calculated from those of the component motions by Eqs. (12d) and (12e), respectively.

The addition of three or more simple harmonic motions of the same frequency will likewise give rise to a resultant motion of the same type, since the motions can be added successively, each time giving an equation of the form of Eq. (12f). Unless considerable accuracy is desired, it is usually more convenient to use the graphical method described in the following section. A knowledge of the resultant phase constant θ, given by Eq. (12e), is not of interest unless it is needed in combining the resultant motion with still another.

The resultant amplitude A depends, according to Eq. (12d), upon the amplitudes a_1 and a_2 of the component motions and upon their difference of phase $\delta = \alpha_1 - \alpha_2$. When we bring together two beams of light, as is done in the Michelson interferometer (Sec. 13.8), the intensity of the light at any point will be proportional to the square of the resultant amplitude. By Eq. (12d) we have, in the case where $a_1 = a_2$,

$$I \approx A^2 = 2a^2(1 + \cos \delta) = 4a^2 \cos^2 \frac{\delta}{2} \qquad (12g)$$

If the phase difference is such that $\delta = 0, 2\pi, 4\pi, \ldots$, this gives $4a^2$, or 4 times the intensity of either beam. If $\delta = \pi, 3\pi, 5\pi, \ldots$, the intensity is zero. For intermediate values, the intensity varies between these limits according to the square of the cosine. These modifications of intensity obtained by combining waves are referred to as *interference* effects, and we shall discuss in the next chapter several ways in which they can be brought about and used experimentally.

12.2 VECTOR ADDITION OF AMPLITUDES

A very simple geometrical construction can be used to find the resultant amplitude and phase constant of the combined motion in the above case of two simple harmonic motions along the same line. If we represent the amplitudes a_1 and a_2 by vectors making angles α_1 and α_2 with the x axis,* as in Fig. 12A(a), the resultant amplitude A

* Here we depart from the usual convention of measuring positive angles in the counterclockwise direction, because it is customary in optics to represent an advance of phase by a clockwise rotation of the amplitude vector.

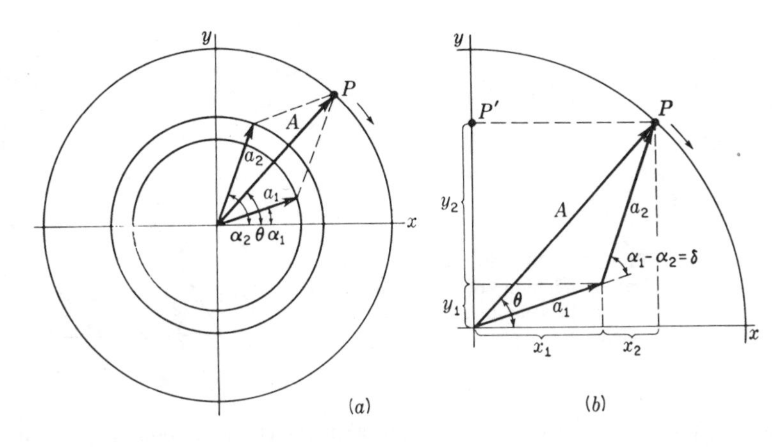

FIGURE 12A
Graphical composition of two waves of the same frequency, but different amplitude and phase.

is the vector sum of a_1 and a_2 and makes an angle θ with that axis. To prove this proposition, we first note from Fig. 12A(b) that in the triangle formed by a_1, a_2, and A the law of cosines gives

$$A^2 = a_1{}^2 + a_2{}^2 - 2a_1a_2 \cos\left[\pi - (\alpha_1 - \alpha_2)\right] \qquad \text{(12h)}$$

which readily reduces to Eq. (12d). Furthermore, Eq. (12e) is obtained at once from the fact that the tangent of the angle θ is the quotient of the sum of the projections of a_1 and a_2 on the y axis by the sum of their projections on the x axis.

That the resultant motion is also simple harmonic can be concluded if we remember that this type of motion may be represented as the projection on one of the coordinate axes of a point moving with uniform circular motion. Figure 12A is drawn for the time $t = 0$, and as time progresses, the displacements y_1 and y_2 will be given by the vertical components of the vectors a_1 and a_2, if the latter revolve clockwise with the same angular velocity ω. The resultant, A, will then have the same angular velocity, and the projection P' of its terminus P will undergo the resultant motion. If one imagines the vector triangle in part (*b*) of the figure to revolve as a rigid frame, it will be seen that the motion of P' will agree with Eq. (12f).

The graphical method is particularly useful where we have more than two motions to compound. Figure 12B shows the result of adding five motions of equal amplitudes a and having equal phase differences δ. Clearly the intensity $I = A^2$ can here vary between zero and $25a^2$, according to the phase difference δ. This is the problem which arises in finding the intensity pattern from a diffraction grating, as discussed in Chap. 17. The five equal amplitudes shown in the figure might be contributed by five apertures of a grating, an instrument which has as its primary purpose the introduction of an equal phase difference in the light from each successive pair of apertures. It will be noted that as Fig. 12B is drawn, the vibrations, starting with that at the origin, lag successively farther *behind* in phase.

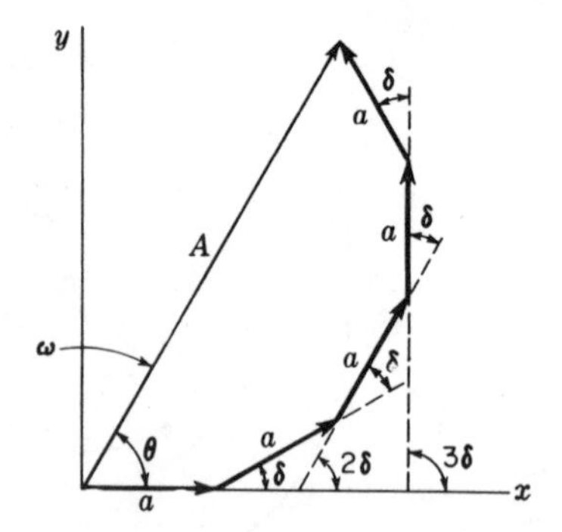

FIGURE 12B
Vector addition of five amplitudes having the same magnitude and phase difference δ.

Either the trigonometric or graphical methods for compounding vibrations may be used to find the resultant of any number of motions with given amplitudes and phases. It is even possible, as we shall see, to apply these methods to the addition of infinitesimal vibrations, so that the summations become integrations. In such cases, and especially if the amplitudes of the individual contributions vary, it is simpler to use a method of adding the amplitudes as complex numbers. We shall take up this method in Sec. 14.8, where it first becomes necessary.

12.3 SUPERPOSITION OF TWO WAVE TRAINS OF THE SAME FREQUENCY

From the preceding section we can conclude directly that the result of superimposing two trains of sine waves of the same frequency and traveling along the same line is to produce another sine wave of that frequency but having a new amplitude which is determined for given values of a_1 and a_2 by the phase difference δ between the motions imparted to any particle by the two waves. As an example, let us find the resultant wave produced by two waves of equal frequency and amplitude traveling in the same direction $+x$, but with one a distance Δ ahead of the other. The equations of the two waves, in the form of Eq. (11y), will be

$$y_1 = a \sin (\omega t - kx) \tag{12i}$$

$$y_2 = a \sin [\omega t - k(x + \Delta)] \tag{12j}$$

By the principle of superposition, the resultant displacement is the sum of the separate ones, so that

$$y = y_1 + y_2 = a\{\sin (\omega t - kx) + \sin [\omega t - k(x + \Delta)]\}$$

Applying the trigonometric formula

$$\sin A + \sin B = 2 \sin \tfrac{1}{2}(A + B) \cos \tfrac{1}{2}(A - B) \tag{12k}$$

we find

$$y = 2a \cos \frac{k\Delta}{2} \sin \left[\omega t - k\left(x + \frac{\Delta}{2}\right)\right] \tag{12l}$$

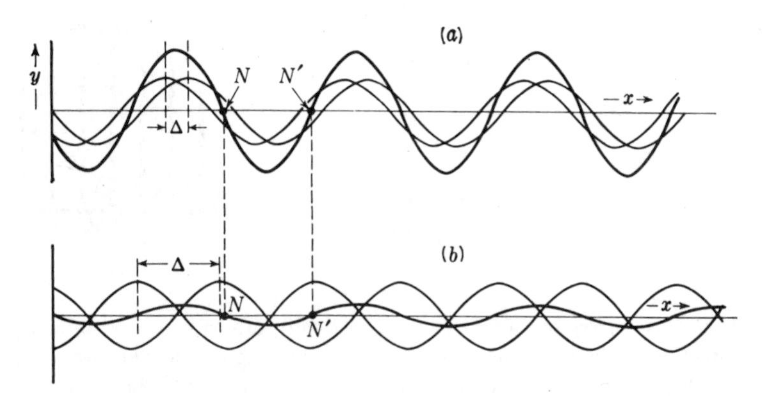

FIGURE 12C
Superposition of two wave trains (*a*) almost in phase and (*b*) almost 180° out of phase.

This corresponds to a new wave of the same frequency but with the amplitude $2a$ cos $(k\Delta/2) = 2a \cos(\pi\Delta/\lambda)$. When Δ is a small fraction of a wavelength, this amplitude will be nearly $2a$, while if Δ is in the neighborhood of $\frac{1}{2}\lambda$, it will be practically zero. These cases are illustrated in Fig. 12C, where the waves represented by Eqs. (12i) and (12j) (light curves) and (12l) (heavy curve) are plotted at the time $t = 0$. In these figures it will be noted that the algebraic sum of the ordinates of the light curves at any value of x equals the ordinate of the heavy curve. The student may easily verify by such graphical construction the facts that the two amplitudes need not necessarily be equal to obtain a sine wave as the resultant and that the addition of any number of waves of the same frequency and wavelength also gives a similar result. In any case, the resultant wave form will have a constant amplitude, since the component waves and their resultant all move with the same velocity and maintain the same relative position. The true state of affairs may be pictured by having all the waves in Fig. 12C move toward the right with a given velocity.

The formation of *standing waves* in a vibrating cord, giving rise to nodes and loops, is an example of the superposition of two wave trains of the same frequency and amplitude but traveling in *opposite* directions. A wave in a cord is reflected from the end, and the direct and reflected waves must be added to obtain the resultant motion of the cord. Two such waves can be represented by the equations

$$y_1 = a \sin(\omega t - kx) \qquad y_2 = a \sin(\omega t + kx)$$

By addition one obtains, in the same manner as for Eq. (12l),

$$y = 2a \cos(-kx) \sin \omega t$$

which represents the standing waves. For any value of x we have simple harmonic motion, whose amplitude varies with x between the limits $2a$ when $kx = 0, \pi, 2\pi, 3\pi, \ldots$ and zero when $kx = \pi/2, 3\pi/2, 5\pi/2, \ldots$. The latter positions correspond to the nodes and are separated by a distance $\Delta x = \pi/k = \lambda/2$. Figure 12C may also serve to illus-

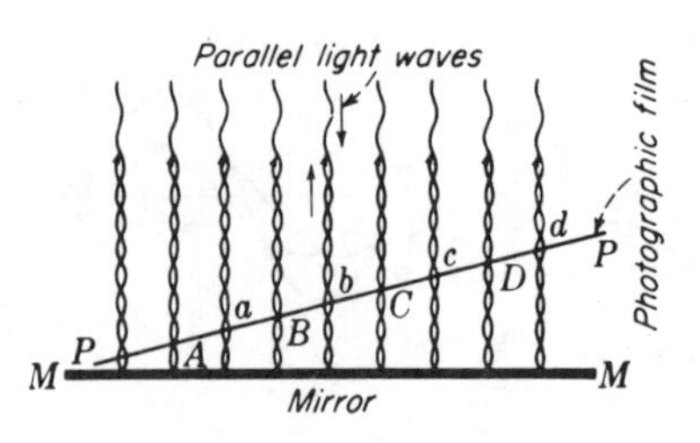

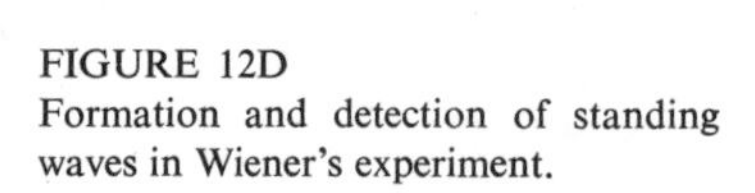
FIGURE 12D
Formation and detection of standing waves in Wiener's experiment.

trate this case if one pictures the two lightly drawn waves as moving in opposite directions. The resultant curve, instead of moving unchanged toward the right, now oscillates between a straight-line position when $\omega t = \pi/2, 3\pi/2, 5\pi/2, \ldots$ and a sine curve of amplitude $2a$ when $\omega t = 0, \pi, 2\pi, \ldots$. At the nodes, such as N and N' in the figure, the resultant displacement is zero at all times.

The standing waves produced by reflecting light at normal incidence from a polished mirror can be observed by means of an experiment due to Wiener,* which is illustrated in Fig. 12D. A specially prepared photographic film only one-thirtieth of a wavelength thick is placed in an inclined position in front of the reflecting surface so that it will cross the nodes and loops successively, as at A, a, B, b, C, c, D, $d, \ldots$. The light will affect the plate only where there is an appreciable amount of vibration and not at all at the nodes. As expected, the developed plate shows a system of dark bands separated by lines of no blackening where it crossed the nodes. Decreasing the angle of inclination of the plate with the reflecting surface causes the bands to move farther apart, since a smaller number of nodal planes are cut in a given distance. Measuring these bands establishes an important fact: the standing waves have a node at the reflecting surface. The phase relations of the direct and reflected waves at this point are therefore such that they continuously annul each other. This is analogous to the reflection of the waves in a rope from a fixed end. Other similar experiments performed by Wiener will be discussed in Sec. 25.12.

12.4 SUPERPOSITION OF MANY WAVES WITH RANDOM PHASES

Suppose that we now consider a large number of wave trains of the same frequency and amplitude to be traveling in the same direction and specify that the amount by which each train is ahead or behind any other is a matter of pure chance. From what has been said above, we can conclude that the resultant wave will be another sine wave of the same frequency, and it becomes of interest to inquire into the amplitude and intensity of this wave. Let the individual amplitudes be a, and let there be n wave trains superimposed. The amplitude of the resultant wave will be the amplitude of motion of a particle undergoing n simple harmonic motions at once, each of amplitude a. If these motions were all in the same phase, the resultant amplitude would be na and

* O. Wiener, *Ann. Phys.*, **40**:203 (1890).

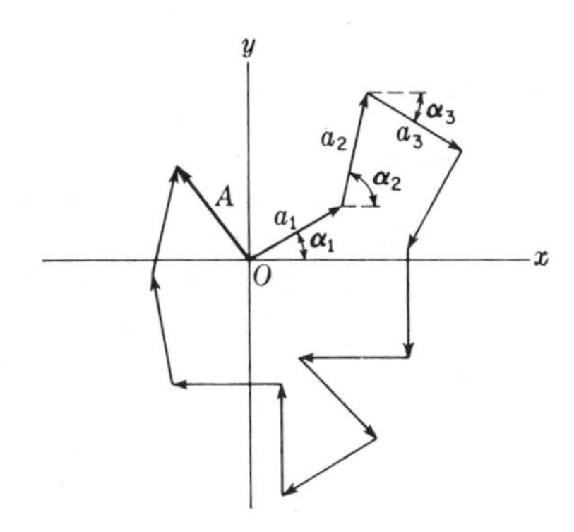

FIGURE 12E
The resultant of 12 amplitude vectors drawn with phases at random.

the intensity n^2a^2, or n^2 times that of one wave. In the case we are considering, however, the phases are distributed purely at random. With the graphical method of compounding amplitudes (Sec. 12.2), one would now obtain a picture like Fig. 12E. The phases $\alpha_1, \alpha_2, \ldots$ take perfectly arbitrary values between 0 and 2π. The intensity due to the superposition of such waves will now be determined by the square of the resultant A. To find A^2, we must square the sum of the projections of all vectors a on the x axis and add the square of the corresponding sum for the y axis. The sum of the x projections is

$$a(\cos\alpha_1 + \cos\alpha_2 + \cos\alpha_3 + \cdots + \cos\alpha_n)$$

When the quantity in parentheses is squared, we obtain terms of the form $\cos^2\alpha_1$ and others of the form $2\cos\alpha_1\cos\alpha_2$. When n is large, one might expect the latter terms to cancel out, because they take both positive and negative values. In any *one* arrangement of the vectors this is far from true, however, and in fact the sum of these cross-product terms actually increases approximately in proportion to their number. Thus we do not obtain a definite result with one given array of randomly distributed waves. In computing the intensity in any physical problem, we are always presented with a large number of such arrays, and we wish to find their average effect. In this case it is safe to conclude that the cross-product terms will average to zero, and we have only the $\cos^2\alpha$ terms to consider. Similarly, for the y projections of the vectors one obtains $\sin^2\alpha$ terms, and the terms like $2\sin\alpha_1\sin\alpha_2$ cancel. Therefore we have

$$I \approx A^2 = a^2(\cos^2\alpha_1 + \cos^2\alpha_2 + \cos^2\alpha_3 + \cdots + \cos^2\alpha_n) + a^2(\sin^2\alpha_1 + \sin^2\alpha_2 + \sin^2\alpha_3 + \cdots + \sin^2\alpha_n)$$

Now since $\sin^2\alpha_k + \cos^2\alpha_k = 1$, we find at once that

● $$I \approx a^2 \times n$$

Thus the average intensity resulting from the superposition of n waves with random phases is just n times that due to a single wave. This means that the amplitude A in Fig. 12E, instead of averaging to zero when a large number of vectors a are repeatedly added in random directions, must actually increase in length as n increases, being proportional to $\sqrt{n}$.

The above considerations can be used to explain why when a large number of violins in an orchestra are playing the same note, interference between the sound waves need not be considered. Owing to the random condition of phases, 100 violins would give about 100 times the intensity due to one alone. The atoms in a sodium flame are emitting light without any systematic relation of phases, and furthermore each is shifting its phase many million times per second. Thus we may safely conclude that the observed intensity is exactly that due to one atom multiplied by the number of atoms. This discussion assumes that *stimulated emission* as found to occur in lasers does not occur here to any large extent. See Chap. 30.

12.5 COMPLEX WAVES

The waves we have considered so far have been of the simple type in which the displacements at any instant are represented by a sine curve. As we have seen, superposition of any number of such waves having the same frequency but arbitrary amplitudes and phases still gives rise to a resultant wave of the same type. However, if only two waves having appreciably different frequencies are superimposed, the resulting wave is complex; i.e., the motion of one particle is no longer simple harmonic, and the wave contour is not a sine curve. The analytical treatment of such waves will be referred to in the following section, and here we shall consider only some of their more qualitative aspects.

It is instructive to examine the results of adding graphically two or more waves traveling along the same line and having various relative frequencies, amplitudes, and phases. The wavelengths are determined by the frequencies according to the relation $\nu\lambda = v$, so that greater frequency means shorter wavelength, and vice versa. Figure 12F illustrates the addition for a number of cases, the resultant curves in each case being obtained, according to the principle of superposition, by merely adding algebraically the displacements due to the individual waves at every point. Figure 12F(*a*) illustrates the case, mentioned in Sec. 12.3, of the addition of two waves of the same frequency but different amplitudes. The resultant amplitude depends on the phase difference, which in the figure is taken as zero. Other phase differences would be represented by shifting one of the component waves laterally with respect to the other and will give a smaller amplitude for the resultant sine wave, its smallest value being the difference in the amplitudes of the components. In (*b*) three waves of different frequencies, amplitudes, and phases are added, giving a complex wave as the resultant, which is evidently very different from a sine curve. In (*c*) and (*d*), where two waves of the same amplitude but frequencies in the ratio 2:1 are added, it is seen that changing the phase difference may produce a resultant of very different form. If these represent sound waves, the eardrum would actually vibrate in a manner represented by the resultant in each case, yet the ear mechanism would respond to two frequencies and these would be heard and interpreted as the two original frequencies regardless of their phase difference. If the resultant wave forms represent visible light, the eye would similarly receive the sensation of a mixture of two colors, which would be the same regardless of the phase difference. Finally (*e*) shows the effect of adding a wave of very high frequency to one of very low frequency and (*f*) the effect

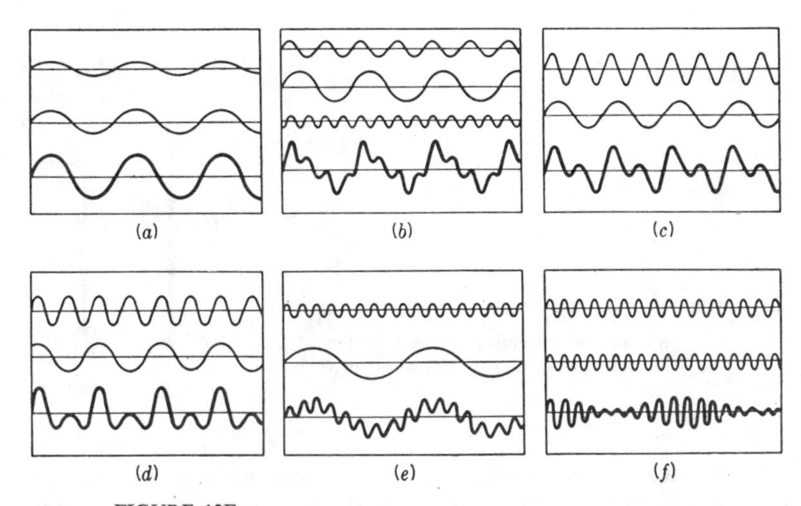

FIGURE 12F
Superposition of two or more waves traveling in the same direction with different relative frequencies, amplitudes, and phases.

of adding two of nearly the same frequency. In the latter case, the resultant wave divides up into groups, which in sound produce the well-known phenomenon of beats. In any of the above cases, if the component waves all travel with the same velocity, the resultant wave form will evidently move with this velocity, keeping its contour unchanged.

Experimental illustrations of the superposition of waves are easily accomplished with the apparatus shown in Fig. 12G. Two small mirrors, M_1 and M_2, are cemented to thin strips of spring steel which are clamped vertically and illuminated by a narrow beam of light. Such a beam is conveniently produced by the concentrated-arc lamp described in Sec. 21.2. An image of this source S is focused on the screen by the lens L. The beam is reflected in succession from the two mirrors, and if one of them is set vibrating, the reflected beam will vibrate up and down with SHM. If now this beam on its way to the screen is reflected from a rotating mirror, the spot of light will trace out a sine wave form which will appear continuous by virtue of the persistence of vision. When both M_1 and M_2 are set vibrating at once, the resultant wave form is the superposition of that produced by each separately. In this way all the curves of Fig. 12F may be produced by using two or more strips of suitable frequencies. The frequencies may be easily altered by changing the free length of the strips above the clamps.

Since for visible light the frequency determines the color, complex waves of light are produced when beams of light of different colors are used. The "impure" colors which are not found in the spectrum will therefore have waves of a complex form. White light, which since Newton's original experiments with prisms we usually speak of as composed of a mixture of all colors, is the extreme example of the superposition of a great number of waves having frequencies differing by only infinitesimal

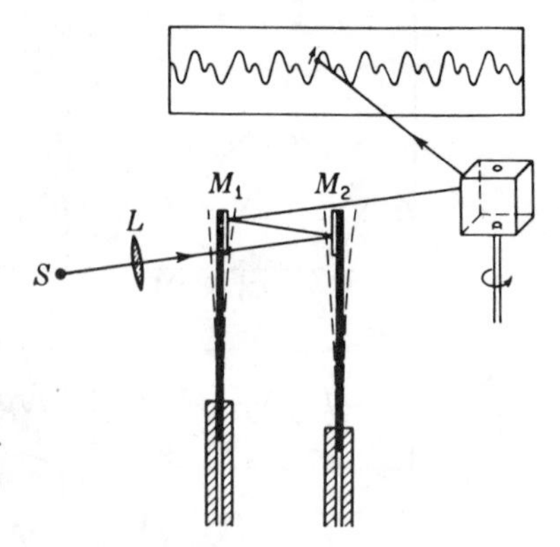

FIGURE 12G
Mechanical and optical arrangement for illustrating the superposition of two waves.

amounts. We shall discuss the resultant wave form for white light in the following section. It was mentioned in the preceding chapter that even the most monochromatic light we can produce in the laboratory still has a finite spread of frequencies. The question of the actual wave forms in such cases, and of how they can be described mathematically, should therefore be considered.

12.6 FOURIER ANALYSIS

Since we can build up a wave of very complex form by the superposition of a number of simple waves, it is of interest to ask to what extent the converse process may be accomplished—that of decomposing a complex wave into a number of simple ones. According to a theorem due to Fourier, any *periodic* function can be represented as the sum of a number (possibly infinite) of sine and cosine functions. By a periodic function we mean one that repeats itself exactly in successive equal intervals, such as the lower curve in Fig. 12F(b). The wave is given by an equation of the type

$$y = a_0 + a_1 \sin \omega t + a_2 \sin 2\omega t + a_3 \sin 3\omega t + \cdots$$
$$+ a'_1 \cos \omega t + a'_2 \cos 2\omega t + a'_3 \cos 3\omega t + \cdots \qquad (12m)$$

This is known as a *Fourier series* and contains, besides the constant term a_0, a series of terms having amplitudes a_1 $a_2, \ldots,$ a'_1, $a'_2, \ldots$ and angular frequencies ω, 2ω, $3\omega, \ldots$. Therefore the resultant wave is regarded as being built up of a number of waves whose wavelengths are as $1 : \frac{1}{2} : \frac{1}{3} : \frac{1}{4} : \cdots$. In the case of sound, these represent the fundamental note and its various harmonics. The evaluation of the amplitude coefficients a_i for a given wave form can be carried out by a straightforward mathematical process for some fairly simple wave forms but in general is a difficult matter. Usually one must have recourse to one of the various forms of *harmonic analyzer*, a mechanical or electronic device for determining the amplitudes and phases of the fundamental and its harmonics.*

* For a detailed account of mechanical harmonic analyzers, see D. C. Miller, "The Science of Musical Sounds," The Macmillan Company, New York, 1922.

Fourier analysis is frequently used today in studying light waves because it is impossible to observe directly the form of a light wave. It is in the investigation of the quality of light and sound that the Fourier analysis of waves has been most used. However, it is important for us to understand the principles of the method, because, as we shall see, a grating or a prism essentially performs a Fourier analysis of the incident light, revealing the various component frequencies which it contains and which appears as *spectral lines.*

Fourier analysis is not limited to waves of a periodic character. The upper part of Fig. 12H shows three types of waves which are not periodic, because, instead of repeating their contour indefinitely, the waves have zero displacement beyond a certain finite range. These *wave packets* cannot be represented by Fourier series; instead *Fourier integrals* must be used, in which the component waves differ only by infinitesimal increments of wavelength. By suitably distributing the amplitudes for the various components, any arbitrary wave form can be expressed by such an integral.* The three lower curves in Fig. 12H represent qualitatively the frequency distribution of the amplitudes which will produce the corresponding wave groups shown above. That is, the upper curves represent the actual wave contour of the group, and this contour can be synthesized by adding up a very large (strictly, an infinite) number of wave trains, each of frequency differing only infinitesimally from the next. The curves shown immediately below each group show the necessary amplitudes of the components of each frequency, in order that their superposition will produce the wave form indicated above. They represent the so-called *Fourier transforms* of the corresponding wave functions.

Curve (*a*) shows the typical wave packet discussed before, and has the Fourier transform (*b*) corresponding to a single spectral line of finite width. The group shown in (*c*) would be produced by passing perfectly monochromatic light through a shutter which is opened for an extremely short time. It is worth remarking here that the corresponding amplitude distribution, shown in curve (*d*), is exactly that obtained for the Fraunhofer diffraction by a single slit, as will be described in Sec. 15.3. Another interesting case, shown in curve (*e*), is that of a single *pulse*, such as the sound pulse sent out by a pistol shot or (better) by the discharge of a spark. The form of such a pulse may resemble that shown, and when a Fourier analysis is made, it yields the broad distribution of wavelengths shown in curve (*f*). For light, such a distribution is called a *continuous spectrum* and is obtained with sources of white light such as an incandescent solid. The distribution of intensity in different wavelengths, which is proportional to the square of the ordinates in the curve, is determined by the exact shape of the pulse. This view of the nature of white light is one which has been emphasized by Gouy and others,† and raises the question whether Newton's experiments on refraction by prisms, which are usually said to prove the composite nature of white

* For a brief discussion of these integrals, and for other references, see J. A. Stratton, "Electromagnetic Theory," pp. 285–292, McGraw-Hill Book Company, New York, 1941. See also J. W. Goodman, "Introduction to Fourier Optics," McGraw-Hill Book Company, New York, 1968, and R. C. Jennison, "Fourier Transforms and Convolutions for the Experimentalist," Pergamon Press, Oxford, England, 1965.

† The reader will find the more detailed discussion of the various representations of white light given in R. W. Wood, "Physical Optics," paperback, Dover Publications, Inc., New York, 1968, of interest in this connection.

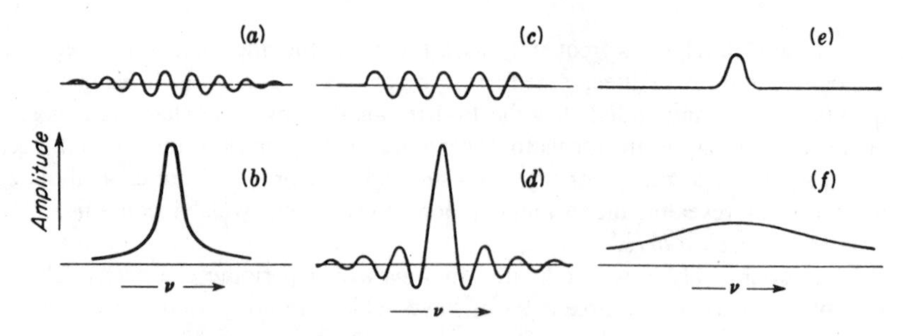

FIGURE 12H
Distribution of amplitudes of different frequencies for various types of wave disturbance of finite length.

light, were of much significance in this respect. Since white light may be regarded as consisting merely of a succession of random pulses, of which the prism performs a Fourier analysis, the view that the colors are manufactured by the prism, which was held by Newton's predecessors, may be regarded as equally correct.

12.7 GROUP VELOCITY

It will be readily seen that if all the component simple waves making up a group travel with the same velocity, the group will move with this velocity and maintain its form unchanged. If, however, the velocities vary with wavelength, this is no longer true, and the group will change its form as it progresses. This situation exists for water waves, and if one watches the individual waves in the group sent out by dropping a stone in still water, they will be found to be moving faster than the group as a whole, dying out at the front of the group and reappearing at the back. Hence in this case the group velocity is less than the wave velocity, a relation which always holds when the velocity of longer waves is greater than that of shorter ones. It is important to establish a relation between the group velocity and wave velocity, and this can easily be done by considering the groups formed by superimposing two waves of slightly different wavelength, such as those already discussed and illustrated in Fig. 12F (*f*). We shall suppose that the two waves have equal amplitudes but slightly different wavelengths, λ and λ', and slightly different velocities, v and v'. The primed quantities in each case will be taken as the larger. Then the propagation numbers and angular frequencies will also differ, such that $k > k'$ and $\omega > \omega'$. The resultant wave will be given by the sum

$$y = a \sin(\omega t - kx) + a \sin(\omega' t - k'x)$$

Again applying the trigonometric relation of Eq. (12k), this equation becomes

$$y = 2a \sin\left(\frac{\omega + \omega'}{2}t - \frac{k + k'}{2}x\right) \cos\left(\frac{\omega - \omega'}{2}t - \frac{k - k'}{2}x\right) \tag{12n}$$

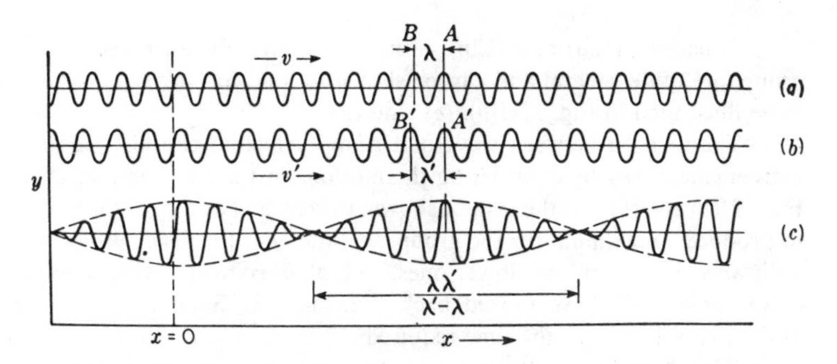

FIGURE 12I
Groups and group velocity of two waves of slightly different wavelength and frequency.

In Figs. 12I(*a*) and (*b*) the two waves are plotted separately, while (*c*) gives their sum, represented by this equation with $t = 0$. The resultant waves have the average wavelength of the two, but the amplitude is modulated to form groups. The individual waves, having the average of the two k's, correspond to variations of the sine factor in Eq. (12n), and according to Eq. (11z), their phase velocity is the quotient of the multipliers of t and x

$$v = \frac{\omega + \omega'}{k + k'} \approx \frac{\omega}{k}$$

That is, the velocity is essentially that of either of the component waves, since these velocities are very nearly the same. The envelope of modulation, indicated by the broken curves shown in Fig. 12I, is given by the cosine factor. This has a much smaller propagation number, equal to the difference of the separate ones, and a correspondingly greater wavelength. The velocity of the groups is

$$u = \frac{\omega - \omega'}{k - k'} \approx \frac{d\omega}{dk} \tag{12o}$$

Since no limit has been set on the smallness of the differences, they may be treated as infinitesimals and the approximate equality becomes exact. Then, since $\omega = vk$, we find for the relation between the *group velocity u* and the *wave velocity v*

$$u = v + k\frac{dv}{dk}$$

If the variable is changed to λ, through $k = 2\pi/\lambda$, one obtains the useful form

● $$u = v - \lambda\frac{dv}{d\lambda} \tag{12p}$$

It should be emphasized that λ here represents the actual wavelength *in the medium*. For light, this will not in most problems be the ordinary wavelength in air (see Sec. 23.7).

Equations (12o) and (12p), although derived for an especially simple type of group, are quite general and can be shown to hold for any group whatever, e.g., the three illustrated in Fig. 12H(*a*), (*c*), and (*e*).

The relation between wave and group velocities can also be derived in a less mathematical way by considering the motions of the two component wave trains in Fig. 12I(*a*) and (*b*). At the instant shown, the crests A and A' of the two trains coincide to produce a maximum for the group. A little later the faster waves will have gained a distance $\lambda' - \lambda$ on the slower ones, so that B' coincides with B and the maximum of the group will have moved back a distance λ. Since the difference in velocity of the two trains is dv, the time required for this is $d\lambda/dv$. But in this time both wave trains have been moving to the right, the upper one moving a distance $v\ d\lambda/dv$. The net displacement of the maximum of the group is thus $v(d\lambda/dv) - \lambda$ in the time $d\lambda/dv$, so that we obtain, for the group velocity,

$$u = \frac{v(d\lambda/dv) - \lambda}{d\lambda/dv} = v - \lambda \frac{dv}{d\lambda}$$

in agreement with Eq. (12p).

A picture of the groups formed by two waves of slightly different frequency may easily be produced with the apparatus described in Sec. 12.5. It is merely necessary to adjust the two vibrating strips until the frequencies differ by only a few vibrations per second. See Fig. 12G.

The group velocity is the important one for light, since it is the only velocity which we can observe experimentally. We know of no means of following the progress of an individual wave in a group of light waves; instead, we are obliged to measure the rate at which a wave train of finite length conveys the *energy*, a quantity which can be observed. The wave and group velocities become the same in a medium having no dispersion, i.e., in which $dv/d\lambda = 0$, so that waves of all lengths travel with the same speed. This is accurately true for light traveling in a vacuum, so that there is no difference between group and wave velocities in this case.

12.8 GRAPHICAL RELATION BETWEEN WAVE AND GROUP VELOCITY

A very simple geometrical construction by which we can determine the group velocity from a curve of the wave velocity against wavelength is based upon the graphical interpretation of Eq. (12p). As an example, the curve of Fig. 12J represents the variation of the wave velocity with λ for water waves in deep water (gravity waves) and is drawn according to the theoretical equation $v = \text{const} \times \sqrt{\lambda}$. At a certain wavelength λ_1, the waves have a velocity v, and the slope of the curve at the corresponding point P gives $dv/d\lambda$. The line PR, drawn tangent to the curve at this point, intersects the v axis at the point R, the ordinate of which is the group velocity u for waves of wavelength in the neighborhood of λ_1. This is evident from the fact that PQ equals $\lambda_1\ dv/d\lambda$, that is, the abscissa of P multiplied by the slope of PR. Hence QS, which is drawn equal to RO, represents the difference $v - \lambda\ dv/d\lambda$, and this is just the value of u, by Eq. (12p). In the particular example chosen here, it will be left as a problem

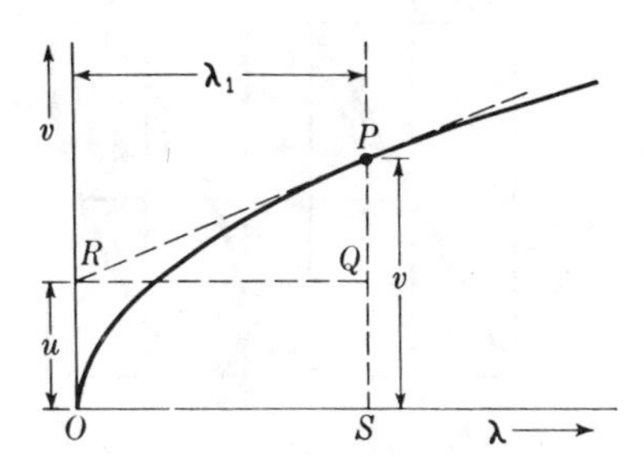

FIGURE 12J
Graphical determination of group velocity from a wave-velocity curve.

for the student to prove that $u = \frac{1}{2}v$ for any value of λ. In water waves of this type, the individual waves therefore move with twice the velocity with which the group as a whole progresses.

12.9 ADDITION OF SIMPLE HARMONIC MOTIONS AT RIGHT ANGLES

Consider the effect when two sine waves of the same frequency but having displacements in two perpendicular directions act simultaneously at a point. Choosing the directions as y and z, we may express the two component motions as

$$y = a_1 \sin(\omega t - \alpha_1) \qquad \text{and} \qquad z = a_2 \sin(\omega t - \alpha_2) \tag{12q}$$

These are to be added, according to the principle of superposition, to find the path of the resultant motion. One does this by eliminating t from the two equations, obtaining

$$\frac{y}{a_1} = \sin \omega t \cos \alpha_1 - \cos \omega t \sin \alpha_1 \tag{12r}$$

$$\frac{z}{a_2} = \sin \omega t \cos \alpha_2 - \cos \omega t \sin \alpha_2 \tag{12s}$$

Multiplying Eq. (12r) by $\sin \alpha_2$ and Eq. (12s) by $\sin \alpha_1$ and subtracting the first equation from the second gives

$$-\frac{y}{a_1} \sin \alpha_2 + \frac{z}{a_2} \sin \alpha_1 = \sin \omega t(\cos \alpha_2 \sin \alpha_1 - \cos \alpha_1 \sin \alpha_2) \tag{12t}$$

Similarly, multiplying Eq (12r) by $\cos \alpha_2$ and Eq. (12s) by $\cos \alpha_1$, and subtracting the second from the first, we obtain

$$\frac{y}{a_1} \cos \alpha_2 - \frac{z}{a_2} \cos \alpha_1 = \cos \omega t(\cos \alpha_2 \sin \alpha_1 - \cos \alpha_1 \sin \alpha_2) \tag{12u}$$

We can now eliminate t from Eqs. (12t) and (12u) by squaring and adding these equations. This gives

$$\sin^2(\alpha_1 - \alpha_2) = \frac{y^2}{a_1^{\,2}} + \frac{z^2}{a_2^{\,2}} - \frac{2yz}{a_1 a_2} \cos(\alpha_1 - \alpha_2) \tag{12v}$$

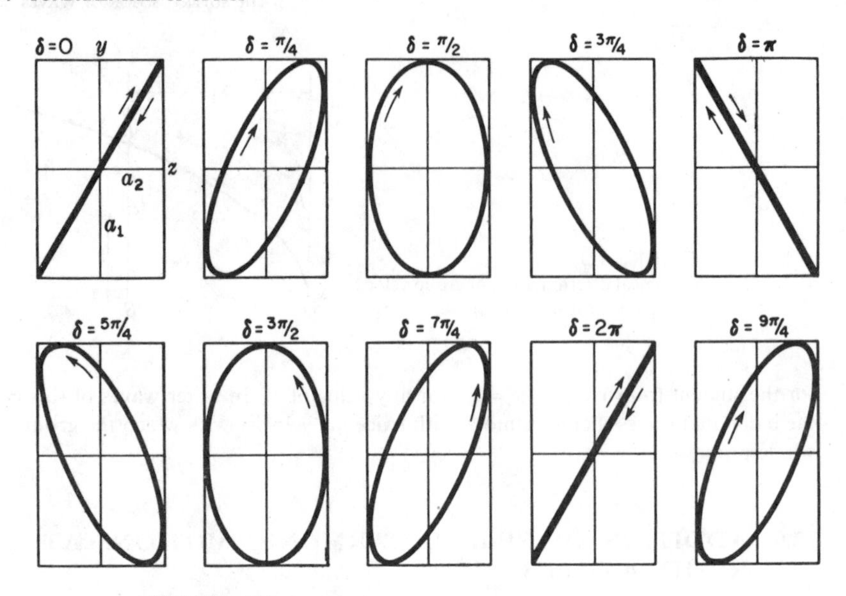

FIGURE 12K
Composition at right angles of two simple harmonic motions of the same frequency but different phase.

as the equation for the resultant path. In Fig. 12K the heavy curves are graphs of this equation for various values of the phase difference $\delta = \alpha_1 - \alpha_2$. Except for the special cases where they degenerate into straight lines, these curves are all ellipses. The principal axes of the ellipse are in general inclined to the y and z axes but coincide with them when $\delta = \pi/2, 3\pi/2, 5\pi/2, \ldots$, as can readily be seen from Eq. (12v). In this case

$$\frac{y^2}{a_1{}^2} + \frac{z^2}{a_2{}^2} = 1$$

which is the equation of an ellipse with semiaxes a_1 and a_2, coinciding with the y and z axes, respectively. When $\delta = 0, 2\pi, 4\pi, \ldots$, we have

$$y = \frac{a_1}{a_2} z$$

representing a straight line passing through the origin, with a slope a_1/a_2. If $\delta = \pi, 3\pi, 5\pi, \ldots$,

$$y = -\frac{a_1}{a_2} z$$

a straight line with the same slope, but of opposite sign.

That the two cases $\delta = \pi/2$ and $\delta = 3\pi/2$, although giving the same path, are physically different can be seen by graphical constructions such as those of Fig. 12L. In both parts of the figure the motion in the y direction is in the same phase, the point

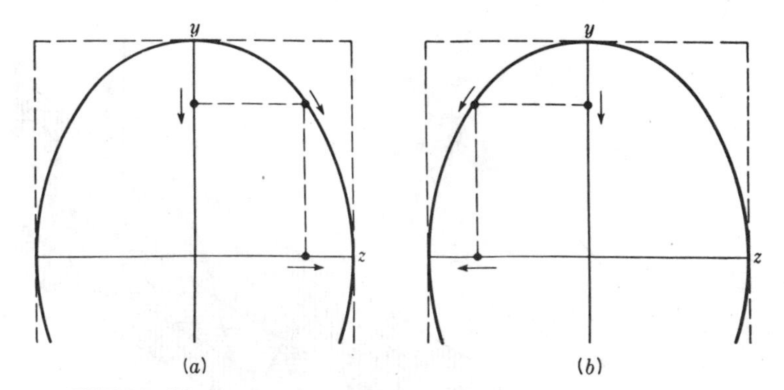

FIGURE 12L
Graphical composition of motions in which y is (a) one-quarter and (b) three-quarters of a period ahead of z.

having executed one-eighth of a vibration beyond its extreme positive displacement. The z motion in part (a) lacks one-eighth of a vibration of reaching this extreme position, while in part (b) it lacks five-eighths. Consideration of the directions of the individual motions, and of that of their resultant, will show that the latter corresponds to the indications of the curved arrows. In the two cases the ellipse is traversed in opposite senses.

Light can be produced for which the form of vibration is an ellipse of any desired eccentricity. The so-called *plane-polarized* light (Chap. 24) approximates a sine wave lying in a plane—say the xy plane of Fig. 12M—and the displacements are linear displacements in the y direction. If one combines a beam of this light with another consisting of plane-polarized waves lying in the xz plane (dotted curve) and having a constant phase difference with the first, the resultant motion at any value of x will be a certain ellipse in the yz plane. Such light is said to be *elliptically polarized* and may readily be produced by various means (Chap. 27). A special case occurs when the amplitudes a_1 and a_2 of the two waves are equal and the phase difference is an odd multiple of $\pi/2$. The vibration form is then a circle, and the light is said to be *circularly polarized.* When the direction of rotation is clockwise ($\delta = \pi/2, 5\pi/2, \ldots$) looking opposite to the direction in which the light is traveling, the light is called *right* circularly polarized, while if the rotation is counterclockwise ($\delta = 3\pi/2, 7\pi/2, \ldots$) it is called *left* circularly polarized.

The various types of motion shown in Fig. 12K can readily be demonstrated with the apparatus described in Sec. 12.5. For this purpose, the two strips are arranged to vibrate at right angles to each other, and the rotating mirror is eliminated. Then one strip imparts a horizontal vibration to the spot of light, and the other a vertical vibration. When both are actuated simultaneously, the spot will trace out an ellipse. This will remain fixed if the two strips are tuned to exactly the same frequency. If they are only slightly detuned, the figure will progress through the forms corresponding

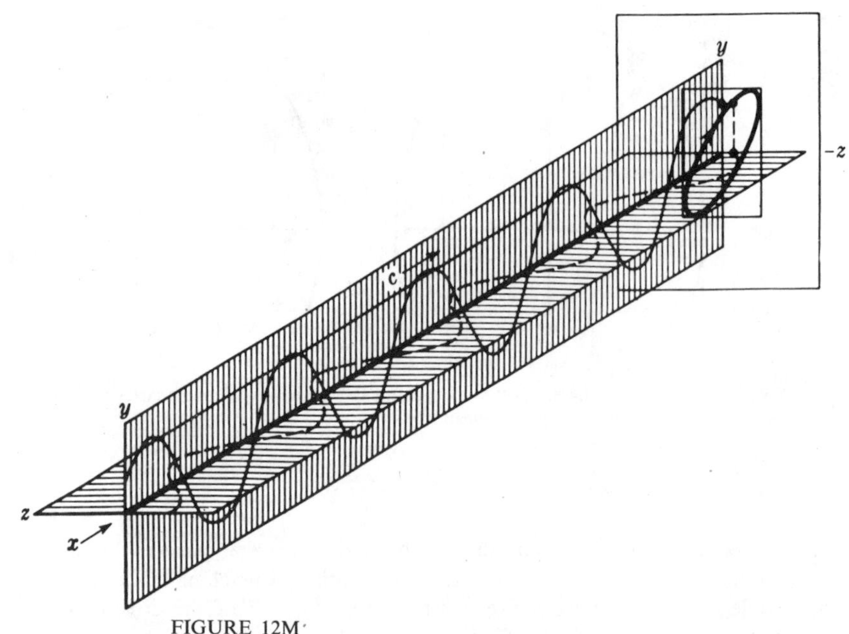

FIGURE 12M
Composition of two sine waves at right angles.

to all possible values of the phase difference, passing in succession a sequence like that shown in Fig. 12K.

PROBLEMS

12.1 Two waves traveling together along the same line are given by $y_1 = 5 \sin(\omega t + \pi/2)$ and $y_2 = 7 \sin(\omega t + \pi/3)$. Find (*a*) the resultant amplitude, (*b*) the initial phase angle of the resultant, and (*c*) the resultant equation of motion.

Ans. (*a*) 11.60, (*b*) 72.4°, (*c*) $y = 11.60 \sin(\omega t + 72.4°)$

12.2 Two waves traveling together along the same line are represented by

$$y_1 = 25 \sin(\omega t - \pi/4) \qquad \text{and} \qquad y_2 = 15 \sin(\omega t - \pi/6)$$

Find (*a*) the resultant amplitude, (*b*) the initial phase angle of the resultant, and (*c*) the resultant equation for the sum of the two motions.

12.3 Three simple harmonic motions are given by $y_1 = 2 \sin(\omega t - 30°)$, $y_2 = 5 \sin(\omega t + 30°)$, and $y_3 = 4 \sin(\omega t + 90°)$. If they are added together, find (*a*) the resultant amplitude, (*b*) the initial phase angle of the resultant, and (*c*) the resultant equation of motion.

12.4 Six simple harmonic motions of the same amplitude and period but differing from the next by +16° are to be added vectorily as shown in Fig. 12B. If each has an amplitude of 5.0 cm, find (*a*) the resultant amplitude and (*b*) the initial phase angle of the resultant with respect to the first. *Ans.* (*a*) 23.09 cm, (*b*) 48.0°

12.5 Two waves having amplitudes of 5 and 8 units and equal frequencies come together at a point in space. If they meet with a phase difference of $5\pi/8$ rad, find the resultant intensity relative to the sum of the two separate intensities.

12.6 Calculate the vibration energy resulting from the superposition of six waves having equal amplitudes of 5 units and initial phase angles of 0, 36, 72, 108, 144, and 180°. Is the resultant energy increased or decreased if the first and sixth waves are removed?

12.7 Compound graphically two waves having wavelengths in the ratio 3:2 and amplitudes 1:2, respectively. Assume that they start in phase.

12.8 Compound graphically two waves having wavelengths in the ratio 4:3 and amplitudes in the ratio 2:3, respectively. Assume that they start in phase.

12.9 Two sources vibrating according to the equations $y_1 = 4 \sin 2\pi t$ and $y_2 = 3 \sin 2\pi t$ send out waves in all directions with a velocity of 2.40 m/s. Find the equation of motion of a particle 5 m from the first source and 3 m from the second. *Note:* $\omega = 2\pi$ rad/s. *Ans.* $y = 6.08 \sin (2\pi t - 25.3°)$

12.10 Standing waves are produced by the superposition of two waves,

$$y_1 = 7 \sin 2\pi \left(\frac{t}{T} - \frac{2x}{\pi}\right) \quad \text{and} \quad y_2 = 7 \sin 2\pi \left(\frac{t}{T} - \frac{2x}{\pi}\right)$$

traveling in opposite directions. Find (*a*) the amplitude, (*b*) the wavelength λ, (*c*) the length of one loop, (*d*) the velocity of the waves, and (*e*) the period.

12.11 If Weiner's experiment is performed with yellow light, $\lambda = 5800 \times 10^{-5}$ cm and the photographic film is inclined at 0.250° with the mirror, find the distance between successive dark bands on the developed film.

12.12 Four equal sources emit waves of the same frequency and amplitude and phases differing by either 0 or π rad. Assuming that each possible combination of phases is equally probable (there are 16 of them) show that the average intensity is just 4 times that of any one of the waves. Remember that the intensity due to each combination is given by the square of the resultant amplitude.

Ans. $++++(16)$, $----(16)$, $---+(4)$, $+++-(4)$, $--+-(4)$, $++-+(4)$, $-+--(4)$, $+-++(4)$, $+---(4)$, $-+++(4)$, $--++(0)$, $++--(0)$, $-++-(0)$, $+--+(0)$, $-+-+(0)$, $+-+-(0)$; sum = 64; average = 4

12.13 Prove that for water waves controlled by gravity the group velocity equals half the wave velocity.

12.14 Calculate the wave and group velocities of water waves at (*a*) $\lambda = 2$ cm, (*b*) $\lambda = 8.0$ cm, and (*c*) $\lambda = 20.0$ cm. The wave velocity of short waves such as these are given by

$$v = \sqrt{\frac{\lambda}{2\pi}\left(g + \frac{4\pi^2 T}{\lambda^2 d}\right)}$$

where λ is the wavelength in meters, T is the surface tension in newtons per meter, which at room temperature is 0.073 N/m, g is the acceleration due to gravity, 9.80 m/s^2, and d is the density of the liquid in kilograms per cubic meter.

12.15 The phase velocity of waves in a certain medium is represented by $v = C_1 + C_2\lambda$, where the C's are constants. What is the value of the group velocity?

Ans. $u = C_1$

12.16 Two simple harmonic motions at right angles are represented by $y = 3 \sin 2\pi t$ and $z = 5 \sin (2\pi t - 3\pi/4)$. Find the equation for the resultant path, and plot this path by the method indicated in Fig. 12L. Verify at least two points on this path by substitution in the resultant equation.

12.17 How must the equation for the y motion in Prob. 12.16 be modified to yield an ellipse having its major axis coincident with z to yield a counterclockwise rotation?

12.18 For the type of waves described in Prob. 12.14, (*a*) find the exact value of the wavelength for which the wave and group velocities become equal and (*b*) find their velocity. (*c*) Plot a graph of v versus λ from 0 to 8.0 cm.

13
INTERFERENCE OF TWO BEAMS OF LIGHT

It was stated at the beginning of the last chapter that two beams of light can be made to cross each other without either one producing any modification of the other after it passes beyond the region of crossing. In this sense the two beams do not interfere with each other. However, in the region of crossing, where both beams are acting at once, we are led to expect from the considerations of the preceding chapter that the resultant amplitude and intensity may be very different from the sum of those contributed by the two beams acting separately. This modification of intensity obtained by the superposition of two or more beams of light we call *interference*. If the resultant intensity is zero or in general less than we expect from the separate intensities, we have *destructive* interference, while if it is greater, we have *constructive* interference. The phenomenon in its simpler aspects is rather difficult to observe, because of the very short wavelength of light, and therefore was not recognized as such before 1800, when the corpuscular theory of light was predominant. The first man successfully to demonstrate the interference of light, and thus establish its wave character, was Thomas Young. In order to understand his crucial experiment performed in 1801, we must first consider the application to light of an important principle which holds for any type of wave motion.

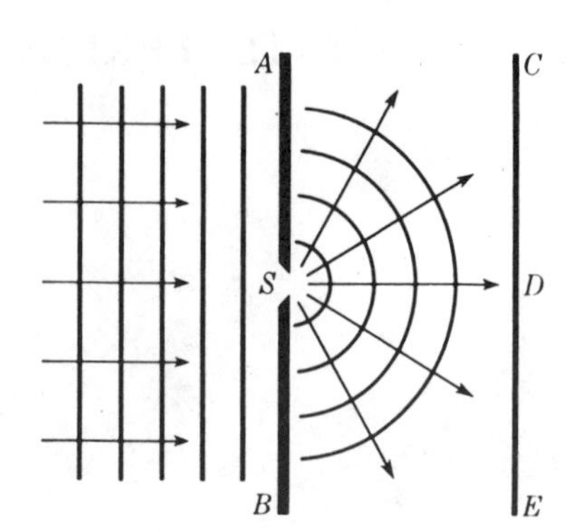

FIGURE 13A
Diffraction of waves passing through a small aperture.

13.1 HUYGENS' PRINCIPLE

When waves pass through an aperture or past the edge of an obstacle, they always spread to some extent into the region which is not directly exposed to the oncoming waves. This phenomenon is called *diffraction.* In order to explain this bending of light, Huygens nearly three centuries ago proposed the rule that *each point on a wave front may be regarded as a new source of waves.** This principle has very far-reaching applications and will be used later in discussing the diffraction of light, but we shall consider here only a very simple proof of its correctness. In Fig. 13A let a set of plane waves approach the barrier AB from the left, and let the barrier contain an opening S of width somewhat smaller than the wavelength. At all points except S the waves will be either reflected or absorbed, but S will be free to produce a disturbance behind the screen. It is found experimentally, in agreement with the above principle, that the waves spread out from S in the form of semicircles.

Huygens' principle as shown in Fig. 13A can be illustrated very successfully with water waves. An arc lamp on the floor, with a glass-bottomed tray or tank above it, will cast shadows of waves on a white ceiling. A vibrating strip of metal or a wire fastened to one prong of a tuning fork of low frequency will serve as a source of waves at one end of the tray. If an electrically driven tuning fork is used, the waves can be made apparently to stand still by placing a slotted disk on the shaft of a motor in front of the arc lamp. The disk is set rotating with the same frequency as the tuning fork to give the stroboscopic effect. This experiment can be performed for a fairly large audience and is well worth doing. Descriptions of diffraction experiments in light will be given in Chap. 15.

If the experiment in Fig. 13A is performed with light, one would naturally expect, from the fact that light generally travels in straight lines, that merely a narrow patch of light would appear at D. However, if the slit is made very narrow, an ap-

* The "waves" envisioned by Huygens were not continuous trains but a series of random pulses. Furthermore, he supposed the secondary waves to be effective only at the point of tangency to their common envelope, thus denying the possibility of diffraction. The correct application of the principle was first made by Fresnel, more than a century later.

FIGURE 13B
Photograph of the diffraction of light from a slit of width 0.001 mm.

preciable broadening of this patch is observed, its breadth increasing as the slit is narrowed further. This is remarkable evidence that light does not always travel in straight lines and that waves on passing through a narrow opening spread out into a continuous fan of light rays. When the screen *CE* is replaced by a photographic plate, a picture like the one shown in Fig. 13B is obtained. The light is most intense in the forward direction, but its intensity decreases slowly as the angle increases. If the slit is small compared with the wavelength of light, the intensity does not come to zero even when the angle of observation becomes 90°. While this brief introduction to Huygens' principle will be sufficient for understanding the interference phenomena we are to discuss, we shall return in Chaps. 15 and 18 to a more detailed consideration of diffraction at a single opening.

13.2 YOUNG'S EXPERIMENT

The original experiment performed by Young is shown schematically in Fig. 13C. Sunlight was first allowed to pass through a pinhole *S* and then, at a considerable distance away, through two pinholes S_1 and S_2. The two sets of spherical waves emerging from the two holes interfered with each other in such a way as to form a symmetrical pattern of varying intensity on the screen *AC*. Since this early experiment was performed, it has been found convenient to replace the pinholes by narrow slits and to use a source giving monochromatic light, i.e., light of a single wavelength. In place of spherical wave fronts we now have cylindrical wave fronts, represented equally well in two dimensions by the same Fig. 13C. If the circular lines represent crests of waves, the intersections of any two lines represent the arrival at those points of two waves with the same phase or with phases differing by a multiple of 2π. Such points are therefore those of maximum disturbance or brightness. A close examination of the light on the screen will reveal evenly spaced light and dark bands or fringes, similar to those shown in Fig. 13D. Such photographs are obtained by replacing the screen *AC* of Fig. 13C by a photographic plate.

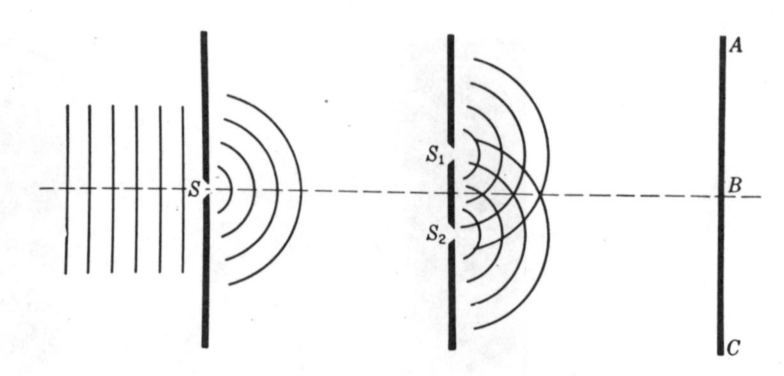

FIGURE 13C
Experimental arrangement for Young's double-slit experiment.

A very simple demonstration of Young's experiment can be accomplished in the laboratory or lecture room by setting up a single-filament lamp L (Fig. 13E) at the front of the room. The straight vertical filament S acts as the source and first slit. Double slits for each observer can be easily made from small photographic plates about 1 to 2 in. square. The slits are made in the photographic emulsion by drawing the point of a penknife across the plate, guided by a straightedge. The plates need not be developed or blackened but can be used as they are. The lamp is now viewed by holding the double slit D close to the eye E and looking at the lamp filament. If the slits are close together, for example, 0.2 mm apart, they give widely spaced fringes, whereas slits farther apart, for example, 1.0 mm, give very narrow fringes. A piece of red glass F placed adjacent to and above another of green glass in front of the lamp will show that the red waves produce wider fringes than the green, which we shall see is due to their greater wavelength.

Frequently one wishes to perform accurate experiments by using more nearly monochromatic light than that obtained by white light and a red or green glass filter. Perhaps the most convenient method is to use the sodium arc now available on the market or a mercury arc plus a filter to isolate the green line, $\lambda 5461$. A suitable filter consists of a combination of didymium glass, to absorb the yellow lines, and a light yellow glass, to absorb the blue and violet lines.

FIGURE 13D
Interference fringes produced by a double slit using the arrangement shown in Fig. 13C.

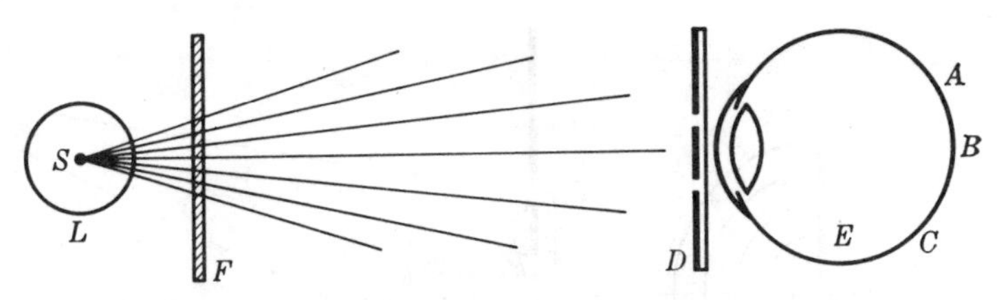

FIGURE 13E
Simple method for observing interference fringes.

13.3 INTERFERENCE FRINGES FROM A DOUBLE SOURCE

We shall now derive an equation for the intensity at any point P on the screen (Fig. 13F) and investigate the spacing of the interference fringes. Two waves arrive at P, having traversed different distances S_2P and S_1P. Hence they are superimposed with a phase difference given by

$$\delta = \frac{2\pi}{\lambda}\,\Delta = \frac{2\pi}{\lambda}\,(S_2P - S_1P) \tag{13a}$$

It is assumed that the waves start out from S_1 and S_2 in the same phase, because these slits were taken to be equidistant from the source slit S. Furthermore, the amplitudes are practically the same if (as is usually the case) S_1 and S_2 are of equal width and very close together. The problem of finding the resultant intensity at P therefore reduces to that discussed in Sec. 12.1, where we considered the addition of two simple harmonic motions of the same frequency and amplitude, but of phase difference δ. The intensity was given by Eq. (12g) as

$$I \approx A^2 = 4a^2 \cos^2 \frac{\delta}{2} \tag{13b}$$

where a is the amplitude of the separate waves and A that of their resultant.

It now remains to evaluate the phase difference in terms of the distance x on the screen from the central point P_0, the separation d of the two slits, and the distance D from the slits to the screen. The corresponding path difference is the distance S_2A in Fig. 13F, where the dashed line S_1A has been drawn to make S_1 and A equidistant from P. As Young's experiment is usually performed, D is some thousand times larger than d or x. Hence the angles θ and θ' are very small and practically equal. Under these conditions, S_1AS_2 may be regarded as a right triangle, and the path difference becomes $d \sin \theta' \approx d \sin \theta$. To the same approximation, we may set the sine of the angle equal to the tangent, so that $\sin \theta \approx x/D$. With these assumptions, we obtain

$$\Delta = d \sin \theta = d\,\frac{x}{D} \tag{13c}$$

This is the value of the path difference to be substituted in Eq. (13a) to obtain the phase difference δ. Now Eq. (13b) for the intensity has maximum values equal to

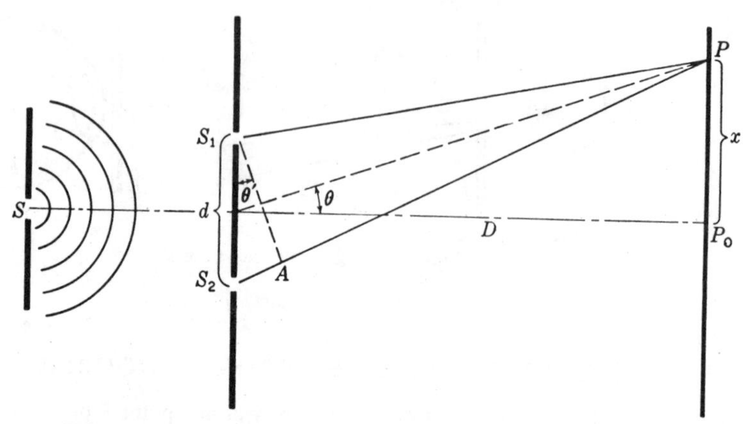

FIGURE 13F
Path difference in Young's experiment.

$4a^2$ whenever δ is an integral multiple of 2π, and according to Eq. (13a) this will occur when the path difference is an integral multiple of λ. Hence we have

$$\frac{xd}{D} = 0, \lambda, 2\lambda, 3\lambda, \ldots = m\lambda$$

or

$$x = m\lambda\,\frac{D}{d} \qquad \textit{Bright fringes} \qquad (13d)$$

The minimum value of the intensity is zero, and this occurs when $\delta = \pi, 3\pi, 5\pi, \ldots$. For these points

$$\frac{xd}{D} = \frac{\lambda}{2}, \frac{3\lambda}{2}, \frac{5\lambda}{2}, \ldots = \left(m + \frac{1}{2}\right)\lambda$$

or

$$x = \left(m + \frac{1}{2}\right)\lambda\,\frac{D}{d} \qquad \textit{Dark fringes} \qquad (13e)$$

The whole number m, which characterizes a particular bright fringe, is called the *order* of interference. Thus the fringes with $m = 0, 1, 2, \ldots$ are called the *zero, first, second,* etc., *orders.*

According to these equations the distance on the screen between two successive fringes, which is obtained by changing m by unity in either Eq. (13d) or (13e), is constant and equal to $\lambda D/d$. Not only is this equality of spacing verified by measurement of an interference pattern such as Fig. 13D, but one also finds by experiment that its magnitude *is directly proportional to the slit-screen distance D, inversely proportional to the separation of the slits d, and directly proportional to the wavelength λ.* Knowledge of the spacing of these fringes thus gives us a direct determination of λ in terms of known quantities.

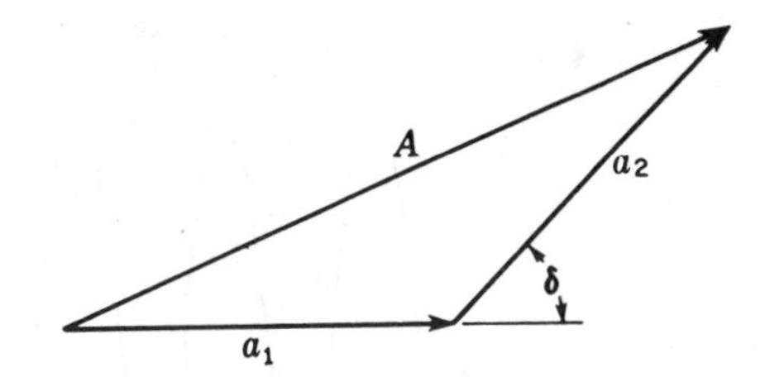

FIGURE 13G
The composition of two waves of the same frequency and amplitude but different phase.

These maxima and minima of intensity exist throughout the space behind the slits. A lens is not required to produce them, although they are usually so fine that a magnifier or eyepiece must be used to see them. Because of the approximations made in deriving Eq. (13c), careful measurements would show that, particularly in the region near the slits, the fringe spacing departs from the simple linear dependence required by Eq. (13d). A section of the fringe system in the plane of the paper of Fig. 13C, instead of consisting of a system of straight lines radiating from the midpoint between the slits, is actually a set of hyperbolas. The hyperbola, being the curve for which the difference in the distance from two fixed points is constant, obviously fits the condition for a given fringe, namely, the constancy of the path difference. Although this deviation from linearity may become important with sound and other waves, it is usually negligible when the wavelengths are as short as those of light.

13.4 INTENSITY DISTRIBUTION IN THE FRINGE SYSTEM

To find the intensity on the screen at points between the maxima, we may apply the vector method of compounding amplitudes described in Sec. 12.2 and illustrated for the present case in Fig. 13G. For the maxima, the angle δ is zero, and the component amplitudes a_1 and a_2 are parallel, so that if they are equal, the resultant $A = 2a$. For the minima, a_1 and a_2 are in opposite directions, and $A = 0$. In general, for any value of δ, A is the closing side of the triangle. The value of A^2, which measures the intensity, is then given by Eq. (13*b*) and varies according to $\cos^2 (\delta/2)$. In Fig. 13H the solid curve represents a plot of the intensity against the phase difference.

In concluding our discussion of these fringes, one question of fundamental importance should be considered. If the two beams of light arrive at a point on the screen exactly out of phase, they interfere destructively and the resultant intensity is zero. One may well ask what becomes of the *energy* of the two beams, since the law of conservation of energy tells us that energy cannot be destroyed. The answer to this question is that the energy which apparently disappears at the minima actually is still present at the maxima, where the intensity is greater than would be produced by the two beams acting separately. In other words, the energy is not destroyed but merely redistributed in the interference pattern. The *average* intensity on the screen is exactly that which would exist in the absence of interference. Thus, as shown in Fig. 13H, the intensity in the interference pattern varies between $4a^2$ and zero. Now each beam acting separately would contribute a^2, and so without interference we would have a uniform intensity of $2a^2$, as indicated by the broken line. To obtain

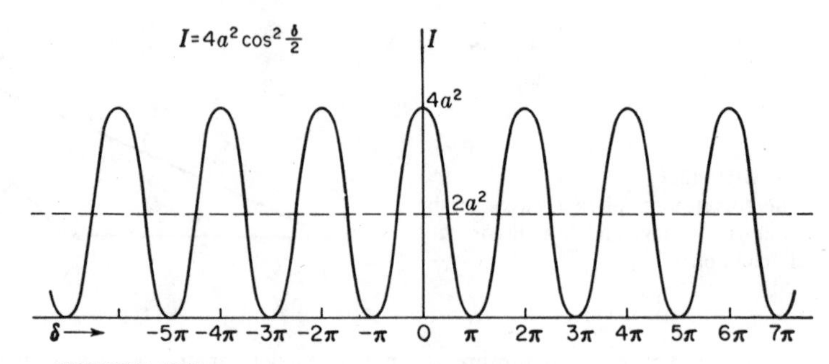

FIGURE 13H
Intensity distribution for the interference fringes from two waves of the same frequency.

the average intensity on the screen for n fringes, we note that the average value of the square of the cosine is $\frac{1}{2}$. This gives, by Eq. (13b), $I \approx 2a^2$, justifying the statement made above, and it shows that no violation of the law of conservation of energy is involved in the interference phenomenon.

13.5 FRESNEL'S BIPRISM*

Soon after the double-slit experiment was performed by Young, the objection was raised that the bright fringes he observed were probably due to some complicated modification of the light by the edges of the slits and not to true interference. Thus the wave theory of light was still questioned. Not many years passed, however, before Fresnel brought forward several new experiments in which the interference of two beams of light was proved in a manner not open to the above objection. One of these, the Fresnel biprism experiment, will be described in some detail.

A schematic diagram of the biprism experiment is shown in Fig. 13I. The thin double prism P refracts the light from the slit sources S into two overlapping beams ac and be. If screens M and N are placed as shown in the figure, interference fringes are observed only in the region bc. When the screen ae is replaced by a photographic plate, a picture like the upper one in Fig. 13J is obtained. The closely spaced fringes in the center of the photograph are due to interference, while the wide fringes at the edge of the pattern are due to diffraction. These wider bands are produced by the vertices of the two prisms, each of which acts as a straightedge, giving a pattern which will be discussed in detail in Chap. 18. When the screens M and N are removed from

* Augustin Fresnel (1788–1827). Most notable French contributor to the theory of light. Trained as an engineer, he became interested in light, and in 1814–1815 he rediscovered Young's principle of interference and extended it to complicated cases of diffraction. His mathematical investigations gave the wave theory a sound foundation.

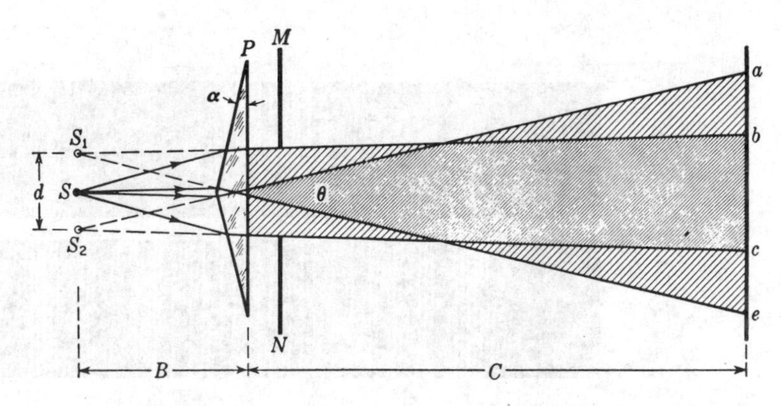

FIGURE 13I
Diagram of Fresnel's biprism experiment.

the light path, the two beams will overlap over the whole region *ae*. The lower photograph in Fig. 13J shows for this case the equally spaced interference fringes superimposed on the diffraction pattern of a wide aperture. (For the diffraction pattern above, without the interference fringes, see lowest figures in Fig. 18U.) With such an experiment Fresnel was able to produce interference without relying upon diffraction to bring the interfering beams together.

Just as in Young's double-slit experiment, the wavelength of light can be determined from measurements of the interference fringes produced by the biprism. Calling B and C the distances of the source and screen, respectively, from the prism P, d the distance between the virtual images S_1 and S_2, and Δx the distance between the successive fringes on the screen, the wavelength of the light is given from Eq. (13d) as

$$\lambda = \frac{\Delta x \, d}{B + C} \tag{13f}$$

Thus the virtual images S_1 and S_2 act like the two slit sources in Young's experiment.

In order to find d, the linear separation of the virtual sources, one can measure their angular separation θ on a spectrometer and assume, to sufficient accuracy, that $d = B\theta$. If the parallel light from the collimator covers both halves of the biprism, two images of the slit are produced and the angle θ between these is easily measured with the telescope. An even simpler measurement of this angle can be made by holding the prism close to one eye and viewing a round frosted light bulb. At a certain distance from the light the two images can be brought to the point where their inner edges just touch. The diameter of the bulb divided by the distance from the bulb to the prism then gives θ directly.

Fresnel biprisms are easily made from a small piece of glass, such as half a microscope slide, by beveling about $\frac{1}{8}$ to $\frac{1}{4}$ in. on one side. This requires very little grinding with ordinary abrasive materials and polishing with rouge, since the angle required is only about 1°.

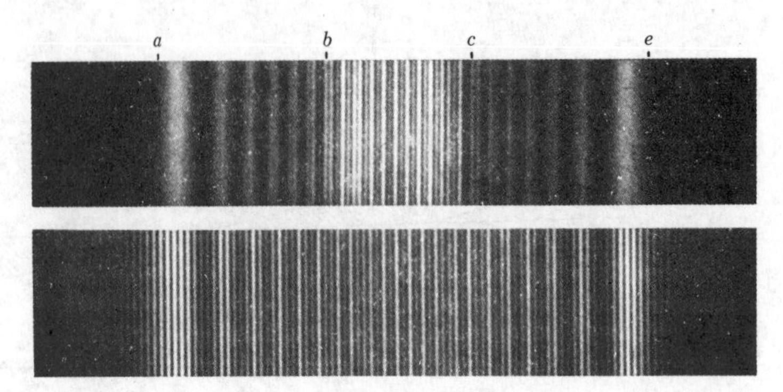

FIGURE 13J
Interference and diffraction fringes produced in the Fresnel biprism experiment.

13.6 OTHER APPARATUS DEPENDING ON DIVISION OF THE WAVE FRONT

Two beams can be brought together in other ways to produce interference. In the arrangement known as *Fresnel's mirrors*, light from a slit is reflected in two plane mirrors slightly inclined to each other. The mirrors produce two virtual images of the slit, as shown in Fig. 13K. They act in every respect like the images formed by the biprism, and interference fringes are observed in the region bc, where the reflected beams overlap. The symbols in this diagram correspond to those in Fig. 13I, and Eq. (13f) is again applicable. It will be noted that the angle 2θ subtended at the point of intersection M by the two sources is twice the angle between the mirrors.

The Fresnel double-mirror experiment is usually performed on an optical bench, with the light reflected from the mirrors at nearly grazing angles. Two pieces of ordinary plate glass about 2 in. square make a very good double mirror. One plate should have an adjusting screw for changing the angle θ and the other a screw for making the edges of the two mirrors parallel.

An even simpler device, shown in Fig. 13L, produces interference between the light reflected in one long mirror and the light coming directly from the source without reflection. In this arrangement, known as *Lloyd's mirror*, the quantitative relations are similar to those in the foregoing cases, with the slit and its virtual image constituting the double source. An important feature of the Lloyd's-mirror experiment lies in the fact that when the screen is placed in contact with the end of the mirror (in the position MN, Fig. 13L), the edge O of the reflecting surface comes at the center of a *dark* fringe, instead of a bright one as might be expected. This means that one of the two beams has undergone a phase change of π. Since the direct beam could not change phase, this experimental observation is interpreted to mean that the reflected light has changed phase at reflection. Two photographs of the Lloyd's-mirror fringes taken in this way are reproduced in Fig. 13M, one taken with visible light and the other with X rays.

If the light from source S_1 in Fig. 13L is allowed to enter the end of the glass

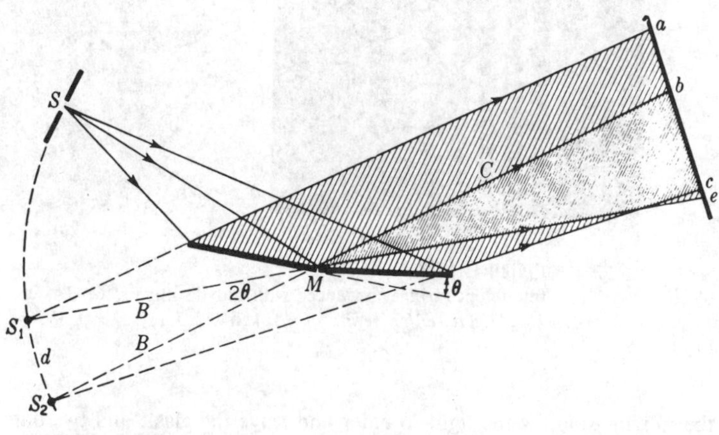

FIGURE 13K
Geometry of Fresnel's mirrors.

plate by moving the latter up, and to be internally reflected from the upper glass surface, fringes will again be observed in the interval OP, with a dark fringe at O. This shows that there is again a phase change of π at reflection. As will be shown in Chap. 25, this is not a contradiction of the discussion of phase change given in Sec. 14.1. In this instance the light is incident at an angle greater than the critical angle for total reflection.

Lloyd's mirror is readily set up for demonstration purposes as follows. A carbon arc, followed by a colored glass filter and a narrow slit, serves as a source. A strip of ordinary plate glass 1 to 2 in. wide and 1 ft or more long makes an excellent mirror. A magnifying glass focused on the far end of the mirror enables one to observe the fringes shown in Fig. 13M. Internal fringes can be observed by polishing the ends of

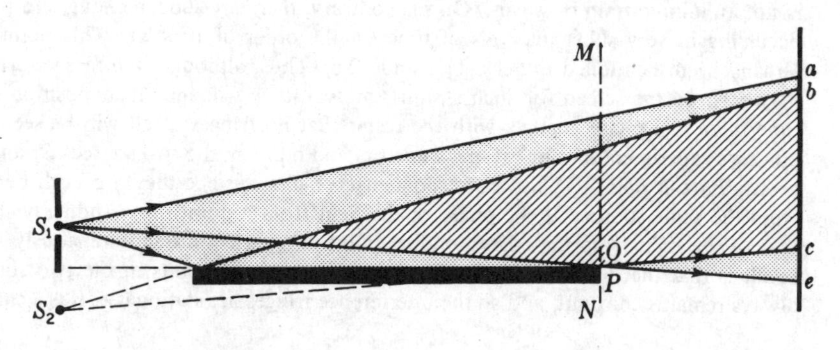

FIGURE 13L
Lloyd's mirror.

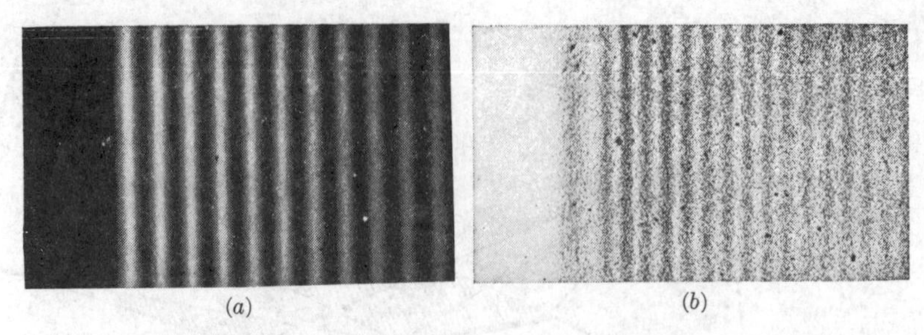

(a) (b)

FIGURE 13M
Interference fringes produced with Lloyd's mirror. (a) Taken with visible light, $\lambda = 4358$ Å. (*After White.*) (b) Taken with X rays, $\lambda = 8.33$ Å. (*After Kellstrom.*)

the mirror to allow the light to enter and leave the glass, and by roughening one of the glass faces with coarse emery.

Other ways exist* for dividing the wave front into two segments and subsequently recombining these at a small angle with each other. For example, one can cut a lens into two halves on a plane through the lens axis and separate the parts slightly, to form two closely adjacent real images of a slit. The images produced in this device, known as *Billet's split lens*, act like the two slits in Young's experiment. A single lens followed by a biplate (two plane-parallel plates at a slight angle) will accomplish the same result.

13.7 COHERENT SOURCES

It will be noticed that the various methods of demonstrating interference so far discussed have one important feature in common: the two interfering beams are always derived from the same source of light. We find by experiment that it is impossible to obtain interference fringes from two separate sources, such as two lamp filaments set side by side. This failure is caused by the fact that the light from any one source is not an infinite train of waves. On the contrary, there are sudden changes in phase occurring in very short intervals of time (of the order of 10^{-8} s). This point has already been mentioned in Secs. 11.1 and 12.6. Thus, although interference fringes may exist on the screen for such a short interval, they will shift their position each time there is a phase change, with the result that no fringes at all will be seen. In Young's experiment and in Fresnel's mirrors and biprism, the two sources S_1 and S_2 always have a point-to-point correspondence of phase, since they are both derived from the same source. If the phase of the light from a point in S_1 suddenly shifts, that of the light from the corresponding point in S_2 will shift simultaneously. The result is that the *difference* in phase between any pair of points in the two sources always remains constant, and so the interference fringes are stationary. It is a charac-

* Good descriptions will be found in T. Preston, "Theory of Light," 5th ed., chap. 7, The Macmillan Company, New York, 1928.

teristic of any interference experiment with light that the sources must have this point-to-point phase relation, and sources that have this relation are called *coherent sources*.

While special arrangements are necessary for producing coherent sources of light, the same is not true of *microwaves*, which are radio waves of a few centimeters wavelength. These are produced by an oscillator which emits a continuous wave, the phase of which remains constant over a time long compared with the duration of an observation. Two independent microwave sources of the same frequency are therefore coherent and can be used to demonstrate interference. Because of the convenient magnitude of their wavelength, microwaves are used to illustrate many common optical interference and diffraction effects.*

If in Young's experiment the source slit S (Fig. 13C) is made too wide or the angle between the rays which leave it too large, the double slit no longer represents two coherent sources and the interference fringes disappear. This subject will be discussed in more detail in Chap. 16.

13.8 DIVISION OF AMPLITUDE. MICHELSON† INTERFEROMETER

Interference apparatus may be conveniently divided into two main classes, those based on *division of wave front* and those based on *division of amplitude*. The previous examples all belong to the former class, in which the wave front is divided laterally into segments by mirrors or diaphragms. It is also possible to divide a wave by partial reflection, the two resulting wave fronts maintaining the original width but having reduced amplitudes. The Michelson interferometer is an important example of this second class. Here the two beams obtained by amplitude division are sent in quite different directions against plane mirrors, whence they are brought together again to form interference fringes. The arrangement is shown schematically in Fig. 13N. The main optical parts consist of two highly polished plane mirrors M_1 and M_2 and two plane-parallel plates of glass G_1 and G_2. Sometimes the rear side of the plate G_1 is lightly silvered (shown by the heavy line in the figure) so that the light coming from the source S is divided into (1) a reflected and (2) a transmitted beam of equal intensity. The light reflected normally from mirror M_1 passes through G_1 a third time and reaches the eye as shown. The light reflected from the mirror M_2 passes back through G_2 for the second time, is reflected from the surface of G_1 and into the

* The technique of such experiments is discussed by G. F. Hull, Jr., *Am. J. Phys.*, **17**:599 (1949).

† A. A. Michelson (1852–1931). American physicist of genius. He early became interested in the velocity of light and began experiments while an instructor in physics and chemistry at the Naval Academy, from which he graduated in 1873. It is related that the superintendent of the Academy asked young Michelson why he wasted his time on such useless experiments. Years later Michelson was awarded the Nobel prize (1907) for his work on light. Much of his work on the speed of light (Sec. 19.3) was done during 10 years spent at the Case Institute of Technology. During the latter part of his life he was professor of physics at the University of Chicago, where many of his famous experiments on the interference of light were done.

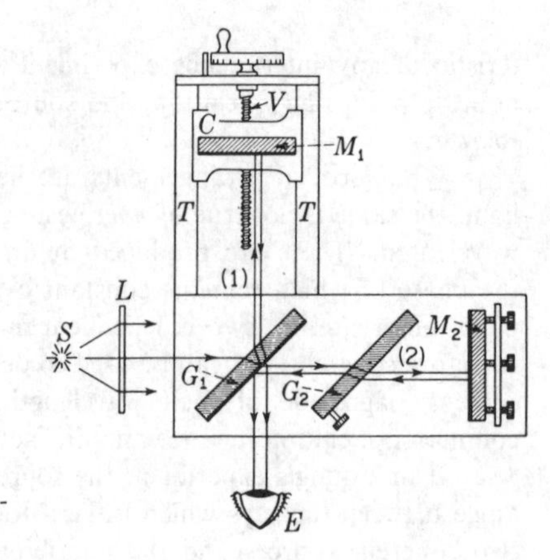

FIGURE 13N
Diagram of the Michelson interferometer.

eye. The purpose of the plate G_2, called the *compensating plate*, is to render the path *in glass* of the two rays equal. This is not essential for producing fringes in monochromatic light, but it is indispensable when white light is used (Sec. 13.11). The mirror M_1 is mounted on a carriage C and can be moved along the well-machined ways or tracks T. This slow and accurately controlled motion is accomplished by means of the screw V, which is calibrated to show the exact distance the mirror has been moved. To obtain fringes, the mirrors M_1 and M_2 are made exactly perpendicular to each other by means of screws shown on mirror M_2.

Even when the above adjustments have been made, fringes will not be seen unless two important requirements are fulfilled. First, the light must originate from an *extended* source. A point source or a slit source, as used in the methods previously described, will not produce the desired system of fringes in this case. The reason for this will appear when we consider the origin of the fringes. Second, the light must in general be *monochromatic*, or nearly so. Especially is this true if the distances of M_1 and M_2 from G_1 are appreciably different.

An extended source suitable for use with a Michelson interferometer may be obtained in any one of several ways. A sodium flame or a mercury arc, if large enough, may be used without the screen L shown in Fig. 13N. If the source is small, a ground-glass screen or a lens at L will extend the field of view. Looking at the mirror M_1 through the plate G_1, one then sees the whole mirror filled with light. In order to obtain the fringes, the next step is to measure the distances of M_1 and M_2 to the back surface of G_1 roughly with a millimeter scale and to move M_1 until they are the same to within a few millimeters. The mirror M_2 is now adjusted to be perpendicular to M_1 by observing the images of a common pin, or any sharp point, placed between the source and G_1. Two pairs of images will be seen, one coming from reflection at the front surface of G_1 and the other from reflection at its back surface. When the tilting screws on M_2 are turned until one pair of images falls exactly on the other, the interference fringes should appear. When they first appear, the fringes will not be clear unless the eye is focused on or near the back mirror M_1, so the observer should look constantly at this mirror while searching for the fringes.

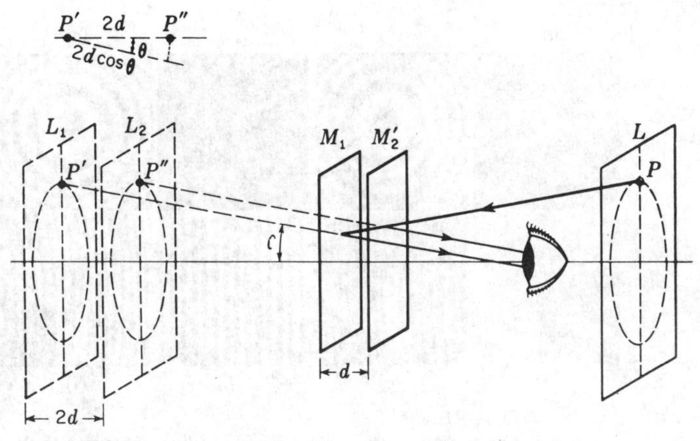

FIGURE 13O
Formation of circular fringes in the Michelson interferometer.

When they have been found, the adjusting screws should be turned in such a way as to continually increase the width of the fringes, and finally a set of concentric circular fringes will be obtained. M_2 is then exactly perpendicular to M_1 if the latter is at an angle of 45° with G_1.

13.9 CIRCULAR FRINGES

These are produced with monochromatic light when the mirrors are in exact adjustment and are the ones used in most kinds of measurement with the interferometer. Their origin can be understood by reference to the diagram of Fig. 13O. Here the real mirror M_2 has been replaced by its virtual image M_2' formed by reflection in G_1. M_2' is then parallel to M_1. Owing to the several reflections in the real interferometer, we may now think of the extended source as being at L, behind the observer, and as forming two virtual images L_1 and L_2 in M_1 and M_2'. These virtual sources are coherent in that the phases of corresponding points in the two are exactly the same at all instants. If d is the separation M_1M_2', the virtual sources will be separated by $2d$. When d is exactly an integral number of half wavelengths, i.e., the path difference $2d$ equal to an integral number of whole wavelengths, all rays of light reflected normal to the mirrors will be in phase. Rays of light reflected at an angle, however, will in general not be in phase. The path difference between the two rays coming to the eye from corresponding points P' and P'' is $2d \cos \theta$, as shown in the figure. The angle θ is necessarily the same for the two rays when M_1 is parallel to M_2' so that the rays are parallel. Hence when the eye is focused to receive parallel rays (a small telescope is more satisfactory here, especially for large values of d) the rays will reinforce each other to produce maxima for those angles θ satisfying the relation

$$2d \cos \theta = m\lambda \qquad \text{(13g)}$$

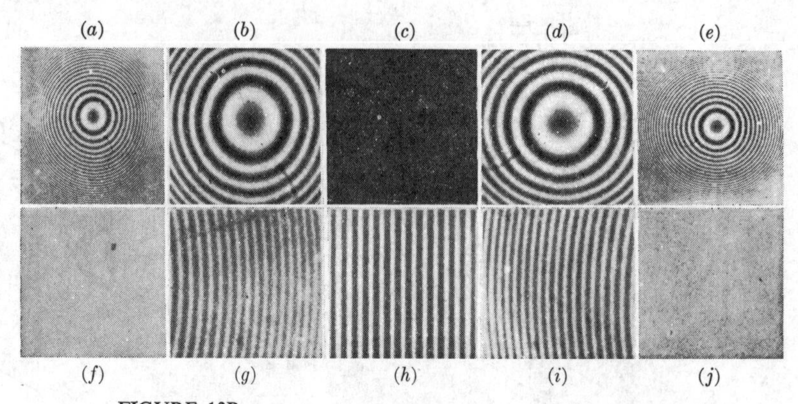

FIGURE 13P
Appearance of the various types of fringes observed in the Michelson interferometer. *Upper row*, circular fringes. *Lower row*, localized fringes. Path difference increases outward, in both directions, from the center.

Since for a given m, λ, and d the angle θ is constant, the maxima will lie in the form of circles about the foot of the perpendicular from the eye to the mirrors. By expanding the cosine, it can be shown from Eq. (13g) that the radii of the rings are proportional to the square roots of integers, as in the case of Newton's rings (Sec. 14.5). The intensity distribution across the rings follows Eq. (13b), in which the phase difference is given by

$$\delta = \frac{2\pi}{\lambda} 2d \cos \theta$$

Fringes of this kind, where parallel beams are brought to interference with a phase difference determined by the angle of inclination θ, are often referred to as *fringes of equal inclination*. In contrast to the type to be described in the next section, this type may remain visible over very large path differences. The eventual limitation on the path difference will be discussed in Sec. 13.12.

The upper part of Fig. 13P shows how the circular fringes look under different conditions. Starting with M_1 a few centimeters beyond M_2', the fringe system will have the general appearance shown in (*a*) with the rings very closely spaced. If M_1 is now moved slowly toward M_2' so that d is decreased, Eq. (13g) shows that a given ring, characterized by a given value of the order m, must decrease its radius because the product $2d \cos \theta$ must remain constant. The rings therefore shrink and vanish at the center, a ring disappearing each time $2d$ decreases by λ, or d by $\lambda/2$. This follows from the fact that at the center $\cos \theta = 1$, so that Eq. (13g) becomes

$$2d = m\lambda \qquad (13h)$$

To change m by unity, d must change by $\lambda/2$. Now as M_1 approaches M_2' the rings become more widely spaced, as indicated in Fig. 13P(*b*), until finally we reach a critical position where the central fringe has spread out to cover the whole field

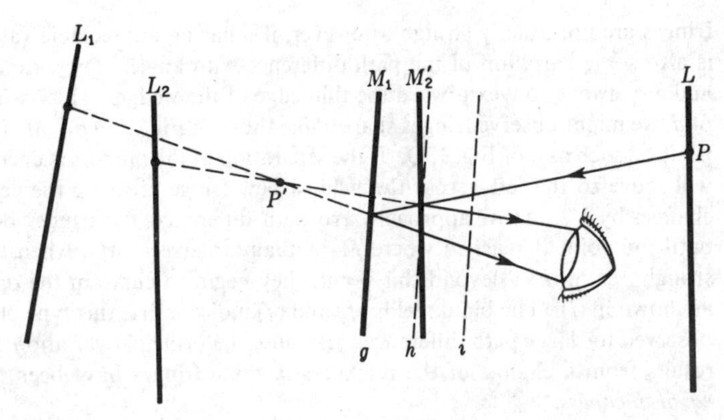

FIGURE 13Q
The formation of fringes with inclined mirrors in the Michelson interferometer.

of view, as shown in (*c*). This happens when M_1 and M_2' are exactly coincident, for it is clear that under these conditions the path difference is zero for all angles of incidence. If the mirror is moved still farther, it effectively passes through M_2', and new widely spaced fringes appear, growing out from the center. These will gradually become more closely spaced as the path difference increases, as indicated in (*d*) and (*e*) of the figure.

13.10 LOCALIZED FRINGES

If the mirrors M_2' and M_1 are not exactly parallel, fringes will still be seen with monochromatic light for path differences not exceeding a few millimeters. In this case the space between the mirrors is wedge-shaped, as indicated in Fig. 13Q. The two rays* reaching the eye from a point P on the source are now no longer parallel, but appear to diverge from a point P' near the mirrors. For various positions of P on the extended source, it can be shown† that the path difference between the two rays remains constant but that the distance of P' from the mirrors changes. If the angle between the mirrors is not too small, however, the latter distance is never great, and hence, in order to see these fringes clearly, the eye must be focused on or near the rear mirror M_1. The localized fringes are practically straight because the variation of the path difference across the field of view is now due primarily to the variation of the thickness of the "air film" between the mirrors. With a wedge-shaped film, the locus of points of equal thickness is a straight line parallel to the edge of the wedge. The

* When the term "ray" is used, here and elsewhere in discussing interference phenomena, it merely indicates the direction of the *perpendicular to a wave front* and is in no way to suggest an infinitesimally narrow pencil of light.

† R. W. Ditchburn, "Light," 2d ed., paperback, John Wiley and Sons, Inc., New York, 1963.

fringes are not exactly straight, however, if d has an appreciable value, because there is also some variation of the path difference with angle. They are in general curved and are always convex toward the thin edge of the wedge. Thus, with a certain value of d, we might observe fringes shaped like those of Fig. 13P(g). M_1 could then be in a position such as g of Fig. 13Q. If the separation of the mirrors is decreased, the fringes will move to the left across the field, a new fringe crossing the center each time d changes by $\lambda/2$. As we approach zero path difference, the fringes become straighter, until the point is reached where M_1 actually intersects M_2', when they are perfectly straight, as in (h). Beyond this point, they begin to curve in the opposite direction, as shown in (i). The blank fields (f) and (j) indicate that this type of fringe cannot be observed for large path differences. Because the principal variation of path difference results from a change of the thickness d, these fringes have been termed *fringes of equal thickness.*

13.11 WHITE-LIGHT FRINGES

If a source of white light is used, no fringes will be seen at all except for a path difference so small that it does not exceed a few wavelengths. In observing these fringes, the mirrors are tilted slightly as for localized fringes, and the position of M_1 is found where it intersects M_2'. With white light there will then be observed a central dark fringe, bordered on either side by 8 or 10 colored fringes. This position is often rather troublesome to find using white light only. It is best located approximately beforehand by finding the place where the localized fringes in monochromatic light become straight. Then a very *slow* motion of M_1 through this region, using white light, will bring these fringes into view.

The fact that only a few fringes are observed with white light is easily accounted for when we remember that such light contains all wavelengths between 400 and 750 nm. The fringes for a given color are more widely spaced the greater the wavelength. Thus the fringes in different colors will only coincide for $d = 0$, as indicated in Fig. 13R. The solid curve represents the intensity distribution in the fringes for green light, and the broken curve that for red light. Clearly, only the central fringe will be uncolored, and the fringes of different colors will begin to separate at once on either side, producing various impure colors which are not the saturated spectral colors. After 8 or 10 fringes, so many colors are present at a given point that the resultant color is essentially white. Interference is still occurring in this region, however, because a spectroscope will show a continuous spectrum with dark bands at those wavelengths for which the condition for destructive interference is fulfilled. White-light fringes are also observed in all the other methods of producing interference described above, if white light is substituted for monochromatic light. They are particularly important in the Michelson interferometer, where they may be used to locate the position of zero path difference, as we shall see in Sec. 13.13.

An excellent reproduction in color of these white-light fringes is given in one of Michelson's books.* The fringes in three different colors are also shown separately

* A. A. Michelson, "Light Waves and Their Uses," plate II, University of Chicago Press, Chicago, 1906.

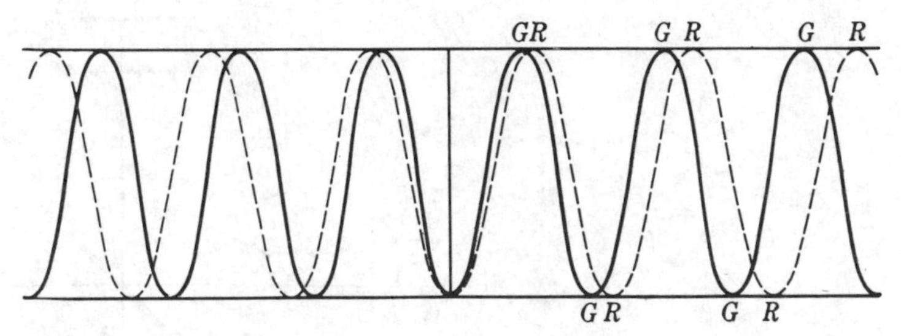

FIGURE 13R
The formation of white-light fringes with a dark fringe at the center.

and a study of these in connection with the white-light fringes is instructive as showing the origin of the various impure colors in the latter.

It was stated above that the central fringe in the white-light system, i.e., that corresponding to zero path difference, is black when observed in the Michelson interferometer. One would ordinarily expect this fringe to be white, since the two beams should be in phase with each other for any wavelength at this point, and in fact this is the case in the fringes formed with the other arrangements, such as the biprism. In the present case, however, it will be seen by referring to Fig. 13N that while ray 1 undergoes an internal reflection in the plate G_1, ray 2 undergoes an external reflection, with a consequent change of phase [see Eq. (14d)]. Hence the central fringe is black if the back surface of G_1 is unsilvered. If it is silvered, the conditions are different and the central fringe may be white.

13.12 VISIBILITY OF THE FRINGES

There are three principal types of measurement that can be made with the interferometer: (1) width and fine structure of spectrum lines, (2) lengths or displacements in terms of wavelengths of light, and (3) refractive indices. As explained in the preceding section, when a certain spread of wavelengths is present in the light source, the fringes become indistinct and eventually disappear as the path difference is increased. With white light they become invisible when d is only a few wavelengths, whereas the circular fringes obtained with the light of a single spectrum line can still be seen after the mirror has been moved several centimeters. Since no line is perfectly sharp, however, the different component wavelengths produce fringes of slightly different spacing, and hence there is a limit to the usable path difference even in this case. For the measurements of length to be described below, Michelson tested the lines from various sources and concluded that a certain red line in the spectrum of cadmium was the most satisfactory. He measured the *visibility*, defined as

$$V = \frac{I_{\text{max}} - I_{\text{min}}}{I_{\text{max}} + I_{\text{min}}} \tag{13i}$$

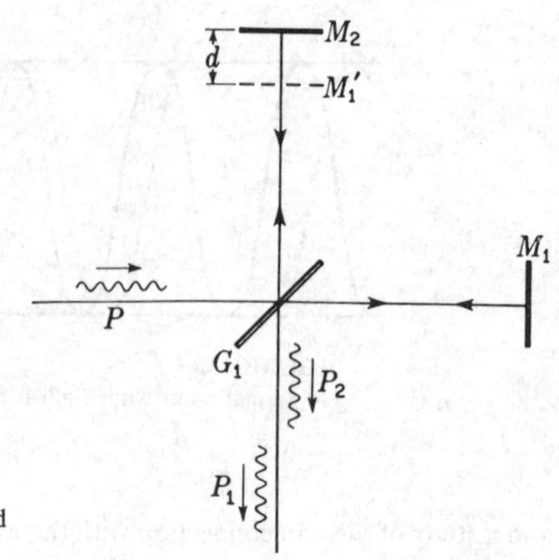

FIGURE 13S
Limiting path difference as determined by the length of wave packets.

where I_{max} and I_{min} are the intensities at the maxima and minima of the fringe pattern. The more slowly V decreases with increasing path difference, the sharper the line. With the red cadmium line, it dropped to 0.5 at a path difference of some 10 cm, or at $d = 5$ cm.

With certain lines, the visibility does not decrease uniformly but fluctuates with more or less regularity. This behavior indicates that the line has a fine structure, consisting of two or more lines very close together. Thus it is found that with sodium light the fringes become alternately sharp and diffuse, as the fringes from the two D lines get in and out of step. The number of fringes between two successive positions of maximum visibility is about 1000, indicating that the wavelengths of the components differ by approximately 1 part in 1000. In more complicated cases, the separation and intensities of the components could be determined by a Fourier analysis of the visibility curves.* Since this method of inferring the structure of lines has now been superseded by more direct methods, to be described in the following chapter, it will not be discussed in any detail here.

An alternative way of interpreting the eventual vanishing of interference at large path differences is instructive to consider at this point. In Sec. 12.6 it was indicated that a finite spread of wavelengths corresponds to wave packets of limited length, this length decreasing as the spread becomes greater. Thus, when the two beams in the interferometer traverse distances that differ by more than the length of the individual packets, these can no longer overlap and no interference is possible. The situation upon complete disappearance of the fringes is shown schematically in Fig. 13S. The original wave packet P has its amplitude divided at G_1 so that two similar packets are produced, P_1 traveling to M_1 and P_2 to M_2. When the beams are reunited, P_2 lags a distance $2d$ behind P_1. Evidently a measurement of this limiting path difference gives a direct determination of the length of the wave packets. This

* A. A. Michelson, "Studies in Optics," chap. 4, University of Chicago Press, Chicago, 1927.

interpretation of the cessation of interference seems at first sight to conflict with the one given above. A consideration of the principle of Fourier analysis shows, however, that mathematically the two are entirely equivalent and are merely alternative ways of representing the same phenomenon.

13.13 INTERFEROMETRIC MEASUREMENTS OF LENGTH

The principal advantage of Michelson's form of interferometer over the earlier methods of producing interference lies in the fact that the two beams are here widely separated and the path difference can be varied at will by moving the mirror or by introducing a refracting material in one of the beams. Corresponding to these two ways of changing the optical path, there are two other important applications of the interferometer. Accurate measurements of distance in terms of the wavelength of light will be discussed in this section, while interferometric determinations of refractive indices are described in Sec. 13.15.

When the mirror M_1 of Fig. 13N is moved slowly from one position to another, counting the number of fringes in monochromatic light which cross the center of the field of view will give a measure of the distance the mirror has moved in terms of λ, since by Eq. (13h) we have, for the position d_1 corresponding to the bright fringe of order m_1,

$$2d_1 = m_1\lambda$$

and for d_2, giving a bright fringe of order m_2,

$$2d_2 = m_2\lambda$$

Subtracting these two equations, we find

$$d_1 - d_2 = (m_1 - m_2)\frac{\lambda}{2} \qquad (13j)$$

Hence the distance moved equals the number of fringes counted, multiplied by a half wavelength. Of course, the distance measured need not correspond to an integral number of half wavelengths. Fractional parts of a whole fringe displacement can easily be estimated to one-tenth of a fringe, and, with care, to one-fiftieth. The latter figure then gives the distance to an accuracy of one-hundredth wavelength, or 5×10^{-7} cm for green light.

A small Michelson interferometer in which a microscope is attached to the moving carriage carrying M_1 is frequently used in the laboratory for measuring the wavelength of light. The microscope is focused on a fine glass scale, and the number of fringes, $m_1 - m_2$, crossing the mirror between two readings d_1 and d_2 on the scale gives λ, by Eq. (13j). The bending of a beam, or even of a brick wall, under pressure from the hand can be made visible and measured by attaching M_1 directly to the beam or wall.

The most important measurement made with the interferometer was the comparison of the standard meter in Paris with the wavelengths of intense red, green, and blue lines of cadmium by Michelson and Benoit. For reasons discussed in the last section, it would be impossible to count directly the number of fringes for a displacement of the movable mirror from one end of the standard meter to the other.

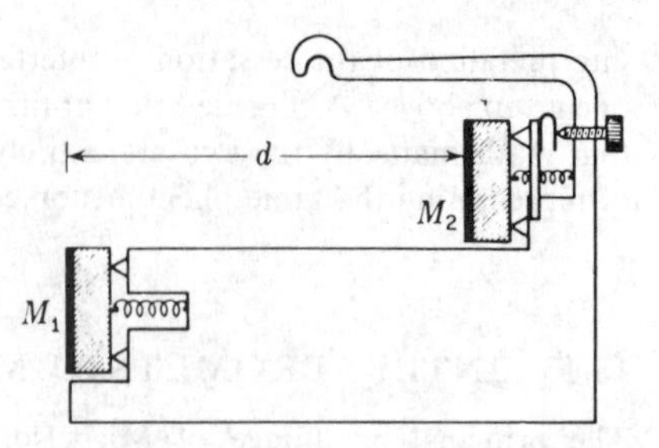

FIGURE 13T
One of the nine etalons used by Michelson in accurately comparing the wavelength of light with the standard meter.

Instead, nine intermediate standards (etalons) were used, of the form shown in Fig. 13T, each approximately twice the length of the other. The two shortest etalons were first mounted in an interferometer of special design (Fig. 13U), with a field of view covering the four mirrors, M_1, M_2, M_1', and M_2'. With the aid of the white light fringes the distances of M, M_1, and M_1' from the eye were made equal, as shown in the figure. Substituting the light of one of the cadmium lines for white light, M was then moved slowly from A to B, counting the number of fringes passing the cross hair. The count was continued until M reached the position B, which was exactly coplanar with M_2, as judged by the appearance of the white-light fringes in the upper mirror of the shorter etalon. The fraction of a cadmium fringe in excess of an integral number required to reach this position was determined, giving the distance M_1M_2 in terms of wavelengths. The shorter etalon was then moved through its own length, without counting fringes, until the white-light fringes reappeared in M_1. Finally M was moved to C, when the white-light fringes appeared in M_2' as well as in M_2. The additional displacement necessary to make M coplanar with M_2 was measured in terms of cadmium fringes, thus giving the exact number of wavelengths in the longer etalon. This was in turn compared with the length of a third etalon of approximately twice the length of the second, by the same process.

The length of the largest etalon was about 10.0 cm. This was finally compared with the prototype meter by alternately centering the white-light fringes in its upper and lower mirrors, each time the etalon was moved through its own length. Ten such steps brought a marker on the side of the etalon nearly into coincidence with the second fiducial mark on the meter, and the slight difference was evaluated by counting cadmium fringes. The 10 steps involve an accumulated error which does not enter in the intercomparison of the etalons, but nevertheless this was smaller than the uncertainty in setting on the end marks.

The final results were, for the three cadmium lines:

Red line	1 m = 1,553,163.5λ	or	λ = 6438.4722 Å
Green line	1 m = 1,966,249.7λ	or	λ = 5085.8240 Å
Blue line	1 m = 2,083,372.1λ	or	λ = 4799.9107 Å

Not only has the standard meter been determined in terms of what we now believe to be an invariable unit, the wavelength of light, but we have also obtained absolute determinations of the wavelength of three spectrum lines, the red line of

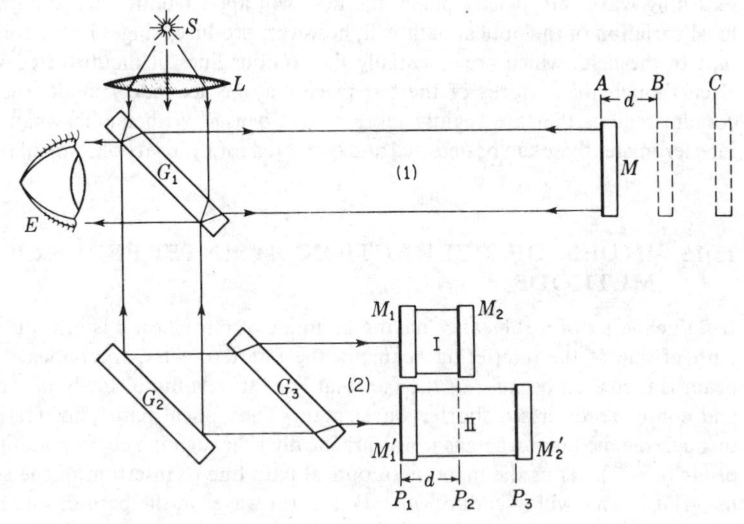

FIGURE 13U
Special Michelson interferometer used in accurately comparing the wavelength of light with the standard meter.

which is at present the primary standard in spectroscopy. More recent measurements on the orange line of the krypton spectrum have been made (see Sec. 14.11). It now is internationally agreed that in dry atmospheric air at 15°C and a pressure of 760 mmHg the orange line of krypton has a wavelength

$$\lambda_0 = 6057.80211 \text{ Å}$$

This is the wavelength the General Conference on Weights and Measures in Paris used in adopting on Oct. 14, 1960, as the international legal standard of length, the following definition of the standard meter:

$$1 \textit{ meter} = 1{,}650{,}763.73 \textit{ wavelengths (orange light of krypton)}$$

13.14 TWYMAN AND GREEN INTERFEROMETER

If a Michelson interferometer is illuminated with strictly parallel monochromatic light, produced by a point source at the principal focus of a well-corrected lens, it becomes a very powerful instrument for testing the perfection of optical parts such as prisms and lenses. The piece to be tested is placed in one of the light beams, and the mirror behind it is so chosen that the reflected waves, after traversing the test piece a second time, again become plane. These waves are then brought to interference with the plane waves from the other arm of the interferometer by another lens, at the focus of which the eye is placed. If the prism or lens is optically perfect, so that the returning waves are strictly plane, the field will appear uniformly illuminated. Any local variation of the optical path will, however, produce fringes in the corresponding part of the field, which are essentially the contour lines of the distorted wave front. Even though the surfaces of the test piece may be accurately made, the glass may contain regions that are slightly more or less dense. With the Twyman and Green interferometer these can be detected and corrected for by local polishing of the surface.*

13.15 INDEX OF REFRACTION BY INTERFERENCE METHODS

If a thickness t of a substance having an index of refraction n is introduced into the path of one of the interfering beams in the interferometer, the optical path in this beam is increased because of the fact that light travels more slowly in the substance and consequently has a shorter wavelength. The optical path [Eq. (1t)] is now nt through the medium, whereas it was practically t through the corresponding thickness of air ($n = 1$). Thus the increase in optical path due to insertion of the substance is $(n - 1)t$.† This will introduce $(n - 1)t/\lambda$ extra waves in the path of one beam; so if we call Δm the number of fringes by which the fringe system is displaced when the substance is placed in the beam, we have

$$(n - 1)t = (\Delta m)\lambda \qquad (13k)$$

In principle a measurement of Δm, t, and λ thus gives a determination of n.

In practice, the insertion of a plate of glass in one of the beams produces a discontinuous shift of the fringes so that the number Δm cannot be counted. With monochromatic fringes it is impossible to tell which fringe in the displaced set corresponds to one in the original set. With white light, the displacement in the fringes of different colors is very different because of the variation of n with wavelength, and the fringes disappear entirely. This illustrates the necessity of the compensating plate G_2 in Michelson's interferometer if white-light fringes are to be observed. If the plate of glass is very thin, these fringes may still be visible, and this affords a method of measuring n for very thin films. For thicker pieces, a practicable method is to use two plates of identical thickness, one in each beam, and to turn one gradually about a vertical axis, counting the number of monochromatic fringes for a given angle of rotation. This angle then corresponds to a certain known increase in effective thickness.

For the measurement of the index of refraction of gases, which can be introduced gradually into the light path by allowing the gas to flow into an evacuated tube, the interference method is the most practicable one. Several forms of refractometers have been devised especially for this purpose, of which we shall describe three, the Jamin, the Mach-Zehnder, and the Rayleigh refractometers.

Jamin's refractometer is shown schematically in Fig. 13V(*a*). Monochromatic

* For a more complete description of the use of this instrument, see F. Twyman, "Prism and Lens Making," 2d ed., chap. 12, Hilger and Watts, London, 1952.

† In the Michelson interferometer, where the beam traverses the substance twice in its back-and-forth path, t is *twice* the actual thickness.

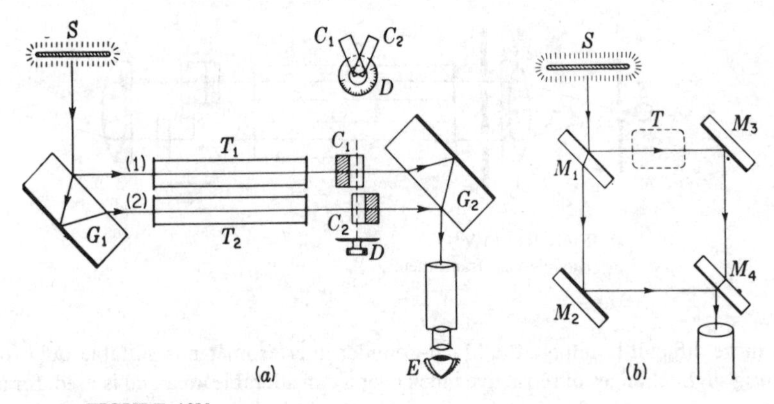

FIGURE 13V
(*a*) The Jamin and (*b*) the Mach-Zehnder interferometer.

light from a broad source S is broken into two parallel beams 1 and 2 by reflection at the two parallel faces of a thick plate of glass G_1. These two rays pass through to another identical plate of glass G_2 to recombine after reflection, forming interference fringes known as Brewster's fringes (see Sec. 14.11). If now the plates are parallel, the light paths will be identical. Suppose as an experiment we wish to measure the index of refraction of a certain gas at different temperatures and pressures. Two similar evacuated tubes T_1 and T_2 of equal length are placed in the two parallel beams. Gas is slowly admitted to tube T_2. If the number of fringes Δm crossing the field is counted while the gas reaches the desired pressure and temperature, the value of n can be found by applying Eq. (13k). It is found experimentally that at a given temperature the value $n - 1$ is directly proportional to the pressure. This is a special case of the *Lorenz-Lorentz* law*, according to which

$$\frac{n^2 - 1}{n^2 + 2} = (n - 1)\frac{n + 1}{n^2 + 2} = \text{const} \times \rho \tag{13l}$$

Here ρ is the density of the gas. When n is very nearly unity, the factor $(n + 1)/(n^2 + 2)$ is nearly constant, as required by the above experimental observation.

The interferometer devised by Mach and Zehnder, and shown in Fig. 13V(*b*), has a similar arrangement of light paths, but they may be much farther apart. The role of the two glass blocks in the Jamin instrument is here taken by two pairs of mirrors, the pair M_1 and M_2 functioning like G_1, and the pair M_3 and M_4 like G_2. The first surface of M_1 and the second surface of M_4 are half-silvered. Although it is

* H. A. Lorentz (1853–1928). For many years professor of mathematical physics at the University of Leyden, Holland. Awarded the Nobel prize (1902) for his work on the relations between light, magnetism, and matter, he also contributed notably to other fields of physics. Gifted with a charming personality and kindly disposition, he traveled a great deal, and was widely known and liked. By a strange coincidence L. Lorenz of Copenhagen derived the above law from the elastic-solid theory only a few months before Lorentz obtained it from the electromagnetic theory.

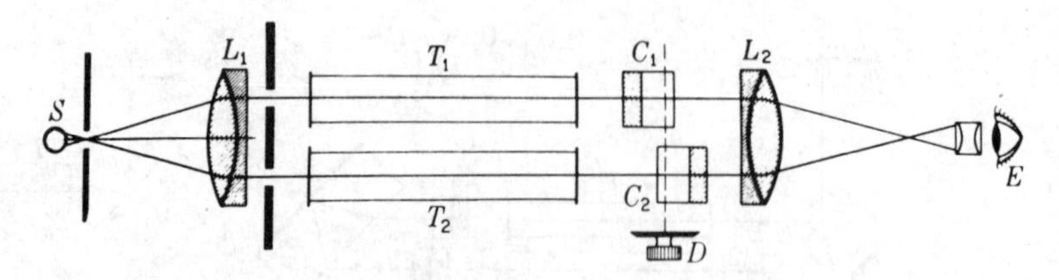

FIGURE 13W
Rayleigh's refractometer.

more difficult to adjust, the Mach-Zehnder interferometer is suitable only for studying slight changes of refractive index over a considerable area and is used, for example, in measuring the flow patterns in wind tunnels (see also Sec. 28.14). Contrary to the situation in the Michelson interferometer, the light traverses a region such as T in the figure in only one direction, a fact which simplifies the study of local changes of optical path in that region.

The purpose of the compensating plates C_1 and C_2 in Figs. 13V(*a*) and 13W is to speed up the measurement of refractive index. As the two plates, of equal thickness, are rotated together by the single knob attached to the dial D, one light path is shortened and the other lengthened. The device can therefore compensate for the path difference in the two tubes. The dial, if previously calibrated by counting fringes, can be made to read the index of refraction directly. The sensitivity of this device can be varied at will, a high sensitivity being obtained when the angle between the two plates is small and a low sensitivity when the angle is large.

In Rayleigh's* refractometer (Fig. 13W) monochromatic light from a linear source S is made parallel by a lens L_1 and split into two beams by a fairly wide double slit. After passing through two exactly similar tubes and the compensating plates, these are brought to interfere by the lens L_2. This form of refractometer is often used to measure slight differences in refractive index of liquids and solutions.

PROBLEMS

13.1 Young's experiment is performed with orange light from a krypton arc. If the fringes are measured with a micrometer eyepiece at a distance 100 cm from the double slit, it is found that 25 of them occupy a distance of 12.87 mm between centers. Find the distance between the centers of the two slits. *Ans.* 1.1297 mm

13.2 A double slit with a separation of 0.250 mm between centers is illuminated with green light from a cadmium-arc lamp. How far behind the slits must one go to measure the fringe separation and find it to be 0.80 mm between centers?

* Lord Rayleigh (third Baron) (1842–1919). Professor of physics at Cambridge University and the Royal Institution of Great Britain. Gifted with great mathematical ability and physical insight, he made important contributions to many fields of physics. His works on sound and on the scattering of light (Sec. 22.9) are the best known. He was a Nobel prize winner in 1904.

13.3 When a thin film of transparent plastic is placed over one of the slits in Young's double-slit experiment, the central bright fringe, of the white-light fringe system, is displaced by 4.50 fringes. The refractive index of the material is 1.480, and the effective wavelength of the light is 5500 Å. (*a*) By how much does the film increase the optical path? (*b*) What is the thickness of the film? (*c*) What would probably be observed if a piece of the material 1.0 mm thick were used? (*d*) Why?

13.4 Lloyd's-mirror experiment is readily demonstrated with microwaves, using as a reflector a sheet of metal lying flat on the table. If the source has a frequency of 12,000 MHz and is located 10.0 cm above the sheet-metal surface, find the height above the surface of the first two maxima 3.0 m from the source.

Ans. (*a*) 18.750 cm, (b) 56.25 cm

Note: A phase change of π occurs upon reflection; see Sec. 13.6.

13.5 A Fresnel biprism is to be constructed for use on an optical bench with the slit and the observing screen 180.0 cm apart. The biprism is to be 60.0 cm from the slit. Find the angle between the two refracting surfaces of the biprism if the glass has a refractive index $n = 1.520$, sodium yellow light is to be used, and the fringes are to be 1.0 mm apart.

13.6 A Fresnel biprism of index 1.7320 and with apex angles of 0.850° is used to form interference fringes. Find the fringe separation for red light of wavelength 6563 Å when the distance between the slit and the prism is 25.0 cm and that between the prism and the screen is 75.0 cm.

13.7 What must be the angle in degrees between the two Fresnel mirrors in order to produce sodium light fringes 1.0 mm apart if the slit is 40.0 cm from the mirror intersection and the screen is 150.0 cm from the slit? Assume $\lambda = 5.893 \times 10^{-5}$ cm.

Ans. 0.06331°

13.8 How far must the movable mirror of a Michelson interferometer be displaced for 2500 fringes of the red cadmium line to cross the center of the field of view?

13.9 If the mirror of a Michelson interferometer is moved 1.0 mm, how many fringes of the blue cadmium line will be counted crossing the field of view?

13.10 Find the angular radius of the tenth bright fringe in a Michelson interferometer when the central-path difference ($2d$) is (*a*) 1.50 mm and (*b*) 1.5 cm. Assume the orange light of a krypton arc is used and that the interferometer is adjusted in each case so that the first bright fringe forms a maximum at the center of the pattern.

Ans. (*a*) 4.885°, (*b*) 1.542°

14
INTERFERENCE INVOLVING MULTIPLE REFLECTIONS

Some of the most beautiful effects of interference result from the multiple reflection of light between the two surfaces of a thin film of transparent material. These effects require no special apparatus for their production or observation and are familiar to anyone who has noticed the colors shown by thin films of oil on water, by soap bubbles, or by cracks in a piece of glass. We begin our investigation of this class of interference by considering the somewhat idealized case of reflection and refraction from the boundary separating different optical media. In Fig. 14A(*a*) a ray of light in air or vacuum incident on a plane surface of a transparent medium like water is indicated by *a*. The reflected and refracted rays are indicated by *ar* and *at*, respectively.

A question of particular interest from the standpoint of physical optics is that of a possible abrupt *change of phase* of waves when they are reflected from a boundary. For a given boundary the result will differ, as we shall now show, according to whether the waves approach from the side of higher velocity or from that of lower velocity. Thus, let the symbol *a* in the left-hand part of Fig. 14A represent the amplitude (not the intensity) of a set of waves striking the surface, let *r* be the fraction of the amplitude reflected, and let *t* be the fraction transmitted. The amplitudes of the two sets of waves will then be *ar* and *at*, as shown. Now, following a treatment given by

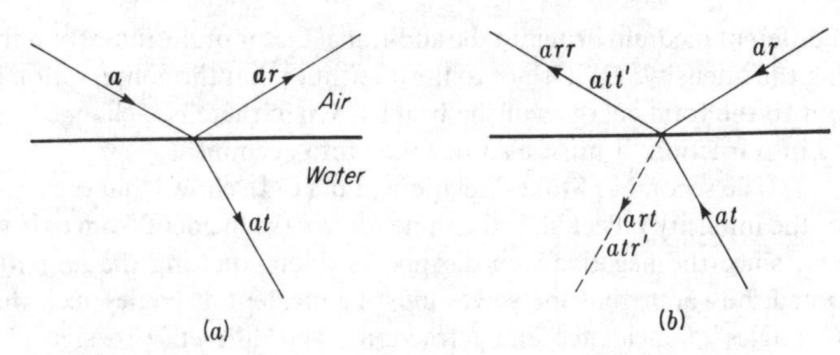

FIGURE 14A
Stokes' treatment of reflection.

Stokes*, imagine the two sets reversed in direction, as in part (*b*) of the figure. Provided there is no dissipation of energy by absorption, a wave motion is a strictly reversible phenomenon. It must conform to the law of mechanics known as the *principle of reversibility*, according to which the result of an instantaneous reversal of all the velocities in a dynamic system is to cause the system to retrace its whole previous motion. That the paths of light rays are in conformity with this principle has already been stated in Sec. 1.8. The two reversed trains, of amplitude *ar* and *at*, should accordingly have as their net effect after striking the surface a wave in air equal in amplitude to the incident wave in part (*a*) but traveling in the opposite direction. The wave of amplitude *ar* gives a reflected wave of amplitude *arr* and a refracted wave of amplitude *art*. If we call r' and t' the fractions of the amplitude reflected and refracted when the reversed wave *at* strikes the boundary from below, this contributes amplitudes att' and atr' to the two waves, as indicated. Now, since the resultant effect must consist only of a wave in air of amplitude a, we have

$$att' + arr = a \qquad (14a)$$

and

$$art + atr' = 0 \qquad (14b)$$

The second equation states that the two incident waves shall produce no net disturbance on the water side of the boundary. From Eq. (14a) we obtain

$$tt' = 1 - r^2 \qquad (14c)$$

and from Eq. (14b)

$$r' = -r \qquad (14d)$$

It might at first appear that Eq. (14c) could be carried further by using the fact that intensities are proportional to squares of amplitudes and by writing, by conservation of energy, $r^2 + t^2 = 1$. This would immediately yield $t = t'$. The result is not correct, however, for two reasons: (1) although the proportionality of intensity with square of amplitude holds for light traveling in a single medium, passage into

* Sir George Stokes (1819–1903), versatile mathematician and physicist of Pembroke College, Cambridge, and pioneer in the study of the interaction of light with matter. He is known for his laws of fluorescence (Sec. 22.6) and of the rate of fall of spheres in viscous fluids. The treatment referred to here was given in his "Mathematical and Physical Papers," vol. 2, pp. 89ff., especially p. 91.

a different medium brings in the additional factor of the index of refraction in determining the intensity; (2) it is not to the intensities that the conservation law is to be applied but to the total energies of the beams. When there is a change in width of the beam, as in refraction, it must also be taken into account.

The second of Stokes' relations, Eq. (14d), shows that the *reflectance*, or fraction of the intensity reflected, is the same for a wave incident from either side of the boundary, since the negative sign disappears upon squaring the amplitudes. It should be noted, however, that the waves must be incident at angles such that they correspond to angles of incidence and refraction. The difference in sign of the amplitudes in Eq. (14d) indicates a difference of phase of π between the two cases, since a reversal of sign means a displacement in the opposite sense. If there is no phase change on reflection from above, there must be a phase change of π on reflection from below; or correspondingly, if there is no change on reflection from below, there must be a change of π on reflection from above.

The principle of reversibility as applied to light waves is often useful in optical problems; for example, it proves at once the interchangeability of object and image. The conclusion reached above about the change of phase is not dependent on the applicability of the principle, i.e., on the absence of absorption, but holds for reflection from any boundary. It is a matter of experimental observation that in the reflection of light under the above conditions, the phase change of π occurs when the light strikes the boundary from the side of higher velocity,* so that the second of the two alternatives mentioned is the correct one in this case. A change of phase of the same type is encountered in the reflection of simple mechanical waves, such as transverse waves in a rope. Reflection with change of phase where the velocity decreases in crossing the boundary corresponds to the reflection of waves from a fixed end of a rope. Here the elastic reaction of the fixed end of the rope immediately produces a reflected train of opposite phase traveling back along the rope. The case where the velocity increases in crossing the boundary has its parallel in reflection from a free end of a rope. The end of the rope undergoes a displacement of twice the amount it would have if the rope were continuous, and it immediately starts a wave in the reverse direction having the same phase as the incident wave.

14.1 REFLECTION FROM A PLANE-PARALLEL FILM

Let a ray of light from a source S be incident on the surface of such a film at A (Fig. 14B). Part of this will be reflected as ray 1 and part refracted in the direction AF. Upon arrival at F, part of the latter will be reflected to B and part refracted toward H. At B the ray FB will be again divided. A continuation of this process yields two sets of parallel rays, one on each side of the film. In each of these sets, of course, the intensity decreases rapidly from one ray to the next. If the set of parallel reflected rays is now collected by a lens and focused at the point P, each ray will have traveled a different distance, and the phase relations may be such as to produce destructive or constructive interference at that point. It is such interference that produces the

* See the discussion in Sec. 13.6 in connection with Lloyd's mirror.

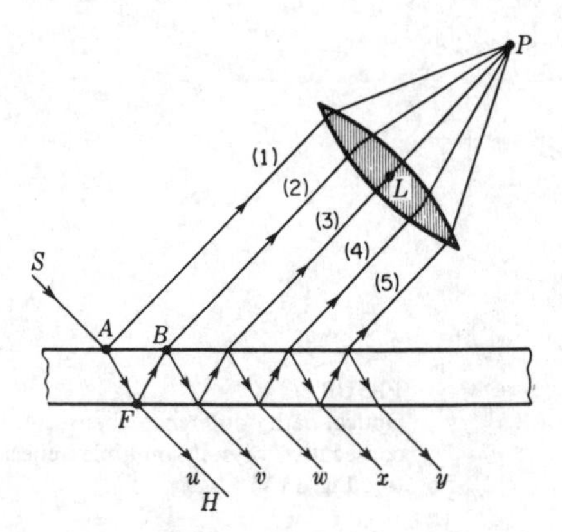

FIGURE 14B
Multiple reflections in a plane-parallel film.

colors of thin films when they are viewed by the naked eye. In such a case L is the lens of the eye, and P lies on the retina.

In order to find the phase difference between these rays, we must first evaluate the difference in the optical path traversed by a pair of successive rays, such as rays 1 and 2. In Fig. 14C let d be the thickness of the film, n its index of refraction, λ the wavelength of the light, and ϕ and ϕ' the angles of incidence and refraction. If BD is perpendicular to ray 1, the optical paths from D and B to the focus of the lens will be equal. Starting at A, ray 2 has the path AFB in the film and ray 1 the path AD in air. The difference in these optical paths is given by

$$\Delta = n(AFB) - AD$$

If BF is extended to intersect the perpendicular line AE at G, $AF = GF$ because of the equality of the angles of incidence and reflection at the lower surface. Thus we have

$$\Delta = n(GB) - AD = n(GC + CB) - AD$$

Now AC is drawn perpendicular to FB; so the broken lines AC and DB represent two successive positions of a wave front reflected from the lower surface. The optical paths must be the same by any ray drawn between two wave fronts; so we may write

$$n(CB) = AD$$

The path difference then reduces to

$$\Delta = n(GC) = n(2d \cos \phi') \qquad (14e)$$

If this path difference is a whole number of wavelengths, we might expect rays 1 and 2 to arrive at the focus of the lens in phase with each other and produce a maximum of intensity. However, we must take account of the fact that ray 1 undergoes a phase change of π at reflection, while ray 2 does not, since it is internally reflected. The condition

$$2nd \cos \phi' = m\lambda \qquad \textit{Minima} \qquad (14f)$$

then becomes a condition for *destructive* interference as far as rays 1 and 2 are concerned. As before, $m = 0, 1, 2, \ldots$ is the order of interference.

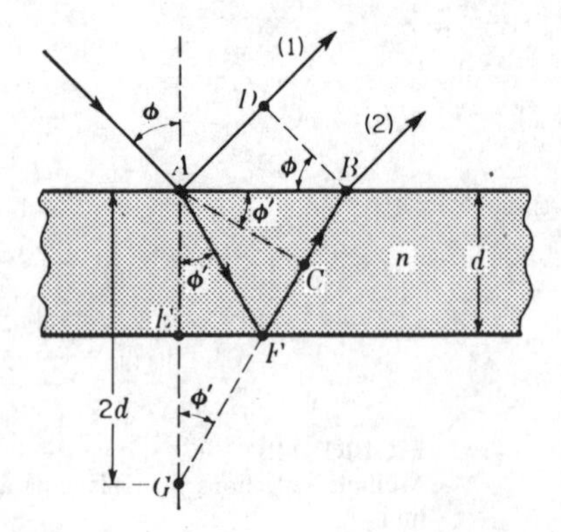

FIGURE 14C
Optical-path difference between two consecutive rays in multiple reflection (see Fig. 14A).

Next we examine the phases of the remaining rays, 3, 4, 5, ... Since the geometry is the same, the path difference between rays 3 and 2 will also be given by Eq. (14e), but here there are only internal reflections involved, so that if Eq. (14f) is fulfilled, ray 3 will be in the same phase as ray 2. The same holds for all succeeding pairs, and so we conclude that under these conditions rays 1 and 2 will be out of phase, but rays 2, 3, 4, ..., will be in phase with each other. On the other hand, if conditions are such that

$$2nd \cos \phi' = (m + \tfrac{1}{2})\lambda \qquad \textit{Maxima} \qquad (14g)$$

ray 2 will be in phase with 1, but 3, 5, 7, ... will be out of phase with 2, 4, 6, ... Since 2 is more intense than 3, 4 more intense than 5, etc., these pairs cannot cancel each other, and since the stronger series combines with 1, the strongest of all, there will be a maximum of intensity.

For the minima of intensity, ray 2 is out of phase with ray 1, but 1 has a considerably greater amplitude than 2, so that these two will not completely annul each other. We can now prove that the addition of 3, 4, 5, ..., which are all in phase with 2, will give a net amplitude just sufficient to make up the difference and to produce complete darkness at the minima. Using a for the amplitude of the incident wave, r for the fraction of this reflected, and t or t' for the fraction transmitted in going from rare to dense or dense to rare, as was done in Stokes' treatment of reflection, Fig. 14D is constructed and the amplitudes labeled as shown. In accordance with Eq. (14d), we have taken the fraction reflected internally and externally to be the same. Adding the amplitudes of all the reflected rays but the first on the upper side of the film, we obtain the resultant amplitude,

$$A = atrt' + atr^3t' + atr^5t' + atr^7t' + \cdots$$
$$= atrt' \, (1 + r^2 + r^4 + r^6 + \cdots)$$

Since r is necessarily less than 1, the geometrical series in parentheses has a finite sum equal to $1/(1 - r^2)$, giving

$$A = atrt' \frac{1}{1 - r^2}$$

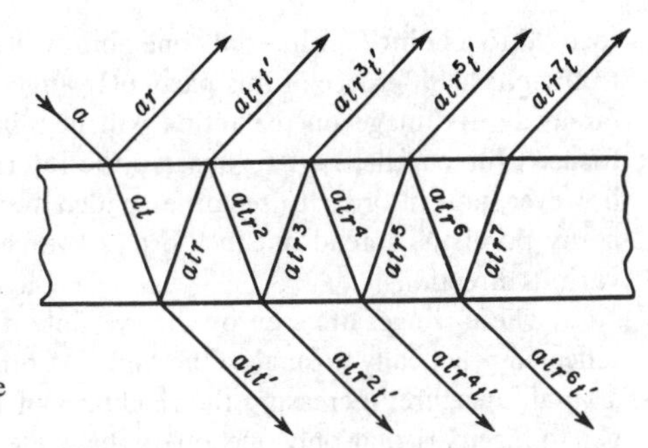

FIGURE 14D
Amplitudes of successive rays in multiple reflection.

But from Stokes' treatment, Eq. (14c), $tt' = 1 - r^2$; so we obtain finally

$$A = ar \tag{14h}$$

This is just equal to the amplitude of the first reflected ray, so we conclude that under the conditions of Eq. (14f) there will be complete destructive interference.

14.2 FRINGES OF EQUAL INCLINATION

If the image of an *extended* source reflected in a thin plane-parallel film is examined, it will be found to be crossed by a system of distinct interference fringes, provided the source emits monochromatic light and provided the film is sufficiently thin. Each bright fringe corresponds to a particular path difference giving an integral value of m in Eq. (14g). For any fringe, the value of ϕ is fixed; so the fringe will have the form of the arc of a circle whose center is at the foot of the perpendicular drawn from the eye to the plane of the film. Evidently we are here concerned with fringes of equal inclination, and the equation for the path difference has the same form as for the circular fringes in the Michelson interferometer (Sec. 13.9).

Note that if m is the order of interference for light incident on the film at $\phi = 0°$, Eq. (14f) gives

$$m = \frac{2nd}{\lambda}$$

which would be a dark fringe. Since the path difference for the first, second, and third, etc., bright fringes will be at progressively larger angles of ϕ and ϕ' [Eq. (14g)], the successive path differences, $2nd \cos \phi'$, will be successively shorter and bright-iight fringes will be at angles where $2nd \cos \phi'$ is equal $(m - \frac{1}{2})\lambda$, $(m - \frac{3}{2})\lambda$, $(m - \frac{5}{2})\lambda$, etc.

The necessity of using an extended source will become clear upon consideration of Fig. 14B. If a very distant point source S is used, the parallel rays will necessarily reach the eye at only one angle (that required by the law of reflection) and will be focused to a point P. Thus only one point will be seen, either bright or dark, according to the phase difference at this particular angle. It is true that if the source is not very far away, its image on the retina will be slightly blurred, because the eye must be focused for parallel rays to observe the interference. The area illuminated is small, however, and in order to see an extended system of fringes, we must obviously have many points S, spread out in a broad source so that the light reaches the eye from various directions.

These fringes are seen by the eye only if the film is very thin, unless the light is reflected practically normal to the film. At other angles, since the pupil of the eye has a small aperture, increasing the thickness of the film will cause the reflected rays to get so far apart that only one enters the eye at a time. Obviously no interference can occur under these conditions. Using a telescope of large aperture, the lens may include enough rays for the fringes to be visible with thick plates, but unless viewed nearly normal to the plate, they will be so finely spaced as to be invisible. The fringes seen with thick plates near normal incidence are often called *Haidinger* fringes*.

14.3 INTERFERENCE IN THE TRANSMITTED LIGHT

The rays emerging from the lower side of the film, shown in Fig. 14B and 14D, can also be brought together with a lens and caused to interfere. Here, however, there are no phase changes at reflection for any of the rays, and the relations are such that Eq. (14f) now becomes the condition for maxima and Eq. (14g) the condition for minima. For maxima, the rays u v, w, ... of Fig. 14B are all in phase, while for minima, v, x, ... are out of phase with u, w, ... When the reflectance r^2 has a low value, as with the surfaces of unsilvered glass, the amplitude of u is much the greatest in the series and the minima are not by any means black. Figure 14E shows quantitative curves for the intensity transmitted I_T and reflected I_R plotted in this instance for $r = 0.2$ according to Eqs. (14n) and (14o), ahead. The corresponding reflectance of 4 percent is close to that of glass at normal incidence. The abscissas δ in the figure represent the phase difference between successive rays in the transmitted set or between all but the first pair in the reflected set, which by Eq. (14e) is

$$\delta = k\Delta = \frac{2\pi}{\lambda}\Delta = \frac{4\pi}{\lambda}nd\cos\phi' \qquad (14i)$$

It will be noted that the curve for I_R looks very much like the $\cos^2$ contour obtained from the interference of two beams. It is not exactly the same, however, and the resemblance holds only when the reflectance is small. Then the first two reflected beams are so much stronger than the rest that the latter have little effect. The important changes that come in at higher values of the reflectance will be discussed in Sec. 14.7.

* W. K. von Haidinger (1795–1871). Austrian mineralogist and geologist, for 17 years director of the Imperial Geological Institute in Vienna.

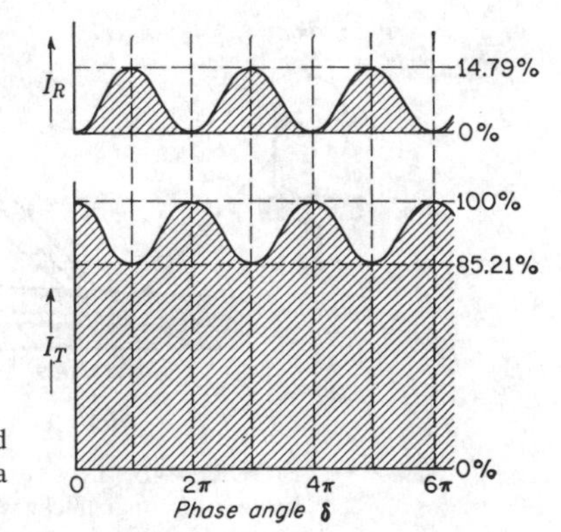

FIGURE 14E
Intensity contours of the reflected and transmitted fringes from a film having a reflectance of 4 percent.

14.4 FRINGES OF EQUAL THICKNESS

If the film is not plane-parallel, so that the surfaces make an appreciable angle with each other as in Fig. 14F(*a*), the interfering rays do not enter the eye parallel to each other but appear to diverge from a point near the film. The resulting fringes resemble the localized fringes in the Michelson interferometer and appear to be formed in the film itself. If the two surfaces are plane, so that the film is wedge-shaped, the fringes will be practically straight, following the lines of equal thickness. In this case the path difference for a given pair of rays is practically that given by Eq. (14e). Provided that observations are made almost normal to the film, the factor cos ϕ' may be considered equal to 1, and the condition for bright fringes becomes

$$2nd = (m + \tfrac{1}{2})\lambda \qquad (14j)$$

In going from one fringe to the next m increases by 1, and this requires the optical thickness of the film nd to change by $\lambda/2$.

Fringes formed in thin films are easily shown in the laboratory or lecture room by using two pieces of ordinary plate glass. If they are laid together with a thin strip of paper along one edge, we obtain a wedge-shaped film of air between the plates. When a sodium flame or arc is viewed as in Fig. 14F, yellow fringes are clearly seen. If a carbon arc and filter are used, the fringes may be projected on a screen with a lens. On viewing the reflected image of a monochromatic source, one will find it to be crossed by more or less straight fringes, such as those in Fig. 14F(*b*).

This class of fringes has important practical applications in the testing of optical surfaces for planeness. If an air film is formed between two surfaces, one of which is perfectly plane and the other not, the fringes will be irregular in shape. Any fringe is characterized by a particular value of m in Eq. (14j), and hence will follow those parts of the film where d is constant. That is, the fringes form the equivalent of *contour lines* for the uneven surface. The contour interval is $\lambda/2$, since for air $n = 1$, and going from one fringe to the next corresponds to increasing d by this amount. The standard method of producing optically plane surfaces uses repeated observation of the fringes

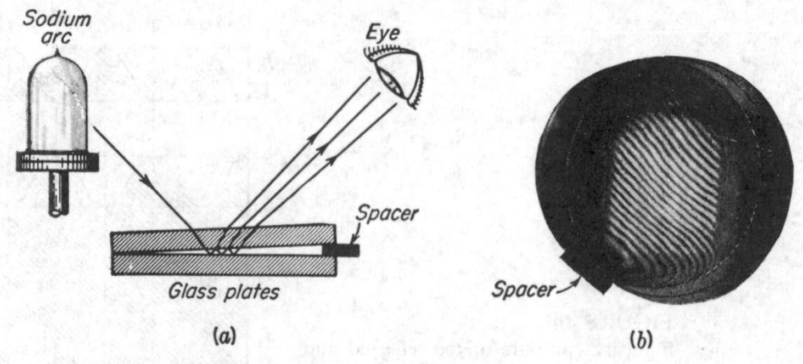

FIGURE 14F
Fringes of equal thickness: (*a*) method of visual observation; (*b*) photograph taken with a camera focused on the plates.

formed between the working surface and an optical flat, the polishing being continued until the fringes are straight. In Fig. 14F(*b*) it will be noticed that there is considerable distortion of one of the plates near the bottom.

14.5 NEWTON'S RINGS

If the fringes of equal thickness are produced in the air film between a convex surface of a long-focus lens and a plane glass surface, the contour lines will be circular. The ring-shaped fringes thus produced were studied in detail by Newton,* although he was not able to explain them correctly. For purposes of measurement, the observations are usually made at normal incidence by an arrangement such as that in Fig. 14G, where the glass plate G reflects the light down on the plates. After reflection, it is transmitted by G and observed in the low-power microscope T. Under these conditions the positions of the maxima are given by Eq. (14j), where d is the thickness of the air film. Now if we designate by R the radius of curvature of the surface A and assume that A and B are just touching at the center, the value of d for any ring of radius r is the sagitta of the arc, given by

$$d = \frac{r^2}{2R} \qquad (14k)$$

Substitution of this value in Eq. (14j) will then give a relation between the radii of the rings and the wavelength of the light. For quantitative work, one may not assume

* Sir Isaac Newton (1642–1727). Besides laying foundations of the science of mechanics, Newton devoted considerable time to the study of light and embodied the results in his famous "Opticks." It seems strange that one of the most striking demonstrations of the interference of light, Newton's rings, should be credited to the chief proponent of the corpuscular theory of light. Newton's advocacy of the corpuscular theory was not so uncompromising as it is generally represented. This is evident to anyone consulting his original writings. The original discovery of Newton's rings is now attributed to Robert Hooke.

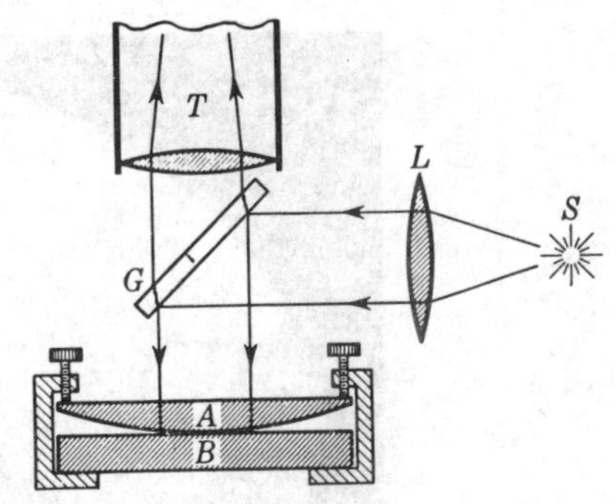

FIGURE 14G
Experimental arrangement used in viewing and measuring Newton's rings.

the plates to barely touch at the point of contact, since there will always be either some dust particles or distortion by pressure. Such disturbances will merely add a small constant to Eq.(14k), however, and their effect can be eliminated by measuring the diameters of at least two rings.

Because the ring diameters depend on wavelength, white light will produce only a few colored rings near the point of contact. With monochromatic light, however, an extensive fringe system such as that shown in Fig. 14H is observed. When the contact is perfect, the central spot is black. This is direct evidence of the relative phase change of π between the two types of reflection, air-to-glass and glass-to-air, mentioned in Sec. 14.1. If there were no such phase change, the rays reflected from the two surfaces in contact should be in the same phase and produce a bright spot at the center. In an interesting modification of the experiment, due to Thomas Young, the lower plate has a higher index of refraction than the lens, and the film between is filled with an oil of intermediate index. Then both reflections are at "rare-to-dense" surfaces, no relative phase change occurs, and the central fringe of the reflected system is bright. The experiment does not tell us at which surface the phase change in the ordinary arrangement occurs, but it is now definitely known (Sec. 25.4) that it occurs at the lower (air-to-glass) surface.

A ring system is also observed in the light transmitted by the Newton's-ring plates. These rings are exactly complementary to the reflected ring system, so that the center spot is now bright. The contrast between bright and dark rings is small, for reasons already discussed in Sec. 14.3.

14.6 NONREFLECTING FILMS

A simple and very important application of the principles of interference in thin films has been the production of *coated* surfaces. If a film of a transparent substance of refractive index n' is deposited on glass of a larger index n to a thickness of one-quarter of the wavelength of light in the film, so that

$$d = \frac{\lambda}{4n'}$$

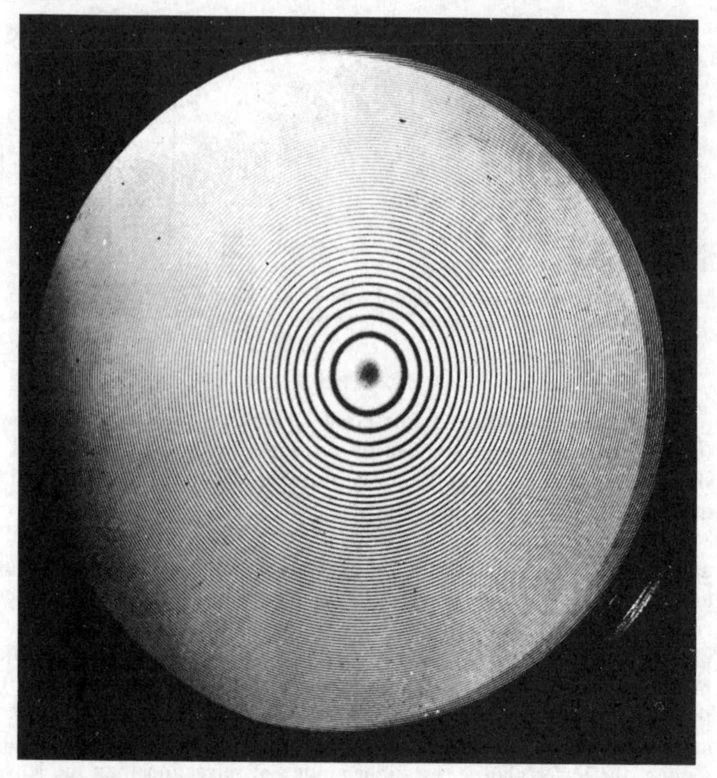

FIGURE 14H
Newton's rings. (*By permission from Bausch & Lomb Incorporated.*)

the light reflected at normal incidence is almost completely suppressed by interference. This corresponds to the condition $m = 0$ in Eq. (14g), which here becomes a condition for *minima* because the reflections at both surfaces are "rare-to-dense." The waves reflected from the lower surface have an extra path of one-half wavelength over those from the upper surface, and the two, combined with the weaker waves from multiple reflections, therefore interfere destructively. For the destruction to be complete, however, the fraction of the amplitude reflected at each of the two surfaces must be exactly the same, since this specification is made in proving the relation of Eq. (14h). It will be true for a film in contact with a medium of higher index only if the index of the film obeys the relation

$$n' = \sqrt{n}$$

This can be proved from Eq. (25*e*) of Chap. 25 by substituting n' for the refractive index of the upper surface and n/n' for that of the lower. Similar considerations will show that such a film will give zero reflection from the glass side as well as from the air side. Of course no light is destroyed by a nonreflecting film; there is merely a redistribution such that a decrease of reflection carries with it a corresponding increase of transmission.

The practical importance of these films is that by their use one can greatly reduce the loss of light by reflection at the various surfaces of a system of lenses or prisms. Stray light reaching the image as a result of these reflections is also largely eliminated, with a resulting increase in contrast. Almost all optical parts of high quality are now coated to reduce reflection. The coatings were first made by depositing several monomolecular layers of an organic substance on glass plates. More durable ones are now made by evaporating calcium or magnesium fluoride on the surface in vacuum or by chemical treatment with acids which leave a thin layer of silica on the surface of the glass. Properly coated lenses have a purplish hue by reflected light. This is a consequence of the fact that the condition for destructive interference can be fulfilled for only one wavelength, which is usually chosen to be one near the middle of the visible spectrum. The reflection of red and violet light is then somewhat larger. Furthermore, coating materials of sufficient durability have too high a refractive index to fulfill the condition stated above. Considerable improvement in these respects can be achieved by using two or more superimposed layers, and such films are capable of reducing the total reflected light to one-tenth of its value for the uncoated glass. This refers, of course, to light incident perpendicularly on the surface. At other angles, the path difference will change because of the factor $\cos \phi'$ in Eq. (14e). Since, however, the cosine does not change rapidly in the neighborhood of 0°, the reflection remains low over a fairly large range of angles about the normal. The multiple films, now called *multilayers*, may also be used, with suitable thickness, to accomplish the opposite purpose, namely to increase the reflectance. They may be used, for example, as beam-splitting mirrors to divide a beam of light into two parts of a given intensity ratio. The division can thus be accomplished without the losses of energy by absorption that are inherent in the transmission through, and reflection from, a thin metallic film.

14.7 SHARPNESS OF THE FRINGES

As the reflectance of the surfaces is increased, either by the above method or by lightly silvering them, the fringes due to multiple reflections become much narrower. The striking changes that occur are shown in Fig. 14I, which is plotted for $r^2 = 0.04$, 0.50, and 0.80 according to the theoretical equations to be derived below. The curve labeled 4% is just that for unsilvered glass which was given in Fig. 14E. Since, in the absence of any absorption, the intensity transmitted must be just the complement of that reflected, the same plot will represent the contour of either set. One is obtained from the other by merely turning the figure upside down or by inverting the scale of ordinates, as shown by the down arrow at the right in Fig. 14I.

In order to understand the reason for the narrowness of the transmitted fringes when the reflectance is high, we use the graphical method of compounding amplitudes already discussed in Secs. 12.2 and 13.4. Referring back to Fig. 14D, we notice that the amplitudes of the transmitted rays are given by att', $att'r^2$, $att'r^4$, ..., or in general for the mth ray by $att'r^{2m}$. We thus have to find the resultant of an infinite

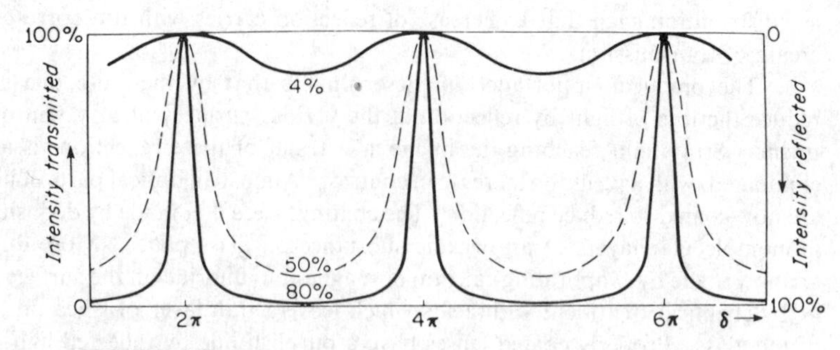

FIGURE 14I
Intensity contours of fringes due to multiple reflections, showing how the sharpness depends on reflectance.

number of amplitudes which decrease in magnitude more rapidly the smaller the fraction r. In Fig. 14J(a) the magnitudes of the amplitudes of the first 10 transmitted rays are drawn to scale for the 50 and 80 percent cases in Fig. 14I, that is, essentially for $r = 0.7$ and 0.9. Starting at any principal maximum, with $\delta = 2\pi m$, these individual amplitudes will all be in phase with each other, so the vectors are all drawn parallel to give a resultant that has been made equal for the two cases. If we now go slightly to one side of the maximum, where the phase difference introduced between successive rays is $\pi/10$, each of the individual vectors must be drawn making an angle of $\pi/10$ with the preceding one and the resultant found by joining the tail of the first to the head of the last. The result is shown in diagram (b). It will be seen that in the case $r = 0.9$, in which the individual amplitudes are much more nearly equal to each other, the resultant R is already considerably less than in the other case. In diagram (c), where the phase has changed by $\pi/5$, this effect is much more pronounced; the

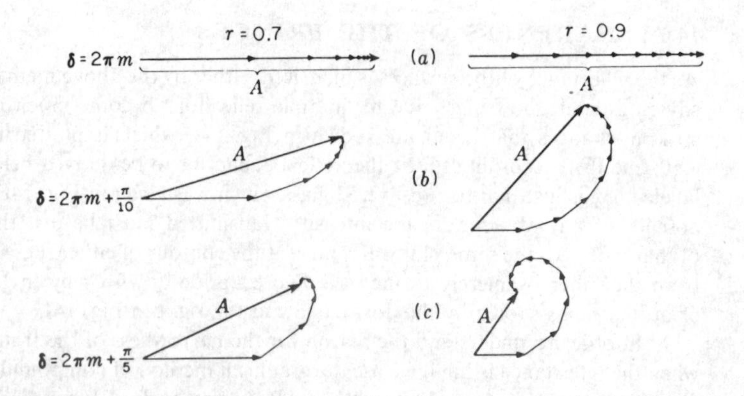

FIGURE 14J
Graphical composition of amplitudes for the first 10 multiply reflected rays, with two difference reflectances.

resultant has fallen to a considerably smaller value in the right-hand picture. Although a correct picture would include an infinite number of vectors, the later ones will have vanishing amplitudes, and we would reach a result similar to that found with the first 10.

These qualitative considerations can be made more precise by deriving an exact equation for the intensity. To accomplish this, we must find an expression for the resultant amplitude A, the square of which determines the intensity. Now A is the vector sum of an infinite series of diminishing amplitudes having a certain phase difference δ given by Eq. (14i). Here we can apply the standard method of adding vectors by first finding the sum of the horizontal components, then that of the vertical components, squaring each sum, and adding to get A^2. In doing this, however, the use of trigonometric functions as in Sec. 12.1 becomes too cumbersome. Hence an alternative way of compounding vibrations, which is mathematically simpler for complicated cases, will be used.

14.8 METHOD OF COMPLEX AMPLITUDES

In place of using the sine or the cosine to represent a simple harmonic wave, one may write the equation in the exponential form*

$$y = ae^{i(\omega t - kx)} = ae^{i\omega t}e^{-i\delta}$$

where $\delta = kx$ and is constant at a particular point in space. The presence of $i = \sqrt{-1}$ in this equation makes the quantities complex. We can nevertheless use this representation and at the end of the problem take either the real (cosine) or the imaginary (sine) part of the resulting expression. The time-varying factor exp $(i\omega t)$ is of no importance in combining waves of the same frequency, since the amplitudes and relative phases are independent of time. The other factor, a exp $(-i\delta)$, is called the *complex amplitude*. It is a complex number whose modulus a is the real amplitude and whose argument δ gives the phase relative to some standard phase. The negative sign merely indicates that the phase is behind the standard phase. In general, the vector $\mathbf{a}$ is given by

$$\mathbf{a} = ae^{i\delta} = x + iy = a(\cos\delta + i\sin\delta)$$

Then it will be seen that

$$a = \sqrt{x^2 + y^2} \qquad \tan\delta = \frac{y}{x}$$

Thus if $\mathbf{a}$ is represented as in Fig. 14K, plotting horizontally its real part and vertically its imaginary part, it will have the magnitude a and will make the angle δ with the x axis, as we require for vector addition.

The advantage of using complex amplitudes lies in the fact that the algebraic

* For the mathematical background of this method, see E. T. Whittaker and G. N. Watson, "Modern Analysis," chap. 1, Cambridge University Press, New York, 1935.

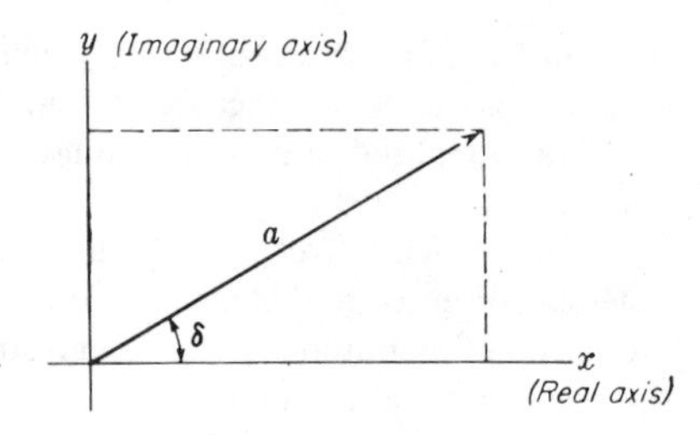

FIGURE 14K
Representation of a vector in the complex plane.

addition of two or more is equivalent to vector addition of the real amplitudes. Thus for two such quantities

$$Ae^{i\theta} = a_1 e^{i\delta_1} + a_2 e^{i\delta_2}$$

so that if

$$x_1 + x_2 = a_1 \cos \delta_1 + a_2 \cos \delta_2 = X$$

and

$$y_1 + y_2 = a_1 \sin \delta_1 + a_2 \sin \delta_2 = Y$$

it will be found that our previous Eqs. (12d) and (12e) require that

$$A^2 = X^2 + Y^2 \qquad \tan \theta = \frac{Y}{X} \tag{14l}$$

Thus, to get a vector sum, we need only obtain the *algebraic* sums $X = \Sigma x_i$ and $Y = \Sigma y_i$ of the real and imaginary parts, respectively, of the complex amplitudes. In obtaining the resultant intensity as proportional to the square of the real amplitude, we multiply the resultant complex amplitude by its complex conjugate, which is the same expression with i replaced by $-i$ throughout. The justification for this procedure follows from the relations

$$\begin{aligned} (X + iY)(X - iY) &= X^2 + Y^2 = A^2 \\ Ae^{i\theta}Ae^{-i\theta} &= A^2 \end{aligned} \tag{14m}$$

14.9 DERIVATION OF THE INTENSITY FUNCTION

For the fringe system formed by the transmitted light, the sum of the complex amplitudes is (see Fig. 14D)

$$\begin{aligned} Ae^{i\theta} &= att' + att'r^2e^{i\delta} + att'r^4e^{i2\delta} + \cdots \\ &= a(1 - r^2)(1 + r^2e^{i\delta} + r^4e^{i2\delta} + \cdots) \end{aligned}$$

where $1 - r^2$ has been substituted for tt', according to Stokes' relation, Eq. (14c). The infinite geometric series in the second parentheses has the common ratio $r^2 \exp i\delta$, and has a finite sum because $r^2 < 1$. Summing the series, one obtains

$$Ae^{i\theta} = \frac{a(1 - r^2)}{1 - r^2e^{i\delta}}$$

By Eq. (14m), the intensity is the product of this quantity by its complex conjugate, which yields

$$I_T \approx \frac{a(1 - r^2)}{1 - r^2e^{i\delta}} \frac{a(1 - r^2)}{1 - r^2e^{-i\delta}} = \frac{a^2(1 - r^2)^2}{1 - r^2(e^{i\delta} + e^{-i\delta}) + r^4}$$

Since $(e^{i\delta} + e^{-i\delta})/2 = \cos \delta$, and $a^2 \approx I_0$, the intensity of the incident beam, we obtain the result, in terms of real quantities only, as

$$I_T = I_0 \frac{(1 - r^2)^2}{1 - 2r^2 \cos \delta + r^4} = \frac{I_0}{1 + [4r^2/(1 - r^2)^2] \sin^2 (\delta/2)} \tag{14n}$$

The main features of the intensity contours in Fig. 14I can be read from this equation. Thus at the maxima, where $\delta = 2\pi m$, we have $\sin^2 (\delta/2) = 0$ and $I_T = I_0$. When the reflectance r^2 is large, approaching unity, the quantity $4r^2/(1 - r^2)^2$ will also be large, and even a small departure of δ from its value for the maximum will result in a rapid drop of the intensity.

For the reflected fringes it is not necessary to carry through the summation, since we know from the conservation of energy that if no energy is lost through absorption,

$$I_R + I_T = 1 \tag{14o}$$

The reflected fringes are complementary to the transmitted ones, and for high reflectances become narrow dark fringes. These can be used to make more precise the study of the contour of surfaces.* If there is appreciable absorption on transmission through the surfaces, as will be the case if they are lightly silvered, one can no longer assume that Stokes' relations or Eq. (14o) hold. Going back to the derivation of Eq. (14n), it will be found that in this case the expression for I_T must be multiplied by $(tt')^2/(1 - r^2)^2$. Here tt' and r^2 are essentially the fractions of the intensity transmitted and reflected, respectively, by a single surface. Where the surfaces are metallized, there will be slight differences between t and t', as well as small phase changes upon reflection. The transmitted fringes may still be represented by Eq. (14n), however, with an overall reduction of intensity and a correction to δ which merely changes slightly the effective thickness of the plate.

14.10 FABRY-PEROT INTERFEROMETER

This instrument utilizes the fringes produced in the transmitted light after multiple reflection in the air film between two plane plates thinly silvered on the inner surfaces (Fig. 14L). Since the separation d between the reflecting surfaces is usually fairly large (from 0.1 to 10 cm) and observations are made near the normal direction, the fringes come under the class of fringes of equal inclination (Sec. 14.2). To observe the fringes, the light from a broad source (S_1S_2) of monochromatic light is allowed to traverse the interferometer plates E_1E_2. Since any ray incident on the first silvered surface is broken by reflection into a series of *parallel* transmitted rays, it is essential to use a lens L, which may be the lens of the eye, to bring these parallel rays together

* S. Tolansky, "Multiple-Beam Interferometry," Oxford University Press, New York, 1948.

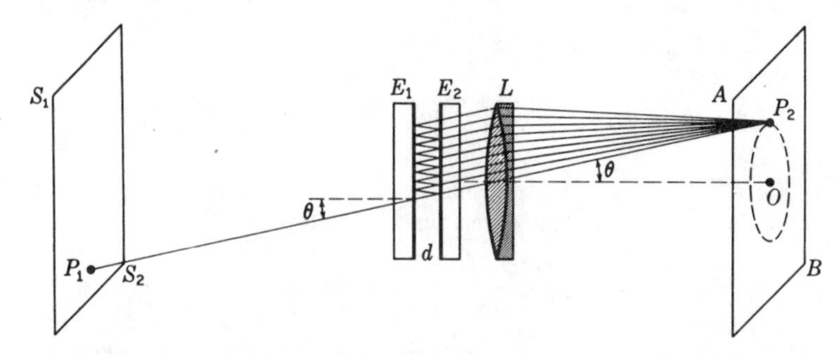

FIGURE 14L
Fabry-Perot interferometer E_1E_2 set up to show the formation of circular interference fringes from multiple reflections.

for interference. In Fig. 14L a ray from the point P_1 on the source is incident at the angle θ, producing a series of parallel rays at the same angle, which are brought together at the point P_2 on the screen AB. It is to be noted that P_2 is *not* an image of P_1. The condition for reinforcement of the transmitted rays is given by Eq. (14f) with $n = 1$ for air, and $\phi' = \theta$, so that

$$2d \cos \theta = m\lambda \qquad \textit{Maxima} \tag{14p}$$

This condition will be fulfilled by all points on a circle through P_2 with their center at O, the intersection of the axis of the lens with the screen AB. When the angle θ is decreased, the cosine will increase until another maximum is reached for which m is greater by 1, 2, . . . , so that we have for the maxima a series of concentric rings on the screen with O as their center. Since Eq. (14p) is the same as Eq. (13g) for the Michelson interferometer, the spacing of the rings is the same as for the circular fringes in that instrument and they will change in the same way with change in the distance d. In the actual interferometer one plate is fixed, while the other may be moved toward or away from it on a carriage riding on accurately machined ways by a slow-motion screw.

14.11 BREWSTER'S* FRINGES

In a single Fabry-Perot interferometer it is not practicable to observe white-light fringes, since the condition of zero path difference occurs only when the two silvered surfaces are brought into direct contact. By the use of two interferometers in series, however, it is possible to obtain interference in white light, and the resulting fringes have had important applications. The two plane-parallel "air plates" are adjusted

* Sir David Brewster (1781–1868). Professor of physics at St. Andrew's, and later principal of the University of Edinburgh. Educated for the church, he became interested in light through repeating Newton's experiments on diffraction. He made important discoveries in double refraction and in spectrum analysis. Oddly enough, he opposed the wave theory of light in spite of the great advances in that theory made during his lifetime.

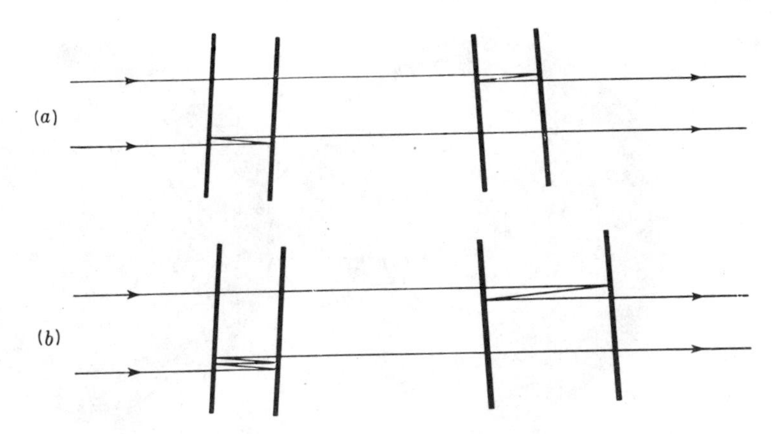

FIGURE 14M
Light paths for the formation of Brewster's fringes. (*a*) With two plates of equal thickness. (*b*) With one plate twice as thick as the other. The inclination of the two plates is exaggerated.

to exactly the same thickness, or else one to some exact multiple of the other, and the two interferometers are inclined to each other at an angle of 1 or 2°. A ray that bisects the angle between the normals to the two sets of plates can then be split into two, each of which after two or more reflections emerges, having traversed the same path. In Fig. 14M these two paths are drawn as separate for the sake of clarity, though actually the two interfering beams are derived from the same incident ray and are superimposed when they leave the system. The reader is referred to Fig. 13V, where the formation of Brewster's fringes by two thick glass plates in Jamin's interferometer is illustrated. A ray incident at any other angle than that mentioned above will give a path difference between the two emerging ones which increases with the angle, so that a system of straight fringes is produced.

The usefulness of Brewster's fringes lies chiefly in the fact that when they appear, the ratio of the two interferometer spacings is very exactly a whole number. Thus, in the redetermination of the length of the standard meter in terms of the wavelength of the red cadmium line, a series of interferometers was made, each having twice the length of the preceding, and these were intercompared using Brewster's fringes. The number of wavelengths in the longest, which was approximately 1 m long, could be found in a few hours by this method. It should finally be emphasized that this type of fringe results from the interference of only *two* beams and therefore cannot be made very narrow, as can the usual fringes due to multiple reflections.

14.12 CHROMATIC RESOLVING POWER

The great advantage of the Fabry-Perot interferometer over the Michelson instrument lies in the sharpness of the fringes. Thus it is able to reveal directly those details of fine structure and line width that previously could only be inferred from the behavior

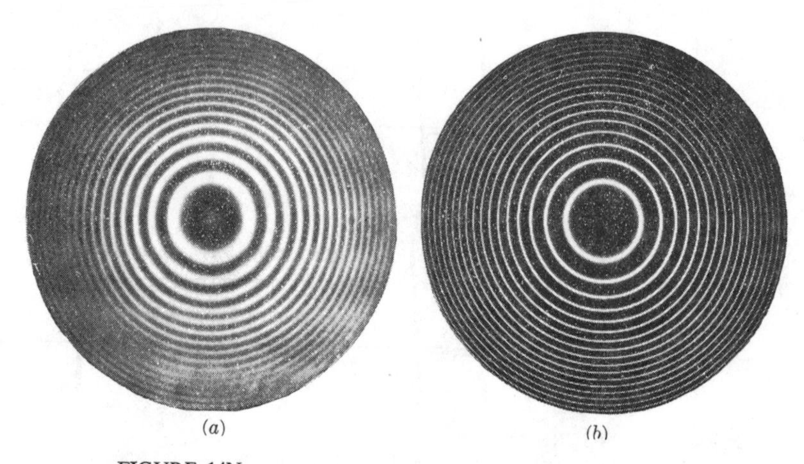

FIGURE 14N
Comparison of the types of fringes produced with (*a*) the Michelson interferometer and (*b*) the Fabry-Perot interferometer with surfaces of reflectance 0.8.

of the visibility curves. The difference in the appearance of the fringes for the two instruments is illustrated in Fig. 14N, where the circular fringes produced by a single spectral line are compared. If a second line were present, it would merely reduce the visibility in (*a*) but would show as a separate set of rings in (*b*). As will appear later, this fact also permits more exact intercomparisons of wavelength.

It is important to know how close together two wavelengths may be and still be distinguished as separate rings. The ability of any type of spectroscope to discriminate wavelengths is expressed as the ratio $\lambda/\Delta\lambda$, where λ is the mean wavelength of a barely resolved pair and $\Delta\lambda$ is the wavelength difference between the components. This ratio is called the *chromatic resolving power* of the instrument at that wavelength. In the present case, it is convenient to say that the fringes formed by λ and $\lambda + \Delta\lambda$ are just resolved when the intensity contours of the two in a particular order lie in the relative positions shown in Fig. 14O(*a*). If the separation $\Delta\theta$ is such as to make the curves cross at the half-intensity point, $I_T = 0.5I_0$, there will be a central dip of 17 percent in the sum of the two, as shown in (*b*) of the figure. The eye can then easily recognize the presence of two lines.

In order to find the $\Delta\lambda$ corresponding to this separation, we note first that in going from the maximum to the halfway point the phase difference in either pattern must change by the amount necessary to make the second term in the denominator of Eq. (14n) equal to unity. This requires that

$$\sin^2 \frac{\delta}{2} = \frac{(1 - r^2)^2}{4r^2}$$

If the fringes are reasonably sharp, the change of $\delta/2$ from a multiple of π will be small. Then the sine may be set equal to the angle, and if we denote by $\Delta\delta$ the change

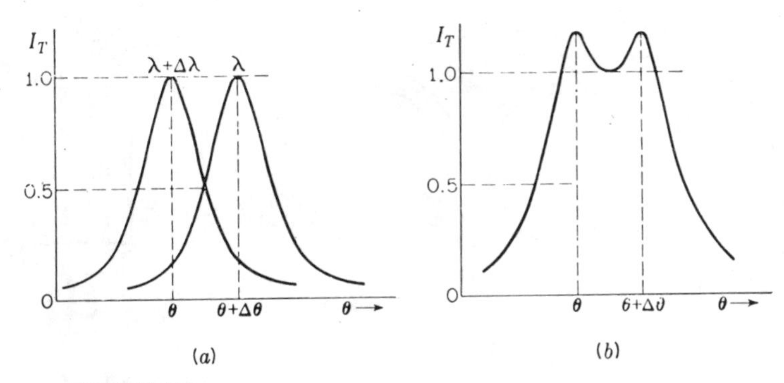

FIGURE 14O
Intensity contour of two Fabry-Perot fringes that are just resolved: (*a*) shown separately; (*b*) added, to give the observed effect.

in going from one maximum to the position of the other, we have

$$(\sin \tfrac{1}{2})\left(\frac{\Delta\delta}{2}\right) \approx \frac{\Delta\delta}{4} = \frac{1 - r^2}{2r} \qquad \text{(14q)}$$

Now the relation between an angular change $\Delta\theta$ and a phase change $\Delta\delta$ can be found by differentiating Eq. (14i), setting $\phi' = \theta$ and $n = 1$.

$$\Delta\delta = -\frac{4\pi d}{\lambda} \sin\theta\, \Delta\theta \qquad \text{(14r)}$$

Furthermore, if the maximum for $\lambda + \Delta\lambda$ is to occur at this same angular separation $\Delta\theta$, Eq. (14p) requires that

$$-2d \sin\theta\, \Delta\theta = m\, \Delta\lambda \qquad \text{(14s)}$$

The combination of Eqs. (14q) to (14s) yields, for the chromatic resolving power,

$$\frac{\lambda}{\Delta\lambda} = m \frac{\pi r}{1 - r^2} \qquad \text{(14t)}$$

It thus depends on two quantities, the order m, which may be taken as $2d/\lambda$, and the reflectance r^2 of the surfaces. If the latter is close to unity, very large resolving powers are obtained. For example, with $r^2 = 0.9$ the second factor in Eq. (14t) becomes 30, and, with a plate separation d of only 1 cm, the resolving power at $\lambda 5000$ becomes 1.20×10^6. The components of a doublet only 0.0042 Å wide could be seen as separate.

14.13 COMPARISON OF WAVELENGTHS WITH THE INTERFEROMETER

The ratio of the wavelengths of two lines which are not very close together, e.g., the yellow mercury lines, is sometimes measured in the laboratory with the form of interferometer in which one mirror is movable. The method is based on observation

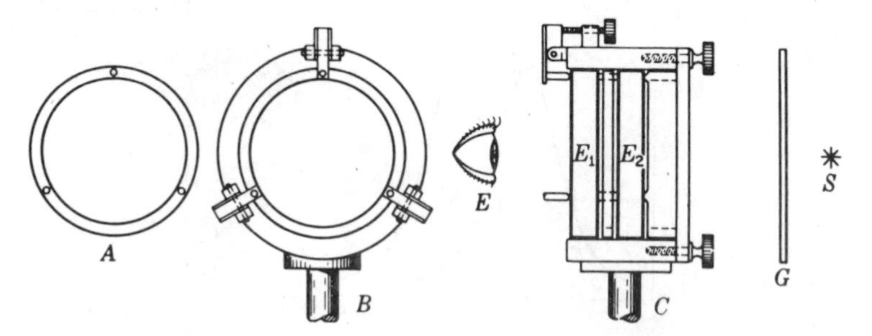

FIGURE 14P
Mechanical details of a Fabry-Perot etalon, showing spacer ring, adjusting screws, and springs.

of the positions of coincidence and discordance of the fringes formed by the two wavelengths, a method already mentioned in Sec. 13.12. When one starts with the two mirrors nearly in contact, the ring system due to the two wavelengths practically coincide. As d is increased, they gradually separate and the maximum discordance occurs when the rings of one set are halfway between those of the other set. Confining our attention to the rings at the center ($\cos\theta = 1$), we can write from Eq. (14p)

$$2d_1 = m_1\lambda = (m_1 + \tfrac{1}{2})\lambda' \tag{14u}$$

where, of course, $\lambda > \lambda'$. From this,

$$m_1(\lambda - \lambda') = \frac{2d_1}{\lambda}(\lambda - \lambda') = \frac{\lambda'}{2} \qquad \text{and} \qquad \lambda - \lambda' = \frac{\lambda\lambda'}{4d_1} = \frac{\lambda^2}{4d_1}$$

if the difference between λ and λ' is small. On displacing the mirror still farther, the rings will presently coincide and then separate out again. At the next discordance

$$2d_2 = m_2\lambda = (m_2 + 1\tfrac{1}{2})\lambda' \tag{14v}$$

Subtracting Eq. (14u) from Eq. (14v), we obtain

$$2(d_2 - d_1) = (m_2 - m_1)\lambda = (m_2 - m_1)\lambda' + \lambda'$$

whence, assuming λ approximately equal to λ', we find

$$\lambda - \lambda' = \frac{\lambda^2}{2(d_2 - d_1)} \tag{14w}$$

We can determine $d_2 - d_1$ either directly from the scale or by counting the number of fringes of the known wavelength λ between discordances.

For the most accurate work, the above method is replaced by one in which the fringe systems of the lines are photographed simultaneously with a fixed separation d of the plates. For this purpose the plates are held rigidly in place by quartz or invar spacers. A pair of Fabry-Perot plates thus mounted is called an *etalon* (Fig. 14P). The etalon can be used to determine accurately the relative wavelengths of several spectral lines from a single photographic exposure. If it were mounted with a lens,

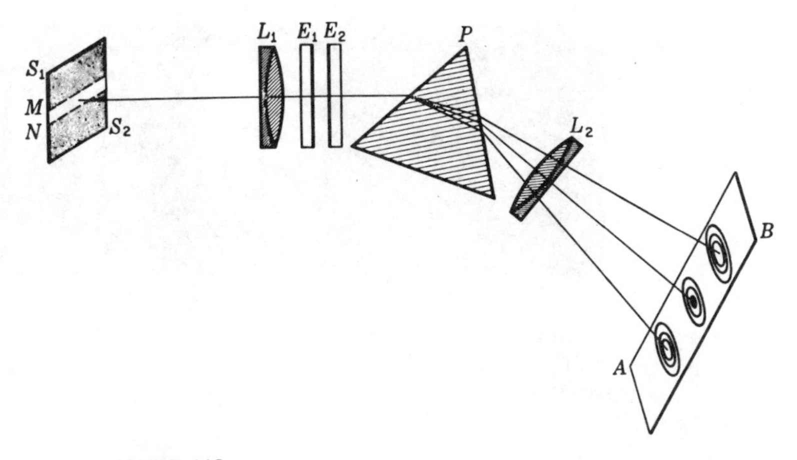

FIGURE 14Q
Fabry-Perot etalon-and-prism arrangement for separating the ring systems produced by different spectrum lines.

as in Fig. 14L, the light containing several wavelengths, the fringe systems of the various wavelengths would be concentric with O and would be confused with each other. However, they can be separated by inserting a prism between the etalon and the lens L. The experimental arrangement is then similar to that shown in Fig. 14Q. A photograph of the visible spectrum of mercury taken in this way is shown in the upper part of Fig. 14R. It will be seen that the fringes of the green and yellow lines

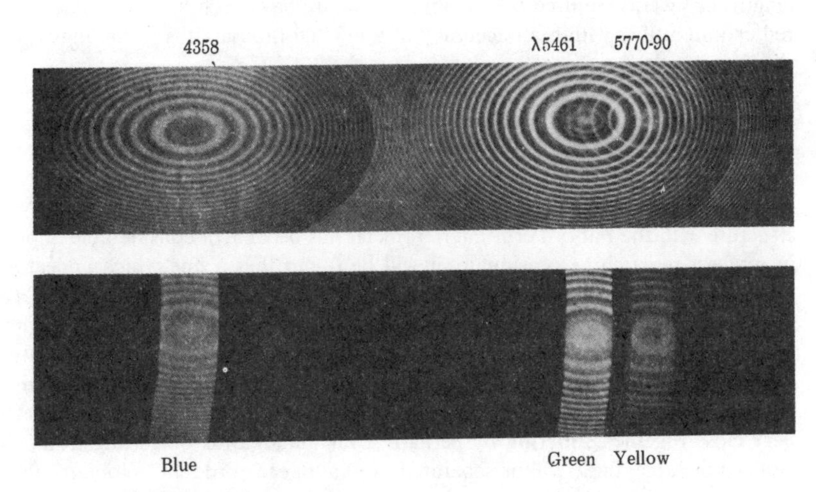

FIGURE 14R
Interference rings of the visible mercury spectrum taken with the Fabry-Perot etalon shown in Fig. 14P.

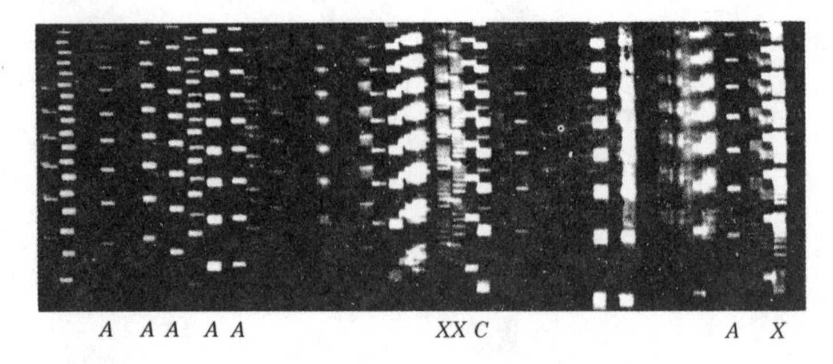

FIGURE 14S
Interference patterns of the lanthanum spectrum taken with a Fabry-Perot etalon; d = 5 mm. (*Courtesy of O. E. Anderson.*)

still overlap. To overcome this, it is merely necessary to use an illuminated slit (MN of Fig. 14Q) of the proper width as the source. When the interferometer is in a collimated beam of parallel light, as it is here, each point on the extended source corresponds to a given point in the ring system. Therefore only vertical sections of the ring system are obtained, as shown in the lower part of Fig. 14R, and these no longer overlap. When the spectrum is very rich in lines, as in Fig. 14S, the source slit must be made rather narrow. In this photograph only sections of the upper half of the fringe systems appear. Measurements of the radii of the rings in a photograph of this type permit very accurate comparisons of wavelengths. The determination of the correct values of m in the different systems and of the exact value of d is a rather involved process which we shall not discuss here.* By this method the wavelengths of several hundred lines from the iron arc have been measured relative to the red cadmium line within an accuracy of a few ten-thousandths of an angstrom.

14.14 STUDY OF HYPERFINE STRUCTURE AND OF LINE SHAPE

Because of its bearing on the properties of atomic nuclei, the investigation of hyperfine structure with the Fabry-Perot interferometer has become of considerable importance in modern research. Occasionally it will be found that a line which appears sharp and single in an ordinary spectroscope will yield ring systems consisting of two or more sets. Examples are found in the lines marked X in the lanthanum spectrum (Fig. 14S). Lines like the one marked C are broadened but not resolved into their components. Those marked A are sharp to a greater or lesser extent. These multiple ring systems arise from the fact that the line is actually a group of lines of wavelengths very close together, differing by perhaps a few hundredths of an angstrom. If d is sufficiently large, these will be separated, so that in each order m we obtain effectively

* See W. E. Williams, "Applications of Interferometry," pp. 83–88, Methuen and Co., Ltd., London, 1930, for a description of this method.

a short spectrum very powerfully resolved. Any given fringe of a wavelength λ_1 is formed at such an angle that

$$2d \cos \theta_1 = m\lambda_1 \qquad \text{(14x)}$$

The next fringe farther out for this same wavelength has

$$2d \cos \theta_2 = (m - 1)\lambda_1 \qquad \text{(14y)}$$

Suppose now that λ_1 has a component line λ_2 which is very near λ_1, so that we may write $\lambda_2 = \lambda_1 - \Delta\lambda$. Suppose also that $\Delta\lambda$ is such that this component, in order m, falls on the order $m - 1$ of λ_1. Then

$$2d \cos \theta_2 = m(\lambda_1 - \Delta\lambda) \qquad \text{(14z)}$$

Equating the right-hand members of Eqs. (14y) and (14z) gives

$$\lambda_1 = m\,\Delta\lambda$$

Substituting the value of m from Eq. (14x) and solving for $\Delta\lambda$, we have

● $$\Delta\lambda = \frac{\lambda_1^{\,2}}{2d \cos \theta_1} \approx \frac{\lambda_1^{\,2}}{2d} \qquad \text{(14za)}$$

if θ is nearly zero. This interval $\Delta\lambda$, called the *spectral range*, is defined as the change in wavelength necessary to shift the ring system by the distance of consecutive orders. We see that it is constant, independent of m. When d and λ are known, the wavelength difference of component lines lying in this small range can be evaluated.*

The equation for the separation of orders becomes still simpler when expressed in terms of frequency. Since the frequencies of light are awkwardly large numbers, spectroscopists commonly use an equivalent quantity called the *wave number*. This is the number of waves per centimeter path in vacuum, and varies from roughly 15,000 to 25,000 cm^{-1} in going from red to violet. Denoting wave number by σ, we have

$$\sigma = \frac{1}{\lambda} = \frac{k}{2\pi}$$

To find the wave-number difference $\Delta\sigma$ corresponding to the $\Delta\lambda$ in Eq. (14za), we can differentiate the above equation to obtain

$$\Delta\sigma = -\frac{\Delta\lambda}{\lambda_2}$$

Substitution in Eq. (14za) then yields

● $$\Delta\sigma = -\frac{1}{2d} \qquad \text{(14zb)}$$

Hence, if d is expressed in centimeters, $1/2d$ gives the wave-number difference, which is seen to be independent of the order (neglecting the variation of θ) and of wavelength as well.

The study of the width and shape of individual spectrum lines, even though they may have no hyperfine structure, is of interest because it can give us information

* For a good account of the methods see K. W. Meissner, *J. Opt. Soc. Am.*, **31**:405 (1941).

on the conditions of temperature, pressure, etc., in the light source. If the interferometer has a high resolving power, the fringes will have a contour corresponding closely to that of the line itself. The small width which is inherent in the instrument can be determined by observations with an extremely small etalon spacer, and appropriate corrections made.

The difficult adjustment of the Fabry-Perot interferometer lies in the attainment of accurate parallelism of the silvered surfaces. This operation is usually accomplished by the use of screws and springs, which hold the plates against the spacer rings shown in Fig. 14P. A brass ring A with three quartz or invar pins constitutes the spacer. A source of light such as a mercury arc is set up with a sheet of ground glass G on one side of the etalon, and then viewed from the opposite side as shown at E. With the eye focused for infinity, a system of rings will be seen with the reflected image of the pupil of the eye at its center. As the eye is moved up and down or from side to side, the ring system will also move along with the image of the eye. If the rings on moving up expand in size, the plates are farther apart at the top than at the bottom. Tightening the top screw will then depress the corresponding separator pin enough to produce the required change in alignment. When the plates are properly adjusted, and if they are exactly plane, the rings will remain the same size as the eye is moved to any point of the field of view.

Sometimes it is convenient to place the etalon in front of the slit of a spectrograph rather than in front of the prism. In such cases the light incident on the etalon need not be parallel. A lens, however, must follow the etalon and must always be set with the slit at its focal plane. This lens selects parallel rays from the etalon and focuses interference rings on the slit. Both these methods are used in practice.

14.15 OTHER INTERFERENCE SPECTROSCOPES

When the light is monochromatic, or nearly so, it is not necessary that the material between the highly reflecting surfaces be air. A single accurately plane-parallel glass plate having its surfaces lightly silvered will function as a Fabry-Perot etalon. The use of two such plates with thicknesses in the ratio of whole numbers will result in the suppression of several of the maxima produced by the thicker plate, since any light getting through the system at a particular angle must satisfy Eq.(14p) for both plates. This arrangement, known as the *compound interferometer*, gives the resolving power of the thicker plate and the free wavelength range, Eq. (14za), of the thinner one.

The spacing of the fringes of equal inclination becomes extremely small when θ departs much from 0°. It opens out again, however, near grazing incidence. The *Lummer-Gehrcke plate* makes use of the first few maxima near $\theta = 90°$. In order to get an appreciable amount of light to enter the plate, it is necessary to introduce it by a total-reflection prism cemented on one end. It then undergoes multiple internal reflections very near the critical angle, and the beams emerging at a grazing angle are brought to interference by a lens. High reflectance and resolving power are thus obtained with unsilvered surfaces.

Because of its flexibility, the Fabry-Perot interferometer has for research purposes largely replaced such instruments having a fixed spacing of the surfaces. For special purposes, however, they may be valuable.*

14.16 CHANNELED SPECTRA. INTERFERENCE FILTER

Our discussion of the Fabry-Perot interferometer was concerned primarily with the dependence of the intensity on plate separation and on angle for a single wavelength, or perhaps for two or more wavelengths close together. If the instrument is placed in a parallel beam of white light, interference will also occur for all the monochromatic components of such light, but this will not manifest itself until the transmitted beam is dispersed by an auxiliary spectroscope. One then observes a series of bright fringes in the spectrum, each formed by a wavelength somewhat different from the next. The maxima will occur, according to Eq. (14p), at wavelengths given by

$$\lambda = \frac{2d \cos \theta}{m} \qquad (14zc)$$

where m is any whole number. If d is a separation of a few millimeters, there will be very many narrow fringes (more than 12,000 through the visible spectrum when $d = 5$ mm), and high dispersion is necessary in order to separate them. Such fringes are referred to as a *channeled spectrum* or as Edser-Butler bands and have been used, for example, in the calibration of spectroscopes for the infrared and in accurate measurements of wavelengths of the absorption lines in the solar spectrum.

An application of these fringes having considerable practical importance uses the situation where d is extremely small, so that only one or two maxima occur within the visible range of wavelengths. With white light incident, only one or two narrow bands of wavelength will then be transmitted, the rest of the light being reflected. The pair of semitransparent metallic films thus can act as a *filter* passing nearly monochromatic light. The curves of transmitted energy against wavelength resemble those of Fig. 14I, since, according to Eq. (14i), the phase difference δ is inversely proportional to wavelength for a given separation d.

For the maxima to be widely separated, m must be a small number. This is attained only by having the reflecting surfaces very close together. If one wishes to have the maximum for $m = 2$ occur at a given wavelength λ, the metal films would have to be a distance λ apart. The maximum $m = 1$ will then appear at a wavelength of 2λ. Such minute separations can be attained, however, with modern techniques of evaporation in vacuum. A semitransparent metal film is first evaporated on a plate of glass. Next, a thin layer of some dielectric material such as cryolite ($3NaF \cdot AlF_3$) is evaporated on top of this, and then the dielectric layer is in turn coated with another similar film of metal. Finally another plate of glass is placed over the films for mechanical protection. The completed filter then has the cross

* For a more detailed description of these and other similar instruments, see A. C. Candler, "Modern Interferometers," Hilger and Watts, London, 1951.

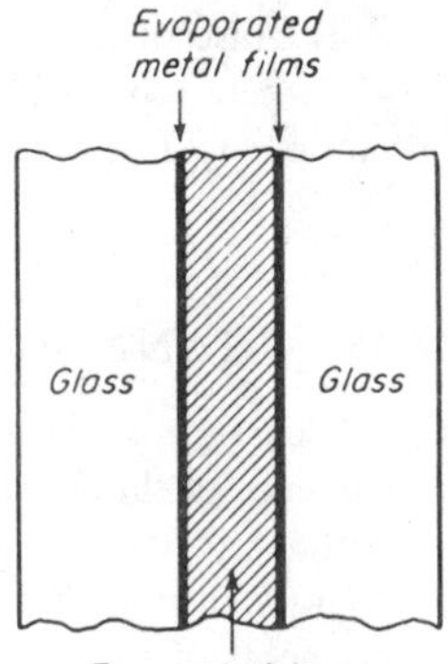

FIGURE 14T
Cross section of an interference filter.

section shown schematically in Fig. 14T, where the thickness of the films is greatly exaggerated relative to that of the glass plates. Since the path difference is now in the dielectric of index n, the wavelengths of maximum transmission for normal incidence are given by

$$\lambda = \frac{2nd}{m} \tag{14zd}$$

If there are two maxima in the visible spectrum, one of them can easily be eliminated by using colored glass for the protecting cover plate. Interference filters are now made which transmit a band of wavelengths of width (at half transmission) only 15 Å, with the maximum lying at any desired wavelength. The transmission at the maximum can be as high as 45 percent. It is very difficult to obtain combinations of colored glass or gelatin filters which will accomplish this purpose. Furthermore, since the interference filter reflects rather than absorbs the unwanted wavelengths, there is no trouble with its overheating.

PROBLEMS

14.1 A transparent film has a thickness of 0.003250 cm, and a refractive index of 1.4000. Find (*a*) the order of interference m at $\theta = 0°$ and (*b*) the first four angles at which red light of wavelength 6500 Å will form bright-light fringes.

Ans. (*a*) $m = 100$, (*b*) 5.73, 9.94, 12.84, and 15.20°

14.2 A thin film has a thickness of 0.04650 cm and a refractive index of 1.5230. Find the angle ϕ at which the dark fringe 122.5 will be observed if monochromatic light 6560 Å is used as an extended source.

14.3 In an experiment involving Newton's rings, the diameters of the fifth and fifteenth bright rings formed by sodium yellow light are measured to be 2.303 and 4.134 mm, respectively. Calculate the radius of curvature of the convex glass surface.

14.4 Three convex spherical glass surfaces have large radii of 200.0, 300.0, and 400.0 cm, respectively. When they are brought into contact in pairs and an extended source of red light with a wavelength of 6500 Å is used, find (*a*) the path difference d and (*b*) the radii r of the twentieth bright ring for each of the three combinations. See Fig. P14.4.
Ans. (*a*) $d = 6.338 \times 10^{-3}$ mm, (*b*) $r_1 = 3.900$ mm, $r_2 = 4.111$ mm, $r_3 = 4.661$ mm

14.5 Three spherical glass surfaces of unknown radius are brought into contact in pairs, and each pair used to form Newton's rings. The diameters of the twenty-fifth bright fringe for the three possible combinations are 8.696, 9.444, and 10.2680 mm, respectively. Find (*a*) the path difference d and (*b*) the radii of the three glass surfaces. Assume the mercury green light used has a wavelength of 5461 Å. See Fig. P14.4.

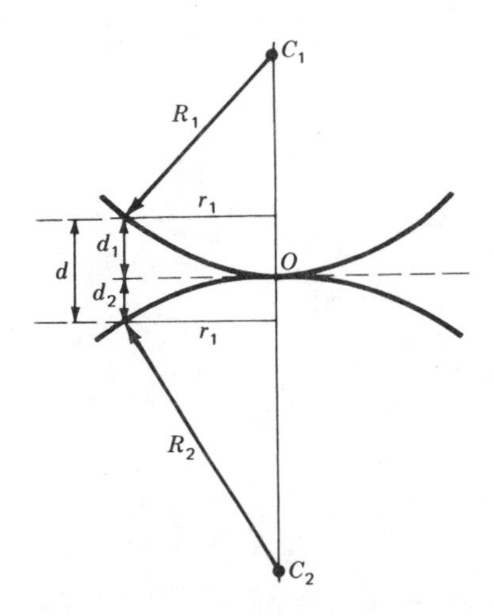

FIGURE P14.4

14.6 A glass lens of index 1.5630 is to be nonreflecting on both surfaces. What should be (*a*) the refractive index of a surface coating material and (*b*) its thickness for green light of wavelength 5500 Å, to produce 0 percent reflectance?

14.7 Using vector diagrams, find the resultant amplitude and intensity in the interference pattern from a Fabry-Perot interferometer having a reflectance of 80.0 percent when the phase difference between successive rays is (*a*) 0°, (*b*) 15.0°, and (*c*) 30.0° (see Figs. 14D and 14J). Use the first six transmitted rays only. Assign unity to the amplitude of the first transmitted ray. Make a drawing.
Ans. (*a*) $A = 2.587$, $A^2 = 6.693$, (*b*) $A = 2.403$, $A^2 = 5.776$, (*c*) $A = 1.948$, $A^2 = 3.793$; see Fig. P14.7.

14.8 The plates of a Fabry-Perot interferometer have a reflectance amplitude of $r = 0.90$. Calculate the minimum (*a*) resolving power and (*b*) plate separation to resolve the two components of the H_α line of the hydrogen spectrum, which is a doublet with a separation of 0.1360 Å.

14.9 The method of coincidences of Fabry-Perot rings is used to compare two wavelengths, one of which is 5460.740 Å, and the other slightly shorter. If coincidences occur at plate separations of 0.652, 1.827, and 3.002 mm, find (*a*) the wavelength difference and (*b*) the wavelength of λ'.

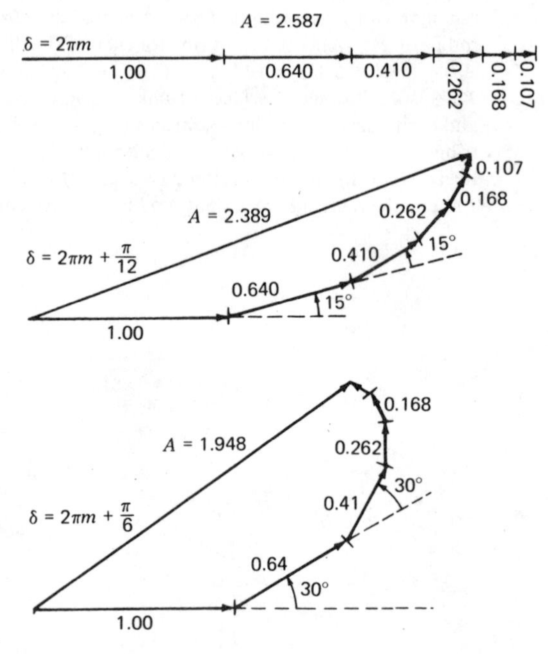

FIGURE P14.7
Graphical addition of amplitudes for Prob. P14.7.

14.10 In making a photograph of a Fabry-Perot pattern using mercury light of wavelength 5460.740 Å, the separation of the plates was 6.280 mm. If a lens with a focal length of 120.0 cm is used, find (*a*) the order of interference for the central spot and (*b*) the order of the sixth ring out from the center. (*c*) What is the wavelength separation of orders and (*d*) the linear diameter of the sixth ring?

Ans. (*a*) 23000.5, (*b*) 22994.5, (*c*) 0.237418 Å, (*d*) 5.5029 cm

15

FRAUNHOFER DIFFRACTION BY A SINGLE OPENING

When a beam of light passes through a narrow slit, it spreads out to a certain extent into the region of the geometrical shadow. This effect, already noted and illustrated at the beginning of Chap. 13, Fig. 13B, is one of the simplest examples of *diffraction*, i.e., of the failure of light to travel in straight lines. It can be satisfactorily explained only by assuming a wave character for light, and in this chapter we shall investigate quantitatively the *diffraction pattern*, or distribution of intensity of the light behind the aperture, using the principles of wave motion already discussed.

15.1 FRESNEL AND FRAUNHOFER DIFFRACTION

Diffraction phenomena are conveniently divided into two general classes, (1) those in which the source of light and the screen on which the pattern is observed are effectively at infinite distances from the aperture causing the diffraction and (2) those in which either the source or the screen, or both, are at finite distances from the aperture. The phenomena coming under class (1) are called, for historical reasons, *Fraunhofer diffraction*, and those coming under class (2) *Fresnel diffraction*. Fraunhofer diffraction is much simpler to treat theoretically. It is easily observed in practice by rendering

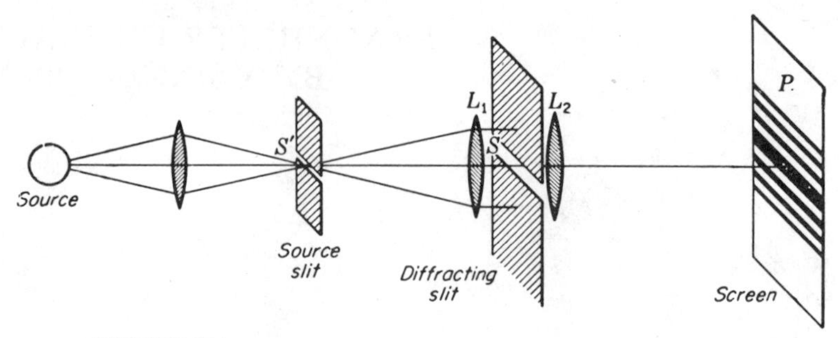

FIGURE 15A
Experimental arrangement for obtaining the diffraction pattern of a single slit; Fraunhofer diffraction.

the light from a source parallel with a lens and focusing it on a screen with another lens placed behind the aperture, an arrangement which effectively removes the source and screen to infinity. In the observation of Fresnel diffraction, on the other hand, no lenses are necessary, but here the wave fronts are divergent instead of plane, and the theoretical treatment is consequently more complex. Only Fraunhofer diffraction will be considered in this chapter, and Fresnel diffraction in Chap. 18.

15.2 DIFFRACTION BY A SINGLE SLIT

A slit is a rectangular aperture of length large compared to its breadth. Consider a slit S to be set up as in Fig. 15A, with its long dimension perpendicular to the plane of the page, and to be illuminated by parallel monochromatic light from the narrow slit S', at the principal focus of the lens L_1. The light focused by another lens L_2 on a screen or photographic plate P at its principal focus will form a diffraction pattern, as indicated schematically. Figure 15B(b) and (c) shows two actual photographs, taken with different exposure times, of such a pattern, using violet light of wavelength 4358 Å. The distance $S'L_1$ was 25.0 cm, and L_2P was 100 cm. The width of the slit S was 0.090 mm, and of S', 0.10 mm. If S' was widened to more than about 0.3 mm, the details of the pattern began to be lost. On the original plate, the half width d of the central maximum was 4.84 mm. It is important to notice that the width of the central maximum is *twice* as great as that of the fainter side maxima. That this effect comes under the heading of diffraction as previously defined is clear when we note that the strip drawn in Fig. 15B(a) is the width of the geometical image of the slit S', or practically that which would be obtained by removing the second slit and using the whole aperture of the lens. This pattern can easily be observed by ruling a single transparent line on a photographic plate and using it in front of the eye as explained in Sec. 13.2, Fig. 13E.

The explanation of the single-slit pattern lies in the interference of the Huygens secondary wavelets which can be thought of as sent out from every point on the wave

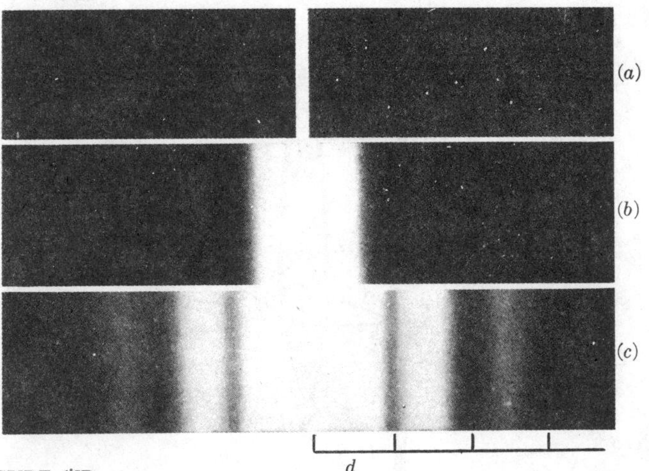

FIGURE 15B
Photographs of the single-slit diffraction pattern.

front at the instant that it occupies the plane of the slit. To a first approximation, one may consider these wavelets to be uniform spherical waves, the emission of which stops abruptly at the edges of the slit. The results obtained in this way, although they give a fairly accurate account of the observed facts, are subject to certain modifications in the light of the more rigorous theory.

Figure 15C represents a section of a slit of width b, illuminated by parallel light from the left. Let ds be an element of width of the wave front in the plane of the slit, at a distance s from the center O, which we shall call the origin. The parts of each secondary wave which travel normal to the plane of the slit will be focused at P_0, while those which travel at any angle θ will reach P. Considering first the wavelet emitted by the element ds situated at the origin, its amplitude will be directly proportional to the length ds and inversely proportional to the distance x. At P it will produce an infinitesimal displacement which, for a spherical wave, may be expressed as

$$dy_0 = \frac{a\,ds}{x} \sin(\omega t - kx)$$

As the position of ds is varied, the displacement it produces will vary in phase because of the different path length to P. When it is at a distance s below the origin, the contribution will be

$$dy_s = \frac{a\,ds}{x} \sin[\omega t - k(x + \Delta)]$$

$$= \frac{a\,ds}{x} \sin(\omega t - kx - ks\sin\theta) \qquad \text{(15a)}$$

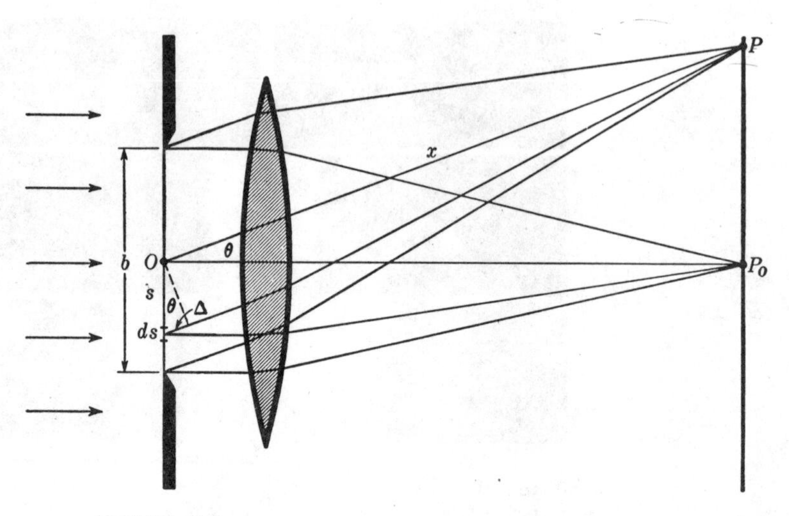

FIGURE 15C
Geometrical construction for investigating the intensity in the single-slit diffraction pattern.

We now wish to sum the effects of all elements from one edge of the slit to the other. This can be done by integrating Eq. (15a) from $s = -b/2$ to $b/2$. The simplest way* is to integrate the contributions from pairs of elements symmetrically placed at s and $-s$, each contribution being

$$dy = dy_{-s} + dy_s$$
$$= \frac{a\,ds}{x}[\sin(\omega t - kx - ks\sin\theta) + \sin(\omega t - kx + ks\sin\theta)]$$

By the identity $\sin\alpha + \sin\beta = 2\cos\frac{1}{2}(\alpha - \beta)\sin\frac{1}{2}(\alpha + \beta)$, we have

$$dy = \frac{a\,ds}{x}[2\cos(ks\sin\theta)\sin(\omega t - kx)]$$

which must be integrated from $s = 0$ to $b/2$. In doing so, x may be regarded as constant, insofar as it affects the amplitude. Thus

$$y = \frac{2a}{x}\sin(\omega t - kx)\int_0^{b/2}\cos(ks\sin\theta)\,ds$$
$$= \frac{2a}{x}\left[\frac{\sin(ks\sin\theta)}{k\sin\theta}\right]_0^{b/2}\sin(\omega t - kx)$$
$$= \frac{ab}{x}\frac{\sin(\frac{1}{2}kb\sin\theta)}{\frac{1}{2}kb\sin\theta}\sin(\omega t - kx) \tag{15b}$$

* The method of complex amplitudes (Sec. 14.8) starts with $(ab/x)\int\exp(iks\sin\theta)\,ds$, and yields the real amplitude upon multiplication of the result by its complex conjugate. No simplification results from using the method here.

The resultant vibration will therefore be a simple harmonic one, the amplitude of which varies with the position of P, since the latter is determined by θ. We may represent its amplitude as

$$A = A_0 \frac{\sin \beta}{\beta} \tag{15c}$$

where $\beta = \frac{1}{2}kb \sin \theta = (\pi b \sin \theta)/\lambda$ and $A_0 = ab/x$. The quantity β is a convenient variable, which signifies one-half the phase difference between the contributions coming from opposite edges of the slit. The intensity on the screen is then

$$I \approx A^2 = {A_0}^2 \frac{\sin^2 \beta}{\beta^2} \tag{15d}$$

If the light, instead of being incident on the slit perpendicular to its plane, makes an angle i, a little consideration will show that it is merely necessary to replace the above expression for β by the more general expression

$$\beta = \frac{\pi b(\sin i + \sin \theta)}{\lambda} \tag{15e}$$

15.3 FURTHER INVESTIGATION OF THE SINGLE-SLIT DIFFRACTION PATTERN

In Fig. 15D(*a*) graphs are shown of Eq. (15c) for the *amplitude* (dotted curve) and Eq. (15*d*) for the *intensity*, taking the constant A_0 in each case as unity. The intensity curve will be seen to have the form required by the experimental result in Fig. 15B. The maximum intensity of the strong central band comes at the point P_0 of Fig. 15C, where evidently all the secondary wavelets will arrive in phase because the path difference $\Delta = 0$. For this point $\beta = 0$, and although the quotient $(\sin \beta)/\beta$ becomes indeterminate for $\beta = 0$, it will be remembered that $\sin \beta$ approaches β for small angles and is equal to it when β vanishes. Hence for $\beta = 0$, $(\sin \beta)/\beta = 1$. We now see the significance of the constant A_0. Since for $\beta = 0$, $A = A_0$, it represents the amplitude when all the wavelets arrive in phase. ${A_0}^2$ is then the value of the maximum intensity, at the center of the pattern. From this *principal maximum* the intensity falls to zero at $\beta = \pm\pi$, then passes through several *secondary maxima*, with equally spaced points of zero intensity at $\beta = \pm\pi, \pm 2\pi, \pm 3\pi, \ldots$, or in general $\beta = m\pi$. The secondary maxima do not fall halfway between these points, but are displaced toward the center of the pattern by an amount which decreases with increasing m. The exact values of β for these maxima can be found by differentiating Eq. (15c) with respect to β and equating to zero. This yields the condition

$$\tan \beta = \beta$$

The values of β satisfying this relation are easily found graphically as the intersections of the curve $y = \tan \beta$ and the straight line $y = \beta$. In Fig. 15D(b) these points of intersection lie directly below the corresponding secondary maxima.

The intensities of the secondary maxima can be calculated to a very close approximation by finding the values of $(\sin^2 \beta)/\beta^2$ at the halfway positions, i.e.,

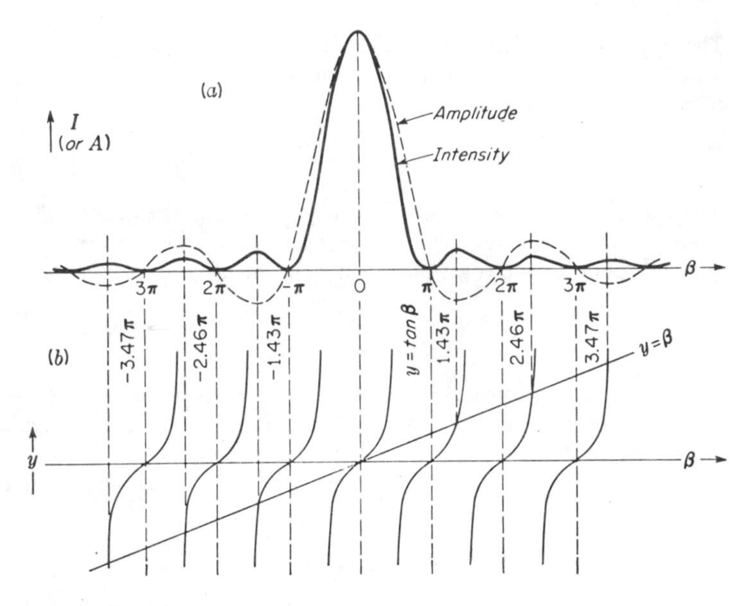

FIGURE 15D
Amplitude and intensity contours for Fraunhofer diffraction of a single slit, showing positions of maxima and minima.

where $\beta = 3\pi/2, 5\pi/2, 7\pi/2, \ldots$. This gives $4/9\pi^2$, $4/25\pi^2$, $4/49\pi^2, \ldots$, or 1/22.2, 1/61.7, 1/121, ..., of the intensity of the principal maximum. Reference to Table 15A ahead are the exact values of the intensity for every 15° intervals for the central maximum. These values are useful in plotting graphs. The first secondary maximum is only 4.72 percent the intensity of the central maximum, while the second and third secondary maxima are only 1.65 and 0.83 percent respectively.

A very clear idea of the origin of the single-slit pattern is obtained by the following simple treatment. Consider the light from the slit of Fig. 15E coming to the point P_1 on the screen, this point being just one wavelength farther from the upper

Table 15A VALUES FOR CENTRAL MAXIMUM FOR FRAUNHOFER DIFFRACTION OF A SINGLE SLIT

β deg	rad	$\sin\beta$	A^2	β deg	rad	$\sin\beta$	A^2
0	0	0	1	105	1.8326	0.9659	0.2778
15	0.2618	0.2588	0.9774	120	2.0944	0.8660	0.1710
30	0.5236	0.5000	0.9119	135	2.3562	0.7071	0.0901
45	0.7854	0.7071	0.8106	150	2.6180	0.5000	0.0365
60	1.0472	0.8660	0.6839	165	2.8798	0.2588	0.0081
75	1.3090	0.9659	0.5445	180	3.1416	0	0
90	1.5708	1.0000	0.4053	195	3.4034	0.2588	0.0058

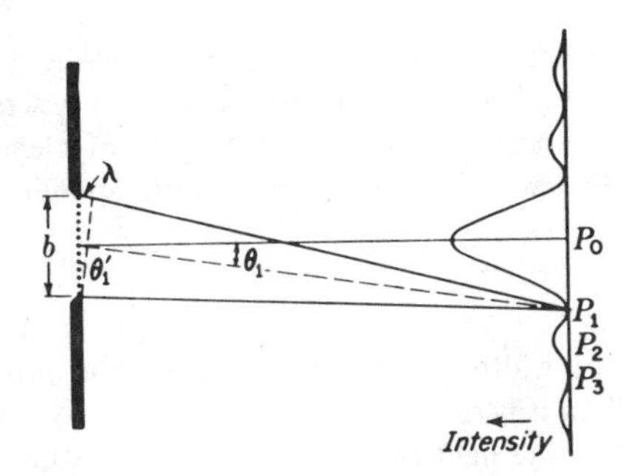

FIGURE 15E
Angle of the first minimum of the single-slit diffraction pattern.

edge of the slit than from the lower. The secondary wavelet from the point in the slit adjacent to the upper edge will travel approximately $\lambda/2$ farther than that from the point at the center, and so these two will produce vibrations with a phase difference of π and will give a resultant displacement of zero at P_1. Similarly the wavelet from the next point below the upper edge will cancel that from the next point below the center, and we can continue this pairing off to include all points in the wave front, so that the resultant effect at P_1 is zero. At P_3 the path difference is 2λ, and if we divide the slit into four parts, the pairing of points again gives zero resultant, since the parts cancel in pairs. For the point P_2, on the other hand, the path difference is $3\lambda/2$, and we divide the slit into thirds, two of which will cancel, leaving one third to account for the intensity at this point. The resultant amplitude at P_2 is, of course, not even approximately one-third that at P_0, because the phases of the wavelets from the remaining third are not by any means equal.

The above method, though instructive, is not exact if the screen is at a finite distance from the slit. As Fig. 15E is drawn, the shorter broken line is drawn to cut off equal distances on the rays to P_1. It will be seen from this that the path difference to P_1 between the light coming from the upper edge and that from the center is slightly greater than $\lambda/2$ and that between the center and lower edge slightly less than $\lambda/2$. Hence the resultant intensity will not be zero at P_1 and P_3, but it will be more nearly so the greater the distance between slit and screen or the narrower the slit. This corresponds to the transition from Fresnel diffraction to Fraunhofer diffraction. Obviously, with the relative dimensions shown in the figure, the geometrical shadow of the slit would considerably widen the central maximum as drawn. Just as was true with Young's experiment (Sec. 13.3), when the screen is at infinity, the relations become simpler. Then the two angles θ_1 and θ_1' in Fig. 15E become exactly equal, i.e., the two broken lines are perpendicular to each other, and $\lambda = b \sin \theta_1$ for the first minimum corresponding to $\beta = \pi$. This gives

$$\sin \theta_1 = \frac{\lambda}{b} \qquad (15f)$$

In practice θ_1 is usually a very small angle, so we may put the sine equal to the angle. Then

$$\theta_1 = \frac{\lambda}{b} \qquad (15g)$$

a relation which shows at once how the dimensions of the pattern vary with λ and b. The *linear* width of the pattern on a screen will be proportional to the slit-screen distance, which is the focal length f of a lens placed close to the slit. The linear distance d between successive minima corresponding to the angular separation $\theta_1 = \lambda/b$ is thus

$$d = \frac{f\lambda}{b}$$

The width of the pattern increases in proportion to the wavelength, so that for red light it is roughly twice as wide as for violet light, the slit width, etc., being the same. If white light is used, the central maximum is white in the middle but reddish on its outer edge, shading into a purple and other impure colors farther out.

The angular width of the pattern for a given wavelength is inversely proportional to the slit width b, so that as b is made larger, the pattern shrinks rapidly to a smaller scale. In photographing Fig. 15B, if the slit S had been 9.0 mm wide, the whole visible pattern (of five maxima) would have been included in a width of 0.24 mm on the original plate instead of 2.4 cm. This fact (that when the width of the aperture is large compared to a wavelength, the diffraction is practically negligible) led the early investigators to conclude that light travels in straight lines and that it could not be a wave motion. Sound waves will be diffracted through large angles in passing through an aperture of ordinary size, such as an open window.

15.4 GRAPHICAL TREATMENT OF AMPLITUDES. THE VIBRATION CURVE

The addition of the amplitude contributions from all the secondary wavelets originating in the slit can be carried out by a graphical method based on the vector addition of amplitudes discussed in Sec. 12.2. It will be worthwhile to consider this method in some detail, because it may be applied to advantage in other more complicated cases to be treated in later chapters, and because it gives a very clear physical picture of the origin of the diffraction pattern. Let us divide the width of the slit into a fairly large number of equal parts, say nine. The amplitude r contributed at a point on the screen by any one of these parts will be the same, since they are of equal width. The phases of these contributions will differ, however, for any point except that lying on the axis, i.e., on the normal to the slit at its center (P_0, Fig. 15E). For a point off the axis, each of the nine segments will contribute vibrations differing in phase, because the segments are at different average distances from the point. Furthermore the difference in phase δ between the contributions from adjacent segments will be constant, since each element is on the average the same amount farther away (or nearer) than its neighbor.

Now, using the fact that the resultant amplitude and phase may be found by the vector addition of the individual amplitudes making angles with each other equal to the phase difference, a vector diagram like that shown in Fig. 15F(b) may be drawn. Each of the nine equal amplitudes a is inclined a tan angle δ with the preceding one, and their vector sum A is the resultant amplitude required. Now suppose that instead of dividing the slit into nine elements, we had divided it into many thousand or, in the

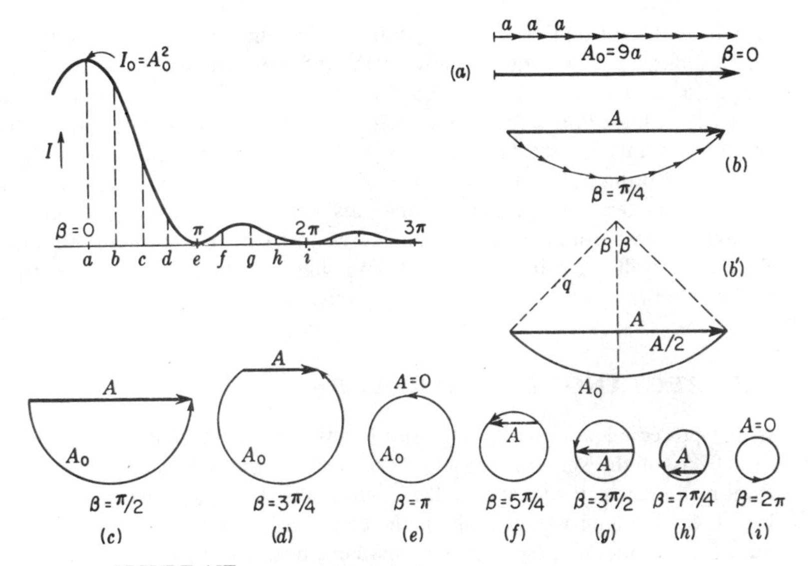

FIGURE 15F
Graphical treatment of amplitudes in single-slit diffraction.

limit, an infinite number of equal elements. The vectors a would become shorter, but at the same time δ would decrease in the same proportion, so that in the limit our vector diagram would approach the arc of a circle, shown as in (b'). The resultant amplitude A is still the same and equal to the length of the chord of this arc. Such a continuous curve, representing the addition of infinitesimal amplitudes, we shall refer to as a *vibration curve*.

To show that this method is in agreement with our previous result, we note that the length of the arc is just the amplitude A_0 obtained when all of the component vibrations are in phase, as in (a) of the figure. Introducing a phase difference between the components does not alter their individual amplitudes or the algebraic sum of these. Hence the ratio of the resultant amplitude A at any point in the screen to A_0, that on the axis, is the ratio of the chord to the arc of the circle. Since β stands for half the phase difference from opposite edges of the slit, the angle subtended by the arc is just 2β, because the first and last vectors a will have a phase difference of 2β. In Fig. 15F(b'), the radius of the arc is called q, and a perpendicular has been dropped from the center on the chord A. From the geometry of the figure, we have

$$\sin \beta = \frac{A/2}{q} \qquad A = 2q \sin \beta$$

and hence

$$\frac{A}{A_0} = \frac{\text{chord}}{\text{arc}} = \frac{2q \sin \beta}{q \times 2\beta} = \frac{\sin \beta}{\beta}$$

in agreement with Eq. (15c).

As we go out from the center of the diffraction pattern, the length of the arc remains constant and equal to A_0, but its curvature increases owing to the larger phase difference δ introduced between the infinitesimal component vectors a. The vibration curve thus winds up on itself as β increased. The successive diagrams (*a*) to (*i*) in Fig. 15F are drawn for the indicated values of β at intervals of $\pi/4$, and the corresponding points are similarly lettered on the intensity diagram. A study of these figures will bring out clearly the cause of the variations in intensity occurring in the single-slit pattern. In particular, one sees that the asymmetry of the secondary maxima follows from the fact that the radius of the circle is shrinking with increasing β. Thus A will reach its maximum length slightly before the condition represented in Fig. 15F(*g*).

15.5 RECTANGULAR APERTURE

In the preceding sections the intensity function for a slit was derived by summing the effects of the spherical wavelets originating from a linear section of the wave front by a plane perpendicular to the length of the slit, i.e., by the plane of the page in Fig. 15C. Nothing was said about the contributions from parts of the wave front out of this plane. A more thorough mathematical investigation, involving a double integration over both dimensions of the wave front,* shows, however, that the above result is correct when the slit is very long compared to its width. The complete treatment gives, for a slit of width b and length l, the following expression for the intensity:

$$I \approx b^2 l^2 \frac{\sin^2 \beta}{\beta^2} \frac{\sin^2 \gamma}{\gamma^2} \tag{15h}$$

where $\beta = (\pi b \sin \theta)/\lambda$, as before, and $\gamma = (\pi l \sin \Omega)/\lambda$. The angles θ and Ω are measured from the normal to the aperture at its center, in planes through the normal parallel to the sides b and l, respectively. The diffraction pattern given by Eq. (15h) when b and l are comparable with each other is illustrated in Fig. 15G. The dimensions of the aperture are shown by the white rectangle in the lower left-hand part of the figure. The intensity in the pattern is concentrated principally in two directions coinciding with the sides of the aperture, and in each of these directions it corresponds to the simple pattern for a slit width equal to the width of the aperture in that direction. Owing to the inverse proportionality between the slit width and the scale of the pattern, the fringes are more closely spaced in the direction of the longer dimension of the aperture. In addition to these patterns there are other faint maxima, as shown in the figure. This diffraction pattern can easily be observed by illuminating a small rectangular aperture with monochromatic light from a source which is effectively a *point*, the disposition of the lenses and the distance of the source and screen being similar to those described for observation of the slit pattern (Sec. 15.2). The cross formed by the brightest spots in the photograph is the one always observed when a bright street light is seen through woven fabric.

Now for a slit having l very large, the factor $(\sin^2 \gamma)/\gamma^2$ in Eq. (15h) is zero for

* See R. W. Wood, "Physical Optics," 2d ed., pp. 195–202, The Macmillan Company, New York, 1921; reprinted (paperback) by Dover Publications, Inc., New York, 1968.

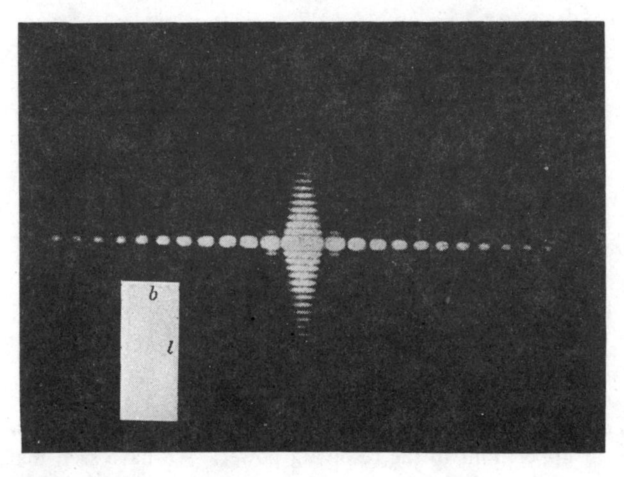

FIGURE 15G
Diffraction pattern from a rectangular opening.

all values of Ω except extremely small ones. This means that the diffraction pattern will be limited to a line on the screen perpendicular to the slit and will resemble a section of the central horizontal line of bright spots in Fig. 15G. We do not ordinarily observe such a line pattern in diffraction by a slit, because its observation requires the use of a *point source*. In Fig. 15A the primary source was a slit S', with its long dimension perpendicular to the page. In this case, each point of the source slit forms a line pattern, but these fall adjacent to each other on the screen, adding up to give a pattern like Fig. 15B. If we were to use a slit source with the rectangular aperture of Fig. 15G, the slit being parallel to the side l, the result would be the summation of a number of such patterns, one above the other, and would be identical with Fig. 15B.

These considerations can easily be extended to cover the effect of widening the primary slit. With a slit of finite width, each line element parallel to the length of the slit forms a pattern like Fig. 15B. The resultant pattern is equivalent to a set of such patterns displaced laterally with respect to each other. If the slit is too wide, the single-slit pattern will therefore be lost. No great change will occur until the patterns from the two edges of the slit are displaced about one-fourth of the distance from the central maximum to the first minimum. This condition will hold when the width of the primary slit subtends an angle of $\frac{1}{4}(\lambda/b)$ at the first lens, as can be seen by reference to Fig. 15H below.

15.6 RESOLVING POWER WITH A RECTANGULAR APERTURE

By the resolving power of an optical instrument we mean its ability to produce separate images of objects very close together. Using the laws of geometrical optics, one designs a telescope or a microscope to give an image of a point source which is as small as

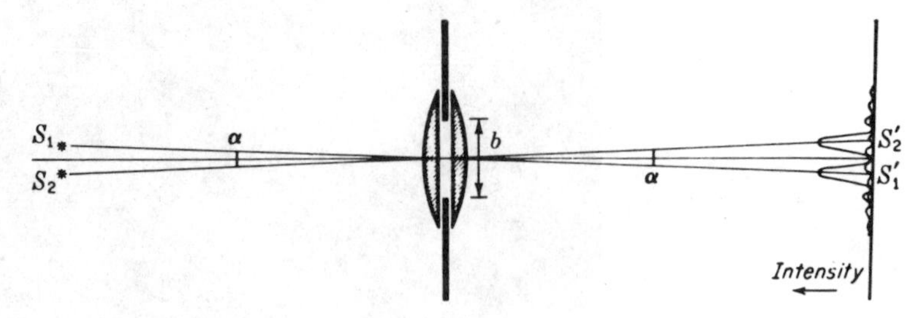

FIGURE 15H
Diffraction images of two slit sources formed by a rectangular aperture.

possible. However, in the final analysis, it is the diffraction pattern that sets a theoretical upper limit to the resolving power. We have seen that whenever parallel light passes through any aperture, it cannot be focused to a point image but instead gives a diffraction pattern in which the central maximum has a certain finite width, inversely proportional to the width of the aperture. The images of two objects will evidently not be resolved if their separation is much less than the width of the central diffraction maximum. The aperture here involved is usually that of the objective lens of the telescope or microscope and is therefore circular. Diffraction by a circular aperture will be considered below in Sec. 15.8, and here we shall treat the somewhat simpler case of a rectangular aperture.

Figure 15H shows two plano-convex lenses (equivalent to a single double-convex lens) limited by a rectangular aperture of vertical dimension b. Two narrow slit sources S_1 and S_2 perpendicular to the plane of the figure form real images S'_1 and S'_2 on a screen. Each image consists of a single-slit diffraction pattern for which the intensity distribution is plotted in a vertical direction. The angular separation α of the central maxima is equal to the angular separation of the sources, and with the value shown in the figure is adequate to give separate images. The condition illustrated is that in which each principal maximum falls exactly on the second minimum of the adjacent pattern. This is the smallest possible value of α which will give zero intensity between the two strong maxima in the resultant pattern. The angular separation from the center to the second minimum in either pattern then corresponds to $\beta = 2\pi$ (see Fig. 15D), or $\sin \theta \approx \theta = 2\lambda/b = 2\theta_1$. As α is made smaller than this, and the two images move closer together, the intensity between the maxima will rise, until finally no minimum remains at the center. Figure 15I illustrates this by showing the resultant curve (heavy line) for four different values of α. In each case the resultant pattern has been obtained by merely adding the intensities due to the separate patterns (dotted and light curves), as was done in the case of the Fabry-Perot fringes (Sec. 14.12).

Inspection of this figure shows that it would be impossible to resolve the two images if the maxima were much closer than $\alpha = \theta_1$, corresponding to $\beta = \pi$. At this separation the maximum of one pattern falls exactly on the first minimum of the other, so that the intensities of the maxima in the resultant pattern are equal to those of the separate maxima. The calculations are therefore simpler than for Fabry-Perot

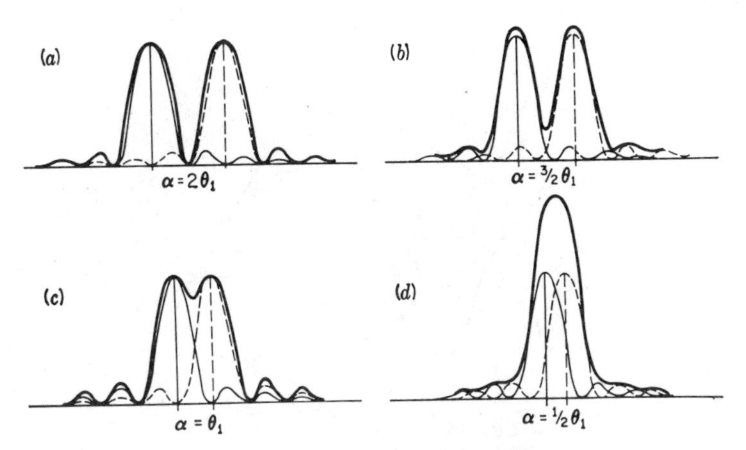

FIGURE 15I
Diffraction images of two slit sources: (*a*) and (*b*) well resolved; (*c*) just resolved; (*d*) not resolved.

fringes, where at no point does the intensity actually become zero. To find the intensity at the center of the resultant minimum for diffraction fringes separated by θ_1, we note that the curves cross at $\beta = \pi/2$ for either pattern and

$$\frac{\sin^2 \beta}{\beta^2} = \frac{4}{\pi^2} = 0.4053$$

the intensity of either relative to the maximum. The sum of the contributions at this point is therefore 0.8106, which shows that the intensity of the resultant pattern drops almost to four-fifths of its maximum value. This change of intensity is easily visible to the eye, and in fact a considerably smaller change could be seen, or at least detected with a sensitive intensity-measuring instrument such as a microphotometer. However, the depth of the minimum changes very rapidly with separation in this region, and in view of the simplicity of the relations in this particular case, it was decided by Rayleigh to arbitrarily fix the separation $\alpha = \theta_1 = \lambda/b$ as the criterion for resolution of two diffraction patterns. This quite arbitrary choice is known as *Rayleigh's criterion.* The angle θ_1 is sometimes called the *resolving power* of the aperture b, although the ability to resolve increases as θ_1 becomes smaller. A more appropriate designation for θ_1 is the *minimum angle of resolution.*

15.7 CHROMATIC RESOLVING POWER OF A PRISM

An example of the use of this criterion for the resolving power of a rectangular aperture is found in the prism spectroscope, if we assume that the face of the prism limits the refracted beam to a rectangular section. Thus, in Fig. 15J, the minimum angle $\Delta\delta$ between two parallel beams which give rise to images on the limit of resolution is such that $\Delta\delta = \theta_1 = \lambda/b$, where b is the width of the emerging beam. The

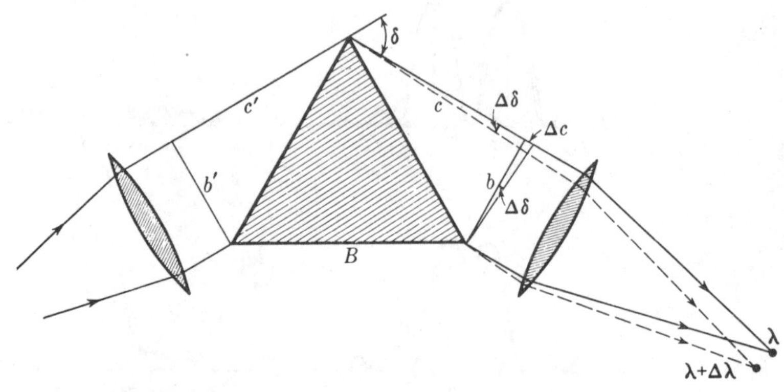

FIGURE 15J
Resolving power of a prism.

two beams giving these images differ in wavelength by a small increment $\Delta\lambda$, which is negative because the smaller wavelengths are deviated through greater angles. The wavelength increment is more useful than the increment of angle, and is the quantity that enters in the chromatic resolving power $\lambda/\Delta\lambda$ (Sec. 14.12). To evaluate this for the prism, we first note that since any optical path between two successive positions b' and b of the wave front must be the same, we can write

$$c + c' = nB \tag{15i}$$

Here n is the refractive index of the prism for the wavelength λ, and B the length of the base of the prism. Now, if the wavelength is decreased by $\Delta\lambda$, the optical path through the base of the prism becomes $(n + \Delta n)B$ and the emergent wave front must turn through an angle $\Delta\delta \doteq \lambda/b$ for the image it forms to be just resolved. Since, from the figure, $\Delta\delta = (\Delta c)/b$, this amount of turning increases the length of the upper ray by $\Delta c = \lambda$. It is immaterial whether we measure Δc along the rays λ or $\lambda + \Delta\lambda$, because only a difference of the second order is involved. Then we have

$$c + c' + \lambda = (n + \Delta n)B$$

and, subtracting Eq. (15i),

$$\lambda = B\,\Delta n$$

The desired result is now obtained by dividing by $\Delta\lambda$ and substituting the derivative $dn/d\lambda$ for the ratio of small increments.

$$\frac{\lambda}{\Delta\lambda} = B\frac{dn}{d\lambda} \tag{15j}$$

It is not difficult to show that this expression also equals the product of the angular dispersion and the width b of the emergent beam. Furthermore, we find that Eq. (15j) can still be applied when the beam does not fill the prism, in which case B must be the difference in the extreme paths through the prism, and when there are two or more prisms in tandem, when B is the sum of the bases.

15.8 CIRCULAR APERTURE

The diffraction pattern formed by plane waves from a point source passing through a circular aperture is of considerable importance as applied to the resolving power of telescopes and other optical instruments. Unfortunately it is also a problem of considerable difficulty, since it requires a double integration over the surface of the aperture similar to that mentioned in Sec. 15.5 for a rectangular aperture. The problem was first solved by Airy* in 1835, and the solution is obtained in terms of Bessel functions of order unity. These must be calculated from series expansions, and the most convenient way to express the results for our purpose will be to quote the actual figures obtained in this way (Table 15B).

The diffraction pattern as illustrated in Fig. 15K(*a*) consists of a bright central disk, known as *Airy's disk*, surrounded by a number of fainter rings. Neither the disk nor the rings are sharply limited but shade gradually off at the edges, being separated by circles of zero intensity. The intensity distribution is very much the same as that which would be obtained with the single-slit pattern illustrated in Fig. 15E by rotating it about an axis in the direction of the light and passing through the principal maximum. The dimensions of the pattern are, however, appreciably different from those in a single-slit pattern for a slit of width equal to the diameter of the circular aperture. For the single-slit pattern, the angular separation θ of the minima from the center was found in Sec. 15.3 to be given by $\sin\theta \approx \theta = m\lambda/b$, where m is any whole number, starting with unity. The dark circles separating the bright ones in the pattern from a circular aperture can be expressed by a similar formula if θ is now the angular semidiameter of the circle, but in this case the numbers m are not integers. Their numerical values as calculated by Lommel† are given in Table 15B, which also includes the values of m for the maxima of the bright rings and data on their intensities.

* Sir George Airy (1801–1892). Astronomer Royal of England from 1835 to 1881. Also known for his work on the aberration of light (Sec. 19.11). For details of the solution here referred to, see T. Preston, "Theory of Light," 5th ed., pp. 324–327, Macmillan & Co., Ltd., London, 1928.

† E. V. Lommel, *Abh. Bayer. Akad. Wiss.*, **15**:531 (1886).

Table 15B

Ring	Circular aperture			Single slit	
	m	I_{max}	I_{total}	m	I_{max}
Central maximum	0	1	1	0	1
First dark	1.220			1.000	
Second bright	1.635	0.01750	0.084	1.430	0.0472
Second dark	2.233			2.000	
Third bright	2.679	0.00416	0.033	2.459	0.0165
Third dark	3.238			3.000	
Fourth bright	3.699	0.00160	0.018	3.471	0.0083
Fourth dark	4.241			4.000	
Fifth bright	4.710	0.00078	0.011	4.477	0.0050
Fifth dark	5.243			5.000	

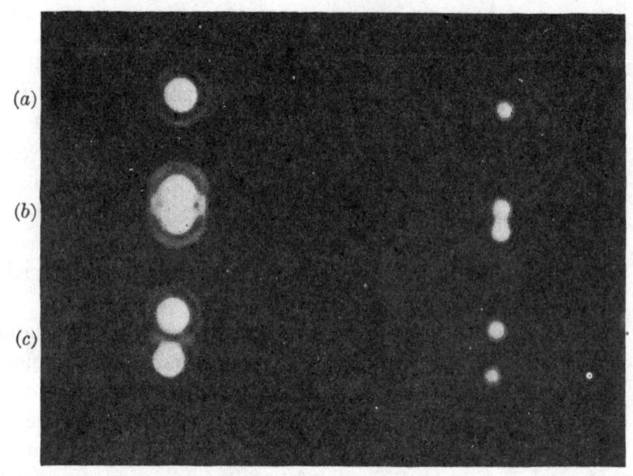

FIGURE 15K
Photographs of diffraction images of point sources taken with a circular aperture: (*a*) one source; (*b*) two sources just resolved; (*c*) two sources completely resolved.

The column headed I_{max} gives the relative intensities of the maxima, while that headed I_{total} is the total amount of light in the ring, relative to that of the central disk. For comparison, the values of m and I_{max} for the straight bands of the single-slit pattern are also included.

15.9 RESOLVING POWER OF A TELESCOPE

To give an idea of the linear size of the above diffraction pattern, let us calculate the radius of the first dark ring in the image formed in the focal plane of an ordinary field glass. The diameter of the objective is 4 cm and its focal length 30.0 cm. White light has an effective wavelength of 5.6×10^{-5} cm, so that the angular radius of this ring is $\theta = 1.220(5.6 \times 10^{-5})/4 = 1.71 \times 10^{-5}$ rad. The linear radius is this angle multiplied by the focal length and therefore amounts to $30 \times 1.71 \times 10^{-5} = 0.000512$ cm, or almost exactly 0.005 mm. The central disk for this telescope is then 0.01 mm in diameter when the object is a point source such as a star.

Extending Rayleigh's criterion for the resolution of diffraction patterns (Sec. 15.6) to the circular aperture we say that two patterns are resolved when the central maximum of one falls on the first dark ring of the other. The resultant pattern in this condition is shown in Fig. 15K(*b*). The minimum angle of resolution for a telescope is therefore

$$\theta_1 = 1.220 \frac{\lambda}{D} \qquad (15k)$$

where D is the diameter of the circular aperture which limits the beam forming the primary image, or usually that of the objective. For the example chosen above, the angle calculated is just this limiting angle, so that the smallest angular separation of a double star which could be theoretically resolved by this telescope is 1.71×10^{-5} rad, or 3.52 seconds of arc. Since the minimum angle is inversely proportional to D, we see that the aperture necessary to resolve two sources 1 second apart is 3.52 times as great as in the example, or that

$$\text{Minimum angle of resolution in seconds } \theta_1 = \frac{14.1}{D} \qquad (15l)$$

D being the aperture of the objective in centimeters. For the largest refracting telescope in existence, that at the Yerkes Observatory, $D = 40$ in. and $\theta_1 = 14$ seconds. This may be compared with the minimum angle of resolution for the eye, the pupil of which has a diameter of about 3.0 mm. We find $\theta_1 = 47$ seconds of arc.* Actually the eye of the average person is not able to resolve objects less than about 1 minute apart, and the limit is therefore effectively determined by optical defects in the eye or by the structure of the retina.

With a given objective in a telescope, the angular size of the image as seen by the eye is determined by the magnification of the eyepiece. However, increasing the size of the image by increasing the power of the eyepiece does not increase the amount of detail that can be seen, since it is impossible by magnification to bring out detail which is not originally present in the primary image. Each point in an object becomes a small circular diffraction pattern or disk in the image, so that if an eyepiece of very high power is used, the image appears blurred and no greater detail is seen. Thus diffraction by the objective is the one factor that limits the resolving power of a telescope.

The diffraction pattern of a circular aperture, as well as the resolving power of a telescope, can be demonstrated by an experimental arrangement similar to that shown in Fig. 15H. The point sources at S_1 and S_2 consist of a sodium or mercury arc and a screen with several pinholes about 0.35 mm in diameter and spaced from 2.0 to 10.0 mm apart. These may be viewed with one of three small holes 1.0, 2.0, and 4.0 mm in diameter, mounted in front of the objective lens to show how an increasing aperture affects the resolution. Under these circumstances the intensity will not be sufficient to show anything but the central disks. In order to observe the subsidiary diffraction rings, the best source to use is the concentrated-arc lamp (Sec. 21.2) or a laser.

The theoretical resolving power of a telescope will be realized only if the lenses are geometrically perfect and if the magnification is at least equal to the so-called normal magnification (Sec. 10.13). To prove the latter statement, we note that two diffraction disks which are on the limit of resolution in the focal plane of the objective

* It might at first appear that the wavelength to be used in this calculation would be that in the vitreous humor of the eye. It is true that the dimensions of the diffraction pattern are smaller on this account, but the separation of two images is also decreased in the same proportion by refraction of the rays as they enter the eye.

must subtend at the eye an angle of at least $\theta = 1.22\lambda/d_e$ in order to be resolved by the eye. Here d_e represents the diameter of the eye pupil. According to Eq. (10k), the magnification

$$M = \frac{\theta'}{\theta} = \frac{D}{d}$$

where D is the diameter of the entrance pupil (objective) and d that of the exit pupil. At the normal magnification, d is made equal to d_e, so that the normal magnification becomes

$$\frac{D}{d_e} \equiv \frac{1.22\lambda/d_e}{1.22\lambda/D} = \frac{\theta_1'}{\theta_1}$$

Hence, if the diameter d of the exit pupil is made larger than d_e, that of the eye pupil, we have $\theta' < \theta_1'$ and the images will cease to be resolved by the eye even though they are resolved in the focal plane of the objective. In other words, any magnification that is less than the normal one corresponds to an exit pupil larger than d_e, and will not afford the resolution that the instrument could give.

15.10 RESOLVING POWER OF A MICROSCOPE

In this case the same principles are applicable. The conditions are, however, different from those for a telescope, in which we were chiefly interested in the smallest permissible angular separation of two objects at a large, and usually unknown, distance. With a microscope the object is very close to the objective, and the latter subtends a large angle $2i$ at the object plane, as shown in Fig. 15L. Here we wish primarily to know the smallest distance between two points O and O' in the object which will produce images I and I' that are just resolved. Each image consists of a disk and a system of rings, as explained above, and the angular separation of two disks when they are on the limit of resolution is $\alpha = \theta_1 = 1.22\lambda/D$. When this condition holds, the wave from O' diffracted to I has zero intensity (first dark ring), and the extreme rays $O'BI$ and $O'AI$ differ in path by 1.22λ. From the insert in Fig. 15L, we see that $O'B$ is longer than OB or OA by $s \sin i$, and $O'A$ shorter by the same amount. The path difference of the extreme rays from O' is thus $2s \sin i$, and upon equating this to 1.22λ, we obtain

$$s = \frac{1.22\lambda}{2 \sin i} \qquad \text{(15m)}$$

In this derivation, we have assumed that the points O and O' were *self-luminous* objects, such that the light given out by each has no constant phase relative to that from the other. Actually the objects used in microscopes are not self-luminous but are illuminated with light from a condenser. In this case it is impossible to have the light scattered by two points on the object entirely independent in phase. This greatly complicates the problem, since the resolving power is found to depend somewhat on the mode of illumination of the object. Abbe investigated this problem in detail and concluded that a good working rule for calculating the resolving power was given by

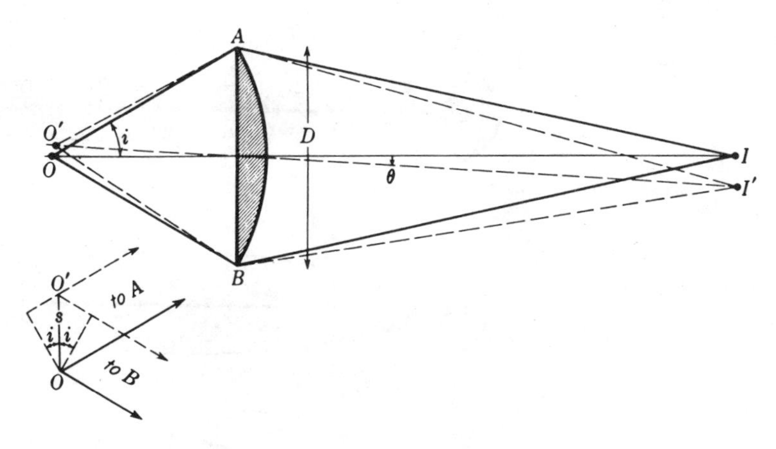

FIGURE 15L
Resolving power of a microscope.

Eq. (15m), omitting the factor 1.22. In microscopes of high magnifying power, the space between the object and the objective is filled with an oil. Besides decreasing the amount of light lost by reflection at the first lens surface, this increases the resolving power, because when refraction of the rays emerging from the cover glass is eliminated, the objective receives a wider cone of light from the condenser. Equation (15m) must then be further modified by substitution of $2ns \sin i$ for the optical path difference, where n is the refractive index of the oil. The result of making these two changes is

$$s = \frac{\lambda}{2n \sin i} \tag{15n}$$

The product $n \sin i$ is characteristic of a particular objective, and was called by Abbe the *numerical aperture*. In practice the largest value of the numerical aperture obtainable is about 1.6. With white light of effective wavelength 5.6×10^{-5} cm, Eq. (15n) gives $s = 1.8 \times 10^{-5}$ cm. The use of ultraviolet light, with its smaller value of λ, has recently been applied to still further increase the resolving power. This necessitates the use of photography in examining the image.

One of the most remarkable steps in the improvement of microscopic resolution has been the development of the *electron microscope*. As will be explained in Sec. 33.4, electrons behave like waves whose wavelength depends on the voltage through which they have been accelerated. Between 100 and 10,000 V, λ varies from 0.122 to 0.0122 nm; i.e., it lies in the region of a fraction of an angstrom unit. This is more than a thousand times smaller than for visible light. It is possible by means of electric and magnetic fields to focus the electrons emitted from, or transmitted through, the various parts of an object, and in this way details not very much larger than the wavelength of the electrons can be photographed. The numerical aperture of electron

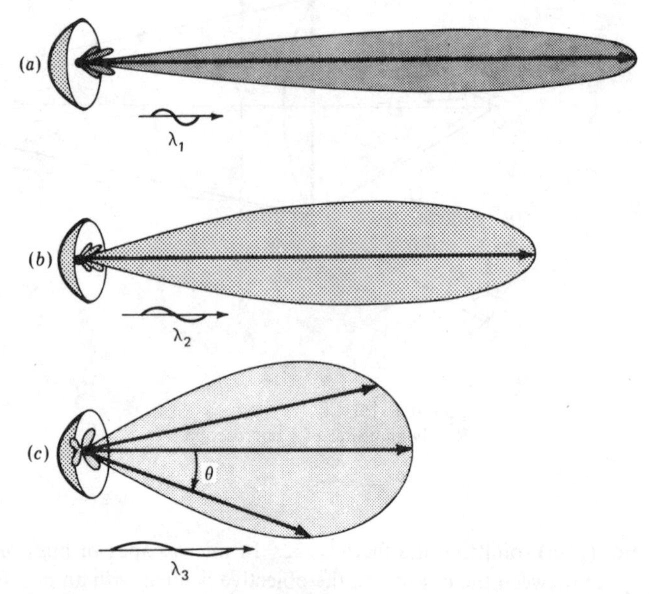

FIGURE 15M
Polar diagrams of the diffraction patterns for waves of different wavelength from the same parabolic reflector.

microscopes is still much smaller than that of optical instruments, but further developments in this large and growing field of *electron optics* are to be anticipated.*

15.11 DIFFRACTION PATTERNS WITH SOUND AND MICROWAVES

The principles of light diffraction from slits, rectangular apertures, and circular openings apply equally well to sound waves and microwaves. A radio loudspeaker with a circular aperture, for example, will form diffraction patterns determined by its diameter and frequencies emitted and give rise to marked changes in sound quality at different points in a closed room or outdoors. As another example the microwaves from a parabolic reflector radiate outward as a single-aperture diffraction pattern, with a central maximum in the forward direction, as shown in Fig. 15M.

It is customary with sound and microwaves to plot radiation diffraction patterns using *polar coordinates* in place of the *rectangular coordinates* used with light waves. Plotted as a polar diagram, the intensity radiated in different directions from a source

* See, for example, V. K. Zworykin, G. A. Morton, and others, "Electron Optics and the Electron Microscope," John Wiley and Sons, Inc., New York, 1945; also V. K. Zworykin, C. A. Morton, and others, "Television in Science and Industry," John Wiley and Sons, Inc., New York, 1958.

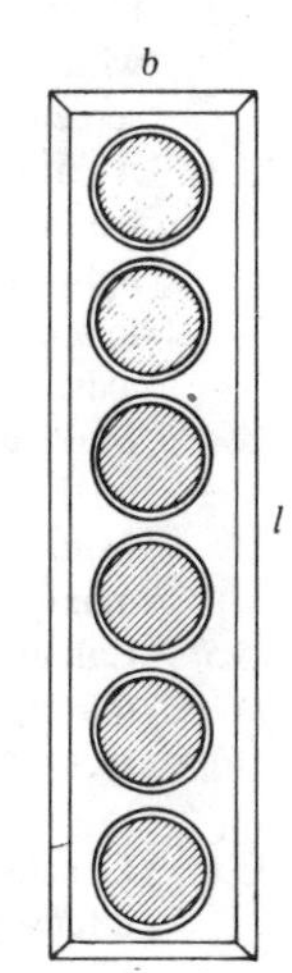

FIGURE 15N
Loudspeaker array for selectively directing sounds to an audience by diffraction.

is called a *lobe diagram*. The length of any arrow at any angle θ is drawn proportional to the relative intensity radiated in that direction. The lobes are then the envelopes of these arrow tips.

The shorter the wavelength and the greater the aperture of any given source of waves, the narrower the lobe pattern. Short waves from a point source at the focus of a given parabolic mirror may produce a very narrow central lobe, as shown in Fig. 15M(*a*), while longer waves will produce proportionately wider beams as shown in diagrams (*b*) and (*c*).

It is quite common today to use loudspeaker arrays with public-address systems, to direct sound in specified directions. An array like that shown in Fig. 15N, with several speakers electrically connected so they vibrate in unison, acts as though the entire rectangular opening sent out plane waves in the forward direction. The three-dimensional diffraction pattern is such that the central lobe is narrow vertically and wide horizontally, thereby directing the sound energy at the spread-out audience. Compare the rectangular source and the shape of the central beam in Fig. 15G. Parabolic reflectors for microwave patterns that are wide horizontally and narrow vertically radiate central beams that are narrow horizontally and wide vertically, thereby permitting reflected beams from distant objects to be located horizontally with precision but vertically with less accuracy.

PROBLEMS

15.1 Parallel light of wavelength 6563 Å is incident normally on a slit 0.3850 mm wide. A lens with a focal length of 50.0 cm is located just behind the slit bringing the diffraction pattern to focus on a white screen. Find the distance from the center of the principle maximum to (*a*) the first minimum and (*b*) the fifth minimum.

Ans. (*a*) 0.852 mm, (*b*) 4.261 mm

15.2 Plane waves of blue light, $\lambda = 4340$ Å, fall on a single slit, then pass through a lens with a focal length of 85.0 cm. If the central band of the diffraction pattern on the screen has a width of 2.450 mm, find the width of the single slit. *Ans.* 0.3011 mm

15.3 A parallel beam of white light falls normally on a single slit 0.320 mm wide, and 1 m behind this slit a small spectroscope is used to explore the spectrum of the diffracted light. Predict if you can what will be seen in the spectroscope if the slit is displaced in a direction perpendicular to the diffracting slit by a distance of 1.250 cm from the axis.

15.4 Make an accurate plot of the intensity in the Fraunhofer diffraction pattern of a slit in the region of the second subsidiary maximum, $\beta = 2\pi$ to $\beta = 3\pi$. Determine from your graph the figures given in Table 15A for the position and intensity of this maximum.

15.5 Calculate the approximate intensity of (*a*) the first and (*b*) the second weak maxima that appear along the diagonal $\beta/\gamma = l/b$ in the Fraunhofer diffraction pattern of a rectangular aperture of width b and height l.

Ans. (*a*) I/I_0 0.2227%, (*b*) I/I_0 0.02716%

15.6 Considering the criterion for the resolution of two diffraction patterns of unequal intensity to be that the drop in intensity between the two maxima shall be 20 percent of the weaker one, find the angular separation required when the intensities are in the ratio 3:1. Express your result in terms of β, the angle required when the intensities are equal. This problem is best solved graphically, using two plots that can be superimposed with a variable displacement.

15.7 From the refractive indices of borosilicate crown glass given in Table 23B, calculate the chromatic resolving power of a 70° prism of this material if the width of the sides is 5.0 cm. Make the calculation for wavelengths (*a*) 5338 Å and (*b*) 4861 Å.

Ans. (*a*) 3.16×10^3, (*b*) 4.13×10^3

15.8 A spectrum line at wavelength 3034 Å is known to be a close doublet. The wavelength difference between the two components is known to be 0.0860 Å. A crystalline quartz prism spectrograph is to be used to photograph this doublet. Such a prism is nearly always made so the refractive index is n_0 of Table 26A. Find (*a*) the dispersion of the quartz prism at $\lambda = 3034.4$ Å and (*b*) the minimum base length for the prism if it is just able to resolve the doublet. Determine the dispersion from a graph of n plotted against λ in the region of 3034.4 Å.

15.9 Carry through the differentiation of Eq. (15*c*) and prove that $\tan \beta = \beta$ is the condition for maxima (see Sec. 15.3).

15.10 Find the diameter of the Airy disk in the focal plane of a refracting telescope having an objective with a focal length of 1.0 m and a diameter of 10.0 cm. Assume the effective wavelength is 5.50×10^{-5} cm. *Ans.* 0.01342 mm

15.11 What is the maximum permissible width of the slit source according to the criterion stated at the end of Sec. 15.5 under the following circumstances: source to diffraction slit = 30.0 cm, width of diffraction slit = 0.40 mm, wavelength of light 5.0×10^{-5} cm?

15.12 The objective of a telescope has a diameter of 12.0 cm. At what distance would two small green objects 30.0 cm apart be barely resolved by the telescope, assuming the resolution to be limited by diffraction by the objective only? Assume $\lambda = 5.40 \times 10^{-5}$ cm.

15.13 A source producing underwater sound waves for submarine detection has a circular aperture 60.0 cm in diameter emitting waves with a frequency of 40.0 kHz. At some distance from this source the intensity pattern will be that of a Fraunhofer pattern from a circular aperture. (*a*) Find the angular spread of the central lobe pattern.

(*b*) Find the angular spread if the frequency is changed to 4.0 kHz. Assume the speed of the sound to be 1.50 km/s. *Ans.* (*a*) 8.74°, (*b*) 99.4°

15.14 A parabolic radar reflector 6.50 m in diameter emits microwaves with a frequency of 6.0×10^{10} Hz. At some distance from this source the lobe pattern is that of Fraunhofer diffraction. Find the angular width of the central lobe if the wave velocity is 3.0×10^{10} cm/s.

15.15 The loudspeaker array in a public-address system consists of six circular speakers each 25.0 cm in diameter and arranged as shown in Fig. 15N. The box in which they are mounted has the inside dimensions of 25.0 cm × 150.0 cm. Assuming Fraunhofer diffraction, find the horizontal and vertical spread of the central lobe pattern for sound waves of frequency (*a*) 5 kHz, (*b*) 1 kHz, and (*c*) 200 Hz. Assume the speed of sound to be 300 m/s.

16

THE DOUBLE SLIT

The interference of light from two narrow slits close together, first demonstrated by Young, has already been discussed (Sec. 13.2) as a simple example of the interference of two beams of light. In our discussion of the experiment, the slits were assumed to have widths not much greater than a wavelength of light, so that the central maximum in the diffraction pattern from each slit separately was wide enough to occupy a large angle behind the screen (Figs. 13A and 13B). It is important to understand the modifications of the interference pattern which occur when the width of the individual slits is made greater, until it becomes comparable with the distance between them. This corresponds more nearly to the actual conditions under which the experiment is usually performed. In this chapter we shall discuss the *Fraunhofer diffraction by a double slit*, and some of its applications.

16.1 QUALITATIVE ASPECTS OF THE PATTERN

In Fig. 16A(*b*) and (*c*) photographs are shown of the patterns obtained from two different double slits in which the widths of the individual slits were equal in each pair but the two pairs were different. Figure 16B shows the experimental arrangement for

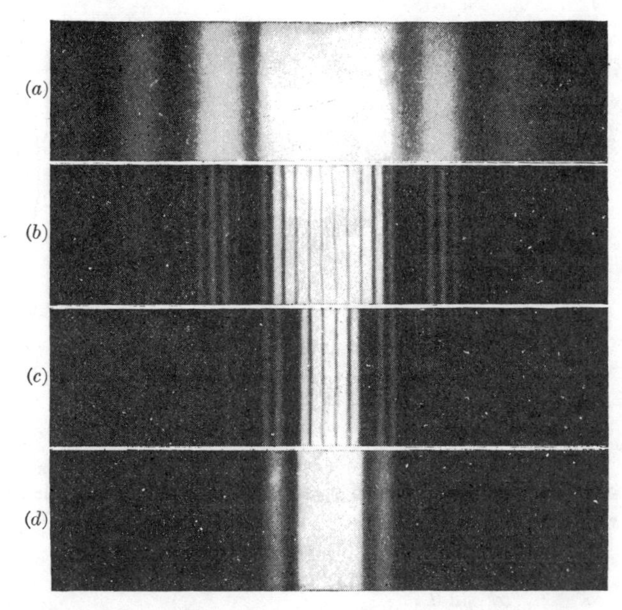

FIGURE 16A
Diffraction patterns from (*a*) a single narrow slit, (*b*) two narrow slits, (*c*) two wider slits, (*d*) one wider slit.

photographing these patterns; the *slit width b* of each slit was greater for Fig. 16A(*c*) than for Fig. 16A(*b*), but the distance between centers $d = b + c$, or the *separation* of the slits, was the same in the two cases. In the central part of Fig. 16A(*b*) are seen a number of interference maxima of approximately uniform intensity, resembling the interference fringes described in Chap. 13 and shown in Fig. 13D. The intensities of these maxima are not actually constant, however, but fall off slowly to zero on either side and then reappear with low intensity two or three times before becoming too faint to observe without difficulty. The same changes are seen to occur much more rapidly in Fig. 16A(*c*), which was taken with the slit widths b somewhat larger.

16.2 DERIVATION OF THE EQUATION FOR THE INTENSITY

Following the same procedure as that used for the single slit in Sec. 15.2, it is merely necessary to change the limits of integration in Eq. (15b) to include the two portions of the wave front transmitted by the double slit.* Thus if we have, as in Fig. 16B, two equal slits of width b, separated by an opaque space of width c, the origin may

* The result of this derivation is obviously a special case of the general formula for N slits, which will be obtained by the method of complex amplitudes in the following chapter.

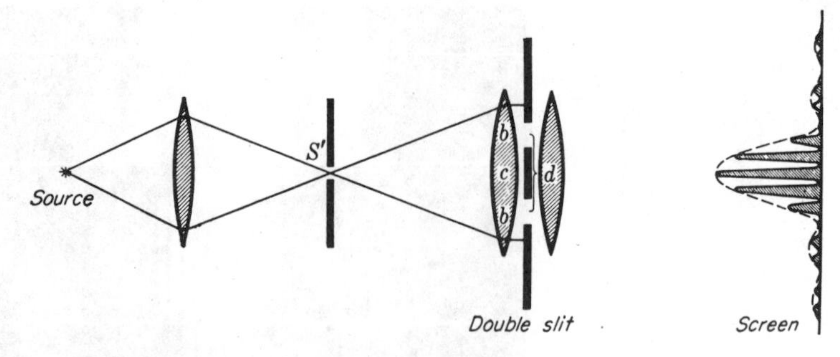

FIGURE 16B
Apparatus for observing Fraunhofer diffraction from a double slit. Drawn for $2b = c$, that is, $d = 3b$.

be chosen at the center of c, and the integration extended from $s = d/2 - b/2$ to $s = d/2 + b/2$. This gives

$$y = \frac{2a}{xk \sin \theta} \{\sin [\tfrac{1}{2}k(d + b) \sin \theta] - \sin [\tfrac{1}{2}k(d - b) \sin \theta]\} [\sin (\omega t - kx)]$$

The quantity in braces is of the form $\sin (A + B) - \sin (A - B)$, and when it is expanded, we obtain

$$y = \frac{2ba}{x} \frac{\sin \beta}{\beta} \cos \gamma \sin (\omega t - kx) \tag{16a}$$

where, as before,

$$\beta = \tfrac{1}{2}kb \sin \theta = \frac{\pi}{\lambda} b \sin \theta$$

and where

$$\gamma = \tfrac{1}{2}k(b + c) \sin \theta = \frac{\pi}{\lambda} d \sin \theta \tag{16b}$$

The intensity is proportional to the square of the amplitude of Eq. (16a), so that, replacing ba/x by A_0 as before, we have

$$I = 4A_0^2 \frac{\sin^2 \beta}{\beta^2} \cos^2 \gamma \tag{16c}$$

The factor $(\sin^2 \beta)/\beta^2$ in this equation is just that derived for the single slit of width b in the previous chapter [Eq. (15d)]. The second factor $\cos^2 \gamma$ is characteristic of the interference pattern produced by two beams of equal intensity and phase difference δ, as shown in Eq. (13b) of Sec. 13.3. There the resultant intensity was found to be proportional to $\cos^2 (\delta/2)$, so that the expressions correspond if we put $\gamma = \delta/2$. The resultant intensity will be zero when either of the two factors is zero. For the first factor this will occur when $\beta = \pi, 2\pi, 3\pi, \ldots$, and for the second factor when $\gamma = \pi/2, 3\pi/2, 5\pi/2, \ldots$. That the two variables β and γ are not independent will be seen from Fig. 16C. The difference in path from the two edges of a given slit to the screen is, as indicated, $b \sin \theta$. The corresponding phase difference is, by Eq. (15c),

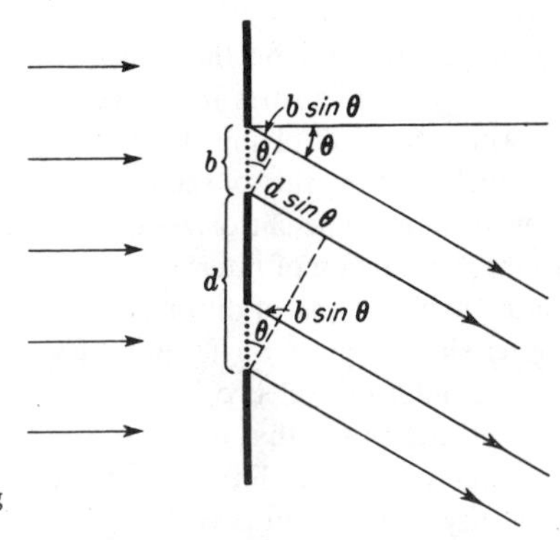

FIGURE 16C
Path differences of parallel rays leaving a double slit.

$(2\pi/\lambda)b \sin \theta$, which equals 2β. The path difference from any two corresponding points in the two slits, as illustrated for the two points at the lower edges of the slits, is $d \sin \theta$, and the phase difference is $\delta = (2\pi/\lambda)d \sin \theta = 2\gamma$. Therefore, in terms of the dimensions of the slits,

$$\frac{\delta}{2\beta} = \frac{\gamma}{\beta} = \frac{d}{b} \qquad \text{(16d)}$$

16.3 COMPARISON OF THE SINGLE-SLIT AND DOUBLE-SLIT PATTERNS

It is instructive to compare the double-slit pattern with that given by a single slit of width equal to that of either of the two slits. This amounts to comparing the effect obtained with the two slits in the arrangement shown in Fig. 16B with that obtained when one of the slits is entirely blocked off with an opaque screen. If this is done, the corresponding single-slit diffraction patterns are observed, and they are related to the double-slit patterns as shown in Fig. 16A(*a*) and (*d*). It will be seen that the intensities of the interference fringes in the double-slit pattern correspond to the intensity of the single-slit pattern at any point. If one or other of the two slits is covered, we obtain exactly the same single-slit pattern in the same position, while if both slits are uncovered, the pattern, instead of being a single-slit one with twice the intensity, breaks up into the narrow maxima and minima called *interference fringes*. The intensity at the maximum of these fringes is 4 times the intensity of either single-slit pattern at that point, while it is zero at the minima (see Sec. 13.4).

16.4 DISTINCTION BETWEEN INTERFERENCE AND DIFFRACTION

One is quite justified in explaining the above results by saying that the light from the two slits undergoes interference to produce fringes of the type obtained with two beams but that the intensities of these fringes are limited by the amount of light arriving at the given point on the screen by virtue of the diffraction occurring at each slit. The relative intensities in the resultant pattern as given by Eq. (16c) are just those obtained by multiplying the intensity function for the interference pattern from two infinitely narrow slits of separation d [Eq. (13b)] by the intensity function for diffraction from a single slit of width b [Eq. (15d)]. Thus, the result may be regarded as due to the joint action of interference between the rays coming from corresponding points in the two slits and of diffraction, which determines the amount of light emerging from either slit at a given angle. But diffraction is merely the result of the interference of all the secondary wavelets originating from the different elements of the wave front. Hence it is proper to say that the whole pattern is an interference pattern. It is just as correct to refer to it as a diffraction pattern, since, as we saw from the derivation of the intensity function in Sec. 16.2, it is obtained by direct summing the effects of all of the elements of the exposed part of the wave front. However, if we reserve the term *interference* for those cases in which a modification of amplitude is produced by the superposition of a finite (usually small) number of beams, and *diffraction* for those in which the amplitude is determined by an integration over the infinitesimal elements of the wave front, the double-slit pattern can be said to be due to a combination of interference and diffraction. Interference of the beams from the two slits produces the narrow maxima and minima given by the $\cos^2 \gamma$ factor, and diffraction, represented by $(\sin^2 \beta)/\beta^2$, modulates the intensities of these interference fringes. The student should not be misled by this statement into thinking that diffraction is anything other than a rather complicated case of interference.

16.5 POSITIONS OF THE MAXIMA AND MINIMA. MISSING ORDERS

As shown in Sec. 16.2, the intensity will be zero wherever $\gamma = \pi/2, 3\pi/2, 5\pi/2, \ldots$ and also when $\beta = \pi, 2\pi, 3\pi, \ldots$. The first of these two sets are the minima for the interference pattern, and since by definition $\gamma = (\pi/\lambda)d \sin \theta$, they occur at angles θ such that

$$d \sin \theta = \frac{\lambda}{2}, \frac{3\lambda}{2}, \frac{5\lambda}{2}, \ldots = (m + \tfrac{1}{2})\lambda \qquad \textit{Minima} \qquad (16e)$$

m being any whole number starting with zero. The second series of minima are those for the diffraction pattern, and these, since $\beta = (\pi/\lambda)a \sin \theta$, occur where

$$b \sin \theta = \lambda, 2\lambda, 3\lambda, \ldots = p\lambda \qquad \textit{Minima} \qquad (16f)$$

the smallest value of p being 1. The exact positions of the *maxima* are not given by any simple relation, but their approximate positions can be found by neglecting the variation of the factor $(\sin^2 \beta)/\beta^2$, a justified assumption only when the slits are very narrow and when the maxima near the center of the pattern are considered [Fig. 16A(b)]. The positions of the maxima will then be determined solely by the $\cos^2 \gamma$ factor, which has maxima for $\gamma = 0, \pi, 2\pi, \ldots$, that is, for

$$d \sin \theta = 0, \lambda, 2\lambda, 3\lambda, \ldots = m\lambda \qquad \textit{Maxima} \qquad (16g)$$

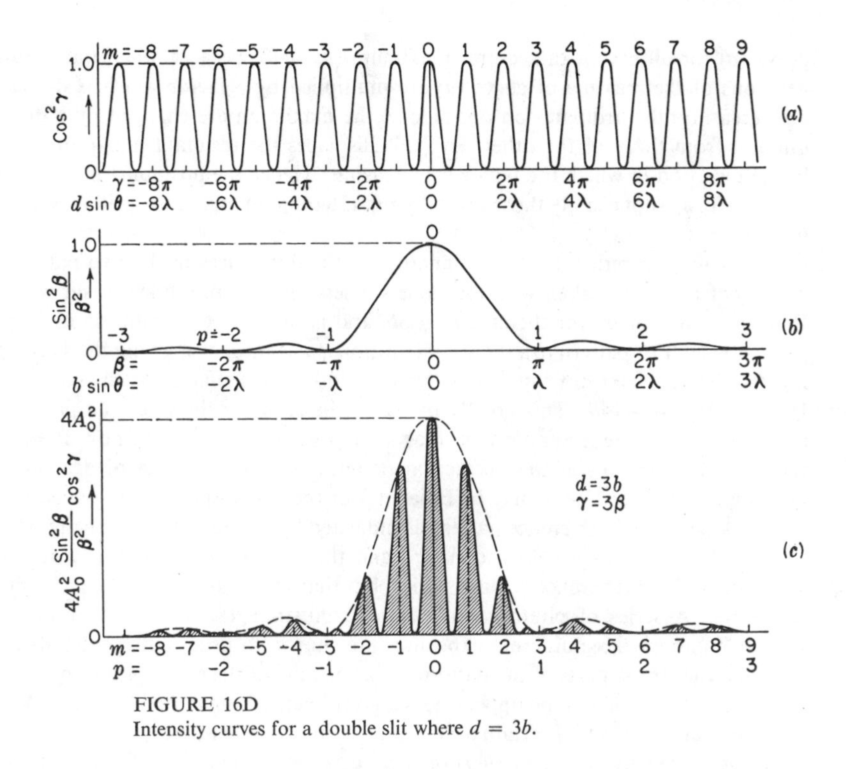

FIGURE 16D
Intensity curves for a double slit where $d = 3b$.

The whole number m represents physically the number of wavelengths in the path difference from corresponding points in the two slits (see Fig. 16C) and represents the *order* of interference.

Figure 16D(a) is a plot of the factor $\cos^2 \gamma$, and here the values of the order, of half the phase difference $\gamma = \delta/2$, and of the path difference are indicated for the various maxima. These are all of equal intensity and equidistant on a scale of $d \sin \theta$, or practically on a scale of θ, since when θ is small, $\sin \theta \approx \theta$ and the maxima occur at angles $\theta = 0, \lambda/d, 2\lambda/d, \ldots$. With a finite slit width b the variation of the factor $(\sin^2 \beta)/\beta^2$ must be taken into account. This factor alone gives just the single-slit pattern discussed in the last chapter, and is plotted in Fig. 16D(b). The complete double-slit pattern as given by Eq. (16c) is the product of these two factors and therefore is obtained by multiplying the ordinates of curve (a) by those of curve (b) and the constant $4A_0{}^2$. This pattern is shown in Fig. 16D(c). The result will depend on the relative scale of the abscissas β and γ, which in the figure are chosen so that for a given abscissa $\gamma = 3\beta$. But the relation between β and γ for a given angle θ is determined, according to Eq. (16d), by the ratio of the slit width to the slit separation. Hence if $d = 3b$, the two curves (a) and (b) are plotted to the same scale of θ. For the particular case of two slits of width b separated by an opaque space of width $c = 2b$, the curve (c), which is the product of (a) and (b), then gives the resultant pattern. The positions of the maxima in this curve are slightly different from those in curve (a) for all except the central maximum ($m = 0$), because when the ordinates near one of the maxima of curve (a) are multiplied by a factor which is decreasing or increasing, the ordinates on one side of the maximum are changed by a different amount from those of the other, and this displaces the resultant maximum slightly in the direction in which the factor is increasing. Hence the positions of the maxima in curve (c) are not exactly those given by Eq. (16g) but in most cases will be very close to them.

Let us now return to the explanation of the differences in the two patterns (b) and (c) of Fig. 16A, taken with the same slit separation d but different slit widths b. Pattern (c) was taken for the case $d = 3b$, and is seen to agree with the description given above. For pattern (b), the slit separation d was the same, giving the same spacing for the interference fringes, but the slit width b was smaller, such that $d = 6b$. In Fig. 13D, $d = 14b$. This greatly increases the scale for the single-slit pattern relative to the interference pattern, so that many interference maxima now fall within the central maximum of the diffraction pattern. Hence the effect of decreasing b, keeping d unchanged, is merely to broaden out the single-slit pattern, which acts as an envelope of the interference pattern as indicated by the dotted curve of Fig. 16D(c).

If the slit-width b is kept constant and the separation of the slits d is varied, the scale of the interference pattern varies, but that of the diffraction pattern remains the same. A series of photographs taken to illustrate this is shown in Fig. 16E. For each pattern three different exposures are shown, to bring out the details of the faint and the strong parts of the pattern. The maxima of the curves are labeled by the order m, and underneath the upper one is a given scale of angular positions θ. A study of these figures shows that certain orders are missing, or at least reduced to two maxima of very low intensity. These *missing orders* occur where the condition for a maximum of the interference, Eq. (16g), and for a minimum of the diffraction, Eq. (16f), are both fulfilled for the same value of θ, that is for

$$d \sin \theta = m\lambda \qquad \text{and} \qquad b \sin \theta = p\lambda$$

so that

$$\frac{d}{b} = \frac{m}{p} \tag{16h}$$

Since m and p are both integers, d/b must be in the ratio of two integers if we are to have missing orders. This ratio determines the orders which are missing, in such a way that when $d/b = 2$, orders 2, 4, 6, . . . are missing; when $d/b = 3$, orders 3, 6, 9, . . . are missing; etc. When $d/b = 1$, the two slits exactly join, and all orders should be missing. However, the two faint maxima into which each order is split can then be shown to correspond exactly to the subsidiary maxima of a single-slit pattern of width $2b$.

Our physical picture of the cause of missing orders is as follows. Considering, for example, the missing order $m = +3$ in Fig. 16D(c), this point on the screen is just three wavelengths farther from the center of one slit than from the center of the other. Hence we might expect the waves from the two slits to arrive in phase and to produce a maximum. However, this point is at the same time one wavelength farther from the edge of one slit than from the other edge of that slit. Addition of the second-

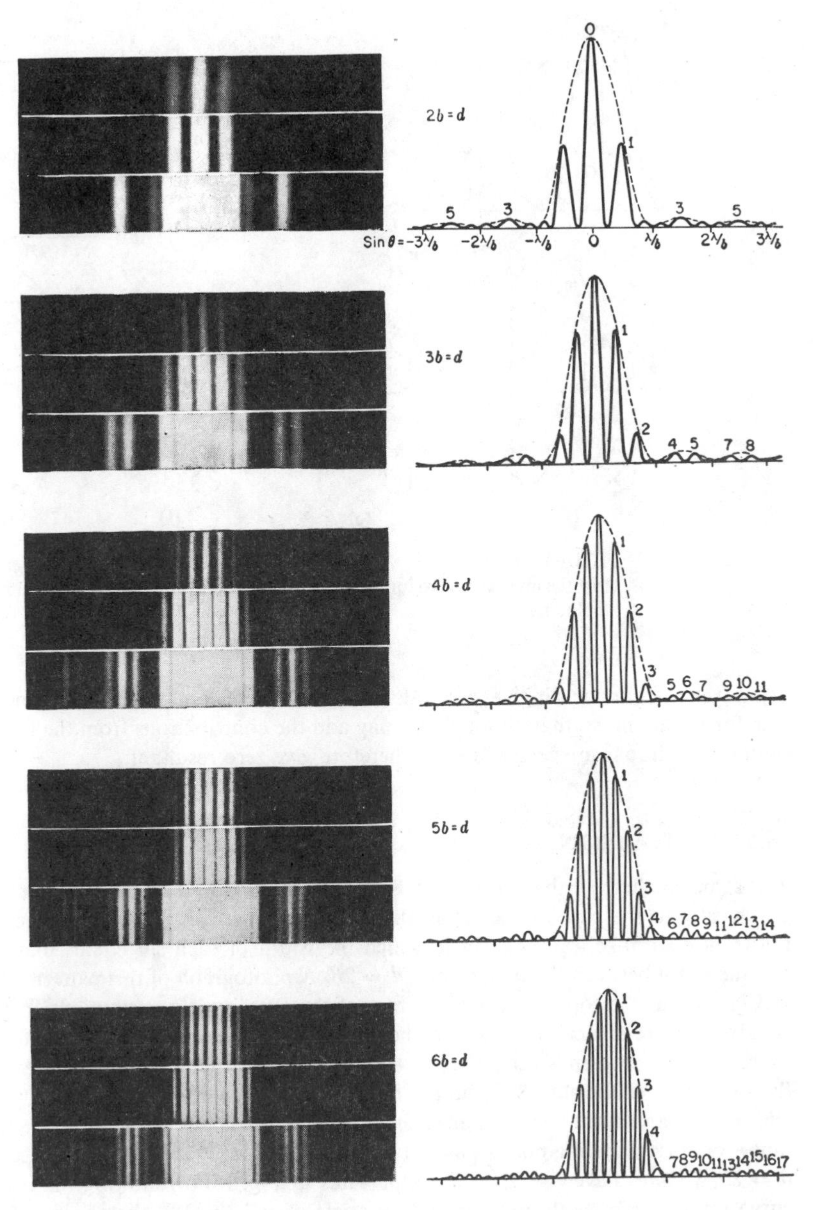

FIGURE 16E
Photographs and intensity curves for double-slit diffraction patterns.

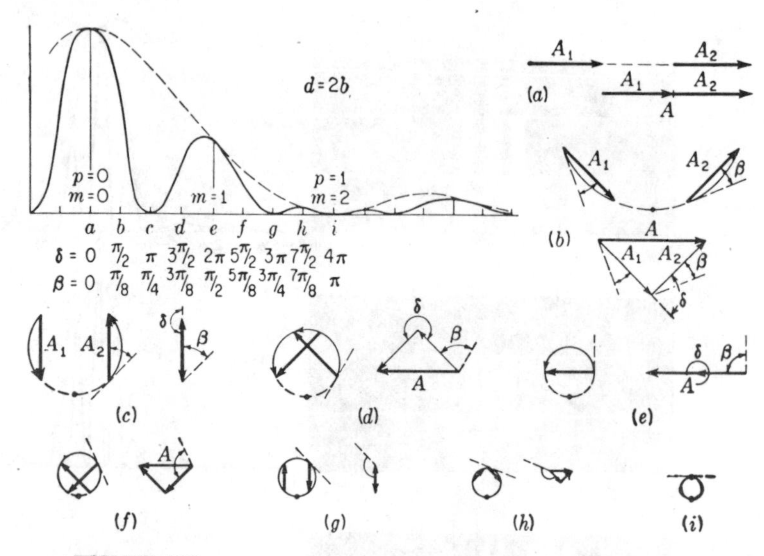

FIGURE 16F
How the intensity curve for a double slit is obtained by the graphical addition of amplitudes.

ary wavelets from one slit gives zero intensity under these conditions. The same holds true for either slit, so that although we may add the contributions from the two slits, both contributions are zero and must therefore give zero resultant.

16.6 VIBRATION CURVE

The same method as that applied in Sec. 15.4 for finding the resultant amplitude graphically in the case of the single slit is applicable to the present problem. For illustration we take a double slit in which the width of each slit equals that of the opaque space between the two, so that $d = 2b$. A photograph of this pattern appears in Fig. 16E at the top. A vector diagram of the amplitude contributions from one slit gives the arc of a circle, as before, the difference between the slopes of the tangents to the arc at the two ends being the phase difference 2β between the contributions from the two edges of the slit. Such an arc must now be drawn for each of the two slits, and the two arcs must be related in such a way that the phases (slopes of the tangents) differ for corresponding points on the two slits by 2γ, or δ. In the present case, since $d = 2b$, we must have $\gamma = 2\beta$ or $\delta = 4\beta$. Thus in Fig. 16F(*b*) showing the vibration curve for $\beta = \pi/8$, both arcs subtend an angle of $\pi/4$ $(= 2\beta)$, the phase difference for the two edges of each slit, and the arcs are separated by $\pi/4$ so that corresponding points on the two arcs differ by $\pi/2(= \delta)$. Now the resultant contributions from the two slits are represented in amplitude and phase by the chords of these two arcs, that is by A_1 and A_2. Diagrams (*a*) to (*i*) give the construction for the points similarly labeled on the intensity curve. The intensity, it will be remembered, is found as the square of the resultant amplitude A, which is the vector sum of A_1 and A_2.

In the example chosen, the slits are relatively wide compared with their separation, and as the phase difference increases the curvature of the individual arcs of the vibration curve increases rapidly, so that the vectors A_1 and A_2 decrease rapidly in length. For narrower slits we obtain a greater number of interference fringes within the central diffraction maximum, because the lengths of the arcs are smaller relative to the radius of curvature of the circle. A_1 and A_2 then decrease in length more slowly with increasing β, and the intensities of the maxima do not fall off so rapidly. In the limit where the slit width a approaches zero, A_1 and A_2 remain constant, and the variation of the resultant intensity is merely due to the change in phase angle between them.

16.7 EFFECT OF FINITE WIDTH OF SOURCE SLIT

A simplification which was made in the above treatment, and which never holds exactly in practice, is the assumption that the source slit (S' of Fig. 16B) is of negligible width. This is necessary in order that the lens shall furnish a single train of plane waves falling on the double slit. Otherwise there will be different sets of waves approaching at slightly different angles, these originating from different points in the source slit. They will produce sets of fringes which are slightly shifted with respect to each other, as illustrated in Fig. 16G(*a*). In the figure the interference maxima are for simplicity drawn with uniform intensity, neglecting the effects of diffraction. Let P and P' be two narrow lines acting as sources. These may be two narrow slits, or, better, two lamp filaments, since we assume no coherence between them. If the positions of the central maxima of the interference patterns produced by these are Q and Q', the fringe displacement QQ' will subtend the same angle α at the double slit as the source slits do. If this angle is a small fraction of the angular separation λ/d of the successive fringes in either pattern, the resultant intensity distribution will still resemble a true $\cos^2 \gamma$ curve, although the intensity will not fall to zero at the minima. The relative positions of the two patterns, and the sum of the two, in this state are illustrated in Fig. 16G, curves (*b*). Curves (*c*) and (*d*) show the effect of increasing the separation PP'. For (*d*) the fringes are completely out of step, and the resultant intensity shows no fluctuations whatever. At a point such as Q the maximum of one pattern then coincides with the next minimum of the other, so that the path difference $P'AQ - PAQ = \lambda/2$. In other words, P' is just a half wavelength farther from A than is P. If the intensity of one set of fringes is given by $4A^2 \cos^2 (\delta/2)$ or $2A^2(1 + \cos \delta)$, that of the other is

$$2A^2[1 + \cos (\delta + \pi)] = 2A^2(1 - \cos \delta)$$

The sum is therefore constant and equal to $4A^2$, so that the fringes entirely disappear. The condition for this disappearance of fringes is $\alpha = \lambda/2d$. If PP' is still further increased, the fringes will reappear, becoming sharp again when α equals the fringe

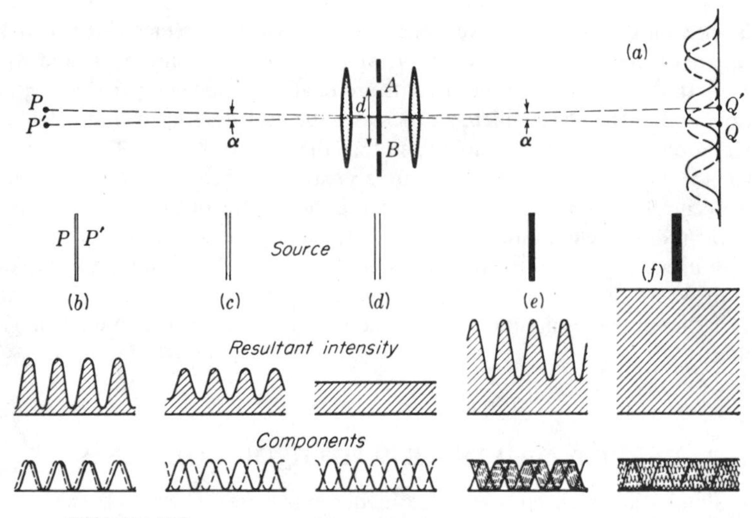

FIGURE 16G
Effect of a double source and of a wide source on the double-slit interference fringes.

distance λ/d, disappear when the fringes are again out of register, etc. In general, the condition for disappearance is

$$\alpha = \frac{\lambda}{2d}, \frac{3\lambda}{2d}, \frac{5\lambda}{2d}, \ldots \qquad \textit{Disappearance of fringes with double source} \qquad (16i)$$

where α is the angle subtended by the two sources at the double slit.

Next let us consider the effect when the source, instead of consisting of two separate sources, consists of a uniformly bright strip of width PP'. Each line element of this strip will produce its own set of interference fringes, and the resultant pattern will be the sum of a large number of these, displaced by infinitesimal amounts with respect to each other. Figure 16G(*e*) illustrates this for $\alpha = \lambda/2d$, that is, for a slit of width such that the extreme points acting alone would give complete disappearance of fringes as in (*d*). The resultant curve now shows strong fluctuations, and the slit must be still further widened to make the intensity uniform. The first complete disappearance will come when the range covered by the component fringes extends over a whole fringe width, instead of one-half, as above. This case is shown in Fig. 16G(*f*), for a slit of width subtending an angle $\alpha = \lambda/d$. Widening the slit still further will cause the fringes to reappear, although they never become perfectly distinct again, with zero intensity between fringes. At $\alpha = 2\lambda/d$ they again disappear completely, and the general condition is

$$\alpha = \frac{\lambda}{d}, \frac{2\lambda}{d}, \frac{3\lambda}{d}, \ldots \qquad \textit{Disappearance of fringes with slit source} \qquad (16j)$$

It is of practical importance, in observing double-slit fringes experimentally, to know how wide the source slit may be made in order to obtain intense fringes without seriously impairing the definition of the fringes. The exact value will depend on our criterion for clear fringes, but a good working rule is to permit a maximum discordance of the fringes of about one-quarter of that for the first disappearance. If f' is the focal length of the first lens, this corresponds to a *maximum permissible width* of the source slit

$$PP' = f'\alpha = \frac{f'\lambda}{4d} \tag{16k}$$

16.8 MICHELSON'S STELLAR INTERFEROMETER

As shown in Sec. 15.9, the smallest angular separation that two point sources may have in order to produce images which are recognizable as separate, in the focal plane of a telescope, is $\alpha = \theta_1 = 1.22\lambda/D$. In this equation [Eq. (15k)] D is the diameter of the objective of the telescope. Suppose that the objective is covered by a screen pierced with two parallel slits of separation almost equal to the diameter of the objective. A separation of $d = D/1.22$ would be a convenient value. If the telescope is now pointed at a double star and the slits are turned so as to be perpendicular to the line joining the two stars, interference fringes due to the double slit will in general be observed. However, according to Eq. (16i), if the angular separation of the two stars happens to be $\alpha = \lambda/2d$, the condition for the first disappearance, no fringes will be seen. Those from one star completely mask those from the other. Hence one could infer from the nonappearance of the fringes that the star was double with an angular separation $\lambda/2d$ or some multiple of this. (The multiples could be ruled out by direct observation without the double slit.) But this separation is only half as great as the minimum angle of resolution of the whole objective $1.22\lambda/D$, which in this case equals λ/d. In this connection it is instructive to compare, as in Fig. 16H, the dimensions of the diffraction pattern due to a *rectangular* aperture of width b with the interference pattern due to two narrow slits whose separation d is equal to b. The central maximum is only half as wide in the second case. Hence it is sometimes said that the resolving power of a telescope can be increased twofold by placing a double slit over the objective. This statement needs two important qualifications, however. (1) The stars are not "resolved" in the sense of producing separate images, but their existence is merely inferred from the behavior of the fringes. (2) A partial blurring of the fringes, without complete disappearance, can be observed for separations much less than $\lambda/2d$, showing the existence of two stars, and from this point of view the minimum resolvable separation is considerably smaller than that indicated by the above statement. In practice it is about one-tenth of this.

The actual measurement of the separation of a given close double star is made by having the slit separation d adjustable. The separation is increased until the fringes first disappear; then, by measuring d, the angular separation is obtained as $\alpha = \lambda/2d$. The effective wavelength λ of the starlight must, of course, be also estimated or measured. Separations of double stars are not often determined by this method, since observations on the doppler effect (Sec. 11.10) afford an even more sensitive means

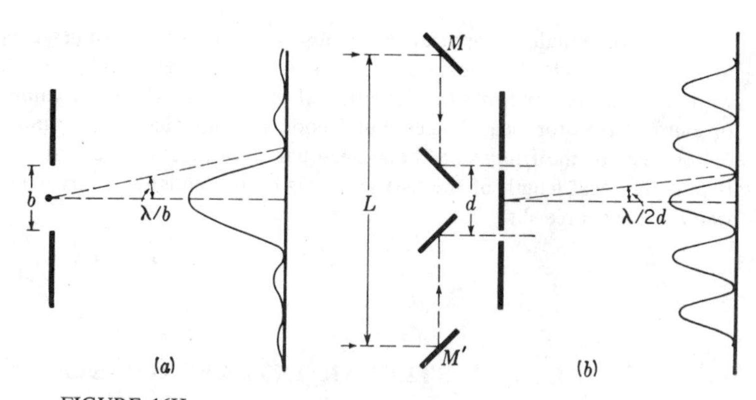

FIGURE 16H
Fraunhofer pattern from (*a*) a rectangular aperture and (*b*) double slit of separation equal to the width of the aperture in (*a*). In (*b*) are shown the four auxiliary mirrors used in the actual stellar interferometer.

of detection and measurement. On the other hand, the method of double-slit interference was, until recently,* the only way of measuring the diameter of the disk of a single star, and in 1920 this method was successfully applied by Michelson for this purpose.

From the discussion of the preceding section it will be seen that if a source such as a star disk subtends a finite angle, disappearance of the fringes would be expected from this cause when the separation of the double slit on a telescope is made great enough. Michelson first demonstrated the practicability of this method by measuring the diameters of Jupiter's moons, which subtend an angle of about 1 second. The values of d for the first disappearance are only a few centimeters in this case, and the measurement could be made by a double slit of variable separation over the objective of a telescope. Because the source is a circular disk instead of a rectangle, a correction must be applied to the equation $\alpha = \lambda/d$ for a slit source. This correction can be found by the same method that is used in finding the resolving power of a circular aperture, and gives the same factor. It is found that $\alpha = 1.22\lambda/d$ gives the first disappearance for a disk source. Estimating the angular diameters of the nearer fixed stars of known distance by assuming they are the same size as the sun, one obtains angles less than 0.01 second. The separations of the double slit required to detect a disk of this size are from 6 to 12 m. Clearly no telescope in existence could be used in the way described above for the measurement of star diameters. Another drawback would be that the fringes would be so fine that it would be difficult to separate them.

Since the blurring of the fringes is the result of variations of the phase difference between the light arriving at the two slits from various points on the source, Michelson realized that it is possible to magnify this phase difference without increasing d. This was done by receiving the light from a star on two plane mirrors M and M'

* See R. Hanbury-Brown and R. Q. Twiss, *Nature*, **178**:1447 (1956).

[Fig. 16H(*b*)] and reflecting it into the slits by these and two other mirrors. Then a variation of α in the angle of the incoming rays will cause a difference of path to the two slits of $L\alpha$, where L is the distance MM' between the outer mirrors. The fringes will now disappear when this difference equals 1.22λ, and so the sensitivity is magnified in the ratio L/d. In the actual measurements, M and M' were two 15 cm mirrors mounted on a girder in front of the 100-in. Mt. Wilson reflector so that they could be moved apart symmetrically. In the case of the star Arcturus, for example, the first disappearance of fringes occurred at $L = 7.2$ m, indicating an angular diameter $\alpha = 1.22\lambda/L$ of only 0.02 second. From the known distance of Arcturus, one then finds that its actual diameter is 27 times that of the sun.*

16.9 CORRELATION INTERFEROMETER

Another approach to the determination of stellar diameters has been to measure some parameter related to the phase of the incident light. Consider light from a distant source incident at one aperture of the Michelson stellar interferometer. Since the intensity at any time over a given light field is composed of a finite number of random wave trains, or photons, one would expect fluctuations in *phase*, *intensity*, and *polarization*. An abrupt change in intensity would be related to an abrupt change in the makeup of the photon field at the slit, which in turn would likely produce an abrupt change in *net phase*. Similarly a momentary lull in intensity fluctuation would correlate with a nonchanging phase. Therefore one would expect fluctuations in phase to be correlated with fluctuations in intensity. Moreover, the fluctuations would occur at frequencies much lower than the frequency of the light itself.

This correlation of light field intensity with phase, called the *Hanbury-Brown–Twiss effect*, was established by these scientists† through experiment in 1956. The technique ultimately led to a stellar interferometer which far surpasses the Michelson interferometer in resolving distant sources of finite angular size. Its major advantage is that the intensity correlation is not sensitive to slight variations in displacement of the optical components.

The crucial problem at the time of their experiment was to develop a method of measuring the intensity-fluctuation correlation with temporal resolution small enough to detect the fluctuations. The answer to this problem was to use two separated parabolic reflectors focused on photomultiplier tubes (see Fig. 16I). The outputs were delivered to electronic circuitry which produces an output proportional to the product of the two inputs. This in turn was sent through an integrating, or averaging, circuit. The variation of this output with detector separation is called the *second-order interference function* and displays an interference pattern similar to the Michelson interferometer (*first order interference*). With this technique the separation of the detectors can be greatly extended without having the interference pattern destroyed by slight variations in the mirror positions.

* Details of these measurements will be found in A. A. Michelson, "Studies in Optics," chap. 11, University of Chicago Press, Chicago, 1927.

† R. Hanbury-Brown and R. Q. Twiss, Correlation between Photons in Two Coherent Beams of Light, *Nature*, **127**:27 (1956).

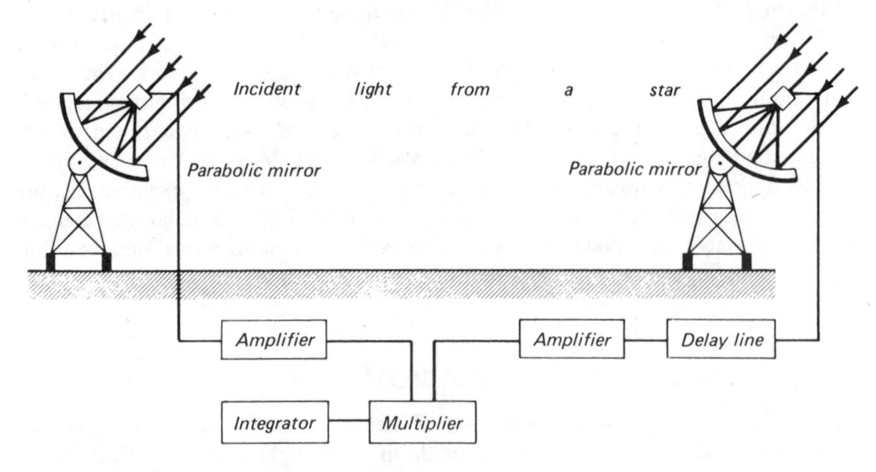

FIGURE 16I
Photoelectric detectors and electronic circuitry for a long-base-line correlation interferometer.

Using ordinary searchlight mirrors to focus starlight onto photomultiplier tubes, Hanbury-Brown and Twiss studied the star Sirius and were able to determine its angular diameter to be 0.0069 second of arc.

Since that time, a correlation interferometer with a base line 188 m long has been built at Narrabri, Australia, where angular diameters as small as 0.0005 second of arc can be measured. This far surpasses the results of the Michelson stellar interferometer.*

16.10 WIDE-ANGLE INTERFERENCE

Nothing has thus far been said about any limit to the angle between the two interfering beams as they leave the source. Consider, for example, the double-slit arrangement shown in Fig. 16J(*a*). The source *S* could be a narrow slit, but to ensure that there is no coherence between the light leaving various points on it, we shall assume that it is a self-luminous object. It is found experimentally that the angle ϕ may be made fairly large without spoiling the interference fringes, provided the width of the source is made correspondingly small. Just how small it must be is seen from the fact that the path difference from the extreme edges of the source to any given point on the screen such as *P* must be less than $\lambda/4$. Now if we call *s* the width of the source,

* For additional reading see W. Martienssen and E. Spiller, Coherence and Fluctuations in Light Beams, *Am. J. Phys.*, **32:** 919 (1964); A. B. Haner and N. R. Isenor, Intensity Correlations from Pseudothermal Light Sources, *Am. J. Phys.*, **38:** 748 (1970); and K. I. Kellermann, Intercontinental Radio Astronomy, *Sci. Am.*, **226:** 72 (1972).

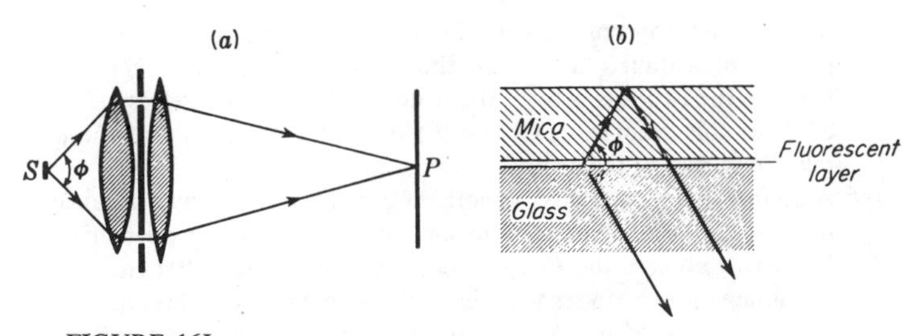

FIGURE 16J
Two methods of investigating wide-angle interference.

the discussion given in Sec. 15.10 shows that this path difference will be $2s \sin(\phi/2)$. Hence, for a divergence of 60°, s cannot exceed one-quarter of a wavelength, or 1.3×10^{-5} cm for green light. If the width is made greater than this, the fringes disappear completely when the path difference is λ, reappear, and then disappear again at 2λ, etc., just as in the stellar interferometer. By using as a source an extremely thin filament, Schrödinger could still detect some interference at an angular divergence ϕ as large as 57°.

An equivalent experiment which permitted using even larger angles of divergence (up to 180°) was performed by Selenyi in 1911. The essential part of his apparatus, shown in Fig. 16J(*b*), was a film of a fluorescent liquid only one-twentieth of a wavelength thick contained between a thin sheet of mica and a plane glass surface. When the film is strongly illuminated, it becomes a secondary source of light having a somewhat longer wavelength than the incident light (see Sec. 22.6). Interference can then be observed in a given direction between the light that comes directly from the film and that which is reflected from the outer surface of the mica. Interesting conclusions about the characteristics of the radiating atoms, in particular whether they radiate as dipoles, quadrupoles, etc., can be drawn from data on the variation of the visibility of the fringes with angle.*

PROBLEMS

16.1 The two slits of a double slit each have a width of 0.140 mm and a distance between centers of 0.840 mm. (*a*) What orders are missing, and (*b*) what is the approximate intensity of orders $m = 0$ to $m = 6$?

Ans. (*a*) 6, 12, 18, 24, . . . , (*b*) $m = 0$, 100%, $m = 1$, 91.2%; $m = 2$, 68.4%; $m = 3$, 40.5%; $m = 4$, 17.1%; $m = 5$, 3.65%; $m = 6$, 0%

16.2 The double slit of Prob. 16.1 is illuminated by parallel light of wavelength 5000 Å, and the pattern is focused on a screen by a lens with a focal length of 50.0 cm. Make a plot of the intensity distribution on the screen similar to Fig. 6D(*c*), using as abscissas the distance in millimeters on the screen. Include the first 12 orders on one side of the central maximum.

* O. Halpern and F. W. Doermann, *Phys. Rev.*, **52**: 937 (1937).

16.3 (*a*) Draw an appropriate vibration curve for the point in a Fraunhofer diffraction pattern of a double slit where the phase difference $\delta = \pi/3$. The opaque space between the two slits is twice the width of the slits themselves. (*b*) What is the value of β for this point? (*c*) Obtain a value for the intensity at the point in question relative to that at the central maximum.

16.4 A double slit has two slits of width 0.650 mm separated by a distance between centers of 2.340 cm. With a mercury arc as a source of light, the green line at $\lambda = 5460.74$ Å is used to observe the Fraunhofer diffraction pattern 100 cm behind the slits. (*a*) Assuming the eye can resolve fringes that subtend 1 minute of arc, what magnification would be required to just resolve the fringes? (*b*) How many fringes could be seen under the central maximum? (*c*) How many under the first side maximum?

Ans. (*a*) 3.1×, (*b*) 71 fringes, (*c*) 35 fringes

16.5 Two double slits are placed on an optical bench. One slit has a spacing of $d_1 =$ 0.250 mm, is illuminated by green light of a mercury arc, $\lambda = 5460.74$ Å, and is used as a double source. The eye located close behind the second double slit, for which $d_2 = 0.750$ mm, sees clear double-slit fringes when observing from the far end of the bench. When the second double slit is moved toward the double-slit source, the fringes completely disappear at a certain point, then appear, then disappear, etc. (*a*) Find the largest distance at which the fringes disappear. (*b*) Find the next largest distance at which they reappear and (*c*) then disappear.

16.6 A double slit with $b = 0.150$ mm, and $d = 0.950$ mm is located between two lenses as shown in Fig. 16G(*a*). The lenses have a focal length of 70 cm. A single adjustable slit is used as a light source at PP', and the green mercury line $\lambda = 5461$ Å illuminates it. According to the usual criterion for clear fringes, how wide should the source slit be made to obtain the best intensity without appreciable sacrifice of clearness?

16.7 Since two equal slits with $d = b$ form a single slit twice the width of either of the slits, prove that Eq. (16c) can be reduced to the equation for the intensity distribution for a single slit of width $2b$.

Ans. Starting with Eq. (16c), we make use of the trigonometric equality that $2 \sin \beta \cos \beta = \sin 2\beta$. Upon substitution, we obtain, $I = 4A_0^2 (\sin^2 2\beta)/4\beta^2$

16.8 If $d = 5b$ for a double slit, determine for Fraunhofer diffraction exactly how much the third-order maximum is shifted from the position given by Eq. (16g) due to modulation by the diffraction envelope. The problem is best solved by plotting exact intensities in the neighborhood of the expected maximum. Express the result as a fraction of the separation of orders.

16.9 With a tungsten lamp with a straight wire filament as a source and a collimating lens of 6.20 cm focal length in front of a double slit, various separations of the double slit are tried, increasing the distance d until the fringes disappear. If this occurs for $d = 0.350$ mm, calculate the filament diameter. Assume $\lambda = 5800$ Å.

16.10 Derive a formula giving the number of interference maxima occurring under the central diffraction maximum of the double-slit pattern in terms of the separation d and the slit width b.

Ans. $N = 2d/b - 1$

17

THE DIFFRACTION GRATING

Any arrangement which is equivalent in its action to a number of parallel equidistant slits of the same width is called a *diffraction grating*. Since the grating is a very powerful instrument for the study of spectra, we shall treat in considerable detail the intensity pattern which it produces. We shall find that the pattern is quite complex in general but that it has a number of features in common with that of the double slit treated in the last chapter. In fact, the latter may be considered as an elementary grating of only two slits. It is, however, of no use as a spectroscope, since in a practical grating many thousands of very fine slits are usually required. The reason for this becomes apparent when we examine the difference between the pattern due to two slits and that due to many slits.

17.1 EFFECT OF INCREASING THE NUMBER OF SLITS

When the intensity pattern due to one, two, three, and more slits of the same width is photographed, a series of pictures like those shown in Fig. 17A(*a*) to (*f*) is obtained. The arrangement of light source, slit, lenses, and recording plate used in taking these pictures was similar to that described in previous chapters, and the light used was the

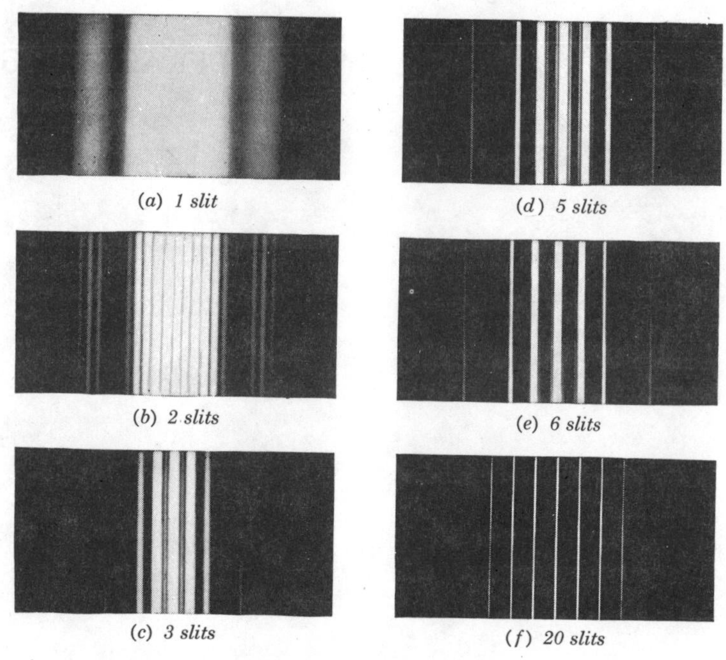

FIGURE 17A
Fraunhofer diffraction patterns for gratings containing different numbers of slits.

blue line from a mercury arc. These patterns therefore are produced by *Fraunhofer diffraction.* In fact, it was because of Fraunhofer's original investigations of the diffraction of parallel light by gratings in 1819 that his name became associated with this type of diffraction. Fraunhofer's first gratings were made by winding fine wires around two parallel screws. Those used in preparing Fig. 17A were made by cutting narrow transparent lines in the gelatin emulsion on a photographic plate, as described in Sec. 13.2.

The most striking modification in the pattern as the number of slits is increased consists of a narrowing of the interference maxima. For two slits these are diffuse, having an intensity which was shown in the last chapter to vary essentially as the square of the cosine. With more slits the sharpness of these *principal maxima* increases rapidly, and in pattern (*f*) of the figure, with 20 slits, they have become narrow lines. Another change, of less importance, which can be seen in patterns (*c*), (*d*), and (*e*) is the appearance of weak *secondary maxima* between the principal maxima, their number increasing with the number of slits. For three slits only one secondary maximum is present, its intensity being 11.1 percent of the principal maximum. Figure 17B shows an intensity curve for this case, plotted according to the theoretical equation (17b) given in the next section. Here the individual slits were assumed very narrow. Actually the intensities of all maxima are governed by the pattern of a single slit of

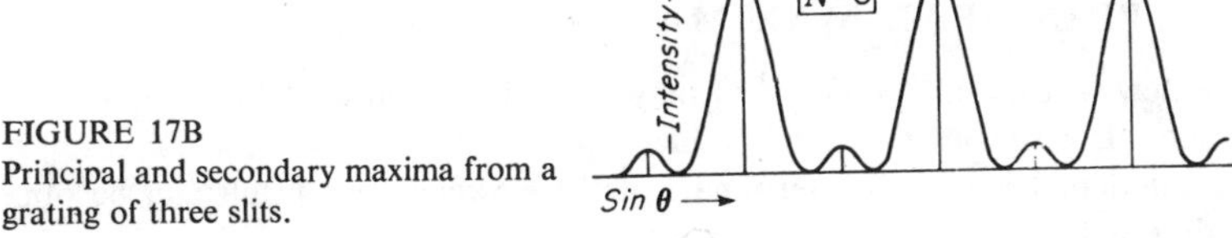

FIGURE 17B
Principal and secondary maxima from a grating of three slits.

width equal to that of any one of the slits used. The width of the intensity envelopes would be identical in the various patterns of Fig. 17A if the slits had been of the same width in all cases. In fact there were slight differences in the slit widths used for some of the patterns.

17.2 INTENSITY DISTRIBUTION FROM AN IDEAL GRATING

The procedure used in Secs. 15.2 and 16.2 for the single and double slits could be used here, performing the integration over the clear aperture of the slits, but it becomes cumbersome. Instead let us apply the more powerful method of adding the complex amplitudes (Sec. 14.8). The situation is simpler than in the case of multiple reflections, because for the grating the amplitudes contributed by the individual slits are all of equal magnitude. We designate this magnitude by a and the number of slits by N. The phase will change by equal amounts δ from one slit to the next; so the resultant complex amplitude is the sum of the series

$$Ae^{i\theta} = a(1 + e^{i\delta} + e^{i2\delta} + e^{i3\delta} + \cdots + e^{i(N-1)\delta}) = a\,\frac{1 - e^{iN\delta}}{1 - e^{i\delta}} \tag{17a}$$

To find the intensity, this expression must be multiplied by its complex conjugate, as in Eq. (14m), giving

$$A^2 = a^2\,\frac{(1 - e^{iN\delta})(1 - e^{-iN\delta})}{(1 - e^{i\delta})(1 - e^{-i\delta})} = a^2\,\frac{1 - \cos N\delta}{1 - \cos\delta}$$

Using the trigonometric relation $1 - \cos\alpha = 2\sin^2(\alpha/2)$, we may then write

$$A^2 = a^2\,\frac{\sin^2(N\delta/2)}{\sin^2(\delta/2)} = a^2\,\frac{\sin^2 N\gamma}{\sin^2\gamma} \tag{17b}$$

where, as in the double slit, $\gamma = \delta/2 = (\pi d \sin\theta)/\lambda$. Now the factor a^2 represents the intensity diffracted by a single slit, and after inserting its value from Eq. (15d) we finally obtain for the intensity in the Fraunhofer pattern of an ideal grating

$$I \approx A^2 = A_0{}^2\,\frac{\sin^2\beta}{\beta^2}\,\frac{\sin^2 N\gamma}{\sin^2\gamma} \tag{17c}$$

Upon substitution of $N = 2$ in this formula, it readily reduces to Eq. (16c) for the double slit.

17.3 PRINCIPAL MAXIMA

The new factor $(\sin^2 N\gamma)/(\sin^2 \gamma)$ may be said to represent the *interference* term for N slits. It possesses maximum values equal to N^2 for $\gamma = 0, \pi, 2\pi, \ldots$. Although the quotient becomes indeterminate at these values, this result can be obtained by noting that

$$\lim_{\gamma \to m\pi} \frac{\sin N\gamma}{\sin \gamma} = \lim_{\gamma \to m\pi} \frac{N \cos N\gamma}{\cos \gamma} = \pm N$$

These maxima correspond in position to those of the double slit, since for the above values of γ

$$d \sin \theta = 0, \lambda, 2\lambda, 3\lambda, \ldots = m\lambda \qquad \textit{Principal maxima} \qquad (17d)$$

They are more intense, however, in the ratio of the square of the number of slits. The relative intensities of the different orders m are in all cases governed by the single-slit diffraction envelope $(\sin^2 \beta)/\beta^2$. Hence the relation between β and γ in terms of slit width and slit separation [Eq. (16d)] remains unchanged, as does the condition for missing orders [Eq. (16h)].

17.4 MINIMA AND SECONDARY MAXIMA

To find the minima of the function $(\sin^2 N\gamma)/(\sin^2 \gamma)$, we note that the numerator becomes zero more often than the denominator, and this occurs at the values $N\gamma = 0, \pi, 2\pi, \ldots$ or, in general, $p\pi$. In the special cases when $p = 0, N, 2N, \ldots$, γ will be $0, \pi, 2\pi, \ldots$; so for these values the denominator will also vanish, and we have the principal maxima described above. The other values of p give zero intensity, since for these the denominator does not vanish at the same time. Hence the condition for a minimum is $\gamma = p\pi/N$, excluding those values of p for which $p = mN$, m being the order. These values of γ correspond to path differences

$$d \sin \theta = \frac{\lambda}{N}, \frac{2\lambda}{N}, \frac{3\lambda}{N}, \ldots, \frac{(N-1)\lambda}{N}, \frac{(N+1)\lambda}{N}, \ldots \qquad \textit{Minima} \qquad (17e)$$

omitting the values $0, N\lambda/N, 2N\lambda/N, \ldots$, for which $d \sin \theta = m\lambda$ and which according to Eq. (17d) represent principal maxima. Between two adjacent principal maxima there will hence be $N - 1$ points of zero intensity. The two minima on either side of a principal maximum are separated by twice the distance of the others.

Between the other minima the intensity rises again, but the secondary maxima thus produced are of much smaller intensity than the principal maxima. Figure 17C shows a plot for six slits of the quantities $\sin^2 N\gamma$ and $\sin^2\gamma$, and also of their quotient, which gives the intensity distribution in the interference pattern. The intensity of the principal maxima is N^2 or 36, so that the lower figure is drawn to a smaller scale. The intensities of the secondary maxima are also shown. These secondary maxima are not of equal intensity but fall off as we go out on either side of each principal maximum. Nor are they in general equally spaced, the lack of equality being due to the fact that the maxima are not quite symmetrical. This lack of symmetry is greatest for the secondary maxima immediately adjacent to the principal maxima,

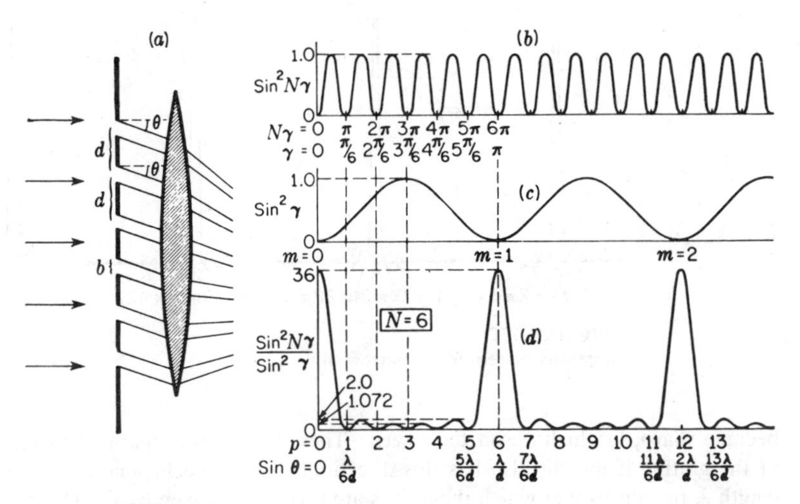

FIGURE 17C
Fraunhofer diffraction by a grating of six very narrow slits and details of the intensity pattern.

and is such that the secondary maxima are slightly shifted toward the adjacent principal maximum.

These features of the secondary maxima show a strong resemblance to those of the secondary maxima in the *single-slit* pattern. Comparison of the central part of the intensity pattern in Fig. 17C(*d*) with Fig. 15D for the single slit will emphasize this resemblance. As the number of slits is increased, the number of secondary maxima is also increased, since it is equal to $N - 2$. At the same time the resemblance of any principal maximum and its adjacent secondary maxima to the single-slit pattern increases. In Fig. 17D is shown the interference curve for $N = 20$, corresponding to the last photograph shown in Fig. 17A. In this case there are 18 secondary maxima between each pair of principal maxima, but only those fairly close to the principal maxima appear with appreciable intensity, and even these are not sufficiently strong to show in the photograph. The agreement with the single-slit pattern is here practically complete. The physical reason for this agreement will be discussed in Sec. 17.10, where it will be shown that the dimensions of the pattern correspond to those from a single "slit" of width equal to that of the entire grating. Even when the number of slits is small, the intensities of the secondary maxima can be computed by summing a number of such single-slit patterns, one for each order.

17.5 FORMATION OF SPECTRA BY A GRATING

The secondary maxima discussed in Sec. 17.4 are of little importance in the production of spectra by a many-lined grating. The principal maxima treated in Sec. 17.3 are called *spectrum lines* because when the primary source of light is a narrow slit they

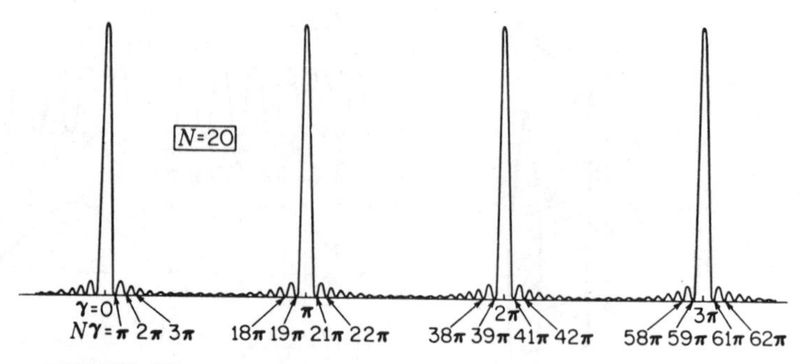

FIGURE 17D
Intensity pattern for 20 narrow slits.

become sharp, bright lines on the screen. These lines will be parallel to the rulings of the grating if the slit also has this direction. For monochromatic light of wavelength λ, the angles θ at which these lines are formed are given by Eq. (17d), which is the ordinary grating equation $d \sin \theta = m\lambda$ commonly given in elementary textbooks. A more general equation includes the possibility of light incident on the grating at any angle i. The equation then becomes

$$d(\sin i + \sin \theta) = m\lambda \qquad \textit{Grating equation} \qquad (17f)$$

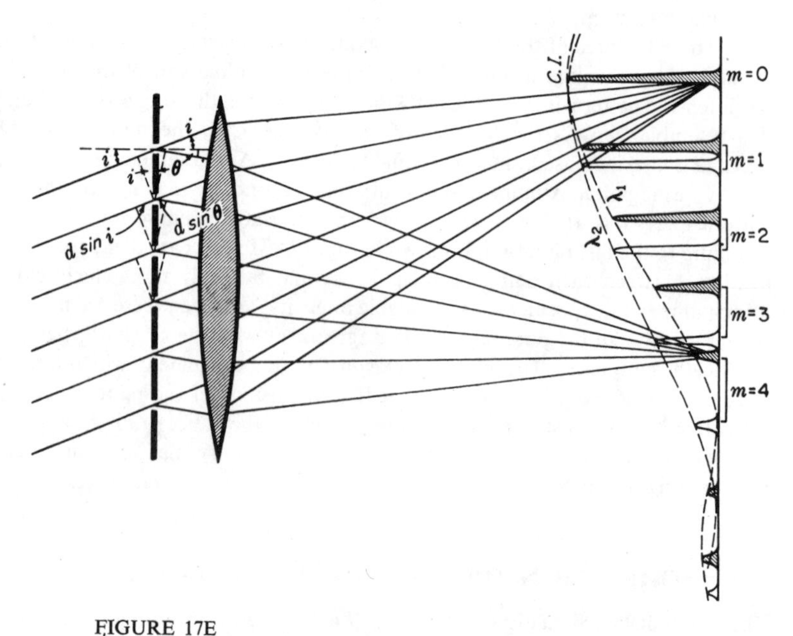

FIGURE 17E
Positions and intensities of the principal maxima from a grating where light containing two wavelengths is incident at an angle i and diffracted at various angles θ.

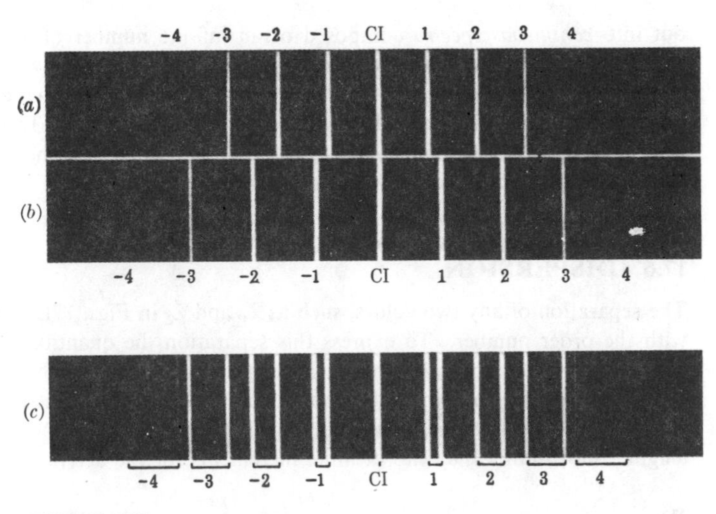

FIGURE 17F
Grating spectra of two wavelengths: (*a*) $\lambda_1 = 4000$ Å; (*b*) $\lambda_2 = 5000$ Å; (*c*) λ_1 and λ_2 together.

since, as will be seen from Fig. 17E, this is the path difference for light passing through adjacent slits. The figure shows the path of the light forming the maxima of order $m = 0$ (called the *central image*), and also $m = 4$ in light of a particular wavelength λ_1. For the central image, Eq. (17f) shows that $\sin\theta = -\sin i$, or $\theta = -i$. The negative sign comes from the fact that we have chosen to call i and θ positive when measured on the same side of the normal; i.e., our convention of signs is such that whenever the rays used *cross over* the line normal to the grating, θ is taken as negative. Those intensity maxima which are shaded show the various orders of the wavelength λ_1. In the case of the fourth order, for example, the path differences indicated are such that $d(\sin i + \sin\theta) = 4\lambda_1$. The intensities of the principal maxima are limited by the diffraction pattern corresponding to a single slit (broken line) and drop to zero at the first minimum of that pattern, which here coincides with the fifth order. The missing orders in this illustration are therefore $m = 5, 10, \ldots$, as would be produced by having $d = 5b$.

Now if the source gives light of another wavelength λ_2 somewhat greater than λ_1, the maxima of the corresponding order m for this wavelength will, according to Eq. (17j), occur at larger angles θ. Since the spectrum lines are narrow, these maxima will in general be entirely separate in each order from those of λ_1 and we have two lines forming a *line spectrum* in each order. These spectra are indicated by brackets in the figure. Both the wavelengths will coincide, however, for the central image, because for this the path difference is zero for any wavelength. A similar set of spectra occurs on the other side of the central image, the shorter wavelength line in each order lying on the side toward the central image. Figure 17F shows actual photographs of grating spectra corresponding to the diagram of Fig. 17E. If the source gives white light, the central image will be white, but for the orders each will be spread out into *continuous spectra* composed of an infinite number of adjacent images of the slit in light of the different wavelengths present. At any given point in such a continuous spectrum, the light will be very nearly monochromatic because of the narrowness of the slit images formed by the grating and lens. The result is in this respect fundamentally different from that with the double slit, where the images were broad and the spectral colors were not separated.

17.6 DISPERSION

The separation of any two colors, such as λ_1 and λ_2 in Figs. 17E and 17F, increases with the order number. To express this separation the quantity frequently used is called the *angular dispersion*, which is defined as the rate of change of angle with change of wavelength. An expression for this quantity is obtained by differentiating Eq. (17f) with respect to λ, remembering that i is a constant independent of wavelength. Substituting the ratio of finite increments for the derivative, one has

$$\frac{\Delta\theta}{\Delta\lambda} = \frac{m}{d \cos \theta} \qquad \textit{Angular dispersion} \qquad (17g)$$

The equation shows in the first place that for a given small wavelength difference $\Delta\lambda$, the angular separation $\Delta\theta$ is directly proportional to the order m. Hence the second-order spectrum is twice as wide as the first order, the third three times as wide as the first, etc. In the second place, $\Delta\theta$ is inversely proportional to the slit separation d, which is usually referred to as the *grating space*. The smaller the grating space, the more widely spread the spectra will be. In the third place, the occurrence of $\cos \theta$ in the denominator means that in a given order m the dispersion will be smallest on the normal, where $\theta = 0$, and will increase slowly as we go out on either side of this. If θ does not become large, $\cos \theta$ will not differ much from unity, and this factor will be of little importance. If we neglect its influence, the different spectral lines in one order will differ in angle by amounts which are directly proportional to their difference in wavelength. Such a spectrum is called a *normal spectrum*, and one of the chief advantages of gratings over prism instruments is this simple linear scale for wavelengths in their spectra.

The *linear dispersion* in the focal plane of the telescope or camera lens is $\Delta l/\Delta\lambda$, where l is the distance along this plane. Its value is usually obtainable by multiplying Eq. (17g) by the focal length of the lens. In some arrangements, however, the photographic plate is turned so the light does not strike it normally, and there is a corresponding increase in linear dispersion. In specifying the dispersion of a spectrograph, it has become customary to quote the *plate factor*, which is the reciprocal of the above quantity and expressed in angstroms per millimeter.

17.7 OVERLAPPING OF ORDERS

If the range of wavelengths is large, e.g., if we observe the whole visible spectrum between 4000 and 7200 Å, considerable overlapping occurs in the higher orders. Suppose, for example, that one observed in the third order a certain red line of wavelength 7000 Å. The angle of diffraction for this line is given by solving for θ the expression

$$d(\sin i + \sin \theta) = 3 \times 70000$$

where d is in angstroms. But at the same angle θ there may occur a green line in the fourth order, of wavelength 5250 Å, since

$$4 \times 5250 = 3 \times 7000$$

Similarly the violet of wavelength 4200 Å will occur in the fifth order at this same place. The general condition for the various wavelengths that can occur at a given angle θ is then

$$d(\sin i + \sin \theta) = \lambda_1 = 2\lambda_2 = 3\lambda_3 = \cdots \tag{17h}$$

where λ_1, λ_2, etc., are the wavelengths in the first, second, etc., orders. For visible light there is no overlapping of the first and second orders, since with $\lambda_1 = 7200$ Å and $\lambda_2 = 4000$ Å the red end of the first order falls just short of the violet end of the second. When photographic observations are made, however, these orders may extend down to 2000 Å in the ultraviolet, and the first two orders do overlap. This difficulty can usually be eliminated by the use of suitable color filters to absorb from the incident light those wavelengths which would overlap the region under study. As an example, a piece of red glass transmitting only wavelengths longer than 6000 Å could be used in the above case to avoid the interfering shorter wavelengths of higher order that might disturb observation of λ7000 and lines in that vicinity.

17.8 WIDTH OF THE PRINCIPAL MAXIMA

It was shown at the beginning of Sec. 17.4 that the first minima on either side of any principal maximum occur where $N\gamma = mN\pi \pm \pi$, or where $\gamma = m\pi \pm (\pi/N)$. When $\gamma = m\pi$, we have the principal maxima, owing to the fact that the phase difference δ or 2γ, in the light from corresponding points of adjacent slits, is given by $2\pi m$, or a whole number of complete vibrations. However, if we change the angle enough to cause a change of $2\pi/N$ in the phase difference, reinforcement no longer occurs, but the light from the various slits now interferes to produce zero intensity. A phase difference of $2\pi/N$ between the maximum and the first minimum means a path difference of λ/N.

To see why this path difference causes zero intensity, consider Fig. 17G(*a*), in which the rays leaving the grating at the angle θ form a principal maximum of order m. For these, the path difference of the rays from two adjacent slits is $m\lambda$, so that all the waves arrive in phase. The path difference of the *extreme* rays is then $Nm\lambda$, since N is always a very large number in any practical case.* Now let us change the angle of diffraction by a small amount $\Delta\theta$, such that the extreme path difference increases by one wavelength and becomes $Nm\lambda + \lambda$ (rays shown by broken lines). This should correspond to the condition for zero intensity, because, as required,

* With a small number of slits, it is necessary to use the true value $(N - 1)m\lambda$, and the subsequent argument must be slightly modified, but it yields the same result [Eq. (17i)].

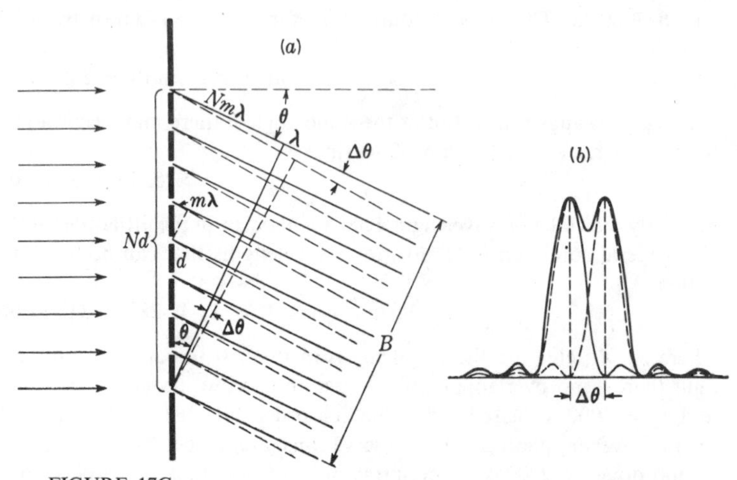

FIGURE 17G
Angular separation of two spectrum lines which are just resolved by a diffraction grating.

the path difference for two adjacent slits has been increased by λ/N. It will be seen that the ray from the top of the grating is now of opposite phase from that at the center, and the effects of these two will cancel. Similarly, the ray from the next slit below the center will annul that from the next slit below the top, etc. The cancellation if continued will yield zero intensity from the whole grating, in entire analogy to the similar process considered in Sec. 15.3 for the single-slit pattern.

Thus the first zero occurs at the small angle $\Delta\theta$ on each side of any principal maximum. From the figure it is seen that

$$\Delta\theta = \frac{\lambda}{B} = \frac{\lambda}{Nd\cos\theta} \qquad \textit{Angular half width of principal maximum} \qquad (17i)$$

It is instructive to note that this is just $1/N$*th of the separation of adjacent orders*, since the latter is represented by the same expression with the path difference $N\lambda$ instead of λ in the numerator.

17.9 RESOLVING POWER

When N is many thousands, as in any useful diffraction grating, the maxima are extremely narrow. The chromatic resolving power $\lambda/\Delta\lambda$ is correspondingly high. To evaluate it, we note first that since the intensity contour is essentially the diffraction pattern of a rectangular aperture, the Rayleigh criterion (Sec. 15.6) can be applied. The images formed in two wavelengths that are barely resolved must be separated by the angle $\Delta\theta$ of Eq. (17i). Consequently the light of wavelength $\lambda + \Delta\lambda$ must form its principal maximum of order m at the same angle as that for the first minimum of wavelength λ in that order [Fig. 17G(b)]. Hence we can equate the extreme path differences in the two cases and obtain

$$mN\lambda + \lambda = mN(\lambda + \Delta\lambda)$$

from which it immediately follows that

$$\frac{\lambda}{\Delta\lambda} = mN \tag{17j}$$

That the resolving power is proportional to the order m is to be understood from the fact that the *width* of a principal maximum, by Eq. (17i), depends on the width B of the emergent beam and does not change much with order, whereas the *separation* of two maxima of different wavelengths increases with the dispersion, which, by Eq. (17g), increases nearly in proportion to the order. Just as for the prism (Sec. 15.7), we have that

Chromatic resolving power = angular dispersion × width of emergent beam since in the present case

$$\frac{\lambda}{\Delta\lambda} = \frac{\Delta\theta}{\Delta\lambda} \times B = \frac{m}{d\cos\theta} \times Nd\cos\theta = mN$$

In a *given order* the resolving power, by Eq. (17j), is proportional to the total number of slits N but is independent of their spacing d. However, at *given angles* of incidence and diffraction it is independent of N also, as can be seen by substituting in Eq. (17j) the value of m from Eq. (17f):

$$\frac{\lambda}{\Delta\lambda} = \frac{d(\sin i + \sin\theta)}{\lambda} N = \frac{W(\sin i + \sin\theta)}{\lambda} \tag{17k}$$

Here $W = Nd$ is the total width of the grating. At a given i and θ, the resolving power is therefore independent of the number of lines ruled in the distance W. A grating with fewer lines gives a higher order at these given angles, however, with consequent overlapping, and would require some auxiliary dispersion to separate these orders, as does the Fabry-Perot interferometer. The method has nevertheless been recently applied with success in the echelle grating to be described later. Theoretically the maximum resolving power obtainable with any grating occurs when $i = \theta = 90°$, and according to Eq. (17k) it equals $2W/\lambda$, or the number of wavelengths in twice the width of the grating. In practice such grazing angles are not usable, however, because of the negligible amount of light. Experimentally one can hope to reach only about two-thirds of the ideal maximum.

17.10 VIBRATION CURVE

Let us now apply the method of compounding the amplitudes vectorially which was used in Sec. 16.6 for two slits and in Sec. 15.4 for one slit. The vibration curve for the contributions from the various infinitesimal elements of a single slit again forms an arc of a circle, but there are now several of these arcs in the curve, corresponding to the several slits of the grating. In Fig. 17H the diagrams corresponding to the

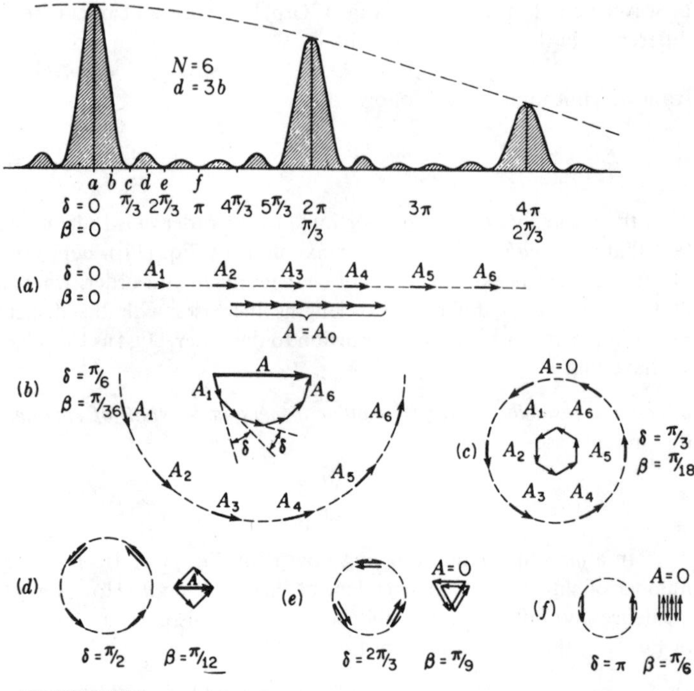

FIGURE 17H
How the intensity curve for a grating of several slits is obtained by the graphical addition of amplitudes.

various points (*a*) to (*f*) of the intensity plot for six slits are shown. For the central maximum the light from all slits, and from all parts of each slit, is in phase, giving a resultant amplitude A which is N times as great as that from one slit, as shown in (*a*) of the figure. Halfway to the first minimum the condition is as shown in (*b*). For this point $\gamma = \pi/12$, so that the phase difference from corresponding points in adjacent slits δ equals $\pi/6$ (compare Fig. 17C). This is also the angle between successive vectors in the series of six resultants A_1 to A_6 which are the chords of six small equal arcs. Just as for the double slit, the final resultant A is obtained by compounding these vectorially, and the intensity is measured by A^2. With increasing angle the individual resultants become slightly smaller in magnitude as β increases, because it is the arc, not the chord, which is constant in length. Their difference is here small, even for point (*f*).

The derivation of the general intensity function for the grating, Eq. (17b), can be very simply done by a geometrical method. In Fig. 17I the six amplitude vectors of Fig. 17H are shown with a phase difference somewhat less than in part (*b*) of Fig 17H. All these have the same magnitude, given by

$$A_n = \frac{\sin \beta}{\beta} A_0 \tag{17l}$$

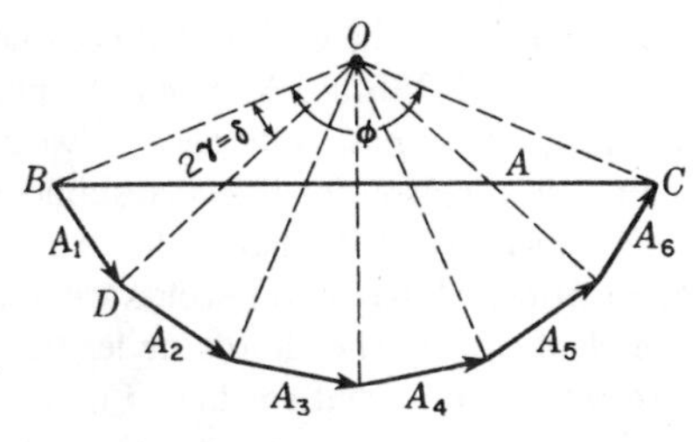

FIGURE 17I
Geometrical derivation of the intensity function for a grating.

since this represents the chord of an arc of length A_0 subtending the angle 2β (see Fig. 15F). Each vector is inclined to the next by the angle $\delta = 2\gamma$, and thus the six form part of a regular polygon. In the figure broken lines are drawn from the ends of each vector to the center O of this polygon. These lines also make the constant angle 2γ with each other. Therefore the total angle subtended at the center is

$$\phi = N\delta = N \times 2\gamma$$

We can now obtain a relation between the resultant amplitude A and the individual ones A_n, which are given by Eq. (17l). By dividing the triangle OBC into two halves with a line from O perpendicular to A, it is seen that

$$A = 2r \sin \frac{\phi}{2}$$

where r represents OB or OC. Similarly, from the triangle OBD as split by a line perpendicular to A_1, we obtain

$$A_n = A_1 = 2r \sin \gamma$$

Dividing the previous equation by this one, we find

$$\frac{A}{A_n} = \frac{2r \sin (\phi/2)}{2r \sin \gamma} = \frac{\sin N\gamma}{\sin \gamma}$$

When we then substitute the value of A_n from Eq. (17l), there results, for the amplitude,

$$A = A_0 \frac{\sin \beta}{\beta} \frac{\sin N\gamma}{\sin \gamma}$$

The square of this, which gives the intensity, is identical with Eq. (17c).

The vibration curve as applied to different numbers of slits helps to understand many features of the intensity patterns. For instance, there is the important question of the narrowness of the principal maxima. The adjacent minimum on one side is reached when the vectors first form a closed polygon, as is (c) of Fig. 17H. It is evident that this will occur for smaller values of δ the larger the number of slits, and this means that the maxima will become sharper. Also one can see at once from the diagram that for this minimum $\delta = 2\pi/N$, or $\gamma = \pi/N$, the condition stated at the beginning of Sec. 17.8. Furthermore, as the number of slits becomes large, the polygon of vectors will rapidly approach the arc of a circle, and the analogy with the pattern due to a single aperture of width equal to that of the grating is thereby seen to be justified. Comparison of Fig. 17H with Fig. 15F for the single slit will show that for large N

the diagrams for the grating will become identical with those for one slit if we replace $N\delta/2$ or $N\gamma$ by β. Since $N\gamma$ is half the phase difference from extreme slits of the grating and β half the phase difference between extreme points in an open aperture, we see the physical reason for the correspondence mentioned in Sec. 17.4.

Finally we note that if the diagrams in Fig. 17H are carried further, the first-order principal maximum occurs when the arc representing each interval d forms one complete circle. The chords under these conditions are all parallel and in the same direction as in (*a*) but smaller in magnitude. The second principal maximum occurs when each arc forms two turns of a circle when the resultant chords again line up. These maxima have no analogue in the pattern for a single slit.

17.11 PRODUCTION OF RULED GRATINGS

Up to this point we have considered the characteristics of an idealized grating consisting of identical and equally spaced slits separated by opaque strips. Actual gratings used in the study of spectra are made by ruling fine grooves with a diamond point either on a plane glass surface to produce a *transmission grating* or more often on a polished metal mirror to produce a *reflection grating*. The transmission grating gives something like our idealized picture, since the grooves scatter the light and are effectively opaque, while the undisturbed parts of the surface transmit regularly and act like slits. The same is true of the reflection grating, except that here the unruled portions reflect regularly, and the grating equation (17f) holds equally well for this case with the same convention of signs for i and θ.

Figure 17J shows microphotographs of the ruled surfaces of two different reflection gratings. The grating shown in (*a*) was ruled lightly, and the grooves are too shallow to obtain maximum brightness. That shown in (*b*) was a high-quality grating having 15,000 lines per inch. One or two vertical cross-rulings have been made to show more clearly the contour of the ruled surface.

Until recently, most gratings were ruled on speculum metal, a very hard alloy of copper and tin. Modern practice, however, is to rule on an evaporated layer of the softer metal aluminum. Not only does this give greater reflection in the ultraviolet, but it causes less wear on the diamond ruling point. The chief requirement for a good grating is that the lines be as nearly equally spaced as possible over the whole ruled surface, which in different gratings varies from 1 to 25 cm in width. This is a difficult requirement to fulfill, and there are very few places in the world where ruling machines of precision adequate for the production of fine gratings have been constructed. After each groove has been ruled, the machine lifts the diamond point and moves the grating forward by a small rotation of the screw which drives the carriage carrying it. To have the spacing of rulings constant, the screw must be of very constant pitch, and it was not until the manufacture of a nearly perfect screw had been achieved by Rowland,* in 1882, that the problem of successfully ruling large gratings was accomplished.

* H. A. Rowland (1848–1901). Professor of physics at the Johns Hopkins University, Baltimore. He is famous for his demonstration of the magnetic effect of a charge in motion, for his measurements of the mechanical equivalent of heat, and for his invention of the concave grating (Sec. 17.15).

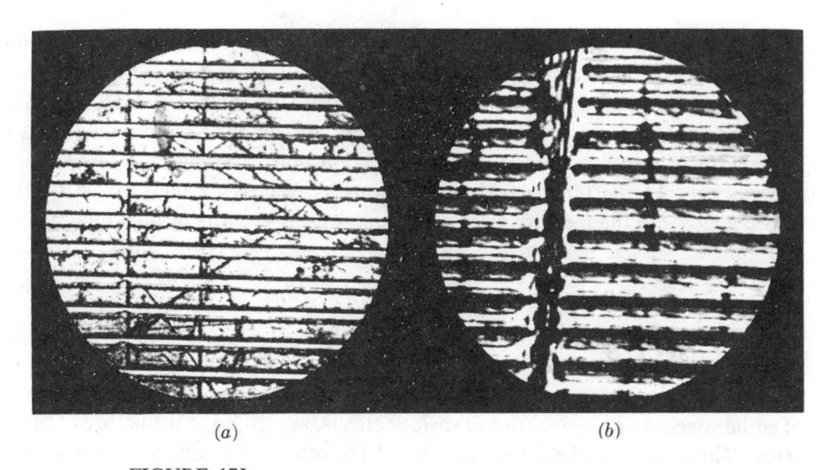

(*a*) (*b*)

FIGURE 17J
Microphotographs of the rulings on reflection gratings: (*a*) light ruling and (*b*) heavy ruling. (*Courtesy of H. D. Babcock, Mt. Wilson Observatory, Pasadena, California.*)

If ruled gratings are used without any auxiliary apparatus to separate the different orders, their overlapping makes it impractical to use values of m above 4 or 5. Hence, to obtain adequate dispersion and resolving power, the grating space must under these circumstances be made very small, and a large number of lines must be ruled. Rowland's engine gave 14,438 per inch, corresponding to $d = 1.693 \times 10^{-4}$ cm, and could produce gratings nearly 15 cm wide. This grating space is about three wavelengths of yellow light, and thus the third order is the highest that can be observed in this color with normal incidence. Correspondingly higher orders can be observed for shorter wavelengths. Even in the first order, however, the dispersion given by such a grating far exceeds that of a prism. From the grating equation one finds that the visible spectrum is spread over an angle of 12°. If it were projected by a lens of 3.0 m focal length, the spectrum would cover a length of about 60.0 cm on the photographic plate. In the second order it would be more than 1.0 m long.

The real advantage of the grating over the prism lies not in its large dispersion, however, but in the high resolving power it affords. One can always increase the linear dispersion by using a camera lens of longer focal length, but beyond a certain minimum set by the graininess of the photographic plate no more detail is revealed thereby. With sufficient dispersion, the final limitation is the chromatic resolving power. A 15 cm Rowland grating in the first order gives $\lambda/\Delta\lambda = 6 \times 14{,}438 \approx 76{,}600$. In the orange region two lines only 0.08 Å apart would be resolved, and with the abovementioned dispersion each line would be only 0.015 mm wide. This separation is only one-eightieth of that of the yellow sodium doublet. A glass prism, even though it had the rather large $dn/d\lambda$ of $-$ 1200 cm^{-1}, would by Eq. (15h) need to have a base 64 cm long to yield the same resolution.

It was first shown by Thorp that fairly good transmission gratings can be made by taking a cast of the ruled surface with some transparent material. Such casts are called *replica gratings*, and may give satisfactory performance where the highest resolving power is not needed. Collodion or cellulose acetate, properly diluted, is poured on the grating surface and dried to a thin, tough film which can easily be detached from the master grating under water. It can then be mounted on a plane glass plate or concave mirror. Some distortion and shrinkage are involved in this process, so that the replica seldom functions as well as the master. With modern improvements in the techniques of plastics, however, replicas of high quality are now being made.

17.12 GHOSTS

In an actual grating the ruled lines will always deviate to some extent from the ideal of equal spacing. This gives rise to various effects, according to the nature of the ruling error. Three types can be distinguished. (1) The error is *perfectly random* in magnitude and direction. In this case the grating will give a continuous spread of light underlying the principal maxima, even when monochromatic light is used. (2) The error *continuously increases* in one direction. This can be shown to give the grating "focal properties." Parallel light after diffraction is no longer parallel but slightly divergent or convergent. (3) The error is *periodic* across the surface of the grating. This is the most common type, since it frequently arises from defects in the driving mechanism of the ruling machine. It gives rise to "ghosts," or false lines, accompanying every principal maximum of the ideal grating. When there is only one period involved in the error, these lines are symmetrical in spacing and intensity about the principal maxima. Such ghosts are called *Rowland ghosts*, and may easily be seen in Fig. 21H(*g*). More troublesome, though of less frequent occurrence, are the *Lyman* ghosts*. These appear when the error involves two periods that are incommensurate with each other or when there is a single error of very short period. Lyman ghosts may occur very far from the principal maximum of the same wavelength.

In recent years the ruling of more perfect gratings has been accomplished through the work of George R. Harrison and George W. Stroke.† These men used ruling engines with the spacings of the rulings being set by a servomechanism, controlled by the automatic counting of interference fringes.

17.13 CONTROL OF THE INTENSITY DISTRIBUTION AMONG ORDERS

The relative intensities of the different orders for a ruled grating do not conform to the term $(\sin^2 \beta)/\beta^2$ derived for the ideal case [Eq. (17c)]. Obviously the light reflected from (or refracted by) the sides of the grooves will produce important modifications. In general there will be no missing orders. The *positions* of the spectral

* Theodore Lyman (1874–1954). For many years director of the Physical Laboratories at Harvard University. Pioneer in the investigation of the far-ultraviolet spectrum.

† See A. R. Ingalls, *Sci. Am.*, **186:** 45 (1952) and J. F. Verrill, *Contemp. Phys.*, **9:** 259 (1968).

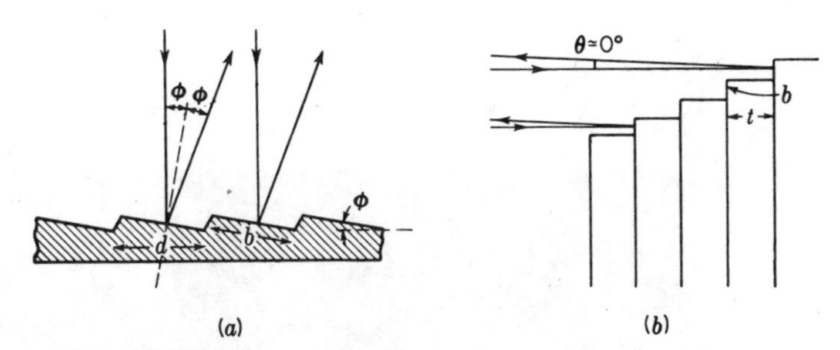

FIGURE 17K
Concentration of light in a particular direction by (*a*) an echelette, or echelle, grating and (*b*) a reflection echelon.

lines are uninfluenced, however, and remain unchanged for any grating of the same grating space *d*. In fact, the only essential requirement for a grating is that it impress on the diffracted wave some periodic variation of either amplitude or phase. The relative intensity of different orders is then determined by the angular distribution of the light diffracted by a single element, of width *d*, on the grating surface. In the ideal grating this corresponds to the diffraction from a single slit. In ruled gratings it will usually be a complex factor, which in the early days of grating manufacture was considered to be largely uncontrollable. More recently, R. W. Wood has been able to produce gratings which concentrate as much as 90 percent of the light of a particular wavelength in a single order on one side. Thus one of the chief disadvantages of gratings compared with prisms—the presence of multiple spectra, none of which is very intense—is overcome.

Wood's first experiments were done with gratings for the infrared, which have a large grating space so that the form of the grooves could be easily governed. These so-called *echelette* gratings had grooves with one optically flat side inclined at such an angle ϕ as to reflect the major portion of the infrared radiation toward the order that was to be bright [Fig. 17K(*a*)]. Of course the light from any one such face is diffracted through an appreciable angle, measured by the ratio of the wavelength to the width *b* of the face. When the ruling of gratings on aluminum was started, it was found possible to control the shape of the finer grooves required for visible and ultraviolet light. By proper shaping and orientation of the diamond ruling point, gratings are now produced which show a *blaze* of light at any desired angle.

Historically, the first application of the principle of concentrating the light in particular orders was made by Michelson in his *echelon* grating [Fig. 17K(*b*)]. This instrument consists of 20 to 30 plane-parallel plates stacked together with a constant offset *b* of about 1.0 mm. The thickness *t* was usually 1.0 cm so that the grating space is very large and concentration occurs in an extremely high order. As used by Michelson, echelons were transmission instruments, but larger path differences and higher orders are afforded by the reflection type first made by Williams.* In either case, the light is concentrated in a direction perpendicular to the fronts of the steps. At most

* W. E. Williams, *Proc. Phys. Soc. (Lond.)* **45**: 699 (1933).

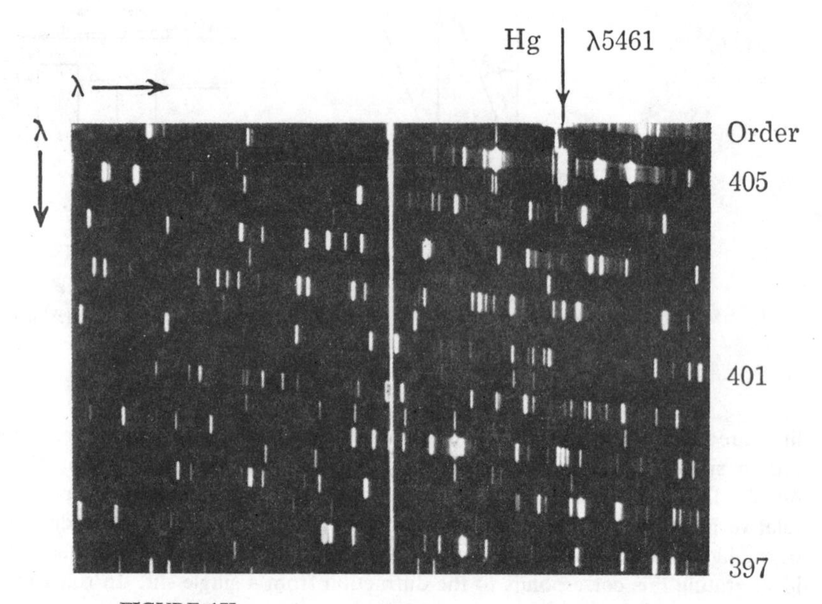

FIGURE 17L
Echellegram of the thorium spectrum. (*Courtesy of Sumner P. Davis, Department of Physics, University of California, Berkeley, California.*)

two orders of a given wavelength appear under the diffraction maximum. These have such large values of m [about $2t/\lambda$ for the reflection type and $(n - 1)t/\lambda$ for the transmission type] that the resolving power mN is very high, even with a relatively small number N of plates. In this respect the instrument is like an interferometer and in the same way requires auxiliary dispersion to separate the lines being studied. Since it has the same defect of lack of flexibility as the Lummer-Gehrcke plate, the echelon is little used nowadays.

A more important type of grating called the *echelle*,* which is intermediate between the echelette and the echelon, has a relatively coarse spacing of the grooves, some 80 to the centimeter. These are shaped as in Fig. 17K(*a*), but with a rather steeper slope. The order numbers for which concentration occurs are in the hundreds, whereas for an echelon they are in the tens of thousands. An echelle must be used in conjunction with another dispersing instrument, usually a prism spectrograph, to separate the various orders. If the dispersion of the echelle is in a direction perpendicular to that of the prism, an extended spectrum is displayed as a series of short strips representing adjacent orders, as shown in Fig. 17L.† This is part of a more extensive

* George R. Harrison, *J. Opt. Soc. Am.*, **39**: 522 (1949); **43**: 853 (1953).

† The separation of orders, in taking the echellegram of Fig. 17L, was accomplished not by a prism but by an ordinary grating. This accounts for the weaker spectra between the orders marked, which occur in the second order and have echelle orders twice as great.

spectrogram, which covers a large wavelength range with a plate factor of only 0.5 Å/mm. Each order contains about 14 Å of the spectrum, the range that is covered by the diffraction envelope of a single groove. This range is sufficient to produce a certain amount of repetition from one order to the next. Thus in Fig. 17L the green mercury line, which has been superimposed as a reference wavelength, appears in the 405th order, and again at the extreme left in the 404th order. The resolving power afforded by the echelle depends only on its total width [Eq. (17k)] and can be some 50 times higher than that of the auxiliary spectrograph. Here it is sufficient to resolve the hyperfine structure of the green line. Besides its high resolution and dispersion, the echelle has the advantages of yielding bright spectra and of registering the spectra in very compact form.

17.14 MEASUREMENT OF WAVELENGTH WITH THE GRATING

Small gratings 2 to 5 cm wide are usually mounted on the prism table of a small spectrometer with collimator and telescope. By measuring the angles of incidence and diffraction for a given spectrum line its wavelength can be calculated from the grating formula [Eq. (17f)]. For this the grating space d must be known, and this is usually furnished with the grating. The first accurate wavelengths were determined by this method, the grating space being found by counting the lines in a given distance with a traveling microscope. Once the absolute wavelength of a single line is known, others can be measured relative to it by using the overlapping of orders. For instance, according to Eq. (17h), a sodium line of wavelength 5890 Å in the third order will coincide with another line of $\lambda = \frac{3}{4} \times 5890 = 4417$ Å in the fourth order. Of course no two lines will exactly coincide in this way, but they may fall close enough together to permit the small difference to be accurately corrected for. This method of comparing wavelengths is not accurate with the arrangement described above because the telescope lens is never perfectly achromatic and the two lines will not be focused in exactly the same plane. To avoid this difficulty Rowland invented the *concave grating*, in which the focusing is done by a concave mirror, upon which the grating itself is ruled.

17.15 CONCAVE GRATING

If the grating, instead of being ruled on a plane surface, is ruled on a concave spherical mirror of metal, it will diffract and focus the light at the same time, thus doing away with the need for lenses. Besides the fact that this eliminates the chromatic aberration mentioned above, it has the great advantage that the grating can be used for regions of the spectrum which are not transmitted by glass lenses, such as the ultraviolet. A mathematical treatment of the action of the concave grating would be out of place here, but we may mention one of the more important results. It is found that if R is the radius of curvature of the spherical surface of the grating, a circle of *diameter R*, that is, radius $r = R/2$, can be drawn tangent to the grating at its midpoint which

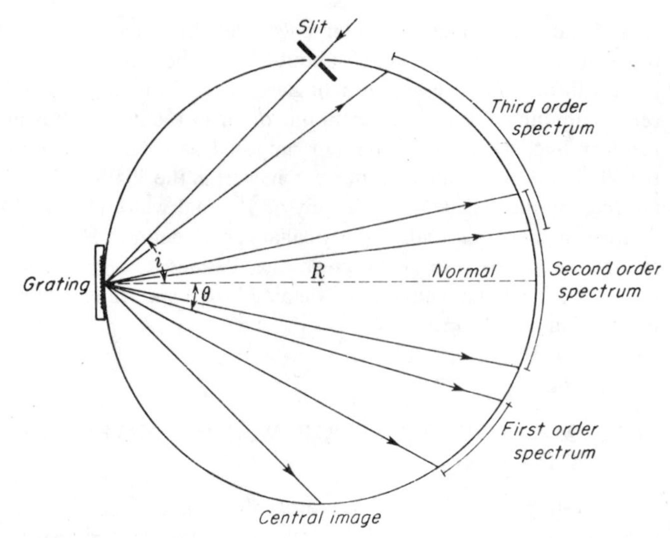

FIGURE 17M
Paschen mounting for a concave grating.

defines the locus of points where the spectrum is in focus, provided the source slit also lies on this circle. This circle is called the *Rowland circle*, and in practically all mountings for concave gratings use is made of this condition for focus. See Fig. 17N(*a*).

17.16 GRATING SPECTROGRAPHS

Figure 17M shows a diagram of a common form of mounting used for large concave gratings, called the *Paschen mounting*. The slit is set up on the Rowland circle, and the light from this strikes the grating, which diffracts it into the spectra of various orders. These spectra will be in focus on the circle, and the photographic plates are mounted in a plateholder which bends them to coincide with this curve. Several orders of a spectrum can be photographed at the same time in this mounting. The ranges covered by the visible spectrum in the first three orders are indicated in Fig. 17M for the value of the grating space mentioned above. In a given order, Eq. (17g) shows that the dispersion is a minimum on the normal to the grating ($\theta = 0$), and increases on both sides of this point. It is practically constant, however, for a considerable region near the normal, because here the cosine is varying slowly. A common value for R is 21 ft, and a concave grating with this radius of curvature is called a *21-ft grating*.

Two other common mountings for concave gratings are the *Rowland mounting* and the *Eagle mounting*, illustrated in Fig. 17N. In the Rowland mounting, which is now mostly of historical interest, the grating G and plateholder P are fixed to opposite ends of a rigid beam of length R. The two ends of this beam rest on swivel trucks

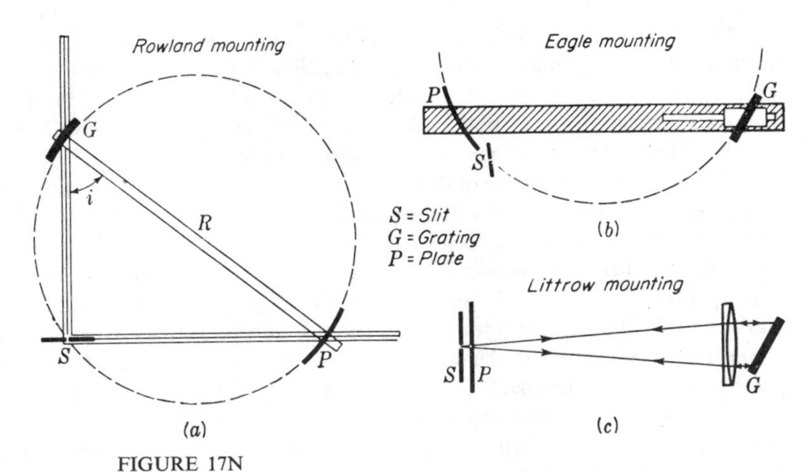

FIGURE 17N
(*a*) One of the earliest and (*b*) one of the commonest forms of concave grating spectrograph. (*c*) Mounting for plane reflection grating.

which are free to move along two tracks at right angles to each other. The slit S is mounted just above the intersection of the two tracks. With this arrangement, the portion of the spectrum reaching the plate may be varied by sliding the beam one way or the other, thus varying the angle of incidence i. It will be seen that this effectively moves S around on the Rowland circle. For any setting the spectrum will be in focus on P, and it will be nearly a normal spectrum (Sec. 17.6) because the angle of diffraction $\theta \approx 0$. The track SP is usually graduated in wavelengths since, as can be easily shown from the grating equation, the wavelength in a given order arriving at P is proportional to the distance SP.

The Eagle mounting, because of its compactness and flexibility, has largely replaced the Rowland and Paschen forms. Here the part of the spectrum is observed which is diffracted back at angles nearly equal to the angle of incidence. The slit S is placed at one end of the plateholder, the latter being pivoted like a gate at S. To observe different portions of the spectrum, the grating is turned about an axis perpendicular to the figure. It must then be moved along horizontal ways, and the plateholder turned, until P and S again lie on the Rowland circle. The instrument can be mounted in a long box or room where the temperature is held constant. Variations of temperature displace the spectrum lines owing to the change of grating space which results from the expansion or contraction of the grating. With a grating of speculum metal it can be shown that a change of temperature of 0.1°C shifts a line of wavelength 5000 Å in any order by 0.013 Å. The Eagle mounting is commonly used in *vacuum spectrographs* for the investigation of ultraviolet spectra in the region below 2000 Å. Since air absorbs these wavelengths, the air must be pumped out of the spectrograph and this compact mounting is convenient for the purpose. The Paschen mounting is also frequently used in vacuum spectrographs with the light incident on the grating at a practically grazing angle. The *Littrow mounting*, also shown in

Fig. 17N, is the only common method of mounting large plane reflection gratings. In principle it is very much like the Eagle mounting, the main difference being that a large achromatic lens renders the incident light parallel and focuses the diffracted light on P, so that it acts as both collimator and telescope lenses at once.

One important drawback of the concave grating as used in the mountings described above is the presence of strong astigmatism. It is least in the Eagle mounting. This defect of the image always occurs when a concave mirror is used off axis. Here it has the consequence that each point on the slit is imaged as two lines, one located on the Rowland circle perpendicular to its plane, the other in this plane and at some distance behind the circle. If the slit is accurately perpendicular to the plane, the sharpness of the spectrum lines is not seriously impaired by astigmatism. Because of the increased length of the lines, however, some loss of intensity is involved. More serious is the fact that it is impossible to study the spectrum of different parts of a source or to separate Fabry-Perot rings by projecting an image on the slit of the spectrograph. For this purpose, a *stigmatic mounting* is required. The commonest of these is the Wadsworth mounting, in which the concave grating is illuminated by parallel light. The light from the slit is rendered parallel by a large concave mirror, and the spectrum is focused at a distance of about one-half the radius of curvature of the grating.

PROBLEMS

17.1 Make a qualitative sketch for the intensity pattern for five equally spaced slits having $d/b = 4$. Label several points on the x axis with the corresponding values of β and γ.

Ans. See Fig. P17.1.

17.2 Make a qualitative sketch for the intensity pattern for seven equally spaced slits having $d/b = 3$. Label points on the x axis with the corresponding values of β and γ.

17.3 Nine coherent sources of microwaves in phase and with a wavelength of 2.50 cm are placed side by side in a straight line, 10.0 cm between centers. Compute (*a*) the angular width of the central maximum. Find the angular separation of (*b*) the principal maxima and (*c*) the subsidiary maxima.

17.4 Light of two wavelengths, $\lambda = 5600$ Å and $\lambda = 5650$ Å, fall normally on a plane transmission grating having 2500 lines per centimeter. The emerging parallel light is focused on a flat screen by a lens of 120 cm focal length. Find the distance on the screen in centimeters between the two spectrum lines (*a*) in the first order and (*b*) in the second order.

17.5 Two spectrum lines at $\lambda = 6200$ Å have a separation of 0.652 Å. Find the minimum number of lines a diffraction grating must have to just resolve this doublet in the second-order spectrum.

17.6 A diffraction grating is ruled with 100,000 lines in a distance of 8.0 cm and used in the first order to study the structure of a spectrum line at $\lambda = 4230$ Å. How does the chromatic resolving power compare with that of a 60.0° glass prism with a base of 8.0 cm and refractive indices 1.5608 at $\lambda = 4010$ Å and 1.5462 at $\lambda = 4450$ Å?

Ans. Grating resolving power = 100,000; prism resolving power = 26,550

17.7 Calculate the dispersion (*a*) in angstroms per degree, (*b*) in degrees per angstrom, and (*c*) in angstroms per millimeter for a grating containing 3000 lines per centimeter when used in the third-order spectrum focused on a screen by a lens with a focal length of 200 cm.

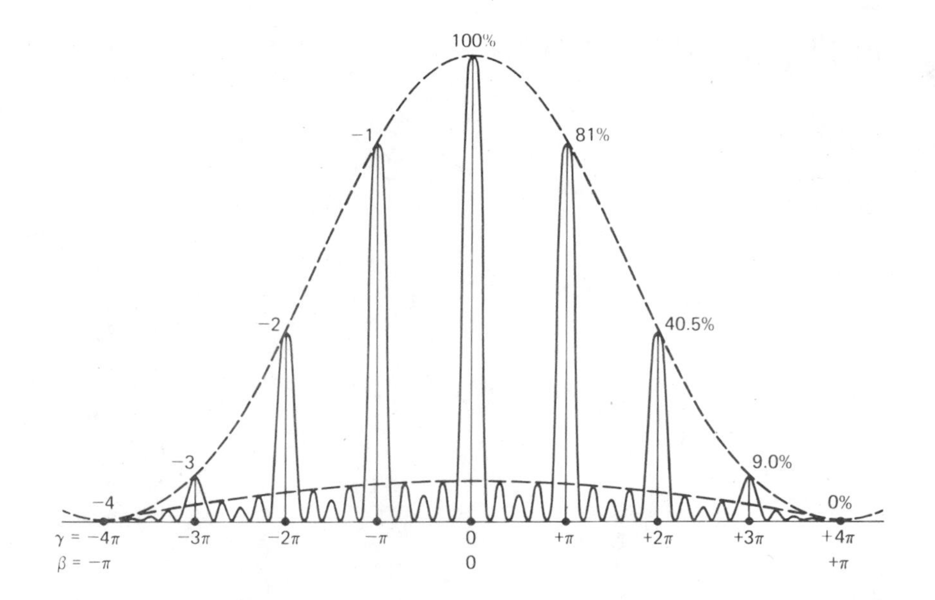

FIGURE P17.1
Intensity graph for a diffraction grating with fine slits, and $d/b = 4$. See Prob. 17.1.

17.8 A group of spectrum lines in the region of 5200 Å is to be studied using a plane grating 15.0 cm wide containing 6000 lines per centimeter and mounted in a Littrow system. Find (*a*) the highest order that can be used, (*b*) the angle of incidence required to observe it, (*c*) the smallest wavelength interval resolved, and (*d*) the plate factor if the lens has a focal length of 2.50 m.

17.9 A diffraction grating containing 5000 lines per centimeter is illuminated at various angles of incidence by light of wavelength 4000 Å. Draw a graph of the deviation of the first-order diffraction beam from the direction of the incident light, using the angle of incidence from 0 to 90° plotted on the x axis.

17.10 Find (*a*) the order number and (*b*) the resolving power for a reflection echelon having 35 plates each 9.0 mm thick if it is used with a cadmium arc to study $\lambda = 5085.82$ Å.

Ans. (*a*) 3.5392×10^4, (*b*) 1.2387×10^6

17.11 An echelette grating has 450 lines per centimeter and is ruled for the concentration of infrared light of wavelength 5.0 μm in the second order. Find (*a*) the angle of the ruled faces to the plane of the grating and (*b*) the angular dispersion at this wavelength, assuming normal incidence. If this grating is illuminated by red light of a helium lamp, (*c*) what order or orders of $\lambda = 6678$ Å will be observed?

17.12 Prove that one can express the resolving power of an echelle grating as $\lambda/\Delta\lambda = (2B/\lambda)[r^2/(1 + r^2)]^{1/2}$, where B is the width of the grating and $r = t/b$ is the ratio of the depth of the steps to their width. It is assumed that the light is incident and diffracted normal to the faces of width b. *Hint:* Use the principle that the resolving power equals the number of wavelengths in the path difference between the rays from opposite edges of the grating.

18
FRESNEL DIFFRACTION

The diffraction effects obtained when either the source of light or the observing screen, or both, are at a finite distance from the diffracting aperture or obstacle come under the classification of *Fresnel diffraction*. These effects are the simplest to observe experimentally, the only apparatus required being a small source of light, the diffracting obstacle, and a screen for observation. In the Fraunhofer effects discussed in the preceding chapters, lenses were required to render the light parallel, and to focus it on the screen. Now, however, we are dealing with the more general case of divergent light which is not altered by any lenses. Since Fresnel diffraction is the easiest to observe, it was historically the first type to be investigated, although its explanation requires much more difficult mathematical theory than that necessary in treating the plane waves of Fraunhofer diffraction. In this chapter we consider some of the simpler cases of Fresnel diffraction, which are amenable to explanation by fairly direct mathematical and graphical methods.

18.1 SHADOWS

One of the greatest difficulties in the early development of the wave theory of light lay in the explanation of the observed fact that light appears to travel in straight lines. Thus if we place an opaque object in the path of the light from a point source, it casts

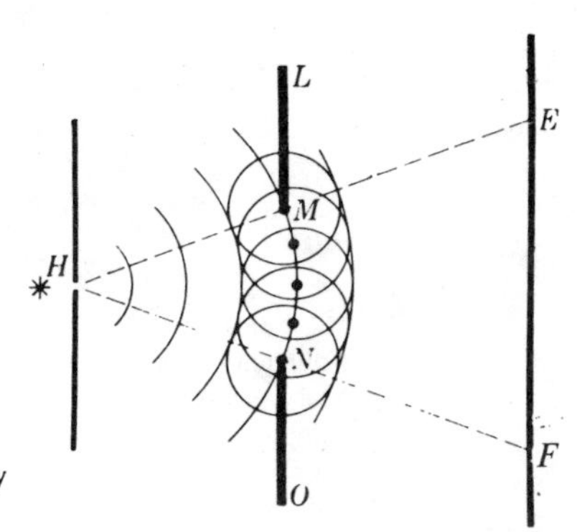

FIGURE 18A
Huygens' principle applied to secondary wavelets from a narrow opening.

a shadow having a fairly sharp outline of the same shape as the object. It is true, however, that the edge of this shadow is not absolutely sharp and that when examined closely it shows a system of dark and light bands in the immediate neighborhood of the edge. In the days of the corpuscular theory of light, attempts were made by Grimaldi and Newton to account for such small effects as due to the deflection of the light corpuscles in passing close to the edge of the obstacle. The correct explanation in terms of the wave theory we owe to the brilliant work of Fresnel. In 1815 he showed not only that the approximately rectilinear propagation of light could be interpreted on the assumption that light is a wave motion but also that in this way the diffraction fringes could in many cases be accounted for in detail.

To bring out the difficulty of explaining shadows by the wave picture, let us consider first the passage of divergent light through an opening in a screen. In Fig. 18A the light originates from a small pinhole H, and a certain portion MN of the divergent wave front is allowed to pass the opening. According to Huygens' principle, we may regard each point on the wave front as a source of secondary wavelets. The envelope of these at a later instant gives a divergent wave with H as its center and included between the lines HE and HF. This wave as it advances will produce strong illumination in the region EF of the screen. But also part of each wavelet will travel into the space behind LM and NO, and hence might be expected to produce some light in the regions of the geometrical shadow outside of E and F. Common experience shows that there is actually no illumination on these parts of the screen, except in the immediate vicinity of E and F. According to Fresnel, this is to be explained by the fact that in the regions well beyond the limits of the geometrical shadow the secondary wavelets arrive with phase relations such that they interfere destructively and produce practically complete darkness.

The secondary wavelets cannot have uniform amplitude in all directions, since if this were so, they would produce an equally strong wave in the backward direction. In Fig. 18A the envelope on the left side of the screen would represent a reverse wave converging toward H. Obviously such a wave does not exist physically, and hence one must assume that the amplitude at the back of a secondary wave is zero. The more exact formulation of Huygens' principle justifies this assumption and also gives

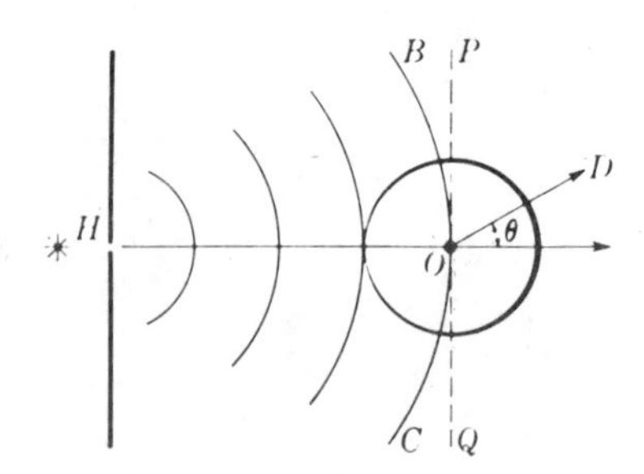

FIGURE 18B
The obliquity factor for Huygens' secondary wavelets.

quantitatively the variation of the amplitude with direction. The so-called *obliquity factor*, as is illustrated in Fig. 18B, requires an amplitude varying as $1 + \cos \theta$, where θ is the angle with the forward direction. At right angles, in the directions P and Q of the figure, the amplitude falls to one-half and the intensity to one-quarter of its maximum value. Another property that the wavelets must be assumed to have, in order to give the correct results, is an advance of phase of one-quarter period ahead of the wave that produces them. The consequences of these two rather unexpected properties and the manner in which they are derived will be discussed later.

18.2 FRESNEL'S HALF-PERIOD ZONES

As an example of Fresnel's approach to diffraction problems, we first consider his method of finding the effect that a slightly divergent spherical wave will produce at a point ahead of the wave. In Fig. 18C let $BCDE$ represent a spherical wave front of monochromatic light traveling toward the right. Every point on this sphere may be thought of as the origin of secondary wavelets, and we wish to find the resultant effect of these at a point P. To do this, we divide the wave front into *zones* by the following construction. Around the point O, which is the foot of the perpendicular from P, we describe a series of circles whose distances from O, measured along the arc, are $s_1, s_2, s_3, \ldots, s_m$ and are such that each circle is a half wavelength farther from P. If the distance $OP = b$, the circles will be at distances $b + \lambda/2$, $b + 2\lambda/2$, $b + 3\lambda/2, \ldots, b + m\lambda/2$ from P.

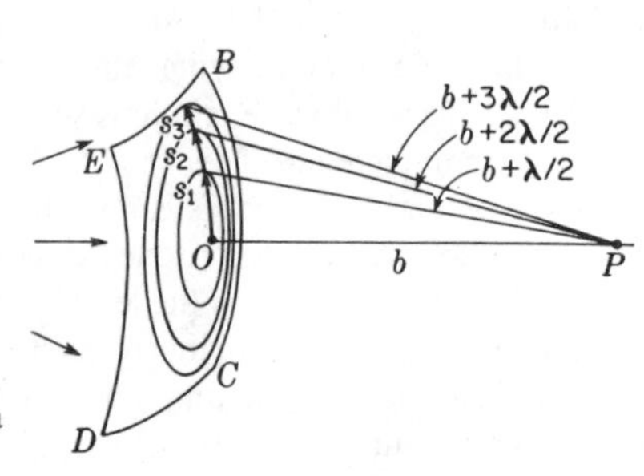

FIGURE 18C
Construction of half-period zones on a spherical wave front.

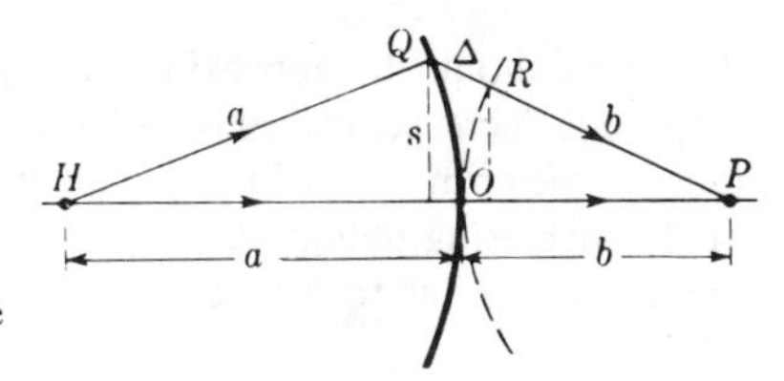

FIGURE 18D
Path difference Δ at a distance s from the pole of a spherical wave.

The areas S_m of the zones, i.e., of the rings between successive circles, are practically equal. In proving this, we refer to Fig. 18D, where a section of the wave spreading out from H is shown with radius a. If a circle of radius b is now drawn (broken circle) with its center at P and tangent to the wave front at its "pole" O, the path HQP is longer than HOP by the segment indicated by Δ. For the borders of the zones, this path difference must be a whole multiple of $\lambda/2$. To evaluate it, we note first that in all optical problems the distance s is small compared with a and b. Then s may be considered as the vertical distance of Q above the axis, and Δ may be equated to the sum of the sagittas of the two arcs OQ and OR. By the sagitta formula we then have

$$\Delta = \frac{s^2}{2a} + \frac{s^2}{2b} = s^2 \frac{a + b}{2ab} \tag{18a}$$

The radii s_m of the Fresnel zones are such that

$$m \frac{\lambda}{2} = s_m{}^2 \frac{a + b}{2ab} \tag{18b}$$

and the area of any one zone becomes

$$S_m = \pi(s_m{}^2 - s_{m-1}{}^2) = \pi \frac{\lambda}{2} \frac{2ab}{a + b} = \frac{a}{a + b} \pi b \lambda \tag{18c}$$

To the approximation considered, it is therefore constant and independent of m. A more exact evaluation would show that the area increases very slowly with m.

By Huygens' principle we now regard every point on the wave as sending out secondary wavelets in the same phase. These will reach P with different phases, since each travels a different distance. The phases of the wavelets from a given zone will not differ by more than π, and since each zone is on the average $\lambda/2$ farther from P, it is clear that the successive zones will produce resultants at P which differ by π. This statement will be examined in more detail in Sec. 18.6. The difference of half a period in the vibrations from successive zones is the origin of the name *half-period zones*. If we represent by A_m the resultant amplitude of the light from the mth zone, the successive values of A_m will have alternating signs because changing the phase by π means reversing the direction of the amplitude vector. When the resultant amplitude due to the whole wave is called A, it may be written as the sum of the series

$$A = A_1 - A_2 + A_3 - A_4 + \cdots + (-1)^{m-1} A_m \tag{18d}$$

Three factors determine the magnitudes of the successive terms in this series: (1) because the area of each zone determines the number of wavelets it contributes, the terms should be approximately equal but should increase slowly; (2) since the amplitude decreases inversely with the average distance from P of the zone, the magnitudes of the terms are reduced by an amount which increases with m; and (3) because of the increasing obliquity, their magnitudes should decrease. Thus we may express the amplitude due to the mth zone as

$$A_m = \text{const}\,\frac{S_m}{d_m}(1 + \cos\theta) \qquad \text{(18e)}$$

where d_m is the average distance to P and θ the angle at which the light leaves the zone. It appears in the form shown because of the obliquity factor assumed in the preceding section. Now an exact calculation of the S_m's shows that the factor b in Eq. (18c) must be replaced by $b + \Delta$, where Δ is the path difference for the middle of the zone. Since at the same time $d_m = b + \Delta$, we find that the ratio S_m/d_m is a constant, independent of m. Therefore we have left only the effect of the obliquity factor $1 + \cos\theta$, which causes the successive terms in Eq. (18d) to decrease very slowly. The decrease is least slow at first, because of the rapid change of θ with m, but the amplitudes soon become nearly equal.

With this knowledge of the variation in magnitude of the terms, we may evaluate the sum of the series by grouping its terms in the following two ways. Supposing m to be odd,

$$A = \frac{A_1}{2} + \left(\frac{A_1}{2} - A_2 + \frac{A_3}{2}\right) + \left(\frac{A_3}{2} - A_4 + \frac{A_5}{2}\right) + \cdots + \frac{A_m}{2}$$

$$= A_1 - \frac{A_2}{2} - \left(\frac{A_2}{2} - A_3 + \frac{A_4}{2}\right) - \left(\frac{A_4}{2} - A_5 + \frac{A_6}{2}\right) - \cdots - \frac{A_{m-1}}{2} + A_m \qquad \text{(18f)}$$

Now since the amplitudes $A_1, A_2, \ldots$ do not decrease at a uniform rate, each one is smaller than the arithmetic mean of the preceding and following ones. Therefore the quantities in parentheses in the above equations are all positive, and the following inequalities must hold:

$$\frac{A_1}{2} + \frac{A_m}{2} < A < A_1 - \frac{A_2}{2} - \frac{A_{m-1}}{2} + A_m$$

Because the amplitudes for any two adjacent zones are very nearly equal, it is possible to equate A_1 to A_2, and A_{m-1} to A_m. The result is

$$A = \frac{A_1}{2} + \frac{A_m}{2} \qquad \text{(18g)}$$

If m is taken to be even, we find by the same method that

$$\frac{A_1}{2} - \frac{A_m}{2} = A$$

Hence the conclusion is that the resultant amplitude at P due to m zones is either half the sum or half the difference of the amplitudes contributed by the first and last zones. If we allow m to become large enough for the entire spherical wave to be divided into

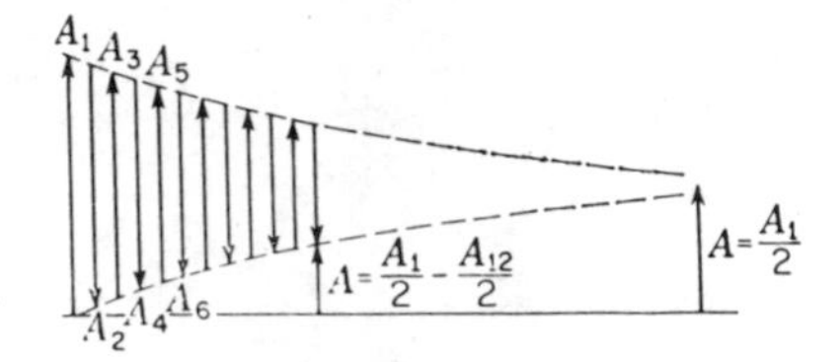

FIGURE 18E
Addition of the amplitudes from half-period zones.

zones, θ approaches 180° for the last zone. Therefore the obliquity factor causes A_m to become negligible, and the amplitude due to the whole wave is just half that due to the first zone acting alone.

Figure 18E shows how these results can be understood from a graphical construction. The vector addition of the amplitudes $A_1, A_2, A_3, \ldots$, which are alternately positive and negative, would normally be performed by drawing them along the same line, but here for clarity they are separated in a horizontal direction. The tail of each vector is put at the same height as the head of the previous one. Then the resultant amplitude A due to any given number of zones will be the height of the final arrowhead above the horizontal base line. In the figure, it is shown for 12 zones and also for a very large number of zones.

18.3 DIFFRACTION BY A CIRCULAR APERTURE

Let us examine the effect upon the intensity at P (Fig. 18C) of blocking off the wave by a screen pierced by a small circular aperture as shown in Fig. 18F. If the radius of the hole $r = OR$ is made equal to the distance s_1 to the outer edge of the first half-period zone,* the amplitude will be A_1 and this is twice the amplitude due to the unscreened wave. Thus the intensity at P is 4 times as great as if the screen were absent. When the radius of the hole is increased until it includes the first two zones, the amplitude is $A_1 - A_2$, or practically zero. The intensity has actually fallen to almost zero as a result of increasing the size of the hole. A further increase of r will cause the intensity to pass through maxima and minima each time the number of zones included becomes odd or even.

The same effect is produced by moving the point of observation P continuously toward or away from the aperture along the perpendicular. This varies the size of the zones, so that if P is originally at a position such that $PR - PO$ of Fig. 18F is $\lambda/2$ (one zone included), moving P toward the screen will increase this path difference to $2\lambda/2$ (two zones), $3\lambda/2$ (three zones), etc. We thus have maxima and minima along the axis of the aperture.

The above considerations give no information about the intensity at points off the axis. A mathematical investigation, which we shall not discuss because of its

* We are here assuming that the radius of curvature of the wave striking the screen is large, so that distances measured along the chord may be taken as equal to those measured along the arc.

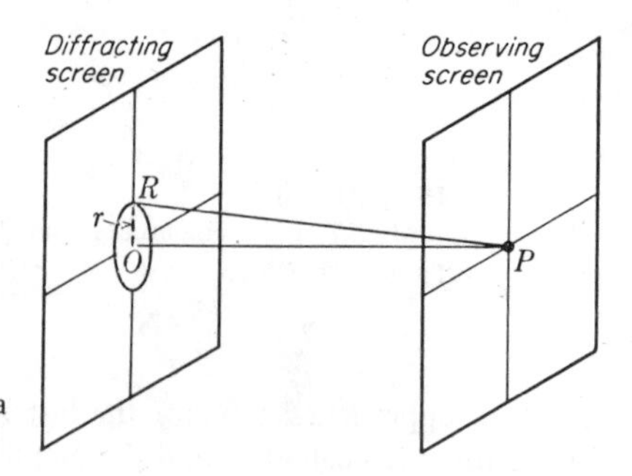

FIGURE 18F
Geometry for light passing through a circular opening.

complexity,* shows that P is surrounded by a system of circular diffraction fringes. Several photographs of these fringes are illustrated in Fig. 18G. These were taken by placing a photographic plate some distance behind circular holes of various sizes, illuminated by monochromatic light from a distant point source. Starting at the upper left of the figures, the holes were of such sizes as to expose one, two, three, etc., zones. The alternation of the center of the pattern from bright to dark illustrates the result obtained above. The large pattern on the right was produced by an aperture containing 71 zones.

18.4 DIFFRACTION BY A CIRCULAR OBSTACLE

When the hole is replaced by a circular disk, Fresnel's method leads to the surprising conclusion that there should be a bright spot in the center of the shadow. For a treatment of this case, it is convenient to start constructing the zones at the edge of the disk. If, in Fig. 18F, $PR = d$, the outer edge of the first zone will be $d + \lambda/2$ from P, of the second $d + 2\lambda/2$, etc. The sum of the series representing the amplitudes from all the zones in this case is, as before, half the amplitude from the first exposed zone. In Fig. 18E it would be obtained by merely omitting the first few vectors. Hence the intensity at P is practically equal to that produced by the unobstructed wave. This holds only for a point on the axis, however, and off the axis the intensity is small, showing faint concentric rings. In Fig. 18H(*a*) and (*b*), which shows photographs of the bright spot, these rings are unduly strengthened relative to the spot by overexposure. In (*c*) the source, instead of being a point, was a photographic negative of a portrait of Woodrow Wilson on a transparent plate, illuminated from behind. The disk acts like a rather crude lens in forming an image, since for every point in the object there is a corresponding bright spot in the image.

The complete investigation of diffraction by a circular obstacle shows that, besides the spot and faint rings in the shadow, there are bright circular fringes bordering the outside of the shadow. These are similar in origin to the diffraction fringes from a straight edge to be investigated in Sec. 18.11.

* See T. Preston, "Theory of Light," 5th ed., pp. 324–327, The Macmillan Company, New York, 1928.

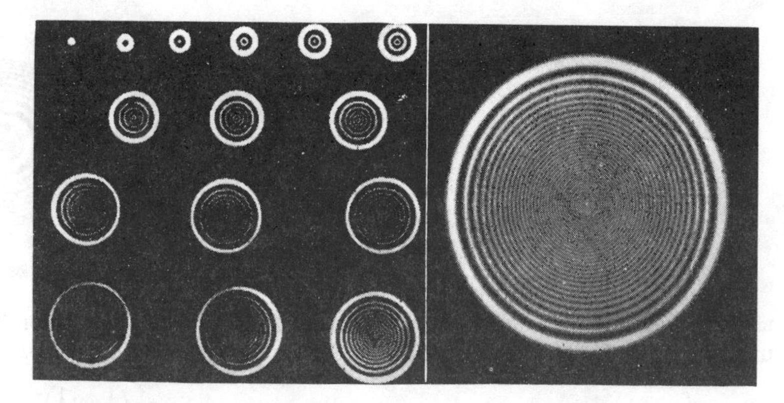

FIGURE 18G
Diffraction of light by small circular openings. (*Courtesy of Hufford.*)

The bright spot in the center of the shadow of a 1-cent piece can be seen by examining the region of the shadow produced by an arc light several meters away, preferably using a magnifying glass. The spot is very tiny in this case and difficult to find. It is easier to see with a smaller object, such as a bearing ball.

18.5 ZONE PLATE

This is a special screen designed to block off the light from every other half-period zone. The result is to remove either all the positive terms in Eq. (18d) or all the negative terms. In either case the amplitude at P (Fig. 18C) will be increased to many times its value in the above cases. A zone plate can easily be made in practice by drawing concentric circles on white paper, with radii proportional to the square roots of whole numbers (see Fig. 18I). Every other zone is then blackened, and the result is photographed on a reduced scale. The negative, when held in the light from a distant point source, produces a large intensity at a point on its axis at a distance corresponding

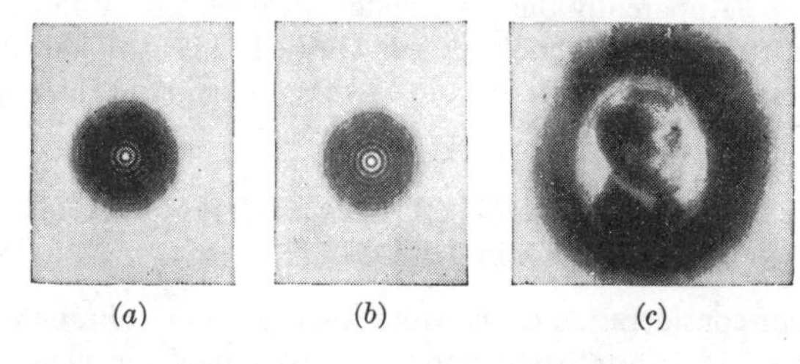

FIGURE 18H
Diffraction by a circular obstacle: (*a*) and (*b*) point source; (*c*) a negative of Woodrow Wilson as a source. (*Courtesy of Hufford.*)

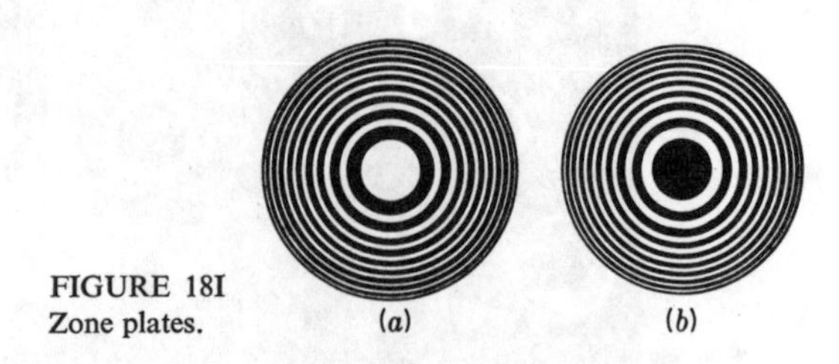

FIGURE 18I
Zone plates.

to the size of the zones and the wavelength of the light used. The relation between these quantities is contained in Eq. (18b), which for the present purpose may be written

$$m\frac{\lambda}{2} = \frac{s_m^2}{2}\left(\frac{1}{a} + \frac{1}{b}\right) \qquad (18h)$$

Hence we see that, for given a, b, and λ, the zones must have $s_m \approx \sqrt{m}$.

The bright spot produced by a zone plate is so intense that the plate acts much like a lens. Thus suppose that the first 10 odd zones are exposed, as in the zone plate of Fig. 18I(*a*). This leaves the amplitudes $A_1, A_3, A_5, \ldots, A_{19}$ (see Fig. 18E), the sum of which is nearly 10 times A_1. The whole wave front gives $\frac{1}{2}A_1$, so that, using only 10 exposed zones, we obtain an amplitude at P which is 20 times as great as when the plate is removed. The intensity is therefore 400 times as great. If the odd zones are covered, the amplitudes A_2 $A_4, A_6, \ldots$ will give the same effect. The object and image distances obey the ordinary lens formula, since, by Eq. (18h),

$$\frac{1}{a} + \frac{1}{b} = \frac{m\lambda}{s_m^2} = \frac{1}{f}$$

the focal length f being the value of b for $a = \infty$, namely,

$$f = \frac{s_m^2}{m\lambda} = \frac{s_1^2}{\lambda} \qquad (18i)$$

There are also fainter images corresponding to focal lengths $f/3, f/5, f/7, \ldots$, because at these distances each zone of the plate includes 3, 5, 7, ... Fresnel zones. When it includes three, for example, the effects of two of them cancel but that of the third is left over.

Apparently the zone plate was invented by Lord Rayleigh as evidenced by an entry in his notebook, dated April 11, 1871: "The experiment of blocking out the odd Huygens zones so as to increase the light at the center succeeded very well. . . ."

18.6 VIBRATION CURVE FOR CIRCULAR DIVISION OF THE WAVE FRONT

Our consideration of the vibration curve in the Fraunhofer diffraction by a single slit (Sec. 15.4) was based upon the division of the plane wave front into infinitesimal elements of area which were actually strips of infinitesimal width parallel to the length of the diffracting slit. The vectors representing the contributions to the amplitude

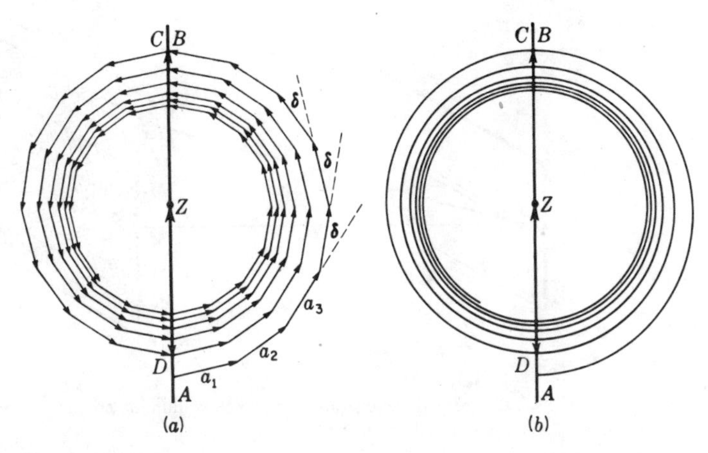

FIGURE 18J
Vibration spiral for Fresnel half-period zones of a circular opening.

from these elements were found to give an arc of a circle. This so-called *strip division* of the wave front is appropriate when the source of light is a narrow slit and the diffracting aperture rectangular. The strip division of a divergent wave front from such a source will be discussed below (Sec. 18.8). The method of dividing the spherical wave from a point source appropriate to any case of diffraction by circular apertures or obstacles involves infinitesimal circular *zones.*

Let us consider first the amplitude diagram when the first half-period zone is divided into eight subzones, each constructed in a manner similar to that used for the half-period zones themselves. We make these subzones by drawing circles on the wave front (Fig. 18C) which are distant

$$b + \frac{1}{8}\frac{\lambda}{2}, \quad b + \frac{2}{8}\frac{\lambda}{2}, \quad \ldots, \quad b + \frac{\lambda}{2}$$

from P. The light arriving at P from various points in the first subzone will not vary in phase by more than $\pi/8$. The resultant of these may be represented by the vector a_1 in Fig. 18J(a). To this is now added a_2, the resultant amplitude due to the second subzone, then a_3 due to the third subzone, etc. The magnitudes of these vectors will decrease very slowly as a result of the obliquity factor. The phase difference δ between each successive one will be constant and equal to $\pi/8$. Addition of all eight subzones yields the vector AB as the resultant amplitude from the first half-period zone. Continuing this process of subzoning to the second half-period zone, we find CD as the resultant for this zone, and AD as that for the sum of the first two zones. These vectors correspond to those of Fig. 18E. Succeeding half-period zones give the rest of the figure, as shown.

The transition to the vibration curve of Fig. 18J(b) results from increasing indefinitely the number of subzones in a given half-period zone. The curve is now a *vibration spiral,* eventually approaching Z when the half-period zones cover the whole

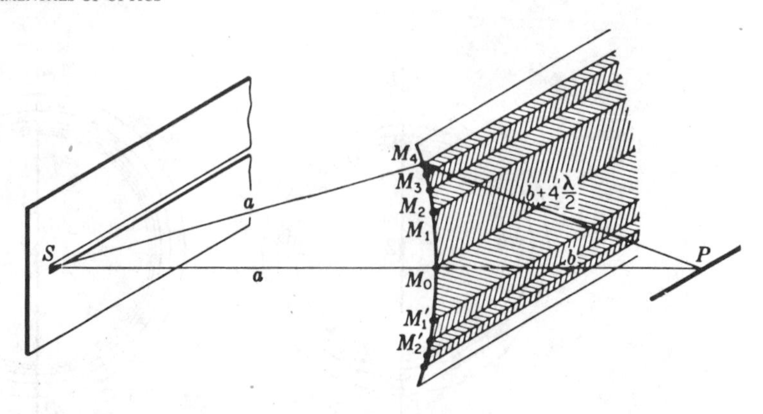

FIGURE 18K
Cylindrical wave from a slit which is illuminated coherently. Half-period strips are marked on the wave front.

spherical wave. Any one turn is very nearly a circle but does not quite close because of the slow decrease in the magnitudes of the individual amplitudes. The significance of the series of decreasing amplitudes, alternating in sign, used in Sec. 18.2 for the half-period zones, becomes clearer when we keep in mind the curve of Fig. 18J(*b*). It has the additional advantage of allowing us to determine directly the resultant amplitude due to any fractional number of zones. It should be mentioned in passing that the resultant amplitude *AZ*, which is just half the amplitude due to the first half-period zone, turns out to be, from this treatment, 90° in phase *behind* the light from the center of the zone system. This cannot be true, since it is impossible to alter the resultant phase of a wave merely by the artifice of dividing it into zones and then recombining their effects. The discrepancy is a defect of Fresnel's theory resulting from the approximations made therein and does not occur in the more rigorous mathematical treatment.

18.7 APERTURES AND OBSTACLES WITH STRAIGHT EDGES

If the configuration of the diffracting screen, instead of having circular symmetry, involves straight edges like those of a slit or wire, it is possible to use as a source a slit rather than a point. The slit is set parallel to these edges, so that the straight diffraction fringes produced by each element of its length are all lined up on the observing screen. A considerable gain of intensity is achieved thereby. In the investigation of such cases, it is possible to regard the wave front as cylindrical, as shown in Fig. 18K. It is true that to produce such a cylindrical envelope to the Huygens wavelets emitted by various points on the slit these must emit *coherently*, and in practice this will not usually be true. Nevertheless, when intensities are added, as is required for noncoherent emission, the resulting pattern is the same as would be produced by a coherent cylindrical wave. In the following treatment of problems involving straight edges, we shall therefore make the simplification of assuming the source slit to be illuminated by a parallel monochromatic beam, so that it emits a truly cylindrical wave.

18.8 STRIP DIVISION OF THE WAVE FRONT

The appropriate method of constructing half-period elements on a cylindrical wave front consists in dividing the latter into strips, the edges of which are successively one-half wavelength farther from the point P (Fig. 18K). Thus the points M_0, $M_1, M_2, \ldots$ on the circular section of the cylindrical wave are at distances $b, b + \lambda/2$, $b + 2\lambda/2, \ldots$ from P. M_0 is on the straight line SP. The half-period strips M_0M_1, $M_1M_2, \ldots$ now stretch along the wave front parallel to the slit. We may call this procedure *strip division* of the wave front.

In the Fresnel zones obtained by circular division, the areas of the zones were very nearly equal. With the present type of division this is by no means true. The areas of the *half-period strips* are proportional to their widths, and these decrease rapidly as we go out along the wave front from M_0. Since this effect is much more pronounced than any variation of the obliquity factor, the latter need not be considered.

The amplitude diagram of Fig. 18L(*a*) is obtained by dividing the strips into substrips in a manner analogous to that described in Sec. 18.6 for circular zones. Dividing the first strip above M_0 into nine parts, we find that the nine amplitude vectors from the substrips extend from O to B, giving a resultant $A_1 = OB$, for the first half-period strip. The second half-period strip similarly gives those between B and C, with a resultant $A_2 = BC$. Since the amplitudes now decrease rapidly, A_2 is considerably smaller than A_1, and their difference in phase is appreciably greater than π. A repetition of this process of subdivision for the succeeding strips on the upper half of the wave gives the more complete diagram of Fig. 18L(*b*). Here the vectors are spiraling in toward Z, so that the resultant for all half-period strips above the pole M_0 becomes OZ.

18.9 VIBRATION CURVE FOR STRIP DIVISION. CORNU'S SPIRAL

When we go over to elementary strips of infinitesimal width, we obtain the vibration curve as a smooth spiral, part of which is shown in Fig. 18M. The complete curve representing the whole wave front would be carried through many more turns, ending at the points Z and Z'. Only the part from O to Z was considered above. The lower half, $Z'O$, arises from the contributions from the half-period strips below M_0.

This curve, called *Cornu's* spiral*, is characterized by the fact that the angle δ

* M. A. Cornu (1841–1902). Professor of experimental physics at the École Polytechnique, Paris.

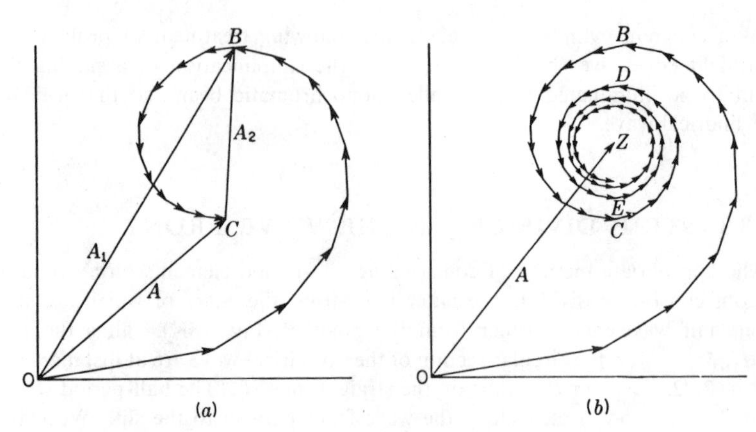

FIGURE 18L
Amplitude diagrams for the formation of Cornu's spiral.

it makes with the x axis is proportional to the *square* of the distance v along the curve from the origin. Remembering that, in a vibration curve, δ represents the phase lag in the light from any element of the wave front, we obtain this definition of the curve by using Eq. (18a) for the path difference, as follows:

$$\delta = \frac{2\pi}{\lambda}\,\Delta = \frac{\pi(a+b)}{ab\lambda}\,s^2 = \frac{\pi}{2}\,v^2 \qquad \text{(18j)}$$

Here we have introduced a new variable for use in plotting Cornu's spiral, namely

$$v = s\sqrt{\frac{2(a+b)}{ab\lambda}} \qquad \text{(18k)}$$

It is defined in such a way as to make it dimensionless, so that the same curve may be used for any problem, regardless of the particular values of a, b, and λ.

18.10 FRESNEL'S INTEGRALS

The x and y coordinates of Cornu's spiral can be expressed quantitatively by two integrals, and a knowledge of them will permit accurate plotting and calculations. They are derived most simply as follows. Since the phase difference δ is the angle determining the slope of the curve at any point (see Fig. 18M), the changes in the coordinates for a given small displacement dv along the spiral are given by

$$dx = dv\cos\delta = \cos\frac{\pi v^2}{2}\,dv \qquad dy = dv\sin\delta = \sin\frac{\pi v^2}{2}\,dv$$

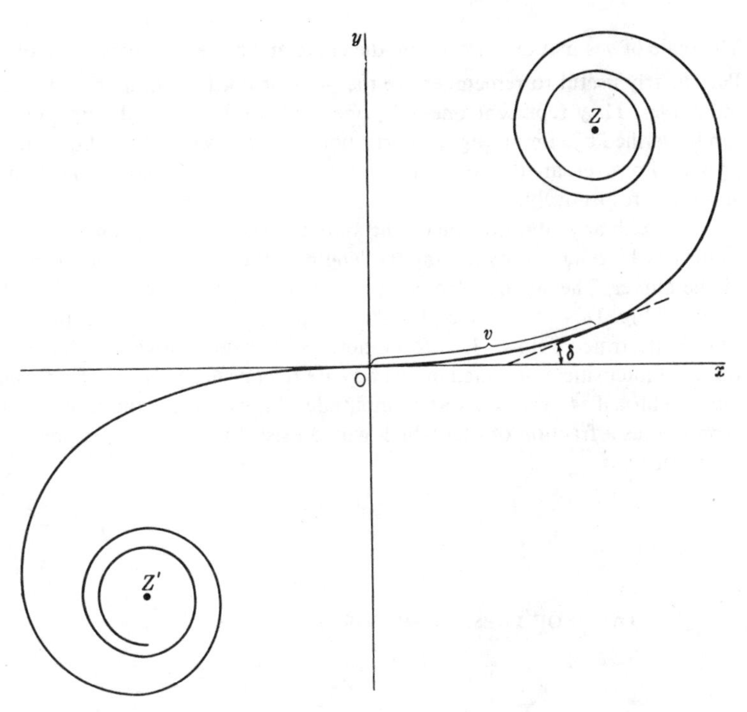

FIGURE 18M
Cornu's spiral drawn to include five half-period zones on either side of the pole.

where the value of δ from Eq. (18j) has been introduced. Thus the coordinates of any point (x,y) on Cornu's spiral become

$$x = \int_0^v \cos \frac{\pi v^2}{2}\, dv \qquad (18l)$$

$$y = \int_0^v \sin \frac{\pi v^2}{2}\, dv \qquad (18m)$$

These are known as *Fresnel's integrals*. They cannot be integrated in closed form but yield infinite series which may be evaluated in several ways.* Although the actual evaluation is too complicated to be given here, we have included a table of the numerical values of the integrals (Table 18A). In Sec. 18.14 the method of using them in accurate computations of diffraction patterns is explained.

Let us first examine some features of the quantitative Cornu's spiral of Fig. 18N, which is a plot of the two Fresnel integrals. The coordinates of any point on the curve give their values for a particular upper limit v in Eqs. (18l) and (18m).

* For the methods of evaluating Fresnel's integrals, see R. W. Wood, "Physical Optics," 2d ed., p. 247, The Macmillan Company, New York, 1921; reprinted (paperback) by Dover Publications, Inc., New York, 1968.

The scale of v is marked directly on the curve and has equal divisions along its length. Particularly useful to remember are the positions of the points $v = 1, \sqrt{2}$, and 2 on the curve. They represent one-half, one, and two half-period strips, respectively, as can be verified by computing the corresponding values of δ from Eq. (18j). More important, however, are the coordinates of the end points Z' and Z. They are $(-\frac{1}{2}, -\frac{1}{2})$ and $(\frac{1}{2}, \frac{1}{2})$, respectively.

As with any vibration curve, the amplitude due to any given portion of the wave front may be obtained by finding the length of the chord of the appropriate segment of the curve. The square of this length then gives the intensity. Thus the Cornu's spiral of Fig. 18N can be used for the graphical solution of diffraction problems, as will be illustrated below. It is to be noted at the start, however, that the numerical values of intensities computed in this way are *relative to a value of 2 for the unobstructed wave*. Thus, if A represents any amplitude obtained from the plot, the intensity I, expressed as a fraction of that which would exist if no screen were present, which we shall call I_0, is

$$\frac{I}{I_0} = \tfrac{1}{2}A^2 \tag{18n}$$

Table 18A TABLE OF FRESNEL INTEGRALS

v	x	y	v	x	y	v	x	y
0.00	0.0000	0.0000	3.00	0.6058	0.4963	5.50	0.4784	0.5537
0.10	0.1000	0.0005	3.10	0.5616	0.5818	5.55	0.4456	0.5181
0.20	0.1999	0.0042	3.20	0.4664	0.5933	5.60	0.4517	0.4700
0.30	0.2994	0.0141	3.30	0.4058	0.5192	5.65	0.4926	0.4441
0.40	0.3975	0.0334	3.40	0.4385	0.4296	5.70	0.5385	0.4595
0.50	0.4923	0.0647	3.50	0.5326	0.4152	5.75	0.5551	0.5049
0.60	0.5811	0.1105	3.60	0.5880	0.4923	5.80	0.5298	0.5461
0.70	0.6597	0.1721	3.70	0.5420	0.5750	5.85	0.4819	0.5513
0.80	0.7230	0.2493	3.80	0.4481	0.5656	5.90	0.4486	0.5163
0.90	0.7648	0.3398	3.90	0.4223	0.4752	5.95	0.4566	0.4688
1.00	0.7799	0.4383	4.00	0.4984	0.4204	6.00	0.4995	0.4470
1.10	0.7638	0.5365	4.10	0.5738	0.4758	6.05	0.5424	0.4689
1.20	0.7154	0.6234	4.20	0.5418	0.5633	6.10	0.5495	0.5165
1.30	0.6386	0.6863	4.30	0.4494	0.5540	6.15	0.5146	0.5496
1.40	0.5431	0.7135	4.40	0.4383	0.4622	6.20	0.4676	0.5398
1.50	0.4453	0.6975	4.50	0.5261	0.4342	6.25	0.4493	0.4954
1.60	0.3655	0.6389	4.60	0.5673	0.5162	6.30	0.4760	0.4555
1.70	0.3238	0.5492	4.70	0.4914	0.5672	6.35	0.5240	0.4560
1.80	0.3336	0.4508	4.80	0.4338	0.4968	6.40	0.5496	0.4965
1.90	0.3944	0.3734	4.90	0.5002	0.4350	6.45	0.5292	0.5398
2.00	0.4882	0.3434	5.00	0.5637	0.4992	6.50	0.4816	0.5454
2.10	0.5815	0.3743	5.05	0.5450	0.5442	6.55	0.4520	0.5078
2.20	0.6363	0.4557	5.10	0.4998	0.5624	6.60	0.4690	0.4631
2.30	0.6266	0.5531	5.15	0.4553	0.5427	6.65	0.5161	0.4549
2.40	0.5550	0.6197	5.20	0.4389	0.4969	6.70	0.5467	0.4915
2.50	0.4574	0.6192	5.25	0.4610	0.4536	6.75	0.5302	0.5362
2.60	0.3890	0.5500	5.30	0.5078	0.4405	6.80	0.4831	0.5436
2.70	0.3925	0.4529	5.35	0.5490	0.4662	6.85	0.4539	0.5060
2.80	0.4675	0.3915	5.40	0.5573	0.5140	6.90	0.4732	0.4624
2.90	0.5624	0.4101	5.45	0.5269	0.5519	6.95	0.5207	0.4591

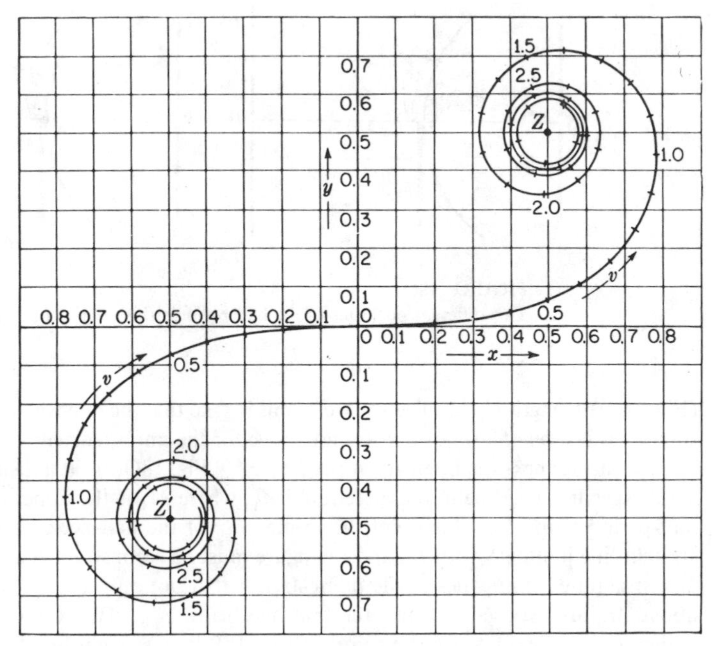

FIGURE 18N
Cornu's spiral; a plot of the Fresnel integrals.

To verify this statement, we note that according to the discussion of Sec. 18.8 a vector drawn from O to Z gives the amplitude due to the upper half of the wave. Similarly, one from Z' to O gives that due to the lower half. Each of these has a magnitude $1/\sqrt{2}$, so that when they are added and the sum is squared to obtain the intensity due to the whole wave, we find that $I_0 = 2$, with the conventional scale of coordinates used in Fig. 18N.*

18.11 THE STRAIGHT EDGE

The investigation of the diffraction by a single screen with a straight edge is perhaps the simplest application of Cornu's spiral. Figure 18O(a) represents a section of such a screen, having its edge parallel to the slit S. In this figure the half-period strips corresponding to the point P being situated on the edge of the geometrical shadow are marked off on the wave front. To find the intensity at P, we note that since the upper half of the wave is effective, the amplitude is a straight line joining 0 and Z

* It will be noticed that the phase of the resultant wave is 45°, or one-eighth period behind that of the wave coming from the center of the zone system (the Huygens' wavelet reaching P from M_0 in Fig. 18K). A similar phase discrepancy, this time of one-quarter period, occurs in the treatment of circular zones in Sec. 18.6. For a discussion of the phase discrepancy in Cornu's spiral, see R. W. Ditchburn, "Light," p. 214, Interscience Publishers, Inc., New York, 1953; 2d ed (paperback), 1963.

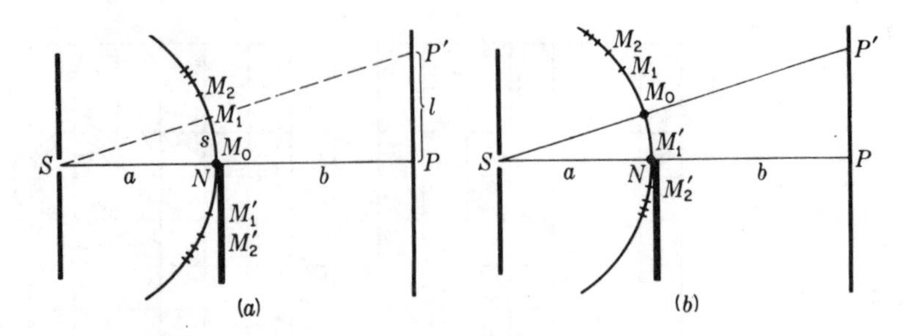

FIGURE 18O
Two different positions of the half-period strips relative to a straight edge N.

(Fig. 18P) of length $1/\sqrt{2}$. The square of this is $\frac{1}{2}$, so that the intensity at the edge of the shadow is just *one-fourth* of that found above for the unobstructed wave.

Consider next the intensity at the point P' [Fig. 18O(*a*)] at a distance l above P. To be specific, let P' lie in the direction SM_1, where M_1 is the upper edge of the first half-period strip. For this point, the center M_0 of the half-period strips lies on the straight line joining S with P', and the figure must be reconstructed as in Fig. 18O(*b*). The straight edge now lies at the point M_1', so that not only all the half-period strips above M_0 are exposed but also the first one below M_0. The resultant amplitude A is therefore represented on the spiral of Fig. 18P by a straight line joining B' and Z. This amplitude is more than twice that at P, and the intensity A^2 more than 4 times as great.

Starting with the point of observation P at the edge of the geometrical shadow (Fig. 18O), where the amplitude is given by OZ, if we move the point steadily upward, the tail of the amplitude vector moves to the left along the spiral, while its head remains fixed at Z. The amplitude will evidently go through a maximum at b', a minimum at c', another maximum at d', etc., approaching finally the value $Z'Z$ for the unobstructed wave. If we go downward from P, into the geometrical shadow, the tail of the vector moves to the right from O, and the amplitude will decrease steadily, approaching zero.

To obtain quantitative values of the intensities from Cornu's spiral, it is only necessary to measure the length A for various values of v. The square of A gives the intensity. Plots of the amplitude and the intensity against v are shown in Figs. 18Q(*a*) and (*b*), respectively. It will be seen that at the point O, which corresponds to the edge of the geometrical shadow, the intensity has fallen to one-fourth that for large negative values of v, where it approaches the value for the unobstructed wave. The other letters correspond with points similarly labeled on the spiral, B', C', D' . . . , representing the exposure of one, two, three, etc., half-period strips below M_0. The maxima and minima of these *diffraction fringes* occur a little before these points are reached. For instance, the first maximum at b' is given when the amplitude vector A has the position shown in Fig. 18P. Photographs of the diffraction pattern from a straight edge are shown in Fig. 18R(*a*) and (*b*). Pattern (*a*) was taken with visible light from a mercury arc, and (*b*) with X rays, $\lambda = 8.33$ Å. Figure 18R(*c*) is a density trace of the photograph (*a*), directly above, and was made with a microphotometer.

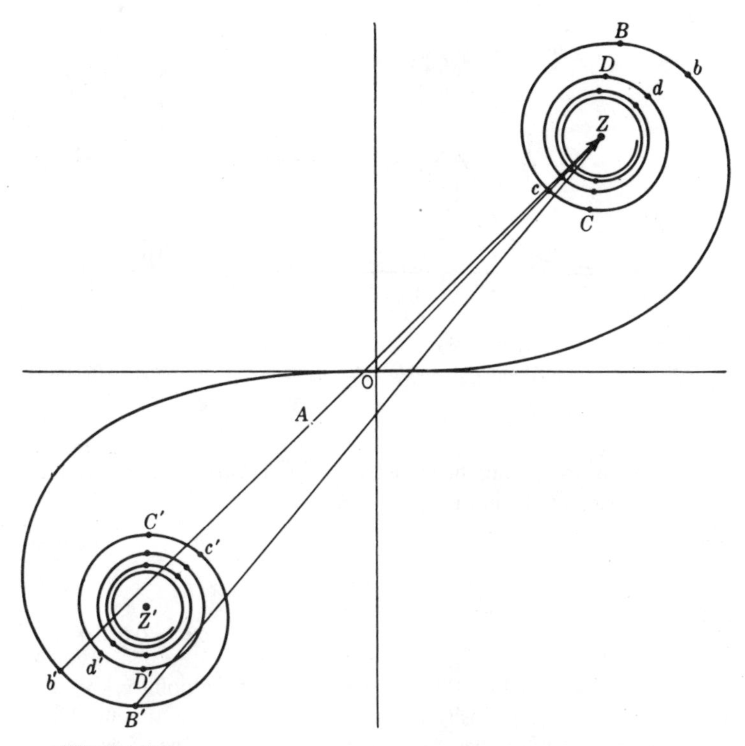

FIGURE 18P
Cornu's spiral, showing resultants for a straight-edge diffraction pattern.

Perhaps the most common observation of the straight-edge pattern, and certainly a very striking one, occurs in viewing a distant street lamp through rain-spattered spectacles. The edge of each drop as it stands on the glass acts like a prism, and refracts into the pupil of the eye rays which otherwise would not enter it. Beyond the edge the field is therefore dark, but the crude outline of the drop is seen as an irregular bright patch bordered by intense diffraction fringes such as those shown in Fig. 18R. The fringes are very clear, and a surprising number can be seen, presumably because of the achromatizing effect of the refraction.

18.12 RECTILINEAR PROPAGATION OF LIGHT

When we investigate the *scale* of the above pattern for a particular case, the reason for the apparently rectilinear propagation of light becomes clear. Let us suppose that in a particular case $a = b = 100$ cm, and $\lambda = 5000$ Å. From Eq. (18k), we then have

$$s = v\sqrt{\frac{ab\lambda}{2(a+b)}} = 0.0354v \qquad \text{cm}$$

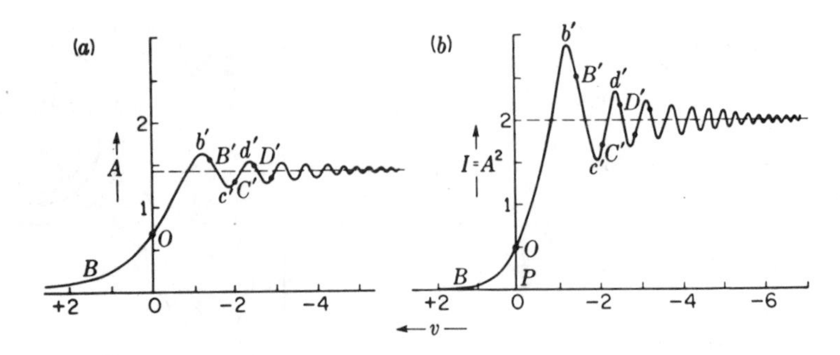

FIGURE 18Q
(*a*) Amplitude and (*b*) intensity contours for Fresnel diffraction at a straight edge.

This is the distance along the wave front [Fig. 18O(*a*)]. To change it to distances l on the screen, we note from the figure that

$$l = \frac{a + b}{a} s = v \sqrt{\frac{b\lambda(a + b)}{2a}} \tag{18o}$$

For the particular case chosen, therefore,

$$l = 2s = 0.0708v \qquad \text{cm}$$

Now in the graph of Fig. 18Q(*b*) the intensity at the point $v = +2$ is only 0.025 or one-eightieth of the intensity if the straight edge were absent. This point has $l =$ 0.142 cm, and therefore lies only 1.42 mm inside the edge of the geometrical shadow.

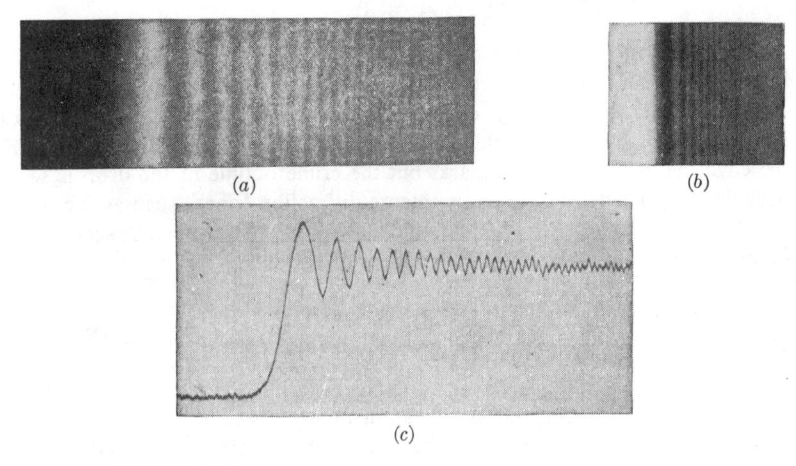

FIGURE 18R
Straight-edge diffraction patterns photographed with (*a*) visible light of wavelength 4300 Å and (*b*) X rays of wavelength 8.33 Å. (*c*) Microphotometer trace of (*a*).

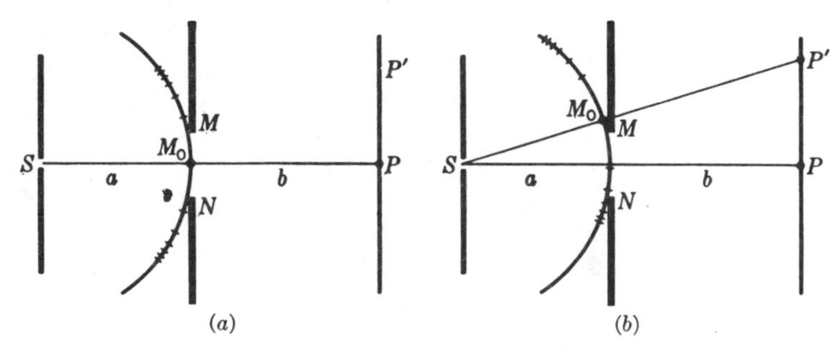

FIGURE 18S
Division of the wave front for Frensel diffraction by a single slit.

The part of the screen below this will lie in practically complete darkness, and this must be due to the destructive interference of the secondary wavelets arriving here from the upper part of the wave.

18.13 SINGLE SLIT

We next consider the Fresnel diffraction of a single slit with sides parallel to a narrow source slit S [Fig. 18S(*a*)]. By the use of Cornu's spiral we wish to determine the distribution of the light on the screen PP'. With the slit located as shown, each side acts like a straight edge to screen off the outer ends of the wave front. We have already seen in Sec. 18.11 how to investigate the pattern from a single straight edge, and the method used there is readily extended to the present case. With the slit in the central position of Fig. 18S(*a*), the only light arriving at P is that due to the wave front in the interval $\Delta s = MN$. In terms of Cornu's spiral we must now determine what length Δv corresponds to the slit width Δs. This is done by Eq. (18k), using Δv for v and Δs for s. Let $a = 100$ cm, $b = 400$ cm, $\lambda = 4000$ Å $= 0.00004$ cm, and the slit width $\Delta s = 0.02$ cm. Substituting in Eq. (18k), we obtain $\Delta v = 0.5$. The resultant amplitude at P is then given by a chord of the spiral, the arc of which has a length $\Delta v = 0.5$. Since the point of observation P is centrally located, this arc will start at $v = -0.25$ and run to $v = +0.25$. This amplitude $A \approx 0.5$ when squared gives the intensity at P.

If we now wish the intensity at P' [Fig. 18S(*b*)], the picture must be revised by redividing the wave front as shown. With the point of observation at P', the same length of wave front, $\Delta s = 0.02$ cm, is exposed, and therefore the same length of the spiral, $\Delta v = 0.5$, is effective. This section on the lower half of the wave front will, however, correspond to a new position of the arc on the lower half of the spiral. Suppose that it is represented by the arc jk in Fig. 18T. The resultant amplitude is proportional to the chord A, and the square of this gives the relative intensity. Thus to get the variation of intensity along the screen of Fig. 18S, we slide a piece of the spiral of *constant* length $\Delta v = 0.5$ to various positions and measure the lengths of the corresponding chords to obtain the amplitudes. In working a specific problem, the

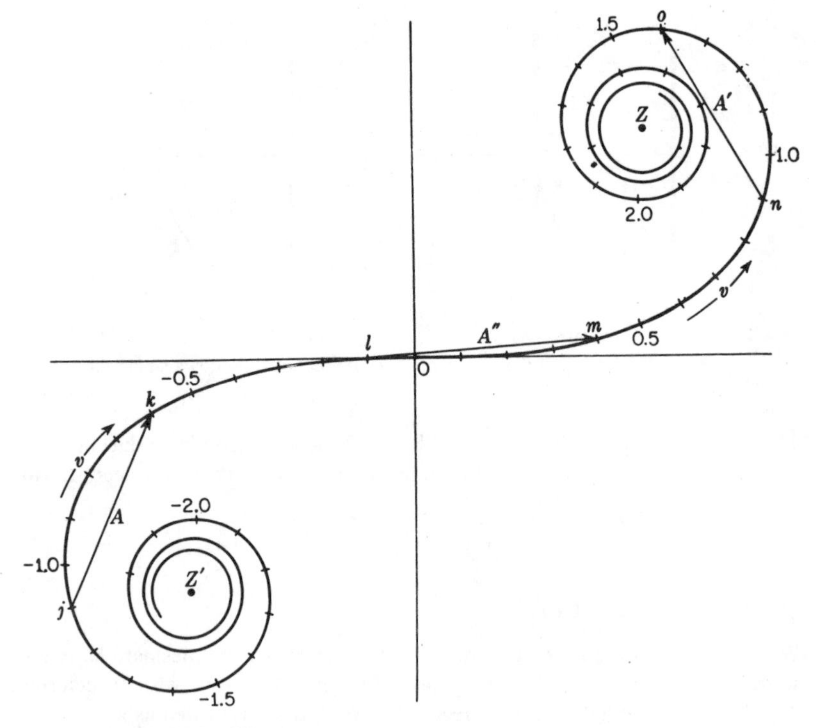

FIGURE 18T
Cornu's spiral, showing the chords of arcs of equal lengths Δv.

student may make a straight scale marked off in units of v to tenths, and measure the chords on an accurate plot such as Fig. 18N, using the scale of v on the spiral to obtain the constant length Δv of the arc. The results should then be tabulated in three columns, giving v, A, and A^2. The value of v to be entered is that for the *central* point of the arc whose chord A is being measured. For example, if the interval from $v = 0.9$ to $v = 1.4$ measured (Fig. 18T), the average value $v = 1.15$ is tabulated against $A = 0.43$.

Photographs of a number of Fresnel diffraction patterns for single slits of different widths are shown in Fig. 18U with the corresponding intensity curves beside them. These curves have been plotted by the use of Cornu's spiral. It is of interest to note in these diagrams the indicated positions of the edges of the geometrical shadow of the slit (indicated on the v axis). Very little light falls outside these points. For a very narrow slit like the first of these where $\Delta v = 1.5$, the pattern greatly resembles the Fraunhofer diffraction pattern for a single slit. The essential difference between the two (compare Fig. 15D) is that here the minima do not come quite to zero except at infinitely large v. The small single-slit pattern at the top was taken with X rays of wavelength 8.33 Å, while the rest were taken with visible light of wavelength 4358 Å. As the slit becomes wider, the fringes go through very rapid changes, approach-

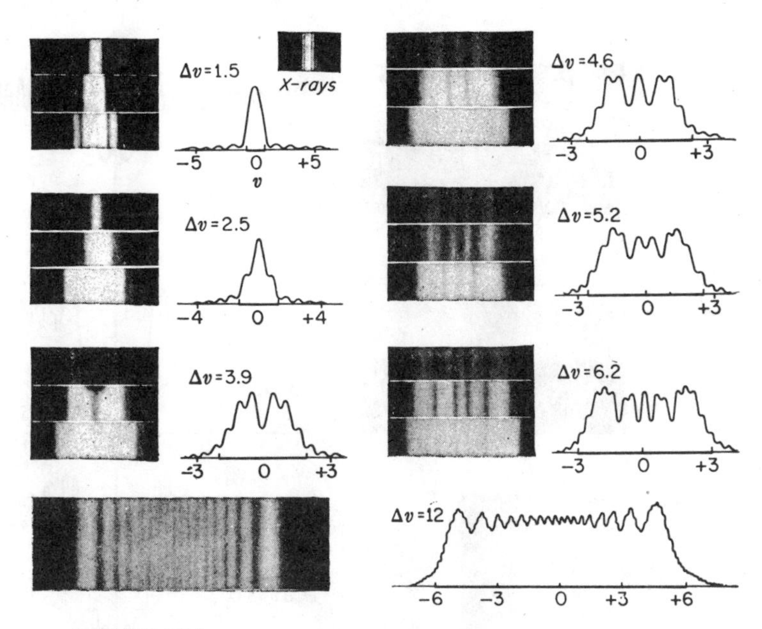

FIGURE 18U
Fresnel diffraction of visible light by single slits of different widths. (*X-ray pattern courtesy of Kellstrom, University of Uppsala, Uppsala, Sweden.*)

ing for a wide slit the general appearance of two opposed straight-edge diffraction patterns. The small closely spaced fringes superimposed on the main fringes at the outer edges of the last figure are clearly seen in the original photograph and may be detected in the reproduction.

18.14 USE OF FRESNEL'S INTEGRALS IN SOLVING DIFFRACTION PROBLEMS

The tabulated values of Fresnel's integrals in Table 18A can be used for higher accuracy than that obtainable with the plotted spiral. For an interval $\Delta v = 0.5$, for example, the two values of x at the ends of this interval are read from the table and subtracted algebraically to obtain Δx, the horizontal component of the amplitude. The corresponding two values of y are also subtracted to obtain Δy, its vertical component. The relative intensity will then be obtained by adding the squares of these quantities, since

$$I \approx A^2 = (\Delta x)^2 + (\Delta y)^2 \tag{18p}$$

The method is accurate but may be tedious, especially if good interpolations are to be made in certain parts of Table 18A. Some problems, such as that of the straight edge,

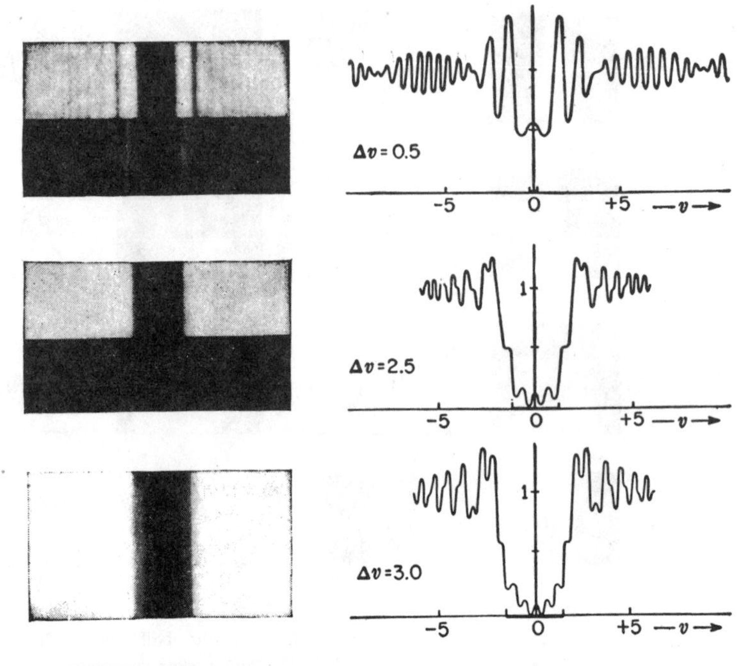

FIGURE 18V
Fresnel diffraction by single opaque strips.

are simplified by the fact that the number of zones on one end of the interval is not limited. The values of both x and y will be $\frac{1}{2}$ at this end. Another example of this type will now be considered.

18.15 DIFFRACTION BY AN OPAQUE STRIP

The shadow cast by a narrow object with parallel sides, such as a wire, can also be studied by the use of Cornu's spiral. In the case of a single slit, treated in Sec. 18.13, it was shown how the resultant diffraction pattern is obtained by sliding a fixed length of the spiral, Δv = const, along the spiral and measuring the chord between the two end points. The rest of the spiral out to infinity, i.e., out to Z or Z' on each side of the element in question, was absent owing to the screening by the two sides of the slit. If now the opening of the slit in Fig. 18S(a) is replaced by an object of the same size and the slit jaws taken away, we have two segments of the spiral to consider. Suppose the obstacle is of such a size that it covers an interval $\Delta v = 0.5$ on the spiral (Fig. 18T). For the position jk the light arriving at the screen will be due to the parts of the spiral from Z' to j and from k to Z. The resultant amplitude due to these two sections is obtained by adding their respective amplitude as vectors. The lower section gives an amplitude represented by a straight line from Z' to j, with the arrowhead at j.

The amplitude for the upper section is represented by a straight line from k to Z with the arrowhead at Z. The vector sum of these two gives the resultant amplitude A, and A^2 gives the intensity for a point v halfway between j and k. Photographs of three diffraction patterns produced by small wires are shown in Fig. 18V, accompanied by the corresponding theoretical curves.

PROBLEMS

18.1 The innermost zone of a zone plate has a diameter of 0.425 mm. (*a*) Find the focal length of the plate when it is used with parallel incident light of wavelength 4471 Å from a helium lamp. (*b*) Find its first subsidiary focal length.

Ans. (*a*) 40.40 cm, (*b*) 13.47 cm

18.2 A zone plate is to be set up on an optical bench, where it will be used as an enlarging lens. Its innermost zone is to be 0.2250 mm in diameter, and monochromatic blue-green light of wavelength 4800 Å from a cadmium arc is to be used. If all enlargements are to be eightfold in diameter, find (*a*) the focal length of the zone plate, (*b*) the object distance, and (*c*) the image distance.

18.3 A parallel beam of microwaves with a wavelength of 1.50 cm passes through a circular adjustable-iris diaphragm. A detector is placed on the axis 2.50 m behind it and the opening gradually increased in diameter. At what diameter would the detector's response reach (*a*) its first maximum, (*b*) its second maximum, and (*c*) its third maximum? (*d*) At the latter radius, give an equation for the positions of the maxima and minima along the axis.

18.4 Using Cornu's spiral, plot a diffraction pattern for a single slit having a width of 0.80 mm. Assume $a = 40.0$ cm, $b = 50.0$ cm, and red light of wavelength 6400 Å. Find (*a*) the value of Δv for use on the spiral and (*b*) plot the graph for intervals of $\Delta v = +0.10$ from $v = -0.10$ to $v = 3.0$.

18.5 A slit is placed at one end of an optical bench and is illuminated with green light of wavelength 5000 Å. A vertical rod 1.60 mm in diameter is mounted 50.0 cm away. Observations of the diffraction around this object are made by mounting a photoelectric cell and narrow slit 50.0 cm behind the rod. What would be (*a*) the value of Δv to be used on Cornu's spiral to represent this opaque object, (*b*) the exact intensity relative to the unobstructed intensity at the edge of the geometrical shadow, and (*c*) the relative intensity at the center of the shadow?

Ans. (*a*) 6.4, (*b*) $0.2282\, I_0$, (*c*) $0.01967\, I_0$

18.6 A slit is placed at one end of an optical bench and illuminated by green light of wavelength 5000 Å. A vertical straightedge is mounted parallel to the slit and 50.0 cm away. Observations of the diffraction pattern produced by the straightedge are made 50.0 cm beyond. What would be the intensity (*a*) 0.40 mm inside the edge of the geometrical shadow of the straightedge at the observation plane and (*b*) 0.40 mm outside the edge?

18.7 A slit is placed at one end of an optical bench and is illuminated by green light of wavelength 5000 Å. A vertical wire 0.40 mm in diameter is mounted 50 cm away. Observations of the diffraction pattern are made 50.0 cm beyond the wire. (*a*) What value of Δv should be used on Cornu's spiral to find the theoretical diffraction pattern? What would be the intensity relative to the unobstructed intensity (*b*) 0.40 mm from the center of the pattern and (*c*) 0.80 mm from the center?

Ans. (*a*) 1.60, (*b*) 26.75%, (*c*) 2.609%

18.8 For the diffraction of light by an opaque strip, investigate by the use of Cornu's spiral (*a*) whether a maximum must necessarily occur at the center of the pattern, as it does in the three cases of Fig. 18V. (*b*) What is the explanation of the beats observed outside the geometrical shadow in the case $v = 0.50$ of Fig. 18V?

18.9 Using Cornu's spiral, investigate the Fresnel diffraction pattern of a double slit. Assume $a = 40.0$ cm, $b = 50.0$ cm, $\lambda = 5625$ Å, and slits 0.1250 cm wide with the opaque interval between them 0.50 mm. Calculate Δv for (*a*) the slit widths and (*b*) the opaque interval. (*c*) Using the values given in Table 18A, calculate the resultant intensity A^2 for intervals $\Delta v = 0.20$ from the center of the pattern to $v = 1.80$. Plot a graph of A^2 against v out to $v = 1.80$ on both sides of the center. From your graph find the value of v for (*d*) the first minimum, (*e*) the first maximum, (*f*) the second minimum, and (*g*) the second maximum.

18.10 From the table of Fresnel integrals, calculate the exact intensity at the points (*a*) $v = +1.50$, (*b*) $v = -1.70$, and (*c*) $v = -1.30$ in the diffraction pattern of a straight edge. *Ans.* (*a*) $0.0210\ I_0$, (*b*) $0.890\ I_0$, (*c*) $1.352\ I_0$

19

THE SPEED OF LIGHT*

In Chap. 1 we observed that light has a finite speed or velocity. There we found that in a vacuum light has its greatest speed and that the generally accepted value is

$$c = 299{,}792.5 \text{ km/s} = 2.997925 \times 10^8 \text{ m/s}$$

We now return to the subject of the *speed of light*, giving a brief history of the subject and seeing what bearing the later experiments have on the theory of relativity.

19.1 RÖMER'S METHOD

Because of the very great speed of light, it is natural that the first successful measurement of its value was an astronomical one, because here very large distances are involved. In 1676 Römer† studied the times of the eclipses of the satellites of the planet Jupiter. Figure 19A(a) shows the orbits of the earth and of Jupiter around the sun S and that of one of the satellites M around Jupiter. The inner satellite has an

* Speed (a scalar) is the magnitude of velocity (a vector).

† Olaf Römer (1644–1710). Danish astronomer. His work on Jupiter's satellites was done in Paris, and later he was made Astronomer Royal of Denmark.

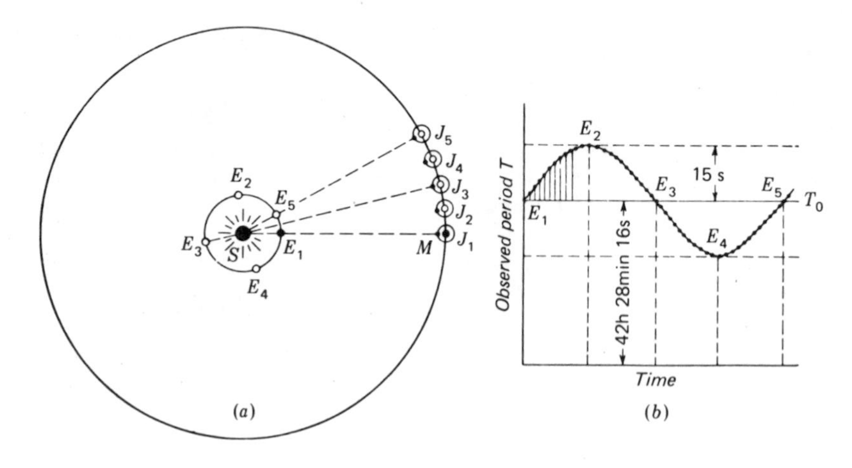

FIGURE 19A
Principle of Römer's astronomical determination of the speed of light from observations on Jupiter's moons.

average period of revolution $T_0 = 42$ h 28 min 16 s, as determined from the average time between two passages into the shadow of the planet. Actually Römer measured the times of *emergence* from the shadow, while the times of transit of the small black spot representing the shadow of the satellite on Jupiter's surface across the median line of the disk can be still more accurately measured.

A long series of observations on the eclipses of the first satellite permitted an accurate evaluation of the average period T_0. Römer found that if an eclipse is observed when the earth is at such a position as E_1 [Fig. 19A(*a*)] with respect to Jupiter J_1 and the time of a later eclipse is predicted by using the average period, it does not in general occur at exactly the predicted time. Specifically, if the predicted eclipse occurred about 3 months later, when the earth and Jupiter were at E_2 and J_2, he found a delay of somewhat more than 10 min. To explain this, he assumed that light travels with a finite velocity from Jupiter to the earth and that since the earth at E_2 is farther away from Jupiter, the observed delay represents the time required for light to travel the additional distance. His measurements gave 11 min as the time for light to go a distance equal to the radius of the earth's orbit. We now know that 8 min 18 s is a more nearly correct figure, and combining this with the average distance to the sun 1.48×10^6 km, we find a speed of about 3.0×10^5 km/s.

It is instructive to inquire how the apparent period of the satellite, i.e, the time between two successive eclipses, is expected to vary throughout a year. If this time could be observed with sufficient accuracy, one would obtain the curve of Fig. 19A(*b*). We may regard the successive eclipses as light signals sent out at regular time intervals of 42 h 28 min 16 s from Jupiter. At all points in its orbit except E_1 and E_3 the earth is changing its distance from Jupiter more or less rapidly. If the distance is increasing, as at E_2, any one signal travels a greater distance than the preceding one and the observed time between them will be increased. Similarly at E_4 it will be decreased. The maximum variation from the average period, about 15 s, is the time for light to cover the distance moved by the earth between two eclipses, which amounts to 4.48 km. At any given position, the total time delay of the eclipse, as observed by Römer, will be obtained by adding the amounts $T - T_0$ [Fig. 19A(*b*)], by which each apparent period is longer than the average. For instance, the delay of an eclipse at E_2, predicted from one at E_1 using the average period, will be the sum of $T - T_0$ for all eclipses between E_1 and E_2.

19.2 BRADLEY'S* METHOD. THE ABERRATION OF LIGHT

Römer's interpretation of the variations in the times of eclipses of Jupiter's satellites was not accepted until an entirely independent determination of the speed of light was made by the English astronomer Bradley in 1727. Bradley discovered an apparent motion of the stars which he explained as due to the motion of the earth in its orbit. This effect, known as *aberration*, is quite distinct from the well-known displacements of the nearer stars known as parallax. Because of parallax, these stars appear to shift slightly relative to the background of distant stars when they are viewed from different points in the earth's orbit, and from these shifts the distances of the stars are computed. Since the apparent displacement of the star is 90° ahead of that of the earth, the effect of parallax is to cause the star which is observed in a direction perpendicular to the plane of the earth's orbit to move in a small circle with a phase differing by $\pi/2$ from the earth's motion. The angular diameters of these circles are very small, being not much over 1 second of arc for the nearest stars. Aberration, which depends on the earth's *velocity*, also causes the stars observed in this direction to appear to move in circles. Here, however, the circles have an angular diameter of about 41 seconds, and they are the same for all stars, whether near or distant. Furthermore, the displacements are always in the direction of the earth's velocity [Fig. 19B(*a*)].

Bradley's explanation of this effect was that the apparent direction of the light reaching the earth from a star is altered by the motion of the earth in its orbit. The observer and his telescope are being carried along with the earth at a velocity of about 29.6 km/s, and if this motion is perpendicular to the direction of the star, the telescope must be tilted slightly toward the direction of motion from the position it would have if the earth were at rest. The reason for this is much the same as that involved when a person walking in the rain must tilt his umbrella forward to keep the rain off his feet. In Fig. 19B(*b*), let the vector v represent the velocity of the telescope relative to a system of coordinates fixed in the solar system, and c that of the light relative to the solar system. We have represented these motions as perpendicular to each other, as would be the case if the star lay in the direction shown in Fig. 19B(*a*). Then the velocity of the light relative to the earth has the direction of c', which is the vector difference between c and v. This is the direction in which the telescope must

* James Bradley (1693–1762). Professor of astronomy at Oxford. He got his ideas about aberration by a chance observation of the changes in the apparent direction of the wind while sailing on the Thames.

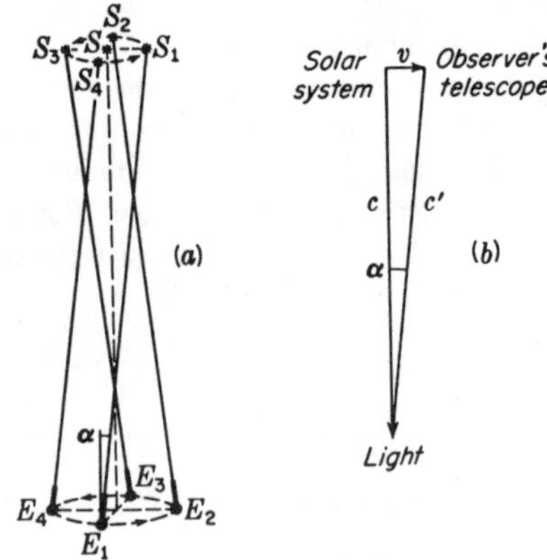

FIGURE 19B
Origin of astronomical aberration when the star is observed perpendicular to the plane of the earth's orbit.

be pointed to observe the star image on the axis of the instrument. We thus see that when the earth is at E_1 the star S has the apparent position S_1, when it is at E_2, the apparent position is S_2, etc. If S were not in a direction perpendicular to the plane of the earth's orbit, the apparent motion would be an ellipse rather than a circle, but the major axis of the ellipse would be equal to the diameter of the circle in the above case.

It will be seen from the figure that the angle α, which is the angular radius of the apparent circular motion, or the major axis of the elliptical one, is given by

$$\tan \alpha = \frac{v}{c} \tag{19a}$$

Recent measurements of this *angle of aberration* give a mean value $\alpha = 20.479''$ $\pm$ 0.008 as the angular radius of the apparent circular orbit. Combining this with the known velocity v of the earth in its orbit, we obtain 299,714 km/s. This value agrees to within its experimental error with the more accurate results obtained by the latest measurements of the speed of light by direct methods, the principles of which we shall now describe.

19.3 MICHELSON'S EXPERIMENTS

The first successful attempts to determine the speed of light, confined to the earth proper, were performed by Fizeau and Foucault in 1849. Their methods and apparatus, described in Sec. 1.2, were improved upon over a period of 80 years by Cornu, Young, Forbes, and Michelson. Of these the latest work by Michelson and his coworkers is considered to be by far the most accurate. Although it now appears that the accuracy of even the best values obtained by Michelson have been surpassed by that of newer methods based on radio-frequency techniques, it will be instructive

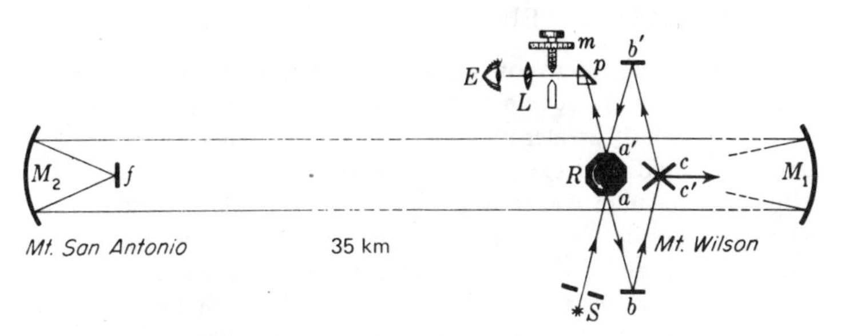

FIGURE 19C
Michelson's arrangement used for determining the speed of light (1926).

to consider, if only briefly, a classical series of measurements he made at the Mt. Wilson Observatory beginning in 1926.

The form of the apparatus Michelson adopted is shown in Fig. 19C. Light from a Sperry arc S passes through a narrow slit and is reflected from one face of the octagonal rotating mirror R. Thence it is reflected from the small fixed mirrors b and c to the large concave mirror M_1 (10-m focus, 60-cm aperture). This gives a parallel beam of light, which travels 35 km from the observing station on Mt. Wilson to a mirror M_2, similar to M_1, on the summit of Mt. San Antonio. M_2 focuses the light on a small plane mirror f, whence it returns to M_1 and, by reflection from c', b', a', and p, to the observing eyepiece L.

Various rotating mirrors, having 8, 12, and 16 sides, were used, and in each case the mirror was driven by an air blast at such a speed that during the time of transit to M_2 and back (0.00023 s) the mirror turned through such an angle that the next face was presented at a'. For an octagonal mirror, the required speed of rotation was about 528 rev/s. The speed was adjusted by a small counterblast of air until the image of the slit was in the same position as when R was at rest. The exact speed of rotation was then found by a stroboscopic comparison with a standard electrically driven tuning fork, which in turn was calibrated with an invar pendulum furnished by the U.S. Coast and Geodetic Survey. This Survey also measured the distance between the mirrors M_1 and M_2 with remarkable accuracy by triangulation from a 40-km base line, the length of which was determined to an estimated error of 1 part in 11 million, or about 3 mm.*

The results of the measurements published in 1926 comprised eight values of the speed of light, each the average of some 200 individual determinations with a given rotating mirror. These varied between the extreme values of 299,756 and 299,803 km/s and yielded the average value of 299,796 $\pm$ 4 km/s. Michelson also made some later measurements with the distant mirror on the summit of a mountain 130 km away, but because of bad atmospheric conditions, they were not considered reliable enough for publication.

* W. Bowie, *Astrophys. J.*, **65**:14 (1927).

19.4 MEASUREMENTS IN A VACUUM

In the preceding discussion we have assumed that the measured velocity in air is equal to that in a vacuum. That is not exactly true, since the index of refraction $n = c/v$ is slightly greater than unity. With white light the effective value of n for air under the conditions existing in Michelson's experiments was 1.000225. Hence the velocity in vacuum $c = nv$ was 67 km/s greater than v, the measured value in air. This correction has been applied in the final results quoted above. A difficulty which becomes important where measurements as accurate as those of Michelson are concerned is the uncertainty of the exact conditions of temperature and pressure of the air in the light path. Since n depends on these conditions, the value of the correction to vacuum also becomes somewhat uncertain.

To eliminate this source of error, Michelson in 1929 undertook a measurement of the velocity in a long evacuated pipe. The optical arrangement was similar to that described above, with suitable modifications for containing the light path in the pipe. The latter was 1.6 km long, and by successive reflections from mirrors mounted at either end the total distance the light traversed before returning to the rotating mirror was about 16 km. A vacuum as low as $\frac{1}{2}$ mmHg could be maintained. This difficult experiment was not completed until after Michelson's death in 1931, but preliminary results were published a year later by his collaborators.* The mean of almost 3000 individual measurements was 299,774 km/s. Because of certain unexplained variations, the accuracy of this result is difficult to assess. It is certainly not as great as that indicated by the computed probable error, and has recently been estimated as ± 11 km/s.

19.5 KERR-CELL METHOD

Determinations by this method have equaled if not surpassed the accuracy of those by the rotating mirror. In 1925 Gaviola devised what amounts to an improvement on Fizeau's toothed-wheel apparatus. It is based on the use of the so-called *electro-optic shutter*. This device is capable of chopping a beam of light several hundred times more rapidly than can be done by a cogwheel. Hence a much shorter base line can be used, and the entire apparatus can be contained in one building so that the atmospheric conditions are accurately known. Figure 19D(a) illustrates the electro-optic shutter, which consists of a Kerr cell K between two crossed nicol prisms N_1 and N_2. K is a small glass container fitted with sealed-in metal electrodes and filled with pure nitrobenzene. Although the operation of this shutter depends on certain properties of polarized light to be discussed later (Chap. 32), all that need be known here in order to understand the method is that no light is transmitted by the system until a high voltage is applied to the electrodes of K. Thus by using an electrical oscillator which delivers a radio-frequency voltage, a light beam can be interrupted at the rate of many millions of times per second.

The first measurements based on this principle used two shutters, one for the

* The final report will be found in A. A. Michelson, F. G. Pease, and F. Pearson, *Astrophys. J.*, **82**:26 (1935).

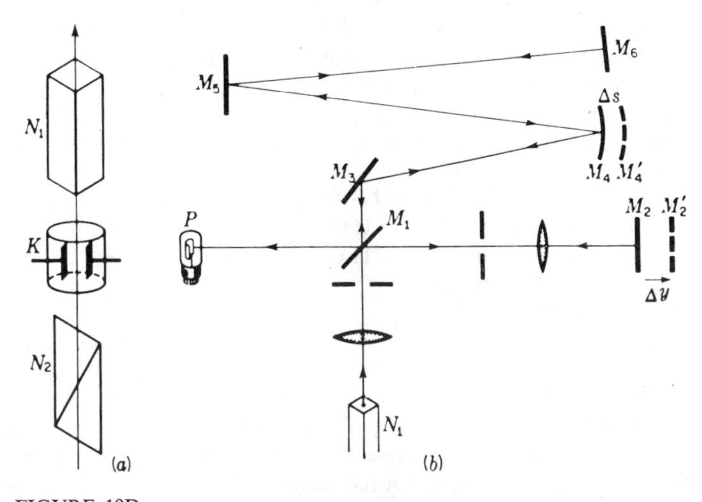

FIGURE 19D
Anderson's method of measuring the speed of light: (*a*) electro-optic shutter and (*b*) the light paths.

outgoing and one for the returning light. Except for the shorter distances, the method closely resembled Fizeau's. Subsequent improvements have led to the apparatus shown in Fig. 19D(*b*), which was used by W. C. Anderson in 1941.* To avoid the difficulty of matching the characteristics of two Kerr cells, he used only one and divided the transmitted light pulses into two beams by means of the half-silvered mirror M_1. One beam traversed the shorter path to M_2 and back through M_1 to the detector P. The other traveled a longer path to M_6 by reflections at M_3, M_4, and M_5, then retraced its course to M_1 which reflected it to P as well. This detector P was a photomultiplier tube, which responded to the sinusoidal modulation of the light waves. One may think of the light wave as the carrier wave, which is amplitude-modulated at the frequency of the oscillator driving the Kerr cell.† The quotient of the wavelength l of the modulation by the period T of the oscillator thus gives the velocity of light.

The accurate measurement of l is based on the following principle. If the longer path exceeds the shorter one by a half-integral multiple of l, the sum of the two modulated waves reaching P will give a constant intensity. The amplifier connected to the photocell was arranged to give zero response under this condition. The adjustment is made by slight motions Δy of the mirror M_2. The extra path beyond M_4 could then be cut out by substituting another mirror M_4' which returned the light directly to M_3. If this extra path (M_4 to M_6 and back) were exactly a whole number

* *J. Opt. Soc. Am.*, **31**:187 (1941).

† Since the shutter transmits at each voltage peak, whether positive or negative, one would expect to use $1/2T$ here. Actually Anderson applied a dc bias to the cell so that each cycle gave a single voltage maximum.

times l, no change in the photocell response would be observed upon cutting it out. As the apparatus was arranged, this was very nearly so, the extra path being about $11l$. By measuring the displacement Δy necessary to reestablish zero response and applying a correction Δs involved in the substitution of M_4', the difference from $11l$ of the measured distance could be exactly determined. Typical results are:

Total path difference = 171.8642 m
Refractive index, air = 1.0002868
$\Delta s = 2.4770$ cm $\quad n = 11.0$
$f = 19.20 \times 10^6$ Hz
$c = 299{,}778$ km/s

The reader will see the resemblance of Anderson's apparatus to a Michelson interferometer for radio waves, since the light pulses have a length essentially equal to the wavelength of the radio waves given by the Kerr-cell oscillator. It is not exactly equal, however, because the speed involved in the experiment is the group velocity of light in air and not the velocity of radio waves. In his final investigation, Anderson made a total of 2895 observations, and the resulting speeds l/T, after correction to vacuum, yielded an average of $299{,}776 \pm 6$ km/s.. The chief source of error was in the difficulty of ensuring that both beams used the same portion of the photoelectric surface. A change in the position of the light spot affects the time of transit of the electrons between the electrodes of the photomultiplier tube. The uncertainty involved here was larger than any errors in the length measurements, and if the frequency of the oscillator were known more accurately than it was, the uncertainty in the final result would be better than 1 part in 1 million.

In the 1951 Kerr-cell determination by Bergstrand (see Table 19A) the last-mentioned difficulty is avoided by using only one beam, and locating the maxima and minima through modulation of the detector in synchronism with the source. The result is indicated to be more than 10 times as accurate as any previous one by optical methods. It disagrees with the concordant values of Anderson and of Michelson, Pease, and Pearson, seeming to show that Michelson's 1926 value was the more nearly correct. It is difficult to understand how the very thorough work in the period 1930–1940 could have been so far in error, but other recent results, to be described below, certainly put the weight of the evidence in favor of the higher value of c.

19.6 SPEED OF RADIO WAVES

The development of modern radar techniques, and especially the interest in their practical application as navigational aids, has led to renewed attempts to improve our knowledge of the speed of light. This speed is of course the same as that for radio waves when both are reduced to vacuum. There are three methods for using microwaves for an accurate measurement of their speed, one of which may easily be performed in vacuum. This is to find the length and resonant frequency of a hollow metal cylinder, or cavity resonator. It is analogous to the common laboratory method for the speed of sound. Measurements of this type were made independently in England, by Essen and Gordon-Smith, and in America, by Bol.* As will be seen from Table 19A, the results agree with each other and with Bergstrand's precise optical value.

The other methods involving radio waves are responsible for the last two entries in our table, and have been developed to a comparable accuracy. The radar method consists in the direct measurement of the time of transit of a signal over a known distance in the open air. The microwave interferometer is the Michelson instrument adapted to radio waves. The speed is found by measuring the wavelength from the motion of a mirror. The details of all the radio methods are interesting and important but must be omitted here as not falling strictly within the scope of optics.

19.7 RATIO OF THE ELECTRICAL UNITS

As we shall find in our consideration of the electromagnetic theory (Chap. 20), c can be found from the ratio of the magnitude of certain units in the electromagnetic and electrostatic systems. Two careful measurements of the ratio have been made and have given results more or less intermediate between the higher and lower values discussed above. Since the accuracy thus far attained is considerably lower than for other methods, these experiments, although they have served to verify the theoretical prediction, have not improved our knowledge of the speed of light.†

19.8 THE SPEED OF LIGHT IN STATIONARY MATTER

A brief description of the early experiments by Foucault in 1850 on the speed of light in stationary matter was given in Chap. 1 (see Fig. 1D).

Much more accurate measurements were made by Michelson in 1885. Using white light, he found for the ratio of the speed in air to that in water a value of 1.330.

* Valuable summaries of the determinations of c, and many original references not given here, will be found in L. Essen, *Nature*, **165:**583 (1950 and K. D. Froome, *Proc. Roy. Soc.* (*Lond.*), **A213:**123 (1952).

† The indirect measurements all antedate the determinations in Table 19A. They have been critically reviewed by R. T. Birge, *Nature*, **134:**771 (1934).

Table 19A RESULTS OF ACCURATE MEASUREMENT OF THE SPEED OF LIGHT

Date	Investigators	Method	Result, m s
1926	Michelson	Rotating mirror	299,796 ± 4
1935	Michelson, Pease, and Pearson	Rotating mirror in vacuum	299,774 ± 11
1940	Hüttel	Kerr cell	299,768 ± 10
1941	Anderson	Kerr cell	299,776 ± 6
1950	Bol	Cavity resonator	299,789.3 ± 0.4
1950	Essen	Cavity resonator	299,792.5 ± 3.0
1951	Bergstrand	Kerr cell	299,793.1 ± 0.2
1951	Alakson	Radar (shoran)	299,794.2 ± 1.9
1951	Froome	Microwave interferometer	299,792.6 ± 0.7

A denser medium, carbon disulfide, gave 1.758. In the latter case he noticed that the final image of the slit was spread out into a short spectrum, which could be explained by the fact that red light travels faster than blue light in the medium. The difference in speed between greenish-blue and reddish-orange light was observed to be 1 or 2 percent.

According to the wave theory of light, the index of refraction of a medium is equal to the ratio of the speed of light in vacuum to that in the medium. If we compare the above figures with the corresponding indices of refraction for white light (water 1.334, carbon disulfide 1.635), we find that while the agreement is within the experimental error for water, the directly measured value is considerably higher than the index of refraction for carbon disulfide.

This discrepancy is readily explained by the fact that the index of refraction represents the ratio of the *wave velocities* in vacuum and in the medium ($n = c/v$), while the direct measurements give the *group velocities*. In a vacuum the two speeds become identical (Sec. 12.7) and equal to c, so that if we call the group velocity in the medium u, the ratios determined by Michelson were values of c/u, rather than c/v. The two velocities u and v are related by the general equation (12p)

$$u = v - \lambda \frac{dv}{d\lambda}$$

The variation of v with λ can be found by studying the change of the index of refraction with color (Sec. 23.2); v is greater for longer wavelengths, so that $dv/d\lambda$ is positive. Therefore u should be less than v, and this is precisely the result obtained above. Using reasonable values for λ and $dv/d\lambda$ for white light, the difference between the two values for carbon disulfide is in agreement with the theory to within the accuracy of the experiments. For water $dv/d\lambda$ is considerably smaller but nevertheless requires the measured value of c/u to be 1.5 percent higher than c/v. That this is not so indicates an appreciable error in Michelson's work. The latest work* on the speed of light in water has given agreement not only on the magnitude of the group velocity but also on its variation with wavelength.

At this point it should be emphasized that all the direct methods for measuring the speed of light that we have described give the group velocity u and not the wave velocity v. Even though it is not evident in the aberration experiment that the wave is divided into groups, it should be obvious that since all natural light consists of wave packets of finite length, any further chopping or modulation is immaterial. In air the difference between u and v is small but nevertheless amounts to 2.2 km/s. Michelson apparently did not apply this correction to his 1926 value, which should therefore have been quoted as 299,798 + 4 km/s.

19.9 SPEED OF LIGHT IN MOVING MATTER

In 1859 Fizeau performed an important experiment to determine whether the speed of light in a material medium is affected by motion of the medium relative to the source and observer. In Fig. 19E the light from S is split into two beams, in much

* R. A. Houstoun, *Proc. R. Soc. Edinb.*, **A62**:58 (1944).

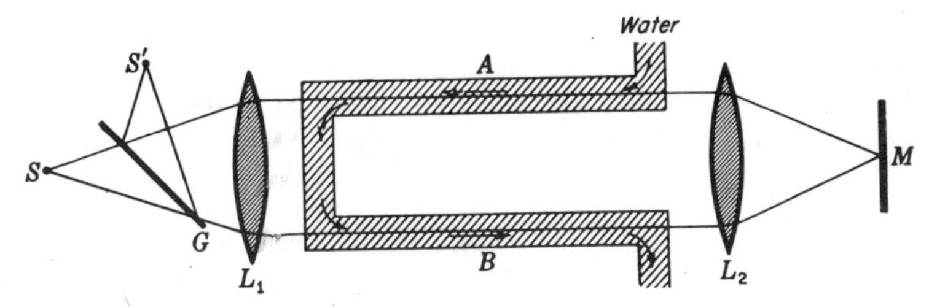

FIGURE 19E
Fizeau's experiment for measuring the speed of light in a moving medium.

the same way as in the Rayleigh refractometer (Sec. 13.15). The beams then pass through the tubes A and B containing water flowing rapidly in opposite directions. On reflection from M, the beams interchange so that when they reach L_1 one has traversed both B and A in the same direction as the flowing water while the other has traversed A and B in the opposite direction to the flow. The lens L_1 then brings the beams together to form interference fringes at S'.

If the light travels more slowly by one route than by the other, its optical path has effectively increased and a displacement of the fringes should occur. Using tubes 150 cm long and a water speed of 700 cm/s, Fizeau found a shift of 0.46 fringe when the direction of flow was reversed. This corresponds to an increase in the speed of light in one tube, and a decrease in the other, of about half of the speed of the water.

This experiment was later repeated by Michelson with improved apparatus consisting essentially of an adaptation of his interferometer to this type of measurement. He observed a shift corresponding to an alteration of the speed of light by 0.434 times the speed of the water.

19.10 FRESNEL DRAGGING COEFFICIENT

The above results were compared with a formula derived by Fresnel in 1818, using the elastic-solid theory of the ether. On the assumption that the density of the ether in the medium is greater than that in vacuum in the ratio n^2, he showed that the ether is effectively dragged along with a moving medium with a speed

$$v' = v\left(1 - \frac{1}{n^2}\right) \qquad (19b)$$

where v is the speed of the medium and n its index of refraction. For water, which has $n = 1.333$ for sodium light, this gives $v' = 0.437v$, in reasonable agreement with Michelson's value for white light quoted in the previous paragraph. The fraction $1 - 1/n^2$ will be referred to as *Fresnel's dragging coefficient*.

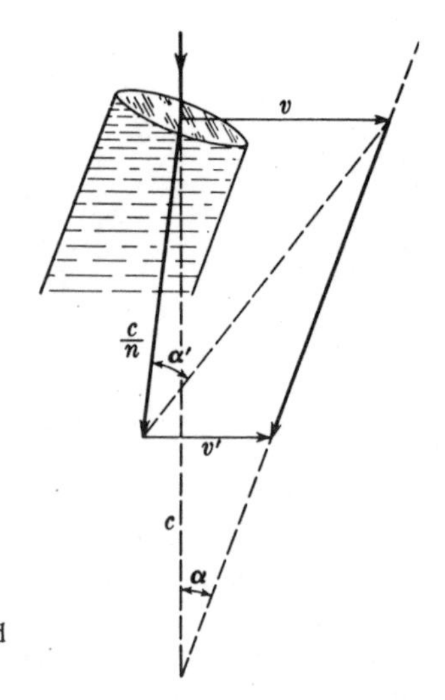

FIGURE 19F
Angle of aberration with a water-filled telescope.

19.11 AIRY'S EXPERIMENT

An entirely different piece of experimental evidence shows that Fresnel's equation must be very nearly correct. In 1872 Airy remeasured the angle of aberration of light (Sec. 19.2), using a telescope filled with water. Upon referring to Fig. 19B(*b*), it will be seen that if the velocity of the light with respect to the solar system is made less by entering water, one would expect the angle of aberration to be increased. Actually the most careful measurements gave the same angle of aberration for a telescope filled with water as for one filled with air.

This negative result can be explained by assuming that the light is carried along by the water in the telescope with the velocity given by Eq. (19b). In Fig. 19F, where the angles are of course greatly exaggerated, the velocity now becomes c/n and is slightly changed in direction by refraction. If one is to observe the ordinary angle of aberration α, it is necessary to add to this velocity the extra component v', representing the velocity with which the light is dragged by the water. From the geometry of this figure it is possible to prove that v' must obey Eq.(19b). The proof will not be given here, however, since a different and simpler explanation is now accepted, based on the theory of relativity (see Sec. 19.15).

19.12 EFFECT OF MOTION OF THE OBSERVER

We have seen that in the phenomenon of aberration the apparent *direction* of the light reaching the observer is altered when he is in motion. One might therefore expect to be able to find an effect of such motion on the *magnitude* of the observed velocity of light. Referring back to Fig. 19B(*b*), we see that the apparent velocity $c' = v/(\sin \alpha)$ is slightly greater than the true velocity $c = v/(\tan \alpha)$. However, α is a

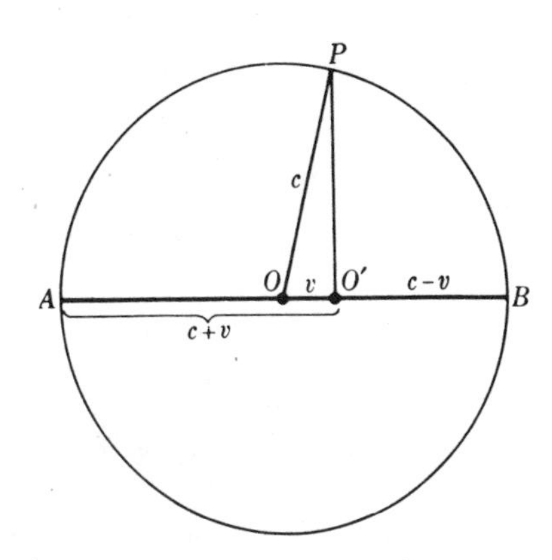

FIGURE 19G
Velocity of light emitted by a moving source.

very small angle, so that the difference between the sine and the tangent is much smaller than the error of measurement of α. A somewhat different experiment embodying the same principle has been devised, which should be sensitive enough to detect this slight change in the apparent velocity if it exists. Before describing this experiment, however, we consider in more detail the effect of motion of the observer on the apparent velocity of light.

In Fig. 19G, let the observer at O be moving toward B with a velocity v. Let an instantaneous flash of light be sent out at O. The wave will spread out in a circle with its center at O, and after 1 s the radius of this circle will be numerically equal to the speed of light c. But during this time the observer will have moved a distance v from O to O'. Hence if the observer were in some way able to follow the progress of the wave, he would find an apparent velocity which would vary with the direction of observation. In the forward direction $O'B$ it would be $c - v$ and in the backward direction $O'A$ it would be $c + v$. At right angles, in the direction $O'P$, he would observe a velocity $\sqrt{c^2 - v^2}$.

It is important to notice that in drawing Fig. 19G we have assumed that the velocity of the light is not affected by the fact that the source was in motion as it emitted the wave. This is to be expected for a wave which is set up in a stationary medium, e.g., a sound wave in the air. The hypothetical medium carrying light waves is the ether, and if v is the velocity with respect to the ether, the same result is expected. For an experiment performed in air, the Fresnel dragging coefficient $1 - 1/n^2$ is so nearly zero that it may be neglected. Thus if the observer were moving with the velocity v of the earth in its orbit, these considerations lead us to expect the changes in the apparent velocity of light described above. Effectively the ether should be moving past the earth with a velocity v, and if any effects on the velocity of light were found, they could be said to be due to an ether wind or *ether drift*. It would not be surprising if this drift did not correspond to the velocity of the earth in its orbit, since we know that the solar system as a whole is moving toward the constellation Hercules with a velocity of 19 km/s and it is more reasonable to expect the ether to be at rest with respect to the system of fixed stars than with respect to our solar system.

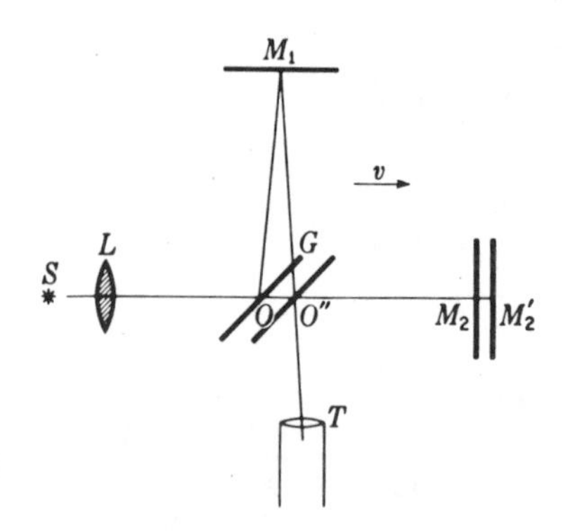

FIGURE 19H
The Michelson interferometer as a test for ether drift.

19.13 THE MICHELSON-MORLEY EXPERIMENT

This experiment, perhaps the most famous of any experiment with light, was undertaken in 1881 to investigate the possible existence of ether drift. In principle it consisted merely of observing whether there was any shift of the fringes in the Michelson interferometer when the instrument was turned through an angle of 90°. Thus in Fig. 19H let us assume that the interferometer is being carried along by the earth in the direction OM_2 with a velocity v with respect to the ether. Let the mirrors M_1 and M_2 be adjusted for parallel light, and let $OM_1 = OM_2 = d$. The light leaving O in the forward direction will be reflected when the mirror is at M_2' and will return when the half-silvered mirror G has moved to O''. Using the expressions for the velocity derived in the previous section, the time required to travel the path $OM_2'O''$ will be

$$T_1 = \frac{d}{c+v} + \frac{d}{c-v} = \frac{2cd}{c^2 - v^2}$$

and the time to travel OM_1O'' will be

$$T_2 = \frac{2d}{\sqrt{c^2 - v^2}}$$

Each of these expressions can be expanded into series, giving

$$T_1 = \frac{2cd}{c^2 - v^2} = \frac{2d}{c}\left(1 + \frac{v^2}{c^2} + \frac{v^4}{c^4} + \cdots\right) \approx \frac{2d}{c}\left(1 + \frac{v^2}{c^2}\right)$$

and

$$T_2 = \frac{2d}{\sqrt{c^2 - v^2}} = \frac{2d}{c}\left(1 + \frac{v^2}{2c^2} + \frac{3v^4}{4c^4} + \cdots\right) \approx \frac{2d}{c}\left(1 + \frac{v^2}{2c^2}\right)$$

Thus the result of the motion of the interferometer is to increase both paths by a slight amount, the increase being twice as large in the direction of motion. The difference in time, which would be zero for a stationary interferometer, now becomes

$$T_1 - T_2 = \frac{2d}{c}\left(1 + \frac{v^2}{c^2}\right) - \frac{2d}{c}\left(1 + \frac{v^2}{2c^2}\right) = d\frac{v^2}{c^3}$$

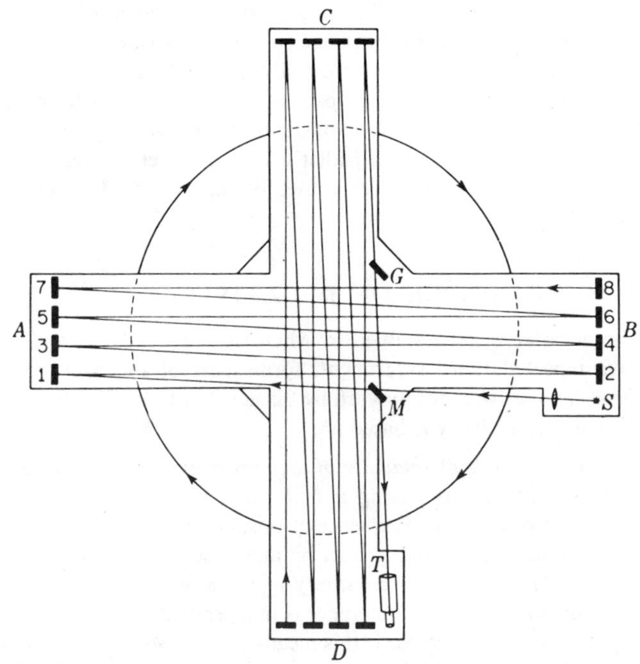

FIGURE 19I
Miller's arrangement of the Michelson-Morley experiment to detect ether drift.

To change this to path difference we multiply by c, obtaining

$$\Delta = d\frac{v^2}{c^2} \tag{19c}$$

If now the interferometer is turned through 90°, the direction of v is unchanged but the two paths in the interferometer will be interchanged. This would introduce a path difference Δ in the opposite sense to that obtained before. Hence we expect a shift corresponding to a change of path of $2dv^2/c^2$.

Michelson and Morely made the distance d large by reflecting the light back and forth between 16 mirrors as illustrated in Fig. 19I. To avoid distortion of the instrument by strains, it was mounted on a large concrete block floating in mercury, and observations were made as it was rotated slowly and continuously about a vertical axis. In one experiment d was 11 m, so that if we take $v = 29.9$ km/s and $c = 300{,}000$ km/s, we find a change in path of 2.2×10^{-5} cm. For light of wavelength 6×10^{-5}, this corresponds to a change of four-tenths of a wavelength, so that the fringes should be displaced by two-fifths of a fringe. Careful observations showed that no shift occurred as great as 10 percent of this predicted value.

This negative result, indicating the absence of an ether drift, was so surprising that the experiment has since been repeated with certain modifications by a number of different investigators. All have confirmed Michelson and Morley in showing that if a real displacement of the fringes exists, it is at most but a small fraction of the expected value. The most extensive series of measurements has been made by D. C. Miller. His apparatus was essentially that of Michelson and Morley (Fig. 19I) but on a larger scale. With a light path of 64 m, Miller thought he had obtained evidence for a small shift of about one-thirtieth of a fringe, varying periodically with sidereal time. The latest analysis of Miller's data, however, makes it probable that the result is not significant, having been caused by slight thermal gradients across the interferometer.*

19.14 PRINCIPLE OF RELATIVITY

The negative result obtained by Michelson and Morley, and by most of those who have repeated their experiment, forms part of the background for the restricted *theory of relativity*, put forward by Einstein† in 1905. The two fundamental postulates on which this theory is based are

1 Principle of relativity of uniform motion The laws of physics are the same for all systems having a uniform motion of translation with respect to one another. As a consequence of this, an observer in any one system cannot detect the motion of that system by any observations confined to the system.

2 Principle of the constancy of the velocity of light The velocity of light in any given frame of reference is independent of the velocity of the source. Combined with principle 1, this means that the velocity of light is independent of the relative velocity of the source and observer.

Returning to our illustration (Fig. 19G) of an observer who sends out a flash of light at O while moving with a velocity v, the above postulates would require that any measurements made by the observer at O' would show that he is the center of the spherical wave. But an observer at rest at O would find that he too is at the center of the wave. The reconciliation of these apparently contradictory statements lies in the fact that the space and time scales for the moving system are different from those for a fixed system. Events separated in space which are simultaneous to an observer at rest do not appear so to one moving with the system.

The first explanation given for the null result of the Michelson-Morley experiment was that the arm of the interferometer that was oriented parallel to the earth's motion was decreased in length because of this motion. The so-called *Fitzgerald-Lorentz contraction* required that, if l_0 is the length of an object at rest, motion parallel to l_0 with a velocity v gives a new length

● $$l = \sqrt{l_0 (1 - v^2/c^2)} \qquad \text{(19d)}$$

This law would satisfy the condition that the difference in path due to ether drift would be just canceled out. Naturally the change in length could not be detected

* R. S. Shankland, S. W. McCuskey, F. C. Leone, and G. Kuerti, *Rev. Mod. Phys.*, **27**:167 (1955).

† Albert Einstein (1879–1955). Formerly director of the Kaiser Wilhelm Institute in Berlin, Einstein in 1935 came to the Institute for Advanced Study at Princeton. Gifted with one of the most brilliant minds of our times, he contributed to many fields of physics besides relativity. Of prime importance was his famous law of the photoelectric effect. He received the Nobel prize in 1921.

by a measuring stick, since the latter would shrink in the same proportion. A contraction of this kind, however, should bring about changes in other physical properties. Many attempts have been made to find evidence for these, but to no avail. According to the first postulate of relativity, they must fail. The ether drift does not exist, nor is there any contraction for an observer moving along with the interferometer.

Starting from the fundamental postulates of the restricted theory, it is possible to show that in a frame of reference that is moving with respect to the observer there should actually be changes in the observed values of length, mass, and time. The mass of a particle becomes

● $$m = m_0 \left(1 - \frac{v^2}{c^2}\right)^{-\frac{1}{2}} \qquad (19e)$$

in which m_0 represents the mass when it is at rest with respect to the observer. If light, which has $v = c$, were regarded as consisting of particles (see Chap. 32), they would have to have zero rest mass since otherwise m becomes infinite. Experimental measurements have been made, mostly with high-speed electrons, which quantitatively verify Eq. (19*e*). Other observable consequences of relativity theory exist, the most striking ones being obtained when it is extended to cover accelerated systems as well as systems in uniform motion.* From this *general theory* of relativity, predictions are made with regard to the deflection of light rays passing close to the sun, and to a decrease in frequency of light emitted by atoms in a strong gravitational field. Accurate measurements of the apparent positions of stars during a total solar eclipse and of the spectra of very dense (white dwarf) stars, have verified these two optical effects.

These experimental proofs of the theory have been sufficiently convincing to lead to the general acceptance of the correctness of the general theory of relativity. While the theory does not directly deny the existence of the ether postulated by Fresnel, it says very definitely that no experiment we can ever perform will prove its existence. For if it were possible to find the motion of a body with respect to the ether, we could regard the ether as a fixed coordinate system with respect to which all motions are to be referred. But it is one of the fundamental consequences of relativity that any coordinate system is equivalent to any other, and no one has any particular claim to finality. Thus, since a fixed ether is apparently not observable, there is no reason for retaining the concept. It cannot be denied, however, that it is historically important and that some of the most important advances in the study of light have come through the assumption of a material ether.

19.15 THE THREE FIRST-ORDER RELATIVITY EFFECTS

There are three optical effects the magnitude of which depends on the first power of v/c. They are

1 The doppler effect

* For a general account of the theory and its consequences, see R. C. Tolman, "Relativity, Thermodynamics and Cosmology," Oxford University Press, New York, 1949. See also Harvey E. White, "Modern College Physics," 6th ed., D. Van Nostrand Co., New York, 1973.

2 The aberration of light
3 The Fresnel dragging coefficient

Equations for these effects have been derived on the basis of classical theory in Secs. 11.10, 19.2, and 19.10. It is characteristic of the theory of relativity that it yields the same results for first-order effects as does the classical theory. Only in second-order effects, which depend on v^2/c^2, do the predictions of the two theories differ. The Michelson-Morley experiment belongs to this class. Even for the first-order effects listed above, the results from the two theories differ in the small terms of the second and higher power of v/c. In the relativity theory, these equations are derived by applying the *Lorentz transformation.* This is a process of translating the description of a motion in terms of one system of coordinates into a description of the same motion in terms of another system which is in uniform motion with respect to the first. Although it is not practicable to give the mathematics of this process here, we shall state the chief results and discuss them briefly.

When the equation for a periodic wave of frequency ν is rewritten in the coordinates of the observer's frame of reference, the frequency assumes a new value given by

$$\nu' = \nu \frac{\sqrt{1 - v^2/c^2}}{1 - v/c} = \nu \left(1 + \frac{v}{c} + \frac{1}{2}\frac{v^2}{c^2} + \frac{1}{2}\frac{v^3}{c^3} + \cdots\right) \qquad \text{(19f)}$$

This is the doppler effect for the source and observer approaching each other with a velocity v along the line joining them. Comparison of the series expansion with our previous Eq. (11z) shows that the prediction from relativity differs from that of the classical theory only in the terms of second and higher orders. Theoretically these arise from the fact that the rate of a moving clock is slower than that of a stationary one. Ives* has given an elegant demonstration of this fact by comparing the frequency of the radiation emitted by hydrogen atoms in a high-speed beam moving first toward the spectroscope, then away from it. In addition to the large first-order shifts of the line toward higher and lower frequencies respectively in these two cases, he observed and measured a small additional shift which was toward higher frequencies in both cases. Since the term in question contains the square of the velocity, it will be the same for either sign of v. This experiment constitutes another verification of the theory of relativity by observation of a second-order effect which does not exist according to the classical theory. It might also be mentioned that relativity predicts a second-order doppler shift even when the source is moving at right angles to the line of sight.

The interpretation of the aberration of light and of Airy's experiment is simpler from the relativistic point of view. According to the second fundamental postulate, the speed of light must always be c to any observer, regardless of his motion. Hence, referring to Fig. 19B(*b*), the observed velocity labeled c' must now be labeled c. The formula for the angle of aberration, instead of being $\tan \alpha = v/c$, then becomes

$$\sin \alpha = \frac{v}{c} \qquad \text{(19g)}$$

* H. E. Ives and A. R. Stilwell, *J. Opt. Soc. Am.*, **28**:215 (1938); **31**:369 (1941).

It is well known that the sine and the tangent differ only in respect to terms of the third and higher orders. Here the angle is so small that in all likelihood the difference will never be detected. In Airy's experiment, the expectation of observing an increase of the angle when the telescope was filled with water arose from the assumption that the water would decrease the velocity of the light with respect to the solar system, in which the ether was regarded as fixed. But according to the point of view of relativity the only "true" speed of light is its velocity in the coordinate system of the observer, and this is inclined at the angle α given by Eq. (19g). Hence reducing the magnitude of this speed by allowing the light to enter water will obviously make no change in its direction.

A positive effect corresponding to Fresnel's ether drag can be observed when the medium is in motion with respect to the observer (Sec. 19.10), but its interpretation by the theory of relativity is entirely different. One result of the Lorentz transformation is that two velocities in coordinate systems that are in relative motion do not add according the methods used in classical mechanics. For example, the resultant of two velocities in the same line is not their arithmetic sum. Let us call V_0 the velocity of light in the coordinate system of a moving medium and v the velocity of this medium in the observer's coordinate system. Then the resultant velocity V of the light with respect to the observer, instead of being merely $V_0 + v$, must be taken as

$$V = \frac{V_0 + v}{1 + (V_0/c)(v/c)} \tag{19h}$$

The student can easily verify the fact that this equation gives the same velocity V for any observer in motion with the velocity v, in the case that $V_0 = c$, that is, in a vacuum. The expression for the Fresnel dragging coefficient follows at once from Eq. (19h) if one neglects second-order terms. Thus the binomial expansion gives

$$V = (V_0 + v)\left(1 - \frac{V_0}{c}\frac{v}{c} - \cdots\right) = V_0 + v - \frac{V_0{}^2 v}{c^2} - \frac{v^2 V_0}{c^2} - \cdots$$

The last term is again a quantity of the second order and is to be neglected. Then we obtain, by substituting n for c/V_0,

$$V = \frac{c}{n} + v\left(1 - \frac{1}{n^2}\right) \tag{19i}$$

The velocity as seen by the observer is changed by the fraction $1 - 1/n^2$, which is just the value required by Eq. (19b). No assumption of any "dragging" is involved in the relativity arguments, nor is the existence of an ether even postulated.

PROBLEMS

19.1 Assuming the speed of light to be 299,793 km/s and the average radius of the earth's orbit around the sun to be 1.49670 × 10^8 km, calculate (*a*) the circumference of the earth's orbit and (*b*) the earth's period in seconds. Calculate (*c*) the earth's average orbital speed in kilometers per second and the maximum angle of aberration of a star in (*d*) degrees and (*e*) seconds of arc. Assume the earth's period to be 365.241 mean solar days.

Ans. (*a*) 9.40404×10^8 km, (*b*) 3.155682×10^7 s, (*c*) 29.80034 km/s, (*d*) 0.00569538°, (*e*) 20.34 seconds of arc

19.2 At the present time it is probably more correct to regard the measurements of astronomical aberration as the determination of the earth's speed than it is the speed of light. Using the value of the angle of aberration given in Sec. 19.2 and Michelson's 1926 value of c, compute the orbital speed of the earth to five figures, (*a*) in kilometers per second and (*b*) in meters per second.

19.3 When Michelson used a 12-sided mirror in his experiment on the speed of light, the image was reflected to its initial position from adjacent faces. Find the distance between the two markers on the two mountain tops, Mt. Wilson and Mt. San Antonio, if the speed of revolution was exactly 352 rev/s. Assume the most probable value of the speed of light to be 299,792.5 km/s.

19.4 In the speed-of-light experiments Michelson, Pease, and Pearson used a long vacuum pipe and a rotating mirror prism with 32 sides. Assuming that the total path the light had to travel was 13.2870 km and that the speed of light is 299,793 km/s, find the speed of rotation of the mirror prism to obtain the first undisplaced image.

Ans. 705.090 rev/s

19.5 If Anderson's Kerr-cell apparatus was arranged so that the total path difference was 171.6985 m and contained 11 wave groups, find (*a*) the length l of one wave group. If the calculated speed is given by lf, find (*b*) the speed c_{air}, (*c*) the speed of light in a vacuum c, and (*d*) the correction from c_{air} to c in kilometers per second. Assume the refractive index of air at the particular time to be 1.0002868 and the frequency of the oscillator to be 19.20 MHz.

19.6 Verify the statement given in Sec. 19.9 that a fringe shift of 0.460 in Fizeau's experiment corresponds to a change in the speed of light by about half the speed of the water flow. Assuming that the effective wavelength of light is 5500 Å and that the refractive index of water is 1.3330, find what fraction it actually gives.

19.7 Carbon disulfide has a refractive index of $n_D = 1.62950$ and a dispersion $dn/d\lambda = -1820\ \text{cm}^{-1}$ at this wavelength. Find (*a*) the ratio of the speed of light in a vacuum to the group velocity in carbon disulfide and (*b*) the exact value of the Fresnel dragging coefficient for this substance. Equation (19b) needs a small correction arising from the fact that for the molecules of moving water the effective frequency is slightly altered by the doppler effect. Prove (*c*) that this can be taken into account by adding a term $-(dn/d\lambda)(\lambda/n)$ to the expression for the dragging coefficient. Here λ is the wavelength in a vacuum. *Hint:* Take the refractive index to vary linearly with frequency and insert the new index, as altered by the doppler effect, in the equation for the velocity of light in the moving medium. *Ans.* (*a*) 1.7367, (*b*) 0.6892

19.8 Suppose a meterstick is moving lengthwise past an observer at 30 percent the speed of light. Find its apparent length in centimeters.

19.9 Find the apparent mass of an electron moving past an observer at one-third the speed of light. Assume the rest mass of the electron to be 9.1096×10^{-31} kg.

19.10 A spaceship with a mass of 6.250×10^6 kg and length of 35.20 m passes the earth with a velocity of 25 percent the speed of light. Find (*a*) its apparent mass and (*b*) its apparent length. *Ans.* (*a*) 6.455×10^6 kg, (*b*) 34.082 m

20

THE ELECTROMAGNETIC CHARACTER OF LIGHT

Our study of the properties of light has thus far led us to the conclusion that light is a wave motion, propagated with an extremely high speed. In the explanation of interference and diffraction it was not necessary to make any assumption about the nature of the displacement y that appears in our wave equations because in these subjects we were concerned only with the interaction of light waves with each other. In the succeeding chapters we are to consider subjects in which the interaction of light with matter plays a part, and here it becomes necessary to specify the physical nature of the quantity y, which is usually termed the *light vector*. Fresnel, who in 1814 first gave the satisfactory explanation of interference and diffraction by the wave theory, imagined the light vector to represent an actual displacement of a material ether, which was conceived as an all-pervading substance of very small density and of high rigidity. This "elastic-solid" theory had considerable success in interpreting optical phenomena and was strongly supported by many leading investigators in the field, such as Lord Kelvin, as late as 1880.

20.1 TRANSVERSE NATURE OF LIGHT VIBRATIONS

The principal objection to the elastic-solid theory lay in the fact that light had been proved to be exclusively a transverse wave motion, i.e., the vibrations are always perpendicular to the direction of motion of the waves. No longitudinal waves of light have ever been detected. The experimental evidence for this comes from the study of the polarization of light (Chap. 24) and is perfectly definite, so that we may here take the fact as established. Now all elastic solids with which we are familiar are capable of transmitting longitudinal as well as transverse waves; in fact, under some circumstances it is impossible to set up a transverse wave without at the same time starting a longitudinal one. Many suggestions were made to overcome this difficulty, but all were highly artificial. Furthermore, the idea of a material ether itself seemed rather forced, inasmuch as its remarkable properties could not be detected by ordinary mechanical experiments.

Thus the time was ripe when Maxwell* proposed a theory which not only *required* the vibrations of light to be strictly transverse but also gave a definite connection between light and electricity. In a paper read before the Royal Society in 1864, entitled A Dynamical Theory of the Electromagnetic Field, Maxwell expressed the results of his theoretical investigations in the form of four fundamental equations which have since become famous as *Maxwell's equations*. They were based on the earlier experimental researches of Oersted, Faraday, and Joseph Henry concerning the relations between electricity and magnetism. They summarize these relations in concise mathematical form, and constitute a starting point for the investigation of all electromagnetic phenomena. We shall show in the following sections how they accounted for the transverse waves of light.

20.2 MAXWELL'S EQUATIONS FOR A VACUUM

The derivation of these equations will not be given here, since it would involve a rather extensive review of the principles of electricity and magnetism.† Instead we shall in this chapter merely state the equations in their simplest form, applicable to empty space, and then prove that they predict the existence of waves having the properties of light waves. The modifications that must be introduced in dealing with different kinds of material media will be considered at the appropriate places in the following chapters.

Maxwell's equations may be written as four vector equations, but for those

* James Clerk Maxwell (1831–1879). Professor of experimental physics at Cambridge University, England. Contributed a paper to the Royal Society at the age of fifteen. Much of his work on the electromagnetic theory was accomplished while an undergraduate at Cambridge. His investigations in many fields of physics bear the stamp of genius. The kinetic theory of gases was given a solid mathematical foundation by Maxwell, whose name is associated with the well-known law of distribution of molecular velocities.

† For a derivation of Maxwell's equations in mks units, see E. Hecht and A. Zajac, "Optics," pp. 29–37, 509, Addison-Wesley Publishing Company, Inc., Reading, Mass.

unfamiliar with vector notation we shall express them by differential equations. In this form the first two equations must be expressed by two sets of three equations each. For a vacuum these become, using a right-handed set of coordinates,

$$\begin{aligned} \frac{1}{c}\frac{\partial E_x}{\partial t} &= \frac{\partial H_z}{\partial y} - \frac{\partial H_y}{\partial z} \\ \frac{1}{c}\frac{\partial E_y}{\partial t} &= \frac{\partial H_x}{\partial z} - \frac{\partial H_z}{\partial x} \\ \frac{1}{c}\frac{\partial E_z}{\partial t} &= \frac{\partial H_y}{\partial x} - \frac{\partial H_x}{\partial_y} \end{aligned} \qquad (20a) \qquad \begin{aligned} -\frac{1}{c}\frac{\partial H_x}{\partial t} &= \frac{\partial E_z}{\partial y} - \frac{\partial E_y}{\partial z} \\ -\frac{1}{c}\frac{\partial H_y}{\partial t} &= \frac{\partial E_x}{\partial z} - \frac{\partial E_z}{\partial x} \\ -\frac{1}{c}\frac{\partial H_z}{\partial t} &= \frac{\partial E_y}{\partial x} - \frac{\partial E_x}{\partial y} \end{aligned} \qquad (20b)$$

The other two equations may be written

$$\frac{\partial E_x}{\partial x} + \frac{\partial E_y}{\partial y} + \frac{\partial E_z}{\partial z} = 0 \qquad (20c) \qquad \frac{\partial H_x}{\partial x} + \frac{\partial H_y}{\partial y} + \frac{\partial H_z}{\partial z} = 0 \qquad (20d)$$

These partial differential equations give the relations in space and time between the vector quantities **E**, the electric field strength, and **H**, the magnetic field strength. Thus E_x, E_y, and E_z are the components of **E** along the three rectangular axes, and H_x, H_y, and H_z those of **H**. The electric field is measured in electrostatic units and the magnetic field in electromagnetic units. The system which uses electrostatic units for all electrical quantities and electromagnetic units for all magnetic ones is known as the *gaussian system* of units. Although not the most convenient one for practical calculations, it is suitable here, and will always be used in what follows. The presence of the important constant c in Eqs. (20a) and (20b) is of course dependent on our choice of units. It represents the ratio of the magnitudes of the electromagnetic and electrostatic units of current.

Equation (20c) merely expresses the fact that no free electric charges exist in a vacuum. The assumption of no free magnetic pole gives rise to Eq. (20d). Equations (20b) express Faraday's law of induced electromotive force. Thus the quantities occurring on the left side of these equations represent the time rate of change of the magnetic field, and the spatial distribution of the resulting electric fields occurs on the right side. These equations do not give directly the magnitude of the emf but only the rates of change of the electric field along the three axes. In particular problems the equations must be integrated to obtain the emf itself.

20.3 DISPLACEMENT CURRENT

Maxwell's principal *new* contribution in giving these equations was the statement of Eqs. (20a). These come from an extension of Ampère's law for the magnetic field due to an electric current. The right-hand members give the distribution of the magnetic field H in space, but the quantities on the left side do not at first sight seem to have anything to do with electric current. They represent the time rate of change of the electric field. But Maxwell regarded this as the equivalent of a current, the *dis-*

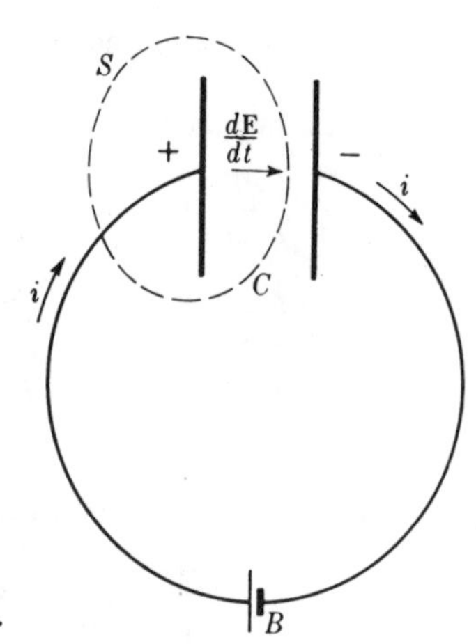

FIGURE 20A
The concept of displacement current.

placement current, which flows as long as the electric field is changing and which produces the same magnetic effects as an ordinary conduction current.

One way of illustrating the equivalence of $\partial \mathbf{E}/\partial t$ to an electric current is shown in Fig. 20A. Imagine an electric capacitor C to be connected to a battery B by conducting wires, the whole apparatus being in a vacuum with a vacuum between the capacitor plates. As the current i flows for an instant, electric charge accumulates on the plates until the capacitor is fully charged to the voltage of the battery. Through the closed surface S, a certain current has been flowing in during this instant, but none has apparently been flowing out. By considerations of continuity, Maxwell was led to assume that as much current should flow out of such a surface as flows in. But no current of the ordinary sort is flowing between the plates of the capacitor. The condition of continuity can be satisfied only by regarding the change of the electric field in this space as the equivalent of a displacement current, the current density j of which is proportional to $\partial E/\partial t$. In our system of units this current is given by $j = 1/4\pi$ times $\partial E/\partial t$. It will be noticed that the displacement current "flows" in a vacuum but stops as soon as $\mathbf{E}$ becomes constant.

One sees at once the analogy between Eqs. (20b) and (20a). By Eqs. (20b) a changing magnetic field produces an emf. This was observed by Faraday and is very simple to verify experimentally. By Eqs. (20a) a changing electric field should produce a magnetic field (*magnetomotive force*). This is a much less familiar idea and cannot be demonstrated by any simple experiment. The reason for the difference is that no substance conducts magnetism as a wire conducts electricity. The peculiarity that some substances possess of being conductors for electricity is the only reason why Eqs. (20b) were discovered before Eqs. (20a). The proof of the correctness of Eqs. (20a) lies in the remarkable success of Maxwell's equations in accounting for phenomena of nature. It should be noted that Maxwell's equations (20a) and (20b) can be written in terms of the displacement current j by replacing the x component $(1/c)(\partial E_x/\partial t)$ by $4\pi j_x$ and the other components by similar expressions.

20.4 THE EQUATIONS FOR PLANE ELECTROMAGNETIC WAVES

Consider the case of plane waves traveling in the x direction, so that the wave fronts are planes parallel to the yz plane. If the vibrations are to be represented by variations of E and H, we see that in any one wave front they must be constant over the whole plane at any instant, and their partial derivatives with respect to y and z must be zero. Therefore Eqs. (20a) to (20d) take the form

$$\frac{1}{c}\frac{\partial E_x}{\partial t} = 0 \qquad \frac{1}{c}\frac{\partial E_y}{\partial t} = -\frac{\partial H_z}{\partial x} \qquad \frac{1}{c}\frac{\partial E_z}{\partial t} = \frac{\partial H_y}{\partial x} \tag{20e}$$

$$-\frac{1}{c}\frac{\partial H_x}{\partial t} = 0 \qquad -\frac{1}{c}\frac{\partial H_y}{\partial t} = -\frac{\partial E_z}{\partial x} \qquad -\frac{1}{c}\frac{\partial H_z}{\partial t} = \frac{\partial E_y}{\partial x} \tag{20f}$$

$$\frac{\partial E_x}{\partial x} = 0 \tag{20g}$$

$$\frac{\partial H_x}{\partial x} = 0 \tag{20h}$$

Considering the first equation of (20e) and Eq. (20g) together, it appears that the longitudinal component E_x is constant in both space and time. Similarly from the top line of Eqs. (20f) and from Eq. (20h), H_x is also constant. These components can therefore have nothing to do with the wave motion but must represent constant fields superimposed on the system of waves. For the waves themselves, we may therefore write

$$E_x = 0 \qquad H_x = 0$$

This means, of course, that the waves are transverse, as stated above.

Of the four remaining equations, we see that the second equation (20e) and the third equation (20f) involve E_y and H_z, while the third equation (20e) and the second equation (20f) involve E_z and H_y. Let us assume, for example, that E_y represents the light vector, so that we are dealing with a plane-polarized wave with vibrations in the y direction. We should then have to put $E_z = H_y = 0$, and consider the two remaining equations

$$\frac{1}{c}\frac{\partial E_y}{\partial t} = -\frac{\partial H_z}{\partial x} \qquad -\frac{1}{c}\frac{\partial H_z}{\partial t} = \frac{\partial E_y}{\partial x} \tag{20i}$$

We now differentiate the first equation with respect to t and the second with respect to x. This gives

$$\frac{1}{c}\frac{\partial^2 E_y}{\partial t^2} = -\frac{\partial^2 H_z}{\partial x\, \partial t} \qquad -\frac{1}{c}\frac{\partial^2 H_z}{\partial t\, \partial x} = \frac{\partial^2 E_y}{\partial x^2}$$

Eliminating the derivatives of H_z, we find

$$\frac{\partial^2 E_y}{\partial t^2} = c^2\frac{\partial^2 E_y}{\partial x^2} \tag{20j}$$

In a similar way, by differentiation of the first equation (20i) with respect to x and the second with respect to t, we find

$$\frac{\partial^2 H_z}{\partial t^2} = c^2 \frac{\partial^2 H_z}{\partial x^2} \tag{20k}$$

Now Eqs. (20j) and (20k) have just the form of the wave equation for plane waves, with E_y and H_z, respectively, playing the part of the displacement y in the two cases. For both, comparison with the wave equation shows that the velocity

$$v = c \tag{20l}$$

Thus we see that two of the four equations in Eqs. (20e) and (20f) predict the existence of a wave of the electric vector, plane-polarized in the xy plane, and an accompanying wave of the magnetic vector, plane-polarized in the xz plane. In the form of Eq. (11a) they would be represented by

$$E_y = f(x \pm ct) \qquad H_z = f(x \pm ct) \tag{20m}$$

The two waves are interdependent, neither can exist without the other. Both are transverse waves, and are propagated in a vacuum with the velocity c, the ratio of the electrical units (Sec. 20.2).

If we had started with the other two equations in Eqs. (20e) and (20f), we would have obtained another pair of waves, plane-polarized with the electric vector in the xz plane. This pair is quite independent of the other and can exist separately from the other pair. A mixture of the two pairs vibrating at right angles to each other and with no constant phase relation between E_y and E_z represents unpolarized light.

20.5 PICTORIAL REPRESENTATION OF AN ELECTROMAGNETIC WAVE

The simplest type of electromagnetic wave is one in which the function f in Eq. (20m) is a sine or cosine. This is a plane-polarized monochromatic plane wave. The three components of **E** and the three of **H** may for such a wave be written

$$\begin{aligned} E_x &= 0 \qquad E_y = A \sin(\omega t - kx) \qquad E_z = 0 \\ H_x &= 0 \qquad H_y = 0 \qquad\qquad\qquad\quad H_z = A \sin(\omega t - kx) \end{aligned} \tag{20n}$$

By substituting the derivatives of these quantities in Eqs. (20a) to (20d), it is easily verified that they represent a solution of Maxwell's equations.

Figure 20B shows a plot of the values of E_y and H_z along the x axis, according to Eq. (20n). In a set of plane waves the values of E_y and H_z at any particular value of x are the same all over the plane x = const; so this figure merely represents the conditions for one particular value of y and z.

Two important points are to be noticed about Fig. 20B. In the first place, the electric and magnetic components of the wave are *in phase* with each other; i.e., when E_y has its maximum value, H_z is also a maximum. The relative directions of these two vectors, as indicated in the figure, agree with Eqs. (20n). The second point to be noted is that the amplitudes of the electric and magnetic vectors are equal. That these

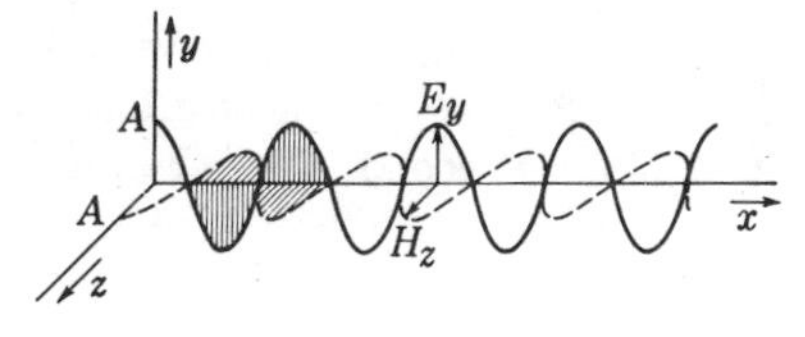

FIGURE 20B
Distribution of the electric and magnetic vectors in a plane-polarized monochromatic wave.

two are numerically equal in the system of units used here is shown by the fact that, in Eqs. (20n), A is the amplitude of each wave.

20.6 LIGHT VECTOR IN AN ELECTROMAGNETIC WAVE

The dual character of the electromagnetic wave raises the question whether it is the electric vector or the magnetic vector which is to be the light vector. This question has little meaning, since we could assume either one to represent the "displacements" we have been using in previous chapters. In every interference or diffraction phenomenon, the electric waves will mutually influence each other in exactly the same way as the magnetic waves. In one respect, however, the electric component plays a dominant part. It will be proved in Sec. 25.12 that it is the electric vector that affects the photographic plate and causes fluorescent effects. Presumably also the electric vector is the one that affects the retina of the eye. In this sense, therefore, the electric wave is the part that really constitutes "light," and the magnetic wave, though no less real, is less important.

20.7 ENERGY AND INTENSITY OF AN ELECTROMAGNETIC WAVE

The intensity of mechanical waves was shown in Sec. 11.3 to be proportional to the square of the amplitude. The same result follows from the electromagnetic equations. It can be shown* that in vacuum the electromagnetic field has an energy density given by

$$\text{Energy per unit volume} = \frac{E^2 + H^2}{8\pi} = \frac{E^2}{4\pi} \tag{20o}$$

where E and H are the instantaneous values of the fields, which here are equal. Half the energy is associated with the electric vector and half with the magnetic vector. The magnitudes of these vectors vary from point to point in any wave; so, in order to obtain the energy in any finite volume, it is necessary to evaluate the average value of E^2 (or H^2). For the plane wave of Eq. (20n), one finds that $E^2 = \frac{1}{2}A^2$, the factor $\frac{1}{2}$ being the average of the square of the sine over all angles. Hence an electromagnetic wave has an energy density $A^2/8\pi$, where A is the amplitude of either the electric or the magnetic component.

* M. V. Klein, "Optics," p. 532, John Wiley and Sons, Inc., New York, 1970.

The intensity of the wave will merely be the product of the above expression by the velocity c, since this represents the volume of the wave that will stream through unit area per second. We therefore have

$$I = \frac{c}{8\pi} A^2 \qquad \text{(20p)}$$

The reader should be reminded that the above statements are applicable only to a wave traveling in vacuum. In matter, not only will the velocity be different, but also the magnitudes of E and H will no longer be equal. Aside from factors of proportionality, however, the intensity is still given by the square of the amplitude of either wave (Sec. 23.9).

20.8 RADIATION FROM AN ACCELERATED CHARGE

A convenient method of representing an electric or magnetic field is by the use of lines of force. These are familiar to anyone who has studied elementary electricity and magnetism. Each line of force indicates the direction of the field at every point along the line, and this is such that a tangent to the line of force at any point gives the direction of the force on a small charge or pole placed at that point. That is, this tangent gives the direction of the electric or magnetic field at that point.

Consider a small positive electric charge at rest at the point A [Fig. 20C(*a*)]. The lines of force are straight lines diverging in every direction from the charge and are uniformly distributed in space. The same picture would hold if the charge were moving in the direction AB with constant velocity, assuming this velocity to be not too large. In these two cases—charge at rest and charge in uniform motion—there is no radiation of electromagnetic waves.

In order to produce electromagnetic radiation, it is necessary to have *acceleration* of the charge. A particularly simple case is represented in Fig. 20C(*b*). Let the charge, originally at rest at A, be accelerated in the direction AC. The acceleration a lasts only until the charge reaches the point B, and from that point on the charge moves with a constant velocity. In this case we can obtain some information about the form of the lines of force radiating from the charge at some later time. Let the time of the acceleration from A to B be Δt, and let the time of the uniform motion from B to C be t. When the charge has reached C, at a time $t + \Delta t$ after it starts, the parts of the original lines of force lying beyond the arc RR', drawn about A with the radius $c(t + \Delta t)$, cannot have been disturbed in any way. This follows from the fact that any electromagnetic disturbance is propagated with the velocity c. At the point C the velocity is uniform, and the lines of force as far as the arc QQ', drawn about B with the radius ct, must be uniform and straight, since the charge has had a uniform velocity during the time t. Consequently we see that in order to have continuous lines of force they must be connected through the region between RR' and QQ' somewhat as shown in the figure. This gives a pronounced "kink" in each line. The exact form of the kink will depend upon the type of acceleration existing between A and B, that is, whether it is uniform or some type of nonuniform acceleration.

What is the significance of such a kink in a line of force? If we select some point

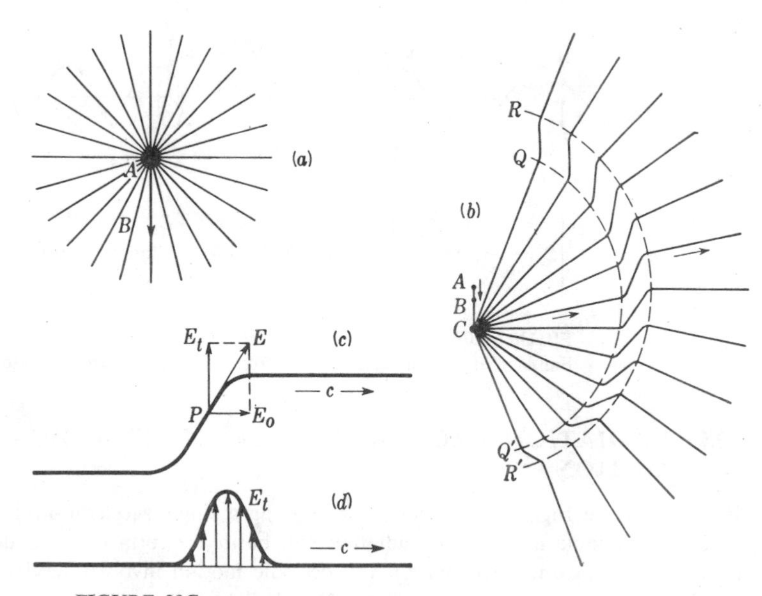

FIGURE 20C
Emission of an electromagnetic pulse from an accelerated charge.

P lying on the kink [Fig. 20C(c)], the vector E drawn tangent to the line at P gives the actual direction of the field at that point. This may be regarded as the resultant of the field E_0, which would be produced by the charge at rest, and a *transverse* field E_t. It is the vector E_t which represents the electric vector of the electromagnetic wave, referred to in the foregoing sections. If we carry out this construction for various points along the kink, we obtain the variations indicated in Fig. 20C(d). This is obviously not a periodic wave form but merely a pulse. There will be a similar pulse of the H vector at right angles to E_t.

Several important features about the production of electromagnetic radiation are illustrated by this example. Most important is the fact that E_t exists only when the charge is *accelerated.* No radiation is produced if there is no acceleration of charge, and, conversely, an accelerated charge will always radiate to a greater or less extent. Also, the example shows how the electric field of the radiation can be transverse to the direction of propagation. The magnitude of the vector E_t obtained by the construction of Fig. 20C(d), i.e., the amplitude of the wave, obviously depends on the steepness of the kink, and this is determined by how rapidly the charge was accelerated from A to B. It can be shown theoretically that the rate of radiation of energy from an accelerated charge is proportional to the square of the acceleration. Finally, we also find that the amplitude of the radiation varies with angle in such a way that it is a maximum in directions perpendicular to the line AC and falls to zero in both directions along AC. The amplitude is easily shown to be proportional to the sine of the angle between AC and the direction considered.

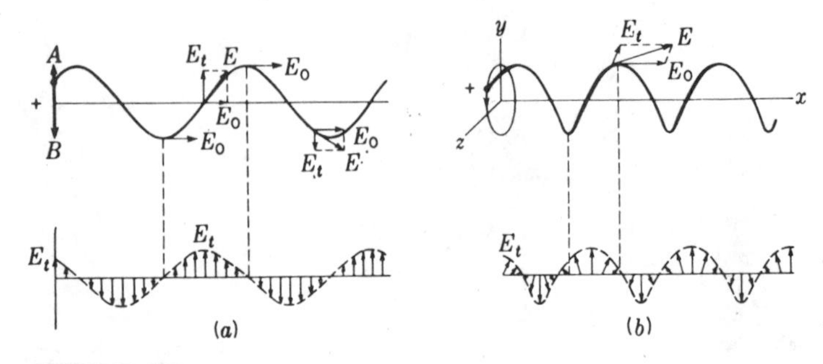

FIGURE 20D
Emission of electromagnetic waves from a charge in periodic motion.

20.9 RADIATION FROM A CHARGE IN PERIODIC MOTION

If the charge in Fig. 20C, instead of undergoing a single acceleration, is caused to execute a periodic motion, the radiation will be in the form of continuous waves instead of a single isolated pulse. Any periodic motion involves accelerations and hence will cause the charge to radiate. We shall here consider only two especially simple cases, that of linear simple periodic motion and that of uniform circular motion. If the positive charge of Fig. 20D(*a*) is moved with simple harmonic motion between the limits A and B, any line of force will be bent into the form of a sine curve. Let the upper curve of Fig. 20D(*a*) represent one such line, say the one running out perpendicular to AB. At the particular instant shown, the electric force E at various points along the line has the direction of the tangent at those points. Resolving it into the undisturbed field E_0 and the transverse component E_t, we find the various values E_t shown just below. These also take the form of a sine curve and represent the variation of the electric vector along the wave sent out. This is a plane-polarized wave.

In part (*b*) of the figure, the positive charge is revolving counterclockwise in a circle, in the yz plane shown in perspective. The same construction now gives values of E_t which are constant in magnitude but vary in direction along the wave. The heads of the arrows lie on a helix similar to that of the line of force, but displaced one-quarter of a wavelength along the direction of propagation, which here is the x axis. This screwlike arrangement of the vectors is characteristic of a circularly polarized wave. It is worth pointing out here that if the radiation along the y or z axes were examined, it would be found to be plane-polarized in the yz plane. Actual observation of these two cases is possible in the Zeeman effect (Sec. 32.1).

20.10 HERTZ'S VERIFICATION OF THE EXISTENCE OF ELECTROMAGNETIC WAVES

We have seen that, starting with a set of equations governing the phenomena of electromagnetism, Maxwell was able to show the possibility of electromagnetic waves and to make definite statements about the production and properties of the waves.

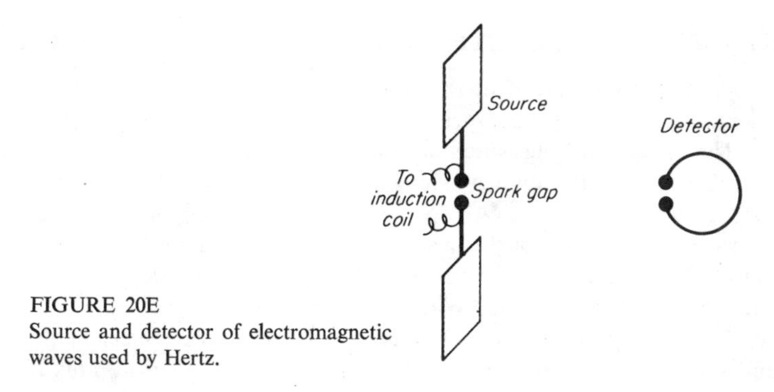

FIGURE 20E
Source and detector of electromagnetic waves used by Hertz.

Thus he could say that they are generated by any accelerated charge, that they are transverse waves, and that they travel with the velocity c in free space. The experimental production and detection of the waves predicted by Maxwell were achieved by Hertz. In 1887 he began a remarkable series of experiments which constitute the first important experiments on radio waves, i.e., electromagnetic waves of long wavelength. The essential features of Hertz's method are illustrated in Fig. 20E. Two plane brass plates are connected to a spark gap and sparks are caused to jump across the gap by charging the plates to high voltage with an induction coil. It is known that the discharge of the plates by the spark is an oscillatory one. Each time the potential difference between the knobs of the gap reaches the point where the air in the gap becomes conducting, a spark passes. This represents a sudden surge of electrons across the gap, and the signs of the charges on the two plates become reversed. But since the air is still conducting, this will produce a return surge, another reversal of sign, and the process repeats until the energy is dissipated as heat by the resistance of the gap. The frequency of these oscillations depends on the inductance and capacity of the circuit. These were very small for Hertz's oscillator, and the frequency correspondingly high. In some of his experiments it reached 10^9 Hz. Thus we have an electric charge undergoing very rapid accelerations, and electromagnetic waves should be radiated.

In Hertz's experiment the presence of electromagnetic waves was detected at some distance from the oscillator by a resonating circuit consisting of a circular wire broken by a very narrow spark gap of adjustable length. The changing magnetic field in the wave induced an alternating emf in the circular wire, whose dimensions were such that the natural frequency of its oscillations was the same as that of the source. Thus the induced oscillations built up by resonance in the detector until they were sufficient to cause sparks to jump the gap.

It was a simple matter to show that the waves were plane-polarized with $\mathbf{E}$ in the y direction and $\mathbf{H}$ in the z direction. If the loop was turned through 90° so that it lay in the xz plane, the sparks ceased. Hertz performed many other experiments with these waves, showing among other things that the waves could be reflected and focused by curved metal reflectors and that they could be refracted in passing through a large 30° prism of pitch. In these respects they therefore showed the same behavior as light waves.

20.11 SPEED OF ELECTROMAGNETIC WAVES IN FREE SPACE

The most convincing proof of the reality of Hertz's electromagnetic waves lay in the demonstration that their speed was that predicted by the theoretical equation (20l). The velocity was measured not directly but indirectly by measuring the wavelength. Then from the known frequency of the oscillations the velocity could be found by the relation $v = \nu\lambda$. To measure the wavelength, standing waves were produced by interference of the direct waves with those reflected from a plane metal reflector. The positions of the nodes could be located by the fact that the detector ceased to spark at these points. With a frequency of 5.5×10^7 Hz, λ was found to be about 5.4 m, which gives v very close to 3×10^8 m/s. The determination could not be made accurately because the oscillations were highly damped, only three or four occurring after each spark, and the wavelength was therefore not accurately defined. Later work by Mercier with undamped waves produced by a vacuum-tube oscillator gave the result 2.9978×10^8 m/s. We have already seen, in Sec. 19.6, how the increased precision obtainable with cavity resonators has added another significant figure to the speed of light.

According to Eq. (20l), this observed speed should equal c, the ratio of the emu to the esu of current. As has been mentioned (Sec. 19.7), this ratio has been accurately measured by different methods, the most recent value being 2.99781×10^8 m/s. But this is just the measured speed of electromagnetic waves and also agrees exactly with the latest measurements of the speed of light by Michelson and others (see Table 19A). For air or other gases at atmospheric pressure, a slight modification in the equations is necessary (Chap. 23), but the predicted speed differs only slightly from that in vacuum.

Hence we are forced to conclude that light consists of electromagnetic waves of extremely short wavelength. Beside the evidence of polarization, which proves that light waves are transverse waves, there is much other evidence of this identity. Spectroscopy has shown that the atoms contain electrons and that by assuming the acceleration of these electrons as they move in orbits around the nucleus one can account for the polarization and intensity of the spectrum lines. Furthermore, as shown in Fig. 11N radio waves, which are obviously electromagnetic in character, join continuously onto the region of infrared light waves. Thus the explanation of light waves as an electromagnetic phenomenon, which in the hands of Maxwell was merely a very elegant theory, has since proved to be a reality, and we accept the electromagnetic character of light as an established fact. In treating the interactions of light with matter we shall therefore use the fact that light consists of oscillations of an electric field at right angles to the direction of propagation of the waves, accompanied by oscillations of the magnetic field, also at right angles to this direction and to the direction of the electric field.

20.12 ČERENKOV RADIATION

It was stated in Sec. 20.8 that an electric charge moving with uniform velocity radiates no energy, but merely carries its electromagnetic field along with it. This is true as long as the charge is traveling in vacuum. If on the other hand, it moves through a

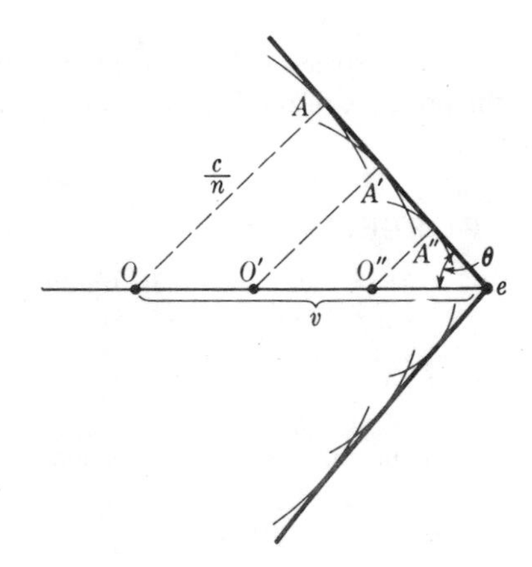

FIGURE 20F
Cross section of the conical wave produced in Čerenkov radiation.

material medium, e.g. as when a high-speed electron or proton enters a piece of glass, it may radiate a small amount of energy even though its velocity is constant. The required condition is that the speed of the charged particle be *greater than the wave velocity c/n* of light in the medium. It then sets up an impulsive wave similar to the shock wave produced by a projectile traveling at a speed greater than that of sound. It is of the same character as the bow wave of a boat, which forms when the boat moves faster than the water waves.

The production of this wave is an excellent illustration of the application of Huygens' principle (Sec. 18.1). In Fig. 20F let e represent an electron moving through glass of index 1.50 with a velocity which is nine-tenths of the velocity of light. (To produce such an electron one would have to accelerate it through a potential difference of some 661 kV.) The disturbances produced when the electron occupied successively the positions O, O', and O'' are represented as secondary wavelets which have radii OA, $O'A'$, and $O''A''$, proportional to the elapsed time and to their velocity c/n. The resulting wave front is the common tangent to these and takes the form of a cone of half angle θ. Since OA is normal to the wave front, it will be seen from the figure that θ is given by

$$\sin \theta = \frac{c}{nv} = \frac{1}{n\beta} \qquad \text{(20q)}$$

where v is the velocity of the charged particle and $\beta = v/c$. If $\beta = 0.9$, as in our example, θ is about 48°. A substantial part of the radiation is in visible light and is detectable by the eye or the photographic plate. Because of dispersion, the variation of n with color, Eq. (20q) is not perfectly exact.* Furthermore, when n is largest (blue light), the cone is narrower and the outer edge of the conical fan of light *rays* will therefore be blue, while its inner edge will be red.

This type of radiation is now commonly observed with the high-speed particles used in nuclear physics. By measuring the angle of the cone, the velocities and energies of the particles can be determined. The light resulting from the passage of

* For the exact equations, see H. Motz and L. I. Schiff, *Am. J. Phys.*, **21**: 258 (1953).

a single particle can be made to register a count with a photomultiplier tube. This is the principle of the Čerenkov counter employed by nuclear physicists.

PROBLEMS

20.1 The waves from a radio transmitter has a frequency of 32.56 MHz and are incident normally on a flat surface of a sheet of metal. The reflected and incident beams set up standing waves that are measured to have nodes 460.3 cm apart. Neglecting the refractive index of air, what does this give for the speed of the waves?

Ans. 299,747 km/s

20.2 Show that Maxwell's equations are satisfied by the solution

$$E_x = A\sin(\omega t + ky) \qquad E_y = 0 \qquad E_z = 0$$
$$H_x = 0 \qquad H_y = 0 \qquad H_z = A\sin(\omega t + ky)$$

(*a*) In which plane is the wave polarized, and (*b*) in which direction does it travel? (*c*) Write down the equations.

20.3 Modify Eqs. (20n) so that they represent (*a*) a plane-polarized wave with the E oscillations in the xy plane but at 30° with the x axis and (*b*) a wave whose oscillations are ellipses in the xy plane (elliptically polarized wave).

20.4 Starting with the following equations, (*a*) make a list of all the partial derivatives occurring in Eqs. (20a) to (20d):

$$E_x = A\sin(\omega t - ky) \qquad H_x = 0$$
$$E_y = 0 \qquad H_y = 0$$
$$E_z = 0 \qquad H_z = A\sin(\omega t - ky)$$

(*b*) Show by direct substitution that these derivatives satisfy

$$\frac{1}{c}\frac{\partial E_x}{\partial t} = \frac{1}{c}\,\omega A\cos(\omega t - ky) = 0 \qquad \frac{\partial E_x}{\partial y} = -kA\cos(\omega t - ky) = 0$$

$$-\frac{1}{c}\frac{\partial H_z}{\partial t} = -\frac{1}{c}\,\omega A\cos(\omega t - ky) = 0 \qquad \frac{\partial H_z}{\partial y} = -kA\cos(\omega t - ky) = 0$$

20.5 (*a*) Prove that the segment of the line of force between Q and R in Fig. 20C(*b*) is a straight line when the acceleration of the charge has been uniform. (*b*) From the slope of this segment show that the ratio E_0/E_t falls off as $1/r$ and hence that at any appreciable distance the transverse component will predominate. *Hint:* Remember that E_0 is given by Coulomb's law.

20.6 The total force F exerted on a charge e that moves in electric and magnetic fields in vacuum is given by

$$F = eE + \frac{evH}{c}$$

where it is assumed that the velocity v is perpendicular to the field H. Find the ratio of an electric force to the magnetic force exerted on an electron in the first Bohr orbit of the hydrogen atom by sunlight which has $E = H = 0.0242$ (gaussian units).

20.7 Calculate the amplitude of the electric field strength of a beam of sunlight, which may be taken as having an intensity of 1.20 kW/m^2.

20.8 (*a*) Show that the amplitude of the electromagnetic wave from an accelerated charge varies as $\sin\theta$, where θ is the angle between the direction of observation and the direction of the acceleration. (*b*) Make a polar plot of the intensity of the *radiation versus angle*.

20.9 Show that the ratio of a charge measured in esu to the same charge measured in emu has the dimensions of a velocity. *Hint:* Start with Coulomb's law in each case.

20.10 Poynting's theorem states that the energy flow in an electromagnetic wave is given by

$$\mathbf{S} = \frac{c}{4\pi}(\mathbf{E} \times \mathbf{H})$$

S is called the Poynting vector, and the expression in parentheses represents the vector product. Show that the conclusions of Secs. 20.5 and 20.7 with regard to the direction and magnitude of the flow relative to the directions and magnitudes of **E** and **H** are in agreement with Poynting's theorem.

20.11 By assuming Einstein's relation between mass and energy and taking the mass equivalent to an electromagnetic wave to move with the velocity c, derive an expression for the pressure that radiation exerts on a perfectly absorbing surface by virtue of its momentum. *Ans.* $p = I/c = A^2/8\pi$

20.12 A beam of protons with an energy of 560 MeV is passed through a sheet of extra dense flint glass, where $n = 1.750$. (*a*) Find the angle between the Čerenkov radiation and the direction of the proton beam inside the glass. (*b*) What is the indicated value of β for these protons? (For MeV see H. E. White, "Modern College Physics," 6th ed., sec. 49.1, D. Van Nostrand Co., New York, 1972.)

21

SOURCES OF LIGHT AND THEIR SPECTRA

Since light is an electromagnetic radiation, we should expect that the emission of light from any source results from the acceleration of electric charges. It is now certain that the electric charges involved in the emission of visible and ultraviolet light are the negative electrons in the outer part of the atom. By assuming that vibratory or orbital motions of these electrons cause radiation, many of the characteristics of different light sources can be explained. It should be emphasized, however, that this concept must not be carried too far. In the interpretation of spectra it fails in several important respects. These all involve the discrete or corpuscular nature of light, which is to be discussed later (Chap. 29). For the present, we shall emphasize only those features which can be explained by the assumption that light consists of electromagnetic waves.

21.1 CLASSIFICATION OF SOURCES

Sources of light which are important for optical and spectroscopic experiments may be divided into two main classes: (1) *thermal* sources, in which the radiation is the result of high temperature, and (2) sources depending on the electrical *discharge through gases*. The sun, with its surface temperature of 5000 to 6000°C, is an important example of the first class, but here must also be included such important sources as tungsten-filament lamps, the various electric arcs at atmospheric pressure, and the flame. Under the second class come high-voltage sparks, the glow discharge in vacuum tubes at low pressure, and certain low-pressure arcs like the mercury arc. The distinction between the two classes is not sharp, and we can go continuously from one to the other, for instance by pumping away the air around an electric arc.

21.2 SOLIDS AT HIGH TEMPERATURE

The majority of practical sources for illuminating purposes use the radiation from a hot solid. In the *tungsten lamp*, the filament is heated to about 2100°C by the dissipation of electric energy due to its resistance. The filament can be run at temperatures as high as 2300°C but will last for only a short period owing to the rapid vaporization of tungsten. In the *carbon arc* in air, the temperature of the positive pole is about 4000°C and that of the negative pole, 3000°C. The positive pole vaporizes and burns away rather rapidly, but it constitutes the brightest thermal source of light available in the laboratory. The heating results chiefly from the bombardment of the positive pole by electrons drawn from the gaseous part of the arc. Relatively little light comes from the gas itself. An interesting type of arc, useful when a very small source of light is needed, is the so-called *concentrated-arc* lamp. A simplified diagram of this device is shown in Fig. 21A(*a*). The cathode consists of a small metal tube packed with zirconium oxide, and the anode consists of a metal plate containing a hole slightly larger than the end of the cathode. Tungsten, tantalum, or molybdenum, because of their high melting points, are used for the metal parts. These are sealed into a glass bulb which is filled with an inert gas like argon to a pressure of nearly 1 atm. The arc runs between the (fused) surface of the zirconium oxide and the surrounding anode, as indicated in part (*b*) of the figure. The tip of the cathode is heated by ion bombardment to 2700°C or higher, giving it a surface brightness almost equal to that in the carbon arc. The light is observed through the hole in the anode, in the direction shown by the arrow in Fig. 21A(*a*). Lamps of this type can be made in which the source is as small as 0.007 cm in diameter. A cheaper way of achieving a source of small dimensions is to use a tungsten lamp with a small spiral filament (automobile headlight bulb), run at a voltage somewhat higher than its rated value. This source does not,

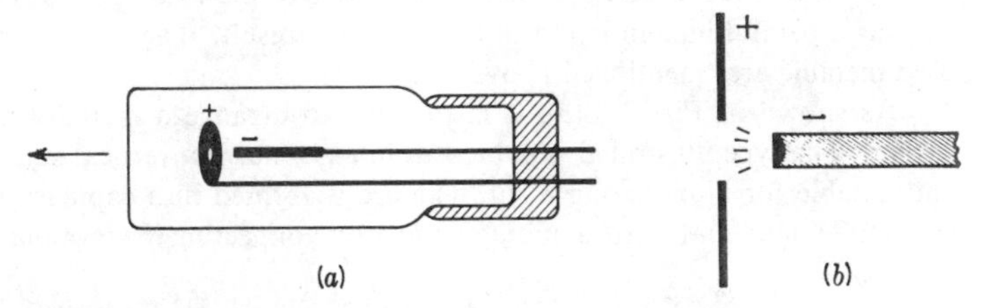

FIGURE 21A
The concentrated arc, a close approximation to a point source.

however, have the smallness and brightness of the concentrated-arc lamp. Other sources of continuous spectra will be considered in Sec. 21.9.

21.3 METALLIC ARCS*

When two metal rods connected to a source of direct current are touched together and drawn apart, a brilliant arc forms between them. A resistance of high current capacity must be connected in series with the circuit and adjusted so that the steady current through the arc is from 3 to 5 A. Higher currents than this will cause excessive heating and melting of the electrodes. A large self-inductance in the circuit will stabilize the arc, and a voltage of 220 is preferable to 110 in this respect. The two poles are held vertically, in line with each other, by clamps with a screw adjustment to vary their separation. In the *iron arc*, the positive pole should be the lower, since then a bead of molten iron oxide collects in the small cavity which soon forms, and this helps the steadiness of the arc. The radiation from an iron, copper, or aluminum arc comes mostly from the gas traversed by the arc, this gas consisting almost entirely of the vapor of the metal. It has been shown that the gas is at a temperature of from 4000 to 7000°C, and it may in cases of very high currents run up to 12,000°C. The equivalent of a metallic arc can be obtained with a carbon arc in which the positive pole has been bored with an axial hole and packed with the salt of a metal, such as calcium fluoride. It is sometimes desirable to run a metallic arc in an atmosphere other than air by enclosing it in an airtight chamber. The arc may then be run at low pressures as well, but this is a difficult procedure.

With the metals of low melting point, the arc may be permanently enclosed in a glass envelope. Of this type are the *mercury arc* and the *sodium arc*, both commonly used in optical laboratories. In the older form of mercury arc, liquid mercury is sealed in a highly evacuated glass container of such a shape that the mercury forms two separate pools. These make electrical connection with two wires sealed through the glass. To start the arc, it is tipped until a thread of mercury connects the two pools for an instant and breaks again. As the arc warms up, the pressure of the mercury vapor increases, and unless a fairly large space is available for cooling and condensation, the arc will go out. With sufficient self-inductance in the circuit, the arc may be run at fairly high temperature and pressure, giving a very intense source. For this purpose the container is made of fused quartz to withstand the higher temperature. Quartz has the advantage that it transmits the ultraviolet light (Sec. 22.3), and quartz arcs are frequently used in spectroscopy and for therapeutic purposes. In using them, great care should be taken not to look at the arc too frequently unless glasses are being worn, as a painful inflammation of the eyes may result. The same is true for the exposed metallic arcs mentioned above.

As shown in Fig. 21B(*a*), it is possible to arrange a mercury arc to be self-starting. The type illustrated provides an intense, narrow vertical source of mercury light suitable for illuminating a slit. The arc is formed in a capillary tube of inside diameter 2.0 mm, and starts a minute or so after connecting the terminals to the 110-V

* These and other sources for use in spectroscopy are well described in G. R. Harrison, R. C. Lord, and J. R. Loofbourow, "Practical Spectroscopy," chap. 8, Prentice-Hall, Inc., Englewood Cliffs, N.J., 1948.

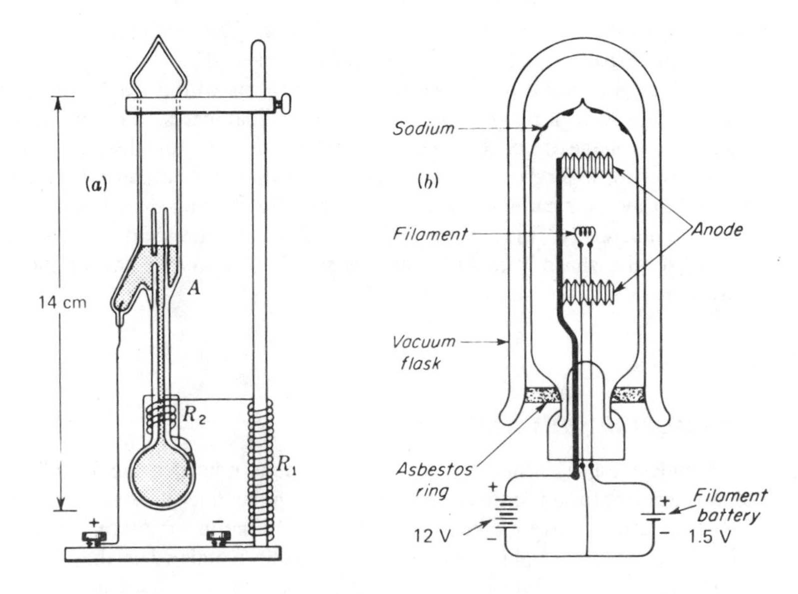

FIGURE 21B
(*a*) Small, self-starting mercury arc. (*b*) Sodium arc.

dc mains. Before this time, the current is limited to about 1.5 A by the resistances R_1 and R_2 of 80 and 7 Ω, respectively. R_2 is wound on the lower part of the capillary and encased in cement so that it heats the mercury at that point until a bubble of vapor is formed and the mercury thread breaks. The resulting arc then generates enough pressure to push the mercury above it up to the point *A*. The arc is then confined to the capillary from *A* to R_2. The current has now fallen to about 1.0 A, owing to the additional resistance of the arc itself.

The sodium arc [Fig. 21B(*b*)] is always contained in a double-walled envelope made of a special glass that is resistant to blackening by hot sodium vapor. The inner envelope contains argon or neon at low pressure and a small amount of metallic sodium. The discharge is initiated in the rare gas by electrons emitted from the coiled filament *F* and is sustained by a relatively small positive potential applied to the anode. Since the space between the double walls is highly evacuated to prevent heat loss, the interior temperature rises rapidly to the point where the sodium melts and vaporizes into the arc. The rare-gas spectrum then fades out, being replaced by radiation from the more easily ionized atoms of sodium. This is nearly all in the yellow sodium doublet, so that the arc yields essentially monochromatic light without the use of filters. The doublet is so narrow (separation 5.97 Å) that for spectroscopy under low dispersion and for interference measurements with small path difference it may be assumed to be a single line with the average wavelength 5892 Å.

Although they are satisfactory sources for use with small gratings and prism spectroscopes, neither of the above arcs yields spectral lines of sufficient sharpness for investigations with very high dispersion. The relatively high pressure, temperature, and current density cause a broadening of the lines. The simplest way to produce sharper lines is to use a discharge through a rare gas with a small admixture of the metal vapor and to limit the current to a few milliamperes. The discharge may be either a low-voltage arc of the type described above or a glow discharge in a vacuum tube (Sec. 21.6). Very convenient sources of this type, not only for mercury and sodium but also for cadmium, zinc, and other low-melting metals, can now be purchased commercially. In fact, the ordinary mercury fluorescent lamp is of the kind required to give sharp lines and would be satisfactory were it not for the coating of fluorescent salt on the inside of the walls.

21.4 BUNSEN FLAME

When sufficient air is admitted at the base of a bunsen burner, the flame is practically colorless, except for a bluish-green cone bounding the inner dark cone of unburnt gas. The temperature above the cone is in the neighborhood of 1800°C, high enough to cause the emission of light from the salts of certain metals when they are introduced into the flame. The color of the flame and its spectrum are characteristic of the metal and do not depend on which salt is used. The chloride is usually most volatile and gives the most intense coloration. The color of the sodium flame is yellow; of strontium, red; of thallium, green; etc. For introducing the salt into the flame, a common method is to use a loop on the end of a platinum wire, which is first dipped in hydrochloric acid and heated until the sodium yellow disappears. Then, while red-hot, it is touched to the powdered salt, melting a small amount which adheres to the wire. When this is again held in the flame, the color is strong but lasts only a short time. A better method is to mix a fine spray of the chloride solution with the gas before it enters the burner. This is best done with the apparatus shown in Fig. 21C, in case air under pressure is available. Air is forced through the atomizer *S*, filling the bottle with a fine spray which is carried into the gas at the base of the burner. This gives a very constant light source, and is convenient for the laboratory study of flame spectra. Unfortunately, it can be used for only a limited number of metals, the suitable ones including lithium, sodium, potassium, rubidium, caesium, magnesium, calcium, strontium, barium, zinc, cadmium, indium, and thallium. Other elements may be used in the hotter oxygas flame or oxyhydrogen flame, but these flames are not as convenient to operate.

21.5 SPARK

By connecting a pair of metal electrodes to the secondary of an induction coil or high-voltage transformer, a series of sparks can be made to jump an air gap of several millimeters. If there is no capacitance in the circuit, the spark is quiet and not very in-

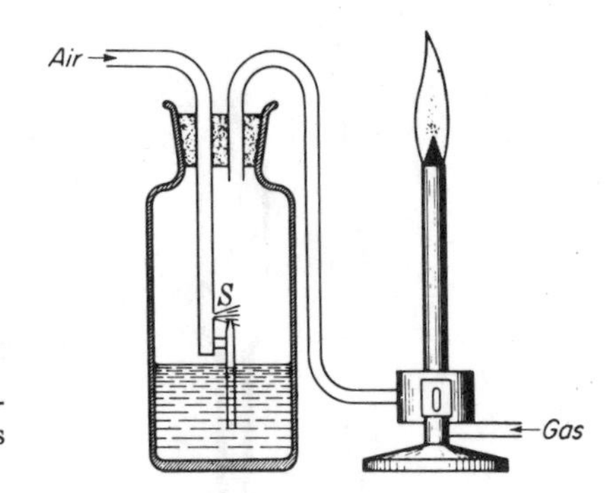

FIGURE 21C
Experimental arrangement for producing spectra by introducing salts of metals into the flame of a bunsen burner.

tense, the radiation coming chiefly from the air in the gap. The spark may be made much more violent and brighter by connecting a capacitor (such as a Leyden jar) in parallel across the gap. We then obtain a *condensed spark*. This is an extremely bright source, the spectrum of which is very rich in lines characteristic of the metal of the electrodes. The condensed spark has the drawbacks not only of noisiness and hazard of electric shock, but also of the considerable breadth of the lines it emits. Nevertheless, it furnishes the most intense excitation available, and is the most efficient source we have for the lines of ionized atoms which have lost one or more electrons. Such lines are usually called high-temperature, or *spark*, *lines*.

21.6 VACUUM TUBE

This common source is familiar because of its application to advertising signs. Neon signs contain pure neon gas at a pressure of about 2 cmHg. Metal electrodes are sealed through the ends of the tube, and an electric current is caused to traverse the gas by connecting the electrodes to a transformer giving a potential of 5000 to 15,000 V. Other colors are produced by introducing a small amount of mercury into a neon or argon tube. The heat of the discharge vaporizes the mercury, and we obtain the characteristic color and spectrum of mercury vapor. If the tube is made of colored glass, certain colors of the mercury light are absorbed and various shades of blue and green may be produced.

In the laboratory, this principle can be used on a smaller scale to excite the characteristic radiations of any gas or vapor. Two common forms of vacuum tube are illustrated in Fig. 21D. Type (*a*) is useful where maximum intensity is not required, e.g., if the tube is to be operated with a small induction coil. The electrodes *E*, *E* are short pieces of aluminum rod, welded to the ends of tungsten wires, the latter being sealed through the glass. The light is most intense in the capillary tube *C*, where the current density is greatest, and it is observed laterally, in the direction indicated by the arrow. Considerably greater intensity can be obtained with the end-on type shown in (*b*). Here the electrodes are of sheet aluminum, rolled up and slipped inside two

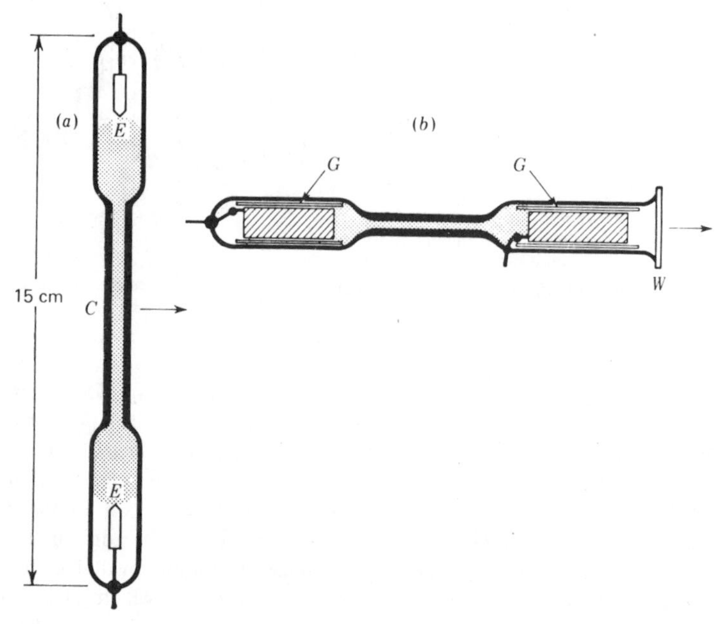

FIGURE 21D
Discharge tubes for obtaining the spectra of gases at low pressure.

loosely fitting inner glass tubes, *G*, *G*. They are fastened to the tungsten leads by wrapping a small strip of aluminum at one end around the wire and pinching it on tightly. The larger area of the electrodes permits the use of greater currents, usually furnished by a transformer, without overheating of the electrodes. The light is observed through a plane glass window *W*, which may be fused directly to the tube. The inner glass tubes serve to prevent the deposition of aluminum on the outer walls of the main tube, which occurs rather rapidly when a tube is used at a low pressure.

The exact pressure at which a vacuum tube should be sealed off varies between about 0.5 and 10 mmHg, according to the gas and to the particular spectrum desired. Only a limited number of gases are suitable for long-continued use in a sealed tube of the above type. Of these, the rare gases neon, helium, and argon are the most satisfactory. Hydrogen, nitrogen, and carbon dioxide tubes will last only a limited time; the gas gradually disappears from the tube, or "cleans up," until a discharge can no longer be maintained. Two processes may be responsible for this. The gas may be decomposed by the discharge and the products deposited on the walls or removed by chemical combination with the metal electrodes. Or, even with a chemically inert gas, a decrease of pressure may be caused by absorption in the above-mentioned metal layers that are "sputtered" on the walls from the electrodes.

21.7 CLASSIFICATION OF SPECTRA

There are two principal classes of spectra, known as *emission spectra* and *absorption spectra.*

Continuous emission spectra	*Continuous absorption spectra*
Line emission spectra	*Line absorption spectra*
Band emission spectra	*Band absorption spectra*

Emission spectra are obtained when the light coming directly from a source is examined with a spectroscope. Absorption spectra are obtained when the light from a source showing a continuous emission spectrum is passed through an absorbing material and thence into the spectroscope. Figures 21G, 21H, and 21J show reproductions of photographed spectra illustrating the three types, both in emission and in absorption. Solids and liquids, with a few rare exceptions,* give only continuous emission and absorption spectra, in which a wide range of wavelengths, without any sharp discontinuities, is covered. Discontinuous spectra (line and band) are obtained with gases. Gases may also, in certain cases, emit or absorb a true continuous spectrum (Sec. 21.9). The three types of emission spectra can easily be observed with a carbon arc. If the spectroscope is pointed at the white-hot pole of the arc, the spectrum is perfectly continuous. If it is pointed at the violet discharge in the gas between the poles, bands in the green and violet are seen and there are always a few lines, like the sodium lines, owing to impurities in the carbons.

21.8 EMITTANCE AND ABSORPTANCE

Although in this chapter we are primarily concerned with various sources of light, and hence with emission, it will be well to state here a very important relation which exists between the emissive and absorptive powers of any surface. A solid, when heated, gives a continuous emission spectrum. The amount of radiation in this spectrum and its distribution in different wavelengths are governed by *Kirchhoff's*† *law* of radiation. This states that the ratio of the radiant emittance to the absorptance is the same for all bodies at a given temperature. As an equation, this law may be written.

$$\frac{W}{a} = \text{const} = W_B \qquad (21a)$$

The quantity W is the total energy radiated per square meter of surface per second, while a represents the fraction of the incident radiation which is not reflected or transmitted by the surface. For the constant representing this ratio, we have used the

* Compounds of some of the rare-earth metals give line spectra superposed on a continuous spectrum when heated to high temperatures. Their absorption spectra, e.g., that of didymium glass, show very narrow regions of absorption, which at liquid-air temperature become sharp absorption lines.

† Gustav Kirchhoff (1824–1887). Professor of physics at Heidelberg and Berlin. Besides discovering some fundamental laws of electricity, he founded (with Bunsen) the science of chemical analysis by spectra.

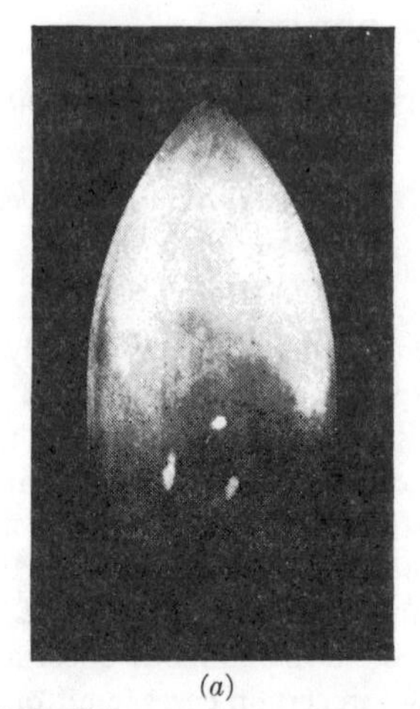

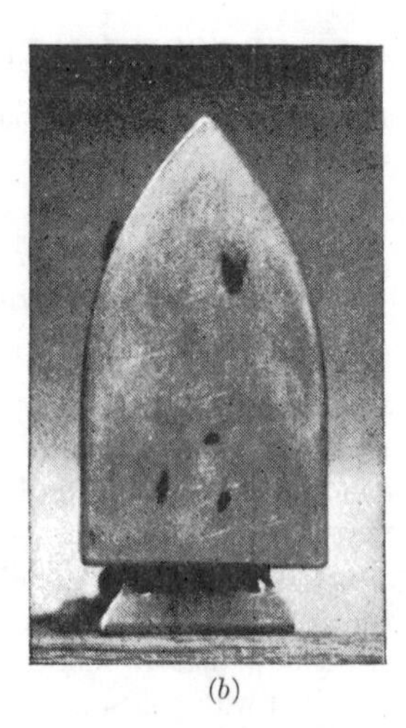

(a) (b)

FIGURE 21E
Photographs of an electric iron, illustrating Kirchoff's law of radiation. (*a*) Taken with infrared-sensitive photo plates, with the iron hot but emitting no visible radiation. (*b*) Taken with ordinary plates and illumination, with the iron at room temperature. For the justification of applying the law at different wavelengths, see text. (*Photographs courtesy of H. D. Babcock.*)

symbol W_B, because it represents the emittance of a *blackbody*. This term specifies a body which is perfectly black, i.e., one which absorbs all the radiation falling on its surface. Hence for such an ideal body, $a_B = 1$, and W_B equals the constant ratio W/a for other bodies.

Kirchhoff's law expresses a very general relation between the emission and absorption of radiation by the surface of different bodies. If the absorptance is high, the emittance must also be high. Here it is essential to realize the difference between the term *absorptance*, which measures the amount of light disappearing at a single reflection, and the *absorption* within the body of the material, as measured by the absorption coefficient α. The latter determines the loss of light upon transmission through the material and has no simple connection with the absorptance of the surface. In the case of metals, for example, a very high absorption coefficient is correlated with a high reflectance. But a high reflectance also means a low absorptance. Thus for metals, and in general for smooth surfaces of pure substances, a high absorption coefficient α necessarily means a low absorptance a.

A blackbody, which is approximated, for example, by a piece of carbon, gives the greatest amount of radiation at a given temperature. Transparent or highly reflecting substances are very poor emitters of visible light, even when raised to high temperatures. Figure 21E shows a practical illustration of the working of Kirchhoff's law. The right-hand picture is a photograph of an ordinary electric iron at room temperature. A few spots of india ink have been made on the surface, and these appear dark since they are regions of high absorptance. The rest of the surface is highly reflecting and hence a poor absorber. The left-hand photograph was taken by the radiation emitted from the iron when heated. The temperature was less than 400°C, so that no visible radiation was emitted. However, with infrared-sensitive photographic plates a successful photograph was obtained, even though the iron was invisible to the eye in the dark. In this picture, it will be seen that the spots which were previously dark (good absorbers) have now become brighter than the surroundings, even though they have the same temperature. Hence they also emit radiation most copiously, as Kirchhoff's law requires. Here we are assuming that the ink spots, because they are black by visible light, are also good absorbers for infrared light. It is in fact essential that W and a refer to the same wavelength or range of wavelengths. For the radiation within a small wavelength interval we may write

$$\frac{W_\lambda}{a_\lambda} = W_{B\lambda} \qquad \text{(21b)}$$

indicating by the subscript the emittance and absorptance at a particular wavelength. This form has important applications to discontinuous spectra (Sec. 21.10).

21.9 CONTINUOUS SPECTRA

The most common sources of continuous emission spectra are solids at high temperature,* and some of these sources were described in Sec. 21.2. Nothing was said there concerning the distribution in different wavelengths of the energy in the continuous spectrum. According to Kirchhoff's law, this depends on the ability of the surface to *absorb* light of different wavelengths. Thus in a piece of china with a red design glazed upon it, the red parts absorb blue and violet light more strongly than red. When the piece is heated to a high temperature in a furnace and withdrawn, the design will appear bluish by the emitted light, since these portions are the best absorbers and emitters for blue. In general, therefore, the reflectance spectrum of such a solid gives a clue to its emission spectrum.

A blackbody, which absorbs all wavelengths completely, is commonly taken as the standard because it constitutes a particularly simple case with which the radiation from other substances may be compared. Figure 21F shows the energy distribution in the radiation from a blackbody at seven different temperatures, and Fig. 21G(*a*) shows photographs of the actual spectra corresponding to these curves.† The curve for 2000 K represents fairly well that for a tungsten filament, while that for 6000 K is close to that of the sun (neglecting the narrow regions of absorption due to the Fraunhofer lines). The area under the curve represents the total energy emitted in all

* A good discussion of the experimental methods employed in this field will be found in W. E. Forsythe (ed.), "The Measurement of Radiant Energy," McGraw-Hill Book Company, New York, 1937.

† In comparing the spectra of Fig. 21G(*a*) with the curves of Fig. 21F it should be borne in mind that photographed spectra do not reproduce the true distribution of intensity in different wavelengths for three reasons: (1) The dispersion of the prism compresses the spectrum at the long-wavelength end. (2) The photographic plate is not equally sensitive to all wavelengths. In particular, the plate used here does not respond at all beyond $\lambda 6600$. (3) The blackening of the plate is not proportional to the intensity.

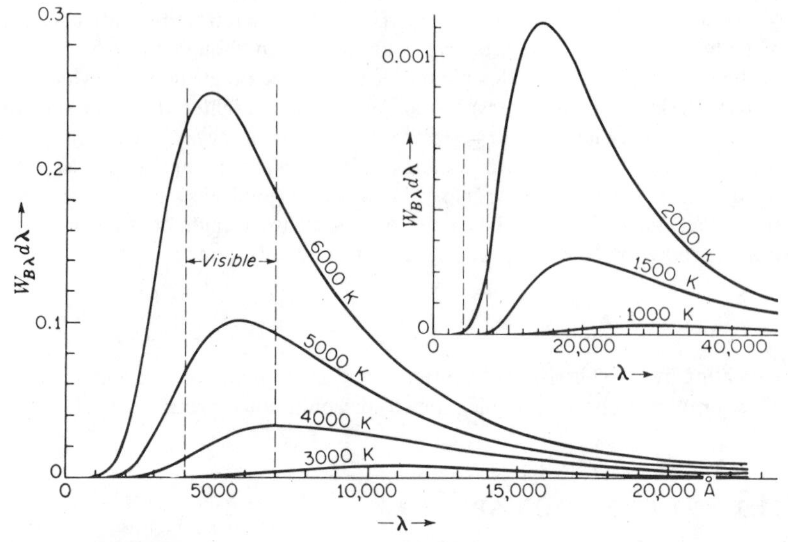

FIGURE 21F
Blackbody radiation curves plotted to scale. Ordinates give the energy in calories per square centimeter per second in a wavelength interval $d\lambda$ of 1 Å. For numerical values, see "Smithsonian Physical Tables," 8th ed., p. 314.

wavelengths, and increases rapidly with the absolute temperature. Calling W_B the total energy emitted from the surface of a blackbody per square meter per second and T the absolute temperature in kelvins, the *Stefan-Boltzmann* law* states that

$$W_B = \sigma T^4 \qquad \text{(21c)}$$

The constant σ has the value 1.3567×10^{-11} kcal/m² s K⁴ or 5.670×10^{-8} J/m² s K⁴. The wavelength of the maximum of each curve λ_{max} depends on the temperature according to *Wien's† displacement law*, which states that

$$\lambda_{max} T = \text{const} = 2.8970 \times 10^{-3} \text{ m K} \qquad \text{(21d)}$$

where λ_{max} is in meters. The shape of the curve itself is given by *Planck's‡ law*, which

* Ludwig Boltzmann (1844–1906). From 1895 to his death by suicide in 1906, professor of physics at Vienna. The law was originally stated by Josef Stefan (1835–1893) and was independently demonstrated theoretically by Boltzmann. The latter is chiefly known for his contributions to the kinetic theory and the second law of thermodynamics.

† Wilhelm Wien (1864–1928). German physicist, awarded the Nobel prize in 1911 for his work in optics and radiation. He also made important discoveries about cathode rays and canal rays.

‡ Max Planck (1858–1947). Professor at the University of Berlin. He was awarded the Nobel prize in 1918 for his derivation of the law of blackbody radiation and other work in thermodynamics.

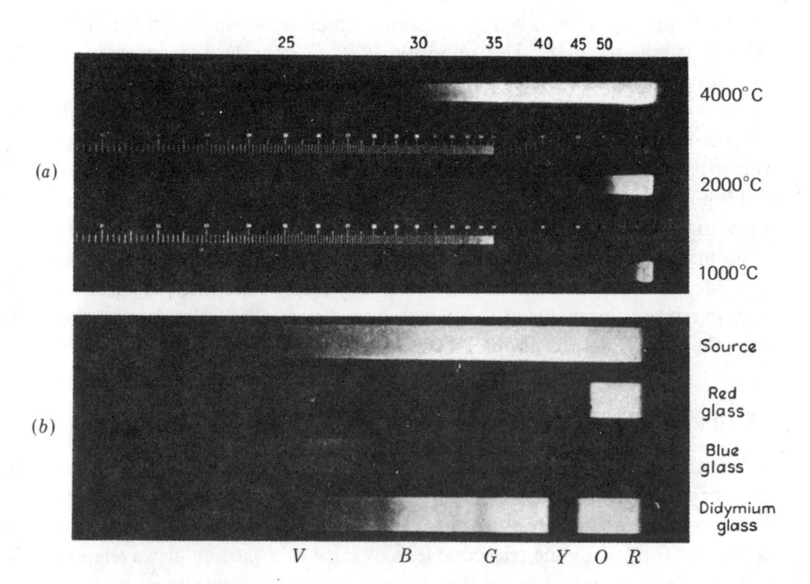

FIGURE 21G
Continuous spectra. (*a*) Continuous emission spectra of a solid at the three temperatures indicated, taken with a quartz spectrograph. The spectra for 1000 and 2000°C were obtained from a tungsten filament. That for 4000°C is from the positive pole of a carbon arc. The wavelength scale is marked in hundreds of angstroms. (*b*) Continuous absorption spectra. The upper spectrum is that of the source alone, extending roughly from 4000 to 6500 Å. The others show the effect on this spectrum of interposing three kinds of colored glass.

may be written

$$W_{B\lambda}\,\Delta\lambda = \frac{hc^3\,\Delta\lambda}{\lambda^5(e^{hc/\lambda kT} - 1)} \tag{21e}$$

where $W_{B\lambda}$ is the energy in the wavelength band between λ and $\lambda + \Delta\lambda$ in joules per second per square meter of surface, c is the velocity of light, λ is the wavelength, T is the absolute temperature, e is the base of the naperian logarithms, k is the *Boltzmann constant* determined from the general gas laws, and h is Planck's constant:

$$h = 6.6262 \times 10^{-34} \text{ J s}$$
$$k = 1.3805 \times 10^{-23} \text{ J/K}$$
$$c = 2.9979 \times 10^{8} \text{ m/s}$$
$$e = 2.7183$$

These constants are of course connected with those in the Stefan-Boltzmann and Wien laws, because Eq. (21c) can be obtained from Eq. (21e) by integrating it from $\lambda = 0$ to $\lambda = \infty$, while Eq. (21d) is obtained if we differentiate Eq. (21e) with respect to λ and equate to zero to obtain the maximum value. These equations apply, of course, only to the radiation from an *ideal* blackbody. This can never be strictly realized experimentally, but it is approximated by a black surface or a hollow cavity with a small opening. The quantity $W_{B\lambda}\, d\lambda$ denotes the emission of unpolarized radiation per square meter per second in all directions in a range $d\lambda$.

A source of a continuous spectrum in the ultraviolet region is sometimes desired for the study of absorption spectra in this region. Hot solids are unsuitable for this purpose, because of the relatively small amount of ultraviolet light they emit, even at the highest temperatures available. It has been found that for this purpose a vacuum-tube discharge through hydrogen gas at 5 to 10 mm pressure is very satisfactory. If a current of a few tenths of an ampere is passed through a tube with a rather wide capillary (5 mm diameter) at 2000 V, a very intense continuous spectrum is obtained. The maximum intensity of this continuum lies in the violet, but it extends far down into the ultraviolet, to about 1700 Å.

21.10 LINE SPECTRA

When the slit of a prism or grating spectroscope is illuminated with the light from a mercury arc, several lines of different color are seen in the eyepiece. Photographs of common line spectra are shown in Fig. 21H(*a*) to (*j*). Each of these lines is an image of the slit formed by the telescope lens by light of a particular wavelength. The different wavelengths are deviated through different angles by the prism or grating; hence the line images are separated. It is important to realize that line spectra derive their name from the fact that a *slit* is customarily used, whose image constitutes the line. If a point, a disk, or any other form of aperture were used in the collimator, the spectrum lines would become points, disks, etc., as the case may be. Frequently, in photographing the spectra from astronomical sources, the collimator is dispensed with entirely, and a prism or grating placed in front of the telescope lens converts the telescope into a spectroscope. In this case, each "line" in the spectrum has the shape of the source. For example, Fig. 21H(*h*) shows the spectrum of the sun at the instant preceding a total eclipse, when the usual dark-line absorption spectrum is replaced by an emission spectrum from the gases of the solar atmosphere, giving the so-called flash spectrum. The chief use of a slit is to produce narrow images, so that the images in different wavelengths do not overlap.

The most intense sources of line spectra are metallic arcs and sparks, although vacuum tubes containing hydrogen or one of the rare gases are very suitable. Flames are often used, because the spectra they give are in general simpler, being not so rich in lines. All common sources of line emission or line absorption spectra are gases. Furthermore, it is now known that only the *individual atoms* give true line spectra. That is, when a molecular compound is used in the source, such as methane gas (CH_4) in a discharge tube, or sodium chloride in a "cored" carbon arc, the lines observed are due to the elements and not to the molecules. For example, methane gives a strong spectrum due to hydrogen, and it is well known that sodium chloride gives the yellow sodium lines. Lines due to carbon and chlorine do not appear with appreciable intensity because these elements are more difficult to excite to emission and their strongest lines lie in the ultraviolet and not in the visible part of the spectrum. In Table 21A are given the wavelengths of the lines in certain commonly used emission spectra.

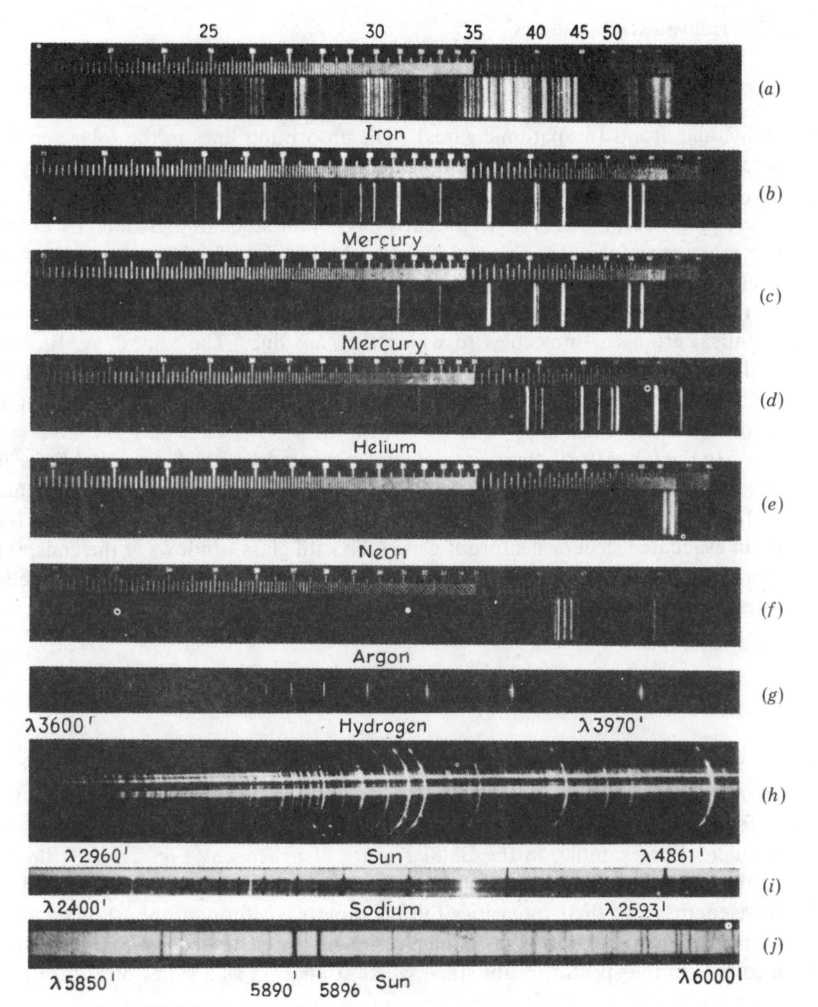

FIGURE 21H
Line spectra. (*a*) Spectrum of the iron arc. The emission spectra (*a*) to (*f*) were all taken with the same quartz spectrograph. Mercury spectrum from an arc enclosed in (*b*) quartz and (*c*) glass. (*d*) Helium in a glass discharge tube. (*e*) Neon in a glass discharge tube. (*f*) Argon in a glass discharge tube. (*g*) Balmer series of hydrogen in the ultraviolet, $\lambda = 3600$ to 4000 Å. This is a grating spectrum. The faint lines on either side of the stronger members are false lines called ghosts (Sec. 17.12). (*h*) Flash spectrum, showing the emission spectrum from the gaseous chromosphere of the sun. This is a grating spectrum taken without a slit at the instant immediately preceding a total eclipse, when the rest of the sun is covered by the moon's disk. The two strongest images, the H and K lines of calcium, show marked prominences, or clouds, of calcium vapor. Other strong lines are due to hydrogen and helium. (*i*) Line absorption spectrum of sodium in the ultraviolet, taken with a grating. The bright lines in the background arise in the source, which here was a carbon arc. Note the slight continuous absorption beyond the series limit. (*j*) Solar spectrum in the neighborhood of the D lines. The two strong lines are absorbed by sodium vapor in the chromosphere and together constitute the first member of the series at 5892 Å shown in (*i*).

Line *absorption* spectra are obtained only with gases ordinarily composed of individual atoms (monatomic gases). The absorption lines in the solar spectrum are due to atoms which exist as such, rather than combined as molecules, only because of the high temperature and low pressure in the "reversing layer" of the sun's atmosphere [Fig. 21H(*h*) and (*j*)]. In the early days of the study of these lines by Fraunhofer, the more prominent ones were designated by letters. The Fraunhofer lines are very useful bench marks in the spectrum, for instance in the measurement and specification of refractive indices. Hence we give here, in Table 21B, their wavelengths and the chemical atoms or molecules to which they are due. The "lines" A, B, and α are really bands, absorbed by the oxygen in the *earth's* atmosphere. It will be seen that b_4 and G are blends of two lines which are not ordinarily resolved but are due to different elements.

In the laboratory, there are only a few substances which are suitable for observing line absorption spectra, because the absorption lines of most monatomic gases lie far in the ultraviolet. The alkali metals are one exception, and if sodium is heated in an evacuated steel or heatproof glass tube with glass windows at the ends, the spectrum of light from a tungsten source viewed through the tube will show the sodium lines in absorption [Fig. 21H(*i*)]. They appear as dark lines against the ordinary continuous emission spectrum.

21.11 SERIES OF SPECTRAL LINES

In the spectra of some elements, lines are observed which obviously belong together to form a *series* in which the spacing and intensities of the lines change in a regular manner. For example, in the Balmer series of hydrogen [Fig. 21H(*g*)] the spacing of the lines decreases steadily as they proceed into the ultraviolet toward shorter wavelengths, and their intensities fall off rapidly. Although only the first four lines lie in the visible region, the Balmer series has been traced by photography to 31 members in the spectra of hot stars, where it appears as a series of absorption lines.

Table 21A WAVELENGTHS, IN ANGSTROMS, OF SOME USEFUL SPECTRAL LINES*

Sodium	Mercury	Helium	Cadmium	Hydrogen
5.889.95 *s*	4046.56 *m*	4387.93 *w*	4678.16 *m*	6562.82 *s*
5895.92 *m*	4077.81 *m*	4437.55 *w*	4799.92 *s*	4861.33 *m*
	4358.35 *s*	4471.48 *s*	5085.82 *s*	4340.46 *w*
	4916.04 *w*	4713.14 *m*	6438.47 *s*	4101.74 *w*
	5460.74 *s*	4921.93 *m*		
	5769.59 *s*	5015.67 *s*		
	5790.65 *s*	5047.74 *w*		
		5875.62 *s*		
		6678.15 *m*		

* *s* = strong, *m* = medium, and *w* = weak.

The absorption spectrum of sodium vapor shows a remarkably long series of lines, each of which is a close doublet [not resolved in Fig. 21H(*i*)], known as the *principal series.* This series also appears in emission from the arc or flame, and the well-known D lines constitute the first doublet of the series. In the sodium spectrum from a flame, about 97 percent of the intensity in this series is in the first member. The emission spectra of the alkalis also show two other series of doublets in the visible region, known as the *sharp* and *diffuse* series. A fourth weak series in the infrared is called the *fundamental* series. The alkaline-earth metals, such as calcium, show two such sets of series—one of single lines, the other of triplets.

A characteristic of any particular series is the approach of the higher series members to a certain limiting wavelength, known as the *limit* or *convergence* of the series. In approaching this limit, the lines crowd closer and closer together, so that there is theoretically an infinite number of lines before the limit is actually reached. Beyond the limit a rather faint continuous spectrum is sometimes observed in emission; in absorption a region of continuous absorption can always be observed if the absorbing vapor is sufficiently dense [Fig. 21H(*i*)]. The series limits furnish the clue to the identification of the type to which the series belongs. Thus the sharp and diffuse series approach the same limit, while the principal series approaches another limit which for the alkalis lies at shorter wavelengths.

21.12 BAND SPECTRA

The most convenient sources of band spectra for laboratory observation are the carbon arc cored with a metallic salt, the vacuum tube, and the flame. Calcium or barium salts are suitable in the arc or flame, and carbon dioxide or nitrogen in a vacuum tube. As observed with a spectroscope of small dispersion, these spectra present a typical appearance which distinguishes them at once from line spectra [Fig. 21I(*a*) to (*d*)]. Many bands are usually observed, each with a sharp edge on one side called the *head.*

Table 21B THE MOST INTENSE FRAUNHOFER LINES

Designation	Origin	Wavelength, Å	Designation	Origin	Wavelength, Å
A	O_2	7594–7621*	b_4	Mg	5167.343
B	O_2	6867–6884*	c	Fe	4957.609
C	H	6562.816	F	H	4861.327
α	O_2	6276–6287*	d	Fe	4668.140
D_1	Na	5895.923	e	Fe	4383.547
D_2	Na	5889.953	G′	H	4340.465
D_3	He	5875.618	G	Fe	4307.906
E_2	Fe	5269.541	G	Ca	4307.741
b_1	Mg	5183.618	g	Ca	4226.728
b_2	Mg	5172.699	h	H	4101.735
b_3	Fe	5168.901	H	Ca^+	3968.468
b_4	Fe	5167.491	K	Ca^+	3933.666

* Band.

From the head, the band shades off gradually on the other side. In some band spectra several closely adjacent bands, overlapping to form *sequences*, will be seen [Fig. 21I(*b*) and (*d*)], while in others the bands are spaced fairly widely, as in Fig. 21I(*c*). When the high dispersion and resolving power of a large grating are used, each band is found to be actually composed of many fine lines, arranged with obvious regularity into series called *branches* of the band. In Fig. 21I(*e*), two branches will be seen starting in opposite directions from a pronounced gap, where no line appears. In (*f*) the band is double, and the two branches of the left-hand member can be seen running side by side.

Various sorts of evidence point to the conclusion that band spectra arise from *molecules*, i.e., combinations of two or more atoms. Thus it is found that while the atomic or line spectrum of calcium is independent of which salt we put in the arc, we obtain different bands by using calcium fluoride, calcium chloride, or calcium bromide. Also, the bands appear in those types of sources where the gas receives less violent treatment. Nitrogen in a vacuum tube subjected to an ordinary uncondensed discharge shows only the band spectrum, whereas if a condensed discharge is used, the line spectrum appears. The most conclusive evidence lies in the fact that the absorption spectrum of a gas which is known to be molecular (O_2, N_2) shows bands but no lines, owing to the absence of any dissociation into atoms. Furthermore, it is found that any simple band spectrum, like those described and illustrated above, is due to a *diatomic* molecule. When calcium fluoride (CaF_2) is put into the arc, the bands observed are due to CaF. The violet bands in the uncored carbon arc are due to CN, the nitrogen coming from the air [Fig. 21I(*e*)]. Carbon dioxide in a vacuum tube gives the spectrum of CO, and there are many other examples of this type of dissociation of the more complex molecules into diatomic ones.

The attempt to interpret the various definite frequencies emitted by the atoms of a gas in producing a line spectrum occupied the best minds in physics during the early part of the twentieth century and eventually had most important consequences. Just as the frequencies of vibration of a violin string give sound waves whose frequencies bear the simple ratio of whole numbers to the fundamental note, it was first supposed that the frequencies of the light in the various spectral lines should bear some definite relation to each other, which would furnish the clue to the modes of vibration of the atom and to its structure. This has proved to be the case, though in a very different way than was at first anticipated. The definite relation of frequencies is actually found in spectral series. However, it will be seen at once that the atomic frequencies do not behave like those of a violin string. In the latter the overtones increase steadily toward an infinite frequency (zero wavelength), while the frequencies in a spectral series approach a definite limiting value. The complete explanation of line spectra has now been obtained by developing an entirely new theory, called the *quantum theory*.* Although this theory is in many respects in direct contradiction to the electromagnetic theory, the latter proved an invaluable guide in attacking such problems as the intensity and polarization of spectral lines. It also gave the first clue

* For an elementary treatment of atomic spectra see H. E. White, "Introduction to Atomic Spectra," McGraw-Hill Book Company, New York, 1934. For a discussion of band spectra, see G. Herzberg, "Molecular Spectra and Molecular Structure," vol. 1, "Diatomic Molecules," D. Van Nostrand Company, Inc., New York, 1950.

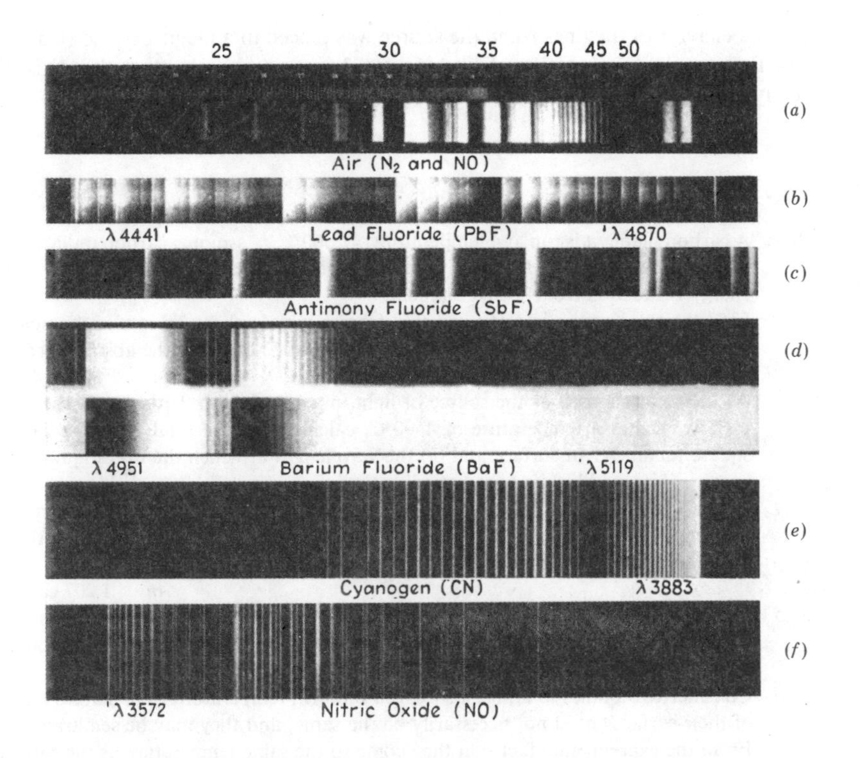

FIGURE 21I
Band spectra. (*a*) Spectrum of a discharge tube containing air at low pressure. Four band systems are present: the γ bands of NO (λ 2300 to 2700 Å), negative nitrogen bands (N_2^+, λ 2900 to 3500 Å), second positive nitrogen bands (N_2, λ 2900 to 5000 Å), and first positive nitrogen bands (N_2, λ 5500 to 7000 Å). (*b*) Spectrum of a high-frequency discharge in lead fluoride vapor. These bands, due to PbF, fall in prominent sequences. (*c*) Spectrum showing part of one band system of SbF, obtained by vaporizing antimony fluoride into active nitrogen. (*b*) and (*c*) were taken with a large quartz spectrograph. (*d*) Emission and absorption band spectra of BaF: emission from a carbon arc cored with BaF_2; absorption of BaF vapor in an evacuated steel furnace. The bands are closely grouped in sequences. Second order of 21-ft grating. (*e*) CN band at λ 3883 from an argon discharge tube containing carbon and nitrogen impurities. Second order of grating. (*f*) Band in the ultraviolet spectrum of NO, obtained from glowing active nitrogen containing a small amount of oxygen. Second order of grating. [(*b*) *and* (*c*) *courtesy of G. D. Rochester.*]

to the behavior of the lines when the source was placed in a magnetic field (Chap. 31). For the complete explanation of line spectra, however, the quantum theory is absolutely essential. We shall return to this subject in Chap. 29.

PROBLEMS

21.1 A carbon filament is run at a temperature of 2500°C. Assuming carbon radiates at this temperature as though it were a blackbody, find the wavelength at which maximum energy is radiated from such a filament.

21.2 Find the total power in watts radiated from a metal sphere 3.0 mm in diameter, the sphere being maintained at a temperature of 2200°C. Assume the absorptance of the surface to be 0.70 and independent of wavelength. *Ans.* 41.97 W

21.3 A carbon arc is used as the source of light in a searchlight. If the tip of the positive carbon reaches a temperature of 4500°C, calculate (*a*) the total power radiated per square millimeter of surface and (*b*) the wavelength at which the maximum radiation occurs. Assume blackbody radiation.

21.4 A small bead of metal is placed in the hollow tip of an iron arc. The bead rises to a temperature of 3027°C where its overall absorptance is 75.0 percent. Find the total heat energy radiated in calories per square millimeter per second. *Ans.* 1.207 cal/s mm^2

21.5 Copper is melted in a furnace; the molten metal surface has an overall absorptance of 82 percent. Calculate the total power radiated per square centimeter in (*a*) joules per second and (*b*) calories per second.

21.6 Consider two bodies in an enclosure at a uniform temperature. The nature and area of their surfaces need not necessarily be the same, and they may be semitransparent. From the experimental fact that they come to the same temperature as the surroundings, show by the energy emitted, absorbed, reflected, and transmitted by each that Kirchhoff's law of radiation must hold.

22
ABSORPTION AND SCATTERING

When a beam of light is passed through matter in the solid, liquid, or gaseous state, its propagation is affected in two important ways: (1) the intensity will always decrease to a greater or lesser extent as the light penetrates farther into the medium, and (2) the velocity will be less in the medium than in free space. The loss of intensity is chiefly due to absorption, although under some circumstances scattering may play an important part. In this chapter we shall discuss the consequences of absorption and scattering, while the effect of the medium on the velocity, which comes under dispersion, we shall consider in the following chapter. The term *absorption* as used in this chapter refers to the decrease of intensity of light as it passes through a substance (Sec. 11.9). It is important to distinguish this definition from that of absorptance, which is given in Sec. 21.8. The two terms refer to different physical quantities, but there are certain relations between them, as we shall now see.

22.1 GENERAL AND SELECTIVE ABSORPTION

A substance is said to show *general absorption* if it reduces the intensity of all wavelengths of light by nearly the same amount. For visible light this means that the transmitted light, as seen by the eye, shows no marked color. There is merely a reduction of the total intensity of the white light, and such substances therefore appear to be gray. No substance is known which absorbs all wavelengths equally, but some, such as suspensions of lamp black or thin semitransparent films of platinum, approach this condition over a fairly wide range of wavelengths.

By *selective absorption* is meant the absorption of certain wavelengths of light in preference to others. Practically all colored substances owe their color to the existence of selective absorption in some part or parts of the visible spectrum. Thus a piece of green glass absorbs completely the red and blue ends of the spectrum, the remaining portion in the transmitted light giving a resultant sensation of green to the eye. The colors of most natural objects such as paints, flowers, etc., are due to selective absorption. These objects are said to show pigment or *body color*, as distinguished from *surface color*, since their color is produced by light which penetrates a certain distance into the substance. Then, by scattering or reflection, it is deviated and escapes from the surface, but only after it has traversed a certain thickness of the medium and has been robbed of the colors which are selectively absorbed. In all such cases the absorptance of the body will be proportional to its true absorption and will depend in the same way upon wavelength. Surface color, on the other hand, has its origin in the process of reflection at the surface itself (Sec. 22.7). Some substances, particularly metals like gold or copper, have a higher reflecting power for some colors than for others and therefore show color by reflected light. The transmitted light here has the complementary color, whereas in body color the color is the same for the transmitted and reflected light. A thin gold foil, for example, looks yellow by reflection and blue-green by transmission. As mentioned in Sec. 21.8, the body absorption of these materials is very high. This causes a high reflectance and a correspondingly low absorptance.

22.2 DISTINCTION BETWEEN ABSORPTION AND SCATTERING

In Fig. 22A let light of intensity I_0 enter a long glass cylinder filled with smoke. The intensity I of the beam emerging from the other end will be less than I_0. For a given density of smoke, experiment shows that I depends on the length d of the column according to the exponential law stated in Sec. 11.9:

● $$I = I_0 e^{-\alpha d} \qquad (22a)$$

Here α is usually called the absorption coefficient, since it is a measure of the rate of loss of light from the direct beam. However, most of the decrease of intensity of I is in this case not due to a real disappearance of the light but results from the fact that some light is *scattered* to one side by the smoke particles and thus removed from the direct beam. Even with a very dilute smoke, a considerable intensity I_s of scattered light may easily be detected by observing the tube from the side in a darkened room. Rays of sunlight seen to cross a room from a window are made visible by the fine dust particles suspended in the air.

True *absorption* represents the actual disappearance of the light, the energy of

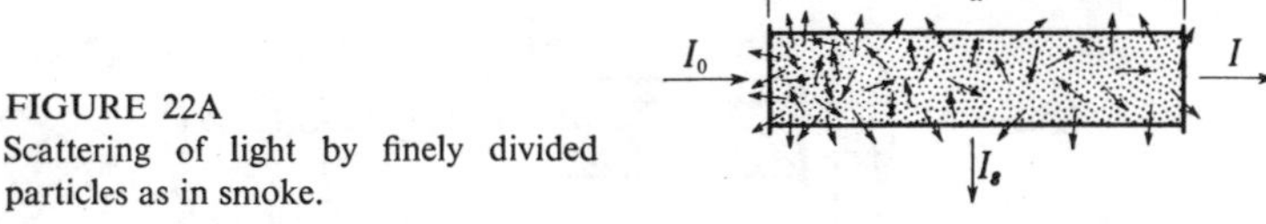

FIGURE 22A
Scattering of light by finely divided particles as in smoke.

which is converted into *heat motion* of the molecules of the absorbing material. This will occur to only a small extent in the above experiment, so that the name "absorption coefficient" for α is not appropriate in this case. In general, we can regard α as being made up of two parts, α_a due to true absorption and α_s due to scattering. Equation (22a) then becomes

$$I = I_0 e^{-(\alpha_a + \alpha_s)d} \qquad (22b)$$

In many cases either α_a or α_s may be negligible with respect to the other, but it is important to realize the existence of these two different processes and the fact that in many cases both may be operating.

22.3 ABSORPTION BY SOLIDS AND LIQUIDS

If monochromatic light is passed through a certain thickness of a solid or of a liquid enclosed in a transparent cell, the intensity of the transmitted light may be much smaller than that of the incident light, owing to absorption. If the wavelength of the incident light is changed, the amount of absorption will also change to a greater or less extent. A simple way of investigating the amount of absorption for a wide range of wavelengths simultaneously is illustrated in Fig. 22B. S_1 is a source which emits a continuous range of wavelengths, such as an ordinary tungsten-filament lamp. The light from this source is rendered parallel by the lens L_1 and traverses a certain thickness of the absorbing medium M. It is then focused by L_2 on the slit S_2 of a prism spectrograph, and the spectrum photographed on the plate P. If M is a "transparent" substance like glass or water, the part of the spectrum on P representing visible wavelengths will be perfectly continuous, as if M were not present. If M is colored, part of the spectrum will be blotted out, corresponding to the wavelengths removed by M, and we call this an *absorption band*. For solids and liquids, these bands are almost always continuous in character, fading off gradually at the ends. Examples of such absorption bands were shown in Fig. 21G(*b*).

Even a substance which is transparent to the visible region will show such selective absorption if the observations are extended far enough into the infrared or the ultraviolet region. Such an extension involves considerable experimental difficulty when a prism spectrograph is used, because the material of the prism and lenses (usually glass) may itself have strong selective absorption in these regions. Thus flint glass cannot be used much beyond 25,000 Å (or 2.5 μm) in the infrared or beyond about 3800 Å in the ultraviolet. Quartz will transmit somewhat farther in the infrared and much farther in the ultraviolet. Table 22A shows the limits of the regions over which various transparent substances used for prisms will transmit an appreciable amount of light.

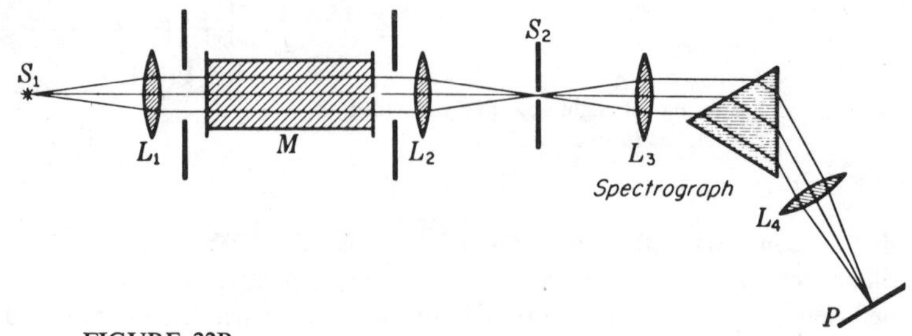

FIGURE 22B
Experimental arrangement for observing the absorption of light by solids, liquids, or gases.

Prisms for infrared investigations are usually of rock salt, while for the ultraviolet quartz is most common. In an ultraviolet spectrograph, there is no advantage in using fluorite unless air is completely removed from the light path, because air begins to absorb strongly below 1850 Å. Also, specially prepared photographic plates must be used below this wavelength, since the gelatin of the emulsion by its absorption renders ordinary plates insensitive below about 2300 Å. In the infrared, photography can now be used as far as 13,000 Å, thanks to innovative methods of sensitizing plates. Beyond this, an instrument based upon measurement of the heat produced, such as a thermopile, is usually used, although as far as 6 μm the *photoconductive cell*, utilizing the change of electrical resistance upon illumination, gives greater sensitivity.

When absorption measurements are extended over the whole electromagnetic spectrum, it is found that no substance exists which does not show strong absorption for some wavelengths. The metals exhibit general absorption, with a very minor dependence on wavelength in most cases. There are exceptions to this, however, as in the case of silver, which has a pronounced "transmission band" near 3160 Å (see Fig. 25N). A film of silver which is opaque to visible light may be almost entirely transparent to ultraviolet light of this wavelength. Dielectric materials, which are poor conductors of electricity, exhibit pronounced selective absorption which is

Table 22A

	Limit of transmission, Å	
Substance	Ultraviolet	Infrared
Crown glass	3500	20,000
Flint glass	3800	25,000
Quartz (SiO_2)	1800	40,000
Fluorite (CaF_2)	1250	95,000
Rock salt (NaCl)	1750	145,000
Sylvin (KCl)	1800	230,000
Lithium fluoride	1100	70,000

most easily studied when scattering is avoided by having them in a homogeneous condition such as that of a single crystal, a liquid, or an amorphous solid. In a general way, it may be said that such substances are more or less transparent to X rays and γ rays, i.e., light waves of wavelength below about 10 Å. Proceeding toward longer wavelengths, we encounter a region of very strong absorption in the extreme ultraviolet, which in some cases may extend to the visible region, or beyond, and in others may stop somewhere in the near ultraviolet (see Table 22A). In the infrared, further absorption bands are encountered, but these eventually give way to almost complete transparency in the region of radio waves. Thus for dielectrics we may usually expect three large regions of transparency, one at the shortest wavelengths, one at intermediate wavelengths (perhaps including the visible), and one at very long wavelengths. The limits of these regions vary a great deal in different substances, and one substance, such as water, may be transparent to the visible but opaque to the near infrared, while another, such as rubber, may be opaque to the visible but transparent to the infrared.

22.4 ABSORPTION BY GASES

The absorption spectra of all gases at ordinary pressures show narrow, dark lines. In certain cases it is also possible to find regions of continuous absorption (Sec. 21.12), but the outstanding characteristic of gaseous spectra is the presence of these sharp lines. If the gas is monatomic like helium or mercury vapor, the spectrum will be a true line spectrum, frequently showing clearly defined series. The number of lines in the absorption spectrum is invariably less than in the emission spectrum. For instance, in the case of the vapors of the alkali metals, only the lines of the principal series are observed under ordinary circumstances [Fig. 21H(*i*)]. The absorption spectrum is therefore simpler than the emission spectrum. If the gas consists of diatomic or polyatomic molecules, the sharp lines form the rotational structure of the absorption bands characteristic of molecules. Here again the absorption spectrum is the simpler, and fewer bands are observed in absorption than in emission from the same gas [Fig. 21I(*d*)].

22.5 RESONANCE AND FLUORESCENCE OF GASES*

Let us consider what happens to the energy of incident light which has been removed by the gas. If true absorption exists, according to the definition of Sec. 22.2, this energy will all be changed into heat and the gas will be somewhat warmed. Unless the pressure is very low, this is generally the case. After an atom or molecule has taken up energy from the light beam, it may collide with another particle, and an increase in the average velocity of the particles is brought about in such collisions. The length

* A comprehensive discussion of the various aspects of this subject is given in A. C. G. Mitchell and M. W. Zemansky, "Resonance Radiation and Excited Atoms," The Macmillan Company, New York, 1934.

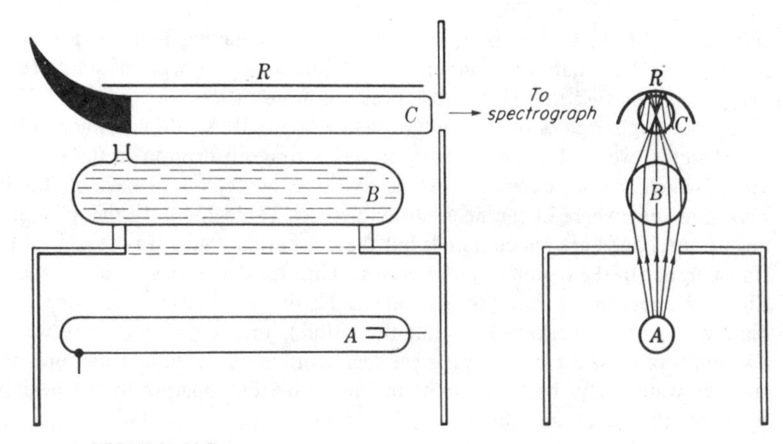

FIGURE 22C
Experimental arrangement for observing the fluorescence of iodine vapor with excitation by monochromatic light.

of time that an energized atom can exist as such before a collision is only about 10^{-7} or 10^{-8} s, and unless a collision occurs before this time, the atom will get rid of its energy as radiation. At low pressures, where the time between the collisions is relatively long, the gas will become a secondary source of radiation, and we do not have true absorption. The reemitted light in such cases usually has the same wavelength as the incident light, and is then termed *resonance radiation*. This radiation was discovered and extensively investigated by R. W. Wood.* The origin of its name is clear, since as has been mentioned the phenomenon is analogous to the resonance of a tuning fork. Under some circumstances the reemitted light may have a longer wavelength than the incident light. This effect is called *fluorescence*. In either resonance or fluorescence, some of the light is removed from the direct beam and dark lines will be produced in the spectrum of the transmitted light. Resonance and fluorescence are not to be classed as scattering. This distinction will be made clear in Sec. 22.12.

Resonance radiation from a gas can readily be demonstrated by the use of a sodium-arc lamp. A small lump of metallic sodium is placed in a glass bulb connected to a vacuum pump. The sodium is distilled from one part of the bulb to another by heating with a bunsen burner, thus liberating the large quantities of hydrogen always contained in this metal. After a high vacuum is attained, the bulb is sealed off and the light of the arc is focused by a lens on the bulb. The bulb must of course be observed from the side in a dark room. On gently warming the sodium with the flame, a cone of yellow light defining the path of the incident light will be seen. At higher

* R. W. Wood (1868–1955). Professor of experimental physics at the Johns Hopkins University. He pioneered in many fields of physical optics and also became one of the most colorful figures in American physics. His discoveries in optics are contained in his excellent text "Physical Optics," 3rd ed., The Macmillan Company, New York, 1934; reprinted (paperback) Dover Publications, Inc., New York, 1968.

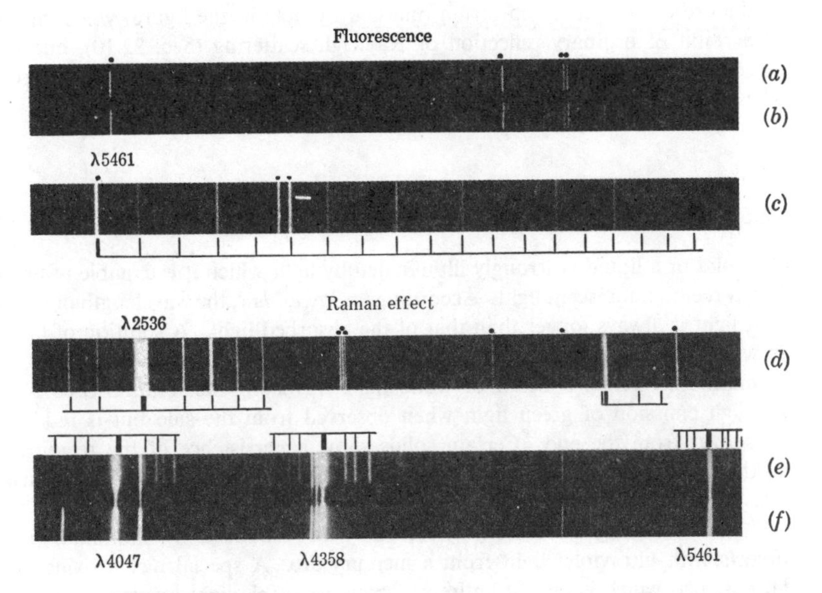

FIGURE 22D
Photographs of (*a*) mercury-arc spectrum; (*b*) fluorescence spectrum of iodine; (*c*) enlarged section of (*b*); (*d*) Raman spectrum of hydrogen (*courtesy of Rasetti*); (*e*) Raman spectrum of liquid carbon tetrachloride (*courtesy of M. Jeppeson*); (*f*) mercury arc.

temperatures, the glowing cone becomes shorter, and eventually is seen merely as a thin bright skin on the inner surface of the glass.

Fluorescence of a gas is most easily shown with iodine vapor, which consists of diatomic molecules, I_2. White light from a carbon arc will produce a greenish cone of light when focused in a bulb containing iodine vapor in vacuum at room temperature. A still more interesting experiment can be performed by using monochromatic light from a mercury arc, as shown in Fig. 22C. The source of light is a long horizontal arc A, which is enclosed in a box with a long slot cut in the top parallel to the arc. Immediately above this is a glass tube B filled with water. This acts as a cylindrical lens to concentrate the light along the axis of tube C, containing the iodine vapor in vacuum. The fluorescent light from the vapor is observed with a spectroscope pointed at the plane window on the end of tube C. The other end is tapered and painted black to prevent reflected light from entering the spectroscope, and a screen with a circular hole placed close to the window helps in this respect. A polished reflector R laid over C increases the intensity of illumination. If B contains a solution of potassium dichromate and neodymium sulfate, only the green line of mercury, $\lambda 5461$, is transmitted. Figure 22D(*b*) and (*c*) were reproduced from a spectrogram taken in this way, though with water in the tube B. Beside the lines of the ordinary mercury spectrum (marked by dots in the figure) which are present as a result of ordinary reflection or Rayleigh scattering (Sec. 22.10), one observes a series of almost equally spaced lines extending toward the red from the green line. These represent the fluorescent light of modified wavelength.

22.6 FLUORESCENCE OF SOLIDS AND LIQUIDS

If a solid or a liquid is strongly illuminated by light which it is capable of absorbing, it may reemit fluorescent light. According to *Stokes' law*, the wavelength of the fluorescent light is always longer than that of the absorbed light. A solution of fluorescein in water will absorb the blue portion of white light and will fluoresce with light of a greenish hue. Thus a beam of white light traversing the solution becomes visible through emission of green light when observed from the side but is reddish when looked at from the end. Certain solids show a persistence of the reemitted light, so that it lasts several seconds or even minutes after the incident light is turned off. This is called *phosphorescence*.

Very striking fluorescent effects may be produced by illuminating various objects with ultraviolet light from a mercury arc. A special nickel oxide glass can be obtained which is almost entirely opaque to visible light but transmits freely the strong group of mercury lines near $\lambda 3650$. If only this light from the arc comes through the glass, many organic as well as inorganic substances are rendered visible almost exclusively by their fluorescent light. Teeth illuminated by ultraviolet light appear unnaturally bright, but artificial teeth look perfectly black. The brilliant red of the ruby gemstone, as another example, is attributed to fluorescence. See Chap. 30.

22.7 SELECTIVE REFLECTION. RESIDUAL RAYS

Substances are said to show selective reflection when certain wavelengths are reflected much more strongly than others. This usually occurs at those wavelengths for which the medium possesses very strong absorption. We are speaking now of dielectric substances, i.e., those which are nonconductors of electricity. The case of metals is rather different and will be considered in Chap. 25. That there is an intimate connection between selective reflection, absorption, and resonance radiation can be seen from an interesting observation made by R. W. Wood with mercury vapor. At a pressure of a small fraction of a millimeter, mercury vapor shows the phenomenon of resonance radiation when illuminated by $\lambda 2536$ from a mercury arc. As the pressure of the vapor is increased, the resonance radiation becomes more and more concentrated toward the surface of the vapor where the incident radiation enters, i.e., on the inner wall of the enclosing vessel. Finally, at a sufficiently high pressure, the secondary radiation ceases to be visible except when viewed at an angle corresponding to the law of reflection. At this angle fully 25 percent of the incident light is reflected in the ordinary way, the remainder having been absorbed and transformed into heat by atomic collisions. However, this high reflection, which is comparable to that of metals in this region, exists only for the particular wavelength λ2536. Other wavelengths are freely transmitted. In this experiment we evidently have a continuous transition from resonance radiation to selective reflection.

A few solids which have strong absorption bands in the visible region also show selective reflection. The dye fuchsine is an example. Such substances have a peculiar metallic sheen by reflected light and are strongly colored. Their color is due to the very high reflection of a certain band of wavelengths—so high that it is frequently termed "metallic" reflection. It is this type of reflection that is responsible for surface color (Sec. 22.1).

The most important application of selective reflection has been its use in locating absorption bands which lie far in the infrared. For example, quartz is found to reflect 80 to 90 percent of radiation having a wavelength of about 8.5 μm, or 85,000 Å. The method of *residual rays* for isolating a narrow band of wavelengths is based upon this fact.* In Fig. 22E, S is a thermal source of radiation, giving a continuous spectrum. After reflection from the four quartz plates Q_1 to Q_4, the radiation is analyzed by means of a wire grating G and thermopile T. It is found to consist almost entirely of the wavelength 8.5 μm. Supposing this wavelength to be 90 percent reflected at each quartz surface and other wavelengths 4 percent reflected, we have, after four reflections, $(0.9)^4 = 0.66$ of the former remaining, but only $(0.04)^4 = 0.0000026$ of the latter. The wavelengths of the residual rays of many substances have been measured in this way. Among the longest wavelengths measured are those from sodium chloride, potassium chloride, and rubidium chloride at 52, 63, and 74 μm, respectively.

22.8 THEORY OF THE CONNECTION BETWEEN ABSORPTION AND REFLECTION

In the electromagnetic theory for the production of resonance radiation it is assumed that light waves are incident upon matter which contains *bound charges* capable of vibrating with a natural frequency equal to that of the impressed wave. Thus a charge e is acted upon by the electric field $\mathbf{E}$ with a force eE, and if $\mathbf{E}$ varies with a frequency exactly matching that with which the charged particle would normally vibrate, a large amplitude may be produced. As a result, the charged particle will reradiate an electromagnetic wave of the same frequency. In a gas at low pressure, where the atoms are relatively far apart, the frequency which can be absorbed will be sharply defined and there will be no systematic relation between the phases of the light reemitted from different particles. The observed intensity from N particles will then be just N times that due to one particle (Sec. 12.4). This is the case with resonance radiation.

If, on the other hand, the particles are close together and interacting strongly with each other, as in a liquid or solid, the absorption will not be limited to a sharply defined frequency but will spread over a considerable range. The result is that the

* For more extensive material on this subject, see R. W. Wood, "Physical Optics," 3d ed., pp. 516–519, The Macmillan Company, New York, 1934; reprinted (paperback) Dover Publications, Inc., New York, 1968.

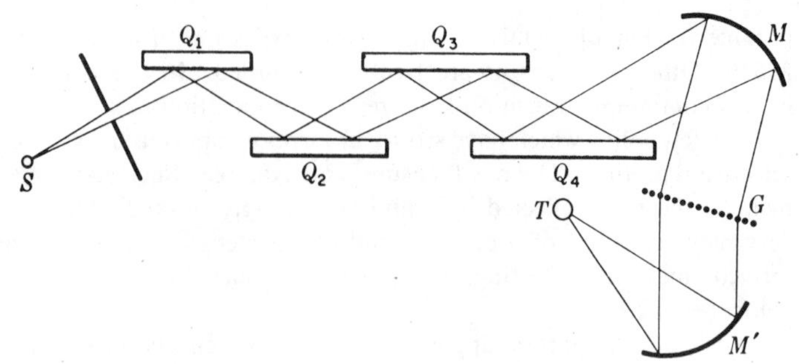

FIGURE 22E
Experimental arrangement for observing residual rays by selective reflection.

phases of the reemitted light from adjacent particles will agree. This will give rise to regular reflection, since the various secondary waves from the atoms in the surface will cooperate to produce a reflected wave front traveling off at an angle equal to the angle of incidence. In fact, this is just the conception used in applying Huygens' principle to prove the law of reflection. Hence selective reflection is also a phenomenon of resonance, and occurs strongly near those wavelengths corresponding to natural frequencies of the bound charges in the substance. The substance will not transmit light of these wavelengths; instead it reflects strongly. True absorption, or the conversion of the light energy into heat, may also occur to a greater or less extent because of the large amplitudes of the vibrating charges which are here involved. If absorption were entirely absent, the reflecting power would be 100 percent at the wavelengths in question.

22.9 SCATTERING BY SMALL PARTICLES

The lateral scattering of a beam of light as it traverses a cloud of fine suspended matter was mentioned in Sec. 22.2. That this phenomenon is closely connected both with reflection and with diffraction can be seen by consideration of Fig. 22F. In (*a*) is shown a parallel beam consisting of plane waves advancing toward the right and striking a small plane reflecting surface. The successive wave fronts drawn are one wavelength apart, so that here the size of the reflector is somewhat greater than a wavelength. The light coming off from the surface of the reflector is produced by the vibration of the electric charges in the surface with a definite phase relation, and the spherical wavelets produced by these vibrations cooperate to produce short segments of plane wave fronts. These are not sharply bounded at their edges by the reflected rays from the edges of the mirror (dotted lines) but spread out somewhat, owing to diffraction. In fact, the distribution of the intensity of the reflected light with angle is just that derived in Sec. 15.2 for the light transmitted by a single slit. The width of the reflector here takes the place of the slit width, so that we shall have greater spreading the smaller the width of the reflector relative to the wavelength.

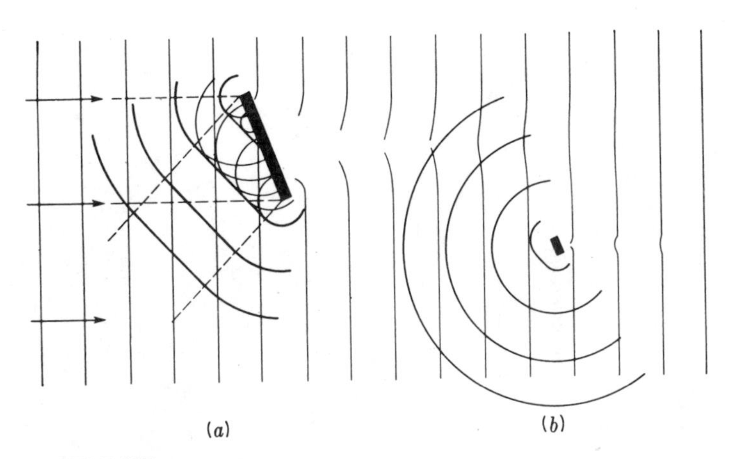

FIGURE 22F
The reflection and diffraction of light by small objects comparable in size with the wavelength of light.

In (*b*) of the figure, the reflector is much smaller than a wavelength, and here the spreading is so great that the reflected waves differ very little from uniform spherical waves. In this case the light taken from the primary beam is said to be scattered, rather than reflected, since the law of reflection has ceased to be applicable. Scattering is therefore a special case of diffraction. The wave scattered from an object much smaller than a wavelength of light will be spherical, regardless of whether or not the object has the plane form assumed in Fig. 22F(*b*). This follows from the fact that there can be no interference between the wavelets emitted by the several points on the surface of the scattering particle, inasmuch as the extreme points are separated by a distance much less than the wavelength.

The first quantitative study of the laws of scattering by small particles was made in 1871 by Rayleigh,* and such scattering is frequently called *Rayleigh scattering.* The mathematical investigation of the problem gave a general law for the intensity of the scattered light, applicable to any particles of index of refraction different from that of the surrounding medium. The only restriction is that the linear dimensions of the particles be considerably smaller than the wavelength. As we might expect, the scattered intensity is found to be proportional to the incident intensity and to the square of the volume of the scattering particle. The most interesting result, however, is the dependence of scattering on wavelength. With a given size of the particles, long waves would be expected to be less effectively scattered than short ones, because the particles present obstructions to the waves which are smaller *compared with the wavelength* for long waves than for short ones. In fact, as will be proved in Sec. 22.13, the intensity is proportional to $1/\lambda^4$:

$$I_s = k \frac{1}{\lambda^4}$$

* Several interesting papers laying the foundation of the theory will be found in "The Scientific Papers of Lord Rayleigh," vols. 1 and 4, Cambridge University Press, New York, 1912.

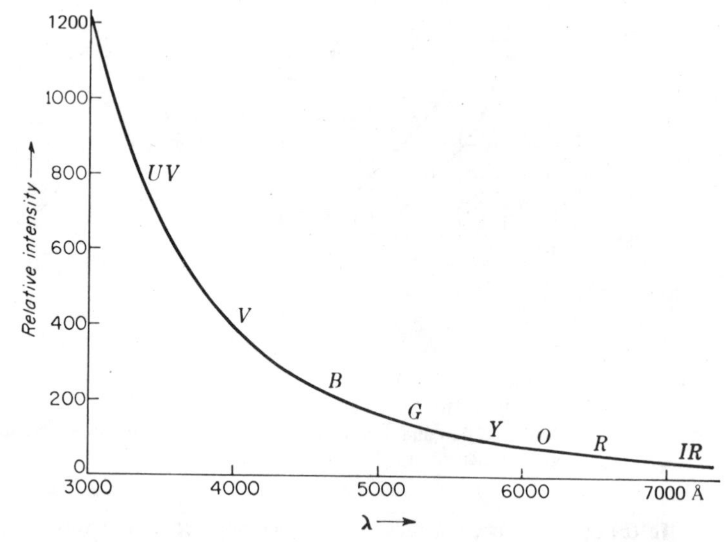

FIGURE 22G
Intensity of scattering versus wavelength according to Rayleigh's law.

Since red light, $\lambda 7200$, has a wavelength 1.8 times as great as violet light, $\lambda 4000$, the law predicts $(1.8)^4$ or 10 times greater scattering for the violet light from particles much smaller than the wavelength of either color. Figure 22G gives a quantitative plot of this relation.

If white light is scattered from sufficiently fine particles, such as those in tobacco smoke, the scattered light always has a bluish color. If the size of the particles is increased until they are no longer small compared with the wavelength, the light becomes white, as a result of ordinary diffuse reflection from the surface of the particles. The blue color seen with very small particles and its dependence on the size of the particles were first studied experimentally by Tyndall,* and his name is often associated with the phenomenon. Chalk dust from an eraser, falling across a beam of light from a carbon arc, will illustrate very effectively the white light scattered by large particles.

22.10 MOLECULAR SCATTERING

If a strong beam of sunlight is caused to traverse a pure liquid which has been carefully prepared to be as free as possible of all suspended particles of dust, etc., observation in a dark room will show that there is a small amount of bluish light scattered

* John Tyndall (1820–1893). British physicist, after 1867 superintendent of the Royal Institution and colleague of Faraday. Tyndall was outstanding for his ability to popularize and clarify physical discoveries.

laterally from the beam. Although some of this light is still due to microscopic particles in suspension, which seem to be almost impossible to eliminate entirely, a certain amount appears to be attributable to the scattering by individual molecules of the liquid. At first sight it is surprising to find that the scattering from liquids is so feeble, in view of the large concentration of molecules present. It is, in fact, much weaker than the scattering from the same number of molecules of a gas. In the latter, the molecules are randomly distributed in space, and in any direction except the forward one the waves scattered by different molecules have perfectly random phases. Thus for N molecules the resultant intensity is just N times that scattered from any individual one (see Sec. 12.4). In a liquid, and even more so in a solid, the spatial distribution has a certain degree of regularity. Furthermore, the forces between molecules act to destroy the independence of phases (Sec. 22.8). The result is that the scattering from liquids and solids in directions other than forward is very weak indeed. The forward-scattered waves are strong and play an essential part in determining the velocity of light in the medium, as we shall see in the following chapter.

Lateral scattering from gases is also weak, but here the weakness is due to the relatively smaller number of scattering centers. When a great thickness of gas is available, however, as in our atmosphere, the scattered light is easily observed. It has been shown by Rayleigh that practically all the light that we see in a clear sky is due to scattering by the molecules of air. If it were not for our atmosphere, the sky would look perfectly black. Actually, molecular scattering causes a considerable amount of light to reach the observer in directions making an angle with that of the direct sunlight, and thus the sky appears bright. Its blue color is the result of the greater scattering of short waves. Rayleigh measured the relative amount of light of different wavelengths in sky light and found rather close agreement with the $1/\lambda^4$ law. The same phenomenon is responsible for the red color of the sun and surrounding sky at sunset. In this case, the scattering removes the blue rays from the direct beam more effectively than the red, and the very great thickness of the atmosphere traversed gives the transmitted light its intense red hue. An experiment demonstrating both the blue of the sky and the red of the sun at sunset is described in Sec. 24.15 and 24.16.

22.11 RAMAN* EFFECT

This is a scattering with change of wavelength somewhat similar to fluorescence. It differs from fluorescence, however, in two important respects. In the first place, the light which is incident on the scattering material must have a wavelength that does not correspond to one of the absorption lines or bands of the material. Otherwise we obtain fluorescence, as in the experiment of Sec. 22.5, where the green line of mercury is absorbed by the iodine vapor. In the second place, the intensity of the light scattered in the Raman effect is much less intense than most fluorescent light. For this reason the Raman effect is rather difficult to detect, and observations must usually be made by photography.

* C. V. Raman (1888–1971). Professor at the University of Calcutta. He was awarded the Nobel prize in 1930 for his work on scattering and for the discovery of the effect that bears his name.

The apparatus illustrated in Fig. 22C is well adapted to observations of the Raman effect.* For this purpose, a liquid or gas which is transparent to the incident light must be used in the tube C. It is convenient to fill tube B with a saturated solution of sodium nitrite, since this absorbs the ultraviolet lines of the mercury arc but transmits the blue-violet line $\lambda 4358$ with great intensity. Figure 22D(e) shows the Raman spectrum of CCl_4. It will be seen that the same pattern of Raman lines is excited by each of the strong mercury lines. Figure 22D(d) illustrates the Raman spectrum of gaseous hydrogen, showing two sets of lines on the side toward the red of the exciting line, which in this case was $\lambda 2536$. Occasionally still fainter lines are seen on the violet side, two of which are visible in (d) and three in (e). This is also sometimes observed in the case of fluorescence. Since the modified light in these lines has a shorter wavelength than the incident light, they represent a violation of Stokes' law (Sec. 22.6) and are called *anti-Stokes* lines.

22.12 THEORY OF SCATTERING

When an electromagnetic wave passes over a small elastically bound charged particle, the particle will be set into motion by the electric field **E**. In Sec. 22.8 we considered the case where the frequency of the wave was equal to the natural frequency of free vibration of the particle. We then obtain resonance and fluorescence under certain conditions, and selective reflection under others. In both cases there may exist a considerable amount of absorption. Scattering, on the other hand, takes place for frequencies not corresponding to the natural frequencies of the particles. The resulting motion of the particles is then one of *forced vibration*. If the particle is bound by a force obeying Hooke's law, this vibration will have the same frequency and direction as that of the electric force in the wave. Its amplitude, however, will be very much smaller than that which would be produced by resonance. Hence the amplitude of the scattered wave will be much less, and this accounts for the relative faintness of molecular scattering. The phase of the forced vibration will differ from that of the incident wave, and this fact is responsible for the difference of the velocity of light in the medium from that in free space. Thus scattering forms the basis of dispersion, which is to be discussed in the following chapter.

The electromagnetic theory is also capable of giving a qualitative picture of the changes of wavelength which occur in the Raman effect and in fluorescence. If the charged oscillator is bound by a force which does not obey Hooke's law, but some more complicated law, it will be capable of reradiating not only the impressed frequency but also various combinations of this frequency with the fundamental and overtone frequencies of the oscillator. For the complete explanation of these phenomena, however, the electromagnetic theory alone is not adequate. It cannot explain the actual magnitudes of the changes in frequency nor the fact that these are predominantly toward lower frequencies. For this, the quantum theory is required.

Rayleigh scattering yields a characteristic distribution of intensity in different

* For a description of the most efficient ways of observing Raman spectra, see G. R. Harrison, R. C. Lord, and J. R. Loofbourow, "Practical Spectroscopy," Prentice-Hall, Inc., Englewood Cliffs, N.J., 1948.

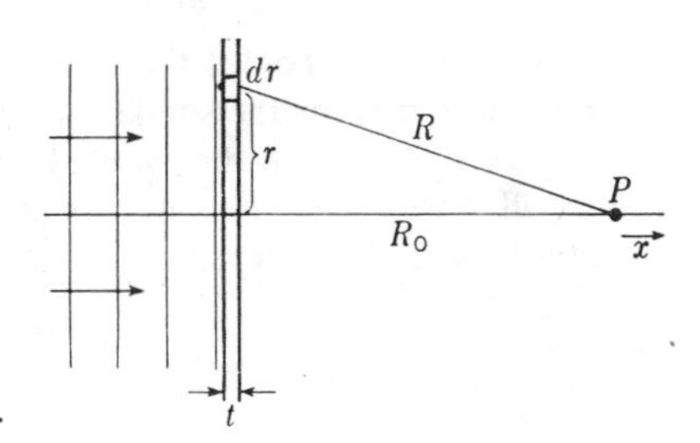

FIGURE 22H
Geometry of scattering by a thin lamina.

directions with respect to that of the primary beam. The scattered light is also strongly polarized. These features are in general agreement with the predictions of the electromagnetic theory. We shall not discuss them, however, until we have taken up the subject of polarization (see Sec. 24.17).

22.13 SCATTERING AND REFRACTIVE INDEX

The fact that the velocity of light in matter differs from that in vacuum is a consequence of scattering. The individual molecules scatter a certain part of the light falling on them, and the resulting scattered waves *interfere* with the primary wave, bringing about a change of phase which is equivalent to an alteration of the wave velocity. This process will be discussed in more detail in the chapter which follows, but here some simplified considerations may be used to show the connection between scattering and refractive index.

In Fig. 22H plane waves are shown striking an infinitely wide sheet of a transparent material, the thickness of which is small compared to the wavelength. Let the electric vector in this incident wave have unit amplitude, so that in the exponential notation (Sec. 14.8) it may be represented at a particular time by $E = e^{ikx}$. If the fraction of the wave that is scattered is small, the disturbance reaching some point P will be essentially the original wave, plus a small contribution due to the light scattered by all the atoms in the thin lamina. To evaluate the latter, we note that its intensity is proportional to the coefficient α_s of Eq. (22b). This measures the fractional decrease of intensity by scattering in traversing the small thickness t, to which the scattered intensity must be proportional. We therefore have

$$-\frac{dI}{I} = \alpha_s t \approx I_s \qquad \text{(22c)}$$

The intensity scattered by a single atom, since there are Nt atoms per unit area of the lamina, becomes

$$I_1 \approx \frac{\alpha_s t}{Nt} = \frac{\alpha_s}{N}$$

and the amplitude

$$E_1 \approx \sqrt{\frac{\alpha_s}{N}}$$

These relations hold if the scattered waves from the different centers are noncoherent, as is true for the smoke particles discussed in Sec. 22.2. The present case of Rayleigh scattering in the forward direction must be taken as *coherent*, however, so that all waves leave the scatterer in phase with each other. Then we must add amplitudes instead of intensities, and the total scattered amplitude

$$E_s \approx Nt\sqrt{\frac{\alpha_s}{N}} = t\sqrt{\alpha_s N}$$

The complex amplitude at P is obtained by integrating this quantity over the surface of the lamina, and adding it to the amplitude of the primary wave. The resultant then becomes

$$E + E_s = e^{ikR_0} + t\sqrt{\alpha_s N}\int_0^\infty \frac{2\pi r\,dr}{R}\,e^{ikR}$$

where the factor $1/R$ enters because of the inverse-square law. Now since $R^2 = R_0{}^2 + r^2$, we have $r\,dr = R\,dR$, and the integral may be written

$$\int_0^\infty \frac{2\pi}{R}\,e^{ikR} r\,dr = 2\pi\int_{R_0}^\infty e^{ikR}\,dR = \frac{2\pi}{ik}\left[e^{ikR}\right]_{R_0}^\infty$$

Since the wave trains always have a finite length, the scattering as $R \to \infty$ can contribute nothing to the coherent wave. Substituting the lower limit of the integral, we find

$$\begin{aligned} E + E_s &= e^{ikR_0} - t\sqrt{\alpha_s N}\,\frac{\lambda}{i}\,e^{ikR_0} \\ &= e^{ikR_0} + t\sqrt{\alpha_s N}\,i\lambda e^{ikR_0} \\ &= e^{ikR_0}(1 + i\lambda t\sqrt{\alpha_s N}) \end{aligned}$$

By our original assumption, the second term in parentheses is small compared with the first. These will be recognized as the first two terms in the expansion of $e^{i\lambda t\sqrt{\alpha_s N}}$, and may here be equated to it, giving

$$E + E_s = \exp ikR_0 \exp(i\lambda t\sqrt{\alpha_s N}) = \exp\left[i(kR_0 + \lambda t\sqrt{\alpha_s N})\right]$$

Thus the phase of the wave at P has been altered by the amount $\lambda t\sqrt{\alpha_s N}$. But we know (Sec. 13.15) that the presence of a lamina of thickness t and refractive index n gives a phase retardation of $(2\pi/\lambda)(n - 1)t$. Hence

$$\lambda t\sqrt{\alpha_s N} = \frac{2\pi}{\lambda}(n - 1)t$$

and finally

$$n - 1 = \frac{\lambda^2}{2\pi}\sqrt{\alpha_s N} \qquad (22d)$$

This important relation contains Rayleigh's law of scattering (Sec. 22.9). Since, by Eq. (22c), I_s is proportional to α_s, this scattered intensity varies as $1/\lambda^4$, assuming n to be independent of wavelength. In our derivation no absorption has been considered, so that the equation is valid only for wavelengths well away from any absorption bands. In the next chapter we shall see how the refractive index behaves as the wavelength approaches that of an absorption band.

PROBLEMS

22.1 A glass tube 3.50 m long contains a gas at normal atmospheric pressure. If the gas under these conditions has an absorption coefficient of 0.1650 m^{-1}, find the relative intensity of transmitted light. *Ans.* 0.561, or 56.1%

22.2 A hollow glass tube 35.0 cm long with end windows contains tiny smoke particles that produce Rayleigh scattering. Under these conditions it transmits 65.0 percent of the light. After precipitation of the smoke particles it transmits 88.0 percent of the light. Calculate the value of (*a*) the scattering coefficient and (*b*) the absorption coefficient.

22.3 A solid plastic rod 60.0 cm long transmits 85.0 percent of the light entering it at one end. When it is subjected to a strong beam of radiation, tiny particles are produced in it that give rise to Rayleigh scattering. Under these modified conditions the rod transmits 55.0 percent of the light. Calculate (*a*) the absorption coefficient and (*b*) the scattering coefficient.

22.4 A certain plastic rod 40.0 cm long has an absorption coefficient of 0.00429 cm^{-1}. If 50.0 percent of the light entering the end of this tube is transmitted, find (*a*) the scattering coefficient and (b) the total coefficient.
Ans. (*a*) 0.01304 cm^{-1}, (*b*) 0.01733 cm^{-1}

22.5 According to the data given in this chapter, are the residual rays of (*a*) rubidium chloride transmitted by rock salt (NaCl) and (*b*) sodium chloride transmitted by quartz?

22.6 The residual rays after five reflections from a certain type of crystal are 4.25×10^6 times more intense than radiation of adjacent wavelengths. Assuming the reflectance at the latter wavelengths to be 4.250 percent, what must be the reflectance at the center of the absorption band?

22.7 Calculate the ratio of the intensities of Rayleigh scattering for the two mercury lines, $\lambda = 2536$ Å in the ultraviolet region of the spectrum and $\lambda = 4916$ Å in the blue-green of the visible region. *Ans.* 14.123

22.8 Photographers know that an orange filter will cut out the bluish haze of scattered light and give better contrast in a landscape photograph. Assuming the spectral composition shown in Fig. 22G, what fraction of the scattered light is removed by a filter that absorbs the light below 5500 Å? The transmission of the camera lens and the film sensitivity limit the normal spectral range of the camera from 3900 from 6200 Å.

23
DISPERSION

The subject of dispersion concerns the speed of light in material substances and its variation with wavelength. Since the speed is c/n, any change in refractive index n entails a corresponding change of speed. We have seen in Sec. 1.4 that the dispersion of color which occurs upon refraction at a boundary between two different substances is direct evidence of the dependence of the n's on wavelength. In fact, measurements of the deviations of several spectral lines by a prism furnish the most accurate means of determining the refractive index, and hence the speed, as a function of wavelength.

23.1 DISPERSION OF A PRISM

When a ray traverses a prism, as shown in Fig. 23A, we can measure with a spectrometer the angles of emergence θ of the various wavelengths. The rate of change, $d\theta/d\lambda$, is called the *angular dispersion* of the prism. It can be conveniently represented as the product of two factors, by writing

$$\frac{d\theta}{d\lambda} = \frac{d\theta}{dn}\frac{dn}{d\lambda} \tag{23a}$$

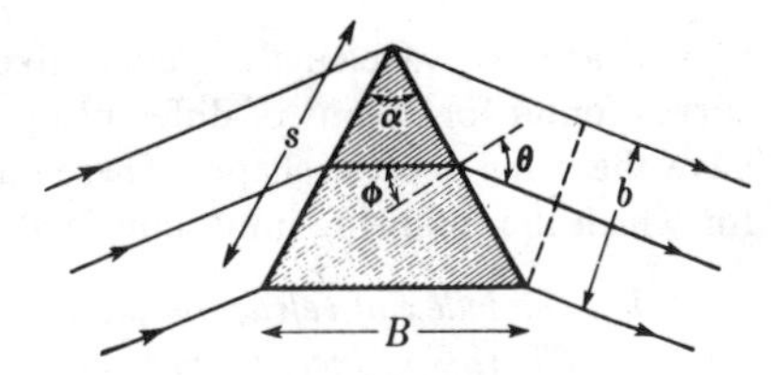

FIGURE 23A
Refraction by a prism at minimum deviation.

The first factor can be evaluated by geometrical considerations alone, while the second is a characteristic property of the prism material, usually referred to simply as its *dispersion*. Before considering the latter quantity, let us evaluate the geometrical factor $d\theta/dn$ for a prism, in the special case of minimum deviation.

For a given angle of incidence on the second face of the prism, we differentiate Snell's law of refraction $n = \sin\theta/\sin\phi$, regarding $\sin\phi$ as a constant, and obtain

$$\frac{d\theta}{dn} = \frac{\sin\phi}{\cos\theta}$$

This is not, however, the value to be used in Eq. (23a), which requires the rate of change of θ for a fixed direction of the rays incident on the *first* face. Because of the symmetry in the case of minimum deviation, it is obvious that equal deviations occur at the two faces, so that the total rate of change will be just twice the above value. We then have

$$\frac{d\theta}{dn} = \frac{2\sin\phi}{\cos\theta} = \frac{2\sin(\alpha/2)}{\cos\theta}$$

where α is the refracting angle of the prism. The result becomes still simpler when expressed in terms of lengths rather than angles. Designating by s, B, and b the lengths shown in Fig. 23A, we write

$$\frac{d\theta}{dn} = \frac{2s\sin(\alpha/2)}{s\cos\theta} = \frac{B}{b} \tag{23b}$$

Hence the required geometrical factor is just the ratio of the base of the prism to the linear aperture of the emergent beam, a quantity not far different from unity. The angular dispersion becomes

$$\frac{d\theta}{d\lambda} = \frac{B}{b}\frac{dn}{d\lambda} \tag{23c}$$

In connection with this equation, it is to be noted that the equation for the chromatic resolving power [Eq. (15j)] follows very simply from it upon the substitution of λ/b for $d\theta$.

23.2 NORMAL DISPERSION

In considering the second factor in Eq. (23a), let us start by reviewing some of the known facts about the variation of n with λ. Measurements for some typical kinds of glass give the results shown in Tables 23A and 23B. If any set of values of n is plotted against wavelength, a curve like one of those in Fig. 23B is obtained. The curves found for prisms of different optical materials will differ in detail but will all have the same general shape. These curves are representative of *normal dispersion,* for which the following important facts are to be noted:

1. *The index of refraction increases as the wavelength decreases.*
2. *The rate of increase becomes greater at shorter wavelengths.*
3. *For different substances the curve at a given wavelength is usually steeper the larger the index of refraction.*
4. *The curve for one substance cannot in general be obtained from that for another substance by a mere change in the scale of the ordinates.*

The first of these facts agrees with the common observation that in refraction by a transparent substance the violet is more deviated than the red. The second fact can also be expressed by saying that the dispersion increases with decreasing wavelength. This follows because the dispersion $dn/d\lambda$ is the slope of the curve (its negative

Table 23A REFRACTIVE INDEX FOR SEVERAL TRANSPARENT SOLIDS

Substance	Color wavelength λ, Å: Violet 4100	Blue 4700	Green 5500	Yellow 5800	Orange 6100	Red 6600
Crown glass	1.5380	1.5310	1.5260	1.5225	1.5216	1.5200
Light flint	1.6040	1.5960	1.5910	1.5875	1.5867	1.5850
Dense flint	1.6980	1.6836	1.6738	1.6670	1.6650	1.6620
Quartz	1.5570	1.5510	1.5468	1.5438	1.5432	1.5420
Diamond	2.4580	2.4439	2.4260	2.4172	2.4150	2.4100
Ice	1.3170	1.3136	1.3110	1.3087	1.3080	1.3060
Strontium titanate ($SrTiO_3$)	2.6310	2.5106	2.4360	2.4170	2.3977	2.3740
Rutile (TiO_2), E ray	3.3408	3.1031	2.9529	2.9180	2.8894	2.8535

Table 23B REFRACTIVE INDICES AND DISPERSIONS FOR SEVERAL COMMON TYPES OF OPTICAL GLASS
Unit of dispersion 1/Å × 10^{-5}

Wavelength λ, Å	Telescope crown n	$-\frac{dn}{d\lambda}$	Borosilicate crown n	$-\frac{dn}{d\lambda}$	Barium flint n	$-\frac{dn}{d\lambda}$	Vitreous quartz n	$-\frac{dn}{d\lambda}$
C 6563	1.52441	0.35	1.50883	0.31	1.58848	0.38	1.45640	0.27
6439	1.52490	0.36	1.50917	0.32	1.58896	0.39	1.45674	0.28
D 5890	1.52704	0.43	1.51124	0.41	1.59144	0.50	1.45845	0.35
5338	1.52989	0.58	1.51386	0.55	1.59463	0.68	1.46067	0.45
5086	1.53146	0.66	1.51534	0.63	1.59644	0.78	1.46191	0.52
F 4861	1.53303	0.78	1.51690	0.72	1.59825	0.89	1.46318	0.60
G′ 4340	1.53790	1.12	1.52136	1.00	1.60367	1.23	1.46690	0.84
H 3988	1.54245	1.39	1.52546	1.26	1.60870	1.72	1.47030	1.12

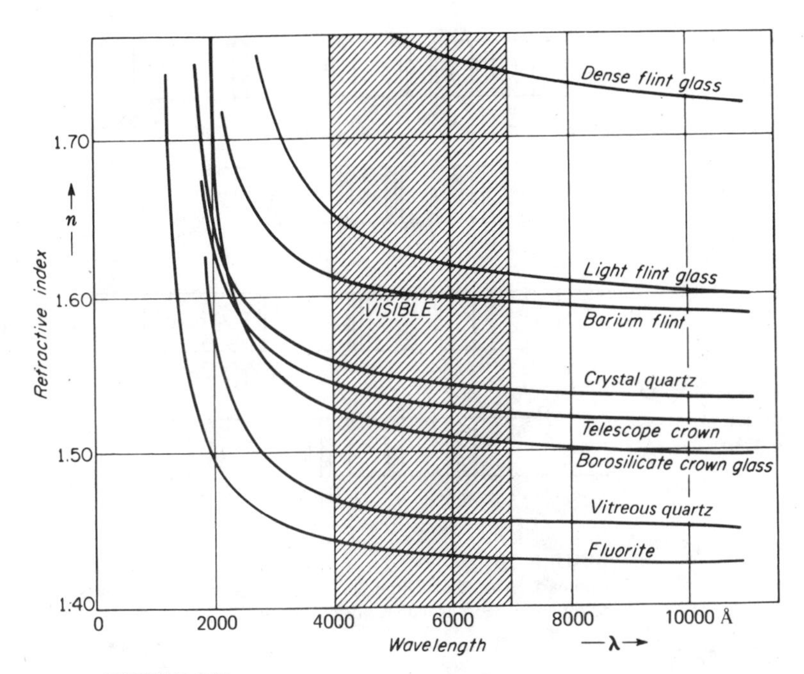

FIGURE 23B
Dispersion curves for several different materials commonly used for lenses and prisms.

sign is usually disregarded), which increases regularly toward smaller λ. An important consequence of this behavior of the dispersion is that in the spectrum formed by a prism the violet end of the spectrum is spread out on a much larger scale than the red end. The spectrum is therefore far from being a normal spectrum (Sec. 17.6). This will be clear from Fig. 23C, in which the spectrum of helium is shown diagrammatically as given by flint- and crown-glass prisms and by a grating used under the proper conditions to give a normal spectrum. In the prism spectra the wavelength scale is compressed toward the red end, as can be seen by comparison with the uniform scale of the normal spectrum.

The third fact stated above requires that for a substance of higher index of refraction, the dispersion $dn/d\lambda$ shall also be greater. Thus, comparing (*a*) and (*b*) in Fig. 23C, the flint glass has the higher index of refraction and gives a longer spectrum because of its greater dispersion. To compare the *relative* spacing of the lines in (*b*) with those in (*a*), the spectrum from crown glass has been enlarged, in (*c*), to have the same overall length between the two lines $\lambda3888$ and $\lambda6678$. When this is done, it is seen that there is not complete agreement with the lines of (*a*). In fact, the spectra from prisms of different substances will never agree exactly in the relative spacing of their spectrum lines. This is a consequence of the fourth of the above facts,

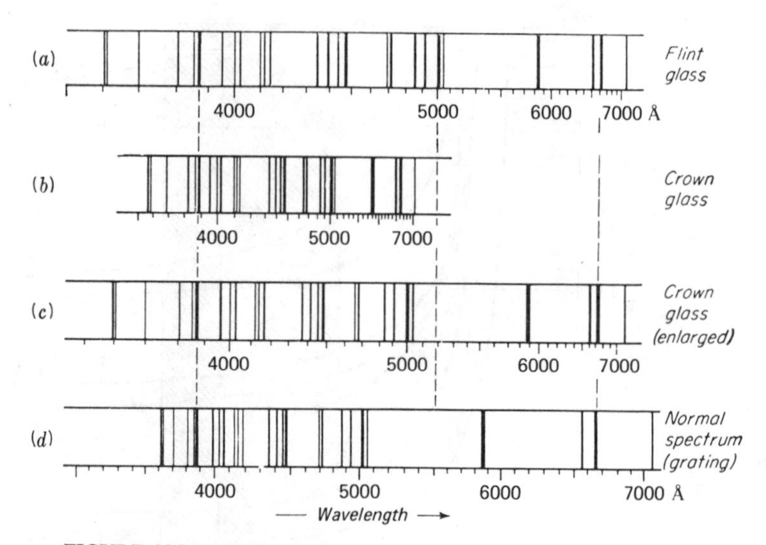

FIGURE 23C
Comparison of the helium spectrum produced by flint-glass and crown-glass prism spectrographs with a normal spectrum.

according to which the shape of the dispersion curve is different for every substance. The curve for flint glass in Fig. 23B has a greater slope at the violet end, relative to that in the red, than does the curve for crown glass. Consequently, the dispersion of different substances is said to be *irrational*, since there is no simple relation between the different curves.

All transparent substances which are not colored show normal dispersion in the visible region. The magnitude of the index of refraction may be quite different in various substances, but its change with wavelength always shows the characteristics described above. In general, the greater the density of the substance the higher its index of refraction and its dispersion. For example, flint glass has a density around 2.8, considerably higher than 2.4 for ordinary crown glass. Water has a smaller n and $dn/d\lambda$, while in a very light substance like air n is practically unity and $dn/d\lambda$ very nearly zero. For air $n = 1.000276$ for red light (Fraunhofer's C line), rising to only 1.000279 for blue light (F line). This rule relating density to index of refraction is only a qualitative one, and many exceptions are known. For instance, ether has a higher index than water (1.36 as compared with 1.33), yet it is less dense, as is shown by the fact that ether floats on the surface of water. Similarly, the correlation of high dispersion with high index is only rough, and there are exceptions to the third rule listed above. Diamond has a density of 3.52 and one of the highest known indices of refraction, varying from 2.4100 for the C line to 2.4354 for the F line. The difference in these values, which is a measure of the dispersion, is only 0.0254, whereas a dense flint glass may give as much as 0.05 for the same quantity.

23.3 CAUCHY'S EQUATION

The first successful attempt to represent the curve of normal dispersion by an equation was made by Cauchy in 1836. His equation may be written

$$n = A + \frac{B}{\lambda^2} + \frac{C}{\lambda^4}$$

where A, B, and C are constants which are characteristic of any one substance. This equation represents the curves in the visible region, such as those shown in Fig. 23B, with considerable accuracy. To find the values of the three constants, it is necessary to know values of n for three different λ's. Then three equations may be set up which, when solved as simultaneous equations, give A, B, and C. For some purposes it is sufficiently accurate to include only the first two terms and the two constants can be found from values of n at only two λ's. The two-constant Cauchy equation is, then,

● $$n = A + \frac{B}{\lambda^2} \tag{23d}$$

from which the dispersion becomes, by differentiation

● $$\frac{dn}{d\lambda} = -\frac{2B}{\lambda^3} \tag{23e}$$

This shows that the dispersion varies approximately as the inverse cube of the wavelength. At 4000 Å it will be about 8 times as large as at 8000 Å. The minus sign corresponds to the usual negative slope of the dispersion curve.

The theoretical reasoning on which Cauchy based his equation was later shown to be false, so that it is to be considered essentially as an empirical equation. Nevertheless it holds very satisfactorily for cases of normal dispersion and is a useful equation from a practical standpoint. We shall show later that it is a special case of a more complete equation which does have a sound theoretical foundation.

23.4 ANOMALOUS DISPERSION

If measurements of the index of refraction of a transparent substance like quartz are extended into the infrared region of the spectrum, the dispersion curve begins to show marked deviations from the Cauchy equation. The deviation is always of the type illustrated in Fig. 23D, where, starting at the point R, the index of refraction is seen to fall off more rapidly than required by a Cauchy equation that represents the values of n for visible light (between P and Q) quite accurately. This equation predicts a very gradual decrease of n for large values of λ (broken curve), the index approaching the limiting value A as λ approaches infinity [Eq. (23d)]. In contrast to this, the measured value of n first decreases more and more rapidly as it approaches a region in the infrared where light ceases to be transmitted at all. This is an absorption band (Sec. 22.3), i.e., a region of selective absorption, the position of which is characteristic of the material. Within the absorption band, n cannot usually be meas-

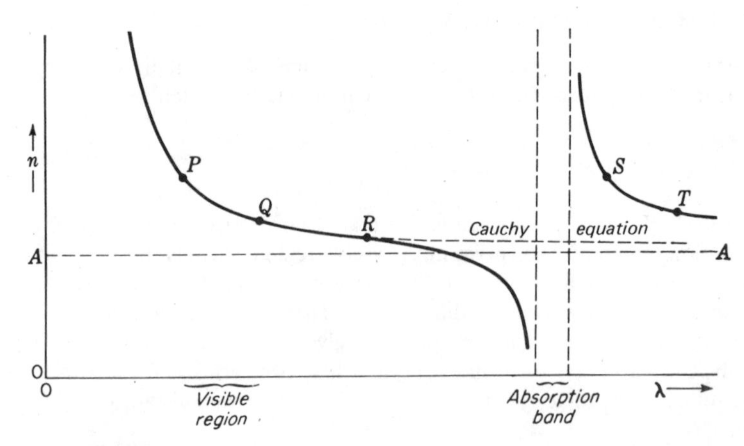

FIGURE 23D
Anomalous dispersion of a transparent substance like quartz in the infrared.

ured because the substance will not transmit radiation of this wavelength. On the long-wavelength side of the absorption band the index is found to be very high, decreasing at first rapidly and then more slowly as we go farther beyond the absorption band. Over the range from S to T, the Cauchy equation will again represent the data, but with different constants. In particular, the constant A will be larger.

The existence of a large discontinuity in the dispersion curve as it crosses an absorption band gives rise to *anomalous dispersion*. The dispersion is anomalous because in this neighborhood the longer wavelengths have a higher value of n and are more refracted than the shorter ones. The phenomenon was discovered with certain substances, such as the dye fuchsin and iodine vapor, whose absorption bands fall in the visible region. A prism formed of such a substance will deviate the red rays more than the violet, giving a spectrum which is very different from that formed by a substance having normal dispersion. When it was later discovered that transparent substances like glass and quartz possess regions of selective absorption in the infrared and ultraviolet, and therefore show anomalous dispersion in these regions, the term "anomalous" was seen to be inappropriate. No substance exists which does not have selective absorption at some wavelengths, and hence the phenomenon, far from being anomalous, is perfectly general. The so-called normal dispersion is found only when we observe those wavelengths which lie between two absorption bands, and fairly far removed from them. Nevertheless, the term "anomalous dispersion" has been retained, although it has little more than historical significance.

A very striking experiment showing the anomalous dispersion of sodium vapor in the neighborhood of the yellow D lines was devised by R. W. Wood in 1904. White light when passed through sodium vapor undergoes strong selective absorption at these lines, which form a close doublet of wavelengths 5890 and 5896 Å. At wave-

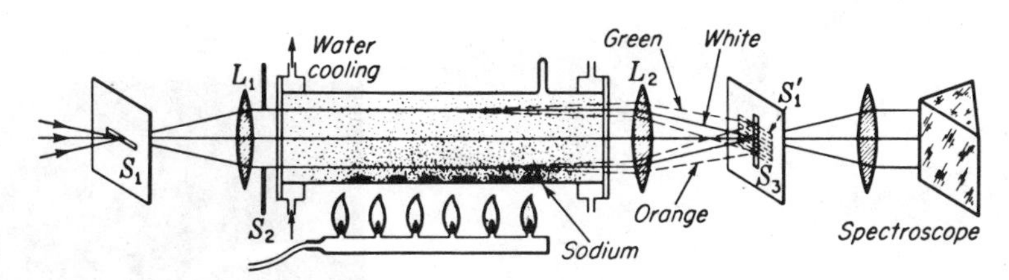

FIGURE 23E
Experimental arrangement for observing the anomalous dispersion of sodium vapor.

lengths far removed from these values, the index of refraction is only very slightly greater than unity, as we expect for a gas. With sodium vapor of appreciable density, the index of refraction in the neighborhood of the D lines passes through a region of anomalous dispersion (strictly speaking, two regions very close together) of the type shown in Fig. 23D. As the D lines are approached from the side of shorter wavelengths, n begins to decrease rapidly, becoming much less than unity as we get very close to them. On the other side, it is at first very high, dropping off rapidly toward unity as λ increases further.

To show this in a direct way, Wood made use of the fact that we can produce the equivalent of a prism of sodium vapor by vaporizing the metal in a partially evacuated tube if the tube is heated from the bottom. The arrangement is shown in Fig. 23E. A number of lumps of metallic sodium are placed along the bottom of a steel tube provided with water-cooled glass windows at the ends and an outlet for pumping. White light from a narrow horizontal slit S_1 is rendered parallel by the lens L_1 and after passing through the tube, forms a horizontal image S_1' on the vertical slit S_3 of an ordinary prism spectroscope. When the sodium tube is cold, S_1' will be a sharp, white image, illuminating one point of the spectroscope slit, and this will be spread out into a narrow horizontal continuous spectrum in the focal plane of the spectroscope camera. If the tube is evacuated to about 2 cm pressure and the sodium is heated by the row of gas burners, it will vaporize slowly, the vapor diffusing upward through the residual gas in the tube. A density gradient is set up, the vapor being densest at the bottom and rarest at the top of the tube. This is equivalent to a prism of vapor, the refracting edge of the prism being perpendicular to the plane of the figure and its thickness increasing downward. This prism will form an anomalous spectrum on S_3, in which the wavelengths shorter than the yellow, i.e., on the green side, are deviated upward ,since their n is less than 1, and the longer ones (on the orange side) will be deviated downward. As a result, we might expect to observe in the spectroscope that the spectrum is deviated upward on the green side of the D lines and downward on the red side. (The directions are actually reversed because the spectroscope inverts the image of the slit.) Three actual photographs of the resulting spectra with different densities of the vapor are shown in Fig. 23F. As a consequence of the inversion mentioned above, the photographs form qualitatively a plot of n against

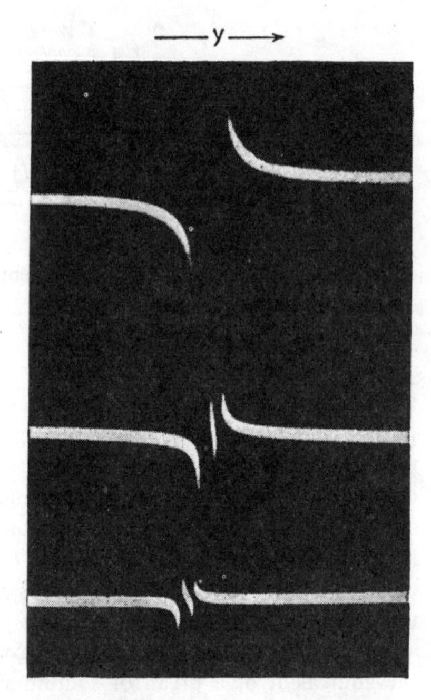

FIGURE 23F
Anomalous dispersion of sodium vapor at three different gas densities. (*Courtesy of Cario.*)

λ, as in Fig. 23D. In the practical performance of this experiment, several refinements are desirable, of which an important one is the introduction of an auxiliary diaphragm S_2 to select that portion of the vapor where the density gradient is most uniform.*

23.5 SELLMEIER'S EQUATION

We have seen that the Cauchy equation is not capable of representing the dispersion curve in a region of anomalous dispersion. The first success in deriving a formula of more general applicability was obtained by postulating a mechanism by which the medium could affect the velocity of the light wave. It was assumed that the medium contains particles bound by elastic forces, so that they are capable of vibrating with a certain definite frequency ν_0. This is the so-called *natural frequency*, i.e., one with which the particles will vibrate in the absence of any periodic force, and is identical with the natural frequency mentioned in Sec. 22.8 in connection with absorption and selective reflection. Passage of the light waves through the medium is then assumed to exert a periodic force on the particles, which causes them to vibrate. If the frequency ν of the light wave does not agree with ν_0, the vibrations will be forced vibrations of relatively small amplitude and of frequency ν. As the frequency of the light

* Further details of the experimental procedure will be found in R. W. Wood, "Physical Optics," 3d ed., pp. 492–496, The Macmillan Company, New York, 1934; reprinted (paperback) Dover Publications, Inc., New York, 1968.

approaches ν_0, the response of the particles will be greater and a very large amplitude will be built up by resonance when $\nu = \nu_0$ exactly. These vibrations will in turn react upon the light wave and alter its velocity. A mathematical investigation of this mechanism was made in 1871 by Sellmeier, who obtained the equation

$$n^2 = 1 + \frac{A\lambda^2}{\lambda^2 - \lambda_0{}^2} \tag{23f}$$

This equation contains two constants, A and λ_0, the latter being related to the natural frequency of the particles by the equation $\nu_0\lambda_0 = c$. Hence λ_0 is the wavelength in vacuum corresponding to ν_0. To allow for the possibility of the existence of several different naturai frequencies, the equation can be written with a series of terms,

$$n^2 = 1 + \frac{A_0\lambda^2}{\lambda^2 - \lambda_0{}^2} + \frac{A_1\lambda^2}{\lambda^2 - \lambda_1{}^2} + \ldots = 1 + \sum_i \frac{A_i\lambda^2}{\lambda^2 - \lambda_i{}^2} \tag{23g}$$

in which $\lambda_0, \lambda_1, \ldots$ correspond to the possible natural frequencies. The constants A_i are proportional to the number of oscillators capable of vibrating with these frequencies.

Figure 23G is a plot of n against λ according to Eq. (23g), asuming two natural frequencies. As λ approaches λ_0 or λ_1, n goes to $-\infty$ or $+\infty$ on the short-wavelength or long-wavelength side, since the denominator of one of the terms in Eq. (23g) goes to zero. Other important characteristics of the curve to be noted are that n approaches unity as λ approaches zero, and that at $\lambda = \infty$, n^2 takes the value $1 + \sum_i A_i$.

Sellmeier's equation represents a great improvement over that of Cauchy and is in fact identical with that derived from the electromagnetic theory with certain simplifying assumptions [see Eq. (23i)]. Not only does it take account of anomalous dispersion, but it also gives a more accurate representation of n in regions far from absorption bands than a Cauchy equation with the same number of constants. That Cauchy's equation is an approximation to Sellmeier's can be seen by writing Eq. (23f) in the form

$$n^2 = 1 + \frac{A}{1 - (\lambda_0{}^2/\lambda^2)}$$

On expanding by the binomial theorem, we find

$$n^2 = 1 + A\left(1 + \frac{\lambda_0{}^2}{\lambda^2} + \frac{\lambda_0{}^4}{\lambda^4} + \ldots\right)$$

For that part of the dispersion curve where λ is considerably greater than λ_0, the higher powers of λ_0/λ will be small and may be neglected. This gives

$$n^2 = 1 + A + A\frac{\lambda_0{}^2}{\lambda^2}$$

Writing M for $1 + A$ and N for $A\lambda_0{}^2$, we have

$$n = (M + N\lambda^{-2})^{1/2}$$

Expanding again gives

$$n = M^{1/2} + \frac{N}{2M^{1/2}\lambda_2} + \frac{N^2}{8M^{3/2}\lambda^4} + \ldots$$

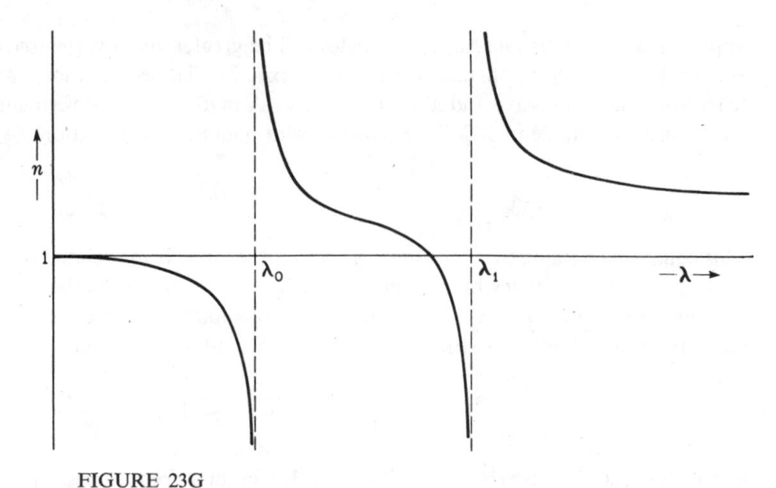

FIGURE 23G
Theoretical dispersion curves given by Sellmeier's equation for a medium having two natural frequencies.

and neglecting higher powers of $1/\lambda$ leads to

$$n = P + \frac{Q}{\lambda^2} + \frac{R}{\lambda^4}$$

This is Cauchy's equation as given in Sec. 23.3.

An instructive experiment to illustrate the origin of dispersion can be performed with a simple pendulum, to the bob of which is attached a light rubber band. If the end of the rubber band is held in the hand and moved to and fro, a periodic force is exerted on the pendulum similar to the action of the light wave on one of the oscillators in the medium. If the frequency motion of the hand is very high compared to the natural frequency of the pendulum, the bob will remain practically motionless. This corresponds to a wave of high frequency and short wavelength, the velocity of which is practically uninfluenced by the presence of the oscillators. In Fig. 23G it will be seen that n approaches unity as λ approaches zero, so the velocity becomes the same as in free space.

If, now, the hand is moved with a frequency only slightly greater than that of the pendulum, it will be found that the pendulum swings 180° out of phase with the motion of the hand. Under these conditions, the rubber band is considerably stretched when the displacements of the hand and bob are in opposite directions and so exerts its maximum force on the hand, tending to pull it back to the central position. This corresponds to an increased restoring force on the "ether" which propagates the wave, and hence to an *increase* in the velocity of the wave. Thus in Fig. 23G, n becomes considerably less than 1 at a wavelength slightly less than λ_0. Finally when the frequency of motion of the hand is made less than the natural frequency, the pendulum will follow the hand, practically in phase with it. In this case the rubber band exerts only small forces on the hand, since the displacements of the pendulum are in the same direction. The forces are less than if the pendulum were at rest, so this corresponds to a decreased restoring force on the ether. The velocity of the wave is therefore *decreased* and n is greater than 1 on the long-wavelength side of λ_0.

The large discontinuity in the dispersion curve at λ_0 is thus seen to be a consequence of the abrupt change of phase by 180° of the oscillator relative to the impressed vibration as the latter passes through resonance frequency. This effect can be demonstrated directly by hanging three pendulums side by side from a horizontal rod clamped at one end. The center pendulum is a heavy one and corresponds to the ether wave while the other two are very light, one being slightly longer and the other slightly shorter than the heavy pendulum. When the center pendulum is set swinging, the two light ones will swing in opposite phases, the shorter one nearly agreeing in phase with the impressed vibration.

23.6 EFFECT OF ABSORPTION ON DISPERSION

Although Sellmeier's equation represents the dispersion curve very successfully in regions not too close to absorption bands, it fails completely at those wavelengths where the medium has appreciable absorption. This can be seen directly from the fact that the curve of Fig. 23G goes to infinity on either side of each λ_i. Not only is this physically impossible, but the form of the curve near λ_i does not agree with experiment. It has been possible to measure the dispersion curve right through an absorption band, although this is a difficult matter because practically all the light is absorbed. By using prisms of very small refracting angle, or thin films of the material with a Michelson interferometer (Sec. 13.15), the indices of refraction of a few dyes, such as cyanine, which have an absorption band in the visible, have been carefully measured. The resulting curve resembles one of those shown by a heavy solid line in Fig. 23H. The true form of the curve in the neighborhood of λ_i is seen to be very different from that required by Sellmeier's equation.

This discrepancy was first shown by Helmholtz* to be due to the fact that Sellmeier's equation takes no account of the absorption of energy of the wave. In the above discussion, and in the suggested mechanical analogy, it was assumed that the oscillator does not experience any frictional resistance to its vibration. Such a resistance is necessary if energy is to be taken continuously from the wave by the oscillator. Helmholtz assumed a frictional force directly proportional to the velocity of the oscillator, and he therefore derived an equation for the index of refraction which takes account of absorption. As a measure of the strength of the absorption, we could use the absorption coefficient α defined in Eq. (11zd), but the equations are simpler when expressed in terms of a constant κ_0, which is related to α as follows:

$$\kappa_0 = \frac{\alpha\lambda}{4\pi} \qquad \text{(23h)}$$

Here λ is the wavelength measured in vacuum. The physical significance of κ_0 is best expressed by the fact that the intensity falls to $1/e^{4\pi\kappa_0}$ of its initial value in going

* H. L. F. von Helmholtz (1821–1894). German physicist who contributed in almost every field of science. His work in physiological optics alone, or in sound, would have made him famous. He is regarded as one of the discoverers of the law of conservation of energy.

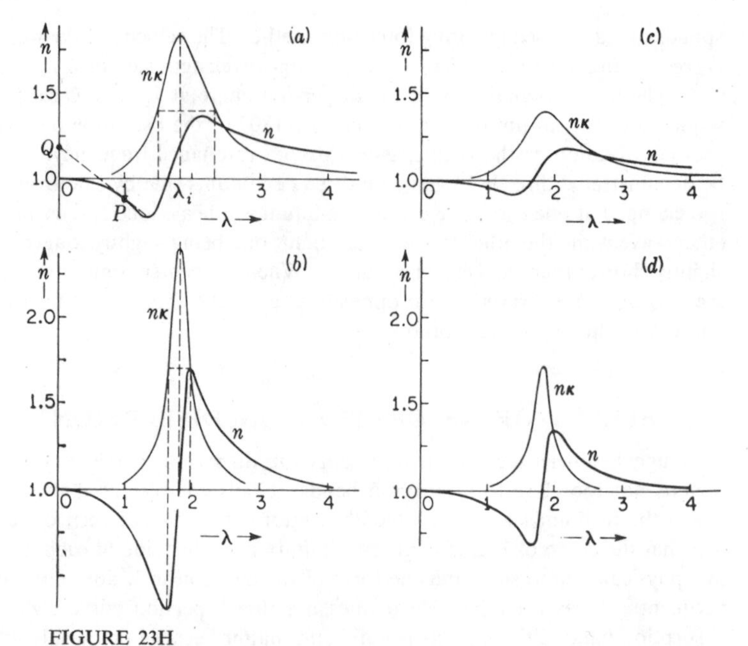

FIGURE 23H
Ideal dispersion curves for an oscillator with different amounts of friction and absorption: (*a*) strong absorption, strong friction; (*b*) strong absorption, weak friction; (*c*) weak absorption, strong friction; (*d*) weak absorption, weak friction.

the distance λ through the medium. The dispersion equations resulting from this purely mechanical theory of Helmholtz may be written

$$
\begin{aligned}
n^2 - \kappa_0{}^2 &= 1 + \sum_i \frac{A_i \lambda^2}{(\lambda^2 - \lambda_i{}^2) + g_i \lambda^2/(\lambda^2 - \lambda_i{}^2)} \\
2n\kappa_0 &= \sum \frac{A_i \sqrt{g_i}\, \lambda^3}{(\lambda^2 - \lambda_i{}^2)^2 + g_i \lambda^2}
\end{aligned}
\tag{23i}
$$

The constant g_i is a measure of the strength of the frictional force. These equations should now hold for all wavelengths, including those within an absorption band. In regions far from absorption bands, κ_0 and g_i are both essentially zero, and the first of the equations reduces to Sellmeier's equation (23g). Figure 23H(*a*) is a plot of n and of $n\kappa_0$, the latter of which by Eq. (23h) is a measure of the absorption coefficient α, for a case of large friction ($g = 1.96 \times 10^{-3}$). It shows quantitatively the course of dispersion and absorption curves through a region of absorption with a maximum at $\lambda_i = 0.1732$ μm. It will be seen that n no longer goes to infinity, as in Fig. 23G, but remains finite at $\lambda = \lambda_i$. The other curves of Fig. 23H are drawn to show the effects of changing both the strength of the absorption and the frictional damping. The former is determined by the total number of oscillators causing the absorption, while the latter depends on the magnitude of the various effects responsible for the breadth of spectral lines. It should be noted in (*b*) and (*d*) that the maxima and minima of the refractive-index curves come exactly at the points where the absorption is half its maximum value.

The pendulum experiments described above may be modified to include the effect of frictional damping and to give some insight into the physical reason for the resulting change in the form of the dispersion curve. Thus if the smaller pendulum, which represents the oscillator, has a wire attached to it which dips in water or oil, we have the desired condition. Two important changes in the response of the pendulum to the impressed vibrations will now be apparent. In the first place, the amplitude will not become nearly as large when the impressed frequency is exactly equal to the natural frequency of the pendulum. With no friction, the amplitude produced by resonance is theoretically infinite (in the final equilibrium state), and the corresponding value of n goes to infinity also. The effect of friction, however, limits this maximum amplitude, and this accounts for the fact that only moderate variations of n are actually observed. In the second place, the change of relative phase between the pendulum and the impressed vibrations when the latter pass through the natural frequency is no longer abrupt but takes place more or less gradually. This accounts for the fact that there is no longer a sharp discontinuity in the dispersion curve, which is rounded off into a continuous curve. The phase change becomes more and more gradual as the friction is made greater, for instance by dipping the wire farther into the water or by using a more viscous liquid.

23.7 WAVE AND GROUP VELOCITY IN THE MEDIUM

In the curves of Figs. 23G and 23H, the abscissas are wavelengths in vacuum $\lambda = c/u$ and the ordinates are the ordinary indices of refraction $n = c/v$, where v is the wave velocity in the medium. For those part of the curve where $n < 1$, the wave velocity is greater than the velocity c of light in vacuum. This is at first sight a contradiction to one of the fundamental results of the theory of relativity, according to which c is the highest attainable velocity. There is actually no contradiction here, however, since relativity applies to the velocity with which *energy* (a light signal) is transmitted, and this is always less than c. Remembering that the energy travels with the group velocity u, we require that it is c/u that shall be greater than unity, rather than c/v. Now u and v are related by Eq. (12p), which may be transformed (see Prob. 23.8) to read

$$\bullet \qquad \frac{c}{u} = n - \lambda \frac{dn}{d\lambda} \tag{23j}$$

where λ is the wavelength in vacuum. Thus the geometrical construction of Sec. 12.8 may also be applied to refractive indices. If in Fig. 23H(*a*) we draw a tangent to the dispersion curve, it will intersect the axis of n at a point Q whose ordinate is c/u. That is, while the ordinate of P is n, or c/v, for that wavelength, the ordinate of Q is the corresponding value of c/u for the same wavelength.

This geometrical construction shows, then, that for any point on the curve where it is descending toward the right, the corresponding c/u is greater than unity, even though n itself may be less than unity. Therefore the group velocity is less than c, and there is no violation of the principle of relativity. An exception to this statement appears to occur in the region within the absorption band, where the curve slopes up steeply to the right. However, in this region we have strong absorption, so that the amplitude of the wave drops practically to zero in a fraction of a wavelength. In this event, the wave velocity and group velocity no longer have any meaning, but other considerations show that in this case also the relativity requirement is fulfilled.

23.8 THE COMPLETE DISPERSION CURVE OF A SUBSTANCE

Although the curve of the refractive index against wavelength is different for every different substance, the curves for all optical media, i.e., substances more or less transparent in the visible region, possess certain general features in common. To illustrate these, let us consider the schematic curve of Fig. 23I, which represents the variation of n from $\lambda = 0$ to several kilometers for an ideal substance. Starting at $\lambda = 0$, the index of refraction is unity, as stated in Sec. 23.5. For the very short waves (γ rays and hard X rays), the index is slightly less than 1. Siegbahn* proved this fact experimentally by refracting X rays through a prism. It was found that the beam was deflected very slightly *away from* the base of the prism, as would be the case if the waves travel faster in the prism than in air. It has also been demonstrated that X rays can be totally reflected from a solid substance by using grazing incidence so that the X rays strike the surface at an angle greater than the critical angle. This property of X rays has been used by A. H. Compton† and others to measure the wavelengths of X rays by diffracting them from an ordinary ruled grating used at grazing incidence.

The first absorption is encountered in the X-ray region at a wavelength depending upon the atomic weight of the heaviest element in the material. For silicon it reaches its maximum at 6.731 Å, and for uranium at 0.1075 Å. This absorption rises rapidly to a maximum, and then falls off sharply at the *K-absorption limit* of the element. It gives rise to a relatively narrow region of strong anomalous dispersion, marked K in Fig. 23I. Beyond this will lie other absorption discontinuities of this element, called $L, M, \ldots$ limits, as well as the $K, L, M, \ldots$ limits of other elements present. Therefore for any actual optical medium there will be many of these sharp discontinuities. For simplicity only three are indicated in the figure.

From the X-ray region the curve descends more rapidly toward longer wavelengths, eventually reaching the broad region λ_1 of strong absorption and anomalous dispersion in the ultraviolet (Sec. 22.3). For most substances this completely covers the region between the soft X rays and the near ultraviolet. The descending course of

* Karl Manne Georg Siegbahn (1886–). Director of the Nobel Institute in Stockholm, Sweden, and winner of the Nobel prize in 1924. He is noted for his fine experimental techniques in the measurement of wavelengths of X rays.

† Arthur H. Compton (1892–1962). Professor of physics at the University of Chicago, and afterward president of Washington University, St. Louis. He received the Nobel prize in 1927, largely for his discovery of the Compton effect in X rays (Sec. 33.2).

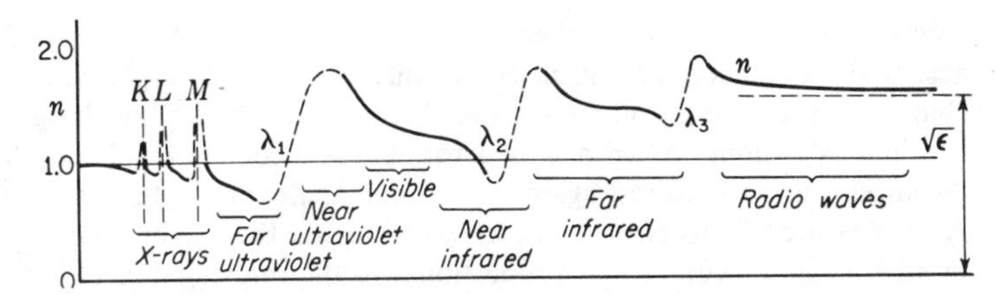

FIGURE 23I
Schematic diagram of a complete dispersion curve for a substance transparent to the visible spectrum.

the curve in the visible region, characteristic of normal dispersion, is seen to be connected with the existence of this ultraviolet absorption. In general the curve will have a steeper slope in the visible region, so that the dispersion $dn/d\lambda$ is greater, the nearer this absorption band lies to the visible. Thus fluorite has a very small dispersion for visible light, quartz somewhat greater, and glass still greater (see Fig. 23B and Table 22A). Dense flint glass, which gives the highest dispersion, frequently has a yellowish color, owing to the fact that the absorption band encroaches slightly on the violet end of the visible spectrum.

Somewhere in the near infrared, the curve begins to descend more steeply, and runs into another absorption band at λ_2. The center of this band is at 8.5 μm for quartz, but the absorption begins to become strong at 4 or 5 μm. Beyond this first absorption band there usually exist one or more others. In passing each of these bands, the index of refraction increases. Thus the index will be higher for certain infrared wavelengths than for any part of the visible. For example, Rubens measured values of n for quartz varying from 2.40 to 2.14 in the region $\lambda = 51$ to 63 μm. An interesting method of isolating radiation of very long wavelengths, called the method of *focal isolation*, is based on this fact. Owing to the high value of n, a convex lens will have a much smaller focal length for these long waves than for the shorter waves, and the latter can be screened off with suitable diaphragms. In this way the longest infrared rays ever measured were isolated by Nichols and Tear (Sec. 11.10).

At wavelengths beyond all the infrared bands, the index decreases slowly and more or less uniformly through the region of radio waves, approaching a certain limiting value for infinitely long waves. There are a few narrow regions of absorption in the radio frequencies, but these are always weak. The limiting value will be shown in the following section to be the square root of ε, the ordinary dielectric constant of the medium.

23.9 THE ELECTROMAGNETIC EQUATIONS FOR TRANSPARENT MEDIA

In Chap. 20 we stated Maxwell's equations as they apply to empty space, and we showed how they predict electromagnetic waves of velocity c. It is now of interest to investigate the characteristics and velocity of such waves in a material substance. For the present we shall consider only nonconducting media, and the more difficult case of conductors will be taken up later in Chap. 25. When a steady electric field acts upon a nonconducting dielectric, there is a small displacement of the bound charges in the atoms, and we say they become *polarized.* The charges do not move continuously along, as in a conductor, but are merely displaced through minute distances and come to rest again in a fashion analogous to the stretching of a spring. As a measure of this *electric displacement* we use the vector quantity **D**,* and since in an isotropic medium it is proportional to the impressed electric field **E**, we may write

$$\mathbf{D} = \varepsilon \mathbf{E} \qquad \text{(23k)}$$

Here ε is the dielectric constant. To apply Maxwell's equations to such a medium it now becomes necessary to replace **E** by **D** wherever it occurs in the equations for empty space [Eqs. (20a) to (20d)]. Hence Maxwell's equations for a nonconducting isotropic medium are written:

$$\frac{\varepsilon}{c}\frac{\partial E_x}{\partial t} = \frac{\partial H_z}{\partial y} - \frac{\partial H_y}{\partial z} \qquad -\frac{1}{c}\frac{\partial H_x}{\partial t} = \frac{\partial E_z}{\partial y} - \frac{\partial E_y}{\partial z}$$

$$\frac{\varepsilon}{c}\frac{\partial E_y}{\partial t} = \frac{\partial H_x}{\partial z} - \frac{\partial H_z}{\partial z} \qquad \text{(23l)} \qquad -\frac{1}{c}\frac{\partial H_y}{\partial t} = \frac{\partial E_x}{\partial z} - \frac{\partial E_z}{\partial x} \qquad \text{(23m)}$$

$$\frac{\varepsilon}{c}\frac{\partial E_z}{\partial t} = \frac{\partial H_y}{\partial x} - \frac{\partial H_x}{\partial y} \qquad -\frac{1}{c}\frac{\partial H_z}{\partial t} = \frac{\partial E_y}{\partial x} - \frac{\partial E_x}{\partial y}$$

$$\varepsilon\left(\frac{\partial E_x}{\partial x} + \frac{\partial E_y}{\partial y} + \frac{\partial E_z}{\partial z}\right) = 0 \qquad \text{(23n)} \qquad \frac{\partial H_x}{\partial x} + \frac{\partial H_y}{\partial y} + \frac{\partial H_z}{\partial z} = 0 \qquad \text{(23o)}$$

If we derive the equations for plane waves as done in Sec. 20.4, starting now with Eqs. (23l) and (23m), we find

$$\frac{\partial^2 H_z}{\partial t^2} = \frac{c^2}{\varepsilon}\frac{\partial^2 H_z}{\partial x^2} \qquad \text{and} \qquad \frac{\partial^2 E_y}{\partial t^2} = \frac{c^2}{\varepsilon}\frac{\partial^2 Ey}{\partial x^2}$$

Comparison with the general wave equation (11b) shows the new velocity to be $c/\sqrt{\varepsilon}$. The index of refraction becomes

$$n = \frac{c}{v} = \sqrt{\varepsilon} \qquad \text{(23p)}$$

The solution of Eqs. (23l) to (23o) for monochromatic plane waves, analogous to Eqs. (20n), is now to be written

$$E_y = A \sin(\omega t - kx) \qquad H_z = \sqrt{\varepsilon}\, A \sin(\omega t - kx)$$

and the magnitude of the electric and magnetic vectors at any instant is such that

$$H_z = \sqrt{\varepsilon}\, E_y$$

Therefore in the usual case $\varepsilon > 1$ the amplitude of the magnetic wave is greater than that of the electric wave in a ratio equal to the index of refraction [Eq. (23p)].

* Strictly speaking, **D** itself is not a direct measure of the displacement of the bound charges. The polarization of the medium is usually written **P**, and **D** depends on **P** by the relation $\mathbf{D} = \mathbf{E} + 4\pi\mathbf{P}$.

The energy carried by electromagnetic waves in dielectric substances may be found by applying the principles of Sec. 20.7, the only change being the replacement of **E** by **D**. The instantaneous energy densities of the above electric and magnetic waves becomes $\varepsilon E_y^2/8\pi$ and $H_z^2/8\pi$, and are therefore again equal. Their sum may be written as $\sqrt{\varepsilon}\, E_y H_z/4\pi$, and when this is multiplied by v from Eq. (23p) to obtain the intensity, one finds

$$I = \frac{c}{\varepsilon}\frac{\varepsilon E_y^{\,2}}{4\pi} = \frac{cn}{4\pi} E_y^{\,2} = \frac{cn}{8\pi} A^2 \qquad (23q)$$

As before, the E_y in this equation represents the root-mean-square (rms) value of the electric vector, since the flow of energy is averaged over a time long compared with the period. The result may also be written $cE_yH_z/4\pi$. In this form it represents the expression of a general law of electromagnetism known as *Poynting's* theorem*, according to which the direction and magnitude of the energy flow are given by the *Poynting vector* $(c/4\pi)\,[\mathbf{E} \times \mathbf{H}]$, the quantity in brackets being the vector product.

Equation (23p) gives very nearly correct values of n for gases, but when we attempt to apply it to denser media, large deviations are found. Thus the dielectric constant for water, measured by placing it between the plates of a condenser charged to a steady potential, is 81, indicating a value of 9 for the index of refraction. For sodium light, the measured index of water is 1.33. For various kinds of glass, ε varies from 4 to 9, which would require n to vary from 2 to 3. This again is higher than the observed values for visible light.

We do not have to look far for the cause of this discrepancy. It lies in the fact that the electric field of a light wave is not a steady field but a rapidly alternating one. For yellow light the frequency is $5 \times 10^{14}\,\text{s}^{-1}$. If the dielectric constant of a substance is measured using an alternating potential on the plates in place of a steady one, the result is found to vary with the frequency. From this we see that the index of refraction must also vary with frequency, or wavelength. As the wavelength becomes very large and approaches infinity, the frequency approaches zero. The limiting case of a steady field thus corresponds to zero frequency, and we are led to expect the index of refraction to approach the square root of the dielectric constant for steady fields. That this is in fact the case is shown by the measurements of the index of refraction of water for electromagnetic waves quoted in Table 23C. The value of $\sqrt{\varepsilon}$ measured for a steady field is shown for comparison. Clearly the value of n approaches exactly the predicted value for infinitely long waves.

23.10 THEORY OF DISPERSION

In order to explain the variation of n (and hence of $\sqrt{\varepsilon}$) with λ by the electromagnetic theory, one must take account of the molecular structure of matter. When an electromagnetic wave is incident on an atom or molecule, the periodic electric force of the wave sets the bound charges into a vibratory motion having the frequency of the wave.

* J. H. Poynting (1852–1914). Professor of physics at the University of Birmingham, England. He is also known for his accurate work on the measurement of the gravitational constant.

The phase of this motion relative to that of the impressed electric force will depend upon the impressed frequency, and will vary with the difference between the impressed frequency and the natural frequency of the bound charges in the way discussed in Secs. 23.5 and 23.6. As the wave traverses the empty space between molecules, it will, of course, have the velocity c, and we must now inquire how it is possible that the presence of the oscillating charges in the molecules produces an effective alteration in the rate at which the wave progresses through the medium.

The clue to the explanation of dispersion lies in the secondary waves which are generated by the induced oscillations of the bound charges. These secondary waves are identical with those which give rise to molecular scattering (Sec. 22.10), as in the explanation of the blue color of the sky. When a light beam traverses a transparent liquid or solid, the amount of light scattered laterally is extremely small, even though the concentration of scattering centers is much greater than that in the air which gives the sky light. This is due to the fact that the scattered wavelets traveling out laterally from the beam have their phases so arranged that there is practically complete destructive interference. But the secondary waves traveling *in the same direction* as the original beam do not cancel out but combine to form sets of waves moving parallel to the original waves. Now the secondary waves must be added to the primary ones according to the principle of superposition, and the results will depend on the phase difference between the two sets. This interference will modify the phase of the primary waves and thus is equivalent to a change in their wave velocity. That is, since the wave velocity is merely the rate at which a condition of equal phase is propagated, an alteration of the phase by interference changes the velocity. We have seen that the phase of the oscillators, and hence of the secondary waves, depends on the impressed frequency, so it becomes clear that the velocity in the medium varies with the frequency of light. This is the physical interpretation of dispersion, expressed in briefest outline.

The foundations for the mathematical treatment of the above mechanism were laid by Rayleigh, who considered the case of mechanical waves, and the theory was later extended to cover the case of electromagnetic waves by Planck, Schuster, and

Table 23C VARIATION OF n WITH λ FOR WATER

Wavelength, cm	Frequency, Hz	n
5.89×10^{-5}	5.1×10^{14}	1.333
12.56×10^{-5}	2.9×10^{14}	1.3210
258×10^{-5}	0.116×10^{14}	1.41
800×10^{-5}	0.0375×10^{14}	1.41
0.40	750×10^{8}	5.3
1.75	171×10^{8}	7.82
8.1	37×10^{8}	8.10
65	4.6×10^{8}	8.88
∞	0×10^{8}	$(9.03 = \sqrt{\varepsilon})$

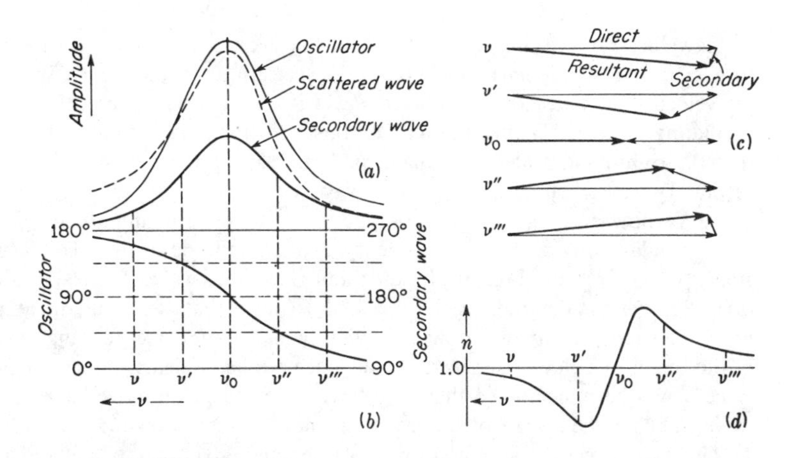

FIGURE 23J
The interpretation of dispersion as the result of interference of the secondary wave with the direct wave.

others. We shall not attempt to give this development here. It leads to dispersion formulas similar to that of Helmhotz [Eq. (23i)]. In fact, there is a close analogy throughout between the electromagnetic and mechanical pictures of the phenomenon. The oscillations of the bound charges must be regarded as damped by a frictional force, just as the particles were in Helmholtz's theory. The nature of the damping forces postulated in electromagnetic theory will be briefly discussed in Sec. 23.11.

To show the relative amplitudes and phases of the incident wave, oscillator, and secondary wave, we consider the schematic diagrams of Fig. 23J. The first curve in (*a*) shows the response of a damped *oscillator* of natural frequency ν_0 to an impressed vibration of frequency ν, the amplitude becoming a maximum when $\nu = \nu_0$. The broken curve shows the amplitude radiated by the oscillator, i.e., of the *scattered wave*. As a consequence of Rayleigh's law that the shorter waves are scattered more effectively, this curve is higher on the sides of higher ν but drops to zero at low frequencies. The third curve gives the amplitude of the *secondary waves* built up from the scattered wavelets. Curve (*b*), in conjunction with the left-hand scale of ordinates, gives the phase difference between the oscillator and the impressed wave. As pointed out in Sec. 23.6, this changes from 0 to 180° in passing through the natural frequency, but not abruptly because of the damping. At ν_0 it is 90° behind that of the impressed wave. Theory shows, furthermore, that the phase of the scattered waves, and therefore of the secondary waves as well, lags 90° behind that of the oscillators.* This is because electromagnetic radiation is proportional to the rate of change of current, or to the acceleration of a charge [see Sec. 20.8 and Fig. 20D(*a*)]. The current itself,

* See, for example, G. P. Harnwell, "Principles of Electricity and Magnetism," 2d ed., pp. 601–602, McGraw-Hill Book Company, 1949.

or the velocity of the charge, has the phase that we attribute to the oscillator. Therefore, since in a simple harmonic motion the acceleration is one-quarter period behind the velocity, the phase of the radiated waves is retarded this much behind that of the oscillating source. Taking account of this additional retardation, it will be seen that the right-hand scale of ordinates in Fig. 23J(*b*) applies to the phase lag of the secondary waves behind the impressed waves.

We now proceed in (*c*) to compound vectorially the amplitudes of the direct and secondary waves. For the frequency ν, the amplitude of the secondary waves is small [curve (*a*)] and lags in phase behind the direct waves by nearly 270° [curve (*b*)]. The vector diagram at the top in (*c*) shows that the resultant amplitude is nearly the same but that the phase is slightly *advanced*, corresponding to a rotation of the vector in a clockwise sense. An advance of phase means an increase in velocity, since it will be remembered that the phase increases as we move backward along a wave. Thus in the dispersion curve (*d*), the index of refraction at ν is slightly less than 1. The second vector diagram, for ν', gives a greater advance of phase, and a considerably smaller resultant amplitude. At $\nu = \nu_0$ there is no change of phase or velocity produced, but merely a reduction of intensity. The energy removed from the resultant forward wave appears in other directions as resonance radiation. Beyond ν_0 there is a retardation instead of an advance of phase, and the velocity of the wave is decreased. Thus it may be seen in a qualitative way how the curve (*d*), having the form for anomalous dispersion, can result from the mechanism described.

23.11 NATURE OF THE VIBRATING PARTICLES AND FRICTIONAL FORCES

In conclusion, we may consider briefly what types of charged particles and damping forces are involved in the various discontinuities of the typical dispersion curve of Fig. 23I. The X-ray absorptions are attributed to the innermost electrons in the atoms, which are assigned to the "shells" *K*, *L*, *M*, etc., of decreasing energy and increasing distance from the nucleus. Because they are deep in the atom, these electrons are shielded from the effects of collisions and electric fields due to neighboring atoms. These two causes of line breadth in spectrum lines are not important for X rays, and the absorption discontinuities are sharp, even in solids. It is only in this region that radiation damping makes any appreciable contribution to the line widths.

The very broad absorption in the far ultraviolet is due to the outer electrons in the atoms and molecules of the material. These are not shielded, and consequently in solids and liquids an extensive region of continuous absorption is produced. For molecular gases the bands may consist of individual rotational lines that are quite sharp, but these are so numerous that they are usually unresolved. In this region the damping due to collisions begins to become more important than that due to radiation, and at still longer wavelengths it usually predominates. The near-infrared absorption bands represent the various natural frequencies of the atoms as a whole, or even of molecules. Since these vibrators are much heavier than electrons, it is clear why they possess lower vibration frequencies. In the far infrared, other molecular vibrations of lower frequency may be involved. Here the frequencies of rotation of molecules as a whole may also play a part, especially in gases.

PROBLEMS

23.1 The refractive indices of a piece of optical glass for the blue and green lines of the mercury spectrum, $\lambda = 4358$ Å and $\lambda = 5461$ Å, are 1.65250 and 1.62450, respectively. Using the two-constant Cauchy equation, calculate values for (*a*) the constants A and B, (*b*) the refractive index for the sodium yellow lines at $\lambda = 5893$ Å, and (*c*) the dispersion at this same wavelength.

Ans. (*a*) $A = 1.57540$ and $B = 1.46431 \times 10^6$ Å^2, (*b*) $n = 1.61757$, (*c*) 1.43104×10^{-5} Å^{-1}

23.2 Using the refractive indices given in Table 23B for borosilicate crown glass, (*a*) find the values of the constants in the three-constant Cauchy equation that fits exactly for wavelengths 4340, 5338, and 6439 Å. (*b*) Using these constants, calculate indices for the other five wavelengths given in the table. (*c*) Compare the observed with the calculated values.

23.3 Using the measured refractive indices for telescopic crown glass given in Table 23B, calculate the values of the constants in the three-term Cauchy equation that fits exactly at wavelengths 6563, 5086, and 3988 Å. (*b*) Compare your calculated values with the measured values at the other five wavelengths given in Table 23B.

23.4 A 50° prism is made of glass for which the constants in the two-term Cauchy equation are $A = 1.53974$ and $B = 4.6528 \times 10^5$ Å^2. Find the angular dispersion in radians per angstrom when the prism is set for minimum deviation for a wavelength of 5500 Å.

Ans. $dn/d\lambda = -5.5932 \times 10^{-6}$ Å^{-1}, $d\theta/dn = 1.12145$, $d\theta/d\lambda = 6.2725 \times 10^{-6}$ rad/Å

23.5 Hartmann worked out an empirical dispersion formula, according to which $n = n_0 + b/(\lambda - \lambda_0)$. (*a*) Find the values of the three constants n_0, b, and λ_0 that fit exactly at wavelengths 6563, 5086, and 3988 Å for telescopic crown glass as given in Table 23B. (*b*) Compare the calculated with the observed values at the five other wavelengths given in the table. (*c*) Compare these values with those calculated using the three-term Cauchy equation. (*d*) Which equation best represents the measured data (see Prob. 23.3)?

23.6 Compare the spectrum formed by a prism having anomalous dispersion in the green part of the spectrum with the spectrum formed by an ordinary piece of glass in the form of a similar prism. Indicate the relative positions of all the colors compared with those produced by normal dispersion.

23.7 Determine from the values of refractive index given in Table 23B a value for (*a*) group velocity and (*b*) wave velocity of violet light $\lambda = 3988$ Å in borosilicate crown glass.

Ans. (*a*) 190,259 km/s, (*b*) 196,526 km/s

23.8 Starting with Eq. (12p) for the relation between group velocity and wave velocity, derive the expression for the group index given by Eq. (23j).

23.9 From the second of Helmholtz's equations (23i) find the relation between the width of the absorption peak at half maximum $n\kappa_0$ and the frictional constant g_i.

23.10 For a particular piece of glass the refractive index for X rays of wavelength 0.70 Å is 1.600×10^{-6} less than unity. At what maximum angle, measured to the surface, must a beam of X rays strike the glass to undergo total reflection? *Ans.* 0.1025°

23.11 According to the electromagnetic theory, the value of A_i is given by

$$A_i = \frac{\lambda_i^2 N_i e_i^2}{\pi c^2 m_i}$$

where N_i represents the number of oscillators per cubic centimeter and e_i and m_i are the charge and mass of one oscillator whose frequency is c/λ_i or ν. Taking the refractive index of air as 1.000279, and assuming only one absorption band in the ultraviolet, calculate the value of e_i/m_i for air. Compare with e/m for the electron.

23.12 (*a*) Use the two-term Cauchy equation which fits the refractive indices of borosilicate crown glass, as given in Table 23B for wavelengths 6563 and 4861 Å, to predict the index for the sodium lines at $\lambda = 5893$ Å. (*b*) Calculate also a value for the dispersion in radians per angstrom for a 60° prism at $\lambda = 5892$ Å.

24

THE POLARIZATION OF LIGHT

From the properties of interference and diffraction we are led to conclude that light is a wave phenomenon, and we utilize these properties to measure the wavelength. These effects tell us nothing about the types of waves with which we are dealing—whether they are longitudinal or transverse, or whether the vibrations are linear, circular, or torsional. The electromagnetic theory, however, specifically requires that the vibrations be traverse, being therefore entirely confined to the plane of the wave front. The most general type of vibration is elliptical, of which linear and circular vibrations are extreme cases. Experiments which bring out these characteristics are those dealing with the polarization of light. Although a longitudinal wave like a sound wave must necessarily be symmetrical about the direction of its propagation, transverse waves may show dissymmetries, and if any beam of light shows such a dissymmetry, we say it is polarized.

The present chapter, by way of introduction to the subject of polarization, gives a brief account of the principal ways of producing plane-polarized light from ordinary unpolarized light. Most of the phenomena to be discussed here will be covered in more detail in later chapters. It will be helpful, however, to have a preliminary acquaintance with the experimental methods and a mental picture of how

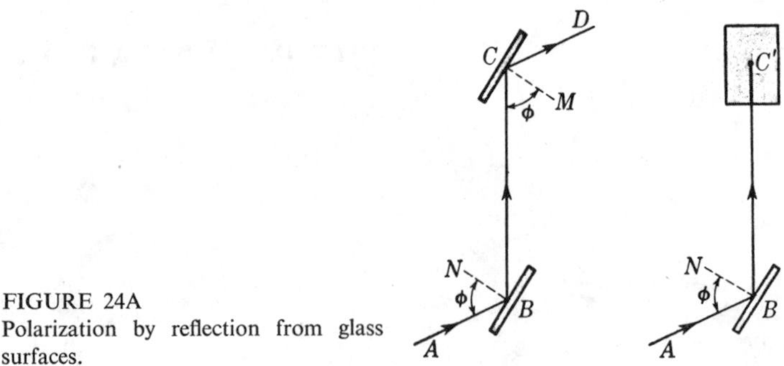

FIGURE 24A
Polarization by reflection from glass surfaces.

the various polarizing devices act to separate ordinary light into its polarized components. The common methods used in producing and demonstrating the polarization of light may be grouped under the following heads: (1) reflection, (2) transmission through a pile of plates, (3) dichroism, (4) double refraction, and (5) scattering.

24.1 POLARIZATION BY REFLECTION

Perhaps the simplest method of polarizing light is the one discovered by Malus in 1808. If a beam of white light is incident at one certain angle on the polished surface of a plate of ordinary glass, it is found upon reflection to be plane-polarized. By plane-polarized is meant that all the light is vibrating parallel to a plane through the axis of the beam (Sec. 11.6). Although this light appears to the eye to be no different from the incident light, its polarization or asymmetry is easily shown by reflection from a second plate of glass as follows. A beam of unpolarized light, AB in Fig. 24A, is incident at an angle of about 57° on the first glass surface at B. This light is again reflected at 57° by a second glass plate C placed parallel to the first as shown at the left. If now the upper plate is rotated about BC as an axis, the intensity of the reflected beam is found to decrease, reaching zero for a rotation of 90°. Rotation about BC keeps the angle of incidence constant. The experiment is best performed with the back surfaces of the glass painted black. The first reflected beam BC' then appears to be cut off and to vanish at C'. As the upper mirror is rotated further about BC the reflected beam CD reappears, increasing in intensity to reach a maximum at 180°. Continued rotation produces zero intensity again at 270°, and a maximum again at 360°, the starting point.

If the angle of incidence on either the lower or upper mirror is not 57°, the twice-reflected beam will go through maxima and minima as before, but the minima will not have zero intensity. In other words there will always be a reflected beam from C. Calling the angle of incidence ϕ in general, the critical value $\bar{\phi}$ which produces a zero minimum for the second reflection is called the *polarizing angle* and varies with the kind of glass used. Before undertaking the explanation of this experiment, it will be worthwhile to consider briefly the accepted ideas concerning the nature of the vibrations in ordinary and polarized light.

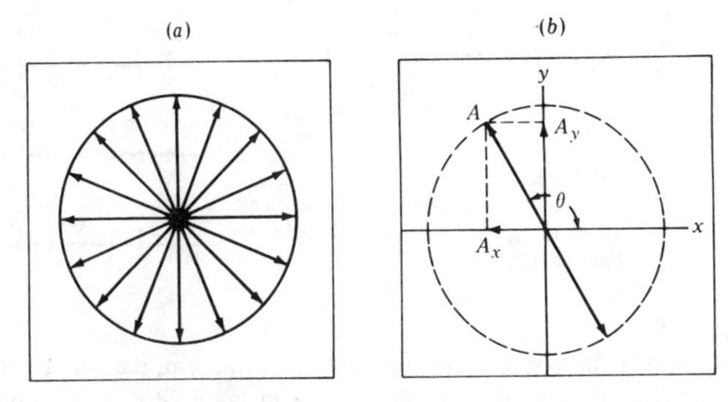

FIGURE 24B
Vibrations in unpolarized light viewed end-on. (*a*) All planes are equally probable. (*b*) Each vibration can be resolved into two components in the x and y directions.

24.2 REPRESENTATION OF THE VIBRATIONS IN LIGHT

According to the electromagnetic theory, any type of light consists of transverse waves, in which the oscillating magnitudes are the electric and magnetic vectors. The question as to which of these is to be chosen as constituting the "vibrations" will be deferred until later (Sec. 25.12), but it is immaterial for our present purpose. Let us assume that in a beam of light traveling toward the observer, along the $+z$ axis in Fig. 24B, the electric vector at some instant is executing a linear vibration with the direction and amplitude indicated. If this vibration continues unchanged, we say that the light is *plane-polarized*, since its vibrations are confined to the plane containing the z axis and oriented at the angle θ. If, on the other hand, the light is *unpolarized*, like most natural light, one may imagine that there are sudden, random changes in θ, occurring in time intervals of the order of 10^{-8} s. Every orientation of A is to be regarded as equally probable, so that, as indicated by the solid circle in Fig. 24B(*a*) the average effect is completely symmetrical about the direction of propagation.

This picture of unpolarized light, although a legitimate one, is oversimplified because if there are fluctuations in orientation, there should be fluctuations in amplitude as well. Furthermore, linear vibrations are a special case of elliptical ones, and there is no reason for this special type to be preferred. Hence a truer picture is one of elliptical vibrations changing frequently in size, eccentricity, and orientation, but confined to the xy plane. This complexity presents little difficulty, however, since because all azimuths are equivalent, the simpler representation in terms of linear vibrations of constant amplitude and rapidly shifting orientation completely describes the facts. Also, since motion in an ellipse can be regarded as made up of two linear motions at right angles (Sec. 12.9), the two descriptions are in fact mathematically the same.

Still another representation of unpolarized light is perhaps the most useful. If we resolve the vibration of Fig. 24B(*b*) into linear components $A_x = A \cos \theta$ and $A_y = A \sin \theta$, they will in general be unequal [see Sec. 24.5 and Eq. (24d)]. But

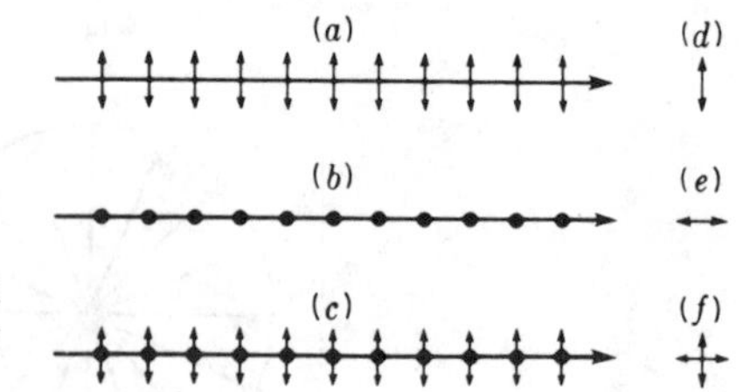

FIGURE 24C
Pictorial representations of side and end views of plane-polarized and ordinary light beams.

when θ is allowed to assume all values at random, the net result is as though we had two vibrations at right angles with equal amplitudes but no coherence of phase. Each is the resultant of a large number of individual vibrations with random phases (Sec. 12.4) and because of this randomness a complete incoherence is produced. Figure 24C shows a common way of picturing these vibrations, parts (*a*) and (*b*) representing the two plane-polarized components, and part (*c*) the two together in an unpolarized beam. Dots represent the end-on view of linear vibrations, and double-pointed arrows represent vibrations confined to the plane of the paper. Thus (*d*), (*e*), and (*f*) of the figure show how the vibrations in (*a*), (*b*), and (*c*) would appear if one were looking along the direction of the rays.

24.3 POLARIZING ANGLE AND BREWSTER'S LAW

Consider unpolarized light to be incident at an angle ϕ on a dielectric like glass, as shown in Fig. 24D(*a*). There will always be a reflected ray OR and a refracted ray OT. An experiment like the one described in Sec. 24.1 and shown in Fig. 24A shows that the reflected ray OR is partially plane-polarized and that only at a certain definite angle, about 57° for ordinary glass, is it plane-polarized. It was Brewster who first discovered that at this polarizing angle $\bar{\phi}$ the reflected and refracted rays are just 90° apart. This remarkable discovery enables one to correlate polarization with the refractive index

$$\frac{\sin \phi}{\sin \phi'} = n \tag{24a}$$

Since at $\bar{\phi}$ the angle $ROT = 90°$, we have $\sin \bar{\phi}' = \cos \bar{\phi}$, giving

$$\frac{\sin \bar{\phi}}{\sin \bar{\phi}'} = \frac{\sin \bar{\phi}}{\cos \bar{\phi}} = n$$

$$n = \tan \bar{\phi} \tag{24b}$$

This is *Brewster's law*, which shows that the angle of incidence for maximum polarization depends only on the refractive index. It therefore varies somewhat with wavelength, but for ordinary glass the dispersion is such that the polarizing angle $\bar{\phi}$ does not change much over the whole visible spectrum. This fact is readily verified by calculating $\bar{\phi}$ for several wavelengths, using the values of n from Table 23B, as suggested in Prob. 24.1 at the end of this chapter.

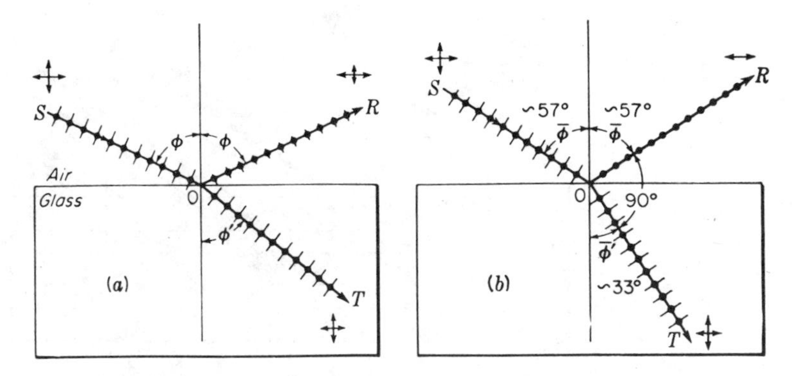

FIGURE 24D
(*a*) Polarization by reflection and refraction. (*b*) Brewster's law for the polarizing angle.

It is not difficult to understand the physical reason why the light vibrating in the place of incidence is not reflected at Brewster's angle. The incident light sets the electrons in the atoms of the material into oscillation, and it is the reradiation from these that generates the reflected beam. When the latter is observed at 90° to the refracted beam, only the vibrations that are perpendicular to the plane of incidence can contribute. Those in the plane of incidence have no component traverse to the 90° direction and hence cannot radiate in that direction. The reason is the same as that which causes the radiation from a horizontal radio-transmitter antenna to drop to zero along the direction of the wires. If the student keeps this picture in mind and remembers that light waves are strictly transverse, he will have no trouble remembering which of the two components is reflected at the polarizing angle.

24.4 POLARIZATION BY A PILE OF PLATES

Upon examining the refracted light in Fig. 24D(*a*) for polarization, it is found to be partially polarized for all angles of incidence, there being no angle at which the light is completely plane-polarized. The action of the reflecting surface may be described somewhat as follows. Let the ordinary incident light be thought of as being made up of two mutually perpendicular plane-polarized beams of light as shown in Sec. 24.2. Of those waves vibrating in the plane of incidence, i.e., in the plane of the page, part are reflected and part refracted for all angles with the single exception of the polarizing angle $\bar{\phi}$, for which all of the light is refracted. Of the waves vibrating perpendicular to the plane of incidence, some of the energy is reflected and the rest refracted for any angle of incidence. Thus the refracted ray always contains some of both planes of polarization. For a single surface of glass with $n = 1.50$, it will be shown later [Sec. 25.1 and Fig. 25B(*a*)] that at the polarizing angle 100 percent of the light vibrating parallel to the plane of incidence is transmitted, whereas for the perpendicular vibrations only 85 percent is transmitted, the other 15 percent being re-

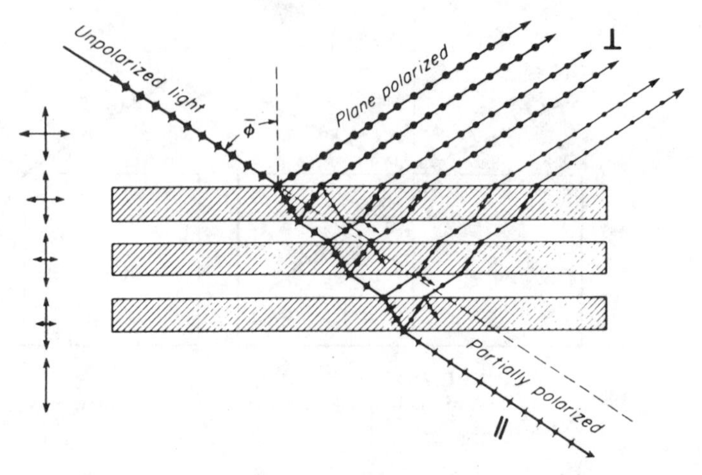

FIGURE 24E
Polarization of light by a pile of glass plates.

flected. Obviously the degree of polarization of the transmitted beam is small for a single surface.

If a beam of ordinary light is incident at the polarizing angle on a pile of plates, as shown in Fig. 24E, some of the vibrations perpendicular to the plane of incidence are reflected at each surface and all those parallel to it are refracted. The net result is that the reflected beams are all plane-polarized in the same plane, and the refracted beam, having lost more and more of its perpendicular vibrations, is partially plane-polarized. The larger the number of surfaces, the more nearly plane-polarized this transmitted beam is. This is illustrated by the vibration figures at the left in Fig. 24E. In a more detailed treatment of polarization by reflection and refraction (see Chap. 25), the polarizing angle for *internal* reflection is shown to correspond exactly to the angle of refraction $\bar{\phi}'$ in Fig. 24D(*b*). This means that light internally reflected at the angle $\bar{\phi}'$ will also be plane-polarized.

The degree of polarization P of the transmitted light can be calculated by summing the intensities of the parallel and perpendicular components. If these intensities are called I_p and I_s, respectively, it has been shown* that

$$P = \frac{I_p - I_s}{I_p + I_s} = \frac{m}{m + [2n^2/(1 - n^2)]} \tag{24c}$$

where m is the number of plates, that is, $2m$ surfaces, and n their refractive index. This equation shows that by the use of enough plates the degree of polarization can be made to approach unity, or ~100 percent. Better methods of producing a wide

* F. Provostaye and P. Desains, *Ann. Chim. Phys.*, **30**:159 (1850). The calculation takes into account not only the ray going directly through but also those internally reflected two or more times (see Fig. 24E). It does not, however, include any effects of absorption, which would increase P somewhat above the value given by Eq. (24c).

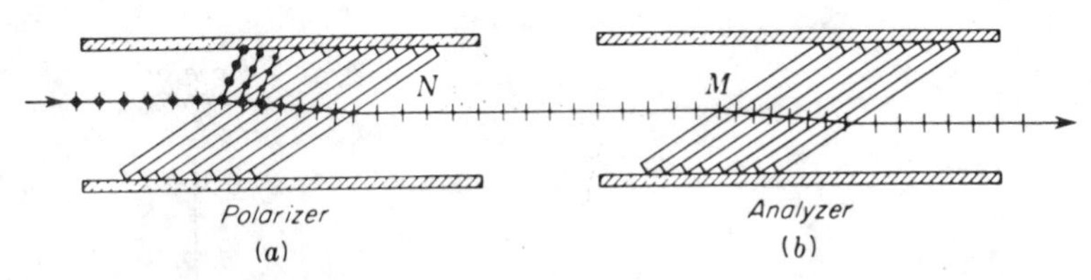

FIGURE 24F
Glass plates mounted at the polarizing angle $\bar{\phi}$.

beam of polarized light now available will be described below. The pile of plates may be used, however, to illustrate a common arrangement in the production and analysis of polarized light.

Figure 24F shows two such piles, the polarizer (*a*) and the analyzer (*b*), with their planes of incidence parallel. The light emerging at *N* is nearly plane-polarized and will be transmitted freely by the analyzer. Rotation of the latter by 90° about the line *NM* as an axis will cause the transmitted light to be nearly extinguished, since the vibrations are now perpendicular to the plane of incidence of the analyzer and will be reflected to the side. A further rotation of 90° will restore the light, and in a complete revolution there will be two maxima and two minima. Any arrangement of polarizer and analyzer in tandem is called a *polariscope* and has numerous uses.

24.5 LAW OF MALUS*

This law tells us how the intensity transmitted by the analyzer varies with the angle that its plane of transmission makes with that of the polarizer. In the case of two piles of plates, the plane of transmission is the plane of incidence, and for the law of Malus to hold we must assume that the transmitted light is completely plane-polarized. A better illustration would be the double reflection experiment of Sec. 24.1, or a combination of two polaroids or nicol prisms (see below), for which the polarization is complete. Then the law of Malus states that the transmitted intensity varies as the *square of the cosine* of the angle between the two planes of transmission.

The proof of the law rests on the simple fact that any plane-polarized vibration—let us say the one produced by our polarizer—can be resolved into two components, one parallel to the transmission plane of the analyzer and the other at right angles to it. Only the first of these gets through. In Fig. 24G, let A represent the amplitude transmitted by the polarizer for which the plane of transmission intersects the plane of the figure in the vertical dashed line. When this light strikes the analyzer, set at the angle θ, one can resolve the incident amplitude into components A_1 and A_2, the latter of which is eliminated in the analyzer. In the pile of plates, it is reflected to one side. The amplitude of the light that passes through the analyzer is therefore

$$A_1 = A \cos \theta \qquad \text{(24d)}$$

* Etienne Louis Malus (1775–1812). French army engineer. His discovery of polarization by reflection was made by accident when looking through a calcite crystal at the light reflected from the windows of the Luxembourg Palace.

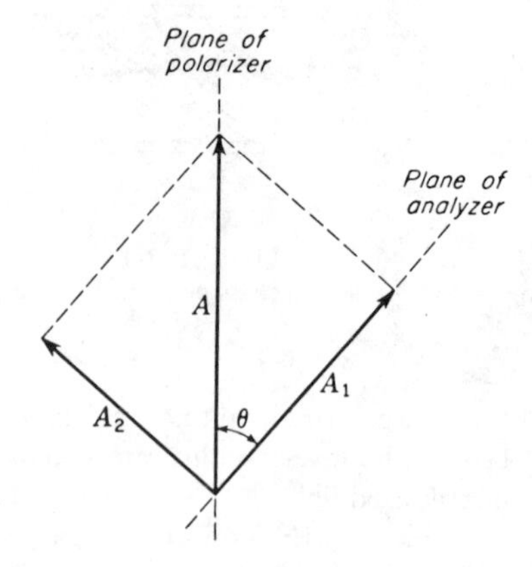

FIGURE 24G
Resolution into components of the amplitude of plane-polarized light.

and its intensity

$$I_1 = A_1{}^2 = A^2 \cos^2 \theta = I_0 \cos^2 \theta \qquad \text{(24e)}$$

Here I_0 signifies the intensity of the incident polarized light. This is, of course, one-half of the intensity of the unpolarized light striking the polarizer, provided one neglects losses of light by absorption in traversing it. There will also be losses in the analyzer. For Polaroids or nicols some light will be removed from the beam by reflection at the surfaces. Although these effects are neglected in deriving Eq. (24e), it will be noticed that they change only the constant in the equation and do not spoil the dependence of the *relative* intensity on $\cos^2 \theta$. Thus Malus' law is rigorously true and applies, for example, to the intensity of the twice-reflected light in the experiment of Sec. 24.1 even though its maximum value is only a small fraction of the original intensity. In such cases, the I_0 in Eq. (24e) is merely the intensity when the analyzer is parallel to the polarizer.

24.6 POLARIZATION BY DICHROIC CRYSTALS

These crystals have the property of selectivity absorbing one of the two rectangular components of ordinary light. Dichroism is exhibited by a number of minerals and by some organic compounds. Perhaps the best known of the mineral crystals is *tourmaline*. When a pencil of ordinary light is sent through a thin slab of tourmaline like T_1, shown in Fig. 24H, the transmitted light is polarized. This can be verified by a second crystal T_2. With T_1 and T_2 parallel to each other the light transmitted by the first crystal is also transmitted by the second. When the second crystal is rotated through 90°, no light gets through. The observed effect is due to a selective absorption by tourmaline of all light rays vibrating in one particular plane (called, for reasons explained below, the O vibrations) but not those vibrating in a plane at right angles (called the E vibrations). Thus in the figures shown, only the E vibrations parallel to the long edges of the crystals are transmitted, so that no light will emerge from the crossed crystals. Since tourmaline crystals are somewhat colored, they are not used in optical instruments as polarizing or analyzing devices.

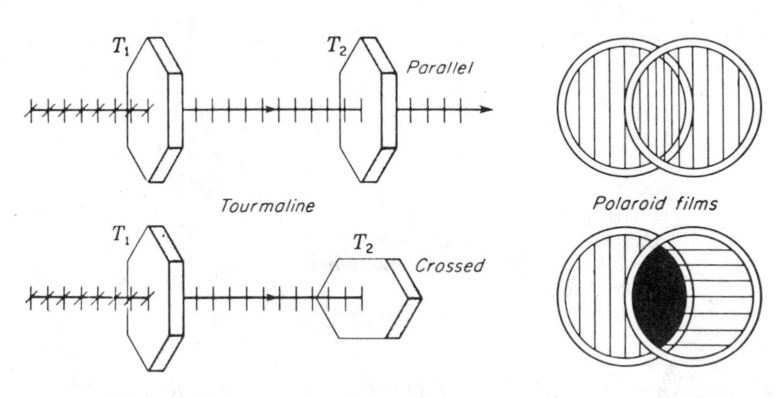

FIGURE 24H
Dichroic crystals and polarizing films in the *parallel* and *crossed* positions.

Attempts to produce polarizing crystals of large aperture were made by Herapath* in 1852. He was successful in producing good but small crystals of the organic compound quinine iodosulfate (now known as herapathite), which completely absorbs one component of polarization and transmits the other with little loss. One variety of Polaroid contains crystals of this substance. Polaroid was invented in 1932 by Land,† and has found uses in many different kinds of optical instruments. These films consist of thin sheets of nitrocellulose packed with ultramicroscopic polarizing crystals with their optic axes all parallel. In more recent developments the lining-up process is accomplished somewhat as follows. Polyvinyl alcohol films are stretched to line up the complex molecules and then are impregnated with iodine. From X-ray diffraction studies of these dichroic films, it can be seen that the iodine is present in polymeric form, i.e., as independent long strings of iodine atoms all lying parallel to the fiber axis, with a periodicity in this direction of about 3.10 Å. Films prepared in this way are called H-Polaroid. Land and Rogers found further that when an oriented transparent film of polyvinyl alcohol is heated in the presence of an active dehydrating catalyst such as hydrogen chloride, the film darkens slightly and becomes strongly dichroic. Such a film becomes very stable and, having no dyestuffs, is not bleached by strong sunlight. This so-called K-Polaroid is very suitable for polarizing uses such as automobile headlights and visors. Polarizing films are usually mounted between two thin plates of optical glass.

24.7 DOUBLE REFRACTION

The production and study of polarized light over a wider range of wavelengths than is afforded by Polaroid use the phenomenon of double refraction in crystals of *calcite* and *quartz*. Both these crystals are transparent to visible as well as ultraviolet light.

* W. B. Herapath, *Phil. Mag.*, **3**:161 (1852).

† A good summary of the development of sheet polarizers is given by E. H. Land, *J. Opt. Soc. Am.*, **41**:957 (1951).

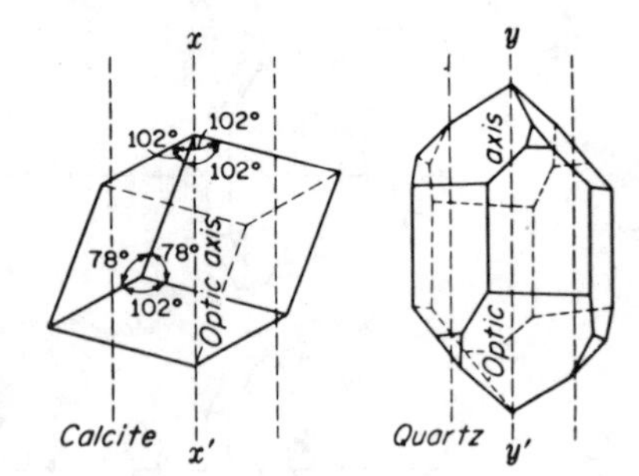

FIGURE 24I
Calcite and quartz crystal forms. The direction of the optic axis is indicated by broken lines.

Calcite, which chemically is calcium carbonate ($CaCO_3$), occurs in nature in a great variety of crystal forms (in the rhombohedral class of the hexagonal system), but it breaks readily into simple cleavage rhombohedrons of the form shown at the left in Fig. 24I. Each face of the crystal is a parallelogram whose angles are 78° 5′ and 101° 55′. If struck a blow with a sharp instrument, each crystal can be made to cleave or break along cleavage planes into two or more smaller crystals which always have faces that are parallelograms with angles shown in Fig. 24J.

Quartz crystals, on the other hand, are found in their natural state to have many different forms, one of the more complicated of which is shown at the right in Fig. 24I. Unlike calcite, quartz crystals will not cleave along crystal planes but will break into irregular pieces when given a sharp blow. Quartz is pure silica (SiO_2). Further details concerning these crystals will be given in this and the following chapters.

When a beam of ordinary unpolarized light is incident on a calcite or quartz crystal, there will be, in addition to the reflected beam, two refracted beams in place of the usual single one observed, for example, in glass. This phenomenon, shown in Fig. 24J for calcite, is called *double refraction*, or *birefringence*. Upon measuring the angles of refraction ϕ' for different angles of incidence ϕ, one finds that Snell's law of refraction

$$\frac{\sin \phi}{\sin \phi'} = n \qquad (24f)$$

holds for one ray but not for the other. The ray for which the law holds is called the *ordinary* or *O ray*, and the other is called the *extraordinary* or *E ray*.

Since the two opposite faces of a calcite crystal are always parallel, the two refracted rays emerge parallel to the incident beam and therefore parallel to each other. Inside the crystal the ordinary ray is always to be found in the plane of incidence. Only for special directions through the crystal is this true for the extraordinary ray. If the incident light is normal to the surface, the extraordinary ray will be refracted at some angle that is not zero and will come out parallel to, but displaced from, the incident beam; the ordinary ray will pass straight through without deviation. A rotation of the orystal about the *O* ray will in this case cause the *E* ray to rotate around the fixed *O* ray.

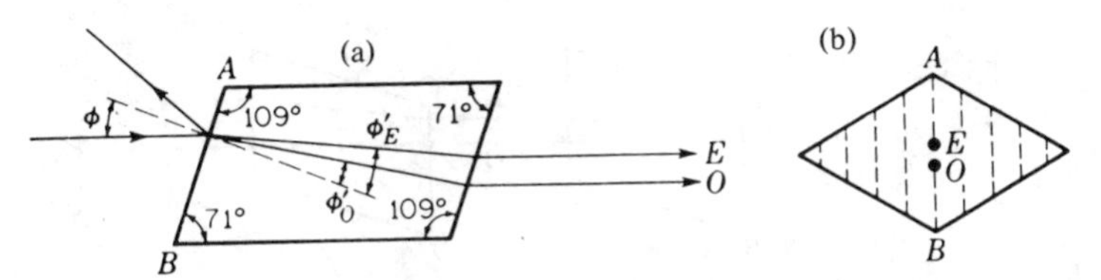

FIGURE 24J
Side and end views of the double refraction of light by a calcite crystal. (a) Cross section of a principal plane. (b) End View.

24.8 OPTIC AXIS

Calcite and quartz are examples of *anisotropic* crystals, or ones in which the physical properties vary with direction. All crystals except those belonging to the cubic system are anisotropic to a greater or less degree. Furthermore, the two examples chosen show the simple type of anisotropy which characterizes *uniaxial* crystals. In these there is a single direction called the *optic axis*, which is an axis of symmetry with respect to both the crystal form and the arrangement of atoms. If any property, such as the heat conductivity, is measured for different directions, it is found to be the same along any line perpendicular to the optic axis. At other angles it changes, reaching a maximum or a minimum along the axis. The directions of the optic axes in calcite and quartz are shown in Fig. 24I.

The double refraction in uniaxial crystals disappears when the light is made to enter the crystal so that it travels along the optic axis. That is, there is no separation of the O and E rays in this case. This is also true in directions at right angles to the axis, but here the O and E rays behave differently in a less obvious respect, namely they differ in velocity. The consequences of the latter difference will be examined in Chap. 27.

The direction of the optic axis in calcite is determined by drawing a line like xx' through a *blunt corner* of the crystal, so that it makes equal angles with all faces. A blunt corner is one where three obtuse face angles come together, and there are only two such corners more or less opposite each other. In quartz the optic axis yy' runs lengthwise of the crystal, its direction being parallel to the six side faces, as shown. It should be emphasized that the optic axis is not a particular line through the crystal but a *direction*. That is, for any given point in the crystal an optic axis may be drawn which will be parallel to that for any other point.

24.9 PRINCIPAL SECTIONS AND PRINCIPAL PLANES

In specifying the positions of crystals, and also the directions of rays and vibrations, it is convenient to use the principal section, made by a plane containing the optic axis and normal to any cleavage face. For a point in calcite, there are therefore three principal sections, one for each pair of opposite crystal faces. A principal section always cuts the surfaces of a calcite crystal in a parallelogram with angles of 71° and 109°, as shown at the left in Fig. 24J. An end view of a principal section cuts the

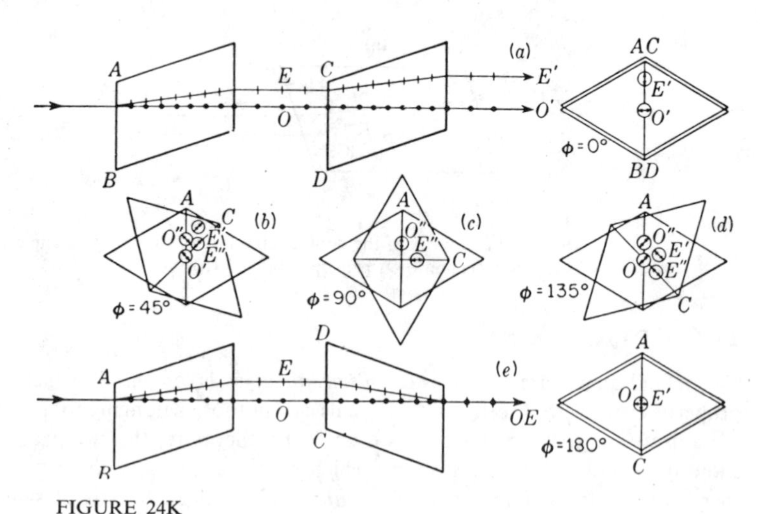

FIGURE 24K
Double refraction and polarization in two calcite crystals with their principal sections making different angles.

surface in a line parallel to AB, shown as a dotted line in the right-hand figure. All other planes through the crystal parallel to the plane represented by AB are also principal sections. These are represented by the other dotted lines.

The principal section, as so defined, does not always suffice in describing the directions of vibrations. Here we make use of the two other planes, the *principal plane of the ordinary ray*, a plane containing the optic axis and the ordinary ray, and the *principal plane of the extraordinary ray*, one containing the optic axis and the extraordinary ray. The ordinary ray always lies in the plane of incidence. This is not generally true for the extraordinary ray. The principal planes of the two refracted rays do not coincide except in special cases. The special cases are those for which the plane of incidence is a principal section as shown in Fig. 24J. Under these conditions the plane of incidence, the principal section, and the principal planes of the O and E rays all coincide.

24.10 POLARIZATION BY DOUBLE REFRACTION

The polarization of light by double refraction in calcite was discovered by Huygens in 1678. He sent a beam of light through two crystals as shown at the top of Fig. 24K. If the principal sections are parallel, the two rays O' and E' are separated by a distance equal to the sum of the two displacements found in each crystal if used separately. Upon rotation of the second crystal each of the two rays O and E is refracted into two parts, making four as shown by an end-on view in (*b*). At the 90°

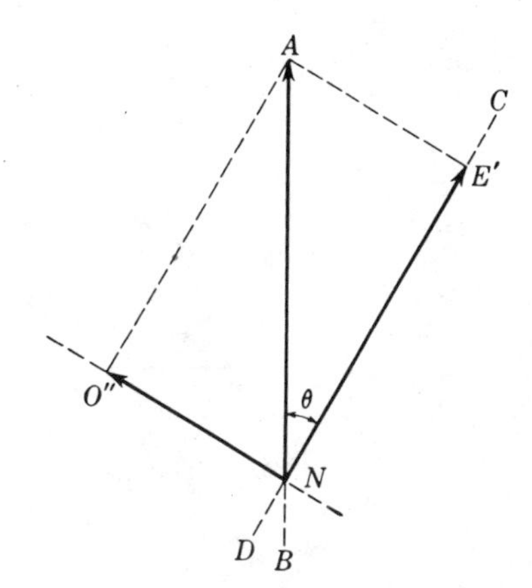

FIGURE 24L
Resolution of polarized light into components by double refraction.

position (*c*) the original O' and E' rays have faded and vanished and the new rays O'' and E'' have reached a maximum of intensity. Further rotation finds the original rays appearing, and eventually, if the crystals are of equal thickness, these rays come together into a single beam in the center for the 180° position shown at the bottom, the rays O'' and E'' having now vanished.

Thus, merely by using two natural crystals of calcite, Huygens was able to demonstrate the polarization of light. The explanation of the movement of the light rays is one simply of deviation by refraction and easily understood. The varying intensity of the spots, however, involves the polarization of the two light beams leaving the first crystal. In brief the explanation is somewhat as follows. Ordinary light upon entering the first calcite crystal is broken up into two plane-polarized rays, one, the O ray, vibrating perpendicular to the principal plane, which is here the same as the principal section, and the other, the E ray, vibrating in the principal section. In other words, the crystal resolves the light into two components by causing one type of vibration to travel one path and the other vibration to travel another path.

Consider more in detail now what happens to one of the plane-polarized beams from the first crystal when it passes through the second crystal oriented at some arbitrary angle θ. Let A in Fig. 24L represent the amplitude of the E ray vibrating parallel to the principal section of the first crystal just as it strikes the face of the second crystal. This second crystal, like the first, transmits light vibrating in its principal section along one path and light vibrating at right angles along another path. The E ray is therefore split up into two components E' with an amplitude $A \cos \theta$ and O'' with an amplitude $A \sin \theta$. These emerge from the second crystal with relative intensities given by $A^2 \cos^2 \theta$ and $A^2 \sin^2 \theta$. At $\theta = 90°$ E' vanishes and O'' reaches a maximum intensity of A^2. At all positions the sum of the two components, $A^2 \sin^2 \theta + A^2 \cos^2 \theta$, is just equal to A^2, the intensity of the incident beam.

The same treatment holds for the splitting up of the O beam from the first crystal into two plane-polarized beams O' and E''.

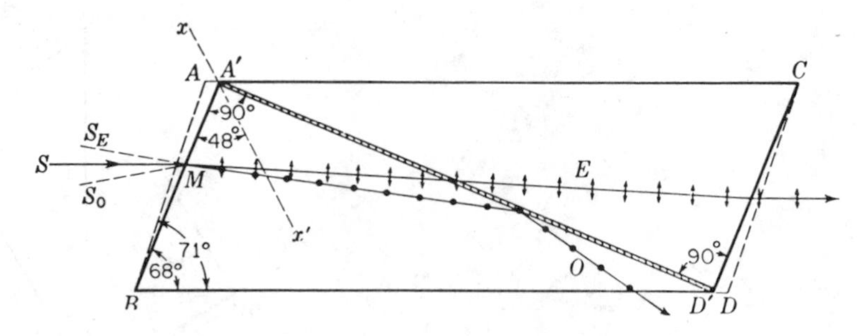

FIGURE 24M
Detailed diagram of a nicol prism, showing how it is made from a calcite crystal.

24.11 NICOL PRISM

This very useful polarizing device is made from a calcite crystal, and derives its name from its inventor.* The nicol prism is made in such a way that it removes one of the two refracted rays by total reflection, as is illustrated in Fig. 24M. There are several different forms of nicol prism,† but we shall describe here one of the commonest ones. First a crystal about 3 times as long as it is wide is taken and the ends cut down from 71° in the principal section to a more acute angle of 68°. The crystal is then cut apart along the plane $A'D'$ perpendicular to both the principal section and the end faces. The two cut surfaces are ground and polished optically flat and then cemented together with canada balsam. Canada balsam is used because it is a clear transparent substance with an index of refraction about midway between the index of the O and E rays. For sodium light,

Index of O ray	$n_O = 1.65836$
Index of canada balsam	$n_B = 1.55$
Index of E ray	$n_E = 1.48641$

Optically the balsam is denser than the calcite for the E ray and less dense for the O ray. The E ray therefore will be refracted into the balsam and on through the calcite crystal, whereas the O ray for large angles of incidence will be totally reflected. The critical angle for total reflection of the O ray at the first calcite to balsam surface is about 69° and corresponds to a limiting angle SMS_0 in Fig. 24M of about 14°. At greater angles than this, some of the O ray will be transmitted. This means that a nicol should not be used in light which is highly convergent or divergent.

The E ray in a nicol also has an angular limit, beyond which it will be totally

* William Nicol (1768–1851). Scotch physicist, who became very skillful in cutting and polishing gems and crystals. He devised his prism in 1828 and reportedly did not himself completely understand how it worked.

† Complete descriptions of polarizing prisms will be found in A. Johannsen, "Manual of Petrographic Methods," 2d ed., pp. 158–164, McGraw-Hill Book Company, New York, 1918.

reflected by the balsam. This is due to the fact that the index of refraction of calcite is different for different directions through the crystal. In the next chapter it will be seen that the index $n_E = 1.486$, as it is usually given, applies only in the special case of light traveling at right angles to the optic axis. Along the optic axis the E ray travels with the same speed as the O ray, and it therefore has the same index of 1.658. For intermediate angles the effective index lies between these two limits 1.486 and 1.658. There will therefore be a maximum angle SMS_E beyond which the balsam will be optically less dense than the calcite, and there will be total reflection of the E vibrations. The prism is so cut that this angle likewise is in the neighborhood of 14°. The direction of the incident light on a nicol therefore is limited on the one side to avoid having the O ray transmitted and on the other side to avoid having the E ray totally reflected. In practice, it is important to keep this limitation in mind.

Polarizing prisms are sometimes made with end faces cut perpendicular to the sides so that the light enters and leaves normal to the surface. The most popular one of this type, the *Glan-Thompson prism*, has an angular tolerance or aperture approaching 40°, hence much larger than that of the nicol. But this prism must be cut with the optic axis parallel to the end faces and is wasteful of calcite, large crystals of which are expensive and difficult to obtain. In another type the halves are held together so that there is a film of air between them instead of balsam. This device, called the *Foucault prism*, will transmit ultraviolet light. It has an angular aperture of only about 8°, however, and some difficulty is experienced with interference occurring in the air film.

24.12 PARALLEL AND CROSSED POLARIZERS

When two nicol prisms are lined up one behind the other, as shown in Fig. 24N, they form a good polariscope (Sec. 24.4). Positions (*a*) and (*c*) are referred to as *parallel polarizers*, and for them the E ray is transmitted. A loss of some 10 percent of the incident light is caused by reflection at the prism faces and absorption in the balsam layer, so that the overall transmission of a nicol for incident unpolarized light is about 40 percent. Position (*b*) in the figure represents one of the two positions called crossed polarizers. Here the E ray from the first nicol becomes an O ray in the second, and is totally reflected to the side. For intermediate angles, the incident E vibrations from the first nicol are broken up into components as shown by the vector diagram in Fig. 24L, where θ is the angle between the principal sections of the two nicols. The E' component is transmitted by the second nicol with the intensity $A^2 \cos^2 \theta$ and the O'' component is totally reflected. Parallel and crossed Polaroid filters are shown in Fig. 24H.

24.13 REFRACTION BY CALCITE PRISMS

Calcite prisms are sometimes cut from crystals for the purpose of illustrating double refraction and dispersion simultaneously as well as single refraction along the optic axis. Two regular prisms of calcite are shown in Fig. 24O, the first cut with the optic

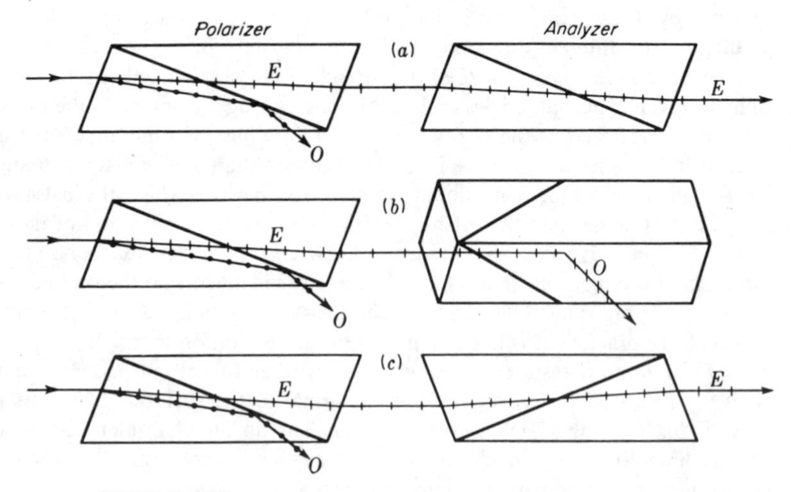

FIGURE 24N
Two nicol prisms mounted as polarizer and analyzer.

axis parallel to the refracting edge A, and the other with the axis also parallel to the base and perpendicular to the refracting edge. In the first prism there is double refraction for all wavelengths and hence two complete spectra of plane-polarized light, one with the electric vector parallel to the plane of incidence and the other with the electric vector perpendicular to it. An interesting demonstration of this polarization is accomplished by inserting a polarizer* into the incident or refracted beams. Upon rotation of the polarizer, first one spectrum is extinguished and then the other.

In the second prism, Fig. 24O(b), only one spectrum is observed, as with glass prisms. Here the light travels along the optic axis, or very nearly so, so that the two spectra are superposed. In this case a polarizer, when rotated, will not affect the

* Although nicol prisms give the most complete polarization of any of the devices commonly found in laboratories, polaroid films or a pile of glass plates mounted as in Fig. 24F are quite suitable for nearly all experimental demonstrations.

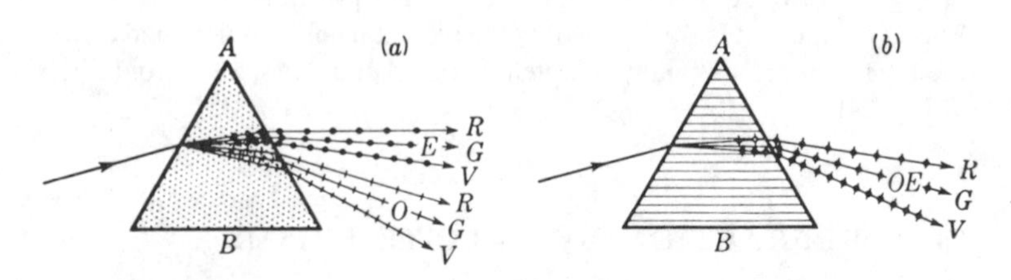

FIGURE 24O
Double and single refraction of white light by prisms cut at different angles from calcite crystals.

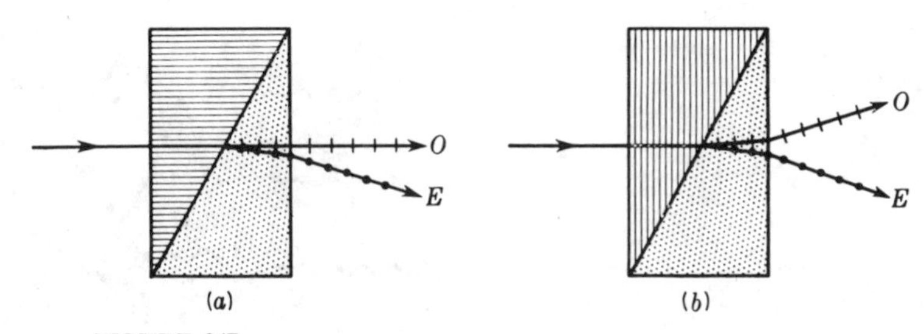

FIGURE 24P
Diagrams of (*a*) Rochon and (*b*) Wollaston prisms made from quartz.

intensity as it does with the first prism. The more detailed treatment of double refraction in Chap. 26 will clarify these experimental observations.

24.14 ROCHON AND WOLLASTON PRISMS

It is often desirable to split a light beam into two plane-polarized components, retaining both of them for a later comparison of their intensities. For this purpose other types of prisms have been designed, the most satisfactory of which are the Rochon and Wollaston prisms. These optical devices, sometimes called double-image prisms, are made of quartz or calcite cut at certain definite angles and cemented together with glycerin or castor oil.

In the Rochon prism [Fig. 24P(*a*)] the light, entering normal to the surface, travels along the optic axis of the first prism and then undergoes double refraction at the boundary of the second prism. The optic axis of the second prism is perpendicular to the plane of the page, as is indicated by the dots. In the Wollaston prism [Fig. 24P(*b*)] the light enters normal to the surface and travels perpendicular to the optic axis until it strikes the second prism, where double refraction takes place. The essential difference between the two is shown in the figures by the directions of the two refracted rays. The Rochon prism transmits the *O* vibrations without deviation, the beam being achromatic. This is frequently desired in optical instruments where only one plane-polarized beam is desired. The *E* beam, which is chromatic, is readily screened off at a sufficiently large distance from the prism.

The Wollaston prism deviates both rays and consequently yields greater separation of the two slightly chromatic beams. It is commonly used where a comparison of intensities is desired. These intensities will of course be equal for unpolarized light but will differ if the light is polarized in any way. It should be noted that in the Rochon prism the light should always enter from the left, in order for it to travel first along the optic axis, as shown in the figure. If it is sent in the other direction, the different wavelengths will emerge vibrating in different planes because of a phenomenon known as rotatory dispersion (see Sec. 28.2). This phenomenon, as well as the directions taken by the doubly refracted beams in quartz, will be treated in detail in the following chapters.

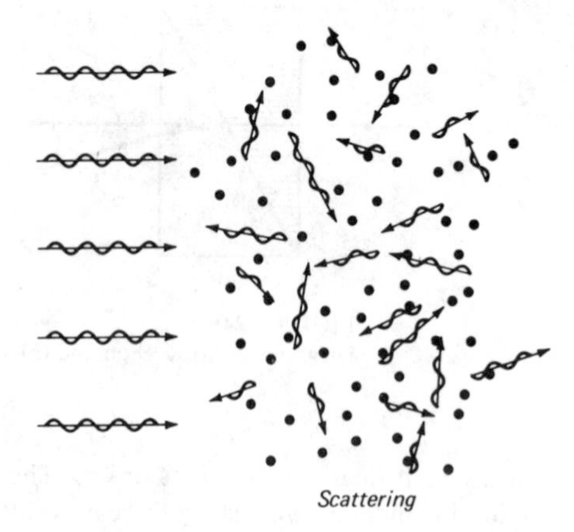

FIGURE 24Q
Light waves scattered by air molecules. *(From H. E. White, "Modern College Physics," 6th ed., D. Van Nostrand Co., New York, 1972. By permission of the publisher.)*

24.15 THE SCATTERING OF LIGHT AND THE BLUE SKY

The scattering of light by small particles of matter is responsible for some of nature's most beautiful phenomena. The blue sky and red sunset are attributed to scattering. As sunlight passes through our atmosphere, a large part of it is absorbed by the air molecules and immediately given out in some new direction (see Sec. 22.9).

The phenomenon of scattering is similar to the action of water waves on floating bodies. When a small stone is dropped into a pond of still water, a small cork floating nearby will bob up and down with the frequency of the passing waves. Light waves are visualized as acting in a similar manner on air molecules, as well as on fine dust and smoke particles. Once a passing light wave sets a molecule or particle into vibration, the wave can be emitted again in some random direction. This is shown schematically in Fig. 24Q. Light waves are shown being scattered in all directions.

It has long been known that short light waves are scattered much more than longer waves. Specifically, scattering is found by experiment to be proportional to the fourth power of the frequency or (what is the same thing) is inversely proportional to the fourth power of the wavelength:

$$\bullet \qquad \text{Scattering} \propto \nu^4 \qquad \text{Scattering} \propto \frac{1}{\lambda^4}$$

This is usually referred to as the *fourth-power law* or the *inverse fourth-power law*. According to these relations, violet light at the short-wavelength end of the spectrum is scattered about 10 times as much as red light at the long-wavelength end. For all six of the spectral colors, violet and blue light are scattered the most, followed by green, yellow, orange, and red. For every red wave ($\lambda = 700$ nm) scattered by sunlight, there are 10 violet waves ($\lambda = 400$ nm):

Red	Orange	Yellow	Green	Blue	Violet
1	2	2.5	3	6	10

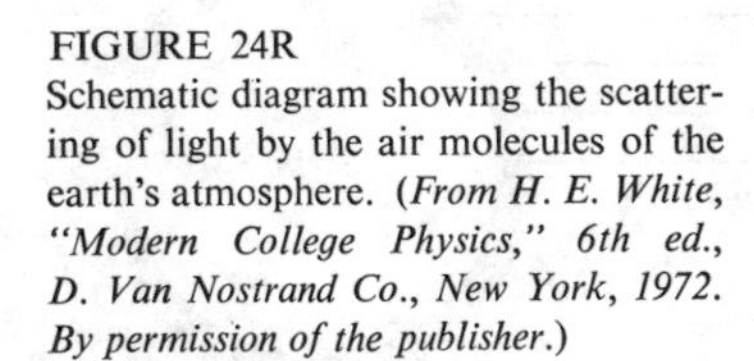

FIGURE 24R
Schematic diagram showing the scattering of light by the air molecules of the earth's atmosphere. (*From H. E. White, "Modern College Physics," 6th ed., D. Van Nostrand Co., New York, 1972. By permission of the publisher.*)

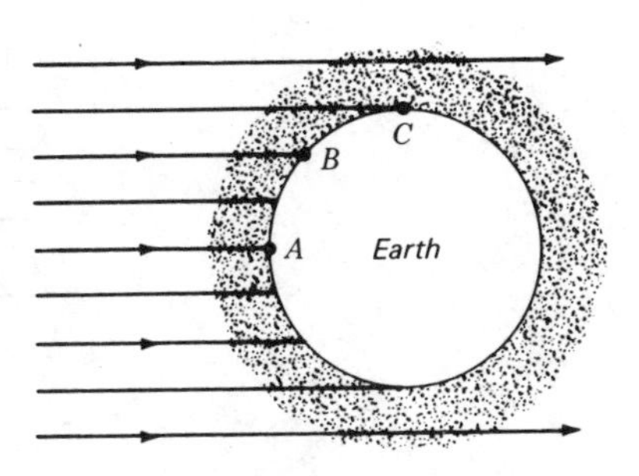

When the sun is brightly shining on a clear day, the whole sky appears to be light blue (see Fig. 24R). This color is a mixture of spectral colors scattered almost entirely by air molecules. It can be demonstrated that spectral colors, mixed in the proportions given by the row of numbers above, will produce the light blue color of the sky. This is most beautifully demonstrated by the sunset experiment in the next section.

24.16 THE RED SUNSET

The sunset on a clear day is never highly colored. To see a highly colored sunset we must have tiny dust and smoke particles in the air. Why this is necessary is shown in Fig. 24S, where a moderate layer of dust and smoke 1 or 2 km thick is spread out over a large area of the earth's surface. Looking straight upward on such a smoky day, an observer will see only a blue sky. The sunlight has traveled the relatively short path of 1 or 2 km through the smoke layer. Since very little if any of the color is scattered, the sun's disk will appear *white* and the surrounding sky *blue*.

As the afternoon wears on and we approach sunset, the direct rays from the sun must travel through an increasing path of dust and smoke. An hour or so before sundown the observer sees rays from the direction of C, and the light path makes a sizeable angle with the horizon. Passing through a longer path than at noon, the blue and violet are scattered out, and the colors coming through to the observer, red, orange, yellow, and green, appear to be light yellow.

Just before sunset, when the observer sees light from the direction D, the rays pass through 10 to 100 km of dust and smoke particles, and all but the red waves of direct sunlight are scattered out. The sun's disk appears red, and much of the immediate surroundings are orange and red. The sky higher up and directly overhead is still light blue. If the dust and smoke layer is very dense, even the red will be scattered in all directions and the deepening red sun will disappear from view before it reaches the horizon.

One of the finest demonstrations in all of science is the scattering of light by fine sulfur particles suspended in water (see Fig. 24T). A parallel beam of white light from a carbon arc and lens L_1 is allowed to pass through a fish tank with all glass sides. The beam then passes through an iris diaphragm, which is imaged on a

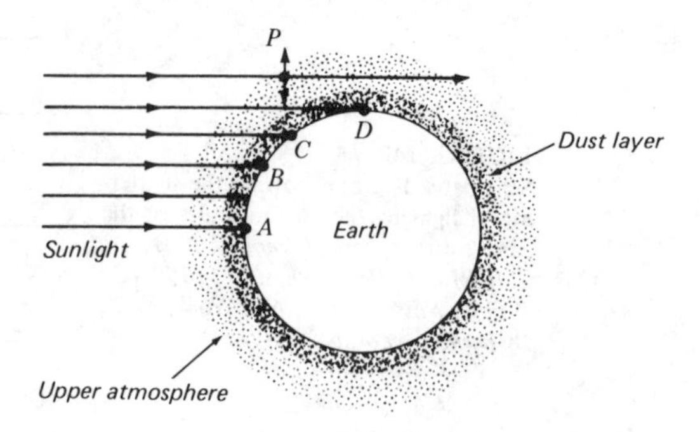

FIGURE 24S
The scattering of light by a layer of dust near the earth's surface causes the sun to turn from white at (A) to yellow at (B), then orange at (C), and finally at (D) to red at sunset. (*From H. E. White, "Modern College Physics," 6th ed., D. Van Nostrand Co., New York, 1972. By permission of the publisher.*)

large screen by a lens L_2. To produce the fine sulfur particles for scattering, about 40 g of photographic fixing powder (*sodium hyposulfite*) is first dissolved in about 7.5 liters of clear distilled water. When one is ready to perform the demonstration for a large or small audience, 1 to 2 ml of concentrated sulfuric acid (previously dissolved in about 100 ml of distilled water) is poured into the tank, and thoroughly stirred.*

The microscopic sulfur particles will begin to form in about 2 min and will be noticed by the pale-blue scattered light from the beam; 2 to 3 min later the beam boundaries should no longer be seen, and the entire tank will be filled with blue light. Light scattered from the central beam is scattered again and again before emerging from the tank. This is called *multiple scattering.*

When scattering first begins to show in the tank, the sun, simulated by the circular image on the large screen, will turn yellow. Later, as more and more scattering takes place, the violet, blue, green, and finally orange will disappear from the direct beam, and the sun will turn from yellow to orange to a beautiful red.

24.17 POLARIZATION BY SCATTERING

If a polarizing plate, like Polaroid, is used to test the blue sky, the light is *partially plane-polarized.* A little exploration will show that maximum polarization occurs at an angle of 90° with the direction of the incoming sunlight and drops to zero at 180° just after the sun goes down. At dusk on a clear day, when the sun has just disappeared over the horizon, one can locate the direction of zero polarization and from it determine the sun's position.

* If more water is needed, use the same proportions of sodium hyposulfite and water as given above. The correct amount of acid to produce the best results is determined by trial.

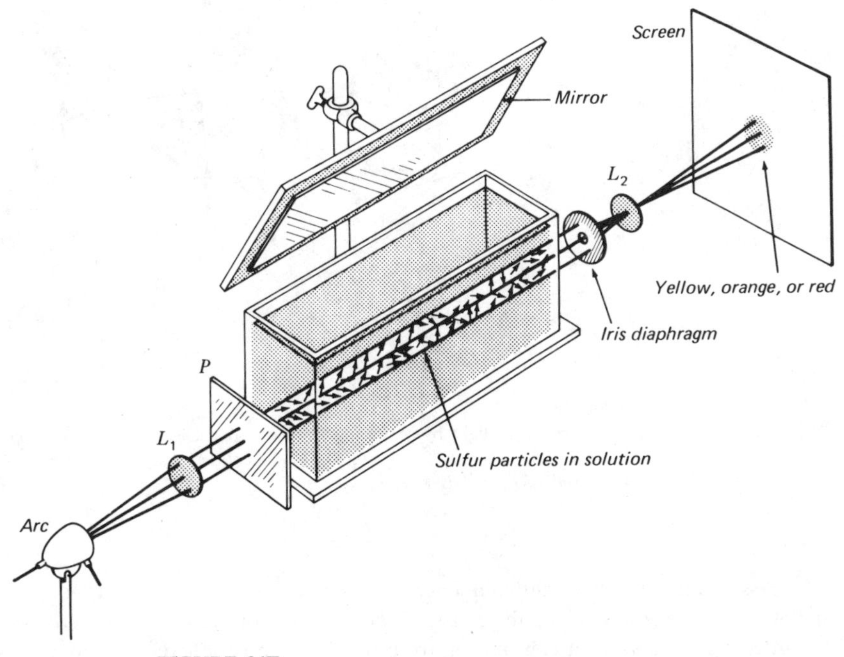

FIGURE 24T
The sunset experiment: demonstration of the scattering and polarization of light by small particles. (*From H. E. White, "Modern College Physics," 6th ed., D. Van Nostrand Co., New York, 1972. By permission of the publisher.*)

The polarization of scattered light can be observed using the fish-tank experiment described in Sec. 24.16. In the early stages of the formation of sulfur particles, one can hold a polarizing plate in front of one eye, and looking at the beam from a 90° angle, rotate the plate and find the scattered light to be nearly 100 percent plane-polarized. Or by placing a polarizing plate in the incident beam, as shown in the figure, and rotating it, observe the beam in the mirror as well as in the tank. It will disappear in the tank, then in the mirror, then in the tank, etc. These experiments are considered proof that light is a transverse wave. Sound waves have a longitudinal character and exhibit none of the above effects.

Consider the scattering of light from a single air molecule P, as shown in Fig. 24U. Suppose ordinary unpolarized light is incident from the left. We assume it is composed of two plane-polarized components, as shown in the diagram. If the incident component is vibrating in the xy plane and is absorbed, it sets the particle vibrating in the y direction. In giving up this energy the same wave can be emitted in any direction except along the y axis. To emit the light in the y direction the wave would have to be longitudinal, and this is forbidden.

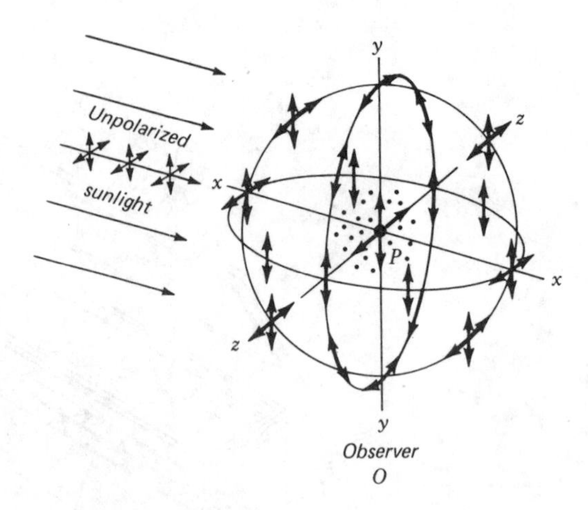

FIGURE 24U
The polarization of light by scattering from fine particles. (*From H. E. White, "Modern College Physics," 6th ed., D. Van Nostrand Co., New York, 1972. By permission of the publisher.*)

Assume the incident light component is vibrating in the xz plane. The particle at P will be set vibrating along the z axis. Reemission is now allowed in all directions except along the z axis. It can be seen from the diagram (a), therefore, why an observer at O looking at the blue sky in a direction making 90° with the sun's rays, will find the blue light *plane-polarized* with its direction of vibration parallel to the z axis. No particle at P can be set vibrating along the x axis, since this would violate the principle that light has no longitudinal component.

Light waves are also well known to be *electromagnetic* in character and as such to have two different components. A single wave has an *electrical component* vibrating in one plane and a *magnetic component* vibrating in a plane at 90° (see Fig. 20B). A number of laboratory experiments on interference show that the *electric component* is responsible for all the known optical effects (see Sec. 25.12).

24.18 THE OPTICAL PROPERTIES OF GEMSTONES

From the earliest times of the ancient emperors of China and India, the czars of Russia, the shahs of Persia, the sheiks of Arabia, and the kings and queens of Europe, gemstones have held a great fascination. *Emeralds*, *rubies*, *sapphires*, and *diamonds* are the most precious of stones, and have served as fine gifts from one wealthy person to another.

Numerous attempts have been made over the centuries to produce synthetic gemstones. Only in recent years have these dreams come true. Not only have our laboratories reproduced nature's products, they have produced many new gems and crystals not found in the earth's crust. Synthetic stones have exactly the same chemical and physical properties of the natural stones and in most instances are more nearly perfect in their crystal formation than their natural counterparts. The principal attraction of well-cut gems is first, their *size*, then their *freedom from flaws*, and finally their *fire* and *luster*.

The first important gemstones to be synthesized in the laboratory belong to the *corundum* family. Corundum is the hexagonal crystal form of α *alumina* (Al_2O_3). Those of high purity are transparent and water-white, and are called *white sapphires*. If a few percent of *chromic oxide* (Cr_2O_3) is added to the growing crystal, one obtains the ruby, a pink or red crystal of great beauty. If other metal oxides, such as *iron* or *titanium*, are added, sapphires of many colors are produced.

Some of the most treasured of gems are the natural star sapphire and star ruby. These natural stones have exactly the same composition as ordinary sapphires and rubies, but they also contain small amounts of *titanium oxide* (TiO_2). These microscopic needles are scattered through the body of the crystal in a symmetrical, three-dimensional array. These stones are usually cut *en cabochon* (dome shaped with a flat base). White light from a point source is reflected by the needles and gives rise to a six-rayed star.

Industrial laboratories have succeeded in duplicating nature's gemstones and have synthesized star rubies and star sapphires. Their synthetic crystals have the same needlelike impurities that produce the six-rayed star effect and have the identical optical properties. The tiger's eye and the cat's eye are similar stones, with all the tiny needles, or hollow tubes, lined up parallel in one direction only.

Emeralds have been synthesized by several laboratories since 1930 and diamonds of small size since 1961. The latter are now produced in sizable quantities and are used in highly specialized machine tools of various kinds.

Pure white, *pale blue*, and *pale yellow diamonds*, up to 1 carat in size, have more recently been produced by the General Electric Research Laboratories (see Fig. 24V). These stones are formed from graphite under extremely high temperatures and pressures. The dispersion of the diamond and the fire and luster of properly cut stones are surpassed by at least two synthetic crystals of large size. These are *strontium titanate* and *rutile*. Refractive indices of diamond and these clear crystals are given in Table 24A. Refractive indices for other wavelengths for rutile (TiO_2) can be calculated from the constants given in the Cauchy equations, Eqs. (26f).

Table 24A REFRACTIVE INDICES FOR THREE GEMSTONES

	Wavelengths λ, Å					
Gemstones	4100	4700	5500	5800	6100	6600
Diamond	2.458	2.444	2.426	2.417	2.415	2.410
$SrTiO_3$	2.613	2.524	2.440	2.417	2.398	2.371
Rutile, *O*	2.975	2.765	2.650	2.621	2.597	2.569
Rutile, *E*	3.330	3.095	2.953	2.917	2.889	2.530

FIGURE 24V
Four of General Electric's famous gem-quality diamonds created in the laboratory from graphite, the soft black substance used in "lead" pencils. The four crystals originally were about 1 carat in weight. After cutting and polishing, each weighs about $\frac{1}{3}$ carat. One is clear, another is light blue, and another is canary yellow. The dark crystal at the bottom is a deep blue in color. (*Courtesy of Herbert M. Strong, General Electric Company, Schenectady, N.Y.*)

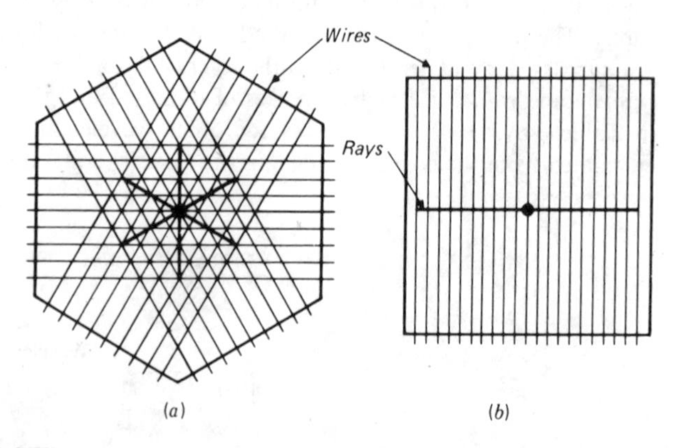

FIGURE 24W
Wire wound around clear plastic sheets for observing the star patterns seen in the gemstones (*a*) star sapphires and star rubies and (*b*) tiger's-eyes and cat's-eyes.

A demonstration of the optical properties of *asteriated*, or star, ruby can be made by winding fine wire around a hexagonal piece of sheet plastic (see Fig. 24W). By observing a point source of white light through the wire mesh one can see the star pattern. Wire wound in one direction around a square plastic sheet will produce the two-rayed star of the tiger's eye and the cat's eye. The overlapping of wires has little effect on the observed pattern.

Gemstone crystals 10 to 20 cm in diameter and over 4000 carats in size are being synthesized by American and foreign laboratories today. These crystals are grown by the ton and are used in various ways. Pink rubies are grown in rods 1 or 2 cm in diameter and are used in lasers for high-quality instruments of many kinds.

PROBLEMS

24.1 Find the variation of the polarization angle throughout the visible spectrum, 4000 to 7200 Å, for the barium flint glass listed in Table 23B. First use the two-term Cauchy equation and the indices for $\lambda = 6563$ Å and $\lambda = 3988$ Å to find the values of A and B, and give the angles for the extreme limits only. Give also the difference between the two angles.

Ans. $A = 1.57664$, $B = 5.0983 \times 10^5$ Å^2, $\bar{\phi}_1 = 57.7757°$, $\bar{\phi}_2 = 58.1310°$, $\Delta\bar{\phi} = 0.3553° = 21'19''$

24.2 Light is reflected from the smooth surface of water at the polarizing angle. Assume $n = 1.3330$. Find (*a*) the angle of incidence and (*b*) the angle of refraction. (*c*) Describe what would be seen if the reflected light were viewed through a calcite crystal which is rotated about the direction of the reflected beam.

24.3 The effective intensity of a source of light is controlled by the use of a polarizer and analyzer by changing the angle θ between their principal sections. To what accuracy in degrees must θ be known to obtain an accuracy of 2 percent in the intensity of the transmitted light at a setting which reduces the maximum intensity to 10 percent?

24.4 A beam of white light is partially polarized by passing it through a single glass plate at the polarizing angle. Assuming 15 percent reflection of the intensity of the s vibrations at each surface, find the degree of polarization (*a*) if multiple reflections within the plate are neglected and (*b*) if internal reflections are taken into account. (*c*) Find the degree of polarization if there are 12 plates. Assume $n = 1.5000$.

Ans. (*a*) 16.11%, (*b*) 14.79%, (*c*) 67.57%

24.5 An ordinary beam of light is sent through three dichroic polarizers, the second of which is oriented at 25° with the first and the third at 50° with the first in the same direction. What intensity gets through the system, relative to that of the incident unpolarized light, (*a*) neglecting light reflected from the six surfaces and (*b*) assuming 4.0 percent of the light reflected at each surface?

24.6 Calculate the relative intensities of the images (*a*) O' and E', and (*b*) O'' and E'' obtained in the two-crystal experiment shown in Fig. 24K when the angle between the principal sections is 60°.

24.7 A crystal is placed in a polariscope, the polarizer and analyzer being parallel. The principal section of the crystal makes an angle of 35° with the planes of transmission of the polarizer and analyzer. Find the ratio of the intensities of the E and O beams (*a*) as they leave the crystal and (*b*) after they leave the analyzer.

Ans. (*a*) 2.040, (*b*) 4.160

24.8 (*a*) Calculate the degree of polarization of the light due to Rayleigh scattering at 70° with the direction of the primary beam. (*b*) Calculate the intensity of this light relative to that scattered straight backward.

24.9 In a Wollaston prism of quartz having refractive angles of 30°, (*a*) what will be the separation of the colors on each side of center? Use the Fraunhofer C to F lines. (*b*) What is the separation of the D light in the two polarized beams? (*c*) What is the ratio of (*b*) to (*a*)? See Table 26A for indices.

25
REFLECTION

Among the subjects touched upon in the last chapter the first to be discussed in greater detail will be related to the sections on polarization by reflection and transmission. There we considered the effects at the particular angle of incidence called the polarizing angle. Going beyond this special case, we shall now investigate the characteristics of the reflected and transmitted light as they depend on wavelength, polarization, and angle of incidence. It will be assumed that the surfaces are optically smooth, which means that any irregularities are small compared with the wavelength. The properties of the reflecting substance play an essential role, and among these absorption is an important one. Metals are in general the best reflectors, a fact which will be found to be related to their ability to conduct electricity, and consequent high absorption. We begin, however, with the simpler case of nonconducting dielectric materials like glass.

25.1 REFLECTION FROM DIELECTRICS

The essential features of reflection from a single glass surface are briefly described as follows. At normal incidence about 4 percent of the intensity of a beam of unpolarized visible light is reflected, and the other 96 percent is transmitted. At other angles of

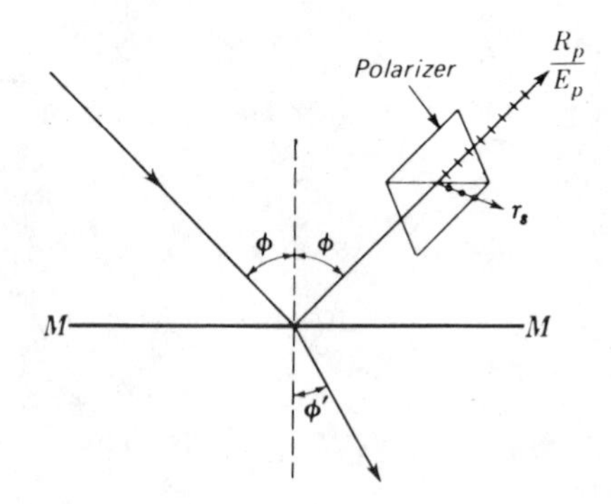

FIGURE 25A
Analysis of the reflected light into its two plane-polarized components.

incidence the reflecting power increases with angle, at first slowly and then more rapidly until at 90°, that is, grazing incidence, all the light is reflected.

It was shown at the beginning of the last chapter that there is one angle of incidence for which the reflected light is completely plane-polarized with its electric vector perpendicular to the plane of incidence. At angles different from this the reflected light is only partially polarized. The relations in this case are most easily described in terms of the reflection of the two plane-polarized components of the incident unpolarized light, the vibrations of which are, respectively, parallel and perpendicular to the plane of incidence. In the laboratory this is usually done by examining the reflected light that passes through a nicol or other polarizer (see Fig. 25A). If the polarizer is oriented with its principal section parallel to the plane of incidence, the p vibrations, i.e., the vibrations parallel to the plane of incidence, can be measured. Rotation of the polarizer through 90° then allows the s vibrations perpendicular to the plane of incidence (s stands for the German *senkrecht*, meaning *perpendicular*) to be measured. The results of such an experiment when plotted against the angle of incidence ϕ are represented by the two solid curves shown in Fig. 25B(a). The ordinates are R_p^2/E_p^2, the fraction of the incident p light reflected, and R_s^2/E_s^2, the corresponding fraction for the s light. These fractions are called the *reflectances* for p and s light. Part (b) of the figure refers to the amplitudes and will be discussed below.

The curves of Fig. 25B are represented very accurately by theoretical equations which were first derived by Fresnel from the elastic-solid theory and are known as *Fresnel's laws of reflection*. For the present we shall merely state them and show their application to the main features from dielectrics. The laws may be written

$$\frac{R_s}{E_s} = -\frac{\sin(\phi - \phi')}{\sin(\phi + \phi')} \qquad \frac{R_p}{E_p} = \frac{\tan(\phi - \phi')}{\tan(\phi + \phi')} \tag{25a}$$

$$\frac{E_s'}{E_s} = \frac{2\sin\phi'\cos\phi}{\sin(\phi + \phi')} \qquad \frac{E_p'}{E_p} = \frac{2\sin\phi'\cos\phi}{\sin(\phi + \phi')\cos(\phi - \phi')} \tag{25b}$$

Here the symbols E, R, and E' mean the amplitudes of the electric vectors in the incident, reflected, and refracted light, respectively, the subscripts denoting the two

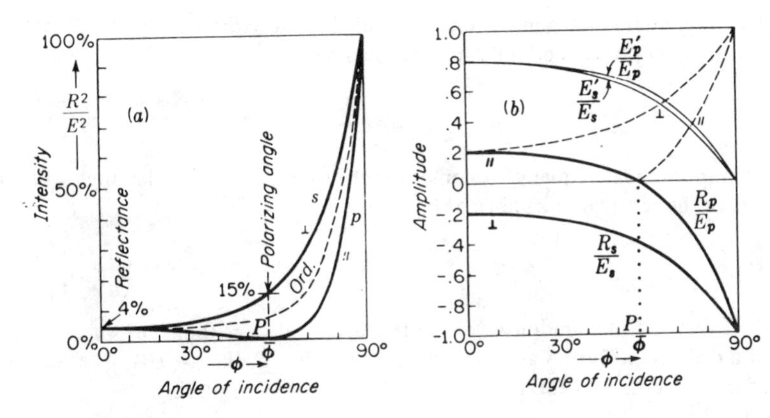

FIGURE 25B
Reflectances and the corresponding amplitudes for a dielectric having $n = 1.50$.

planes of vibration. The angles ϕ and ϕ', following our usual notation, are the angles of incidence and refraction.

The fractional amplitudes given by Eqs (25a) and (25b) are plotted against the angle of incidence in Fig. 25B(*b*), the values of ϕ and ϕ' used in the equations corresponding to the index of refraction 1.50. Solid curves represent the amplitudes, both positive and negative, as they are given by the equations, while the absolute magnitudes of the reflected components are shown by broken curves. The negative signs indicate phase changes of π, which will be discussed below. These, however, are immaterial for the intensities since the latter are dependent on the squares of the amplitudes.

The reflectances are given by

$$\frac{R_s^2}{E_s^2} \quad \text{and} \quad \frac{R_p^2}{E_p^2} \tag{25c}$$

and these are the curves in part (*a*) of the figure. At normal incidence, where $\phi = 0$, the parallel and perpendicular components must be equally reflected because here the plane of incidence is undefined and the two components are not distinguishable. With increasing ϕ R_p^2/E_p^2 drops and R_s^2/E_s^2 rises until at the polarizing angle their values are zero and 15 percent, respectively. At grazing incidence both components are totally reflected. Even an unsilvered glass surface becomes a nearly perfect mirror when the light source is viewed very close to the reflecting plane. It is easily verified that the glaze on a page of this book becomes highly reflecting at a grazing angle.

The value of the reflectance at normal incidence does not follow immediately from Eqs. (25a) by setting $\phi = 0$, since this substitution gives an indeterminate result. It can be evaluated, however, as follows. Since both ϕ and ϕ' become small as we approach perpendicular incidence, we may set the tangents equal to the sines, obtaining

$$\frac{R_p}{E_p} = -\frac{R_s}{E_s} = \frac{\sin(\phi - \phi')}{\sin(\phi + \phi')} = \frac{\sin\phi\cos\phi' - \cos\phi\sin\phi'}{\sin\phi\cos\phi' + \cos\phi\sin\phi'}$$

Dividing numerator and denominator of the last expression by sin ϕ' and replacing sin ϕ/(sin ϕ') by n, we find that it reduces to

$$\frac{R}{E} = \frac{n \cos \phi' - \cos \phi}{n \cos \phi' + \cos \phi} \approx \frac{n - 1}{n + 1} \tag{25d}$$

The approximate equality becomes exact in the limit when the angles become zero. Hence the reflectance at normal incidence is

$$\frac{R^2}{E^2} = \left(\frac{n - 1}{n + 1}\right)^2 \tag{25e}$$

This very useful equation gives the reflectance at $\phi = 0$ for any single clean surface of a dielectric. Thus a glass having $n = 1.50$ has $R^2/E^2 = 0.04$, or exactly 4 percent as indicated in Fig. 25B(a).

25.2 INTENSITIES OF THE TRANSMITTED LIGHT

One might expect that the transmitted intensities would be complementary to the reflected ones, so that the two would add to give the incident intensity, but this is not so. The intensity is defined as the energy crossing unit area per second, and the cross-sectional area of the refracted beam is different from that of the incident and reflected beams except at normal incidence. It is the total energy in these beams that is complementary. There are, however, simple relations between the incident, reflected, and transmitted *amplitudes*, which follow, as we shall show later, from the boundary conditions of electromagnetic theory. These are

$$\frac{E'_s}{E_s} - \frac{R_s}{E_s} = 1 \qquad \text{and} \qquad n\frac{E'_p}{E_p} - \frac{R_p}{E_p} = 1 \tag{25f}$$

In Fig. 25B(b) it will be seen that the curves for E'_s and R_s run parallel to each other. Those for E'_p and R_p are not parallel but become so if the ordinates of the former curve are multiplied by n. Since Eqs. (25f) are simpler than the Fresnel equations (25b), it is sufficient to remember only the former in addition to Eqs. (25a) in order to solve problems involving transmitted amplitudes and intensities.

The fraction of the incident intensity that is transmitted, or the transmittance, when light enters a dielectric of refractive index n is not given directly by the square of the relative amplitude. The intensity in the medium according to Eq. (23g) also contains a factor n, so that the transmittance becomes $n(E'/E)^2$. Now, as stated above, the sum of this and the reflectance $(R/E)^2$ does not equal unity, as can easily be verified from Eqs. (25a) and (25b). The total energy flux in the refracted beam is its intensity times its area, and the latter differs from that of the incident or reflected beams in the ratio (cos ϕ')/(cos ϕ). Conservation of energy is then expressed by the relation

$$\left(\frac{R}{E}\right)^2 + n\left(\frac{E'}{E}\right)^2 \frac{\cos \phi'}{\cos \phi} = 1$$

which applies to either s or p light.

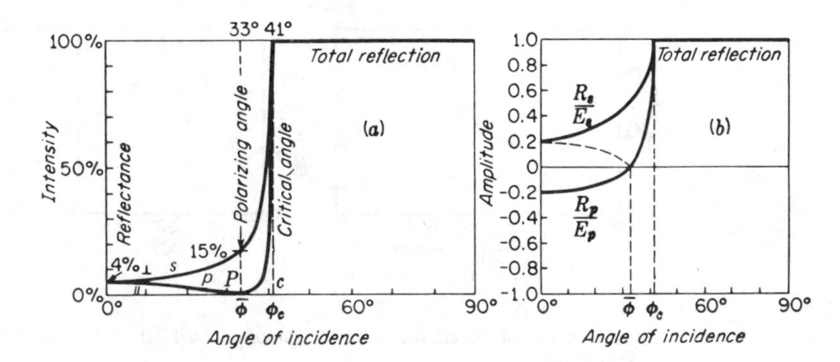

FIGURE 25C
Intensity and amplitude curves for internal reflection at a dielectric boundary, $n = 1.54$.

25.3 INTERNAL REFLECTION

In the above discussion it was assumed that the light strikes the boundary from the side of the optically less dense medium (usually air) so that we were dealing with the so-called rare-to-dense, or *external, reflection.* Fresnel's laws apply equally well to the case of dense-to-rare, or *internal, reflection.* If the same value of n is to be retained for the dense medium, the latter case merely involves the exchange of ϕ and ϕ' in the equations. The resulting curves are plotted in Fig. 25C, the reflectances in (*a*) and the amplitudes in (*b*). Up to the critical angle $\phi_c = 41°$ these resemble the curves for external reflection, starting with $R^2/E^2 = 4$ percent at normal incidence and diverging until the polarizing angle $\bar{\phi}$ is reached. This angle, 33°, corresponds to the angle of refraction at the polarizing angle in the external case, since the angle in the rarer medium (57°) must be such as to render the refracted and reflected rays perpendicular to each other.

At the critical angle the refracted ray leaves at a grazing angle, and the internal reflectance becomes 100 percent just as for external reflection at grazing incidence. When ϕ exceeds the critical angle, Fresnel's equations contain imaginary quantities but as we shall see may still be used. It will be found that the reflection remains total but that there is a continually changing phase shift.

25.4 PHASE CHANGES ON REFLECTION

Returning for the moment to external reflection, where $\phi > \phi'$ through the whole range, we find from Eqs. (25a) that the sign of R_s/E_s is always negative. This means that there is an abrupt change of phase of 180° in the process of reflection. We express it by writing $\delta_s = \pi$. For the p light the sign is positive for small ϕ, indicating no phase change, but when the condition $\phi + \phi' = 90°$ is reached, the tangent in the

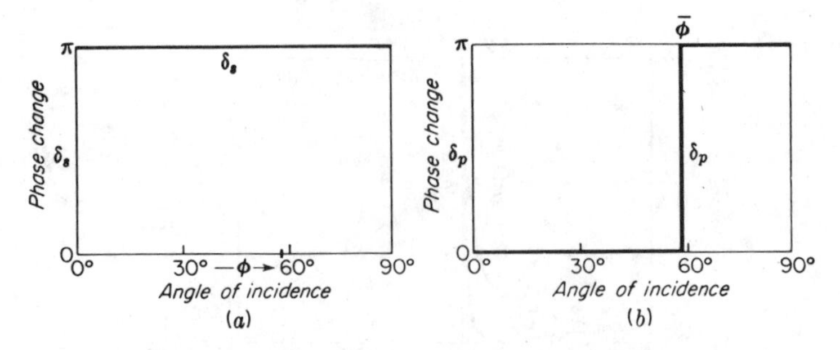

FIGURE 25D
Phase change of the electric vector of plane-polarized light externally reflected from a dielectric.

denominator goes to infinity and changes sign. Thus δ_p changes abruptly from zero to π at the polarizing angle. No real discontinuity is involved, however, because at this angle the amplitude of the p light goes through zero [Fig. 25B(*b*)]. Plots of δ_p and δ_s for the entire range of ϕ are given in Fig. 25D.

The directions in space of the electric vector before and after reflection are shown in Fig. 25E. It is seen that in case (*a*), where δ_p is taken as zero, the incident and reflected vectors are in nearly opposite directions. This apparent contradiction comes from our convention of regarding a displacement as positive or negative according as it is seen looking *against the light* in all cases. If the observer turns from viewing the incident beam to viewing the reflected beam, the rotation occurring in the plane of incidence, he finds that the two arrows maintain the same orientation relative to him. It is unfortunate that this convention gives a phase change for the s light but none for the p light at normal incidence, because at $\phi = 0$ the distinction

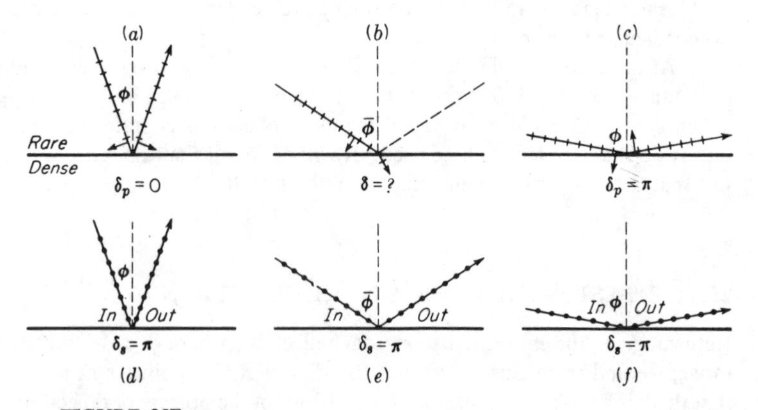

FIGURE 25E
Positions in space of the electric vector just before and just after external reflection from a dielectric.

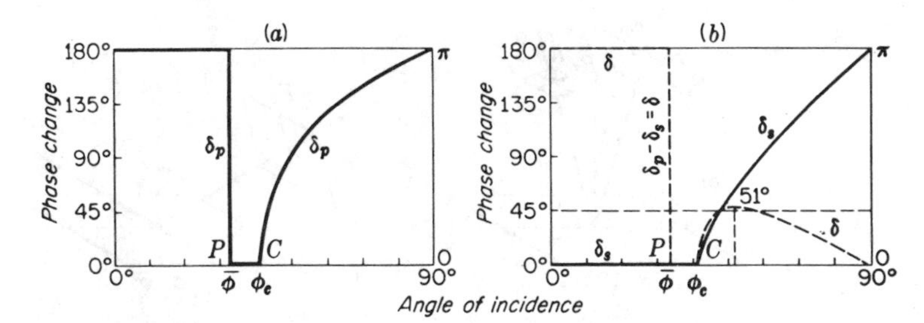

FIGURE 25F
Phase changes of the electric vector for internal reflection from a dielectric, $n = 1.51$.

between s and p vanishes. Using the opposite convention for p would, however, lead to an equally bad inconsistency in case (c) of the figure.

The phase changes that occur in internal reflection are, up to the critical angle, exactly the reverse of those at the corresponding angles for external reflection. This is a necessary consequence of Stokes' relations [Eq. (14d)], according to which there must be a relative difference of π between the two cases. Beyond ϕ_c, in the region of total reflection, Eqs. (25a) lead* to the following expressions for the tangent of half the phase change:

$$\tan\frac{\delta_s}{2} = \frac{\sqrt{n^2 \sin^2 \phi - 1}}{n \cos \phi} \qquad \tan\frac{\delta_p}{2} = n\,\frac{\sqrt{n^2 \sin^2 \phi - 1}}{\cos \phi} \tag{25g}$$

In Fig. 25F are shown separate curves for δ_p and δ_s, and for their difference $\delta = \delta_p - \delta_s$. The δ_p curve rises more steeply than δ_s and at $\phi = 45°$, according to Eqs. (25g), is exactly twice as large. Since the curves come together again at $\phi = 90°$, their difference δ passes through a maximum and decreases to zero. The principle of the Fresnel rhomb (Sec. 25.6) is based on this fact.

25.5 REFLECTION OF PLANE-POLARIZED LIGHT FROM DIELECTRICS

We are now prepared to predict the nature of the reflected light when plane-polarized light is incident on the surface at any angle. Let the light fall on a plate of glass as in Fig. 25G with the plane of vibration making an angle $\psi = 45°$ with the *perpendicular* to the plane of incidence.† This angle we shall call the *azimuth* angle, whether it refers to the light vibrations in the incident, reflected, or refracted beam. The incident light, of amplitude E, can here be resolved into two equal components E_p and E_s and each of these treated separately.

* See, for example, M. Born, "Optik," p. 43, J. Springer, Berlin, 1933.

† It is customary to measure ψ this way because the plane of polarization was originally defined to be at right angles to what we now call the plane of vibration.

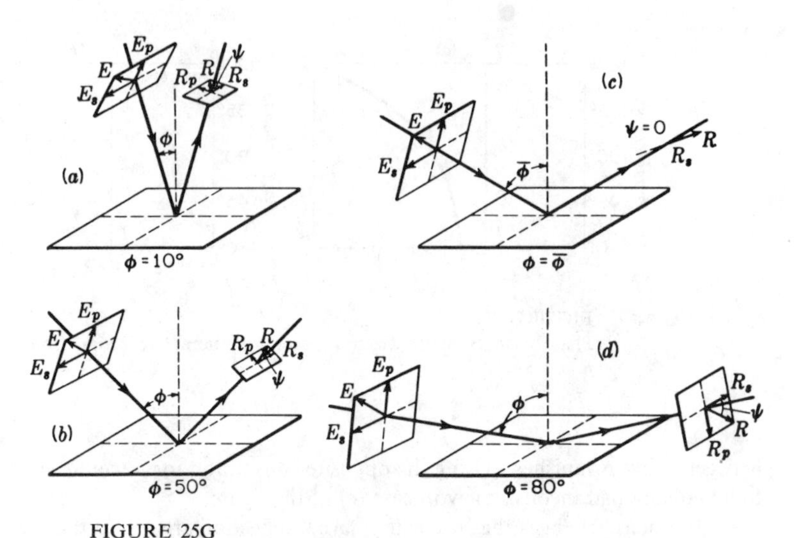

FIGURE 25G
Azimuths and amplitudes of plane-polarized light externally reflected from a glass surface at different angles of incidence.

Consider first the case where the angle of incidence ϕ is small, as in diagram (*a*) of the figure. Referring to Fig. 25B(*b*), the amplitudes of the two reflected components are found to be small and very nearly equal in magnitude. They are also out of phase by 180°. If the angle ϕ is about 10°, the component R_s is just a little larger than R_p. Taking the vector sum of the reflected components, one finds R in the direction shown. In case (*b*) the azimuth of the incident light is again 45°, but the angle of incidence is about 50°. R_p is now quite small and in phase with E_p, whereas R_s is larger than before and still 180° out of phase with E_s. The reflected ray remains plane-polarized, but the plane of vibration has turned farther away from the plane of incidence. When $\phi = \bar{\phi}$, as in (*c*), $R_p = 0$, while R_s is still larger and maintains the same phase. The resultant amplitude has continued to grow and now stands at right angles to the plane of incidence. In diagram (*d*), where the angle ϕ approaches 90° (grazing incidence), the reflected components have increased markedly, approaching in magnitude those of the corresponding components in the incident light. Both these components have now undergone a phase change of 180°, so that the reflected light approaches 100 percent in intensity and the plane of vibration approaches the plane of the incident light.

An equation giving the variation of the plane of vibration of the reflected light with the angle of incidence is readily obtained by dividing the two equations (25a).

$$\frac{R_p}{R_s} = -\frac{E_p \cos(\phi + \phi')}{E_s \cos(\phi - \phi')} \qquad \text{(25h)}$$

This is the tangent of the angle ψ, that is,

$$\frac{R_p}{R_s} = \tan\psi \qquad \text{(25i)}$$

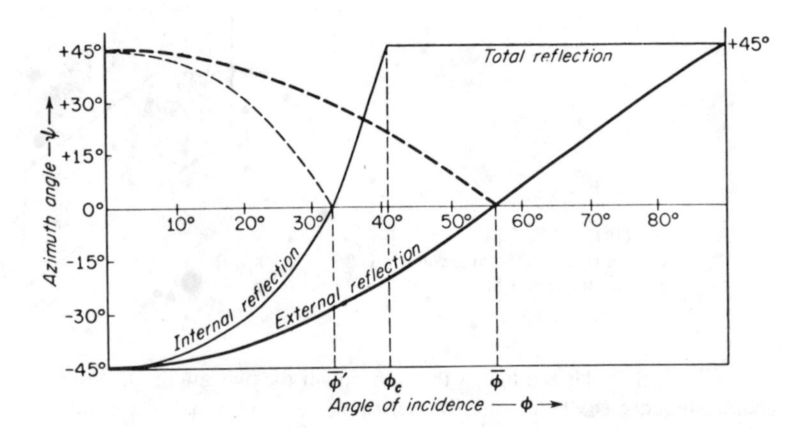

FIGURE 25H
The azimuth angle for plane-polarized light reflected from a dielectric.

since the azimuth ψ is the angle between R and R_s. This angle is plotted in Fig. 25H for the case where the incident light has an azimuth of 45°, making $E_p = E_s$. The heavily drawn curves are for external reflection, while the lighter curves, to be discussed in the following section, are for internal reflection.

25.6 ELLIPTICALLY POLARIZED LIGHT BY INTERNAL REFLECTION

Reference to Fig. 25F(*b*), which gives the phase change for light internally reflected from a glass surface, will show that at an angle of incidence near 50° there is a phase difference of slightly more than 45° between the two components. More exactly, the phase difference for $n = 1.51$ reaches a maximum of 45° 56′ at $\phi = 51°20'$ and is just 45° at the two angles $\phi = 48°37'$ and 54° 37′. This behavior of the phase difference was first tested and confirmed by Fresnel, who constructed a glass rhomb of the form illustrated in Fig. 25I. Plane-polarized light falls normally on the shorter surface of the rhomb with its plane of vibration oriented at 45° to the plane of the paper. It then strikes the first diagonal surface at an (internal) angle of incidence of 54° 37′. Here it is totally reflected with a phase difference of 45° between the two components. Now we have seen in Sec. 12.9 that the result of combining two linear vibrations at right angles is in general an elliptical one, the shape of the ellipse depending on the two amplitudes and on their phase difference δ. Only when δ is some integral multiple of π is the resultant linear, and the light plane-polarized. This situation exists in all cases of external reflection, and in internal reflection up to the critical angle. But in total reflection one obtains elliptically polarized light as the result of a single internal reflection at $\phi > \phi_c$. The systematic study of elliptical and circular polarization will be taken up in Sec. 27.5.

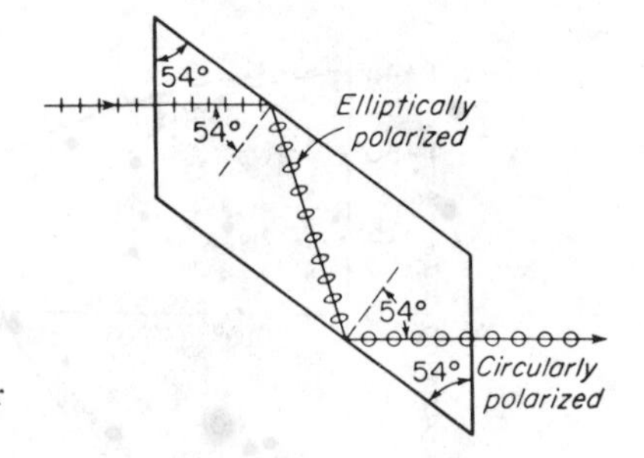

FIGURE 25I
Fresnel rhomb. The angle shown is for glass having $n = 1.51$.

Circularly polarized light will occur only if the two amplitudes are equal and the phase difference is 90°. In the Fresnel rhomb an additional phase difference of 45° is produced by a second internal reflection, so that on emergence the p component is 90° ahead in phase. The device is therefore useful in producing and analyzing circularly polarized light, and, as we shall see later, there are other somewhat more common methods of doing this.

The polarization of the reflected beam when plane-polarized light undergoes a single internal reflection at various angles of incidence is shown in Fig. 25J. The amplitude of the electric vector in the incident and reflected light, and their components, are designated as in Fig. 25G for external reflection. Here, however, they are shown as they would appear to an observer looking against the direction of each beam, with the plane of incidence cutting the plane of the page in a horizontal line. If these diagrams are studied in connection with Figs. 25C, F, and H, their main features should

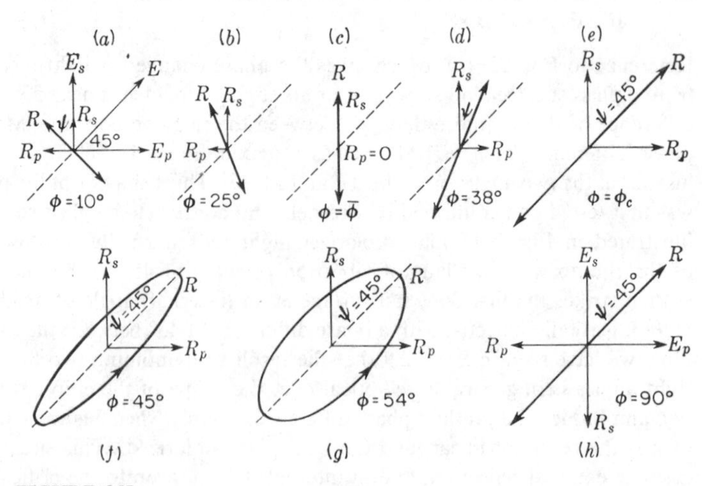

FIGURE 25J
Modes of light vibrations internally reflected in glass at various angles of incidence.

be clear. From $\phi = 0°$ to $\phi = \phi_c$ the reflected light remains plane-polarized but changes its azimuth steadily and increases in intensity. Beyond ϕ_c the vibration opens out into an ellipse having its maximum width at $\phi = 51°$, then narrows again to a linear vibration at 90°.

25.7 PENETRATION INTO THE RARE MEDIUM

One might conclude from the fact that internal reflection beyond the critical angle is total that the amplitude of the light drops discontinuously to zero just at the reflecting surface. According to the boundary conditions of electromagnetic theory this is not possible, however, and furthermore there is experimental evidence that a disturbance capable of producing light exists for a short distance beyond the surface. Suppose that a given surface is totally reflecting an intense beam of light and one brings the edge of a razor blade very close to the surface, or scatters fine particles on it. Observation of the edge of the blade or of the particles with a microscope will show them to be secondary sources of light. In the absence of such foreign matter the electromagnetic theory predicts a disturbance which dies off exponentially beyond the surface* but which involves no net transfer of energy through it. The energy merely oscillates in and out through the surface. The disturbance is periodic in a direction parallel to the surface but not at right angles to it, and hence cannot properly be called a light wave. When the electromagnetic field is distorted by the presence of denser matter sufficiently close to the surface, however, energy is drained off in the form of light.

An instructive experiment to illustrate this penetration was performed by Hall,† who used it for quantitative measurements of the distance of penetration. The apparatus, as shown in Fig. 25K, consists of two total-reflection prisms, one of which has a slightly convex face. If the two prisms are barely in contact at the point C and the angle of incidence exceeds the critical angle, total reflection would send all the light in the direction (b). Actually it is found that in the reflected light there is a dark patch about C and a corresponding bright one in the transmitted light. Photographs of these are shown in the figure. As the angle of incidence is increased further beyond ϕ_c, the size of the patch shrinks, showing that the distance of penetration decreases. At an angle of incidence just below the critical angle (the rays indicated by broken lines), the complete set of Newton's rings in reflection and transmission appears, as illustrated by the ring patterns at the left and right of the figure. Hall used measurements of the diameters of these rings to measure the thickness of the air layers corresponding to different observed diameters of the patch mentioned above. He thus had an accurate measure of the distance of penetration. Both theory and experiment give a decrease of the energy to about $\frac{1}{100}$ in a distance of one wavelength when $\phi = 45°$ and $n = 1.51$. When $\phi = 60°$, it decreases to 1/40,000 in the same distance.

* Quantitative relations are given, for example, in R. W. Ditchburn, "Light," p. 434, Interscience Publishers, Inc., New York, 1953; reprinted (paperback), 1963.

† E. E. Hall, *Phys. Rev.*, **15**:73 (1902). See also K. H. Drexhage, Monomolecular Layers and Light, *Sci. Am.*, **222**:108 (March 1970).

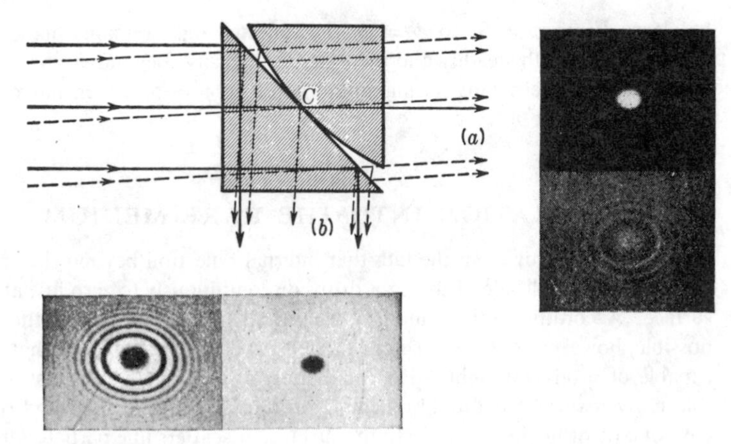

FIGURE 25K
Hall's experiment for measuring the penetration that occurs in total reflection.

25.8 METALLIC REFLECTION

In general, highly polished metallic surfaces have a higher reflectance than dielectrics. At normal incidence, for example, silver and aluminum reflect over 90 percent of all visible light. Experiments show that the reflectance depends not only on the particular metal but on the preparation of the surface and on the wavelength and direction of the incident light. If plane-polarized light is reflected from a metal at other than normal incidence (Fig. 25L), the p and s components of the incident electric vector are reflected with a phase difference and this gives rise to elliptical polarization. It is a general observation for all metals that plane-polarized light is not reflected as plane-polarized light except when it vibrates either in the plane of incidence or perpendicular to it.

In discussing the reflectance of metals it is convenient (just as for dielectrics) to resolve the incident light vector **E** into two components E_p and E_s. Curves for the two reflectances as a function of the angle of incidence are shown in Fig. 25M. These are experimental curves obtained by using white light from a tungsten-filament lamp. A comparison with the corresponding curves for a dielectric [Fig. 25B(*a*)] shows similarities and at the same time striking differences. Metals and dielectrics are similar

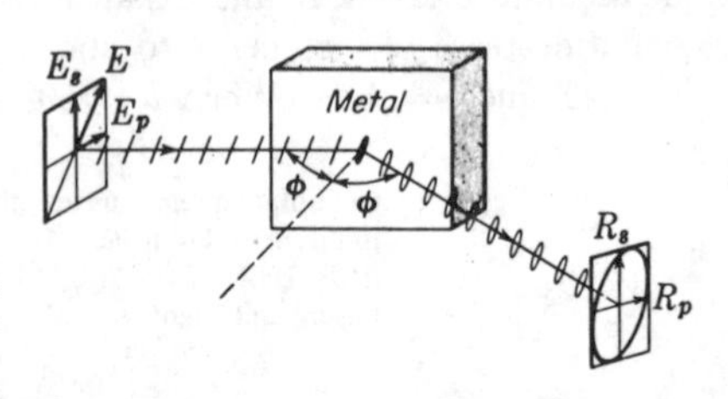

FIGURE 25L
Reflection of plane-polarized light from a metal surface to give elliptically polarized light.

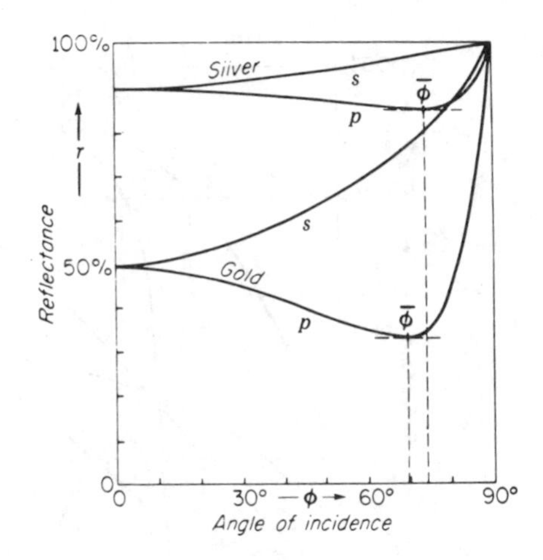

FIGURE 25M
Reflectances for plane-polarized white light from gold and silver mirrors.

in that the values for the p and s components start together at normal incidence, separate, and then come together again at grazing incidence. The essential differences are the much higher reflectance of metals at normal incidence and the relatively high minimum at $\bar{\phi}$. This angle for minimum reflection of E_p is called the *principal angle of incidence*.

The reflectance of a metal usually varies considerably with wavelength. In Fig. 25N this variation is shown for a number of typical metals. In spite of irregularities at the shorter wavelengths, all metals reflect very well in the red and infrared. The face plates on the Apollo space suits worn by the astronauts on the moon were coated with a thin layer of gold. The coating reflected at least 70 percent of the light from the sun. Objects seen through the visor appear light-blue or green in color, but the eyes readily adapt to this color, which soon appears to be practically white. Such face plates are designed to decrease the thermal load on the suit's cooling system by strongly reflecting the infrared radiation from the sun while transmitting sufficient visible light. Gold films deposited on the surface of a sheet of plastic to be used as window shades are to be found on the sunny sides of many houses and office buildings for the same reasons.

Silver and aluminum are of particular importance for general use because they maintain their high reflectance throughout the visible spectrum. The development of methods of depositing metal films by evaporation in vacuum has rendered aluminum the most satisfactory substance for mirrors in optical instruments. This is due chiefly to two factors: (1) aluminum retains its high reflectance in the near ultraviolet as well as in the visible, and (2) the surface does not easily tarnish even after years of exposure to air. It is now standard practice to coat the mirrors of large reflecting telescopes, such as the 200-in. instrument at Mt. Palomar, with evaporated aluminum. A freshly made silver mirror actually has a slightly greater reflectance in the visible, but it soon tarnishes and becomes poorer than aluminum. For the reflecting surfaces of Fabry-Perot etalons, however, silver is preferred for use in visible and infrared

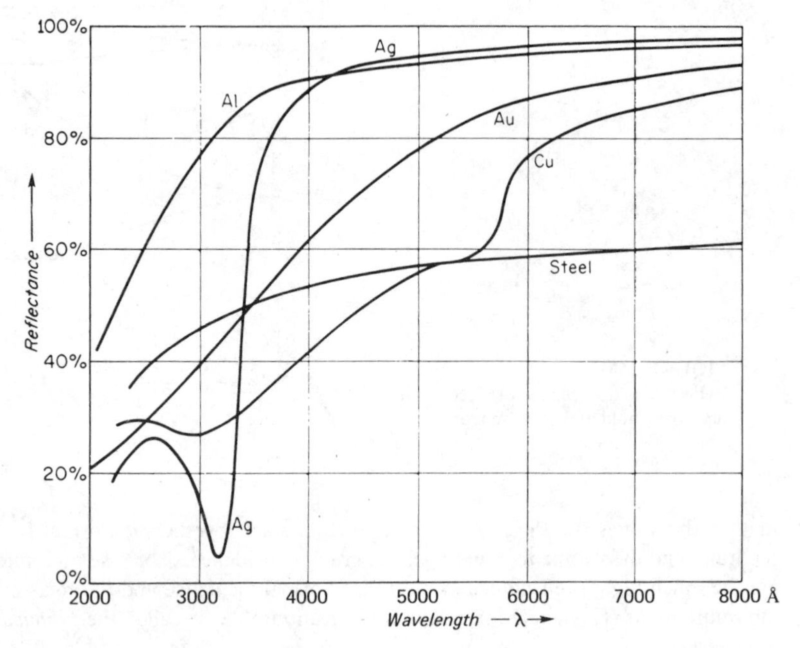

FIGURE 25N
Reflectances at normal incidence of aluminum, silver, gold, copper, and steel.

light. For ultraviolet light aluminum or a mixture of aluminum and magnesium is better.

Silver is exceptional in showing a narrow region of very low reflectance near 3200 Å. The light of this wavelength that is not reflected may be mostly transmitted if the silver film is sufficiently thin. Such a transmission band is also possessed by the alkali metals at still shorter wavelengths.* A sodium film, for example, can be used as an ultraviolet filter, opaque to all wavelengths except those near 1950 Å.

25.9 OPTICAL CONSTANTS OF METALS

The optical properties of a dielectric at a particular wavelength are completely described by one constant, the refractive index n at the wavelength. For a metal, however, another constant must be specified which measures the strength of absorption of light as it enters the metal. Because they contain free electrons, metals have very high absorption, the intensity falling to practically zero in a small fraction of a wavelength.

* For details, see R. W. Wood, "Physical Optics," 3d ed., pp. 558–566, The Macmillan Company, New York, 1934; reprinted (paperback) Dover Publications, Inc., New York, 1968.

An important quantity used in dealing with the optics of metals is the absorption index κ, which is defined in terms of the absorption coefficients κ_0 and α (Sec. 23.6) as

$$\kappa = \frac{\kappa_0}{n} = \frac{\alpha\lambda}{4\pi n} \tag{25j}$$

The determination of n for a dielectric material is usually accomplished by refraction measurements, but it can also be done using the reflected light by finding the polarizing angle and applying Brewster's law. With metals the absorption is so strong that measurements are difficult to do with transmitted light. It has been possible, working with very thin samples, to measure rough values of n and κ, but besides their inaccuracy the results may not be strictly applicable to the metal in bulk. Hence accurate values of the optical constants of metals are determined by investigating the reflected light.

Since there are the two constants to be found, n and κ, two measurements are required. In analogy to the measurement of Brewster's angle for dielectrics, one of these may be the *principal angle of incidence* $\bar{\phi}$, defined above. The other is then the corresponding azimuth, called the *principal azimuth* $\bar{\psi}$. In view of the fact that the light reflected from metals is elliptically polarized, it is not immediately evident what is meant by its azimuth. The definition is made by disregarding the phase difference between the p and s components, which is actually 90° if the light is incident at $\bar{\phi}$, and defining the azimuth in the same way as for dielectrics, namely by the equation

$$\tan \psi = \frac{R_p}{R_s} \tag{25k}$$

The theory shows that to a fair approximation* the two constants can be found from the relations

$$n\sqrt{1 + \kappa^2} = \sin \bar{\phi} \tan \bar{\phi} \qquad \kappa = \tan 2\bar{\psi} \tag{25l}$$

The method of measurement of $\bar{\phi}$ and $\bar{\psi}$ will be briefly described below, after we have considered the variation in the character of the reflected light with the angle of incidence.

The values of the optical constants found in the literature show considerable variations because of different preparation of the surfaces, purity of the samples, and accuracy of the equations used. We quote in Table 25A, however, some typical values, including in the last column the reflectances at normal incidence. It will be seen that there are large variations of n among the metals, those for the better conductors running considerably below unity. These refractive indices cannot be interpreted in the same way as for dielectrics, since here we are dealing with highly damped waves (see Sec. 23.6). The value of κ_0 for copper, for instance, corresponds to the intensity falling to $1/e$ when the light penetrates a depth of only one thirty-third of a vacuum wavelength.

* See H. Geiger and K. Scheel, "Handbuch der Physik," vol. 20, pp. 240–250, Springer-Verlag OHG, Berlin, 1928, which in general follows the work of C. Pfeiffer, "Beiträge zur Kentnisse der Metallreflexion," dissertation, Giessen, 1912.

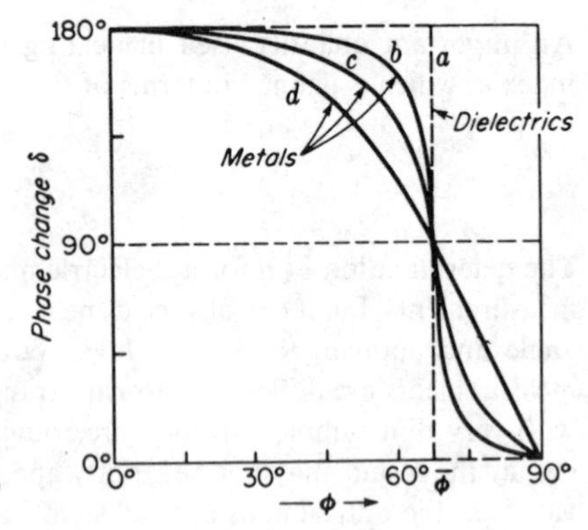

FIGURE 25O
Plots of the phase difference $\delta_p - \delta_s$ for a dielectric, a, and for three metals, b, c, and d, of increasing absorption index κ.

25.10 DESCRIPTION OF THE LIGHT REFLECTED FROM METALS

If plane-polarized light is reflected from a metal, the shape and orientation of the elliptical vibration in the reflected light depend on the orientation of the incident vibration, on the magnitude of the reflected p and s components, and on the phase difference between them. The latter factor has not thus far been discussed, and a quantitative treatment of it would require a more extensive mathematical development than would be profitable to include here. We may, however, consider the main result on the behavior of δ $(= \delta_p - \delta_s)$ as a function of ϕ.

Figure 25O shows plots of the theoretical equations for the phase differences for three different metals b, c, and d, in the order of increasing absorption index κ. It also shows, as the broken line a, the plot for a dielectric, which has $\kappa = 0$. We see that the discontinuous transition of δ from π to zero which occurs at $\bar{\phi}$ for dielectrics becomes for metals a more or less gradual change. We also note that at the principal angle of incidence the value of δ is always exactly 90°.

Knowing the values of R_p/E_p, R_s/E_s, and δ, it is possible to predict the shape of the elliptical vibration reflected at each angle of incidence. Thus suppose, as in Fig.

Table 25A OPTICAL CONSTANTS FOR VARIOUS METALS FOR SODIUM LIGHT, $\lambda = 5893$ Å

Metal	$\bar{\phi}$	$\bar{\psi}$	n	κ	κ_0	r, %
Steel*	77°9′	27°45′	2.485	1.381	3.433	58.4
Cobalt*	78°5′	31°40′	2.120	1.900	4.040	67.5
Copper*	71°34′	39°5′	0.617	4.258	2.630	74.1
Silver*	75°35′	43°47′	0.177	20.554	3.638	95.0
Gold	72°18′	41°39′	0.37	7.62	2.82	85.1
Sodium	71°19′	44°58′	0.005	522.0	2.61	99.7

* Data supplied the authors courtesy of R. S. Minor.

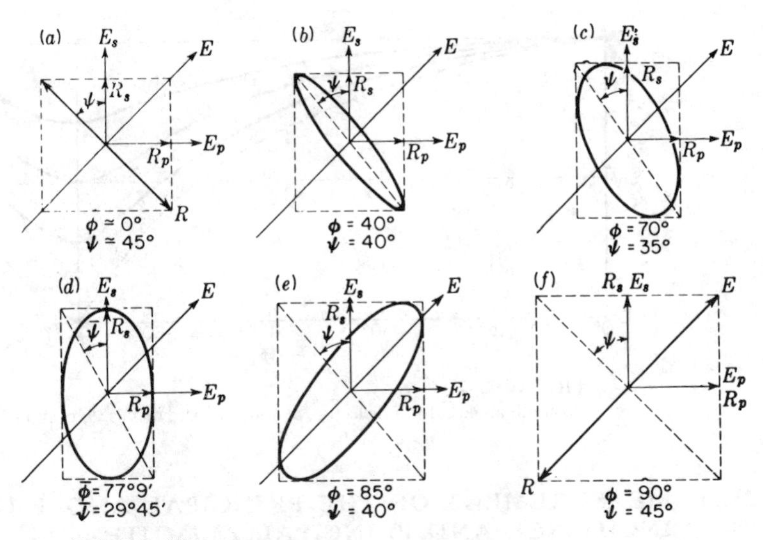

FIGURE 25P
Elliptically polarized light (λ5893) reflected at different angles ϕ from a steel mirror.

25P, that the electric vector in the incident plane-polarized light makes an angle of 45° with the plane of incidence, so that $E_p = E_s$. We have taken the reflecting metal to be steel, which according to Fig. 25N has a reflectance $R^2/E^2 = 0.58$ for sodium light at normal incidence. Hence near normal incidence [case (a) of Fig. 25P] we construct the reflected amplitudes $R_p = R_s = 0.76E_p = 0.76E_s$, since $0.76 = \sqrt{0.58}$. Now because of the phase change of π shown in Fig. 25O, we must shift the p vibration in the reflected light 180° ahead of the s vibration, and the result is a linear vibration of amplitude R in the direction shown. This direction is actually opposite in space to that of E [see Fig. 25G(*a*)]. As the angle of incidence increases from zero, the gradual change in the phase difference causes the vibration to open out into an ellipse which is contained in a rectangle of sides $2R_p$ and $2R_s$. When the angle $\bar{\phi}$ is reached, as in (*d*) of the figure, we obtain an ellipse symmetrical to the axes, and the one that has the least eccentricity. From then on the ellipse becomes slimmer until finally at grazing incidence, as in (*f*), we have a linear vibration of the same amplitude as the incident light, but exactly out of phase with it.

The meaning of the azimuth angle ψ is best seen from Fig. 25P. It is the angle that the diagonal of the rectangle makes with R_s. From the figure we see that ψ first diminishes and then increases again in going from $\phi = 0$ to $\phi = 90°$. The minimum value occurs at $\bar{\phi}$, but it is not zero at this angle as it is for a dielectric. The depth of this minimum becomes less for metals of larger κ. This effect may be seen in Fig. 25Q, where the letters a to d have the same significance as in Fig. 25O. In the figure we have marked the value of the principal azimuth $\bar{\psi}$ for the particular metal c.

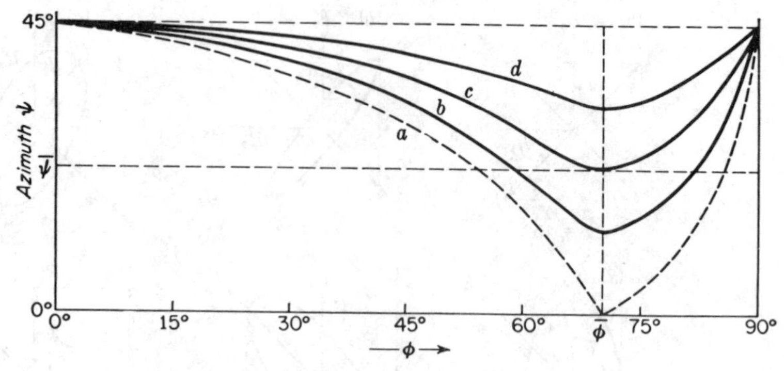

FIGURE 25Q
Azimuth angle $\bar{\psi}$ for a dielectric, a, and for three metals, b, c, and d.

25.11 MEASUREMENT OF THE PRINCIPAL ANGLE OF INCIDENCE AND PRINCIPAL AZIMUTH

The determination of these quantities is a special case of the general problem of the analysis of elliptically polarized light, a problem which will be treated in some detail in Chap. 27. It is not difficult to see, however, with the aid of Fig. 25R in connection with Fig. 25P(d), how measurements of $\bar{\phi}$ and $\bar{\psi}$ might be made. Let the nicol N_1 of Fig. 25R be oriented so that the incident vibrations are at 45° to the plane of incidence. In the reflected beam is placed some type of compensator C, which retards the p vibrations by a quarter of a period, or 90°, with respect to the s vibrations. This could be a Fresnel rhomb (Sec. 25.6) but is more commonly a quarter-wave plate or a Soleil compensator (Secs. 27.2 and 27.4). Now at any angle of incidence other than $\bar{\phi}$ the value of δ is different from 90°, so that the phase difference will not be completely removed by the compensator. The light transmitted by C will still be elliptically polarized and cannot be extinguished by rotation of the analyzer N_2. Various angles of incidence are tried until complete extinction becomes possible, and under this condition the light is incident at $\bar{\phi}$.

The fact that it is possible to obtain complete extinction with a nicol means that the compensator has changed the elliptically polarized reflected light into plane-polarized light. In Fig. 25P(d) the ellipse is converted into a linear motion along the

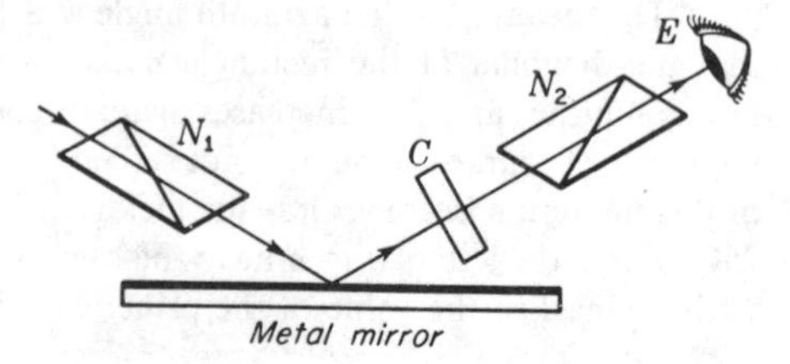

FIGURE 25R
Apparatus for determining the principle angle of incidence and principal azimuth for a metal.

diagonal of the rectangle by removing the phase difference of 90° that exists between the p and s components. It will therefore be seen that when the condition of extinction has been found the plane of transmission of the analyzer makes the angle $\bar{\psi}$ with R_p, namely, with the plane of incidence.

25.12 WIENER'S EXPERIMENTS

In Sec. 12.3 we described a classical experiment in which Wiener demonstrated the formation of standing waves of light by reflection from a silver mirror. The object of this experiment was not only to reveal the standing waves but also to tell whether it is the electric or the magnetic vector that produces the observed effects, and hence is to be identified as the "light vector." According to the electromagnetic theory, the incident and reflected electric vectors are oppositely directed in space for external reflection at normal incidence. With dielectrics the reflected waves have a much smaller amplitude than the incident ones so that the destructive interference is not complete. For metals, however, we should obtain a node of the electric vector at the surface.* As to the magnetic vectors, their relative directions in the incident and reflected light can be found from the fact that **E**, **H**, and the direction of propagation are related according to the right-hand screw rule. The result is shown in Fig. 25S. If the angle of incidence is made to approach zero, we see that the **H**″ and **H** vectors approach the same direction for each polarization. Their superposition should produce an antinode of the standing waves at the surface. Now as was explained before, *Weiner observed a node where the detecting plate touched the surface. This indicated that the electric vector was the important one, at least for photographic action.*

One would expect from theory that the electric vector would be more important than the magnetic one in producing the observed effects of light. Wherever it is a question of the action of light on electrons, the electric fields in the wave exert much greater forces than the magnetic fields. In fact, only 2 years after Wiener's work Drude and Nernst showed that the same result holds when fluorescence instead of photography is used for detection. Later Ives demonstrated it using the photoelectric effect. It is assumed that the electric vector is also responsible for vision.

An even more convincing demonstration, and one which does not depend on the phase changes or on the achievement of perfect contact of the end of the photographic plate with the mirror, was given by Wiener in the following way. Plane-polarized light was reflected at an angle of incidence of exactly 45°. Then the incident and reflected rays are at right angles to each other, and the orientations of the vectors in space are those illustrated in Fig. 25S. We see that for s polarization the electric vectors E_s and R_s vibrate along the same line and can interfere. On the other hand E_p and R_p are perpendicular to each other, and no interference is possible. Exactly the reverse is true for the **H** vectors. The experiment is illustrated schematically

* The values of δ_p and δ_s are not exactly 0 or 180° for metals at normal incidence, although their difference is. The only effect of this, however, is to shift the position of the node so that it does not occur exactly at the surface. With silver, for example, the node is located 0.043λ below the surface.

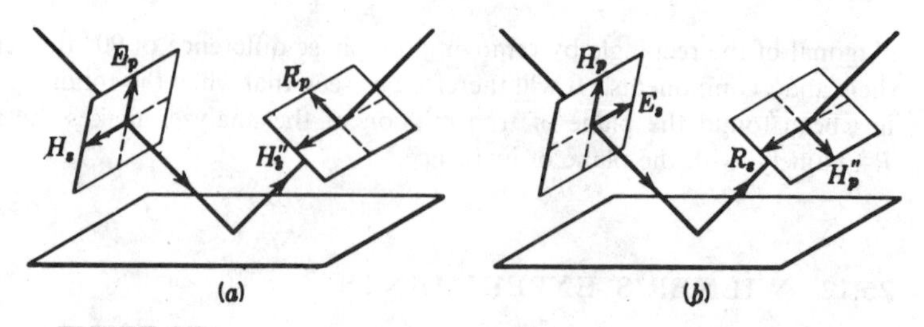

FIGURE 25S
Space relations between the incident and reflected **E** and **H** vectors: (*a*) for *p* polarization and (*b*) for *s* polarization. The angle of incidence is assumed to be less than $\bar{\phi}$.

in Fig. 25T. In part (*a*) the electric vector is perpendicular to the plane of the figure, a condition which could be achieved by a preliminary reflection from a glass plate at Brewster's angle, and interference can occur along the horizontal planes marked by dots. These planes are $1/\sqrt{2}$ times as far apart as for normal incidence. In the figure the phase change of π on reflection is indicated by the change from solid to broken lines, and vice versa. For the corresponding magnetic vectors, as shown in part (*b*), there is no phase change on reflection. On the surface at point *A* the resultant is a linear vibration normal to the surface. Farther up the vibration becomes elliptical, then circular as at *a*, and finally linear again at *B* with horizontal vibrations. The reverse sequence is followed to point *C*, with the points *A*, *B*, and *C* being separated by a distance $\lambda/2$ along the ray. The energy associated with all these vibration forms is the same (Sec. 28.8), and hence, if the magnetic vector were the active one, the test plate should be uniformly blackened. Wiener actually found interference bands in the case illustrated, and uniform blackening when the vibrations in the incident light were turned through 90°.

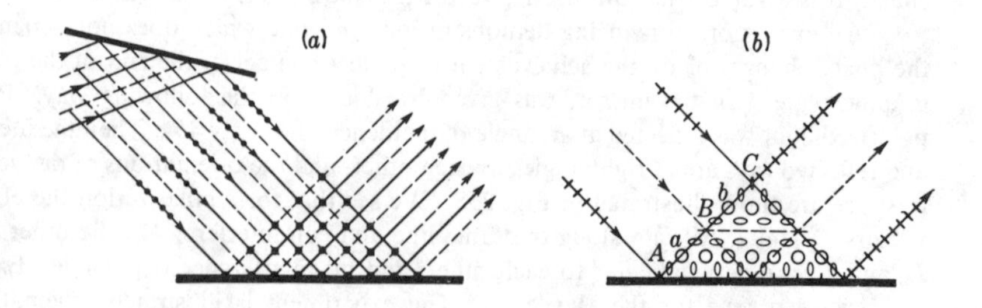

FIGURE 25T
Wiener's experiment at 45° incidence. Interference is observed for the electric vector oriented as in (*a*), while the corresponding magnetic vector (*b*) would show none.

PROBLEMS

25.1 Plot the external reflection intensity curves for red light incident on a transparent crystal of diamond. Use the refractive index given in Table 23A.

25.2 Plot the internal reflectance curves for red light in diamond. Use the refractive index given in Table 23A.

25.3 Compute the reflectance at normal incidence for the following materials: (*a*) diamond, $n = 2.426$; (*b*) quartz, $n = 1.547$; (*c*) rutile, $n = 2.946$; (*d*) crown glass, $n = 1.526$; (*e*) metallic silver, $n = 0.177$, $\kappa_0 = 3.638$; (*f*) steel, $n = 2.485$, $\kappa_0 = 3.433$.
Ans. (*a*) 17.32%, (*b*) 4.61%, (*c*) 24.32%, (*d*) 4.34%, (*e*) 95.16%, (*f*) 58.46%

25.4 (*a*) Derive an equation for the azimuth of the refracted light in a dielectric. Assuming $n = 1.50$, (*b*) make a plot of this angle ψ' against ϕ', similar to that for reflected light in Fig. 25H.

25.5 Plane-polarized light is incident at $\phi = 70°$ on a glass surface, with its electric vector vibrating at 30° with the plane of incidence. Assuming $n = 1.750$, calculate (*a*) the polarizing angle, (*b*) the critical angle, (*c*) the relative magnitudes E_p to E_s, (*d*) the relative magnitudes of R_p and R_s, and (*e*) the azimuth angle ψ.

25.6 Plane-polarized light is internally reflected at $\phi = 45°$ from the hypotenuse of a total-reflection prism made of glass of index 1.650. If the azimuth of the incident light is 45°, calculate (*a*) the phase change of the p and s components and (*b*) the phase difference between the p and s components. (*c*) Plot to scale the elliptical vibrations form as was done in Fig. 25J.

25.7 Unpolarized light strikes a smooth glass surface at an angle of 35°. Assume the glass index to be 1.750. Calculate (*a*) the amplitudes and (*b*) the intensities of the reflected p and s components. (*c*) Find the degree of polarization of the refracted light (see Sec. 24.4).
Ans. (*a*) amplitudes, 0.2055 and 0.3374, (*b*) reflectances, 0.04223 and 0.11384, (*c*) polarization 3.884%

25.8 (*a*) Make a plot of the phase changes on internal reflection in glass having an index of 1.825. Limit the plot to angles between the critical angle and grazing incidence. (*b*) Take the difference $\delta = \delta_p - \delta_s$, and find what two angles could be used in designing a Fresnel rhomb of this glass.

25.9 Explain why in making a Fresnel rhomb described in Sec. 25.6, it is more desirable to choose the angle 54°37′ than the other angle (48°37′), which also gives a phase difference of $\delta = 45°$.

25.10 The optical constants for a given polished metal surface are $n = 2.340$ and $\kappa = 1.176$, for green light. Calculate (*a*) its reflectance at normal incidence, (*b*) its principal angle of incidence, and (*c*) its principal azimuth. *Ans.* (*a*) 50.0%, (*b*) 75.0°, (*c*) 24.81°

25.11 For a certain polished metal surface the principal angle of incidence is measured to be 65.5° and the principal azimuth 38.4°. Determine (*a*) the optical constants of this metal and (*b*) its reflectance at normal incidence.

26
DOUBLE REFRACTION

From the standpoint of physical optics, doubly refracting crystals are classified as either *uniaxial* or *biaxial*. We have seen that in uniaxial crystals the refractive indices, and hence the velocities, of the O and E waves become equal along a unique direction called the optic axis. In biaxial crystals, on the other hand, there are two directions in which the velocity of plane waves is independent of the orientation of the incident vibrations. These two optic axes make a certain angle with each other which is characteristic of the crystal and depends to some extent on the wavelength. Uniaxial crystals may be thought of as a special case of biaxial crystals where the angle between the axes is zero.

26.1 WAVE SURFACES FOR UNIAXIAL CRYSTALS

Uniaxial crystals may be divided into two classes, *negative* and *positive*. In a negative crystal like calcite, the extraordinary index of refraction is less than the ordinary index. In quartz, a positive crystal, the index of the extraordinary ray is greater than that of the ordinary ray. A general treatment of the propagation of light in positive

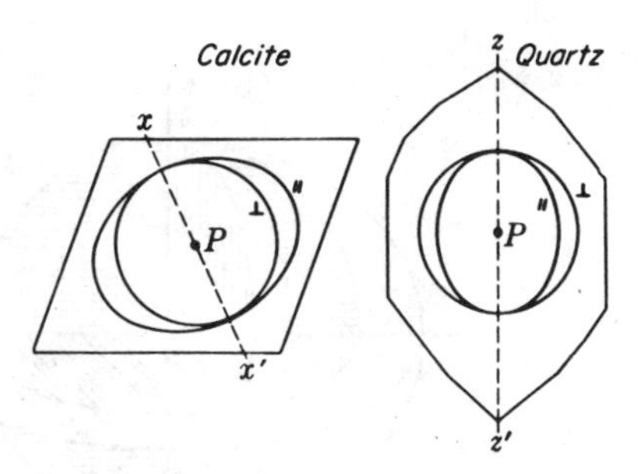

FIGURE 26A
Wave-surface diagrams in calcite and quartz crystals.

and negative crystals is usually given in terms of wave surfaces, which lend themselves so aptly to Huygens' construction.

The wave surface is a wave front (or pair of wave fronts) completely surrounding a point source of monochromatic light. Thus if in one of the crystals of Fig. 26A the source is at P, the circle and ellipse around it represent the traces of the wave fronts, which are the loci of points of *equal phase* in the waves given out by P. If these crystals were isotropic substances such as glass, there would be a single wave surface, which would take the form of a sphere, showing that the velocity of the wave in all directions is the same. In most crystalline substances, however, two wave surfaces are formed, one called the *ordinary wave surface* and the other the *extraordinary wave surface*. In both calcite and quartz the ordinary wave surface is a sphere and the extraordinary wave surface an ellipsoid of revolution. The actual three-dimensional surfaces are obtained by rotating the cross-sectional figures of Fig. 26A about the optic axes which, for reasons to be explained, are labeled xx' and zz'. The circle generates a sphere, and the ellipse an ellipsoid of revolution. The three cross sections of these surfaces are shown in Fig. 26B. The eccentricity of the elliptical sections in these figures is exaggerated, the major and minor axes actually differing by only 11 percent for calcite and 0.6 percent for quartz.

In calcite the ellipsoid touches the enclosed sphere at the two points where the optic axis through P passes through the surfaces. In quartz the sphere and enclosed ellipsoid do not quite touch at the optic axis through P. The fact that they do not touch gives rise to a whole new phenomenon called *optical activity*, a subject which will be treated in detail in Chap. 28. The approach of the two surfaces along the optic axis is so close, however, that for the present they will be assumed to touch as they in fact do in other positive crystals like titanium dioxide, zinc oxide, ice, etc. It should be pointed out that owing to the dispersion of all media the wave surfaces shown apply to one wavelength only. Correspondingly smaller or larger surfaces would be drawn for other wavelengths. Furthermore, it is important to remember that the radii drawn from P are proportional to phase velocities and hence do not measure the rate of propagation of energy. The group velocities, which in dispersive media are usually less than phase velocities (Sec. 23.7), would be represented by proportionately smaller surfaces. They would be the same as the wave surfaces described here only in the case of ideal monochromatic light.

The directions of vibration in the two wave surfaces are indicated in Fig 26A

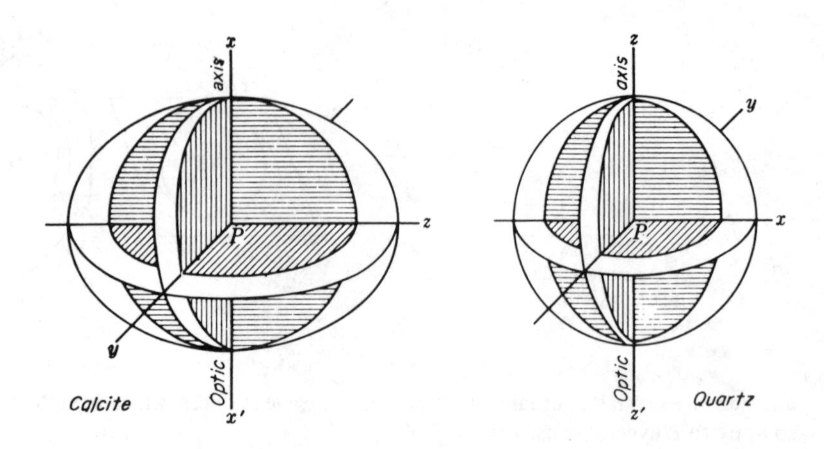

FIGURE 26B
Cross sections of the wave surfaces for calcite and quartz crystals.

by ⊥ for vibrations perpendicular to the page and by ∥ for vibrations in the plane of the page. These will be more exactly specified after we have considered how the wave surfaces may be applied.

26.2 PROPAGATION OF PLANE WAVES IN UNIAXIAL CRYSTALS

The origin of the double refraction of light at a crystal surface is readily explained in terms of the wave surfaces just described. This is accomplished by the use of Huygens' principle of secondary wavelets. Consider, for example, a beam of parallel light incident normally on the surface of a crystal like calcite, whose optic axis makes some arbitrary angle with the crystal surface (see Fig. 26C). The optic axis has the direction shown by the broken lines. According to Huygens' principle, we may now choose points anywhere along the wave front as new point sources of light. Here A, B, and C are chosen just as the wave strikes the crystal boundary. After a short time interval the Huygens wavelets entering the crystal from these points will have the form shown in the figure.

If one now proceeds to find the common tangents to these secondary wavelets, the result is the two plane waves labeled OO' and EE' in the figure. Since the first is tangent to spherical wavelets, it behaves like a wave in an isotropic substance, traveling perpendicular to the surface with a velocity proportional to AA', BB', and CC'. We have seen in the last chapter that the vibrations for this O wave are normal to the principal section. The tangent to the ellipsoidal wavelets represents the wave front for the E vibrations, which take place in the principal section. The E rays, connecting the origins of the wavelets with the points of tangency, diverge from the O rays and are no longer perpendicular to the wave front. They represent the direc-

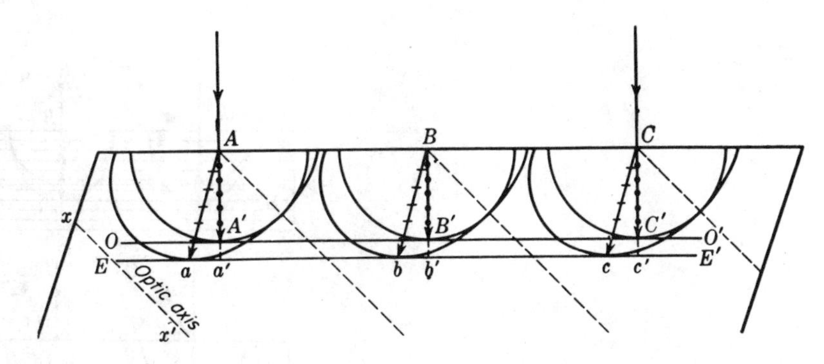

FIGURE 26C
Huygens' construction for a plane wave incident normally on a calcite crystal.

tion in which a narrow beam of light would be refracted, which is the direction in which the energy of the E vibrations is transmitted. Its velocity, proportional to Aa, Bb, or Cc, is called the *ray velocity*. This is greater than the normal velocity, measured by Aa', Bb', or Cc', the velocity with which the wave advances through the crystal in a direction normal to its own plane.

If the normal velocity Aa' is plotted in polar coordinates as a function of the angle between the optic axis and the E-wave normal, we get the dashed ovals of Fig. 26D. These ovals are of course three-dimensional surfaces symmetrical about the optic axis. Now it is seen that the wave surface, i.e., the ellipsoid of revolution, is really a ray-velocity surface. The normal-velocity surface and the ray-velocity surface for the ordinary vibrations are both represented by the same circle or sphere. Hereafter the ellipsoid of revolution will be referred to as the *wave surface* of the E wave and the oval of revolution as the *normal-velocity surface* of the E wave.

In constructing Fig. 26C the optic axis was assumed to be in the plane of the

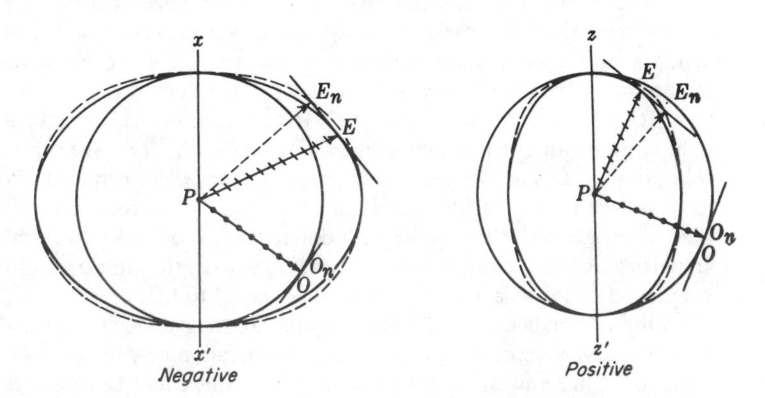

FIGURE 26D
Wave surfaces and normal-velocity surfaces for uniaxial crystals.

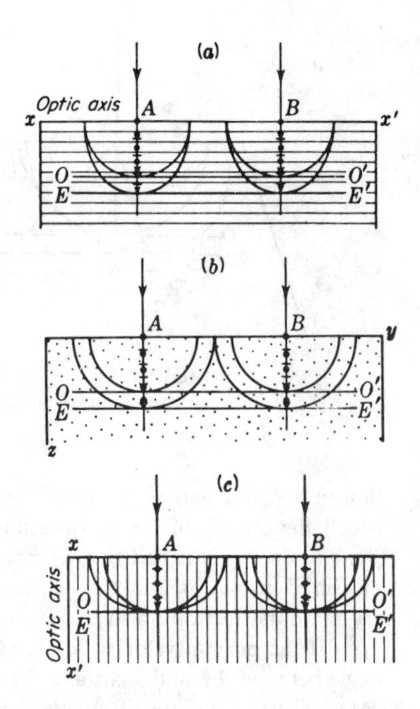

FIGURE 26E
Propagation of normally incident plane waves through calcite crystals cut parallel and perpendicular to the optic axis.

page. In case the optic axis is not in the plane of the page, a plane drawn tangent to the ellipsoidal wavelets will make contact at points in front of or in back of the plane of the page. If the optic axis is either parallel or perpendicular to the surface of the crystal, however, the situation is especially simple. Figure 26E illustrates Huygens' construction in these important cases, where the crystal face is cut (1) parallel to the optic axis as in (*a*) and (*b*), and (2) perpendicular to the optic axis as in (*c*). In both cases the ray velocities are equal to the normal velocities, and there is no double refraction. In case 1, however, the *E* wave travels faster than the *O* wave. When a difference in these velocities exists, we obtain the phenomenon of interference in polarized light, discussed in the following chapter.

It will assist in understanding the rather complex behavior of the velocity of light vibrating in different directions, and described by the wave surface, to note the following facts. The *O* wave, which vibrates everywhere perpendicular to the optic axis, has the same velocity in every direction. The vibrations of the *E* wave make a different angle with the axis for each different ray that is drawn from *P* (Fig. 26D). In particular, for the ray drawn along the optic axis, the vibrations of which are perpendicular to the axis, the velocity becomes equal to that of the *O* ray, which is also vibrating perpendicular to the axis. These facts suggest that the velocity of light is for some reason dependent on the angle of inclination between the vibrations and the optic axis. In terms of the elastic-solid theory, this could be explained by assuming two different coefficients of elasticity for vibrations parallel and perpendicular to the optic axis. In calcite, for example, the restoring force is taken to be greater for the *E* ray

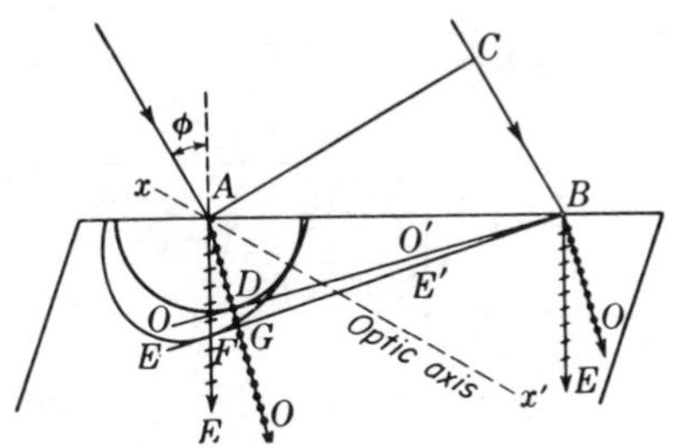

FIGURE 26F
Huygens' construction when the optic axis of a calcite crystal lies in the plane of incidence.

traveling perpendicular to the optic axis (vibrations parallel to the axis) than for the O ray in the same direction (vibrations perpendicular to the axis). Hence the E wave travels faster in this direction.

26.3 PLANE WAVES AT OBLIQUE INCIDENCE

Continuing the study of the double refraction of light in uniaxial crystals, consider the case of a beam of parallel light incident at an angle on the surface of a crystal whose optic axis lies in the plane of incidence and at the same time makes some arbitrary angle with the crystal surface (see Fig. 26F). At the point A where the light first strikes the boundary, the O-wave surface is drawn with such a radius that the ratio CB/AD is equal to the refractive index of the O ray. The ellipsoidal wave surface is then drawn tangent to the circle at the intersection with the optic axis xx'. The points D and F and the new wave fronts DB and FB are located by drawing tangents from the common point B to the circle and ellipse. While the light is traveling from C to B in air, the O vibrations travel from A to D in the crystal and the E vibrations travel from A to F. In the more general case where the optic axis is not in the plane of incidence, the refracted ray will not lie in the same plane. Such cases require three-dimensional figures and cannot be easily shown.

The principles of Huygens' construction are applied to three special cases in Fig. 26G. In (*a*) and (*c*), the optic axis, the plane of incidence, and the E and O principal planes coincide with the plane of the page. In (*b*) the axis is perpendicular to the plane of incidence, and the cross sections of the wave surfaces from A yield two circles. This is a case where the two principal planes defining the directions of vibrations of the O and E rays (Sec. 24.9) are separate from each other and from the principal section.

From geometry it may be shown for the special case of Fig. 26G(*a*), where the optic axis is in the surface as well as in the plane of incidence, that the directions of the refracted rays are given by

$$\frac{n_E}{n_O} = \frac{\tan \phi_E'}{\tan \phi_O'}$$

Here ϕ_E' and ϕ_O' are the angles of refraction, and n_E and n_O are the principal refractive indices.

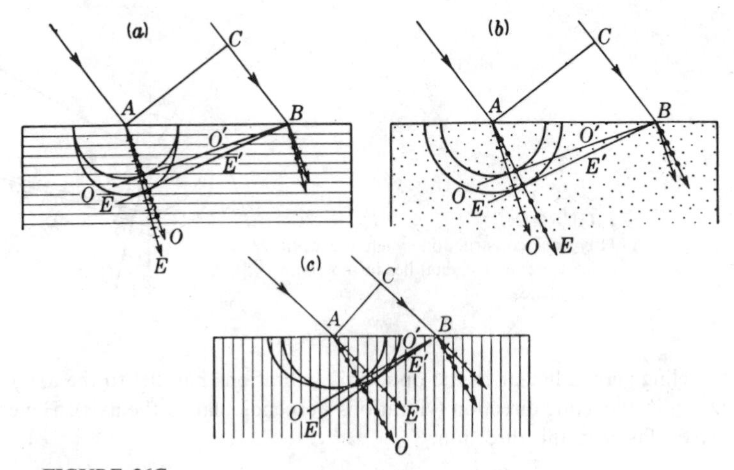

FIGURE 26G
Double refraction for crystals cut with the optic axis parallel and perpendicular to the surface.

26.4 DIRECTION OF THE VIBRATIONS

In crystals the physical nature of the "vibrations" must be specified more closely than merely as the oscillations of the electric (or magnetic) vector used thus far. For reasons to be discussed below, the direction of the electric displacement **D** (Sec. 23.9) is not in general the same as that of the electric field **E**. The application of Maxwell's equations to anisotropic substances along the lines to be outlined in Sec. 26.9 shows that the vibrations lying in the wave front are those of **D**. The vibrations of **E**, however (i.e., of the electric vector—not to be confused with the symbol E for the extraordinary wave) are perpendicular to the ray, and hence inclined to the wave front. Hence the extraordinary wave is a transverse wave for **D** but not for **E**. In our Figs. 26C and 26D, and in what follows, we indicate as the direction of the vibrations that of the electric displacement **D**.

For uniaxial crystals, the directions of vibration of the O and E rays may be specified in terms of the principal planes for these rays defined in Sec. 24.9. The O vibrations are perpendicular to the principal plane of the O ray, which contains this ray and the optic axis. They are also tangent to the O-wave surface. The E vibrations lie in the principal plane of the E ray and are tangent to the E-wave surface. These definitions may seem unnecessarily complicated in cases like Fig. 26C, where the principal section and the two principal planes all coincide with the plane of the figure, but they are essential in the more general case, where all three of these planes are different. Another way of determining the directions of the vibrations, which holds quite generally for all cases including biaxial crystals, is as follows. The electric displacements associated with one ray (the E ray in uniaxial crystals) are in the direction of the projection of the ray on its wave front. Those associated with the other ray may then be found from the fact that for a given direction of the wave normal the two possible directions of **D** are mutually perpendicular. Inspection of our figures will show agreement with these rules in the simple cases we have considered.

26.5 INDICES OF REFRACTION FOR UNIAXIAL CRYSTALS

The index of refraction is usually defined as the ratio of the velocity of light in vacuo to the velocity in the medium in question. In uniaxial crystals there are two principal indices of refraction, one expressing the velocity of the E wave when traveling normal to the optic axis, and the other the velocity of the O wave. These are related to the two elastic coefficients mentioned in Sec. 26.2. In negative crystals, such as calcite, *the principal index for the extraordinary wave is defined as the velocity of light in vacuo divided by the maximum velocity in the crystal.*

$$n_E = \frac{\text{velocity in vacuo}}{\text{maximum velocity of } E \text{ wave}} \tag{26a}$$

The maximum normal velocity, it should be noted, is equal to the maximum ray velocity. The ordinary index is defined as

$$n_O = \frac{\text{velocity in vacuo}}{\text{velocity of } O \text{ wave}} \tag{26b}$$

In positive uniaxial crystals the principal index for the extraordinary wave is defined as

$$n_E = \frac{\text{velocity in vacuo}}{\text{minimum velocity of } E \text{ wave}} \tag{26c}$$

The principal indices for calcite and quartz are given in Table 26A for several wavelengths throughout the visible, ultraviolet, and near-infrared spectrum.

Since the E-wave surface touches the O-wave surface at the optic axis, the ordinary index n_O also gives the velocity of the E wave along the axis. Each pair of values of n_O and n_E for a given wavelength therefore determines the ratio between the major and minor axes of the extraordinary wave surfaces for that wavelength of light.

The principal indices for uniaxial crystals are readily determined experimentally by refracting light through a prism of known angle. If either of the prisms in Fig. 26H is placed on a spectrometer table, two spectra will be observed. For any given wavelength there will be two spectrum lines and hence two angles of minimum deviation. The O and E indices are then calculated in the usual way (Sec. 2.5) by the formula

$$n = \frac{\sin \frac{1}{2}(\alpha + \delta_m)}{\sin \frac{1}{2}\alpha} \tag{26d}$$

where δ_m is the angle of minimum deviation and α is the angle of the prism.

At minimum deviation in prism (*a*) the E ray is traveling essentially perpendicular to the optic axis, the necessary conditions for measuring the principal index n_E. In prism (*b*) it should be noted that the cross section of the wave surface yields two

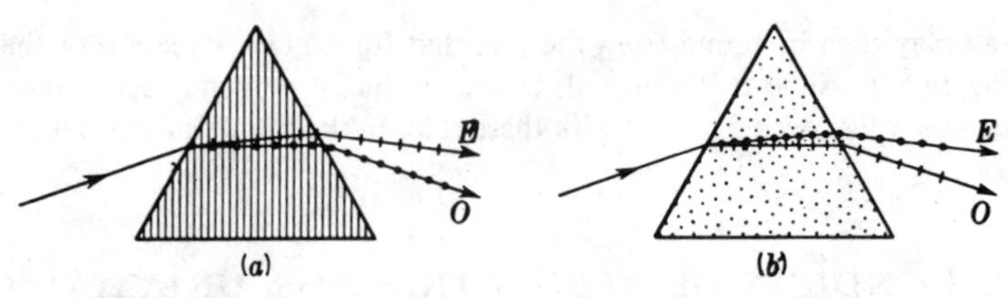

FIGURE 26H
Double refraction in prisms cut from a negative uniaxial crystal.

circles. This means that the velocity of the E ray, as well as that of the O ray, is independent of direction in the plane of the figure and Snell's law of refraction holds for it also.

Two useful relationships for calculating points on an ellipse, plotted in rectangular coordinates, are

$$\frac{x^2}{a^2} + \frac{y^2}{b^2} = 1 \qquad \begin{aligned} x &= a \cos \phi \\ y &= b \sin \phi \end{aligned} \tag{26e}$$

One most interesting uniaxial crystal, called *rutile*, is synthesized TiO_2 (titanium oxide), a clear water-white crystal used in making gemstones that show approximately

Table 26A PRINCIPAL INDICES OF CALCITE AND QUARTZ AT 18°C

Source element	Wavelength, Å	Calcite		Quartz		Fused quartz
		n_O	n_E	n_O	n_E	
Au	2000.60	1.90302	1.57663	1.64927	1.66227	
Cd	2265.03	1.81300	1.54914	1.61818	1.62992	1.52308
Cd	2573.04	1.76048	1.53013	1.59622	1.60714	1.50379
Cd	2748.67	1.74147	1.52267	1.58752	1.59813	1.49617
Sn	3034.12	1.71956	1.51366	1.57695	1.58720	1.48594
Cd	3403.65	1.70080	1.50561	1.56747	1.57738	1.47867
Hg	4046.56	1.68134	1.49694	1.55716	1.56671	1.46968
H_γ	4340.47	1.67552	1.49552	1.55396	1.56340	1.46690
H_β	4861.33	1.66785	1.49076	1.54968	1.55898	1.46318
Hg	5460.72	1.66168	1.48792	1.54617	1.55535	1.46013
Hg	5790.66	1.65906	1.48674	1.54467	1.55379	
Na	5892.90	1.65836	1.48641	1.54425	1.55336	1.45845
H_α	6562.78	1.65438	1.48461	1.54190	1.55093	1.45640
He	7065.20	1.65207	1.48359	1.54049	1.54947	1.45517
K	7664.94	....	....	1.53907	1.54800	
Rb	7947.63	....	....	1.53848	1.54739	1.45340
	8007.00	1.64867	1.48212			
O	8446.70	....	....	1.53752	1.54640	
	9047.0	1.64579	1.48095			
Hg	10140.6	....	....	1.53483	1.54360	
	10417.0	1.64276	1.47982			

6 times the fire of a diamond. The refractive indices given in Table 26B, are calculated from the modified two-term Cauchy equations;

$$\begin{aligned} O \text{ ray:} \quad & n_O^2 = 5.913 + \frac{2.441 \times 10^7}{\lambda^2 - 0.803 \times 10^7} \\ E \text{ ray:} \quad & n_E^2 = 7.197 + \frac{3.322 \times 10^7}{\lambda^2 - 0.843 \times 10^7} \end{aligned} \tag{26f}$$

26.6 WAVE SURFACES IN BIAXIAL CRYSTALS

The majority of crystals occurring in nature are biaxial, possessing two optic axes, or directions of single normal velocity. Double refraction in such crystals, just as in calcite and quartz, is most easily described in terms of wave-surface diagrams and Huygens' principle. Three cross-sectional views of the wave surfaces for a biaxial crystal are given in Fig. 26I. As before, the directions of vibration are shown by dots and lines. Each section cuts the two surfaces in one circle and one ellipse, and these are different in the three sections. The figures are drawn for the case where the semi-axes of the intersections of the wave surface with the coordinate planes are, as shown in the figure, $a = 3$, $b = 2$, and $c = 1$. (Such large differences in a, b, and c are never found in nature.)

Of the three cross sections, the center one (that in the xz plane) is the most interesting, for it contains the four singular points where the outer wave surface (light line) touches the inner surface (heavy line). As shown again in Fig. 26J(a), the rays OR_1 and OR_2 represent directions in which there is but one ray velocity. These are not the optic axes. The optic axes are located by drawing the tangent planes, A_1M_1 and A_2M_2. It is difficult to show in two dimensions that these tangent planes touch the three-dimensional outer surface in circles whose diameters are A_1M_1 and A_2M_2, but such is the case. Since the cross section of one surface is a circle, the lines OA_1 and OA_2 are perpendicular to the tangent planes. They therefore give the same normal velocity for both the ellipse and the circle, so that OA_1 and OA_2 are the optic axes for the point O.

From Fig. 26I it is seen that one can determine the shape of the wave surfaces

Table 26B REFRACTIVE INDICES FOR TiO_2 (RUTILE), FOR SEVERAL OF THE PRINCIPAL FRAUNHOFER LINES

Designation	λ, Å	n_O	n_E
C (H_α)	6561	2.5710	2.8560
D (Na)	5890	2.6131	2.9089
E (Fe)	5270	2.6738	2.9857
F (H_β)	4861	2.7346	3.0631
G′(H_γ)	4340	2.8587	3.2232
H (Ca^+)	3968	3.0128	3.4261

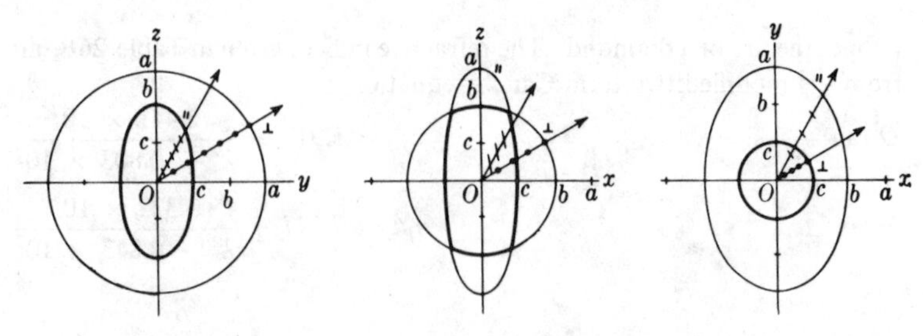

FIGURE 26I
Cross sections of wave surfaces for a biaxial crystal.

by specifying three principal indices of refraction. These are determined by the fact that there are three particular velocities, corresponding to vibrations parallel to x, y, and z, respectively. The elastic-solid theory specified three different coefficients of elasticity for these three types of vibration, which gave rise to these three velocities. If the wave surfaces represent the wave fronts after they have traveled a time of 1 s from the point O, the indices are given by

$$n_a = \frac{V}{a} \qquad n_b = \frac{V}{b} \qquad n_c = \frac{V}{c} \tag{26g}$$

where V is the distance light travels in 1 s in vacuum and a, b, and c are the semiaxes of the elliptic sections of the wave front. Values of n_a, n_b, and n_c are given for several crystals in Table 26C.

The distinction between positive and negative crystals is made according to to whether the angle α of Fig. 26J(*a*) is less or greater than 45°.

The angle α in Fig. 26J(*a*) can be calculated from the geometry of a circle and an ellipse, and is given by the relation,

$$\cos \alpha = \sqrt{\frac{b^2 - c^2}{a^2 - c^2}} \tag{26h}$$

Table 26C PRINCIPAL INDICES OF REFRACTION FOR BIAXIAL CRYSTALS (FOR SODIUM LIGHT)

Crystal and formula	n_a	n_b	n_c	Angle α, deg
Negative crystals:				
Mica $[KH_2Al_3(SO_4)_3]$	1.5601	1.5936	1.5977	71.0
Aragonite $[CaO(CO)_2]$	1.5310	1.6820	1.6860	81.4
Lithargite (PbO)	2.5120	2.6100	2.7100	46.3
Stibnite$(Sb_2S_3)(\lambda 7620)$	3.1940	4.0460	4.3030	80.7
Positive crystals:				
Anhydrite $(CaSO_4)$	1.5690	1.5750	1.6130	22.1
Sulfur (S)	1.9500	2.0430	2.2400	37.3
Topaz $[(2AlO)FSiO_2]$	1.6190	1.6200	1.6270	20.8
Turquoise $(CuO_3 \cdot Al_2O_3 \cdot 2P_2O_5 \cdot 9H_2O)$	1.5200	1.5230	1.5300	33.3

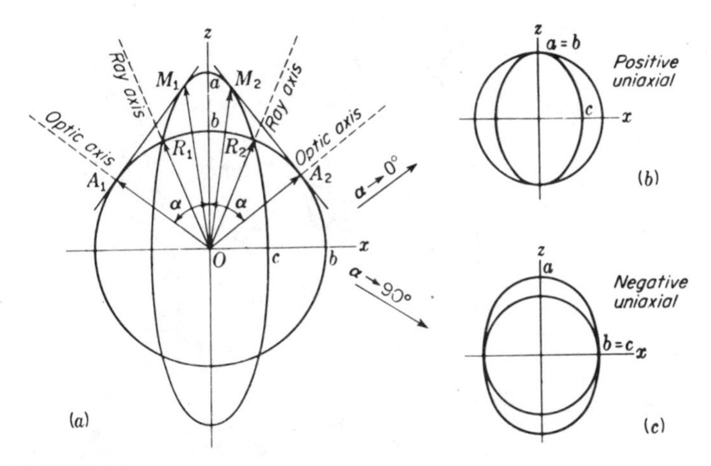

FIGURE 26J
Wave-surface diagram for (*a*) a biaxial crystal and (*b*) and (*c*) the limiting cases of uniaxial crystals.

It can be seen from the diagram that, as a approaches b, α approaches zero and the surface takes the form of a positive uniaxial crystal [Fig. 26J(*b*)]. When, on the other hand, $\alpha = 90°$, we have $b = c$ and the surface is that of a negative uniaxial crystal, as in (*c*) of the figure. In terms of the refractive indices these limiting cases are

$n_a = n_b < n_c$ *Positive uniaxial crystal having* $n_O = n_a$ *or* n_b, $n_E = n_c$

$n_a < n_b = n_c$ *Negative uniaxial crystal having* $n_O = n_b$ *or* n_c, $n_E = n_a$

It is to be noted in Fig. 26I that each coordinate plane contains one circular cross section of the wave surface. This means that one of the two rays refracted into a crystal along any of these planes will obey Snell's law. Prisms can therefore be cut from crystals in such a way as to make use of this fact in determining the principal indices of refraction.

One quadrant of the wave surface for a biaxial crystal is shown in Fig. 26K to illustrate the directions of the electric displacements **D**, that is, the vibrations in the wave fronts, and also to show the normal velocity surface (dashed lines). The outer sheet touches the inner one at only four points, where it forms "dimples." These are located at points such as R_2, where the surface is intersected by the ray axes. Along the x, y, and z axes the ray velocity is equal to the normal velocity. It will be seen that the vibrations in the wave surface wherever it has a circular section are perpendicular to the coordinate plane, for only under these conditions do they maintain a constant angle with the optic axes.

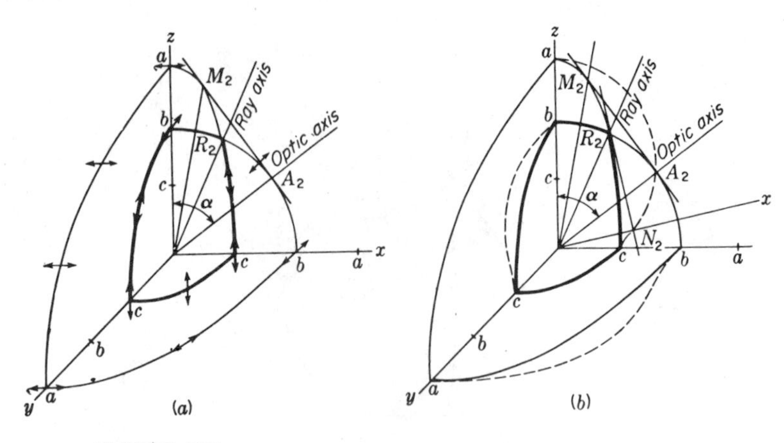

FIGURE 26K
Quadrant cross sections of wave surfaces for a biaxial crystal. Broken lines are normal-velocity surfaces. Arrows show the direction of the electric displacement.

26.7 INTERNAL CONICAL REFRACTION

The investigation of refraction in biaxial crystals follows the same lines as that given in the preceding sections for uniaxial crystals. To treat refraction in the *xz* plane, for example, we may apply Huygens' construction using secondary wavelets of the form shown in Fig. 26J. One finds in general two plane-polarized refracted rays, so that we have double refraction here also. Two special cases arise, however, in which the behavior of a biaxial crystal is different from the simpler uniaxial type. They correspond to the singular case where light is sent along the optic axis of a uniaxial crystal. One of these, *internal conical refraction*, is observed when a beam is directed along one of the *optic* axes inside the crystal. In the other, *external conical refraction*, it is sent along one of the *ray* axes.

Internal conical refraction comes about as follows: It has already been mentioned that the tangent plane A_2M_2 [Figs. 26J(*a*) and 26L(*a*)] makes contact with the three-dimensional wave surface in a circle of diameter A_2M_2. Suppose now that a plane-parallel plate is cut from a crystal so that its surfaces are perpendicular to an optic axis and that the crystal has the thickness OA_2 of Fig. 26L(*a*). Let a ray of unpolarized light be incident normally on the first surface at O. The perpendicular vibrations will travel along the optic axis OA_2 and pass through undeviated. The parallel vibrations will be propagated along OM_2, and will emerge after a second refraction traveling in the same direction as OA_2. Now the incident unpolarized ray contains vibrations in all planes through the ray (Sec. 24.2) and for each particular plane of vibration there is a different direction along which the wave will be propagated with the same normal velocity as along any other ray. In three dimensions these rays will form a cone of light in the crystal spreading out from O. Arriving simultaneously at the second surface A_2M_2, all these waves are refracted parallel to each other to form a circular cylinder. When this hollow beam of light is looked at end-on, the planes of vibration will be as shown in Fig. 26L(*b*).

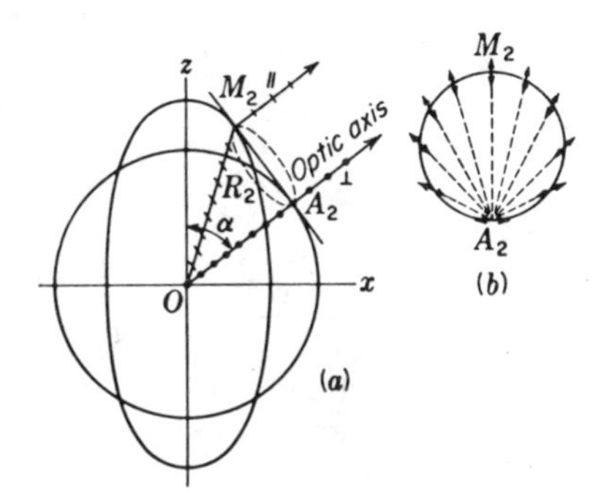

FIGURE 26L
(*a*) Geometry of internal conical refraction. (*b*) End view of internal conically refracted light, showing the directions of vibration.

Internal conical refraction was predicted by Sir William Hamilton and at his suggestion is reputed to have first been verified experimentally by Lloyd in 1833. The observations are usually made now with a parallel crystal plate as shown in Fig. 26M. A beam of light, confined to a narrow pencil by two movable pinholes S_1 and S_2, is incident at just such an angle that the light vibrating perpendicular to the plane of incidence is refracted along the optic axis. When the pinhole S_2 is moved around to vary the angle of incidence, there will be only two refracted rays until the correct direction for internal conical refraction is reached. When this happens, the light spreads out from the two spots near A_2 and M_2 into a ring.*

26.8 EXTERNAL CONICAL REFRACTION

External conical refraction in biaxial crystals deals with the refraction of an external hollow cone of light into a narrow pencil or ray of light inside the crystal (Figs. 26N and 26O). Suppose that a beam of monochromatic light is moving inside a crystal

* A photograph of internal conically refracted light is given in Max Born, "Optik," p. 240, J. Springer, Berlin, 1933.

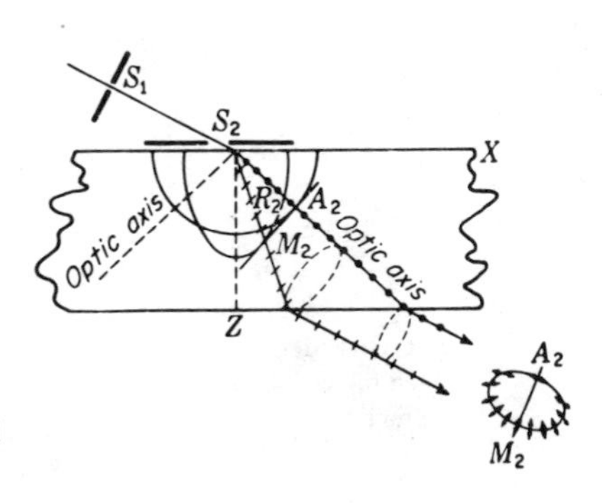

FIGURE 26M
Internal conical refraction in a biaxial crystal plate.

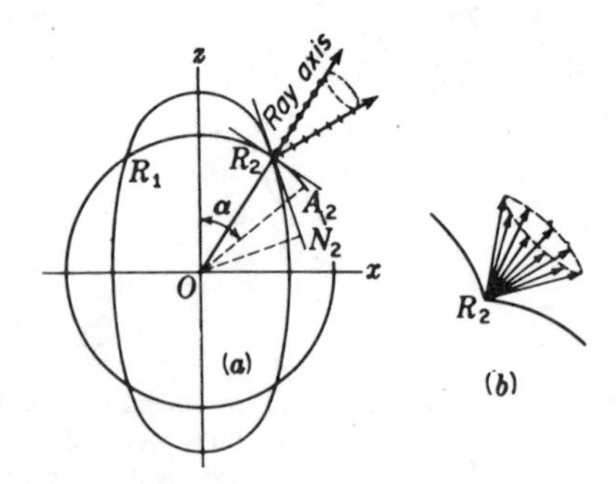

FIGURE 26N
Geometry of external conical refraction.

along the ray axis OR_2. From the diagram in Fig. 26N two tangents can be drawn at the intersection R_2, one to the ellipse and one to the circle.

In the three-dimensional wave surfaces the point R_2 is like a dimple, and there is an infinite number of wave fronts enveloping the obtuse cone. Corresponding to these wave fronts there will be an infinite number of wave normals, each with its own particular direction of vibration [Fig. 26N(*b*)], and these will form an acute-angled cone. When these wave fronts, the energy of each one of which travels along the ray axis, arrive at the crystal surface, they will emerge as a cone of rays, since each wave normal inside corresponds to a refracted ray outside. There is thus a cone of wave normals outside as well as inside. By the principle of the reversibility of light rays a hollow cone of polarized light rays from outside a crystal should unite to form one ray inside traveling along the single-ray axis.

Experimentally a solid cone of converging unpolarized light, somewhat larger than necessary, is made to fall on a crystal plate cut as shown in Fig. 26O. The ray axis is located by moving one of the pinhole apertures S_1 and S_2. From the incident light the crystal picks the hollow cone of rays vibrating in the proper planes that unite to form one ray. The various other rays travel different directions in the crystal and are stopped by the screen S_2. Upon refraction from the second crystal surface a hollow cone of polarized light is observed emerging from S_2. The cone shown in Fig. 26O is not the same as the one shown in Fig. 26N(b), but is the one produced by refraction of the latter.

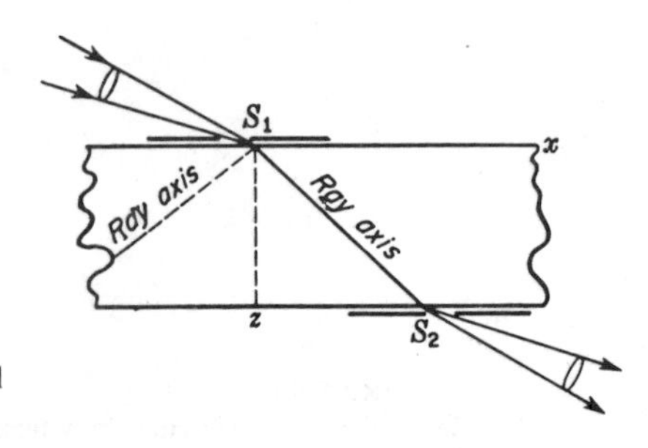

FIGURE 26O
Method of observing external conical refraction.

26.9 THEORY OF DOUBLE REFRACTION

Maxwell's equations for crystalline media have the same form as those given in Sec. 23.9 for transparent media in general, namely,

$$\begin{aligned} \frac{1}{c}\frac{\partial D_x}{\partial t} &= \frac{\partial H_z}{\partial y} - \frac{\partial H_y}{\partial z} & -\frac{1}{c}\frac{\partial H_x}{\partial t} &= \frac{\partial E_z}{\partial y} - \frac{\partial E_y}{\partial z} \\ \frac{\partial D_x}{\partial x} + \frac{\partial D_y}{\partial y} + \frac{\partial D_z}{\partial z} &= 0 & \frac{\partial H_x}{\partial x} + \frac{\partial H_y}{\partial y} + \frac{\partial H_z}{\partial z} &= 0 \end{aligned} \tag{26i}$$

Only in the case of an isotropic substance like glass, however, is it permissible to write for the electric displacement $\mathbf{D} = \varepsilon\mathbf{E}$, as was done in Sec. 23.9. In anisotropic crystals it is found that the measured values of the dielectric constant ε vary with the orientation of the optic axis or axes relative to the electric field $\mathbf{E}$. In the electron theory of dielectric media, the value of the dielectric constant depends on the *polarization* of the atoms under the influence of the electric field. This fact was mentioned in connection with our discussion of dispersion. The effect of the electric field is to produce a slight relative displacement of the positive and negative charges, so that the atom acquires an electric moment. Now the moment generated in a given atom depends on the electric field at that atom, which will be determined in part by the fields from other polarized atoms in its immediate neighborhood. If these other atoms are arranged in a particular way, it is clear that the polarization and the effective dielectric constant should depend on the orientation of the electric vector of the waves. In calcite, for example, the oxygen atoms in the CO_3 group are the most easily polarized, and they exert a strong influence on each other. Under this influence, they are more easily polarized by an electric field parallel to the plane of the group than by one perpendicular to it. At a result, we shall find that the refractive index should be greatest for light having its electric vector perpendicular to the trigonal axis.

The fact that ε varies with direction in these crystals can be shown by the electromagnetic theory to give rise to double refraction. The direction of $\mathbf{D}$ differs from that of $\mathbf{E}$ except in three singular directions, which are mutually perpendicular. The value of ε is a maximum along one of these axes, a minimum along another, and intermediate along the third. Designating them by x, y, and z, we therefore find that, for the three components of $\mathbf{D}$ in Maxwell's equations, we must now write

$$D_x = \varepsilon_x E_x \qquad D_y = \varepsilon_y E_y \qquad D_z = \varepsilon_z E_z \tag{26j}$$

When these values are substituted in Eqs. (26i), and the equation for plane electromagnetic waves derived,* it is found that for any direction of the wave front there are two velocities for vibrations of the vector $\mathbf{D}$ in two mutually perpendicular directions, and this is the fundamental aspect of double refraction.

The most concise way of presenting the results of the electromagnetic theory

* See, for example, P. Drude, "Theory of Optics," English edition, pp. 314–317, Longmans, Green & Co., Inc., New York, 1922.

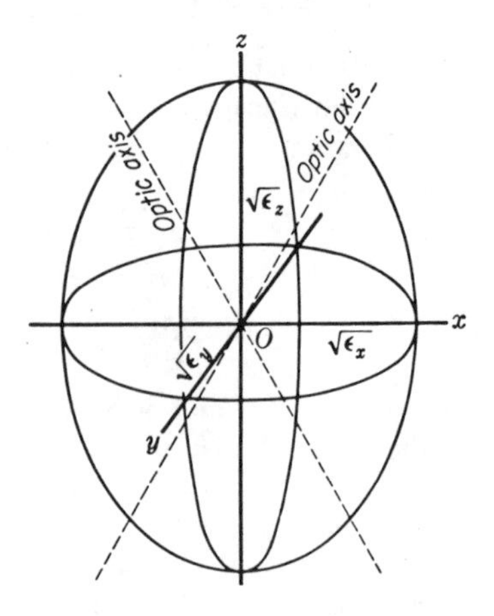

FIGURE 26P
Dielectric ellipsoid for a biaxial crystal.

is by the use of what we may call the *dielectric ellipsoid.* This is an ellipsoid described by the equation

$$\frac{x^2}{\varepsilon_x} + \frac{y^2}{\varepsilon_y} + \frac{z^2}{\varepsilon_z} = 1 \tag{26k}$$

in which ε_x, ε_y, and ε_z are the *principal dielectric constants* of Eqs. (26j). The semiaxes of the ellipsoid are $\sqrt{\varepsilon_x}$, $\sqrt{\varepsilon_y}$, and $\sqrt{\varepsilon_z}$, as shown in Fig. 26P, where we have taken $\varepsilon_x < \varepsilon_y < \varepsilon_z$. From the ellipsoid we may obtain the two velocities and the corresponding directions of vibration for a wave traveling in any arbitrary direction through the crystal, as explained below. This mode of representation was first given by Fresnel in terms of the elastic-solid theory of light. Since, in the older theory, the velocity was dependent on the elasticity and density of the ether, Fresnel's ellipsoid could be either an "ellipsoid of elasticity" or an "ellipsoid of inertia." When it is replaced by a dielectric ellipsoid, Fresnel's results can be translated directly into the terms of the electromagnetic theory.

Suppose now that ordinary light waves vibrating in all planes are moving through the point O in the crystal in every direction and that we wish to determine the double wave surfaces presented in the preceding sections. In Eq. (23p), the velocity of light was given by

$$v = \frac{c}{\sqrt{\varepsilon}} \tag{26l}$$

where c is the velocity in vacuo. We therefore have the relations

$$v_a = \frac{c}{\sqrt{\varepsilon_x}} \qquad v_b = \frac{c}{\sqrt{\varepsilon_y}} \qquad v_c = \frac{c}{\sqrt{\varepsilon_z}}$$

$$n_a = \sqrt{\varepsilon_x} \qquad n_b = \sqrt{\varepsilon_y} \qquad n_c = \sqrt{\varepsilon_z}$$

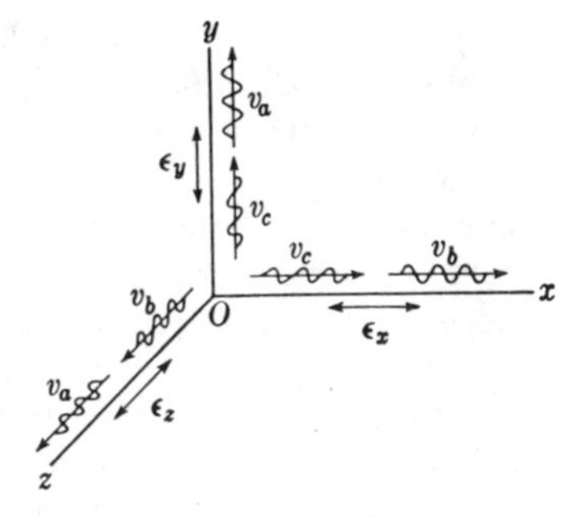

FIGURE 26Q
Correlation of the velocities and directions of vibration in the waves with the directions of the three principal dielectric constants.

where $v_a > v_b > v_c$. Now v_a represents the velocity of waves traveling perpendicular to the x axis with their electric displacements parallel to x. Their velocity is therefore determined by ε_x. The application of this statement to the other directions of vibration and velocities of propagation along the three coordinate axes may be seen by inspection of Fig. 26Q.

Let us now see how the two velocities in any arbitrary direction may be determined by the use of the dielectric ellipsoid. First we note that the velocities along any one of the coordinate axes are inversely proportional to the major and minor axes of the elliptical section of the ellipsoid made by the coordinate plane that is perpendicular to that axis. In the same way, for any other direction of propagation, we pass a plane through O parallel to the plane of the wave. This will cut the ellipsoid in an ellipse with major and minor axes OA and OB, Fig. 26R(a). On the wave normal the distances OM and ON are measured off inversely proportional to OA and OB.

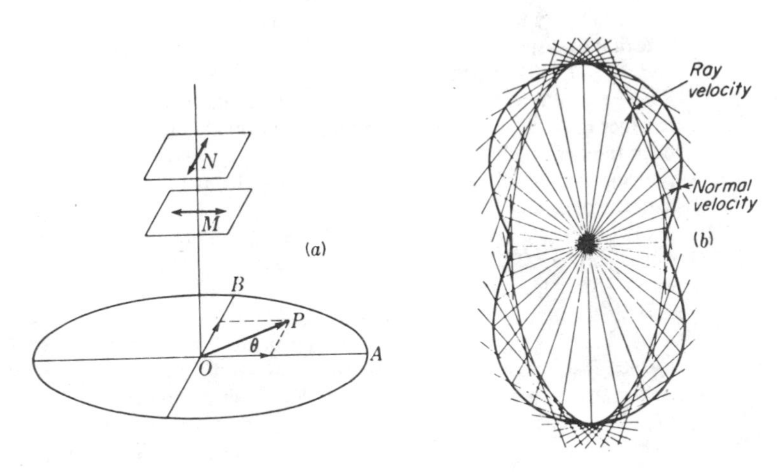

FIGURE 26R
The construction of a normal-velocity surface.

The planes M and N parallel to the initial plane represent a later position of the waves vibrating parallel to the two axes of the ellipse. If we consider a single vibration in plane AOB making an angle θ with OA, the electric vector OP may be resolved into two components $OP \cos \theta$ and $OP \sin \theta$. These components along the major and minor axes travel with the two different velocities. If now the plane AOB is rotated about O in every possible direction the points M and N will trace out the normal-velocity surfaces (dotted lines) shown in Fig. 26K(b).

For every ellipsoid with three different axes there are two planes only for which the cross sections are circles. For these two planes, OA and OB in Fig. 26R(a) will be equal and the planes M and N will coincide. The directions of the normal for these two circular cross sections of the dielectric ellipsoid give the optic axes of the crystal, i.e., the directions of equal *normal* velocity for all planes of vibration. The envelope of all plane waves at the instant they reach the normal-velocity surface is the wave surface previously described in Sec. 26.6. This envelopment, giving a surface of elliptical section, is illustrated in Fig. 26R(b).

The optical properties of doubly refractive crystals are completely determined when one knows the values of the three principal refractive indices and the directions of two of the principal axes. As was mentioned, these can be measured by cutting the crystal into prisms of different orientation. There exist, however, more powerful and more convenient methods based on the interference effects resulting from the difference in velocity of the two polarized components, and these will be discussed in the next chapter.

PROBLEMS

26.1 A light ray strikes the surface of a crystal of ice at grazing incidence in a plane perpendicular to the optic axis. The crystal has been cut so that its axis lies parallel to the surface. Find the separation in millimeters of the O and E rays at the opposite face of the crystal, which is a plane-parallel slab 4.20 mm thick. Assume $n_O = 1.3090$ and $n_E = 1.3104$ for sodium light. *Ans.* 0.01271 mm

26.2 Find by graphical construction how thick a natural calcite crystal would need to be in order for a ray of sodium light incident normally on one cleavage face to emerge from the opposite face as two rays separated by a linear distance of 2.50 mm. In a principal section of calcite, the optic axis can be assumed to make an angle of 45° with the normal.

26.3 A ray of unpolarized light falls on a calcite crystal, the optic axis of which is parallel to the surface. The angle of incidence is 32°, and the plane of incidence coincides with the principal section of the crystal. Find the angles of refraction of the O and E rays for the green mercury line (see Table 26A and footnote Sec. 26.3).

26.4 A 50° prism is made of ammonium phosphate for which $n_O = 1.5250$ and $n_E = 1.4790$. If the prism is cut with its optic axis parallel to its refracting edge, calculate (a) the angles of minimum deviation and (b) their difference.
Ans. (a) $\delta_O = 30.26°$, $\delta_E = 27.37°$, (b) 2.89°

26.5 Draw to scale the two cross sections of the wave surface of Rutile (TiO_2) made by planes (a) parallel to the optic axis and (b) perpendicular to the axis. Indicate the directions of the vibrations on each diagram. (c) Is rutile a positive or negative crystal? Assume the light is for the Fraunhofer F line, $\lambda = 4861$ Å.

26.6 The angle 2α between the optic axes of a biaxial crystal is given by Eq. (26g). The principal indices of refraction for two unidentified crystals are measured and found to be (*a*) for the first, $n_a = 1.6842$, $n_b = 1.6935$, and $n_c = 1.7126$ and (*b*) for the second, $n_a = 2.1547$, $n_b = 2.3282$, and $n_c = 2.4034$. Find the angle α for both crystals, and determine whether they are positive or negative.

Ans. (*a*) 35.24° positive, (*b*) 58.77° negative

26.7 Draw to scale cross sections in the three coordinate planes of the wave surfaces of a biaxial crystal of sulfur. See Table 26C for refractive indices.

26.8 Construct one quadrant of the xz section of the elliptical wave surface for stibnite. From this graphically construct the corresponding normal velocity surface in this same plane [see Fig. 26R(*b*)]. Show the optic axis.

26.9 A crystal of stibnite is cut into a 20° prism with its refracting edge perpendicular to the plane containing the optic axes. The angle of minimum deviation is measured for a ray of sodium light with its vibrations parallel to the refracting edge. What value would be expected according to the refractive indices given in Table 26C?

Ans. $\delta = 69.3°$

26.10 The axis of *single ray velocity* in a biaxial crystal makes an angle β with the z axis, the cosine of which is a/b times as large as the value of $\cos \alpha$. Find the apex angle of the cone of internal conical refraction in a crystal of stibnite, using the refractive indices given in Table 26C.

27

INTERFERENCE OF POLARIZED LIGHT

The first investigation of the interference of polarized light was made by Arago in 1811. Examining the blue light of the sky with a calcite crystal, he observed that when a thin sheet of clear mica was interposed, the ordinary and extraordinary rays became richly colored. This color effect is produced by nearly all crystals and is due in most cases to the interference of polarized light and in a relatively few to optical activity. The latter subject will be treated in detail in the next chapter, and the phenomena attributed to interference will be considered now.

27.1 ELLIPTICALLY AND CIRCULARLY POLARIZED LIGHT

Suppose that plane-polarized light from a nicol prism N_1, as in Fig. 27A, is incident normally on a thin plate of calcite C cut with faces parallel to the optic axis. Making use of the wave-surface diagrams and Huygens' construction as in Fig. 26E(*a*), we can now determine qualitatively the nature of the light emerging from the calcite plate. Upon entering the crystal at P normal to the surface but with the vibrations at an angle with the optic axis, the light will break up into two components E and

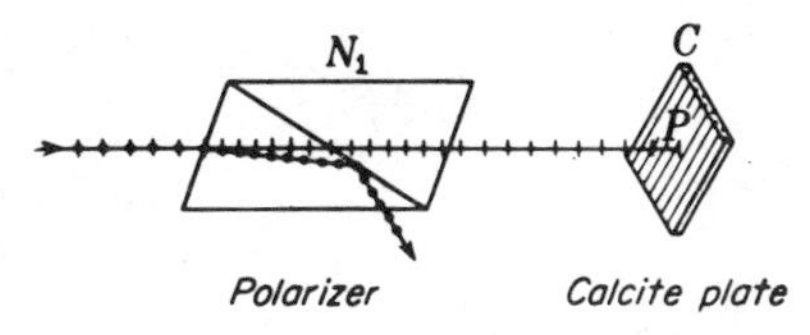

FIGURE 27A
Plane-polarized light incident normally on a crystal plate cut parallel to the optic axis.

O (Fig. 27C). As shown in Fig. 27B, the E wave having vibrations parallel to the optic axis will travel faster than the O wave, but along the same path.

To find just how far the E vibrations get ahead of the O vibrations by the time they have traversed the thickness d of the crystal, we take the difference in the *optical paths* (Sec. 1.5) and then convert this into phase difference. The optical path for the O ray is, according to Eq. (1i), merely $n_O d$, and that for the E ray, $n_E d$. The path difference is therefore

$$\Delta = d(n_O - n_E) \qquad (27a)$$

The corresponding phase difference, by Eq. (13a), is given by $2\pi/\lambda$ times the path difference,

$$\delta = \frac{2\pi}{\lambda} d(n_O - n_E) \qquad (27b)$$

Here d may also represent the distance of penetration into a given crystal, and one sees that the phase difference δ steadily increases in proportion to this distance.

Looking end-on against the light beam, as in Fig. 27C, let the plane-polarized light vibrations from the first nicol N_1 meet the first crystal face, making an angle θ with the principal section. If A is the amplitude of this light, it will be broken up into two components, $E = A \cos \theta$ traveling with the faster velocity v_E and $O = A \sin \theta$ traveling with the slower velocity v_O. After leaving the crystal, the O and E rays will continue in the same straight line and, of course, with their vibrations perpendicular to each other.

Within the crystal there are, at any given point, two vibrations at right angles having the phase difference δ. They are of the same frequency, equal to that of the

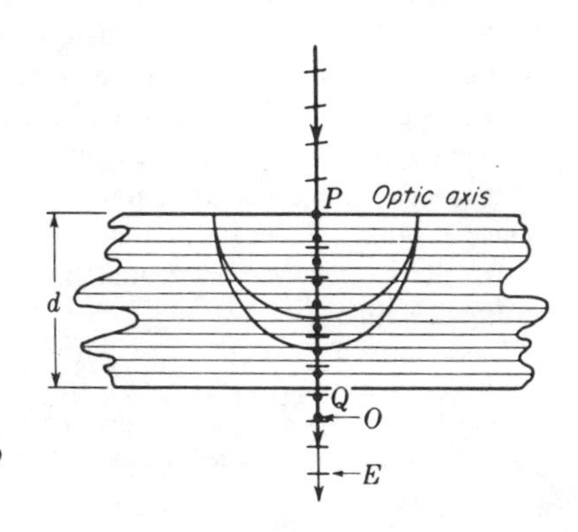

FIGURE 27B
Advance of the E wave ahead of the O wave in a negative crystal plate.

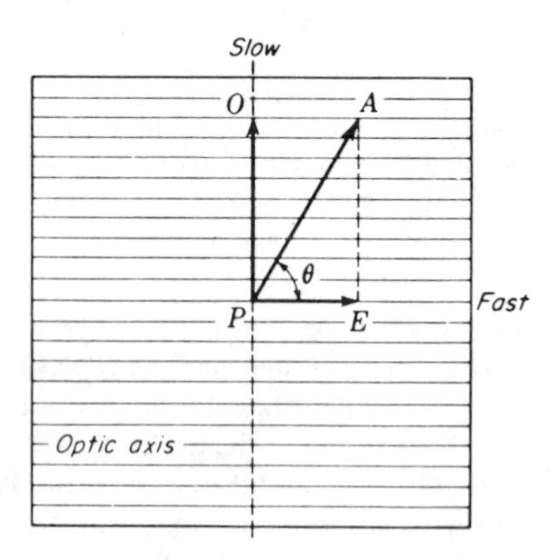

FIGURE 27C
Resolution by the crystal of plane-polarized light incident as in Fig. 27A.

light outside the crystal. The problem of the composition of such vibrations has already been investigated in Sec. 12.9, where it was shown that the resultant motion is one of the various elliptical figures for equal frequencies (Fig. 12K). The vibration is therefore an ellipse, a straight line, or a circle. In fact, at steadily increasing distances through the crystal from P towards Q (Fig. 27B), the vibration forms will progress through a whole sequence of figures like those in Fig. 12K and usually will repeat the sequence many times. It is only when the light emerges from the crystal, however, that the type of vibration may be readily observed. Depending on the thickness of the crystal and on the other quantities in Eq. (27b), this will be some figure enclosed within a rectangle of sides $2A \cos \theta$ and $2A \sin \theta$. For $\delta = 0, 2\pi, 4\pi, \ldots$ the linear incident vibration will emerge unchanged, while for $\delta = \pi, 3\pi, 5\pi$ it, ... will be transformed into another linear vibration making an angle 2θ with its original direction. At all intermediate values of δ the motion is in an ellipse, the shape of which is determined by the particular θ and δ according to the principles explained in Sec. 12.9. Such light is termed *elliptically polarized* light, of which linearly polarized* and circularly polarized light are obviously special cases.

Let us consider for a moment what is meant by the statement that the vibrations in a beam of light are elliptical. Since the "vibration" is actually a periodic variation of the electric field in space, it means that at any given point in an elliptically polarized beam *the terminus of the electric vector moves in an ellipse* in a plane perpendicular to the direction of propagation of the light. The vector thus varies continuously in both direction and magnitude, returning to the original values with the frequency of the wave. At other points along the wave the motion is similar but of different phase, so that the vector is in a different part of the ellipse. In a "snapshot" of the wave, the electric vectors would have a screwlike arrangement like that illustrated in Fig. 20D(*b*).

* The terms plane-polarized and linearly polarized are often used interchangeably. The latter is rather to be preferred when comparisons are made with elliptically polarized light.

For the crystal to produce *circularly polarized* light, two conditions must hold. First, the amplitudes of the O and E rays must be equal. This requires that $\sin\theta = \cos\theta$ or that $\theta = 45°$. Second, the phase difference must be either $\pi/2$ or $3\pi/2$. (Addition of any multiple of 2π to either of these is of no consequence.) The difference between the two cases is one of the direction of rotation in the circle, as was explained in Sec. 12.9 in connection with Fig. 12L. Which value of δ gives right circular polarization and which left will depend on whether the plate is made of a positive or of a negative crystal. With calcite, for example, the E wave travels faster, and if $\delta = \pi/2$, a left-handed rotation is produced, looking against the direction of the light. The directions parallel to and perpendicular to the optic axis in a negative crystal are often called the *fast* and *slow* axes of the plate, as is indicated in Fig. 27C. In a positive crystal these designations relative to the axis are of course interchanged.

27.2 QUARTER- AND HALF-WAVE PLATES

The simplest device for producing and detecting circularly polarized light is known as a quarter-wave plate, or $\lambda/4$ plate. Such plates are usually made of thin sheets of split mica, although they can be of quartz cut parallel to the optic axis. The thickness is adjusted so as to introduce a 90° phase change between the O and E vibrations.* The correct thickness for such plates can in the case of uniaxial crystals be computed by use of Eq. (27b). Since the phase difference δ depends upon the wavelength, the principal indices for yellow sodium light, $\lambda 5893$, are usually used for computing the required thickness for a quarter-wave plate. When a quarter-wave plate is oriented at an angle of 45° with the plane of the incident polarized light, the emerging light is circularly polarized. Reasonably good plates can be made by splitting good clear mica into very thin sheets about 0.035 mm thick. This can be done with a penknife or a needle, using a micrometer caliper for checking the thickness.

Use is often made of plates which introduce a phase difference of 180° between the components and which are therefore called *half-wave* plates. As mentioned in the previous section, the effect of a plate of this kind is merely to alter the direction of vibration of plane-polarized light by the angle 2θ, where θ is the angle between the incident vibrations and the principal section. In certain instruments where it is desired to compare two adjacent fields of light polarized at a certain angle with each other, half the field is covered with a half-wave plate.

* Strictly speaking, mica is a negative biaxial crystal of which there are many different forms. The angle between the optic axes may be almost anything from 0 to 42° depending upon the chemical constitution as well as the crystal structure. The most common mica, muscovite (pale brown in color) has an angle of 42° ($= 180° - 2\alpha$) between the optic axes (see Table 26C). The cleavage plane along which it splits most easily is the yz plane in Figs. 26I and 26J. The difference between the velocities along the x axis is therefore very small. This is an advantage, since then the plates do not have to be made too thin and fragile. Quartz has no natural cleavage planes and has to be cut and the faces polished to optical flatness. Quarter-wave plates can also be made in large sheets of plastic that has been formed by *extrusion*. These sheets are double refracting and by careful control of the thickness can be made to produce a phase difference of $\pi/2$ rad, or any other value.

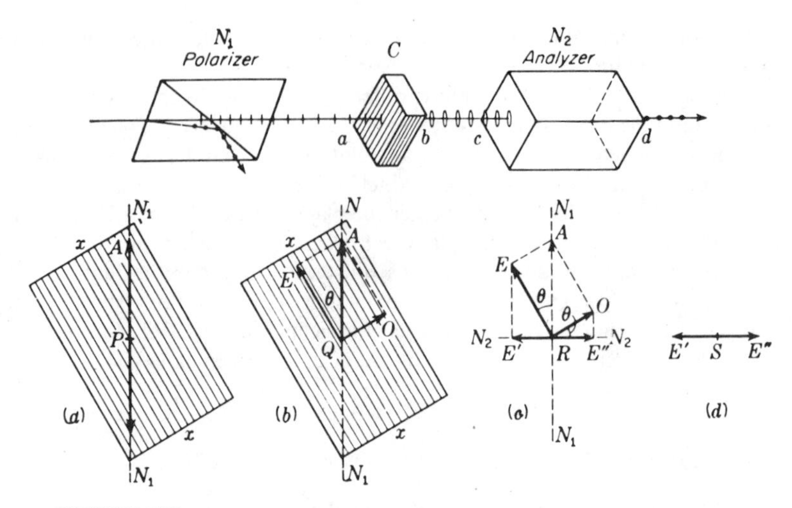

FIGURE 27D
Origin of the components brought to interference by crossed analyzer and polarizer.

27.3 CRYSTAL PLATES BETWEEN CROSSED POLARIZERS

As explained in Sec. 24.12, when the polarizer and analyzer are set at right angles to each other, no light is transmitted by the combination. Suppose now that a crystal plate cut parallel to the optic axis is inserted between crossed polarizers as shown in Fig. 27D. The observed result is that light will now get through the analyzer. One way of interpreting this result is to say that the plane-polarized light entering the crystal at a emerges as elliptically polarized light at b and thus has developed a component parallel to the transmitting plane of the analyzer. This view is correct and very simple: it is just the component A_1 shown in Fig. 27E that is passed by the analyzer, and the corresponding intensity is proportional to $A_1{}^2$. For purposes of computation, however, it is possible to consider the phenomenon as one of *interference* between the two component vibrations emerging from the plate, part of each being transmitted by the analyzer. In Fig. 27D, the four lower diagrams represent end-on views of the light (looking *against* the light) at the four points designated by corresponding letters in the diagram above. In (a) the plane vibration is shown as it arrives at the crystal plate with an amplitude A and making an angle θ with the optic axis. This amplitude is broken up into two components, $E = A \cos \theta$ along the optic axis, and $O = A \sin \theta$ perpendicular to it. One of these components travels faster in the crystal and, upon emerging, will be ahead of the other in phase. In (c) these two components are shown as they arrive at the analyzer N_2, where only the E vibrations parallel to its principal section N_2N_2 can be transmitted. In other words, only the components E' and E'' get through and they are now vibrating in the same plane. These have the magnitudes

$$E' = E \sin \theta = A \cos \theta \sin \theta \qquad (27c)$$

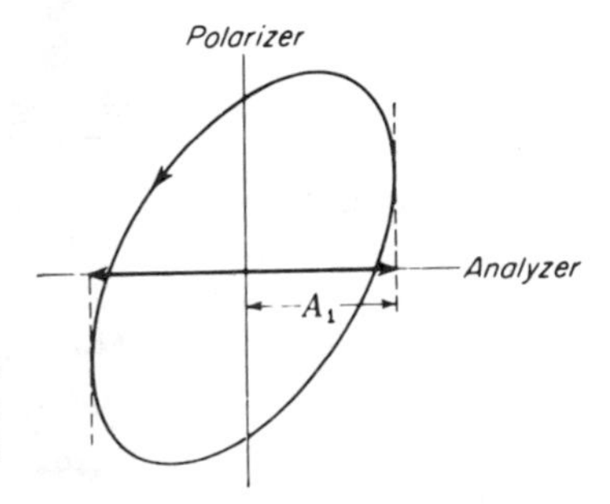

FIGURE 27E
Component of elliptically polarized light transmitted by a crystal plate, as in Fig. 27D, and then by an analyzer that is crossed with the polarizer.

and

$$E'' = O \cos \theta = A \sin \theta \cos \theta \qquad (27d)$$

This result shows that regardless of the angle θ both components E' and E'' transmitted by the analyzer are equal in magnitude when the polarizers are crossed.

These two components are now vibrating in the same plane and have the phase difference given by Eq. (27b). If the thickness of the plate is such that $\delta = 0, 2\pi, 4\pi, \ldots,$ the two *destructively interfere*. (Note that for zero thickness $d = 0$, $\delta = 0$, and the components E' and E'' are in opposite directions and hence cancel.) For all other phase angles the resultant of the two vibrations will be transmitted. To find the amplitude and intensity of this transmitted light the two components are combined as shown in Fig. 12A. The equations for these quantities will be derived in Sec. 27.6.

It should be noted that destructive interference is not produced in front of the analyzer. It is only after the two components are brought into the same plane that interference is brought about. This principle is best expressed by the *Fresnel-Arago laws*, the two most significant of which are

1 *Two rays polarized at right angles do not interfere.*
2 *Two rays polarized at right angles (obtained from the same beam of plane-polarized light) will interfere in the same manner as ordinary light only when brought into the same plane.*

27.4 BABINET COMPENSATOR

Frequently in the study of the optical phenomena a crystal plate of variable thickness is useful in producing or analyzing elliptically polarized light. Such a plate, with faces cut parallel to the optic axis, was first made by Babinet and is called a *Babinet compensator*. It consists of two wedge-shaped prisms of quartz cut at a very small angle as shown in Fig. 27F(*a*). The optic axes are parallel and perpendicular, respectively, to the two refracting edges. If plane-polarized light is incident normally on the compensator with the plane of vibration at some arbitrary angle θ to the optic axis, it will be broken up into two components. The E component, parallel to the optic axis in the first crystal, travels slower (since the compensator is made of quartz)

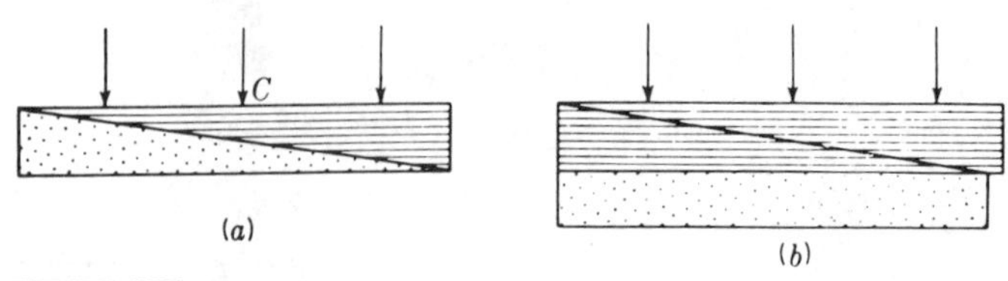

FIGURE 27F
Diagrams of (*a*) Babinet compensator and (*b*) Soleil compensator.

than the O component until it reaches the second crystal. At this point the E vibration becomes the O vibration since it is now perpendicular to the axis. At the same point the O vibration from the first crystal becomes the E vibration in the second. In other words, the two vibrations exchange velocities in passing from one prism to the other. The effect is such that one prism tends to cancel the effect of the other. Along the center at C, where both paths are equal, the cancellation is complete and the effect is that of a plate of zero thickness. On each side of C one vibration will be behind or ahead of the other because of the different path lengths. Thus the effect is that of a plate the thickness of which is zero along the centerline and varies linearly in both directions from this line.

The chief disadvantage of a Babinet compensator is that a specified plate thickness or a certain desired retardation is confined to a narrow region along the plate parallel to the refracting edges of the prisms. A modification which permits a changeable thickness which is the same over a large field consists of two wedges cut and mounted together with axes shown in Fig. 27F(*b*). The effective thickness is varied by a calibrated screw which slides the top prism along the other. By making the prism angles very small a careful adjustment to a $\lambda/4$ or $\lambda/2$ plate thickness is readily made for any color of light. This arrangement is called a *Soleil compensator*.

The properties of a Babinet compensator are well illustrated by the following experiment. Light from a carbon-arc lamp is polarized by a nicol prism N_1 as shown

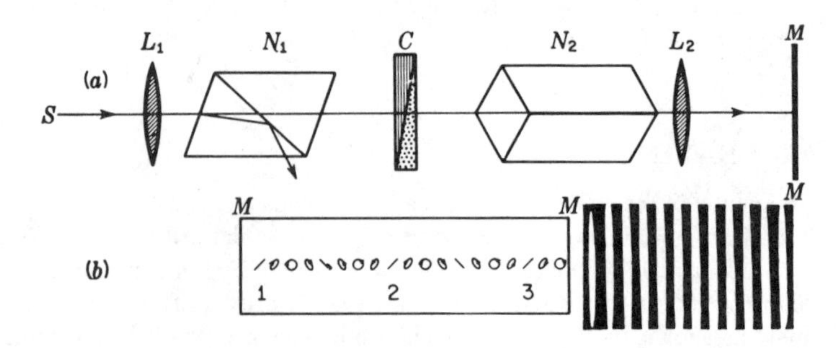

FIGURE 27G
Polarization and light bands produced by a Babinet compensator between crossed nicols.

in Fig. 27G(*a*). The compensator C is oriented at about 45° with N_1 and its image focused on a screen MM by means of a lens L_2. Owing to the effectively variable thickness along the compensator, the light on the screen (with N_2 removed) will be polarized as shown in Fig. 27G(*b*) (see also Sec. 27.1 and Fig. 12K). If a second nicol is inserted at N_2 and oriented perpendicular to one set of plane-polarization regions, e.g., those marked 1, 2, and 3 in the figure, no light will be transmitted at these points. The screen will thus be crossed by a set of equidistant parallel dark bands. With white light the bands are colored and look like Young's double-slit fringes but with the center one black. Best results are obtained with monochromatic light. Piles of glass plates mounted in tubes or Polaroid films may of course be used in place of nicols N_1 and N_2.

27.5 ANALYSIS OF POLARIZED LIGHT

If one has a beam of light that is completely polarized, linearly, elliptically, or circularly, it will appear to the eye no different from ordinary unpolarized light. By using simple auxiliary apparatus, however, its character and vibration form can easily be determined. For this purpose an analyzer, in the form of a nicol or Polaroid, is used in conjunction with either a quarter-wave plate or some form of compensator. For many purposes a quarter-wave plate is adequate, and the compensator need be used only where precise measurements of elliptical polarization are required.

To illustrate the use of the quarter-wave plate, suppose, for example, that it is placed in a beam of circularly polarized light. Regardless of the orientation of the optic axis, the circular vibration is equivalent to two linear and mutually perpendicular ones along the slow and fast axes, 90° out of phase with each other. Upon emerging from the plate these two are in phase and recombine to give plane-polarized light vibrating at 45° with the axes of the plate. The plane of the emergent light depends on the direction of rotation of the incident circularly polarized light. In either of the possible cases it can be completely extinguished by the analyzer. If the light to be studied is elliptically polarized, it will be converted into plane-polarized light only when the fast axis of the quarter-wave plate coincides with either the major or the minor axis of the ellipse. The ratio of these axes can then be found as the tangent of the angle that the plane of transmission of the analyzer makes with the fast axis when extinction has been achieved.

The same information may be found with greater accuracy by means of a Babinet compensator, which has the additional advantage of being usable for any wavelength. We have seen that when the incident light is plane-polarized at 45° to the principal section of one wedge, a dark band occurs at the center. If for some other kind of light the dark band is displaced from this position, a phase difference must exist between the two rectangular components of this light, and this means that it is elliptical to a greater or less degree. Since a phase difference of 2π corresponds to one whole fringe, the actual difference can be found from the fractional fringe displacement. The measurement is done by screwing one wedge over the other until the dark fringe returns to the center, so that the phase difference has been *compensated*.

For details of the use of the compensator, the reader should refer to more advanced texts.*

If the light is not completely polarized but contains some admixture of unpolarized light, it is still possible to completely determine its character by using a quarter-wave plate and an analyzer in the systematic way outlined in Table 27A. The light is first studied with the analyzer alone. If there is no change in intensity on rotating it, the procedure outlined in part *A* of the table is followed. If there is some change of intensity, part *B* is followed. The seven kinds of light which can be identified in this way represent all the possible conditions of polarization. Other more complicated mixtures can be shown to be equivalent to one or another of these.

To specify quantitatively the state of polarization of a beam of light, it is found that just four numbers are required. These *Stokes parameters* can be determined by making four suitably chosen measurements. One of these involves the total intensity, and another requires some phase-shifting device like a quarter-wave plate in conjunction with an analyzer. The remaining two can be made with the analyzer alone.†

27.6 INTERFERENCE WITH WHITE LIGHT

Referring to Eq. (27b), it is observed that the phase difference between the *E* and *O* rays depends upon the wavelength as well as on the thickness of the plate. As for the difference between the principal indices of refraction ($n_O - n_E$), the values given in Table 26A show that there is little change throughout the visible region. As the thickness of a crystal plate increases, the phase difference δ between the *O* and *E* rays for violet light, $\lambda 4000$, increases nearly twice as fast as the phase difference for red light, $\lambda 6500$, since λ occurs in the denominator of the expression for δ. This fact gives rise to the rich colors frequently observed in thin plates of mica, quartz, calcite, etc., cut parallel to the axis and placed between crossed nicols. The reason for the color is that some one or more parts of the continuous visible spectrum are stopped by the second nicol prism.

Suppose that a thin sheet of mica which introduces a phase change for yellow light of 2π rad, that is, a full-wave plate, is introduced between crossed nicols and oriented at an angle $\theta = 45°$. The orange and red wavelengths will undergo a phase change less than 2π and the green, blue, and violet more than 2π. Components of all colors but yellow light will therefore get through the second nicol. With yellow absent, the resultant color will be the mixture of red, orange, green, blue, and violet, giving a purple hue.

If, in the above experiment with a mica sheet, the analyzing nicol is replaced with a thick natural calcite crystal, one obtains the ordinary vibrations O' and O'' as well as the extraordinary ones (Fig. 27H) but in a different position. This *O* beam is also colored and is complementary to the *E* beam containing the components

* M. Born, "Optik," p. 244, J. Springer, Berlin, 1933.

† A summary of the uses of the Stokes parameters, including their application to photons and elementary particles, is given by W. H. McMaster, *Am. J. Phys.*, **22:** 351 (1954).

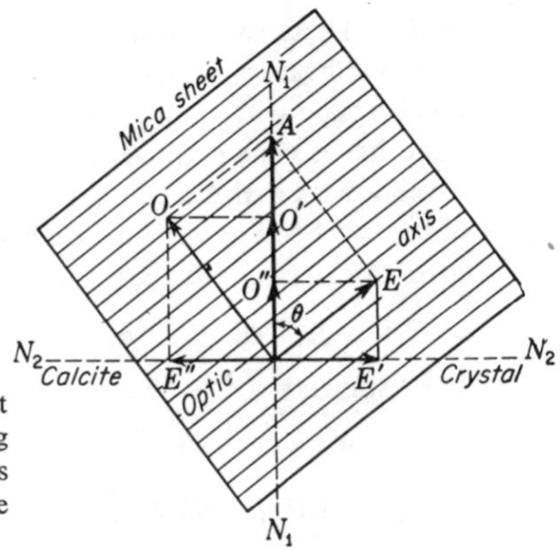

FIGURE 27H
Components of plane-polarized light transmitted by a thin doubly refracting plate and an analyzing crystal. The lines N_1 and N_2 indicate the directions of the O and E vibrations in the calcite.

Table 27A ANALYSIS OF POLARIZED LIGHT

<table>
<tr><td colspan="3">A. No intensity variation with analyzer alone</td></tr>
<tr><td>I. If with $\lambda/4$ plate in front of analyzer</td><td colspan="2">II. If with $\lambda/4$ plate in front of analyzer one finds a maximum, then</td></tr>
<tr><td>1. One has no intensity variation,

one has

natural unpolarized light</td><td>2. If one position of analyzer gives zero intensity,

one has

circularly polarized light</td><td>3. If no position of analyzer gives zero intensity,

one has

mixture of circularly polarized light and unpolarized light</td></tr>
</table>

<table>
<tr><td colspan="4">B. Intensity variation with analyzer alone</td></tr>
<tr><td>I. If one position of analyzer gives</td><td colspan="3">II. If no position of analyzer gives zero intensity</td></tr>
<tr><td rowspan="3">1. Zero intensity,

one has

plane-polarized light</td><td colspan="3">2. Insert a $\lambda/4$ plate in front of analyzer with optic axis parallel to position of maximum intensity</td></tr>
<tr><td rowspan="2">(a) If get zero intensity with analyzer,

one has

elliptically polarized light</td><td colspan="2">(b) If get no zero intensity,</td></tr>
<tr><td>(1) But the same analyzer setting as before gives the maximum intensity,

one has

mixture of plane-polarized light and unpolarized light</td><td>(2) But some other analyzer setting than before gives a maximum intensity,

one has

mixture of elliptically polarized light and plane-polarized light</td></tr>
</table>

E' and E''. These two beams if made to overlap will give white light, since what is absent in one beam is present in the other. A slight increase or decrease in thickness of the mica plate will change the wavelength or color of the light interfering destructively and hence change the color of each transmitted beam.

To show that these two colors are complementary it must be shown that the sum of the two beams gives the original intensity A^2. For the E beam the components E' and E'' must be combined with their proper phase angle difference.

$$\begin{aligned} A_1{}^2 &= E'^2 + E''^2 + 2E'E'' \cos(\delta + \pi) \\ &= (A \sin\theta \cos\theta)^2 + (A \sin\theta \cos\theta)^2 + 2A^2 \sin^2\theta \cos^2\theta \cos(\delta + \pi) \\ &= 2A^2 \sin^2\theta \cos^2\theta\,(1 - \cos\delta) \\ &= 4A^2 \sin^2\theta \cos^2\theta \sin^2\frac{\delta}{2} \end{aligned}$$

where δ is the phase angle difference given by Eq. (27b) and π is added since E' and E'' are oppositely directed when the plate thickness $d = 0$ (Fig. 27H).

Similarly, for the O beam the two components O' and O'' must be combined.

$$\begin{aligned} A_2{}^2 &= O'^2 + O''^2 + 2O'O'' \cos\delta \\ &= (A \cos^2\theta)^2 + (A \sin^2\theta)^2 + 2A^2 \sin^2\theta \cos^2\theta \cos\delta \\ &= A^2\left[\sin^4\theta + \cos^4\theta + 2\sin^2\theta\cos^2\theta\left(1 - 2\sin^2\frac{\delta}{2}\right)\right] \\ &= A^2\left[(\sin^2\theta + \cos^2\theta)^2 - 4\sin^2\theta\cos^2\theta\sin^2\frac{\delta}{2}\right] \\ &= A^2 - 4A^2\sin^2\theta\cos^2\theta\sin^2\frac{\delta}{2} \end{aligned}$$

When added together, the two intensities yield the original one, since

$$A_1{}^2 + A_2{}^2 = A^2$$

Because of the rapid change of δ with wavelength, if a plate several times as thick as the one described above is inserted between crossed nicols, several narrow bands of the visible spectrum will be absent from the transmitted light. This can be shown experimentally with a crystal plate cut parallel to the axis as follows. A calcite plate about 0.01 to 0.03 mm thick or a quartz plate 0.2 to 1.0 mm thick is placed in a beam of plane-polarized light and beyond it a prism spectroscope arrangement as shown in Fig. 27I. With an arc lamp as a source at S, a continuous spectrum will be formed on the screen MM. If the crystal plate axis is oriented at an angle $\theta = 45°$, this light is polarized as shown schematically in the figure. To test this polarization, a second nicol is now inserted between C and S_1. When it is crossed with the polarizer, the intensity will vary sinusoidally through the spectrum, with zeros at those wavelengths for which the light transmitted by C is plane-polarized with vibrations perpendicular to the transmission plane of the second nicol. The thicker the plate the larger the number of dark bands across the spectrum.

With thick plates the combined light of the spectrum will appear white, since

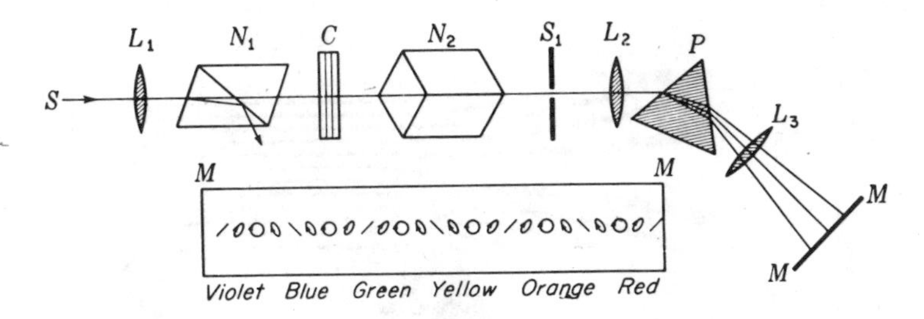

FIGURE 27I
Interference of white light produced by a crystal plate between crossed polarizers.

the large number of very narrow bands removed all along the spectrum affect the eye only as a lowering of the intensity. If a Soleil compensator is used in place of a fixed plate in the above experiment, any desired number of dark fringes can be introduced across the spectrum. A slow continuous change in thickness will cause the bands to move sideways across the spectrum and at the same time to increase or decrease slowly in number.

27.7 POLARIZING MONOCHROMATIC FILTER

The dark bands produced in the spectrum as described above have been used in an ingenious manner by Lyot* to construct a "light filter" which transmits one or more very narrow bands of wavelength. The separation of the bands produced in the spectrum by a single crystal is inversely proportional to the thickness of the crystal. Hence if one uses a thick crystal followed by one exactly half as thick, every other maximum due to the thick one will be suppressed because it will coincide with a minimum for the thinner one. Still another crystal one-quarter as thick as the first will blot out every other maximum that is transmitted by the first two. Thus it will be seen that, by placing in series a number of quartz plates the thicknesses of which vary in the geometrical progression 1:2:4:8 . . . , it is possible to isolate a very few narrow wavelength bands. Then the unwanted ones can be cut out by an ordinary color filter.

In one polarization filter Lyot used six quartz plates varying from 2.221 to 71.080 mm in thickness, with Polaroids between each pair. The optic axes of all the plates were perpendicular to the light beam, and parallel to each other, while the Polaroids were set at 45° with the optic axes. This filter transmitted 13 narrow bands only 2 Å wide. Filters of this type have been very valuable to astronomers, since they permit the study of the solar corona and prominences without the necessity of a total eclipse. By varying the temperature of the filter, it is possible to shift the wavelength of the transmitted bands to the desired value, thanks to the expansion of the plates, and decrease of the refractive indices, with rising temperature.

* B. Lyot, *C. R.*, **197**:1593 (1933).

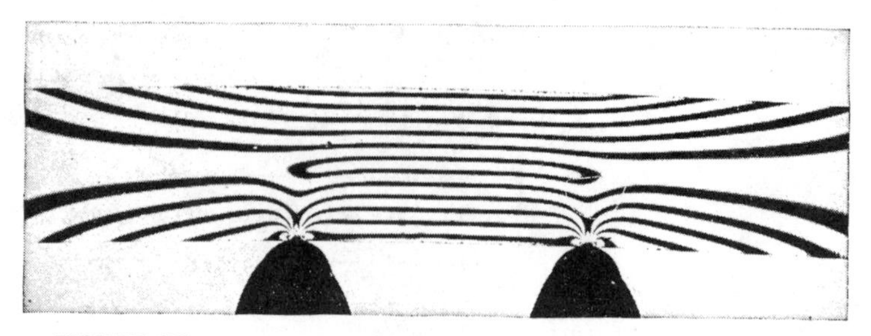

FIGURE 27J
Photoelastic effect in a plastic beam loaded at two points. (*Courtesy of R. W. Clough, Jr.*)

27.8 APPLICATIONS OF INTERFERENCE IN PARALLEL LIGHT

If the source is sufficiently intense, very small amounts of double refraction can be detected by the restoration of light when the sample is placed between crossed polarizers. If a transparent, isotropic substance like glass is subjected to mechanical stress, it becomes weakly doubly refracting with the effective optic axis in the direction of the stress. Glass blowers examine their finished work in a polariscope to check it for proper annealing. Engineers make models of structures out of a transparent plastic in order to study the distribution of stresses when a load is applied. The stresses are revealed by the distribution of light when the model is placed between crossed Polaroids. As a simple example, Fig. 27J shows the interference fringes produced by a rectangular beam when stressed at two points by small rollers. This field of *photoelasticity* is obviously one of great practical importance.*

Many common transparent substances such as silk fibres, white hair, fish scales, etc., possess a small anisotropy which can be detected by examination in polarized light. Such substances are often highly colored when viewed in a polariscope. This fact is made use of in the study of microscopic crystal growths, the color yielding a contrast that permits ready observation of the normally transparent crystals.

These applications are cited merely as examples of the many practical uses of interference in polarized light. Another is discussed in the following section, and still further ones will be found in Chap. 32.

27.9 INTERFERENCE IN HIGHLY CONVERGENT LIGHT

Up to this point in our discussion of interference of polarized light we have considered only uniaxial crystals in parallel beams. In Sec. 27.4 conditions of interference were described in which the thickness of the crystal could be varied continuously, thus altering the phase difference between the O and E rays by any desired amount. A

* Complete descriptions of the methods used are given in M. Frocht, "Photoelasticity," vol. 1, 1941, vol. 2, 1948, John Wiley and Sons, Inc., New York.

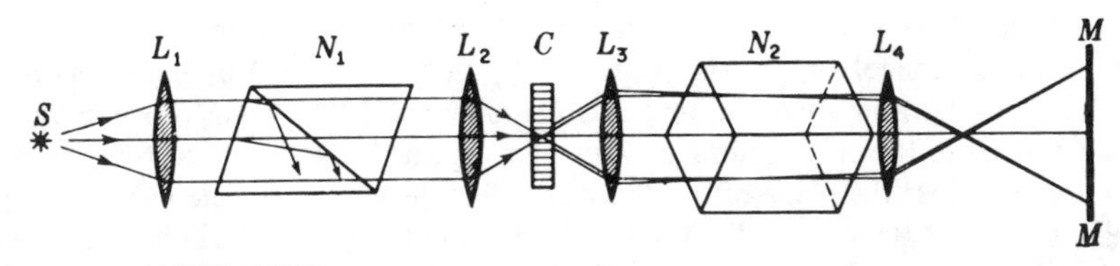

FIGURE 27K
Arrangement for projecting "rings and brushes" obtained by the interference of highly convergent polarized light in birefringent materials.

similar result can be achieved by sending light at different angles through a plate of uniform thickness. In this case a single plane-parallel plate is usually cut with its two faces perpendicular to the optic axis. Experimentally such a plate is inserted between crossed analyzer and polarizer as shown in Fig. 27K. Parallel white light from the polarizer is made highly convergent by one or more short focus lenses at L_2. After passing through the crystal C, the light is made parallel again by a similar lens L_3. Beyond the analyzer N_2 another lens L_4 focuses on the screen MM all parallel rays leaving C. This lens therefore images the secondary focal plane of L_3 on MM.

A detailed diagram of the convergent light passing through a uniaxial crystal is shown in Fig. 27L(*a*). The central ray parallel to the optic axis undergoes no change in phase since both the O and E components travel with the same speed and, in fact, there is no distinction between them. Other rays like P and Q, however, travel a greater distance in the crystal and being at an angle with the optic axis are doubly refracted. As they travel at different speeds, there will be a phase difference between the O and E rays which will increase with increasing angle of incidence. Referring to the end view in Fig. 27L(*b*), all rays entering at points on the circle P, H, Q, G pass through the same thickness of crystal and show the same phase difference upon

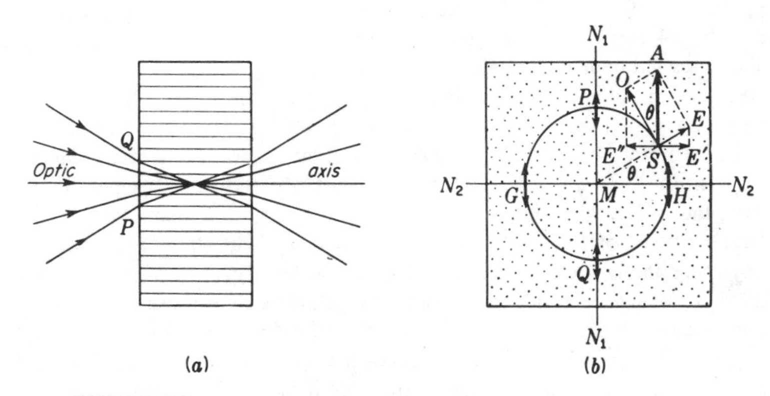

FIGURE 27L
Resolution of O and E components for interference of highly convergent polarized light in birefringent materials.

emergence. The vertical line N_1 represents the plane of vibration of the incident light from the first nicol and N_2 the plane of vibration transmitted by the second.

Consider now any point on the circle of Fig. 27L(*b*) such as S where the light is not normal to the crystal surface. This light will be broken up into two components O and E. Since the plane of incidence contains the optic axis, the refracted rays will also be in this plane. The E vibrations with amplitude $A \sin \theta$ will lie in the plane of incidence, and the O ray vibrations with amplitude $A \cos \theta$ will be perpendicular to it, as shown. Upon arrival at the second nicol N_2, the components E' and E'' will be transmitted and will interfere destructively or otherwise depending on the emergent phase relations. Whatever the phase relations for the point S, they will be the same for all other points on the same circle. For points on some other circle the phase will be different. If the plate is of several millimeters thickness, there will be a number of regularly spaced concentric dark circles where the phase difference is some multiple of 2π, so that destructive interference is produced. Thus the transmitted light will give interference rings, as shown in Fig. 27M(*a*). If white light is used, these fringes will be highly colored because of the various wavelengths present.

The dark cross appearing in these patterns, usually referred to as the "brushes," may be explained by again using Fig. 27L(*b*). As the point S approaches G or H, the components E' and E'' vanish. At these points the vibrations traverse the crystal as pure O vibrations. They therefore undergo no change and are stopped by the analyzer. Similarly the light striking P and Q is transmitted as E vibrations. Hence the intensity along the directions N_1 and N_2, corresponding to the planes of the two nicols, is zero. On each bright fringe it rises steadily to a maximum at 45° to these directions.

If the second nicol is parallel to the first, the interference pattern becomes just the complement, in every respect, of the one described. This pattern is shown in the lower part of (*a*) in Fig. 27M. One sees that this must be true by remembering that any light that is stopped by a crossed nicol will be passed by a parallel one, and vice versa.

It is possible to eliminate the brushes by introducing quarter-wave plates immediately before and behind the crystal. The light traversing the latter is then circularly polarized, and since there is no preferred direction, there can be no brushes. The so-called *optical ring sight* is made in this way, using Polaroids as the polarizing elements. Looking through such a combination one sees white-light interference rings, the center of which is exactly on the foot of the perpendicular from the eye. It can therefore be used as a gun sight of great accuracy and convenience.

In the case where the crystal is cut not perpendicular but parallel to the optic axis, the fringes turn out to be hyperbolic instead of circular. Part (*c*) of the figure shows fringes of this type. Because in this case the phase difference is not small anywhere in the field, white light cannot be used in observing these fringes. The interference figures produced by biaxial crystals, such as those shown in (*d*), are more complicated in their explanation, but the same general principles apply. The two "eyes" show the points of intersection of the optic axes with the surface of the crystal. Such figures are of importance in the identification of mineral specimens, and the

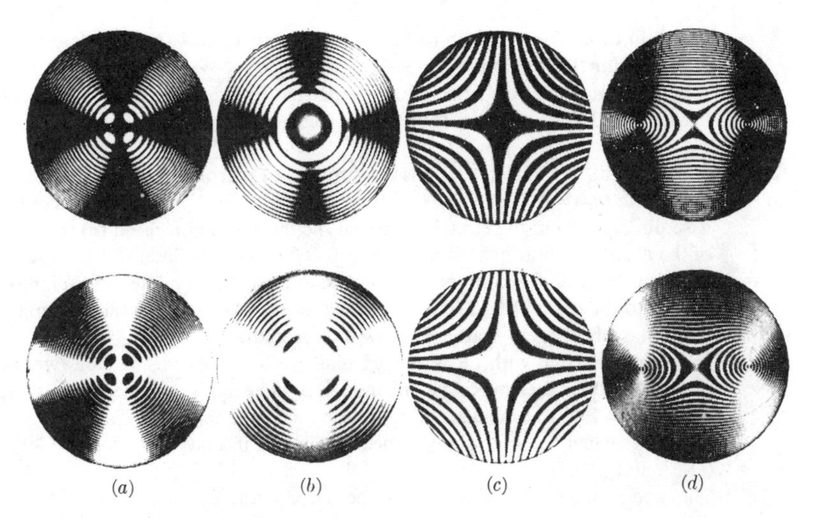

FIGURE 27M
Interference patterns from crystals placed in highly convergent light. *Upper photographs:* crossed polarizers; *lower photographs:* parallel polarizers. (*a*) Calcite cut perpendicular to the optic axis. (*b*) Quartz cut perpendicular. (*c*) Quartz cut parallel. (*d*) Aragonite cut perpendicular to the bisector of the two optic axes.

mineralogist obtains them in a microscope fitted with polarizing attachments.* The bright center of the ring system observed with the uniaxial crystal quartz [photographs (*b*)] will be explained in the next chapter.

PROBLEMS

27.1 A calcite plate cut with its faces parallel to the optic axis is placed between two crossed nicols with its principal section at 35° with the polarizer. Find (*a*) the amplitudes and (*b*) the intensities of the O and E vibrations leaving the calcite. Find also (*c*) the relative amplitudes and (*d*) intensities leaving the analyzer.

Ans. (*a*) 0.819 and 0.574, (*b*) 0.671 and 0.329, (*c*) both equal 0.470, (*d*) both equal 0.221

27.2 A quartz plate 0.850 mm thick is illuminated at normal incidence by green light of wavelength 5461 Å (see Table 26A). The optic axis is parallel to the surface. Find (*a*) the optical paths of the two rays in traveling through the plate and (*b*) the phase difference between these two in degrees.

27.3 It is desired to make a half-wave plate from a biaxial crystal of topaz. From the refractive indices given in Table 26C determine (*a*) in which plane the crystal should

* See A. Johannsen, "Manual of Petrographic Methods," 2d ed., McGraw-Hill Book Company, New York, 1918.

be cut in order for the plate to be least thin and fragile. (*b*) Calculate the required thickness for this section.

27.4 A quarter-wave plate is to be made of quartz. Using the refractive indices for blue light, $\lambda = 4340$ Å, given in Table 26A, calculate the required thickness.

Ans. 0.01149 mm

27.5 Sodium light, $\lambda = 5893$ Å, is passed through a Polaroid and then through a corundum plate ($n_O = 1.768$, $n_E = 1.760$) oriented with its axis at 35° in a counterclockwise direction from the electric vector of the incident light. Find (*a*) the magnitudes of the O and E vibrations. If the plate is 0.160 mm thick, find (*b*) the phase difference between the O and E components in passing through the plate, and (*c*) make a plot similar to the one in Fig. 27E showing the form of vibration of the transmitted light. Draw the vibration to scale and show its direction.

27.6 Into a polariscope with crossed Polaroids are inserted side by side two half-wave plates with their axes making a small angle α. The fields are of equal intensity when the direction of the incident vibrations bisects the angle α. Find the ratio of the intensities when the analyzer is turned through 1° if α has the values (*a*) 30°, (*b*) 10°, (*c*) 5°, and (*d*) 2°.

27.7 The wedge angles of a Babinet compensator made of quartz are 2.75°. Find the distance apart of sodium-light fringes when this device is placed between crossed nicols of a polariscope (see Fig. 27G). *Ans.* 0.674 mm

27.8 When a light beam of unknown polarization is viewed through a nicol prism, its intensity varies upon rotation of the latter but does not go to zero in any position. A quarter-wave plate is inserted in front of the analyzer when set for a maximum intensity, and the *fast axis* is turned parallel to the plane of transmission of the analyzer. A clockwise rotation of the analyzer by 60° will then completely extinguish the light. (*a*) What is the type of polarization? (*b*) Describe quantitatively the mode of vibration.

Ans. (*a*) elliptically polarized light, (*b*) clockwise elliptical vibration with major to minor axis ratio of 1.732

27.9 It is desired to determine the direction of rotation in a beam of circularly polarized light. When a quarter-wave plate is placed in front of the analyzer and the latter set for extinction, the fast axis of the quarter-wave plate lies in such a position that it must be turned 45° clockwise in order to bring that axis in line with the direction of transmission of the analyzer. (*a*) Make a diagram. (*b*) Does the light have right or left circular polarization?

27.10 Devise an arrangement that could be used to produce a beam of elliptically polarized light for which the major axis of the ellipse is horizontal, the ratio of the major to minor axis is 3 : 2, and the direction of rotation is clockwise. Make a scale drawing. Carefully specify each part of the apparatus and its orientation.

28

OPTICAL ACTIVITY AND MODERN WAVE OPTICS

In the preceding chapters on the behavior of polarized light in crystals we have seen that when the light travels along the optic axis there is no double refraction. In this particular direction one expects that any kind of light will be propagated without change. As early as 1811, however, Arago discovered exceptions to this simple rule. He found that certain substances, notably crystalline quartz, will restore the light when placed between crossed nicols even though the optic axis is parallel to the direction of the light. An example of this effect was shown in Fig. 27M(*b*).

28.1 ROTATION OF THE PLANE OF POLARIZATION

When a beam of plane-polarized light is directed along the optic axis of quartz, the plane of polarization turns steadily about the direction of the beam, as shown in Fig. 28A, and emerges vibrating in some other plane than that at which it entered. The amount of this rotation is found experimentally to depend upon the distance traveled in the medium and upon the wavelength of the light. The former fact shows that the action occurs within the medium and not at the surface. This phenomenon of the rotation of the plane of vibration is frequently called *optical activity*, and many

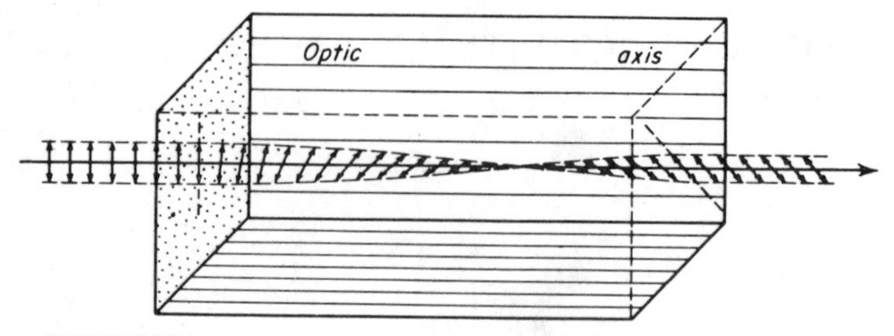

FIGURE 28A
Rotation of the plane of vibration in an optically active substance.

substances are now known to exhibit the effect. Some of these are cinnabar, sodium chlorate, turpentine, sugar crystals, sugar in solution, and strychnine sulfate.

Some quartz crystals and sugar solutions rotate the plane of vibration to the right and some to the left. Substances which rotate to the right are called *dextrorotatory* or *right-handed*, and those which rotate to the left are called *levorotatory* or *left-handed.* Right-handed rotation means that upon looking *against* the oncoming light the plane of vibration is rotated in a clockwise direction. Left-handed substances rotate the light counterclockwise.*

28.2 ROTARY DISPERSION

A striking feature of optical activity is that different colors are rotated by very different amounts. The first accurate measurements of this effect were made by Biot, who found that the rotation is very nearly proportional to the inverse square of the wavelength. In other words, there is a rotatory dispersion, violet light being rotated nearly 4 times as much as red light. This effect is illustrated diagrammatically for quartz in Fig. 28B(*a*). Let plane-polarized white light be incident normally on a quartz plate, and let the direction of its vibration be indicated by *AA*. Upon passing through 1 mm thickness of the crystal, the violet light is rotated about 50°, the red about 15°, and the other colors by intermediate amounts. More exact values for 15 wavelengths throughout the visible and ultraviolet spectrum are given in Table 28A.

This rotation for a 1-mm plate, plotted in Fig. 28B(*b*), is called the *specific rotation.* Careful measurements on quartz and other substances as well show that Biot's inverse-square law is only approximately true. In fact, optical activity is closely enough connected with ordinary dispersion theory for the regular dispersion formulas for refractive index to be applied to rotation. Cauchy's equation (Sec. 23.3), for example, can be used to represent the specific rotation for quartz in the visible region. Thus we have

$$\rho = A + \frac{B}{\lambda^2} \qquad (28a)$$

where A and B are constants to be determined.

* Although the convention used here seems to be the most common, many books will be found which use the opposite convention.

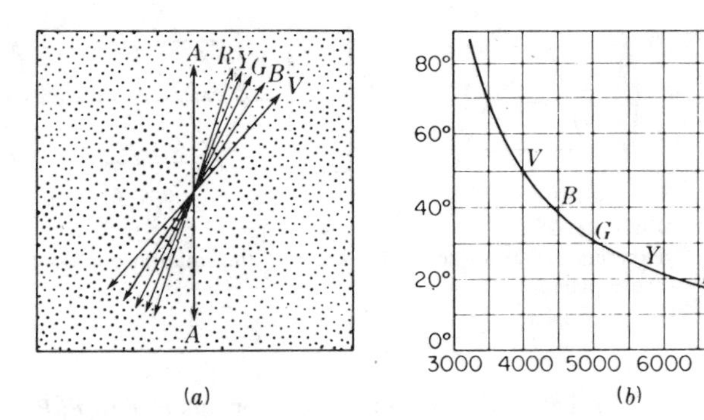

FIGURE 28B
(*a*) Rotation for different colors by a plate of quartz 1 mm thick; (*b*) specific rotation curve.

Experimentally, the phenomenon of rotatory dispersion can be illustrated by inserting a quartz plate between crossed analyzer and polarizer as shown in Fig. 28C. With a monochromatic source at S, some light will get through the analyzer to the screen MM, since in passing through the quartz along the optic axis the plane of vibration has been rotated. This is shown diagrammatically in Fig. 28D(*a*). After the vibration is rotated from the plane AP to the plane A_1P, a certain component $EP = A_1P \sin\theta$ gets through the analyzer N_2. If now the analyzer is made parallel to A_1P, all the light will be transmitted, whereas if it is normal to A_1P, none will be transmitted.

Suppose that white light is used in place of monochromatic light, so that upon passage through the crystal the different colors are rotated by different amounts as shown in Fig. 28D(*b*). The new planes of vibration are RP for the red and VP for the violet. On arriving at N_2 the two horizontal components E_RP to E_VP will get through. Since more violet light is transmitted than red, the image on the screen will be colored. What has happened is that more of the red light has been eliminated in the second nicol. This can be seen by the following modification of the experiment.

Let the analyzer in Fig. 28C be replaced by a calcite crystal. This will transmit in one beam the E vibrations given by the analyzer alone and, in a separate beam,

Table 28A SPECIFIC ROTATION ρ OF PLANE-POLARIZED LIGHT IN QUARTZ

Wavelength, Å	Deg/mm	Wavelength, Å	Deg/mm	Wavelength, Å	Deg/mm
2265.03	201.9	4358.34	41.548	5892.90	21.724
2503.29	153.9	4678.15	35.601	6438.47	18.023
3034.12	95.02	4861.33	32.761	6707.86	16.535
3403.65	72.45	5085.82	29.728	7281.35	13.924
4046.56	48.945	5460.72	25.535	7947.63	11.589

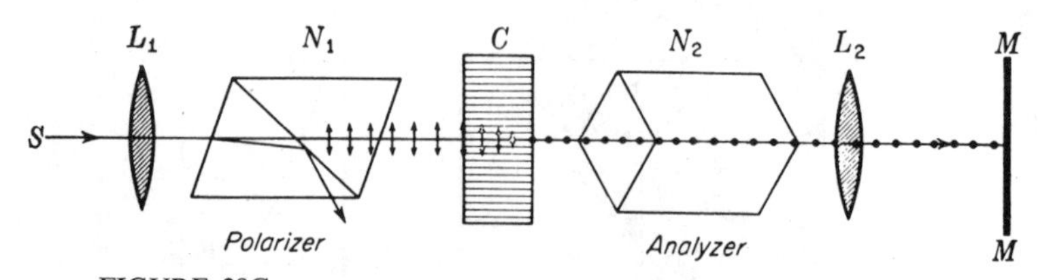

FIGURE 28C
Experimental arrangement for studying the rotation produced by an optically active plate C.

the O vibrations. The E beam will contain the components E_RP to E_VP (see Fig. 28E) and the O beam the components O_RP to O_VP. In other words, what the E beam does not contain, the O beam does. The two images on the screen MM are therefore the *complementary* colors, and if made partly to overlap, the regions of overlapping will be white. This is an excellent method for demonstrating a series of complimentary colors, for if the calcite is turned slowly, varying amounts of the different colors can be thrown into the O or E beams.

Another very striking demonstration of optical activity and rotatory dispersion is achieved by passing plane-polarized light vertically into a clear solution of cane sugar contained in a large glass tube. On observing the tube from the side with a nicol prism, a very fine spiral arrangement of colors, somewhat like a barber-pole, will be seen.

28.3 FRESNEL'S EXPLANATION OF ROTATION

Fresnel proposed an explanation for rotation in crystals like quartz which is based upon the assumption that circularly polarized light is propagated along the optic axis without change. This explanation, while not a theory in the sense of giving the

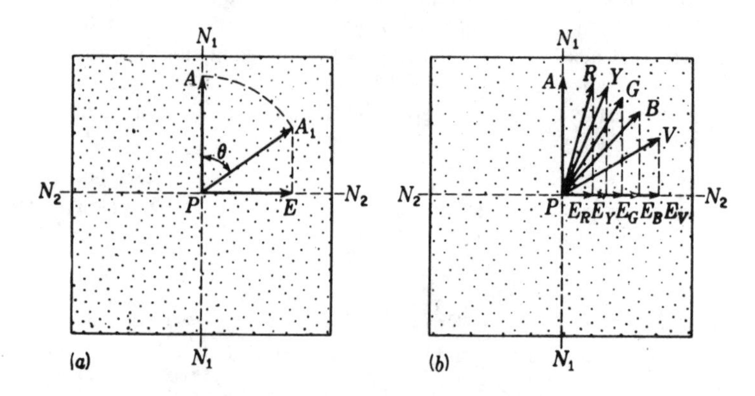

FIGURE 28D
Rotation of white light, showing the various colors transmitted by a crossed analyzer.

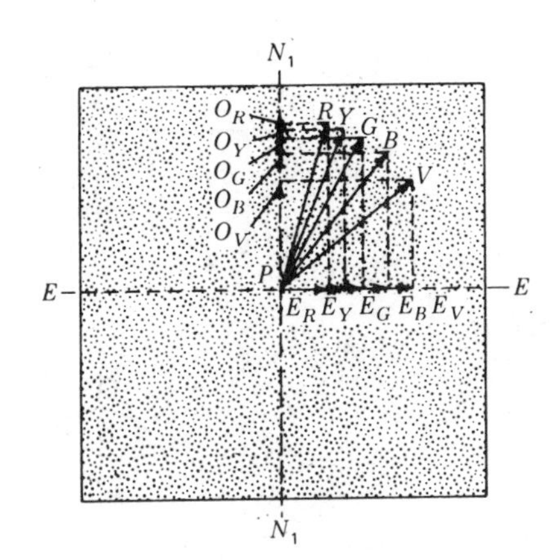

FIGURE 28E
Quartz plate between a polarizer N_1 and a calcite crystal E as analyzer.

basic cause of the phenomenon, nevertheless gives an admirable account of the facts. It is based upon the elementary principle in mechanics that any simple harmonic motion along a straight line can be described as the resultant of two opposite circular motions.

Fresnel's first assumption is that *plane-polarized light entering a crystal along the optic axis is decomposed into two circularly polarized vibrations rotating in opposite directions with the same frequency.* In a crystal like calcite, which is not optically active, these two circular motions R and L travel with the same speed as shown in Fig. 28F(*a*). Since both vibrations arrive simultaneously at any given point along their path, their resultant will be a simple harmonic motion in the plane of the original vibration as indicated in (*b*). Thus, in calcite, a plane-polarized wave along the axis is propagated with its vibrations always in the same plane.

In an optically active crystal, the two circular vibrations move forward with very slightly different velocities. In right-handed quartz, the right-handed or clockwise motion (looking against the light) travels faster, and in left-handed quartz the left-handed or counterclockwise motion travels faster.

Consider now some point Q, in a right-handed crystal, along the path of a plane-polarized incident beam as shown in Fig. 28F(*c*). Let the amplitude and plane of the incident vibration be represented by AP in Fig. 28F(*d*). The right circular component R of this vibration arrives at Q first, and as the wave travels on, the displacement turns through an angle θ before the left-handed component L arrives. At this instant, the two circular motions are in opposite senses with the same frequency, the one starting at R and the other at L. The result is that the point B' vibrates along the fixed line BQ with the same amplitude and frequency as the original vibration AP and this represents the vibration form of the light at Q. Thus in traveling from the crystal face at P to the point Q, the plane of vibration has been rotated through an angle $\theta/2$. It is clear, therefore, that the plane of vibration would under these assumptions rotate continuously as the light penetrates deeper and deeper into the crystal and that the angle of rotation would be proportional to the thickness.

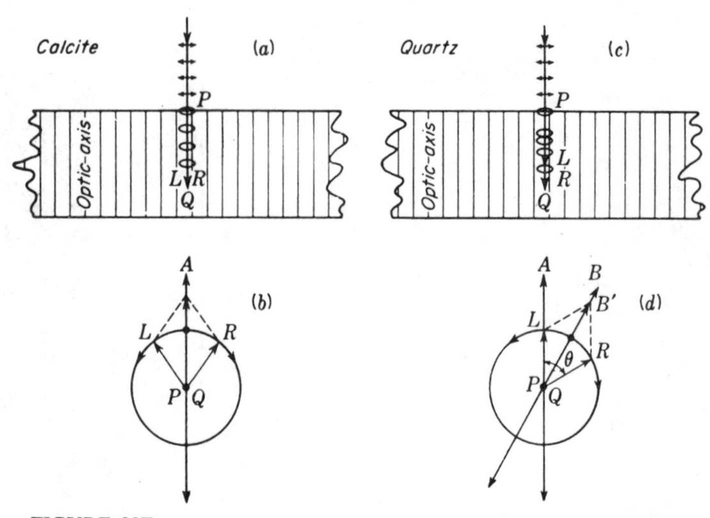

FIGURE 28F
Resolution of plane-polarized light into circularly polarized components.

28.4 DOUBLE REFRACTION IN OPTICALLY ACTIVE CRYSTALS

Since the ability to rotate the plane of polarization is a special property not possessed by many anisotropic crystals, the question arises as to its relation to the ordinary type of double refraction discussed in previous chapters. Optical activity is shown only by a certain type of crystal, but such a crystal will also show double refraction when the light is transmitted in some direction other than that of the axis. Hence the one phenomenon must change continuously into the other as the angle is changed. To understand this, we must realize that Fresnel's two velocities for right and left circular light are actually velocities represented by the wave surfaces introduced in Chap. 26 (Fig. 26B). There it was pointed out that the two sheets of the wave surface in quartz do not touch at the optic axis as they do in calcite. In Fig. 28G the wave surfaces for quartz are shown again. In the equatorial plane linear vibrations O and E, perpendicular and parallel, respectively, to the optic axis, are propagated with different velocities but unchanged in form, as shown. Along the axis z, z', right and left circular vibrations R and L are propagated with slightly different velocities. Along intermediate directions like (*b*) and (*c*) only *elliptical vibrations* of a definite form can be transmitted unchanged.

In calcite the elliptic wave surface gave a measure of the velocity of *plane*-polarized light in the various directions, and the variation of the velocity, represented by the radius vector of the surface, was due to the varying angle that the vibrations made with the optic axis. In quartz or any optically active crystal, each of the two surfaces represents the velocity of various kinds of polarized light, depending on the direction of propagation. For a direction parallel to the axis, the velocity for the outer

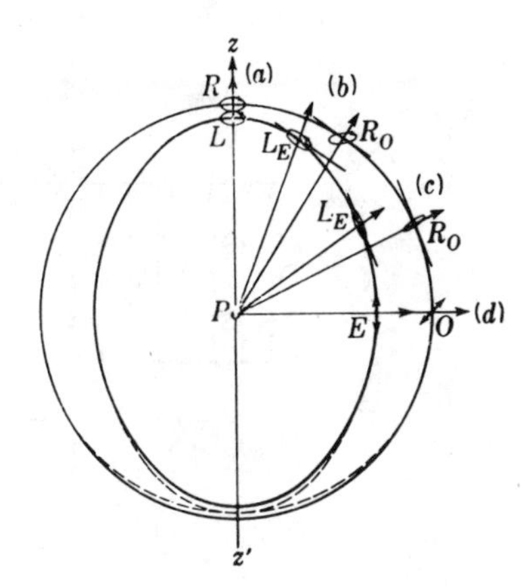

FIGURE 28G
Wave-surface diagram for a right-handed quartz crystal, showing the two vibrations associated with different directions of the wave normal.

surface is that of right circularly polarized light (right-handed quartz), and that for the inner surface is of left circularly polarized light. For directions at an angle with this the velocity is that of two elliptically polarized components. The major axes of the two ellipses are perpendicular to each other, and the ellipses become narrower with increasing angle from the axis, degenerating into lines (plane-polarized light) for a direction at right angles to the axis.

The behavior of plane-polarized light when it enters a crystal traveling either parallel or perpendicular to the optic axis, as in parts (*a*) and (*b*) of Fig. 28H, is easily understood from the above characteristics of the wave surface. In (*a*) the incident linear vibrations are resolved, upon entering the crystal, into two circular vibrations which travel with different velocities. The resultant of these two gives merely a plane vibration which rotates by an amount depending on the thickness of the crystal and the wavelength. In (*b*) the incident vibrations are again linear, but are here parallel to the optic axis so that the light is transmitted as an E beam with the velocity determined by the inner sheet of the wave surface. If the vibrations were perpendicular to the axis, they would travel with the greater velocity of the O beam. In either case the form and direction of vibration would remain unchanged. At other angles for the incident vibrations there would be two linear components moving with different velocities, and these give rise to elliptically polarized light. Hence for light traveling perpendicular to the optic axis quartz behaves precisely as do other uniaxial crystals and gives the interference effects described in the last chapter.

When the axis is not perpendicular to the ray, the effects of optical activity will manifest themselves to a greater or less extent, becoming greatest when the ray moves parallel to the axis. In Fig. 28H(*c*), where the incident vibrations lie in the principal section, they are decomposed upon entering the crystal into two ellipses L_E and R_O of different size. The major axes are at right angles, and the senses of rotation are opposite. In contrast to the case of nonactive crystals, an incident ray vibrating parallel to the principal section is not transmitted as a single E ray, but instead gives two rays of different intensity. We shall see in the following sections that except when the angle between the ray and the axis is very small, the intensity of the ray

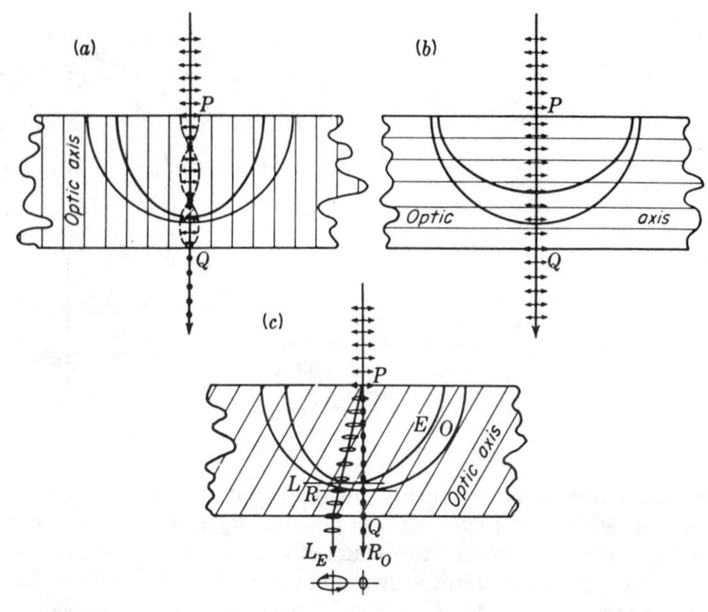

FIGURE 28H
Effects on plane-polarized light passing through quartz crystals cut in three different planes.

marked R_O is very low and L_E is a very slim ellipse. We shall also see that the O wave surface is not strictly spherical, so that R_O is slightly deviated even at normal incidence.

Some biaxial crystals are known to show optical activity. In general the phenomenon is accompanied by double refraction and is somewhat difficult to show. The wave surfaces for such crystals have the same general appearance as those given in Chap. 26 with the exception that the inner and outer surfaces do not quite touch at the ray axes, i.e., at the dimple in the outer surface.

28.5 SHAPE OF THE WAVE SURFACES IN QUARTZ

In order to explain the polarization effects observed when light is sent through quartz crystals, it is found that the usual spherical and ellipsoidal surfaces for nonactive crystals must be assumed to be distorted, ever so slightly, in the neighborhood of the optic axis. The outer surface is bulged and the inner surface flattened as shown in an exaggerated way at the bottom in Fig. 28G. The dotted lines represent a true circle and ellipse, while the solid lines give the actual wave surface. The exact shape of these two surfaces, however, is not so important optically as is the distance between them. Actually the change from circularly polarized light to almost plane-polarized light takes place within a very small angle of the optic axis, so that except for very small angles quartz acts essentially like an ordinary uniaxial crystal. This is due to the fact that the difference in velocity (or difference in the refractive indices) of the two circularly polarized rays R and L moving parallel to the optic axis is small compared to the difference in velocity of the O and E rays moving perpendicularly. This is best seen from the values given for red and violet light in Table 28B.

Along the optic axis the separation of the two surfaces as compared with the radius of a spherical surface is as 1:26,000 for red light and 1:14,000 for violet. Perpendicular to the axis the ratios are 1:170 and 1:160, respectively.

Since there are two velocities for circular vibrations along the optic axis, the angle of rotation of plane-polarized light can be calculated from the refractive indices. The phase difference δ between two waves a given distance apart is given by Eq. (27b) as

$$\delta = \frac{2\pi}{\lambda} d(n_L - n_R) \qquad (28b)$$

where d is the distance traveled in the medium λ the wavelength of the light, and $n_L - n_R$ the difference in refractive index. If the R circular motion gets δ rad ahead of the L, the plane of vibration will be rotated through $\delta/2$ rad [see Fig. 28F(d)].

For a quartz plate 1 mm thick, for example, we get, on substitution in Eq (28b),

$$\delta = \frac{2\pi}{0.000076 \text{ cm}} (0.1 \text{ cm})(0.00006) = 0.5 \text{ rad}$$

This gives a rotation for red light, $\lambda 7600$, of about 14° [see Fig. 28B(b)]. It should be pointed out, however, that accurate differences for $n_L - n_R$ are in practice calculated from the observed rotation.

28.6 FRESNEL'S MULTIPLE PRISM

The first experimental demonstration of double refraction into two circularly polarized rays was made by Fresnel. He reasoned that if two circular components travel with different velocities along the optic axis of quartz, they should, upon emerging obliquely from a crystal surface into the air, be refracted at different angles. Upon failing to observe the effect with a single quartz prism, Fresnel constructed a train of right- and left-handed prisms cut and mounted together, as shown in Fig. 28I. With this prism train two circularly polarized beams were observed, one rotating to the right and the other to the left.

The reason the two rays get farther and farther apart at each oblique surface

Table 28B REFRACTIVE INDICES FOR QUARTZ

Wavelength, Å	n_E	n_O	n_R	n_L
3968	1.56771	1.55815	1.55810	1.55821
7620	1.54811	1.53917	1.53914	1.53920

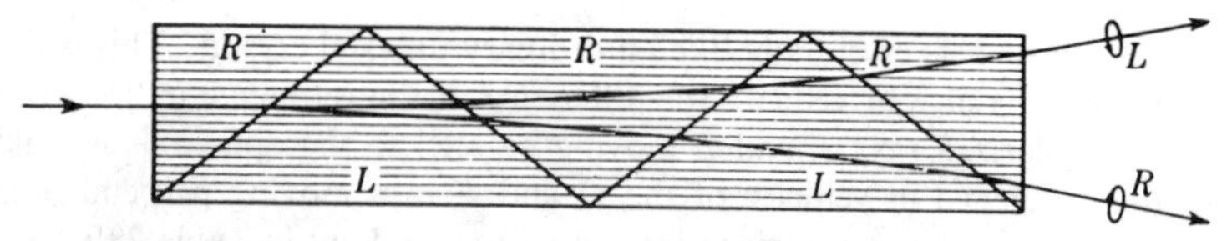

FIGURE 28I
Fresnel multiple prism for demonstrating the two circularly polarized components.

can be explained as follows. With light incident normally on the first crystal surface the two component circular vibrations travel along the optic axis with different speeds. Upon passing through the first oblique boundary, the R motion which was faster in the first prism becomes the slower in the second. The opposite is true for the L motion. By the ordinary law of refraction, then, the one ray is refracted away from the normal to the boundary and the other toward it. At the second boundary the velocities are again interchanged, so that the ray bent toward the normal at the first oblique boundary is now bent away from it. The net result is that the angular separation of the two rays increases at each successive refraction.

If such a prism is available to the student, Fresnel's observations can be repeated by placing it on the table of a small laboratory spectrometer. If the two images in the eyepiece are examined with a nicol prism or other analyzing device, they appear unchanged as the analyzer is rotated. If a quarter-wave plate is inserted in front of the nicol, both circular vibrations become plane-polarized perpendicular to each other. Now the images will disappear alternately with each 90° rotation of the nicol.

28.7 CORNU PRISM

The double refraction into circularly polarized light is appreciable even with a single quartz prism cut with the optic axis parallel to the base as in Fig. 28J(*a*). For sodium light and a 60° prism, the angular separation is only 27 seconds of arc, and hence is greatly exaggerated in the figure. When quartz prisms are to be used in spectrographs, even this small doubling of spectrum lines cannot be tolerated, particularly in the instruments of large dispersion. To overcome this effect, Cornu devised a 60° prism made up of right- and left-handed quartz as shown in Fig. 28J(*b*). Because of the interchange in velocities, light can be transmitted without double refraction if the prism is set at minimum deviation. Practically all 60° quartz prisms used in spectrographs are of this type.

In a Littrow spectrograph only one-half of a Cornu prism is used, and this takes the place of the grating shown in Fig. 17N. In this case the back face AB of, say, the R prism, Fig. 28J(*b*), is made a reflector by depositing silver or aluminum on the surface. By reflecting the light back the half prism is used a second time, giving the same dispersion as the Cornu prism. The R vibrations approaching the mirror become L vibrations after reflection, thus nullifying the double refraction.

Fused quartz prisms and lenses are sometimes used in optical devices, but not

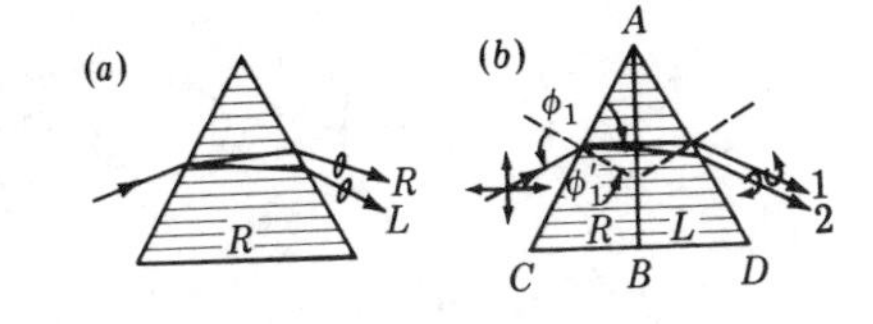

FIGURE 28J
(*a*) Simple crystal quartz prism. (*b*) Cornu prism.

where the best performance is desired. Although fused quartz is transparent and not doubly refracting, manufacturing processes have not yet produced large samples sufficiently free from inhomogeneities to make them useful for precision work.

28.8 VIBRATION FORMS AND INTENSITIES IN ACTIVE CRYSTALS

In Sec. 28.4 the propagation of light in various directions with respect to the optic axis in quartz was briefly described in terms of the wave surface for such a crystal. In right-handed quartz, for example, the outer sheet of the wave surface represents the velocity of a right-handed circular vibration traveling along the axis, of an elliptical vibration at some angle to it, and of a linear vibration in the equatorial plane. Looking against the light from the positions (*a*), (*b*), (*c*), and (*d*) of Fig. 28G, these vibrations would appear as shown in Fig. 28K. All vibrations are confined to planes tangent to the wave surface, with the major axis of each ellipse on the outer sheet perpendicular to the optic axis, and the minor axis of each ellipse on the inner surface also perpendicular to this axis. In left-handed quartz the directions of rotation would be interchanged, but the figures would otherwise remain unchanged.

As has been mentioned, the transition from circularly polarized to essentially linearly polarized light actually occurs within a few degrees of the optic axis.* For example, in quartz·the ratio between the axes of the elliptical vibrations (major to minor axes of each) is already 2.37 for sodium light traveling at 5° to the optic axis. At 10° the ratio has increased to 7.8. These are the ratios used in drawing Fig. 28K(*b*) and (*c*).

When a quartz plate cut perpendicular to the axis is placed in highly convergent light between analyzer and polarizer, so that the light traverses the crystal at various angles to the axis, the interference figures (see Fig. 27M) are much the same as those obtained with a nonactive crystal like calcite. The essential difference is that the center of the pattern, even with crossed analyzer and polarizer, is nearly always bright instead of dark. A rotation of the vibration plane has the result that some light gets through at the center of the otherwise dark brushes. The effect may be seen in both the photographs shown in part (*b*) of Fig. 27M.

The intensities of the two elliptically polarized beams derived from an incident

* Equations giving the difference of velocity as a function of angle are derived in P. Drude, "Theory of Optics," English edition, pp. 408–412, Longmans, Green & Co., Inc., New York, 1922; reprinted (paperback) by Dover Publications, Inc., New York, 1968.

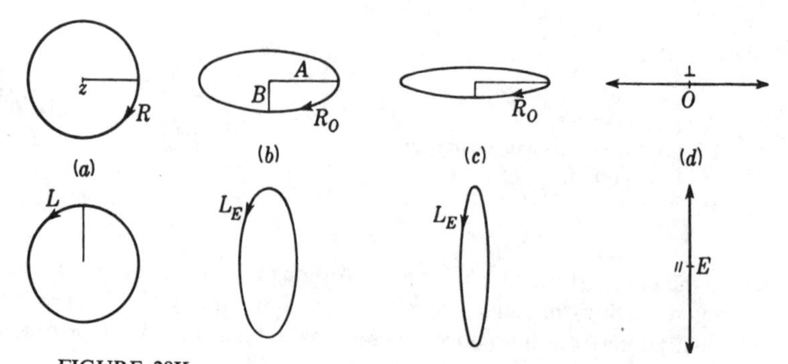

FIGURE 28K
Vibrations for light traveling through an optically active crystal at different angles with the optic axis.

unpolarized beam will always be equal. The two ellipses, such as those in part (*b*) of Fig. 28K, are alike except for their orientation. Remembering that an elliptical vibration can be regarded as made up of two linear ones at right angles and 90° out of phase, we find the corresponding intensity in terms of the major and minor semi-axes A and B as

$$I \approx A^2 + B^2 \qquad \text{(28c)}$$

In the limiting case of circularly polarized light having the radius $B = A$, we therefore have

$$I \approx 2A^2 \qquad \text{(28d)}$$

and for linearly polarized light ($B = 0$) the usual relation

$$I \approx A^2 \qquad \text{(28e)}$$

If each beam is to maintain the same intensity regardless of the eccentricity, the amplitude of the linear vibration must therefore be $\sqrt{2}$ times as large as the radius of the corresponding circular one.

If the incident light is *plane-polarized*, as in the example of Fig. 28H(*c*), the two ellipses are of different size. Now, in order that they may represent components of the original linear vibration, Fig. 28L shows that the minor axis of the large ellipse must equal the major axis of the smaller one. Namely, it is necessary that $B_E - A_O = 0$ for the horizontal components to cancel. Furthermore, since for the vertical ones to add up to the initial linear vibration $A_E + B_O = A$, it follows that $A_E/B_E = A_O/B_O$, and the ellipses have the same shape. The ratio of the corresponding intensities will depend on the actual value of either ratio A/B, and will vary from unity in the direction of the optic axis to zero at right angles to it.

For unpolarized light, which is equivalent to two independent linear vibrations at right angles, each of these will yield two counterrotating ellipses of different size. When one combines the two left-handed ones to obtain another left-handed ellipse, and the two right-handed ones to obtain a right-handed ellipse, it is found that the resultant ellipses have the same size. These are the ones illustrated in Fig. 28K.

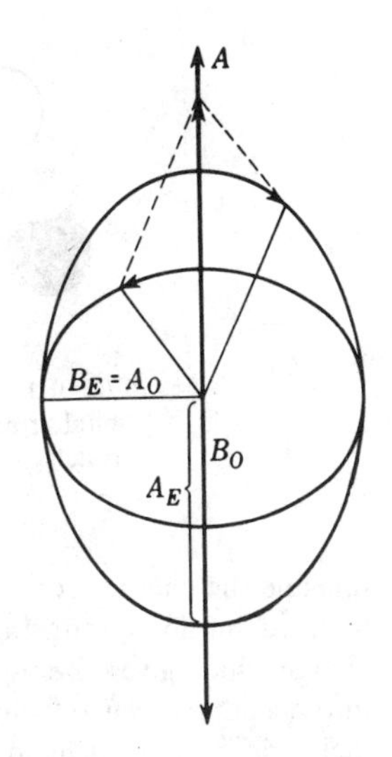

FIGURE 28L
Resolution of a linear harmonic vibration into two similar elliptical vibrations.

28.9 THEORY OF OPTICAL ACTIVITY

The theory of the rotation of plane-polarized light in optically active substances goes back to an early experiment performed by Reusch. He found that when plane-polarized light was incident normally on a pile of thin mica plates cut parallel to the axis and each plate turned through a small angle to the right of the one preceding it, the plane of vibration rotates to the right. The smaller the angle between successive plates, the more nearly the pile imitates rotation along the axis in quartz.

Reusch's experiment thus suggests that optically active crystals are made up of atomic layers which are twisted slightly one from the other. In right-handed crystals the layers are built up clockwise around the optic axis, and in left-handed crystals, counterclockwise. The known structure of crystalline quartz, which chemically is SiO_2, confirms this. Looking along the axis of a quartz-crystal model, one finds columns of silicon and oxygen atoms built up in spirals, as shown in Fig. 28M. These spirals of atoms form planes which give the effect of rotation along the optic axis. From the diagrams of right- and left-handed crystals in Fig. 28N this twisted formation is suggested by the arrangement of the smaller crystal faces. One crystal, in both its gross and atomic structures, is a mirror image of the other. The above analogy to a pile of plates must not be interpreted to mean that the plane of vibration rotates as fast as the atomic layers, as this would preclude any rotatory dispersion.

The electromagnetic theory of optical activity is due mainly to Born and his collaborators, and has been well summarized by Condon.* In an ordinary dielectric an imposed electric field produces a separation of charges and a resultant polarization of the medium in the direction of E (Sec. 23.9). In an optically active substance, we

* E. U. Condon, *Rev. Mod. Phys.*, **9**:432–457 (1937).

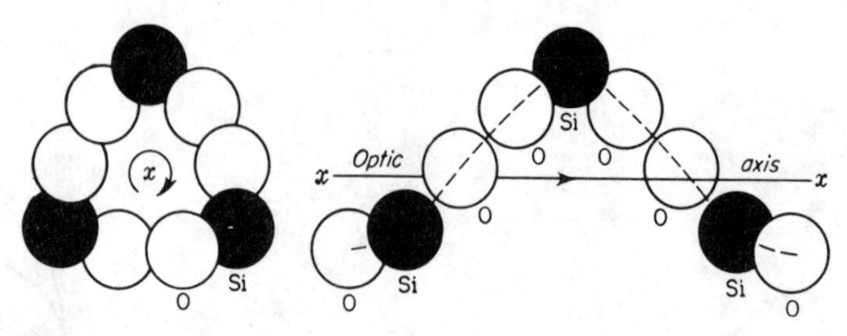

FIGURE 28M
Spiral arrangement of silicon and oxygen atoms along the optic axis in quartz crystals.

imagine the charges constrained to move in helical paths, so that in addition to the forward motion producing ordinary polarization there is a circulatory motion of charge which gives rise to magnetic effects. Drude has shown that this can be taken into account by introducing an additional term in one of Maxwell's equations for a dielectric [in the left-hand member of Eq. (23l)]. The solution of the equations then yields the phenomenon of optical activity. Born assumed that each molecule or crystal unit consists of a set of oscillators coupled together by electric forces. The simplest such unit, according to him, necessarily contains at least four oscillators arranged in a form that does not have symmetry. A tetrahedron, as an example, has symmetry properties, so that any crystal built on this structure will not show optical activity. If, however, the tetrahedron is twisted slightly out of shape, optical activity is a natural result. Born's early theoretical treatments have been applied to quartz by Hylleraas* and found to give excellent agreement with observations. It has since been shown by Condon and others that the assumption of coupled oscillators is not essential, and that the desired results can be obtained with a single-oscillator model.

28.10 ROTATION IN LIQUIDS

The rotation of the plane of vibration by liquids was discovered quite accidentally by Biot in 1811. He found that turpentine behaved like quartz in producing a rotation proportional to the light path through the substance and very nearly proportional to the inverse square of the wavelength. In such cases the rotation is attributable to the molecular structure itself. In fact, most liquids exhibiting rotation are organic compounds involving complex molecules.

Each molecule of a liquid may be thought of as a small crystal with an optic axis along which plane-polarized light is rotated. Since in a liquid the molecules are oriented at random, the observed rotation is an average effect of all the molecules and therefore the same in any direction through the liquid. One might at first sight think that the random orientation of the molecules would cancel out the rotatory effect entirely. But each molecule has a screwlike arrangement of atoms, and a right-handed screw is always right-handed, no matter from which end it is viewed.

* E. A. Hylleraas, *Z. Phys.*, **44**:871 (1927).

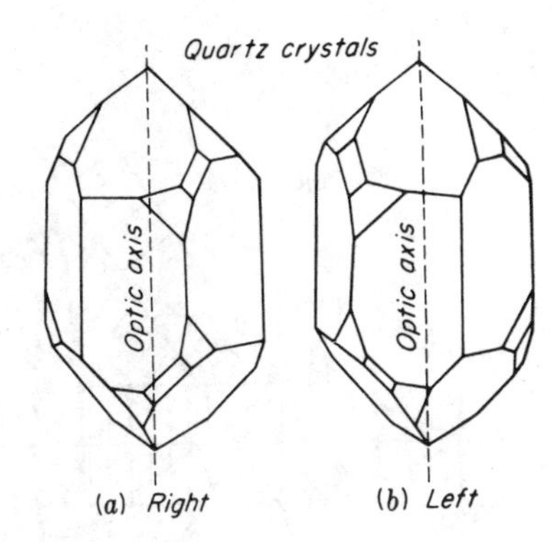

FIGURE 28N
Diagrams of crystal planes in right- and left-handed quartz crystals. Each is a mirror image of the other.

Liquids made up of an optically active substance and an inactive solvent are found to produce a rotation very nearly proportional to the amount of the active substance present. This has led to the very wide use of polarized light in industry as an accurate means of determining the amounts of sugar, an optically active substance, in the presence of nonactive impurities. The *specific rotation* or *rotatory power* is defined as the rotation produced by a 10-cm column of liquid containing 1 g of active substance for every cubic centimeter of solution. This may be written as an equation,

$$[\rho] = \frac{10\theta}{ld} \tag{28f}$$

where $[\rho]$ is the specific rotation, d the number of grams of active substance per cubic centimeter, l the length of the light path in centimeters, and θ the angle of rotation.

In general the rotation in liquids is considerably less than in crystals. For example, 10.0 cm of turpentine rotates sodium light $-37°$. (The minus sign indicates left-handed or counterclockwise rotation, looking against the direction of propagation.) An equal thickness of quartz, on the other hand, rotates sodium light 2172°. It is for this reason that the specific rotation for crystals is taken as the angle for a 1-mm path.

Careful determinations of the rotatory power of an optically active substance in various nonactive solvents have given slightly different results. There is a variation not only with solvents but with the concentration of the active substance. From experiment the rotatory power is found to be adequately given by

$$\rho = L + Md + Nd^2 \tag{28g}$$

where L, M, and N are constants and d is the amount of the active material in solution.

Like crystals, active substances in solution give rise to rotatory dispersion quite similar to that shown for quartz in Fig. 28B(*b*). Just as normal dispersion is a special case of anomalous dispersion observed near absorption bands in ordinary nonactive materials, so normal rotatory dispersion is a special case of anomalous rotatory dispersion known to exist at the absorption bands in optically active substances.

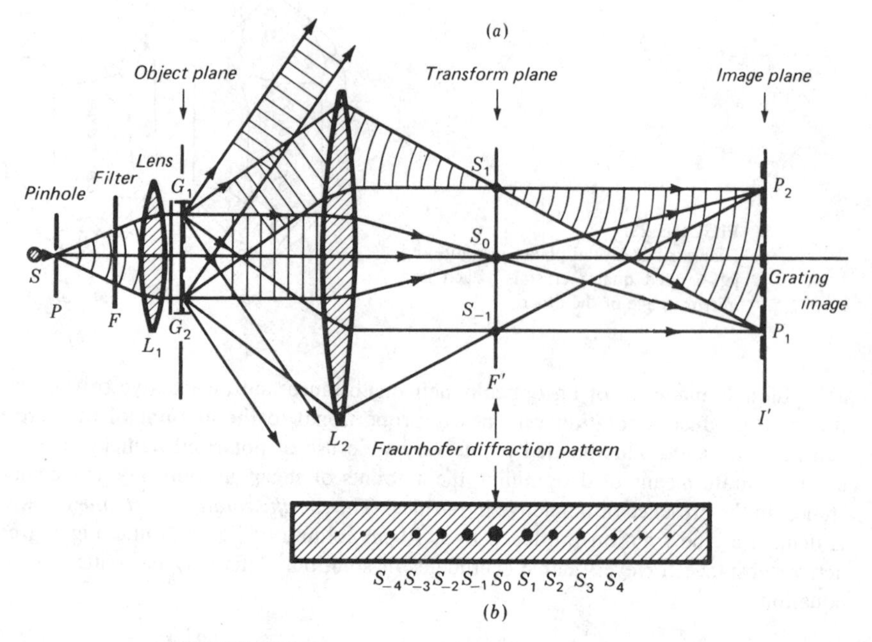

FIGURE 28O
Fraunhofer diffraction from a grating G_1G_2, showing the diffraction image $S_1S_0S_{-1}$ in the plane F' and the grating P_1P_2 in the image plane I'.

28.11 MODERN WAVE OPTICS

The major discoveries attributed to the wave properties of light, *diffraction*, *interference* and *polarization*, were all accounted for over 100 years ago. Up until the beginning of the twentieth century, nearly all optical phenomena had been studied by Fresnel, Fraunhofer, Huygens, Abbe, Airy, Foucault, Young, and a few others. The wave theory, due largely to Fresnel, accounts for all their observations in the minutest detail.

Over the years these fundamental principles have found many practical applications in the development of microscopes, binoculars, periscopes, telescopes, interferometers, etc. (see Chap. 10). Detailed studies of diffraction phenomena have led in recent years to the development of a large number of useful optical instruments. Although an explanation of their basic principles is quite complex, they are best described using the wave picture of light. A brief account of discoveries best explained by use of *quantum theory* and *quantum optics* will be presented in Chaps. 29 to 33.

Consider the diffraction grating experiment diagramed in Fig. 28O. Parallel monochromatic waves from a laser beam (see Chap. 30) or from a strong source S, a pinhole P, a filter F, and a lens L_1 are incident normally on an *object plane* as shown. Acting on these waves the diffraction grating G_1G_2 and lens L_2 produce a sharp, well-defined Fraunhofer diffraction pattern of spots in the *diffraction image plane.* This is the secondary focal plane of the lens L_2, and is sometimes called the *transform plane.* Parallel rays from all open grooves of the grating come to a focus there. Diverging rays from any one groove like G_1, however, come to a focus in the conjugate plane at I, where the real image of the diffraction grating itself is formed.

If the grating spacing in Fig. 28O is of the order of the wavelength of light, only the spot at or near the center images will be formed at F', since higher orders of interference will miss the lens L_2 and be lost. If the grating spacing is of the order of 10 or more wavelengths of light, the diffracted rays come together at points corresponding to different *orders of interference* [see diagram (*b*)]. These orders, given by

$$m = 0, \pm 1, \pm 2, \pm 3, \pm 4, \ldots \tag{28h}$$

correspond to increasing *spatial frequencies* (cycles or lines per centimeter) in the image (or object) plane.

In terms of Fourier components, $m = 0$ produces uniform illumination in the image plane; $m = \pm 1$ sinusoidally modulates this illumination at a fundamental spatial frequency called the *first harmonic*, and characterized by the separation between lines of the grating. $m = \pm 2$ corresponds to the *second harmonic* with twice the spatial frequency in the image plane; $m = \pm 3$ to the *third harmonic*, etc. The addition of each higher Fourier component serves to sharpen up the image (see Secs. 17.1 to 17.3), approaching the detail of the original object.

If we look upon the points $S_0, S_1, S_2, S_3, \ldots$ as though they were point sources of Huygens wavelets, their diffraction pattern P_1P_2 on the image plane is a real image of the diffraction grating G_1G_2. Looking at this in another way, the waves from the lens L_1 are diffracted by the grating and then diffracted again by the lens L_2, for if L_2 were not there, a Fresnel diffraction pattern of the grating would appear on the image plane and a Fraunhofer pattern would be developed at infinity.

These principles were first investigated by Abbe in connection with the theory of the microscope* (see Sec. 15.10). The lens L_2 represents the microscope objective, and the diffraction grating represents the slide specimen being illuminated by the substage lens L_1 and source S. The importance of this study by Abbe was his discovery that a microscope objective of large aperture provides higher resolution than a small one, since it collects higher orders of diffraction from small objects in the specimen. It was previously thought that since the light beam from the substage goes through the central part of the objective lens, the dark space outside the beam but still within the microscope tube is not used and a small objective should be adequate.

28.12 SPATIAL FILTERING

Let us now consider the setup of an optical system composed of two identical lenses spaced twice their focal length apart (see Fig. 28P). Since each lens has primary and secondary focal planes, this divides the system into five equally spaced regions, an

* H. Volkmann, Ernst Abbe and His Work, *Appl. Opt.*, **5**:1720 (1966).

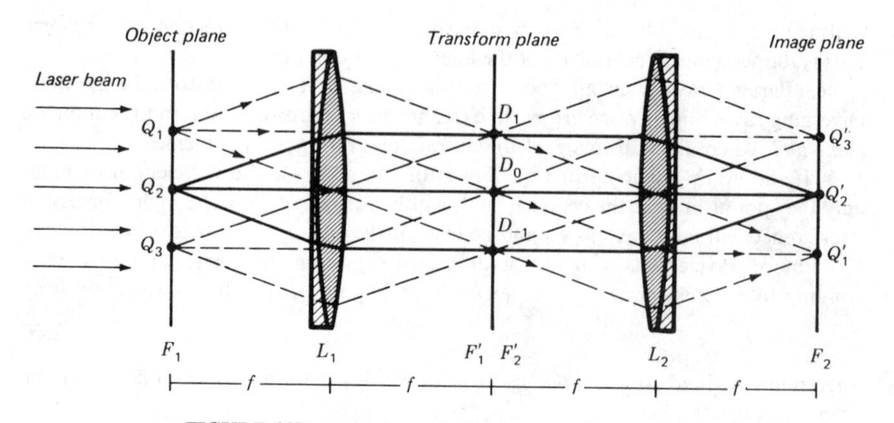

FIGURE 28P
Symmetrical well-corrected lenses form an optical system that permits spatial filtering. This arrangement is called an optical computer.

object plane F_1, a *lens* L_1, a *transform plane* F_1F_2, a *lens* L_2, and a *final image plane* F_2. Parallel rays from a laser are incident from the left.

Diverging bundles of rays from object points Q_1, Q_2, and Q_3 emerge as parallel bundles from L_1, arrive at L_2 as parallel bundles, and upon passing through L_2, converge to real image points Q'_1, Q'_2, and Q'_3, respectively. If Q_1, Q_2, and Q_3 are thought of as grooves of a diffraction grating (see Fig. 28O), parallel bundles of rays from the grating form a Fraunhofer diffraction pattern on the secondary focal plane F'_1 (see Fig. 17C).

Figure 28P is called an *optical computer*. The object is diffracted (scrambled) by the first half of the system and diffracted again (unscrambled) by the second half.* We are now ready to insert masks into the diffraction pattern of the tranform plane and block various features of the object, thereby preventing them from reaching the final image plane. This process is called *spatial filtering*.

As an illustration consider the laboratory demonstration shown in Fig. 28Q, using a laser beam or a point source and two high-quality lenses, each with a focal length of approximately 1 m. With a square wire screen or gauze as an object, the diffraction pattern in the transform plane will be a two-dimensional pattern of equally spaced spots, while the real image in the image plane will be that of the screen, inverted as shown (see Fig. 28R).

We now place a slit in the transform plane and turn it about the very center of the system until it transmits the vertical line of spots. The observer's eye sees the horizontal wires of the screen, with no hint of the vertical wires. By rotating the slit

* In terms of advanced mathematics, the diffraction pattern is the two-dimensional Fourier transform of a two-dimensional object, and the real image is the Fourier transform of the diffraction pattern. Neglecting scale factors, the Fourier transform of a Fourier transform is the original function. Fourier analysis is treated in Sec. 12.6.

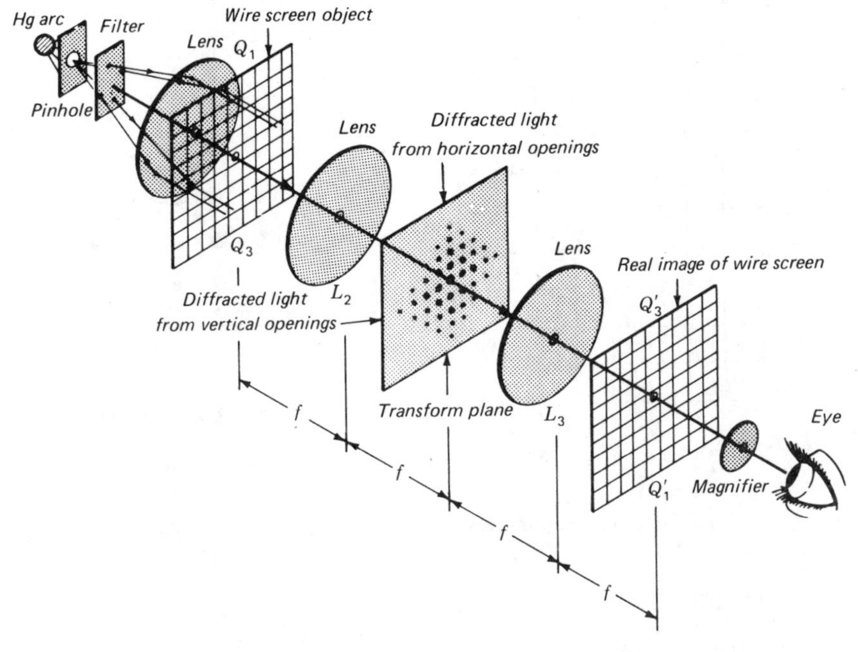

FIGURE 28Q
A laboratory experiment on spatial filtering. An optical computer with a wire screen as an object.

so that only the horizontal row of spots gets through, only the vertical wires are visible. Turning the slit to 45° and at other angles so that other rows of spots get through is a part of the experiment that must be observed to be appreciated.

If a mask with a small round hole in the center is placed over the transform plane, so that only the central spot gets through, the image screen shows only a uniformly illuminated field. If a number of masks are made with small holes in them, to pass certain sets of symmetrically located spots and stop others, some very interesting and informative effects can be observed in the image. For example, with the slit in a horizontal position across the center, the two spots $m = \pm 1$ are covered up, the pattern observed changes to a set of vertical wires with one-half the normal spacing. These experiments will illustrate the relationship between rows and sets of openings in the object, the Fourier components in the transform plane, and what is visible in the final image plane. Modern optical science makes use of these rather sophisticated techniques for intercepting parts of a diffraction pattern of an object to change the character of the image.

An excellent example of spatial filtering is shown in Fig. 28S. Here a photographic montage of the lunar surface consists of many horizontal film strips pieced

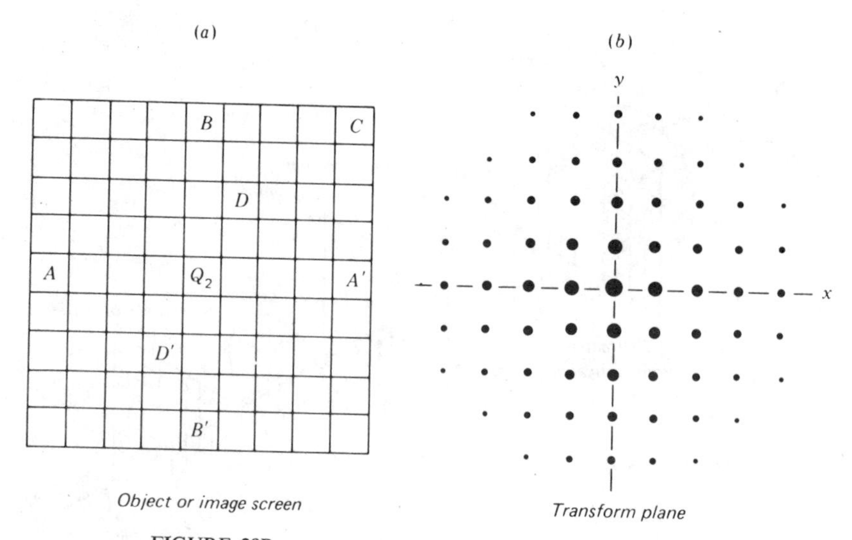

FIGURE 28R
Correlation diagram between the object or image screen and the transform plane for the demonstration shown in Fig. 28Q.

together. The strips were transmitted to the earth by Lunar Orbiter 1 as it circled the moon. This photo was placed in the object plane of Fig. 28P and a photographic plate at the transform plane. When the transform plate was exposed and printed, a photograph like the one reproduced in diagram (*b*) was obtained. The moon as a whole reproduced the mottled diffraction pattern, and the fairly regularly spaced lines between adjacent strips generated the inconspicuous vertical dot pattern.

Two narrow masks, shown in black in photograph (*d*), were then carefully mounted in the transform plane, to intercept and remove the spot pattern, thereby preventing all higher orders from reaching the final image on the photographic plate at F_2. Light rays passing by these masks reached all the points of the final image, thereby revealing a complete picture with little sign of the horizontal lines of the original montage.

The actual photographs in Fig. 28S were made at the Jet Propulsion Laboratory in Pasadena. The light source of the optical computer was composed of a 20× microscope objective and 10-μm pinhole used to filter out random spatial noise from a laser beam emitting light of wavelength 6328 Å (see Fig. 31S). An air-spaced doublet collimates the diverging beam into a 15-cm-diameter parallel beam with $\frac{1}{10}\lambda$ wavefront flatness. The transform lens L_1 in Fig. 28P and retransform lens L_2 are identical and located symmetrically about the transform plane. Aligned as a confocal pair, they image a 10 by 10 cm field from the object plane of L_1 to the image plane of L_2, with a resolution of 100 line pairs per millimeter. These two high-quality lenses are designed for 6328 Å light and to have a flat undistorted image plane. Each lens has five air-spaced elements in a 28-cm-diameter lens cell 63 cm long with a mass of 115 kg.

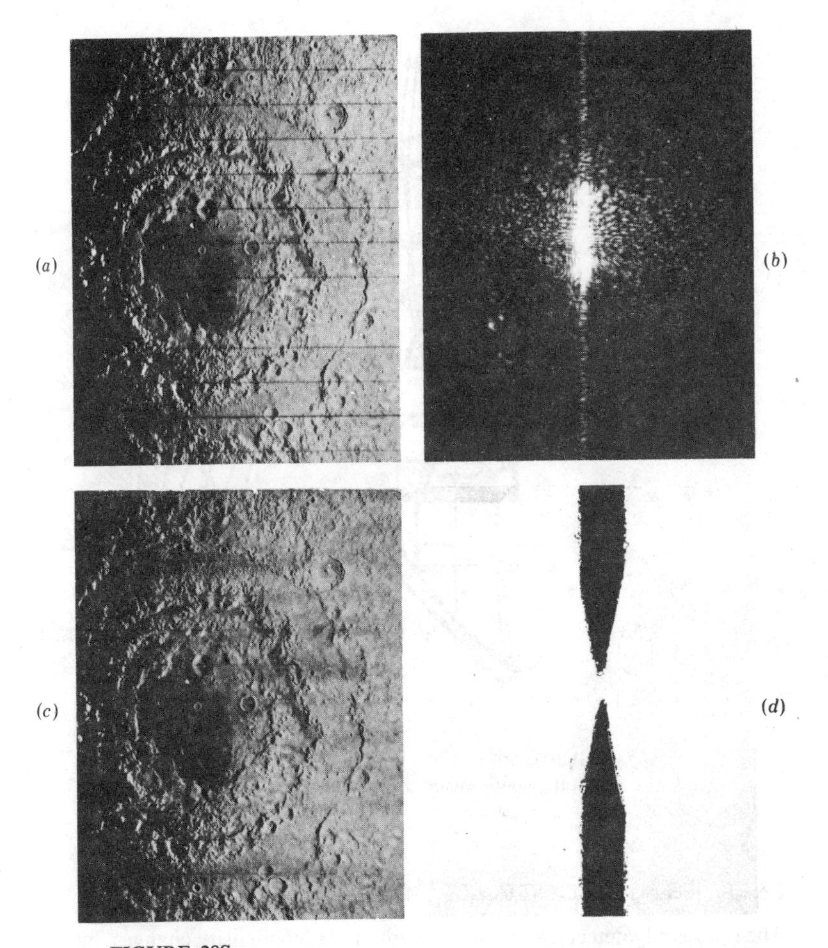

FIGURE 28S
(*a*) Lunar orbiter montage of the moon's surface. (*b*) Fraunhofer diffraction pattern of montage (*a*) made in the transform plane of the optical computer. (*d*) Shape of mask used to filter out the vertical dot pattern in (*b*). (*c*) Image-plane photo made with filter (*d*) in the transform plane, almost eliminating the horizontal lines of the montage. Note the concentric-ring pattern of mountains in (*a*) and (*d*), suggesting the impact of a giant meteor early in the moon's history. (*Courtesy of David Norris and Thomas Bicknell, Jet Propulsion Laboratory, California Institute of Technology.*)

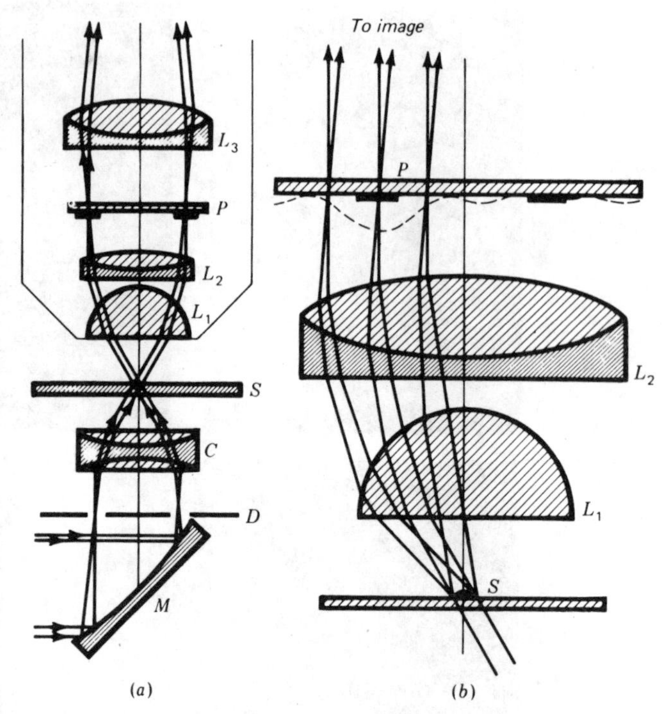

FIGURE 28T
Optical components of the phase-contrast microscope.

28.13 PHASE-CONTRAST MICROSCOPE

The eye readily detects differences in amplitude by intensity changes, but it is not able to see changes in phase directly. Thus, as long as the objects on a microscope slide are colored, opaque, or absorbing, they can be seen in the image. If they are transparent, however, and differ only slightly from their surroundings in refractive index or in thickness, they will ordinarily not be visible. It is nevertheless possible to convert the *phase changes* produced by such objects into *amplitude changes* in the final image. The so-called phase-contrast microscope, devised in 1935 by Zernike*, functions in this way.

Figure 28T shows how this is done. In part (*a*) are shown the two essential additions to an ordinary microscope: the *phase plate P* and an annular diaphragm *D*. The latter is placed in the front focal plane of the substage condenser *C*, and an image

* F. Zernike (1888–1966). Professor of Physics at the University of Groningen, Holland. In 1953 he was awarded the Nobel prize for his discovery of the phase-contrast principle. For more reading, see E. Hecht and A. Zajac, "Optics," pp. 474–478, Addison-Wesley Publishing Company, Inc., Reading, Mass., 1974.

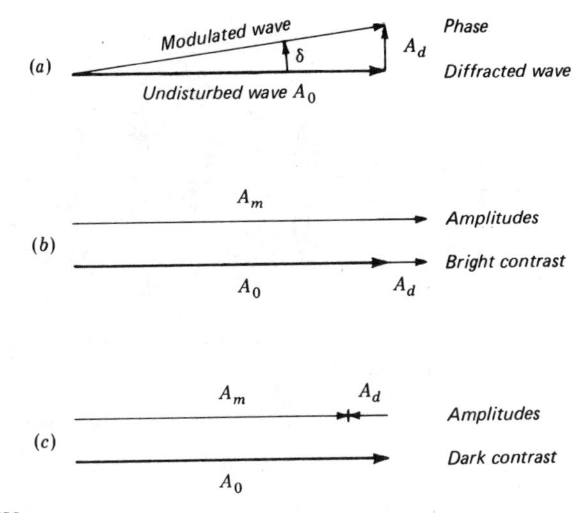

FIGURE 28U
Vector diagrams for the waves at the transform plane of the objective lens in a phase-contrast microscope: (*a*) Relative phases of the waves arriving at the phase plate; and amplitudes for waves leaving the phase plate, (*b*) for bright-contrast, and (*c*) dark-contrast illumination.

of the light source is focused upon D by the concave mirror M. The object on the slide S is therefore illuminated by a hollow cone of parallel light. If there were no diffraction by objects on the slide, this light would be focused again by the first two lenses of the objective O to form an image of D on the phase plate P.

This phase plate is seen to be in the transform plane of the object. Typically it consists of a glass plate upon which is evaporated an annular layer of transparent material to such a thickness that it increases the optical path by one-quarter of a wavelength of green light. The size of this retarding ring is such as to match the image of D.

Let us assume that a small transparent object on the slide retards the phase of the light passing through it by a small angle δ, relative to the phase of the undisturbed light transmitted by the unobstructed parts of the slide [see Fig. 28U(*a*)]. It can easily be shown that a small phase shift of this sort produces a *modulated wave*, which is given by the sum of the *undisturbed wave* and a new *diffracted wave* retarded in phase by approximately $\pi/2$. This *retarded wave* is typically characterized by a varying spatial structure and will therefore form a relatively broad and complex diffraction pattern at the transform plane P. For simplicity this is shown as a single-slit diffraction pattern in Fig. 28T(*b*). Most of the light in this diffracted wave will therefore miss the annular ring. The undisturbed wave is not diffracted and will only pass through the thicker annular layer, where it undergoes a phase retardation of $\pi/2$ with respect to the diffracted light. Therefore the phase plate brings the two into phase, with a resulting increase in the intensity at the corresponding point of the final

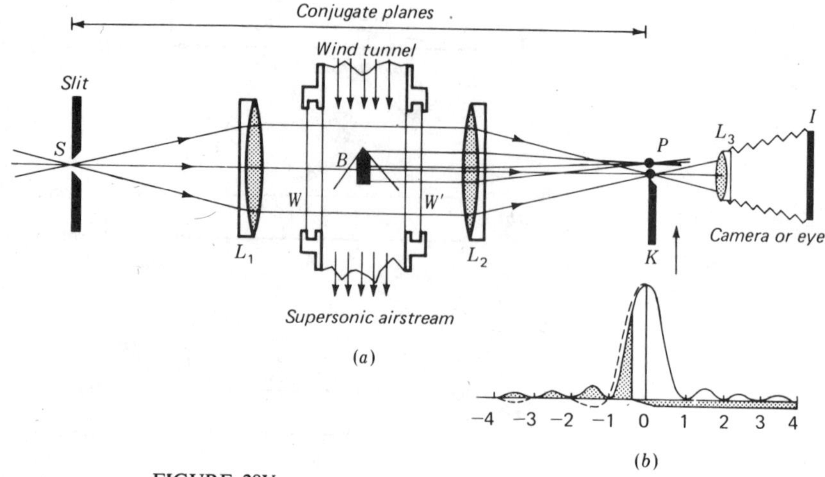

FIGURE 28V
Schlieren optics for ballistics and ultrasonic wind-tunnel studies: (*a*) symmetrical lens system; (*b*) Fraunhofer diffraction pattern for a single slit.

image [see Fig. 28U(*b*)]. The diffracting object is then rendered visible by what is called *negative* or *bright contrast.*

For *positive* or *dark contrast*, the annular phase plate is made thinner so that the direct light is advanced in phase with respect to the diffracted light. The interference at the image is then destructive and the object is dark [see Fig. 28U(*c*)]. For best results a metal film is usually deposited on the annular portion of the phase plate to make it absorbing, since otherwise the undisturbed light is too strong relative to the diffracted light and the destructive interference is not sufficiently complete.

It is thus apparent that by introducing phase changes in the transform plane, i.e., in the back focal plane of the objective, an object which influences the transmitted beam only through changing its optical path can be made visible, provided of course that such an object produces a diffraction pattern.

28.14 SCHLIEREN OPTICS

This is a method developed originally for observing the shock waves that develop around bullets in ballistics and airfoils in jet aircraft when the flight of these objects is at supersonic speeds.

Let us set up a symmetrical lens system for observing a single-slit diffraction pattern as shown in Fig. 28V. With a monochromatic light source in front of the slit Fraunhofer diffraction pattern of the slit is observed in the conjugate plane *P* (see ‹. 15D). Between the two identical lenses we now insert a *wind tunnel*, in the center

FIGURE 28W
A schlieren photograph of the supersonic shock waves around a shuttle configuration (*Courtesy of C. M. Jackson and Roy V. Harris, NASA, Langley Research Center, Hampton, Va.*)

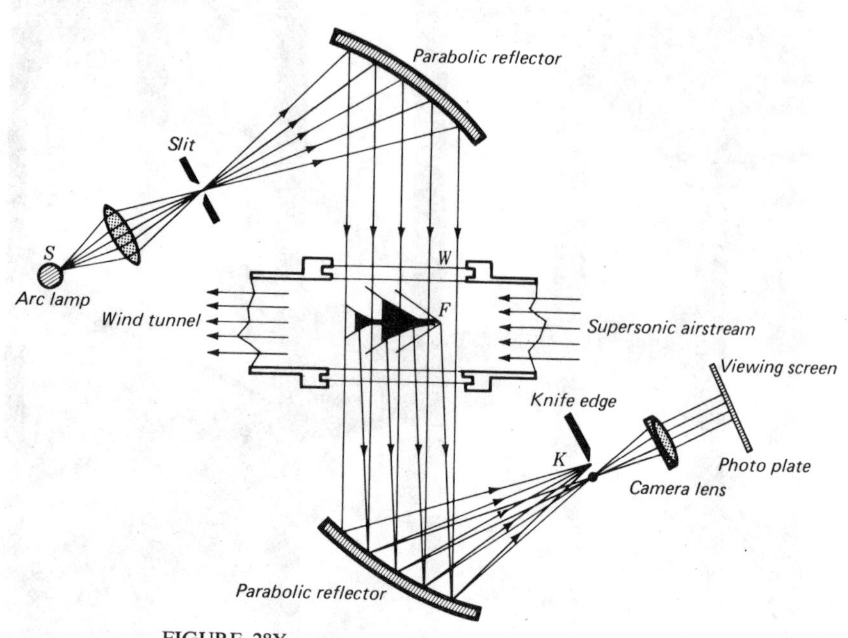

FIGURE 28X
Schlieren optics using concave parabolic mirrors.

of which is mounted a stationary diffracting object like a rifle bullet or the foil of a model jet airplane. As the supersonic airstream passes this object, shock waves develop around it, changing the refractive index of the air according to pressure differences in different areas. These changes of index give rise to diffraction patterns which are formed by L_2 on the plane P.

A knife-edge K parallel to the slit S is mounted in this diffraction image plane and is slowly raised by means of a micrometer screw. As the sharp edge crosses the central intense section of the diffraction pattern [see diagram (*b*)], the lower half of the pattern is blocked off from the camera or eye of the observer. Just before the extinction of the central maximum (the zero-order light) the field of view becomes relatively dark (sometimes called the dark-ground condition), and the shock waves become visible. Phase changes between higher orders of interference on the one side, shown dotted in diagram (*b*), produce constructive and destructive interference patterns (see Fig. 28W).

The lenses and wind-tunnel windows of the schlieren apparatus of Fig. 28V must be of the highest obtainable quality, for any imperfections in the glass surfaces or glass density will be clearly visible in the field of view. Although the lenses may be corrected for chromatic aberration, second-order effects are troublesome, and in recent years front-silvered mirrors have been used (see Fig. 28X).

A mirror-system schlieren apparatus uses large off-axis, precision-made parabolic mirrors, and the light as a parallel beam passes through the wind tunnel normal to the glass plates. These plates are polished flat to less than one wavelength of light and cause little trouble in the final image. Light sources with a broad spectrum permit color film to be used in the camera, with the result that a number of color schlieren systems have been made.

PROBLEMS

28.1 A quartz plate cut perpendicular to the optic axis is to be used to rotate the plane polarized light through an angle of 90°. If the light to be used is green light of wavelength 5461 Å, find its thickness. *Ans.* 3.524 mm

28.2 (*a*) Find the thickness of a quartz plate, cut perpendicular to the optic axis, that will rotate plane-polarized light of wavelength $\lambda = 5086$ Å through an angle of 720°. (*b*) Plot on a full sheet of graph paper the specific rotation for quartz for wavelengths in the range 4000 to 7000 Å [see Fig. 28B(*b*)]. (*c*) Using this graph, find what wavelengths will be missing if plane-polarized light is sent through this crystal and the light is examined with a spectroscope. Assume the polarizer is parallel to the analyzer.

28.3 Calculate the values of A and B in Cauchy's equation for rotary dispersion, using the values given in Table 28A, for $\lambda = 5086$ and 5893 Å.

28.4 Violet light of wavelength 3968 Å is refracted by a 60° quartz prism cut with the optic axis parallel with the base. Find the angle between the right- and left-handed circularly polarized rays refracted at or close to minimum deviation (see Tables 26A and 28A). *Ans.* 32 seconds of arc, or 0.0101°

28.5 A quartz rod 5.639 cm long is cut from a crystal with the ends polished perpendicular to the optic axis. The rod is then placed in a polariscope set with crossed polarizer and analyzer, and white light is sent through the system. The transmitted light is observed in a spectroscope. (*a*) Use a full sheet of graph paper ($8\frac{1}{2}$ in. × 11 in.) and plot a rotation curve for the wavelength range 4000 to 7000 Å. (*b*) What wavelengths as read from this graph will be missing in the spectograph? What is (*c*) the smallest and (*d*) the largest rotation involved in the missing wavelengths?

28.6 Plane-polarized light is incident normally on a plane-parallel plate of quartz cut with its optic axis at an angle of 5° with the normal as shown in Fig. 28H(*c*). (*a*) Using the axis ratio given in Sec. 28.8, make plots of the vibration forms in the two refracted beams L_O and R_E. (*b*) If the plate is of such a thickness as to produce a phase difference of 90° between these beams, find by graphical composition the resultant vibration form of the emergent light.

28.7 In measuring the rotation produced by sugar solutions, the accuracy obtainable by using the ordinary extinction point of an analyzer is not sufficient. Best results are obtained by matching the intensity of two adjacent fields produced by altering the polarizer so that it gives two beams linearly polarized at a small angle α with each other. Investigate the action of such a device by plotting the intensities of the two fields for a complete rotation of the analyzer. Take α 10°.

28.8 What should be the angle α in Prob. 7 to be able to measure the rotation to 1 minute of arc, supposing the eye can detect 2 percent difference of intensity in the two fields? *Ans.* $\alpha = 6.659°$ (see Fig. P28.8)

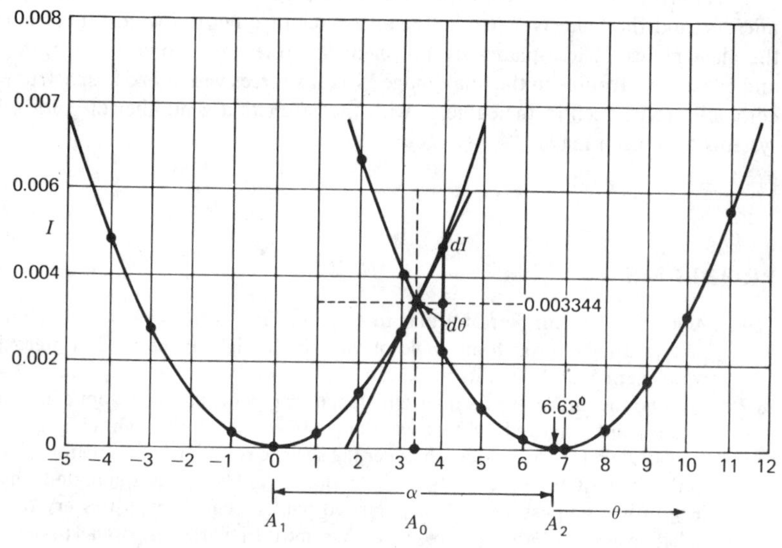

FIGURE P28-8
Detailed drawing for Prob. P28.8.

28.9 An unknown solution is suspected of containing *l*-glucose and does not contain any other optically active substance. If a 15-cm length of this solution rotates sodium light 25.6°, what is the concentration of *l*-glucose? $[\rho] = -51.4°$ for *l*-glucose.
Ans. 33.20 g/l

28.10 A weighed amount of 14.50 g of sucrose is dissolved in water to give 60 cm^3 of solution. When this is placed in a polarimeter tube 15 cm long, it rotates the plane of polarization of sodium light 16.8° to the right. Find the fraction of the sample that is not sucrose. For sucrose $[\rho] = 66.5°$.

28.11 A coarse transmission grating with 40 grooves per centimeter is located in the object plane of an optical computer. Both lenses have a focal length of 100.0 cm. If laser light with a wavelength of 6943 Å is used, find the spot separation in the transform plane.
Ans. 2.777 mm

28.12 A square mesh or gauze containing 30 wires per centimeter each way is located in the object plane of an optical computer. If the lenses have focal lengths of 90.0 cm and laser light of wavelength 6328 Å is used, find the spot separation in the transform plane.
Ans. 1.709 mm

28.13 A Cornu prism with a vertex angle of 60° is made of right- and left-handed quartz crystals. See Fig. 28J. Parallel light of wavelength 3968 Å is incident on the left-hand face so the upper refracted ray travels through the prism exactly along the axes. Assume the refractive indices are those given in Table 28B. Find (*a*) the angle of incidence of the unpolarized light at surface *AC*, and (b) the angle of refraction of the upper ray as it leaves the prism at surface *AD*. Find (*c*) the angle of refraction of the lower ray as it leaves surface *AC*, (*d*) the angle of incidence of the lower ray at surface *AB*, (*e*) the angle of refraction of the lower ray at surface *AB*, and (*f*) the angle of refraction of the lower ray at surface *AD*. Use a nine- or ten-place calculator.

PART THREE

Quantum Optics

29

LIGHT QUANTA AND THEIR ORIGIN

In Chap. 21, Sources of Light and Their Spectra, we observed that solids and gases, when heated to high temperatures, are the chief man-made sources of light. The plasma state of our sun and the distant stars, at high temperatures, are certainly the most prominent light sources in the universe. The fact that so many of the brightest stars emit the same spectra that we observe in our laboratories is direct evidence that light throughout the universe comes from the same chemical elements we find on the earth.

The origin of light from within gas molecules, liquids, and solids is similar in many respects to that within individual atoms. Although the processes are reasonably well understood, many of them are quite complex. We take time and space in this chapter to give only a brief account of present concepts of the origin of light from within atoms, and in the next chapter use these concepts to present the principal features of lasers.

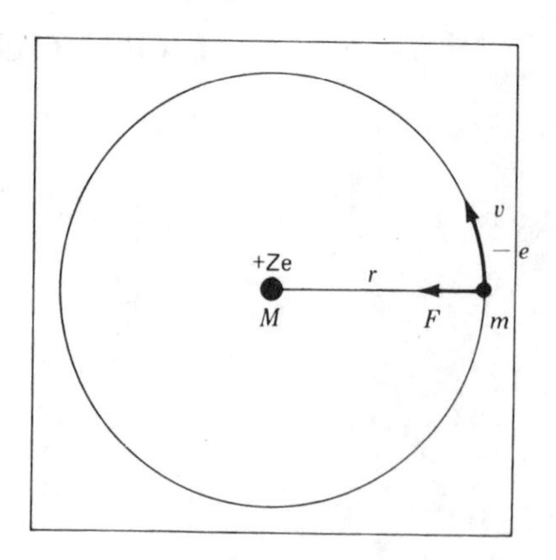

FIGURE 29A
Orbital diagram of the hydrogen atom according to the Bohr theory (1916).

29.1 THE BOHR ATOM

Historically the atomic and molecular structure of nearly all of the known chemical elements was established during the first three decades of the twentieth century. They became known through the establishment of the quantum theory and the various relationships found between the frequencies of light waves they emit (see Figs. 21H and 21J).

The Bohr model of the hydrogen atom is a logical starting point for any systematic presentation of atomic structure* because the energy relations developed in Bohr's theory are basic to an understanding of the quantum theory.

According to Bohr, the hydrogen atom is composed of a single electron of mass m and charge $-e$ rotating like a planet in a circular orbit around a positively charged nucleus of mass M and charge $+Ze$ (see Fig. 29A). Z, the *atomic number*, is equal to unity for hydrogen. According to classical laws of electrodynamics, the electron's motion is governed by the equation

$$m\frac{v^2}{r} = k\frac{Ze^2}{r^2} \tag{29a}$$

Centripetal force = *electrostatic attraction*

Bohr adopted this relation as his *first assumption* and then introduced the *quantum theory*. His *second assumption* says that the angular momentum of the electron mvr must always be equal to a whole number of units of $h/2\pi$

$$mvr = n\hbar \tag{29b}$$

* N. Bohr, *Phil. Mag.*, **26**:1 (1913); L. M. Rutherford, *Phil. Mag.*, **21**:669 (1911). For an elementary account of atomic structure and atomic spectra, see Harvey E. White, "Modern College Physics," 6th ed., D. Van Nostrand Company, New York, 1972. For a detailed account, see Harvey E. White, "Introduction to Atomic Spectra," McGraw-Hill Book Company, New York, 1934.

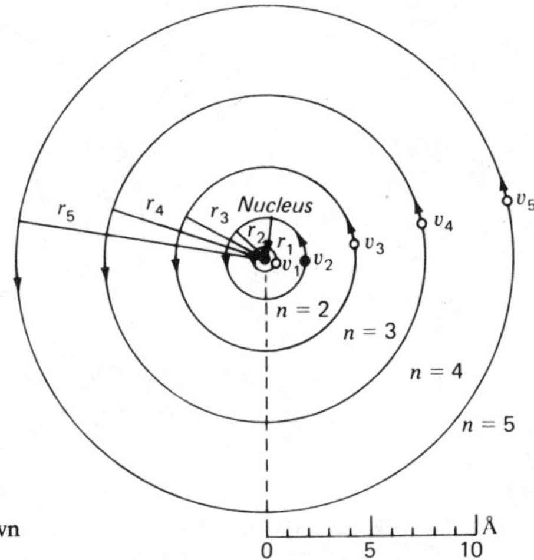

FIGURE 29B
Bohr circular orbits of hydrogen drawn to scale.

where m is the electron mass, $\hbar$ is Planck's constant of action (h divided by 2π), first introduced in 1905 by Max Planck to derive the law of thermal radiation, and n is a whole number called the *principal quantum number*;

$$m = 9.10956 \times 10^{-31} \text{ kg} \qquad e = -1.602192 \times 10^{-19} \text{ C}$$

$$h = 6.62620 \times 10^{-34} \text{ J s} \qquad \hbar = 1.054592 \times 10^{-34} \text{ J s}$$

$$k = 8.98755 \times 10^{9} \text{ N m}^2/\text{C}^2 \qquad Z = 1 = \text{atomic number of hydrogen}$$

$$n = 1, 2, 3, 4, \ldots$$

This means that the electron is not free to move in just any orbit, like a satellite in classical mechanics, but only in certain quantized orbits. By combining Eqs. (29a) and (29b) and solving for the orbit radius we obtain

$$r = n^2 \frac{\hbar^2}{me^2Zk} = n^2(0.529177 \times 10^{-10}) \qquad \text{m} \tag{29c}$$

and by solving for the orbital speed v we obtain

$$v = \frac{1}{n}\frac{e^2Zk}{\hbar} = \frac{1}{n}(2.18768 \times 10^6) \qquad \text{m/s} \tag{29d}$$

A diagram showing the relative sizes of the first five circular orbits is shown in Fig. 29B. Bohr's first success is to be attributed to the fact that with $n = 1$ or 2, Eq. (29c) gives a theoretical size that agrees with previously known values and that Eq. (29d) gives an orbital frequency approximately equal to that of visible light.

Bohr's *final assumption* regarding the hydrogen atom concerns the emission of light. Bohr postulated that light is not emitted by an electron when it is moving in

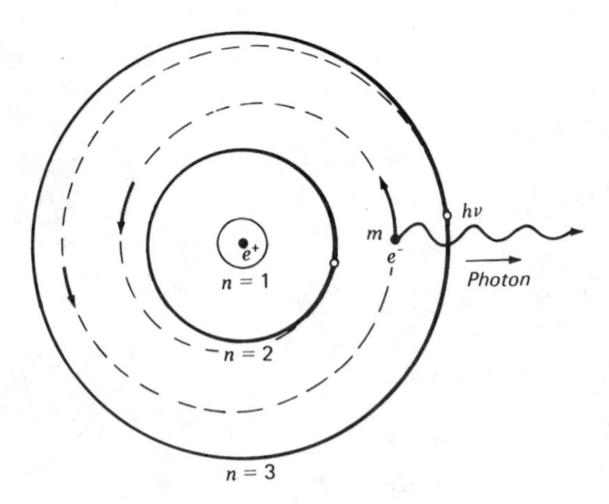

FIGURE 29C
Bohr's quantum theory of the radiation of light from a hydrogen atom.

one of its allowed orbits, as one might well expect classically of an accelerated electric charge, but only when the electron jumps from one orbit to another, as shown schematically in Fig. 29C. The frequency of the emitted light is not given by the orbit frequency of either the initial or final orbit but by a kind of average of the two, given by the simple relation

$$h\nu = E_i - E_f \tag{29e}$$

where E_i is the *total energy in the initial orbit*, E_f is the *total energy in the final orbit*, h is again Planck's constant, and ν is the frequency of the emitted light.

To illustrate, let $E_1, E_2, E_3, E_4, \ldots$ represent the total energy of the electron when it is in orbits $n_1, n_2, n_3, n_4, \ldots$, respectively. When the electron is in orbit $n = 3$, where its energy is E_3, and it jumps to orbit $n = 2$, where its energy is E_2, the energy difference $E_3 - E_2$ is ejected from the atom in the form of a light wave of energy $h\nu$, called a *photon*. This is the origin of light waves from within the atom (see Fig. 29C).

By combining the three equations (29a), (29b), and (29e) and inserting the known values of the atomic constants, Bohr derived the following equation for all of the frequencies of light emitted by free hydrogen atoms:

$$\nu = 3.28984 \times 10^{15} \left(\frac{1}{n_f^2} - \frac{1}{n_i^2} \right) \quad \text{Hz} \tag{29f}$$

where n_i and n_f are the principal quantum numbers of the *initial* and *final orbits*. If we introduce the wave equation

$$c = \nu\lambda \tag{29g}$$

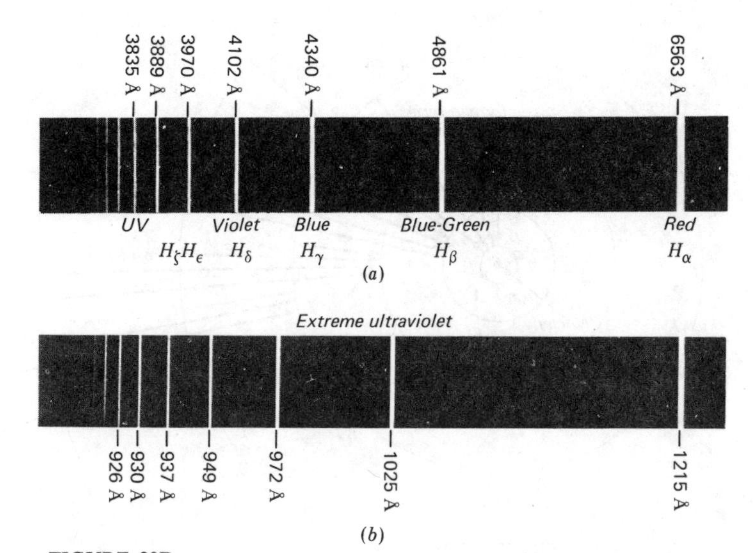

FIGURE 29D
The spectrum of the hydrogen atom: (*a*) the Balmer series and (*b*) the Lyman series.

and replace ν by c/λ, we obtain for the wavelengths of light*

$$\lambda = 911.503 \frac{n_i^2 \times n_f^2}{n_i^2 - n_f^2} \quad \text{Å} \tag{29h}$$

Bohr observed that if $n_f = 2$ and $n_i = 3, 4, 5, 6, \ldots$, this equation gives with great precision the wavelengths of all the Balmer series of hydrogen (see Fig. 29D).

By substituting $n_f = 1$ and $n_i = 2, 3, 4, 5, \ldots$ Bohr predicted a series of lines in the extreme ultraviolet region of the spectrum. These lines were first photographed by T. Lyman at Harvard University, and the wavelengths were found to check exactly with those calculated. This series, now called the *Lyman series*, which can be photographed only in a vacuum spectrograph, is reproduced in Fig. 29D. Observe that the Lyman series arises from electron jumps from any outer orbit directly to the innermost orbit, the *ground state*.

Other hydrogen series with the electron jumping to $n_f = 3, n_f = 4, n_f = 5, \ldots$, found still later in the infrared region of the hydrogen spectrum, were found exactly where they were predicted (see Fig. 29E).

* Due to the relativistic increase in the electrons mass with velocity and the rotation of the electron and proton around their common center of mass, the value 911.267 obtained for Eq. (29h) has been multiplied by a small correction factor 1.000259 to obtain 911.503.

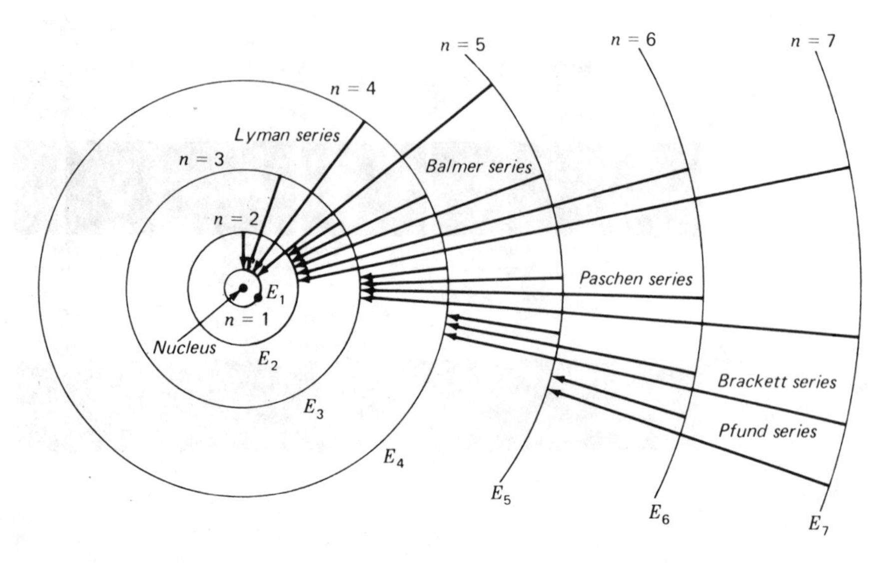

FIGURE 29E
Bohr circular orbits of hydrogen showing the transitions giving rise to the emitted light waves, or photons, of different frequency.

29.2 ENERGY LEVELS

The total energy E_{tot} of the electron in each of the Bohr orbits can be calculated from Bohr's first two postulates, Eqs. (29a) and (29b). Classically the potential energy E_{pot} is electrical in concept and is given by

$$E_{pot} = -k \frac{Ze^2}{r}$$

The kinetic energy, on the other hand, is mechanical, and is given by

$$E_{kin} = \tfrac{1}{2}mv^2 = k \frac{Ze^2}{2r}$$

Adding these two energies, we can eliminate r and v and obtain for the total energy

$$E_{tot} = -\frac{me^4Z^2k^2}{2n^2\hbar^2} \tag{29i}$$

The minus sign, as we might expect, signifies that one must do work on the electron to remove it from the atom. The electron is bound to the atom, and the closer it is to the nucleus the greater the energy that must be supplied to remove it from the atom.

With the exception of the principal quantum number n, all quantities in Eq. (29i) are fixed atomic constants for hydrogen, and we can write

$$E_{tot} = -R\frac{1}{n^2} \tag{29j}$$

where R has the value*

$$R = \frac{me^4Z^2k^2}{2\hbar^2} = 2.179350 \times 10^{-18} \text{ J} \tag{22k}$$

Equation (29j) is an important equation in atomic structure: it gives the energy of the hydrogen atom when it occupies any one of its allowed states. Instead of drawing orbits to scale as in Fig. 29E, it is customary to draw horizontal lines to an energy scale, as shown in Fig. 29F. This is called an *energy level diagram.* The various jumps between orbits can be represented by vertical arrows between the levels.

The importance of this kind of diagram is at least twofold: (1) regardless of the atomic model presented, whether it is an orbital model, a quantum-mechanical wave model, or any other yet to be proposed in the future, it represents with a high degree of precision the stationary energy states of hydrogen; and (2) it represents the well-established law of conservation of energy as applied through Bohr's third postulate, Eq. (29e), that the energy of each radiated photon $h\nu$ is given by the energy difference between two energy levels.

The first line of the Balmer series, $\lambda = 6561$ Å, the red line in Fig. 29D(*a*), corresponds to the short arrow, $n = 3$ to $n = 2$. The second line of the same series is the blue-green line, $\lambda = 4861$ Å, and corresponds to the slightly longer arrow, $n = 4$ to $n = 2$, etc.

● $$E_i - E_f = -R\left(\frac{1}{n_i^2} - \frac{1}{n_f^2}\right) \tag{29l}$$

29.3 BOHR-STONER SCHEME FOR BUILDING UP ATOMS

Bohr and Stoner proposed an extension of the orbital model of hydrogen to include all the chemical elements. As shown by the examples in Fig. 29G, each atom is composed of a positively charged nucleus with a number of electrons around it.

Although the nucleus is a relatively small particle less than 10^{-14} m in diameter, it contains nearly all the mass of the atom, a mass equal in *atomic mass units* to the *atomic weight.* The positive charge carried by the nucleus is equal numerically to the atomic number, and it determines the number of electrons located in orbitals outside.

A helium atom, atomic number $Z = 2$, contains two positive charges on the nucleus and two electrons outside. The lithium atom, atomic number $Z = 3$, contains three positive charges on the nucleus and three electrons outside. A mercury atom, atomic number 80, contains 80 positive charges on the nucleus and 80 electrons outside.

* For correction see footnote on page 615.

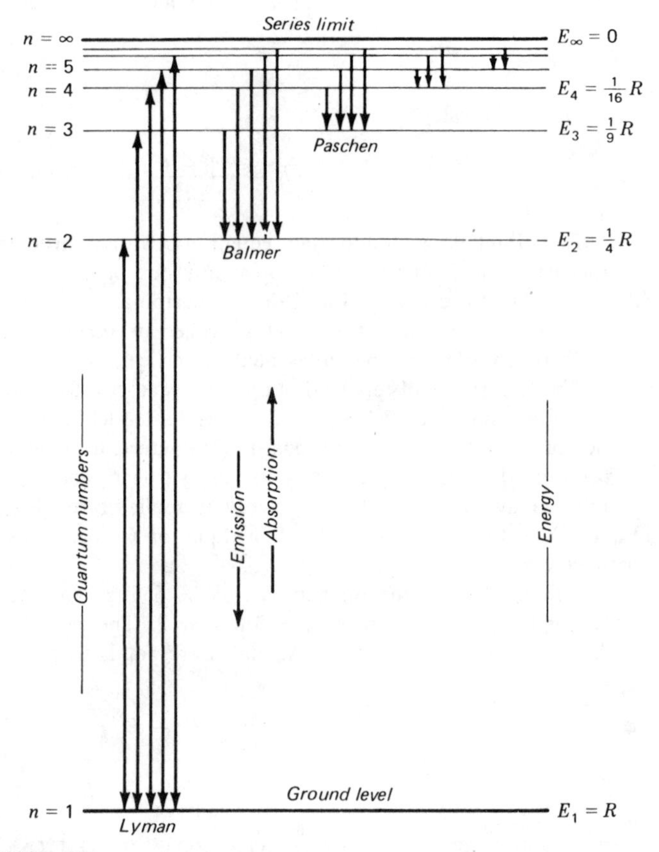

FIGURE 29F
Energy level diagram for the hydrogen atom. Vertical arrows represent electron transitions.

The orbits to which the electrons are confined are Bohr orbits of hydrogen with $n = 1, 2, 3, 4, \ldots$, called *electron shells*. As one goes from element to element in the atomic table, starting with hydrogen, electrons are added one after another, filling first one shell then another. A shell is filled only when it contains a number of electrons given by $2n^2$. To illustrate this, the first shell $n = 1$ is filled when it has two electrons, the second shell $n = 2$ when it has 8 electrons, the third shell $n = 3$ when it has 18 electrons, etc., $2 \times 1^2 = 2$, $2 \times 2^2 = 8$, $2 \times 3^2 = 18$, etc.

Quantum number n	1	2	3	4
Number of electrons	2	8	18	32

Among the heavier elements there are several departures from the order in which the shells are filled, e.g., the mercury atom. The four innermost shells $n = 1, 2, 3$, and

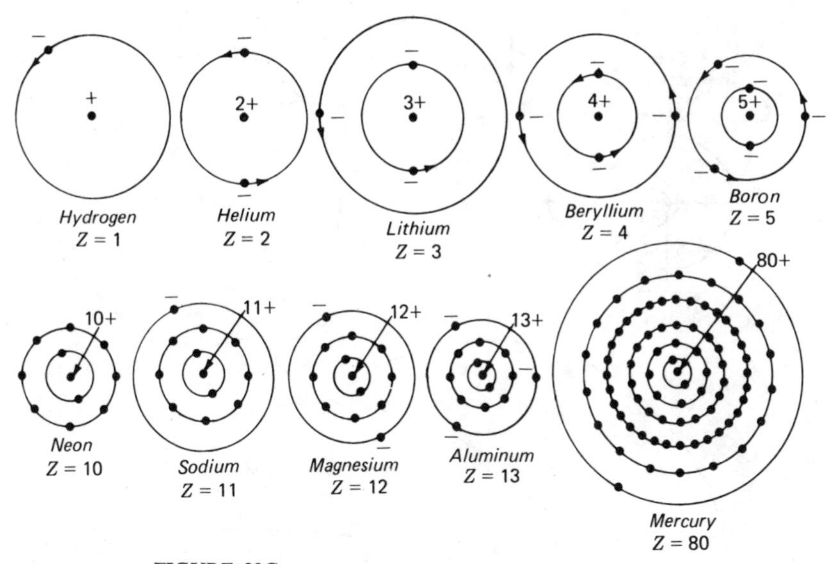

FIGURE 29G
Bohr-Stoner orbital models for the light and heavy atoms of the periodic table of elements.

4 are entirely filled with 2, 8, 18, and 32 electrons, respectively, while the fifth shell contains only 18 electrons and the sixth shell 2 electrons. The reasons for such departures are well understood and are now known to follow another rule.

It is important to note that as the nuclear charge increases and additional electrons are added in outer shells, the inner shells, under the stronger attraction by the nucleus, shrink in size. The net result of this shrinkage is that the heaviest elements in the periodic table are not much larger in diameter than the lighter elements. The schematic diagrams in Fig. 29G are drawn approximately to the same scale.

The experimental confirmation of these upper limits to the allowed number of electrons in each shell is now considered one of the most fundamental principles of nature. A sound theoretical explanation of this principle of atomic structure, first given by W. Pauli in 1925, is commonly referred to as the *Pauli exclusion principle*. For the order in which shells are filled throughout the periodic table see Appendix II.

29.4 ELLIPTICAL ORBITS, OR PENETRATING ORBITALS

Within only a few months after Bohr (in Denmark) published a report telling of his phenomenal success in explaining the hydrogen spectrum with quantized circular orbits, Sommerfeld* (in Germany) extended the theory to include quantized elliptical

* A. Sommerfeld, *Ann. Phys.*, **51**:1 (1916); W. Wilson, *Phil. Mag.*, **29**:795 (1915).

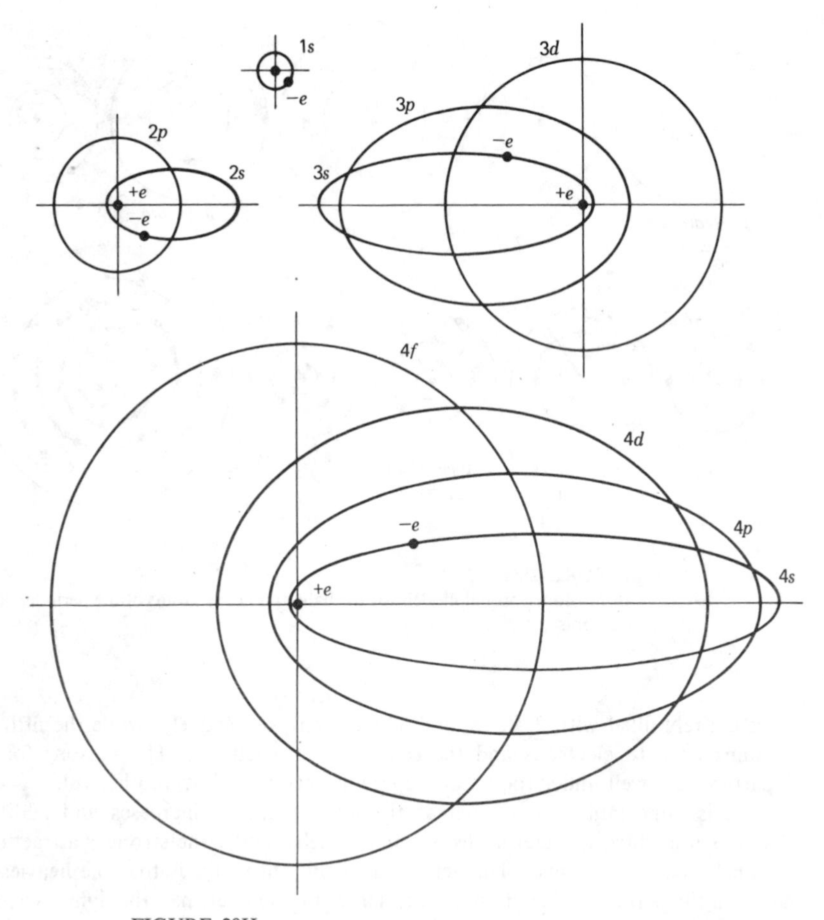

FIGURE 29H
A scale model for the hydrogen atom showing the innermost families of orbits and designations according to the Bohr-Sommerfeld theory.

orbits as well. Because these orbitals played such an important role in later developments in atomic structure, they deserve some attention here.

The net result of Sommerfeld's theory showed that the electron in any one of the allowed energy levels of a hydrogen atom may move in any one of a number of orbits. For each energy level $n = 1$, $n = 2$, $n = 3, \ldots$, as shown in Fig. 29E, there are n possible orbits (see Fig. 29H). For $n = 4$, for example, there are four orbits with designations $l = 3$, $l = 2$, $l = 1$, and $l = 0$. The diameter of the circular orbit given by Bohr's theory is just equal to the major axis of the three elliptical orbits. The minor axes are one-fourth, two-fourths, and three-fourths the major axis.

It is common practice to assign letters l to the quantum numbers as follows:

$l = 0$	$l = 1$	$l = 2$	$l = 3$	$l = 4$	...
s	p	d	f	g	...

According to this system, the circular orbit with $n = 3$ and $l = 2$ is designated $3d$, while the elliptical orbit $n = 2$ and $l = 0$ is designated $2s$, etc. n is the *principal quantum number* and l is the *orbital quantum number*. All orbits having the same value of n have the same total energy, the energy given by Bohr's equation for circular orbits, Eq. (29i).

Each of the allowed orbits of the Bohr-Sommerfeld model of the hydrogen atom becomes a subshell into which electrons are added to build up the elements of the periodic table in the Bohr-Stoner scheme. These subshells are given in Table 29A.

The maximum number of electrons in any one subshell is given by the relation

$$2(2l + 1)$$

This is called the *Pauli exclusion principle*, each subshell being filled when it contains the following number of electrons:

	Subshell l				
	0	1	2	3	4
Designation	s	p	d	f	g
Number of electrons	2	6	10	14	18

A model of the argon atom, atomic number 18, is shown in Fig. 29I. There are 18 protons on the nucleus, and drawn to scale are 18 electrons around it in circular and elliptical orbits. There are 2 electrons each in the $1s$, $2s$, and $3s$ orbits, and 6 electrons each in the $2p$ and $3p$ orbits. Together all these electrons are represented by

$$1s^2\ 2s^2\ 2p^6\ 3s^2\ 3p^6$$

which is called the *complete electron configuration* of the atom.

If argon atoms are excited to emit light, e.g., in an electric discharge in a tube containing argon gas, one of the outer electrons, $3p$ or $3s$, is excited to outer virtual orbits. Upon returning to lower energy states, the atom will emit one or more photons.

Table 29A ELECTRON ORBITAL DESIGNATIONS

	Subshell l				
Shell n	0	1	2	3	4
1	$1s$				
2	$2s$	$2p$			
3	$3s$	$3p$	$3d$		
4	$4s$	$4p$	$4d$	$4f$	
5	$5s$	$5p$	$5d$	$5f$	$5g$

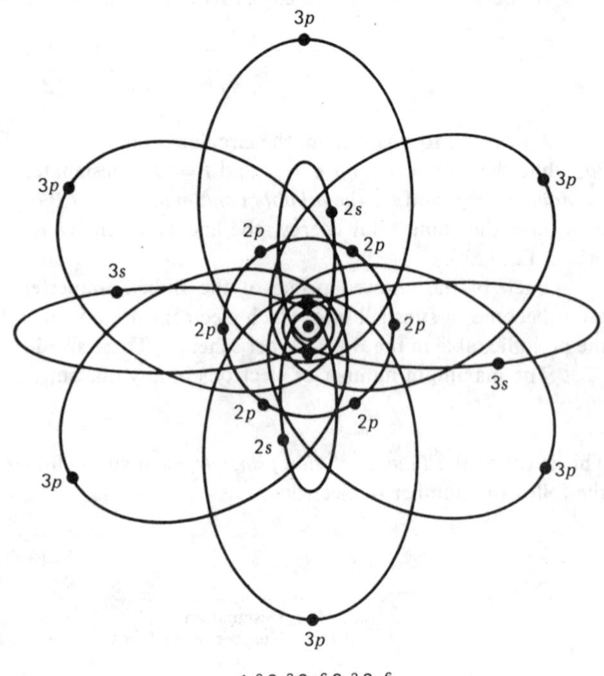

FIGURE 29I
Orbital diagram for an argon atom, Z = 18.

When such diagrams are drawn for atoms of higher atomic number, they become more and more tedious, and a scheme like that shown for cesium in Fig. 29J is frequently drawn. The electron configuration

$$1s^2\ 2s^2\ 2p^6\ 3s^2\ 3p^6\ 3d^{10}\ 4s^2\ 4p^6\ 4d^{10}\ 5s^2\ 5p^6\ 6s$$

shows 54 electrons filling *closed subshells*, with the fifty-fifth, or valence, electron alone in the 6*s* subshell. When cesium atoms are excited in an electric-discharge tube, it is this outer valence electron that jumps from orbit to orbit emitting photons. For the order in which subshells are filled see Appendix II.

29.5 WAVE MECHANICS

In 1924 the French physicist Louis de Broglie* derived an equation predicting that all moving particles have an associated wavelength. A beam of electrons, for example, should, under the proper experimental conditions, act like trains of light waves or a

* L. de Broglie, *Phil. Mag.*, **47**:446 (1924); *Ann. Phys.*, **3**:22 (1925).

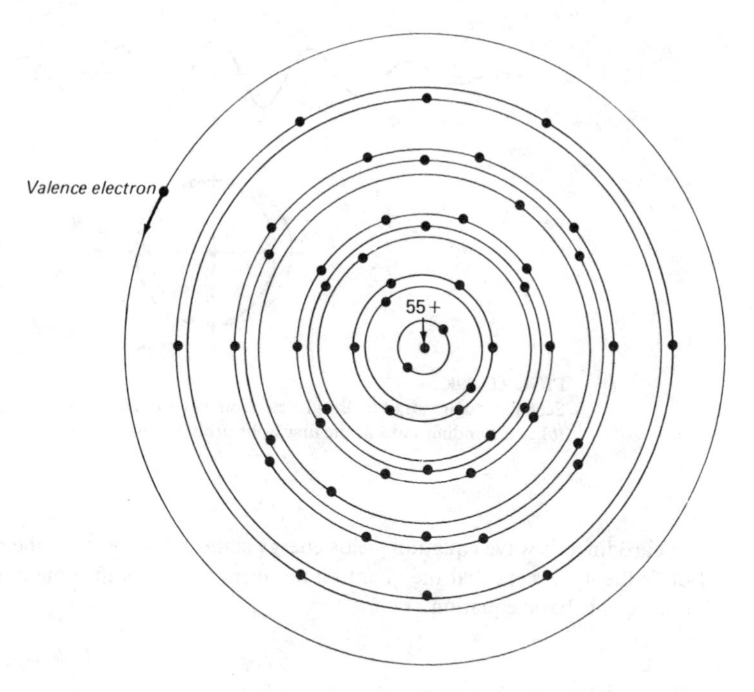

FIGURE 29J
Schematic diagram of electron shells and subshells in the atom of cesium 55.

beam of photons. The wavelength of these particle waves depends upon the mass and velocity of the particles according to the equation

$$\lambda = \frac{h}{mv} \qquad (29\text{m})$$

This is known as *de Broglie's wave equation* [see Fig. 29K(*a*)]. For an electron moving at high speed, as it does in the first Bohr circular orbit of hydrogen, the denominator *mv* is large and the wavelength is just equal to the circumference of the orbit [see Fig. 29K(*b*)].

With the development of matrix mechanics by Heisenberg in 1925 and wave mechanics by Schrödinger* in 1926, the orbital picture of the atom was replaced by one of de Broglie's waves. According to Schrödinger's formulation, the electron energy states in the atom of hydrogen can be described in terms of three-dimensional standing waves called *spherical harmonics.*

* E. Schrödinger, *Ann. Phys.*, **79**:361, 489, and 734 (1926); *Phys. Rev.*, **28**:1047 (1926).

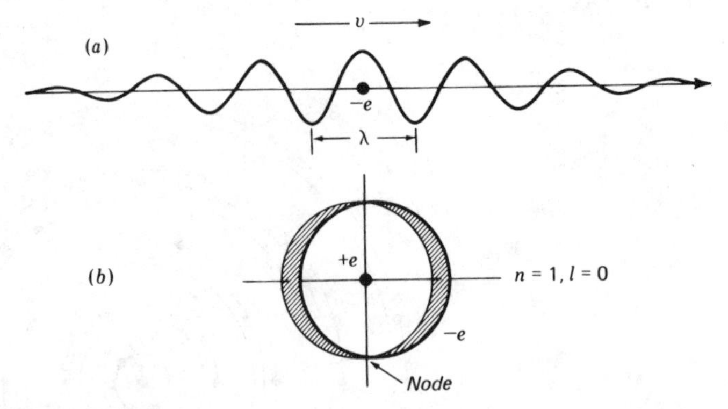

FIGURE 29K
Schematic diagram of de Broglie electron wave, moving in (*a*) a straight line and (*b*) as a standing wave in the first Bohr orbital of hydrogen.

Schrödinger's wave equation yields energy states having exactly the same values as Bohr's theory, except that the quantum numbers n and l both come out as natural solutions of his basic equation

$$\nabla^2\psi + \frac{2m}{\hbar^2}(W - V)\psi = 0 \qquad (29n)$$

where V is the potential energy, W is the total energy (kinetic energy and potential energy), and ψ is called the *wave function* of the electron. It may be thought of as the amplitude of the electron wave, and it is related to the probability density at any point within the atom. This is Schrödinger's wave equation.

Although solutions of this equation will not be given here, pictures representing six states of the hydrogen atom are shown in Fig. 29L for 1*s*, 2*s*, 2*p*, 2*p*, 3*d*, and 4*f* orbitals.* If these pictures were shown to the same scale of dimensions, their sizes should be enlarged as n^2, thereby comparing closely in size to their Bohr-orbit counterparts shown to the same scale in Fig. 29H.

In 1928 Dirac† included the spin of the electron in the Schrödinger wave equation and found similar probability density distributions for hydrogen, with appreciable differences in the angular distributions for the lower states with small n.

The radial distributions of charge density in the Bohr-Stoner scheme of atomic structures come out in such a way that closed shells and subshells form spherical symmetry about the nucleus, while valence electrons in incomplete subshells form an angular distribution similar to electron orbits. Because three-dimensional probability density figures are so hard to draw, it is common practice to represent electron states as orbital diagrams.

* See White, "Introduction to Atomic Spectra," chap. 4.
† P. A. M. Dirac, *Proc. R. Soc.*, **A117**:610 (1928); **A118**:351 (1928).

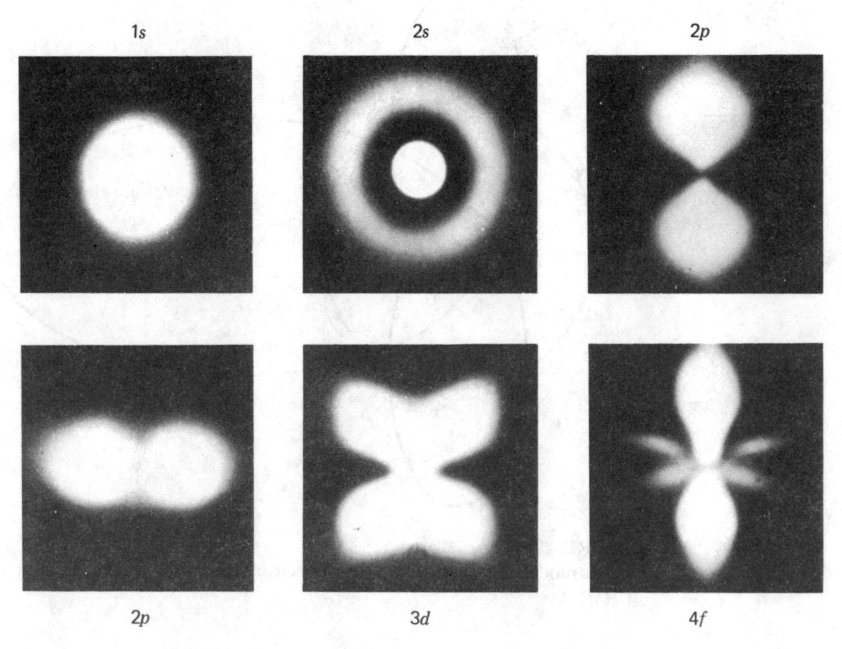

FIGURE 29L
Wave-mechanical pictures of six different states of the hydrogen atom (according to Schrödinger's equations). (*By permission from* H. E. White, *Phys. Rev.*, **37**: *1416* (*1931*).

29.6 THE SPECTRUM OF SODIUM

Except for elements in the first two columns of the periodic table, the spectra of all elements are quite complex [see Fig. 21H(*a*) and (*b*)]. Although their spectra have all been analyzed and converted into the atomic structures of their atoms, they were historically a long time in being thoroughly analyzed.

The spectra of the alkali metals Li, Na, Mg, Ca, Sr, Ba, and Ra are relatively simple compared with those elements near the center of the periodic table. As an example of other than hydrogen atoms we shall consider the structure of the sodium atom, its energy levels, and its observed spectrum. As the eleventh element in the periodic table, with a chemical valence of 1, each sodium atom contains 11 protons in the nucleus and 11 electrons in quantized orbits outside (see Fig. 29M). The 2 electrons in each of the 1*s* and 2*s* subshells, plus the 6 electrons in the 2*p* subshell, all form three closed subshells. As closed subshells, their total angular momenta are all zero; their spins all cancel in pairs, as do their orbital angular momenta.

As far as the electric field outside the core of 10 electrons is concerned, they screen or neutralize approximately 10 of the nuclear charges, and the eleventh, or valence, electron moves in a field that is almost hydrogenlike. It is not surprising, therefore, that the four known series of spectrum lines in sodium, produced by this

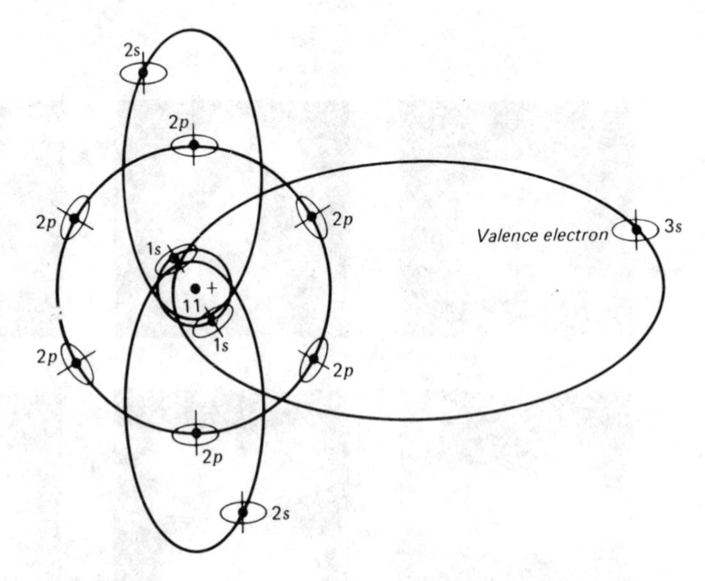

FIGURE 29M
Atomic model of sodium, $Z = 11$. All electrons spin with an angular momentum of $\frac{1}{2}(h/2\pi)$.

one electron jumping from orbit to orbit, are not far different in frequency and wavelength from hydrogen.

The sodium energy level diagram in Fig. 29N shows the normal state, or ground state, as 3^2S and the succeeding excited states as 3^2P, 4^2S, 3^2D, 4^2P, etc. These level designations correspond to the orbit designations 3*s*, 3*p*, 4*s*, 3*d*, 4*p*, etc. The superscript 2 indicates that all levels, with the exception of *S* states, are doublets. This doubling is due to the spinning of the electron and results in the doubling of all lines in all series.

Transitions from the two 3^2P levels to the ground state 3^2S give rise to the most prominent lines, the yellow D lines, of the principal series of sodium. These two particular lines account for the yellow color of all sodium lamps and are called the *resonance lines*. Other lines of this and other series are shown by arrows.

At relatively low temperatures nearly all sodium atoms are in their ground state. As the temperature is raised, more and faster collisions occur between atoms and more have their valence electron bumped into excited states, with the subsequent emission of light.

29.7 RESONANCE RADIATION

A good demonstration of resonance is shown with sound waves, using two tuning forks having exactly the same natural frequency, i.e., the same pitch. Fork *A* is set vibrating for a moment and then stopped. Fork *B* 10 m or more away will then be

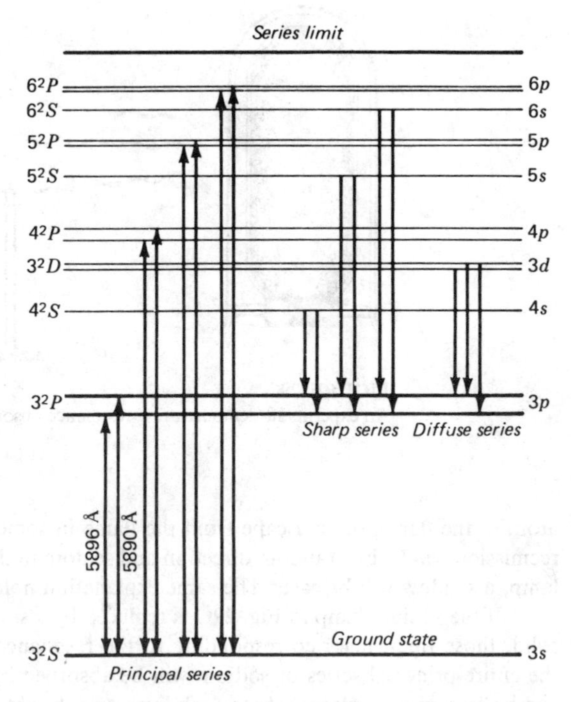

FIGURE 29N
An energy level diagram for the sodium atom, $Z = 11$, showing transitions for first members of the sharp, principal, and diffuse series.

found to be vibrating. Each sound pulse that emerges with each wave from fork A, pushing with just the right frequency on the prongs of fork B, sets it vibrating. If fork B is now stopped, fork A will again be found vibrating as the result of the waves from fork B. Such *resonance absorption* will fail if there is a frequency mismatch between the second fork and the passing waves.

An analogous demonstration of resonance absorption with visible light is shown in Fig. 29O. Light from a sodium lamp, in passing through a sodium flame of a bunsen burner, casts a pronounced dark shadow on a nearby screen. A small piece of asbestos paper soaked in common table salt (NaCl) and placed in an ordinary gas flame can be used to produce an abundance of free sodium atoms.

The atomic process of resonance absorption taking place in this experiment is shown in Fig. 29P. An excited atom in the sodium lamp emits a wave $\lambda = 5890$ Å by the downward transition from the upper of the two 3^2P excited levels to the 3^2S ground state. Coming close to a normal sodium atom in the flame, this wave will be absorbed and will raise the single valence electron to the corresponding 3^2P level. This second atom will in turn emit the same frequency again, to be absorbed by another

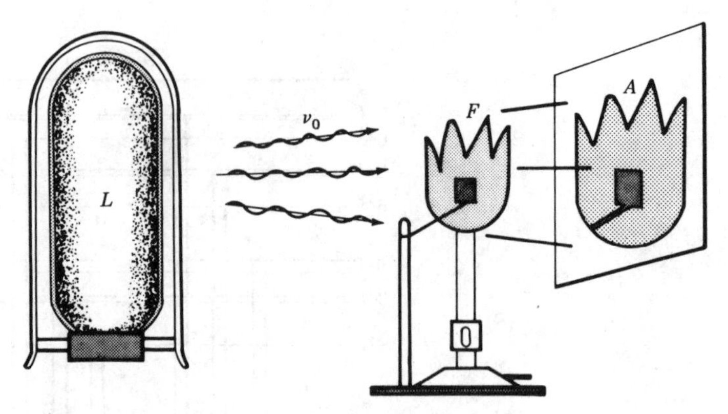

FIGURE 29O
An experiment demonstrating resonance absorption of sodium light.

atom in the flame, or to escape from the flame in some random direction. Because reemission will be in a random direction and seldom in the original direction from the lamp, a shadow will be cast. The same explanation holds for $\lambda = 5896$ Å.

If the sodium lamp in Fig. 29O is replaced by a source of white light from a hot solid, those frequencies corresponding to the resonance lines 5890 and 5896 Å and the entire principal series of sodium will be absorbed by the flame. The absorption can be seen in a spectrograph as dark lines on a bright continuous background [see Fig. 29H(*i*) and (*j*)]. Hence all arrows showing downward transitions to the ground state in Fig. 29N can also have arrowheads at their upper ends, indicating resonance absorption. All absorption lines originate on the ground state only.

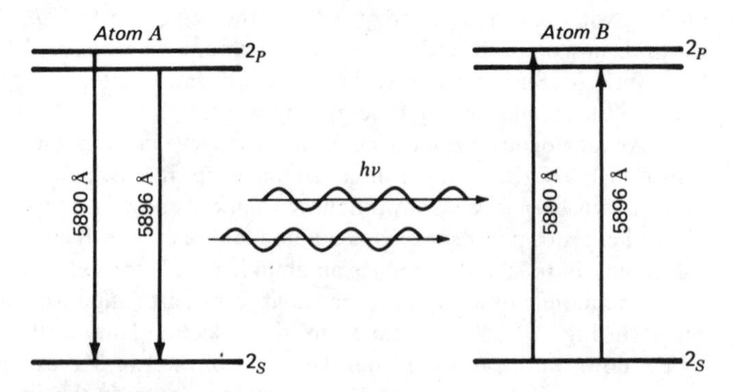

FIGURE 29P
Energy level diagram illustrating light emission and resonance absorption between two sodium atoms.

29.8 METASTABLE STATES

In gases like those found in a bunsen burner or in an electric-discharge tube emitting visible light most of the atoms have their valence electrons in the ground state. When by collision with another particle or atom the valence electron is raised to an excited state, it remains there for approximately 1.6×10^{-8} s before jumping down to a lower level with the emission of a photon.

Transitions to lower states by excited atoms are governed by well-established selection rules, such that not all transitions are allowed. For all atoms with one valence electron, the selection rules are quite simple:

$$\begin{aligned} \Delta n &= 0, \pm 1, \pm 2, \pm 3, \pm 4, \cdots \\ \Delta l &= \pm 1 \qquad \text{only} \end{aligned} \tag{29o}$$

For an extension of the selection rule to atoms with more than one valence electron, like the alkaline earths Be, Mg, Ca, Sr, Ba, and Ra, new sets of rules apply. With two electrons taking part in producing the various energy levels, transitions may occur as one electron jumps from orbit to orbit or two electrons may jump simultaneously, with the emission of a single radiated frequency. Selection rules for two electron systems in general may be written

$$\Delta l_1 = \pm 1 \qquad \text{and} \qquad \Delta l_2 = 0, \pm 2 \tag{29p}$$

If a single electron jumps, the l value of one changes by 1 and the other remains unchanged. If two electrons jump simultaneously, the l value of one changes by 1 and the other by 0 or 2. There are no restrictions on the total quantum number n of either electron. In calcium, for example, the two-electron transition $3d$ to $4p$ and $4s$ to $3d$ gives rise to three groups of lines, called *multiplets* which constitute some of the strongest lines in the visible spectrum.

An examination of the energy level diagram of sodium in Fig. 29N shows that certain transitions, like 3^2D to 3^2S, are forbidden. To arrive at the ground state from 3^2D, an electron cannot jump directly to 3^2S, as this would involve a Δl by 2. The electron can jump from 3^2D to 3^2P, emitting one photon, and then from 3^2P to 3^2S with the emission of a second photon of a different frequency. Both these transitions involve $\Delta l = -1$.

In some atoms it is not possible for an electron to get back to the ground state with the emission of light. Such is the case, for example, in ionized calcium, where the one valence electron in the atom accounts for the observed spectrum (see Fig. 29Q).*

Once an electron finds itself in the 3^2D state, selection rules do not allow it to return to the ground state, with the emission of a photon, and it remains there indefinitely. It can return to the ground state, however, if upon collision with another atom it transfers its excitation energy to the colliding atom. Such impacts are called *collisions of the second kind.* Both the existence of metastable states and the transfer of energy from one atom in a metastable state to another by collision are of importance in lasers.

* For energy level values for most of the elements in the periodic table see R. F. Bacher and S. Goudsmit, "Atomic Energy States," McGraw-Hill Book Company, New York, 1936; reprinted by Greenwood Press, Inc., Westport, Conn., 1969.

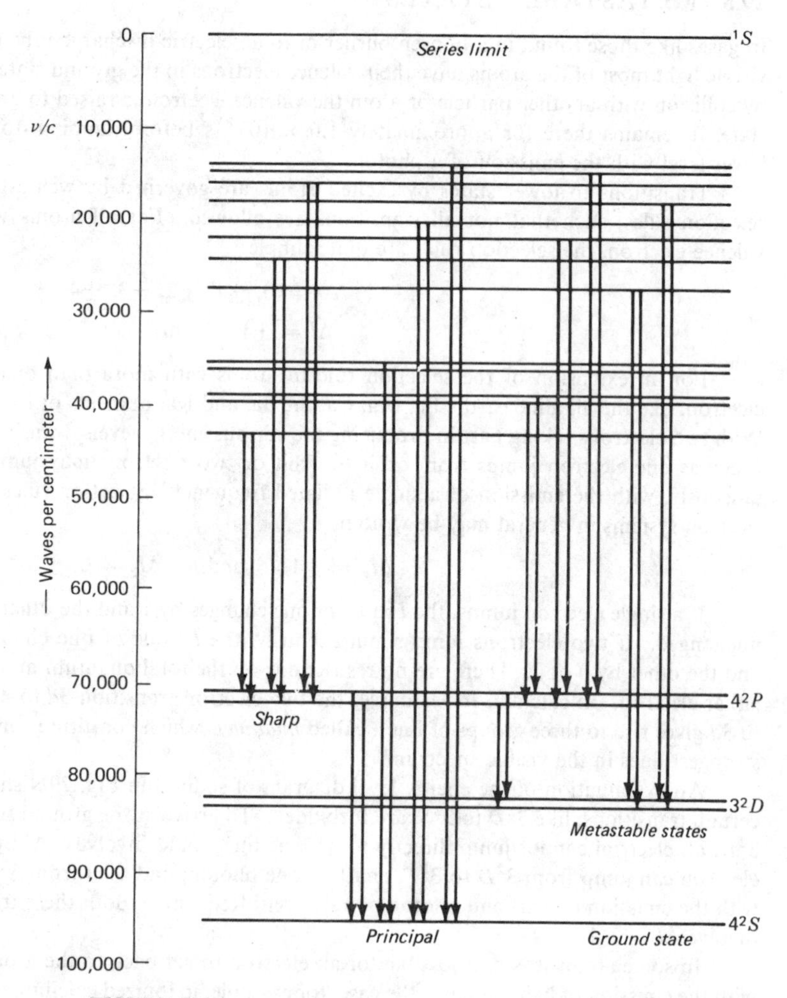

FIGURE 29Q
Energy level diagram of ionized calcium, showing the existence of metastable states.

29.9 OPTICAL PUMPING

Nearly all atoms in solids, liquids, or gases at near absolute zero are in their ground state. As the temperature is raised, by some form of energy input, more and more electrons are bumped into excited states. The populations of electrons in the higher energy levels increases at the expense of those that were in the ground level.

At 5000 K the populations in all states will have increased considerably, with the numbers in higher energy states at values less than those lying deeper. At any constant temperature a steady state will exist, and just as many electrons will be jumping into any level as will be jumping out of that level.

If a metastable state exists, the situation is different. As atoms are excited to higher levels, more and more of them get caught in the metastable level and relatively few of them get out except through mechanical collisions with other atoms. A steady state can exist, however, and just as many will be leaving per second as there are arriving there. The average populations of atoms in the metastable levels may be thousands and even millions of times that of any other, except the ground state. If they exceed the number in the ground state, it is called *population inversion.*

By shining light of a higher energy $h\nu$ than that required to excite an electron from the ground state to a metastable level, atoms can be *pumped* into that state by light absorption. The stronger the light source, the greater the number of electrons getting to the upper levels and then dropping back into the trap. This process is called *optical pumping.*

While the mean life of an electron in most excited states is of the order of 10^{-8} s, the *mean life* of a metastable level can be millions of times longer.

PROBLEMS

29.1 Calculate the orbital frequency of the electron in (*a*) the first, (*b*) the second, and (*c*) the third Bohr circular orbits. (*d*) To what wavelengths in angstroms would such frequencies belong?

Ans. (*a*) 6.760×10^{15} Hz, (*b*) 4.112×10^{14} Hz, (*c*) 8.123×10^{13} Hz, (*d*) 443.5, 7290, and 36,907 Å

29.2 Show that Eq. (29c) is derived from Eqs. (29a) and (29b).

29.3 Show that Eq. (29d) is derived from Eqs. (29a) and (29b).

29.4 Calculate the diameters of (*a*) the tenth, (*b*) the twenty-fifth, and (*c*) the hundredth circular orbits of the hydrogen atom according to Bohr's theory.

Ans. (*a*) 1.0584×10^{-8} m, (*b*) 6.615×10^{-8} m, (*c*) 1.0584×10^{-6} m

29.5 Calculate the wavelengths of (*a*) the fifth, (*b*) the tenth, and (*c*) the fiftieth lines of the Balmer series of hydrogen. (*d*) Find the wavelength of the series limit, i.e., as $n_i \rightarrow \infty$.

29.6 Calculate the wavelengths of (*a*) the first and (*b*) the fifth lines of the Paschen series of hydrogen (see Fig. 29E). (*c*) Find the series limit when $n_i = \infty$.

29.7 Calculate the wavelengths of (*a*) the fourth, (*b*) the tenth, and (*c*) the twentieth lines of the Lyman series of hydrogen. (*d*) Find the wavelength of the series limit, i.e., as $n_i \rightarrow \infty$. *Ans.* (*a*) 949.48 Å, (*b*) 919.10 Å, (*c*) 913.57 Å, (*d*) 911.50 Å

29.8 (*a*) Make a diagram of a zinc atom, atomic number 30, according to the Bohr-Stoner scheme, showing all subshells as circles. (*b*) Write down the complete electron configuration.

29.9 What would be the approximate quantum number n for an orbit of hydrogen to be 1.00 mm in diameter?

29.10 Starting with the first two equations in Sec. 29.2, derive Eq. (29i).

29.11 Show that the magnitude of the kinetic energy $\frac{1}{2}mv^2$ of a Bohr circular orbit is just one-half the magnitude of the potential energy.

30
LASERS

The name *laser* is an acronym of light amplification by stimulated emission of radiation. A laser is a device that produces an intense, concentrated, and highly parallel beam of coherent light. So parallel is the beam from a visible light laser 10 cm in diameter that at the moon's surface 384,000 km away the beam is no more than 5 km wide.

Historically the laser is the outgrowth of the *maser*, a similar device using radio microwaves instead of visible light waves. The first successful maser was built by C. H. Townes* and his associates at Columbia University between 1951 and 1954. During the next 7 years great strides were made in maser technology.

In 1958, A. H. Schawlow and C. H. Townes set forth the principles of the optical maser, or laser. The first successful laser based on these principles was built by T. H. Maiman of the Hughes Aircraft Company Laboratories in the summer of 1960. Extensive research on laser development has been carried on since that time.

* Charles H. Townes (1915–   ), born in Greenville, South Carolina. He received his Ph.D. degree from the California Institute of Technology in 1939, and is now Professor-at-large at the University of California. He is noted for his outstanding work in the development of masers and lasers, for which he received the Nobel prize in physics in 1964.

Because such devices have such widespread use in so many fields of research and development, a brief account of their basic principles will be presented here.

30.1 STIMULATED EMISSION

There are at least 10 basic principles involved in the operation of most lasers:

1 Metastable states
2 Optical pumping
3 Fluorescence
4 Population inversion
5 Resonance
6 Stimulated emission
7 Coherence
8 Polarization
9 Fabry-Perot interferometry
10 Cavity oscillation

While most of these concepts have long been known to science, the principle of coherence accompanying stimulated emission was the key to the realization of maser and laser operation.*

Consider a gas enclosed in a vessel containing free atoms having a number of energy levels, at least one of which is *metastable*. By shining white light into this gas many atoms can be raised, through resonance, from the *ground state* to *excited states*. As the electrons drop back, many of them will become trapped in the metastable state. If the pumping light is intense enough, we may obtain a *population inversion*, i.e., more electrons in the metastable state than in the ground state.

When an electron in one of these metastable states spontaneously jumps to the ground state, as it eventually will, it emits a photon of energy $h\nu$. This is called *fluorescent* or *phosphorescent radiation*. As the photon passes by another nearby atom in the same metastable state, it can, by the principle of *resonance*, immediately *stimulate* that atom to radiate a photon of the exact same frequency and return it to its ground state (see Fig. 30A). Amazingly enough this stimulated photon has exactly the same *frequency*, *direction*, and *polarization* as the primary photon (*spatial coherence*) and exactly the same *phase* and *speed* (*temporal coherence*).

Both of these photons may now be considered primary waves, and upon passing close to other atoms in their metastable states, they stimulate them to emission in the same direction with the same phase. However, transitions from the ground state to the excited state can also be stimulated, thereby absorbing the primary wave. An excess of stimulate emission therefore requires a population inversion, i.e., more atoms in the metastable state than the ground state. Thus if the conditions in the gas are right, a chain reaction can be developed, resulting in high-intensity coherent radiation.

* For a detailed treatment of lasers, see W. V. Smith and P. P. Sorokin, "The Laser," McGraw-Hill Book Company, New York, 1966, and E. Hecht and A. Zajac, "Optics," pp. 481–490, Addison-Wesley Publishing Company, Inc., Reading, Mass., 1974.

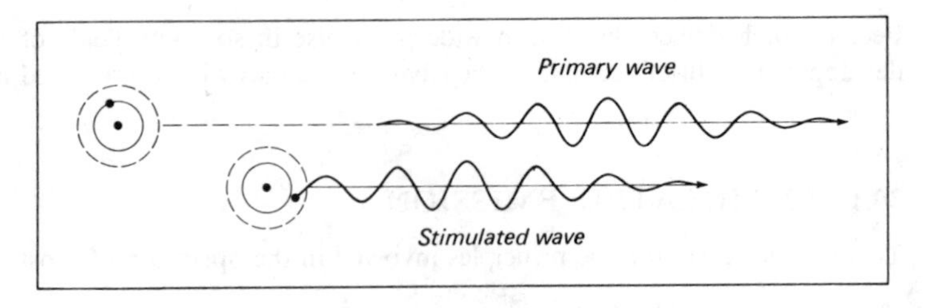

FIGURE 30A
The principle of stimulated emission of light from an atom. Both waves have the same wavelength λ and are in phase and vibrating in parallel planes.

30.2 LASER DESIGN

In order to produce a laser, one must collimate the stimulated emission, and this is done by properly designing a *cavity* in which the waves can be used over and over again. Here in optics the principles of the *Fabry-Perot interferometer* are applied (see Secs. 14.10 and 14.13). Suppose we retain the high reflecting power of the two end mirrors of the etalon and increase the distance between them. Into this cavity we then introduce an appropriate solid, liquid, or gas having metastable states in the atoms or molecules of its structure (see Fig. 30B).

By one means or another we now excite electrons in these atoms or molecules and produce a population inversion. If one or more atoms in a metastable state spontaneously radiate, those photons moving at an appreciable angle to the walls of the cavity, or tube, will escape and be lost. Those emitted parallel to the axis will

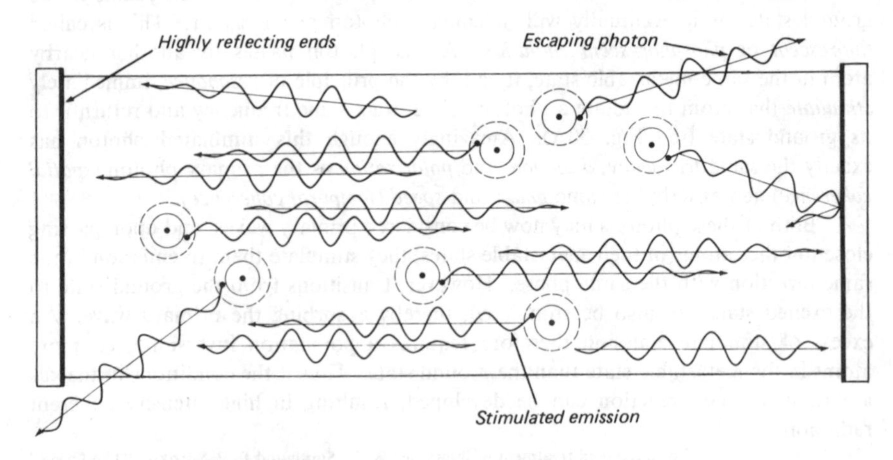

FIGURE 30B
A laser cavity with highly reflecting ends, showing the stimulated emission of light and the escape of some primary photons through the side walls.

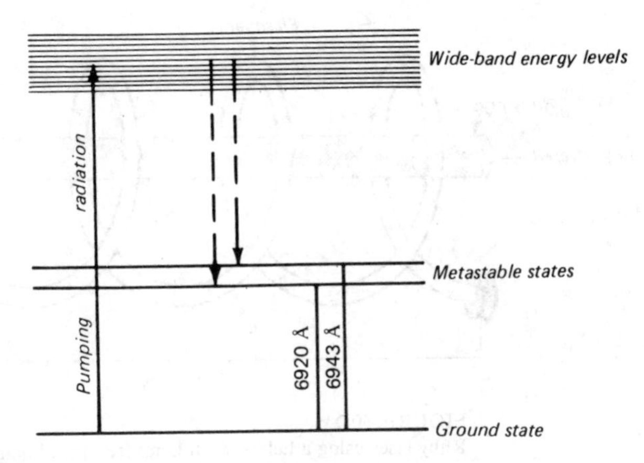

FIGURE 30C
Energy level diagram for a ruby crystal.

reflect back and forth from end to end. Their chance of stimulating emission will now depend upon a high reflectance at the end mirrors and a high population density of metastable atoms within the cavity. If both these conditions are satisfied, the buildup of photons surging back and forth through the cavity can be self-sustaining and the system will oscillate, or *lase*, spontaneously.

30.3 THE RUBY LASER

The first successful laser, developed by Maiman in 1960, used a single crystal of synthetic pink ruby as its resonating cavity. The ruby is primarily a transparent crystal of corundum (Al_2O_3) doped with approximately 0.05 percent of trivalent chromium ions in the form of Cr_2O_3, the latter providing its pink color. The aluminum and oxygen atoms of the corundum are inert; the chromium ions are the active ingredients.

As grown in the laboratory, a ruby crystal is cylindrical in shape. It is cut some 10 cm or so long and the ends polished flat and parallel. (Later beveled at Brewster's angle; see Fig. 30K.) In a typical ruby laser one end is highly reflective (about 96 percent), and the other end is close to half-silvered (about 50 percent).

When white light enters a crystal, strong absorption by the chromium ions in the blue-green part of the spectrum occurs (see Fig. 30C). Light from an intense source surrounding the crystal will therefore raise many electrons to a wide band of levels as shown by the "up" arrow at the left. These electrons quickly drop back, many returning to the ground level. However, some of the electrons drop down to the intermediate levels, not by the emission of photons, but by the conversion of the vibrational energy of the atoms forming the crystal lattice. Once in the intermediate levels, the electrons remain there for several milliseconds (about 10,000 times longer

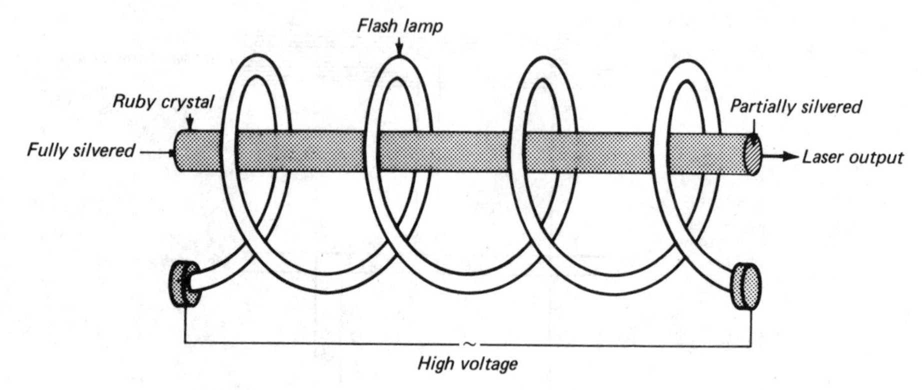

FIGURE 30D
Ruby laser using a helical flash lamp for optical pumping.

than in most excited states), and randomly jump back to the ground level, emitting visible red light. This fluorescent radiation enhances the pink or red color of the ruby and gives it its brilliance.

To greatly increase the electron population in the metastable levels, very intense light sources, as well as light-gathering systems, have been developed. The arrangement used by Maiman is shown in Fig. 30D. A high-intensity helical flash lamp surrounding the ruby provides adequate pumping light to produce a population inversion.

Another effective arrangement is shown in Fig. 30E. By placing a strong pulsed light source at one focus of a cylindrical reflector of elliptical cross section and the ruby rod at the other focus, high efficiency can be realized. A bank of capacitors can be discharged through the lamp for high-intensity pulsed operation.

A number of other pumping light sources of energy have been developed and used successfully; exploding wires, chemical reactions, and concentrated sunlight are but a few.

By pumping from a strong surrounding light source, a large part of the stored energy is converted into a coherent beam. Coherent waves traveling in opposite directions in the ruby crystal set up *standing waves* comparable to a resonating cavity in microwaves. With one end only partially reflecting, part of the internal light is transmitted as an emerging beam. (See Fig. 30F.) For some purposes both end mirrors are fully silvered, and a center spot on one is left clear to transmit part of the light as a narrow emergent beam.

30.4 THE HELIUM-NEON GAS LASER

The first successful gas laser was put into operation by Javan, Bennett, and Harriott in 1961. Since that time many different gas lasers, using gases of many kinds and mixtures, have been put into operation. Because it is inexpensive, unusually stable,

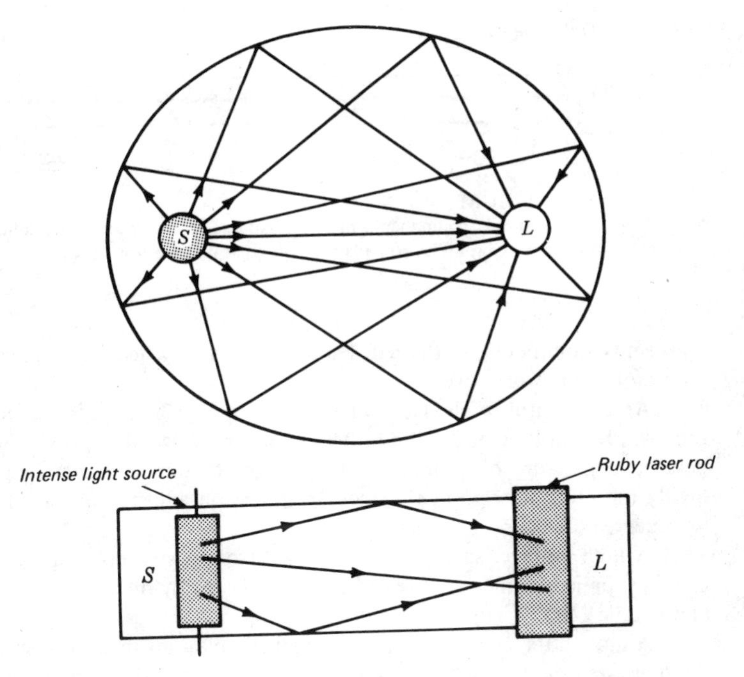

FIGURE 30E
Elliptical reflector for concentrating light from a source S on a laser L.

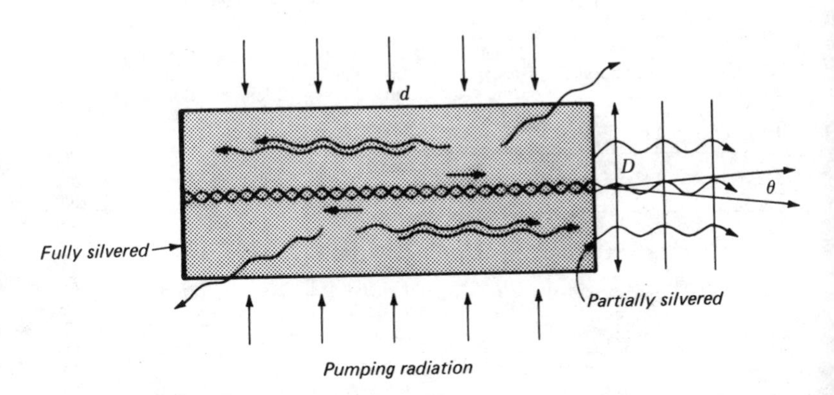

FIGURE 30F
The coherent stimulation of light waves in a solid-state laser such as a ruby crystal. Reflection from the ends sets up standing waves and resonance.

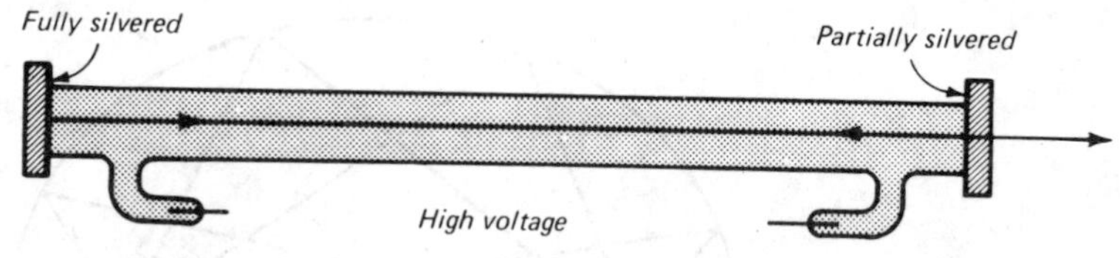

FIGURE 30G
Simple components of a He–Ne gas laser. Micrometer adjusting screws for making the mirror planes highly parallel are not shown.

and emits continuously, the He–Ne laser is widely used in optics and physics laboratories the world over.

An early form of He–Ne laser is shown in Fig. 30G. It is composed of a glass tube nearly 1 m long and contains helium at a pressure of approximately 1 torr and neon at a pressure of approximately $\frac{1}{10}$ torr. (1 torr = 1 mmHg pressure.) The highly reflecting mirrors at the ends are precision-adjusted and made parallel to a high degree of accuracy.

A high voltage, such as that obtained from a step-up transformer or a Tesla coil, is supplied by means of sealed-in electrodes or by metal bands around the ends and middle.

Although there are 10 times as many helium atoms present in such a mixture as there are neon atoms, the orange color of the gaseous discharge is characteristic of neon atoms. The visible spectrum of helium contains strong lines in the red, yellow, green, and blue, so the discharge appears as white light. The spectrum of neon, on the other hand, has so many strong lines in the orange and red and so few in the

Table 30A THE LOWER ENERGY LEVELS, THEIR VALUES IN WAVE NUMBERS, AND DESIGNATIONS FOR HELIUM AND NEON

Element	Electron configuration	Level designation	Energy, cm^{-1}
He	$1s^2$	$^1S_0$	0
He	$1s2s$	$^3S_1$	159,843
		$^1S_0$	166,265
Ne	$2p^6$	$^1S_0$	0
Ne	$2p^53s$	$^3P_2$	134,042
		$^3P_1$	134,460
		$^3P_0$	134,820
		$^1P_1$	135,889
Ne	$2p^53p$	*1*(1)	148,258
		2(3)	149,658
		3(2)	149,825
		4(1)	150,122
		5(2)	150,316
Ne	$2p^53p$	*6*(0)	150,918
		7(1)	150,773
		8(2)	150,856
		9(0)	151,039
		10(0)	152,971
Ne	$2p^54s$	$^3P_2$	158,605
		$^3P_1$	158,797
		$^3P_0$	159,381
		$^1P_1$	159,534
Ne	$2p^55s$	$^3P_2$	165,829
		$^3P_1$	165,913
		$^3P_0$	166,607
		$^1P_1$	166,659

green, blue, and violet, that its gaseous discharge appears to be orange-red [see Fig 21H(*e*) and (*f*)]. The neon spectrum also contains a large number of lines in the near infrared.

All the lower energy levels for helium and neon are given in Table 30A, and an energy level diagram for the same states is given in Fig. 30H.

The normal state of helium is a $^1S_0$ level arising from two valence electrons in $1s$ orbits. The excitation of either one of these electrons to the $2s$ orbit finds the atom in a $^1S_0$ or a $^3S_1$ state, both quite metastable, since transitions to the normal state are forbidden by selection rules [see Eq. (29o)].

Neon, with $Z = 10$, has 10 electrons in the normal state and is represented by the configuration $1s^2\ 2s^2\ 2p^6$. When one of the six $2p$ electrons is excited to the $3s$, $3p$, $3d$, $4s$, $4p$, $4d$, $4f$, $5s$, etc., orbit, *triplet* and *singlet energy levels* arise. A subshell like $2p^5$, lacking only one electron from a closed subshell, behaves as though it were a subshell containing one $2p$ electron. The number and designations of the levels produced are therefore the same as for two electrons, all triplets and singlets.

As free electrons collide with helium atoms during the electric discharge, one of the two bound electrons may be excited to $2s$ orbits, i.e., to the $^3S_1$ or $^1S_0$ states. Since downward transitions are forbidden by radiation selection rules, these are metastable states and the number of excited atoms increases. We therefore have optical pumping, out of the ground state $^1S_0$ and into the metastable states $^3S_1$ and $^1S_0$.

When a metastable helium atom collides with a neon atom in its ground state, there is a high probability that the excitation energy will be transferred to the neon, raising it to one of the $^1P_1$ or $^3P_0$, $^3P_1$, or $^3P_2$ levels of $2p^5 5s$. The small excess energy is converted into kinetic energy of the colliding atoms.

In this process each helium atom returns to the ground state as each colliding neon atom is excited to the upper level of corresponding energy. The probability of a neon atom being raised to the $2p^5 3s$ or $2p^5 3p$ levels by collision is extremely small because of the large energy mismatch. The collision transfer therefore selectively increases the population of the upper levels of neon.

Since selection rules permit transitions from these levels downward to the 10 levels of $2p^5 3p$ and these in turn to the 4 levels of $2p^5 3s$, stimulated emission can speed up the process of lasing. Lasing requires only that the $4s$ and $5s$ levels of neon be more densely populated than the $3p$ levels. Since the $3p$ levels are only sparsely populated, lasing can be initiated without pumping a majority of the atoms out of the ground state.

Light waves emitted within the laser at wavelengths such as 6328, 11,177, and 11,523 Å will occasionally be omitted parallel to the tube axis. Bouncing back and forth between the end mirrors, these waves will stimulate emission of the same frequency from other excited neon atoms, and the initial wave with the stimulated wave will travel parallel to the axis. Most of the amplified radiation emerging from the ends of the He–Ne gas laser are in the near-infrared region of the spectrum, between 10,000 and 35,000 Å, the most intense amplified wavelength in the visible spectrum being the red line at 6328 Å. A photograph of an inexpensive laboratory type of He–Ne gas laser is shown in Fig. 30I. Methods for operating such lasers at one wavelength will be presented in Sec. 30.7.

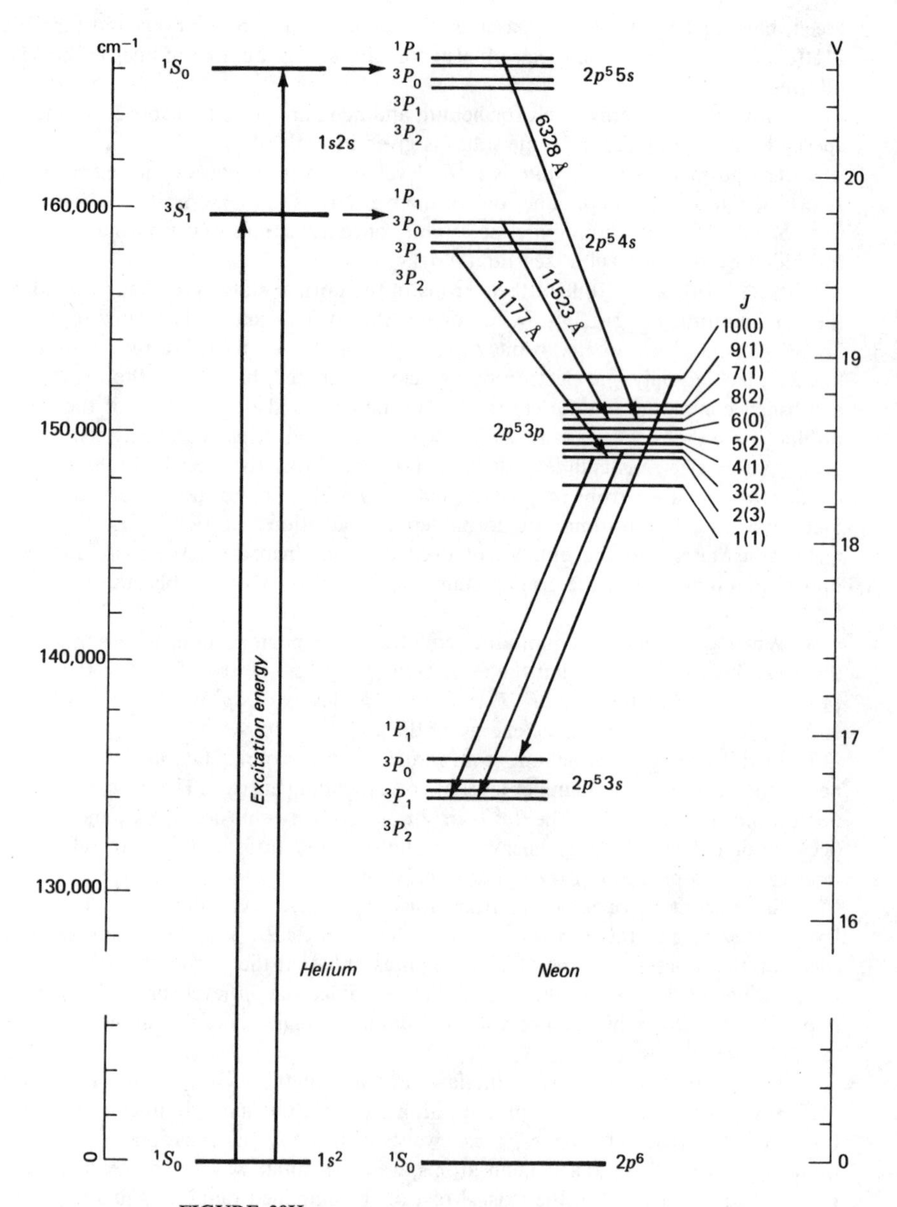

FIGURE 30H
Correlation energy level diagrams for helium and neon atoms involved in the He–Ne gas laser.

FIGURE 30I
Photograph of a He–Ne gas laser of the type used in the elementary and advanced physics laboratories for student experimentation. (*Metrologic Instruments, Inc.*)

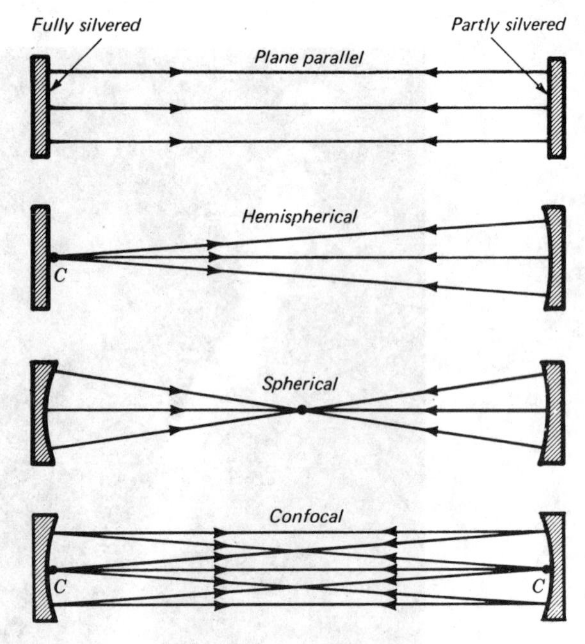

FIGURE 30J
Four types of end mirrors in common use for lasers. (Mirror curvatures are exaggerated.)

30.5 CONCAVE MIRRORS AND BREWSTER'S WINDOWS

A great many improvements have been made in laser technology. One of these is the use of concave mirrors at one or both ends of the resonating cavity, resulting in less sensitivity to misalignment. These mirrors are often separated from the plasma to provide for easy adjustment and to permit insertion of a variety of optical components into the standing-wave section.

Four commonly used configurations are shown in Fig. 30J. The *hemispherical* arrangement at the center, with a concave mirror at one end only, has its center of curvature at the center of the reflecting surface of the plane mirror. The *spherical* mirror arrangement has the two centers of curvature falling together at the center point C of the configuration. The *confocal* arrangement has the two centers at the centers of the opposite mirror faces. One mirror is usually *fully silvered*, and the other is *partially silvered* or fully silvered with a clear spot at its center.

With the end plates of a laser normal to the axis, reflection losses of approximately 4 percent at each of the interfaces are detrimental to coherence. By tilting these plates or beveling the ends of a solid laser to the polarizing angle $\bar{\phi}$, the windows or ends will have a 100 percent transmission for light whose electric vector is parallel to the plane of incidence (see Fig. 30K). The normal component is partially reflected at each interface with each traversal of the laser. The laser beam is thereby polarized,

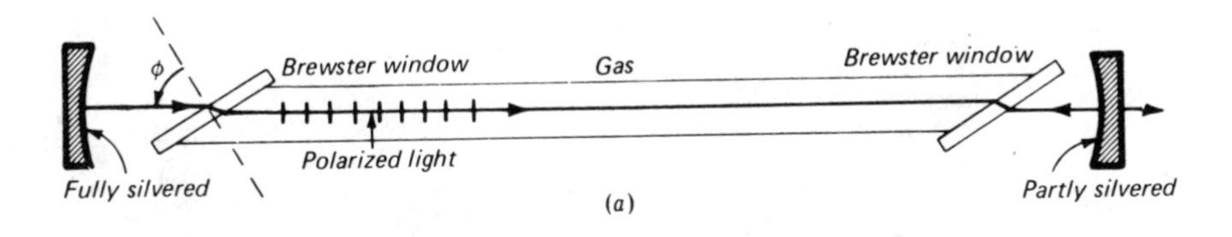

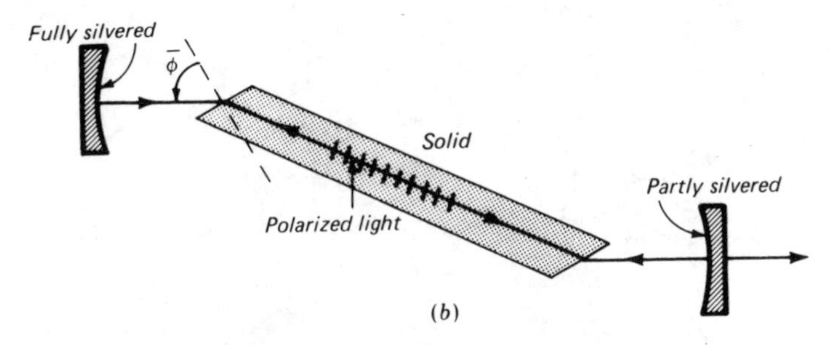

FIGURE 30K
Beveling laser ends at Brewster's polarizing angle eliminates detrimental reflections and at the same time polarizes the light in the plane of incidence: (*a*) gaseous laser and (*b*) solid-state laser like a ruby crystal of index *n*.

as with a pile of plates (see Figs. 24D, 24E, 24F, and 25B). The polarizing angle is given by

$$\tan \bar{\phi} = n \qquad \text{(30a)}$$

where n is the refractive index of the medium. For glass of index 1.50, $\bar{\phi} = 57°$, and at this angle of incidence in the rare medium, the normal component has approximately a 15 percent reflectance in crossing each interface. As stated earlier the plane of polarization of any stimulated photon is exactly the same as that of the stimulating photon.

30.6 THE CARBON DIOXIDE LASER

An example of a high-power molecular-gas laser is one that operates on carbon dioxide gas molecules. This optical device produces a continuous laser beam with a power output of several kilowatts and at the same time maintains a relatively high degree of purity and coherence.

The significance of such laser power can be demonstrated by the experimental fact that a focused beam can cut through a diamond and thick steel plates in a matter of seconds. Furthermore, such lasers generate a wide range of infrared frequencies and are tunable over a range of wavelengths. The beams also have applications in optical communications systems, as well as optical radar, and are well suited for use in terrestrial and extraterrestrial systems, since infrared light is only slightly scattered or absorbed by the atmosphere. (*Scattering is proportional to* ν^4.)

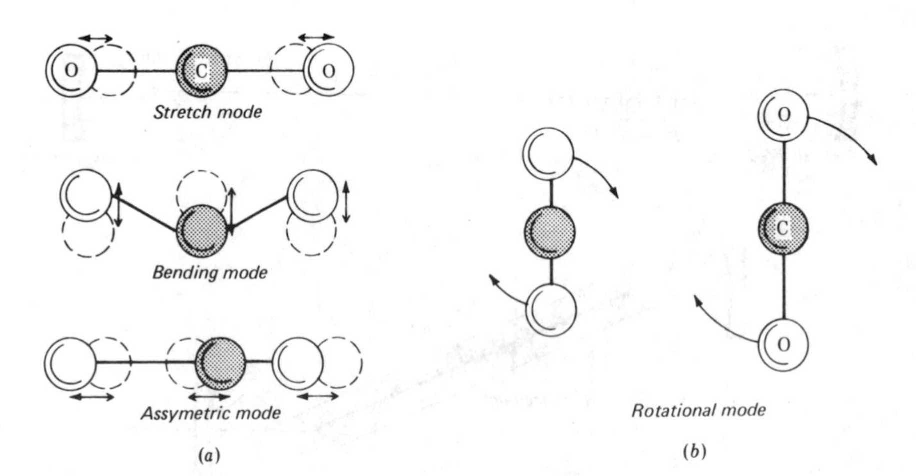

FIGURE 30L
Diagrams showing vibrational and rotational quantized modes of the CO_2 molecule.

The spectra of molecular gases are considerably more complicated than those of many atomic gases. In addition to the electronic energy levels of a free atom, a molecule can have levels arising from quantized vibrations and rotations of the atoms themselves. Thus, for a given electronic configuration in a molecule, there are a number of almost equally spaced vibration levels, and for each vibrational level there are a number of rotational levels. Figure 30L shows these molecular modes in separate diagrams. Note that while vibrating in any one of the three quantum states (*a*), a molecule may take on any one of a number of quantized rotational states like (*b*).

The energy levels for the electron configuration of the ground state are shown in Fig. 30M. The number with which each level is labeled gives the rotational angular momentum in units of $\hbar$. Two of the allowed infrared transitions between rotational levels belonging to different vibrational levels are shown. See the simplified energy level diagram in Fig. 30N.

The addition of nitrogen gas N_2 to the laser cavity results in the selective raising of CO_2 molecules to the desired laser levels. This is similar to the selective transfer of excitation energy from helium to neon atoms in the He–Ne laser (see Fig. 30H).

The high efficiency of the CO_2 laser is attributed largely to the fact that low-lying vibrational and rotational states require little energy for excitation and a good share of this energy is transferred to the laser beam. Whereas it requires some 20 V of energy to excite a helium atom to its first metastable states, only $\frac{1}{3}$ V is required to excite a CO_2 molecule to one of its lower vibrational and rotational levels (see Probs. 30.11 and 30.12 at the end of this chapter):

$$1 \text{ V} = 8065 \text{ cm}^{-1} \tag{30b}$$

$$1 \text{ cm}^{-1} = 1.2399 \times 10^{-4} \text{ V} \tag{30c}$$

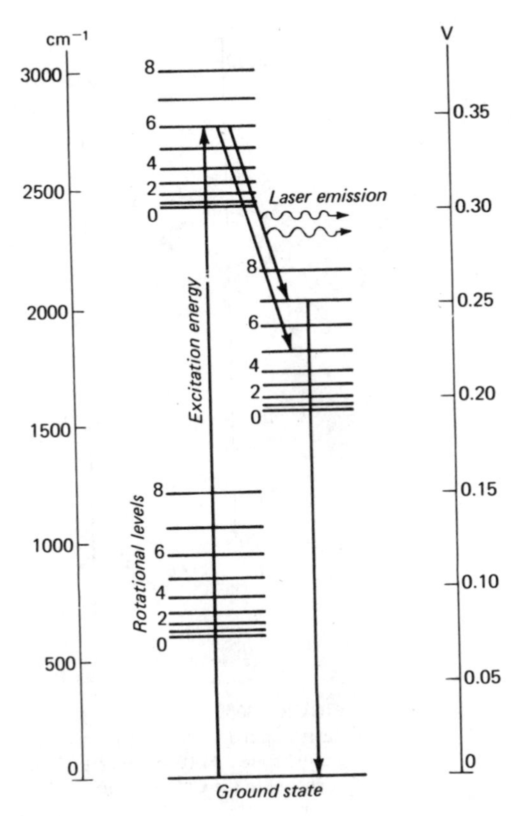

FIGURE 30M
Energy level diagram of a CO_2 molecule, showing three vibration states, each with nine rotation states.

One form of CO_2 laser is shown in Fig. 30O. Because the upper vibrational levels have a relatively long lifetime, one can store energy in a gaseous discharge tube for nearly a millisecond by blocking the path of light within the resonating cavity, thereby preventing laser oscillation.

When the barrier is suddenly removed, output from the laser results in a sudden pulse whose peak power is at least a 1,000 times larger than the average CW (continuous-wave) power. This is called *Q switching* or *Q spoiling*; it can be accomplished by the insertion of any of a variety of elements in the cavity, such as a mechanical chopper, a rotating mirror, a Kerr cell, a Pockels cell, etc. (see Chap. 33).

With a rotating mirror arranged in the position shown in Fig. 30O, an infrared light pulse at 10.6 μm is emitted every time it lines up with the opposite mirror. A 100-W CW laser will produce pulsed power of 100 kW in bursts approximately 150 ns long at a rate of 400 pulses per second.

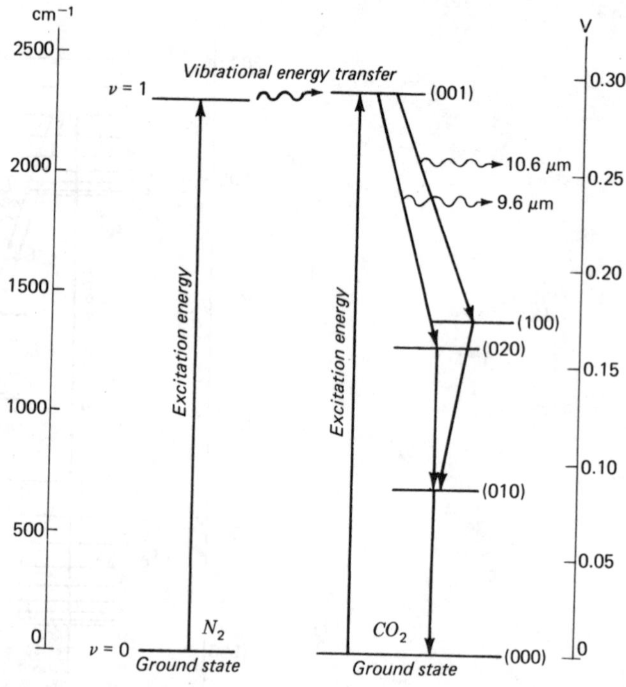

FIGURE 30N
Energy level diagrams comparing N_2 with CO_2. Excitation of nitrogen from the ground state $v = 0$, to the first vibrational excited state $v = 1$, and the transfer of energy to the CO_2 molecule.

30.7 RESONANT CAVITIES

A laser cavity can be operated in a variety of oscillation modes similar to those of a waveguide. As waves travel back and forth between the end mirrors, a distance d apart, standing waves are set up when

$$m = \frac{d}{\lambda/2} \tag{30d}$$

where m is a large integer. The frequency of the oscillation ν_m is given by

$$\nu_m = \frac{mv}{2d} \tag{30e}$$

where v is the speed of the waves in the cavity medium.

The frequency difference between modes is given by

$$\Delta\nu = \frac{v}{2d} \tag{30f}$$

and is the reciprocal of the round-trip time. For a gas laser 1 m long, $\Delta\nu = 150$ MHz.

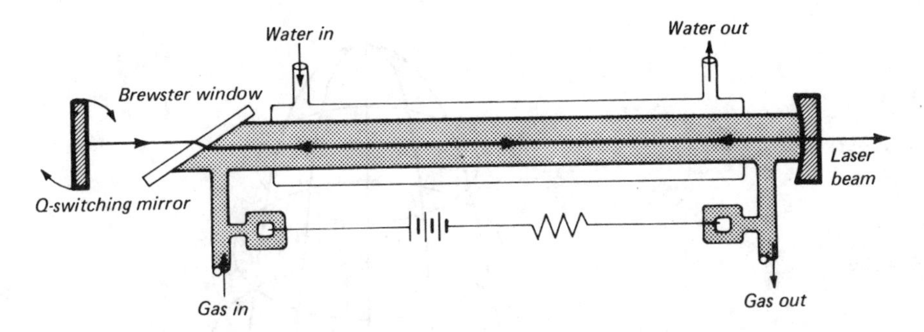

FIGURE 30O
Carbon dioxide laser with water cooling jacket, Brewster's window, and rotating mirror for pulsing the output laser beam.

In a spectrum-rich source single wavelengths can be selected for oscillation by inserting a silvered prism for one of the mirrors, as shown in Fig. 30P. Owing to the dispersion of the prism the optical path can be "tuned" to be collinear for the desired wavelength only. This technique makes use of the Littrow spectrograph, where either a prism or diffraction grating is used as the dispersion unit [see Fig. 17N(c)].

In addition to the longitudinal modes of oscillation, transverse modes can be sustained simultaneously. Since the fields within a gas are nearly normal to the cavity axis, these are known as *transverse electric and magnetic* (TEM_{mn}) *modes*. The subscripts m and n specify the integral number of transverse nodal lines across the emerging beam. In other words, the beam in cross section is segmented into layers.*

The simplest mode, TEM_{00}, is the most widely used, in which the flux density over the beam cross section is approximately gaussian (see Fig. 30Q). There are no phase changes across the beam, as there are in other modes, so the beam is spatially coherent. The angular spread of the emergent beam is limited by diffraction at the exit aperture and to a first approximation (*assuming uniform intensity over the beam*

* For photographs of these mode patterns, see E. Hecht and A. Zajac, "Optics," p. 484.

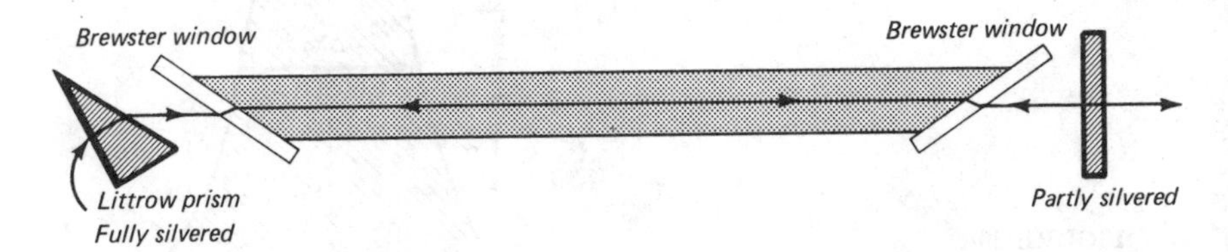

FIGURE 30P
A fully silvered prism at one end of a laser disperses the light so that only one spectrum line is collinear with the laser axis and is amplified by setting up standing waves.

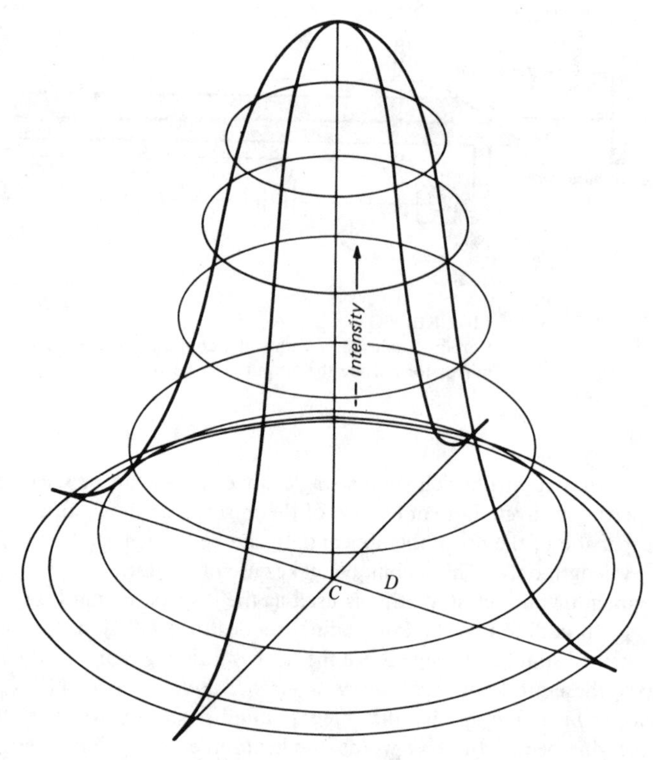

FIGURE 30Q
Gaussian distribution of the light intensity over the beam cross section for a laser oscillating in the TEM_{00} mode.

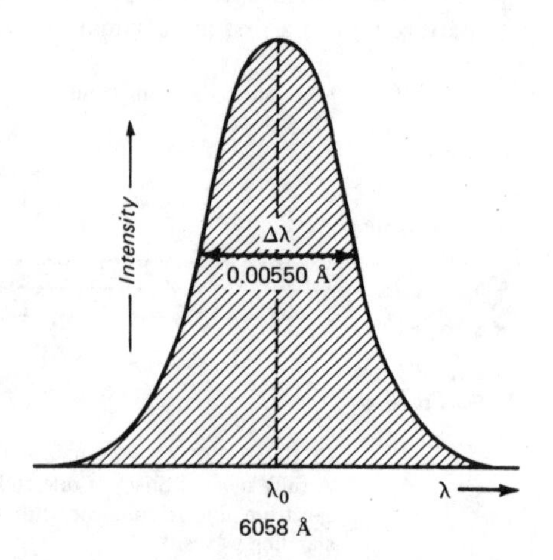

FIGURE 30R
Intensity-wavelength graph of the orange line of the krypton (^{86}Kr) spectrum, $\lambda = 6058$ Å. The line with $\Delta\lambda$ is due largely to doppler broadening.

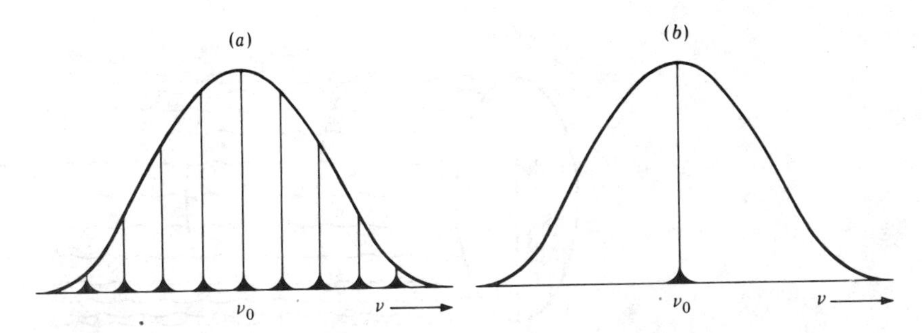

FIGURE 30S
Laser modes for two operational configurations of a CW gas laser showing a gaussian envelope and (*a*) nine resonant frequencies without etalon control and (*b*) a single frequency with etalon control (see Fig. 30T).

cross section of diameter D) is given by the single-aperture diffraction pattern, Eq. (15k),

$$\theta = 2.44 \frac{\lambda}{D} \qquad (30g)$$

where $\theta = 2\theta_1$. See Fig. 30F.

The resonant modes of a laser cavity are much narrower in frequency than the bandwidth of the normal spontaneous atomic transition. Most of the bandwidths of spectrum lines emitted by a discharge tube are due to doppler broadening (see Fig. 30R). Only those modes obeying Eq. (30d) are sustained in a cavity. A single radiation transition within the atom or molecule produces a band of frequencies, out of which the cavity will select and simplify only certain narrow bands. The number of these narrow bands depends upon the wavelength λ and the distance between laser ends D [see Fig. 30S(*a*)].

One method for selecting one narrow band only is shown in Fig. 30T. An etalon with a much shorter length than the laser and with lightly silvered plates is inserted in the laser cavity and fine-tuned by tilting to resonate to the selected frequency ν_0. The next sideband frequency to ν_0 on either side will come at too wide an angle to enter and be amplified by the longer cavity. Hence only ν_0 is sustained by the combination.

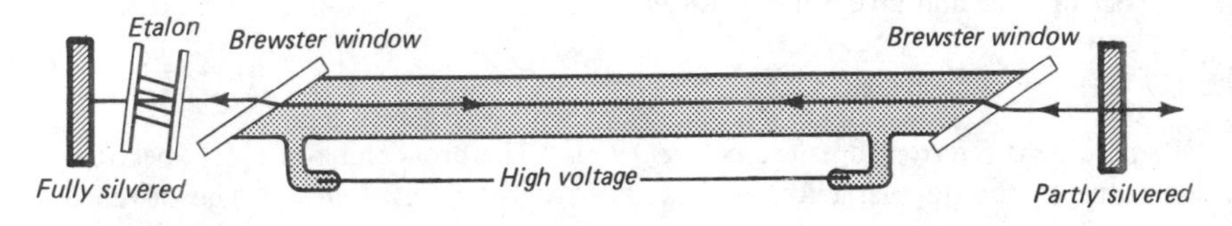

FIGURE 30T
Configuration for etalon control of a single laser oscillation mode.

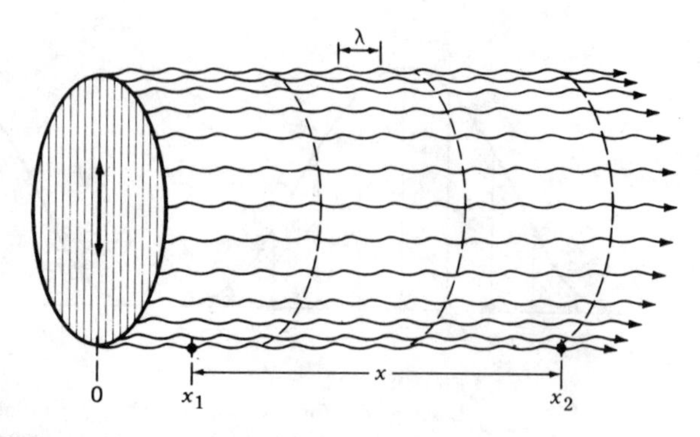

FIGURE 30U
Diagram illustrating plane, monochromatic, and polarized coherent waves emerging from a laser.

30.8 COHERENCE LENGTH

Consider a point source of light emitting an infinitely long monochromatic wave train having spherical or plane wave fronts (see Fig. 30U). Under such ideal conditions the phase difference $\Delta\phi$ between two fixed points x_1 and x_2, spaced any distance apart along any ray, is time-independent. Equivalent to this, the phase difference measured at a single point in space at the beginning and end of a fixed time interval Δt does not change with time t. These are statements of *perfect temporal coherence.*

Alternately, the phase difference for any two fixed points in a plane normal to a ray direction is time-independent. This is a statement of *perfect spatial* or *lateral coherence.*

Since real light sources emit wave trains of finite length and this length is important to the production of interference phenomena of all kinds, we should determine practical values for *coherence length.* The average lifetime of an atom in the state of radiating is approximately 1.6×10^{-8} s. Traveling with the speed of light, each wave train has a length of about 3 m. Whether these waves are *damped* or of *constant amplitude*, a Fourier analysis of the waves leads to a frequency distribution called the *natural breadth of a spectrum line* [see Fig. 30V(a)].

Thermal light sources are composed of atoms randomly emitting wave trains at random times, and their frequencies are altered by thermal motions and by local electric and magnetic fields. The sum of all these effects is to greatly widen each spectrum line and give it a *bandwidth*,

$$\Delta\nu = \frac{1}{\Delta t} \tag{30h}$$

where Δt is referred to as the *coherent time.* The broadening of most spectrum lines is due to the doppler effect and is called *doppler broadening.** The distance light

* See Harvey E. White, "Introduction to Atomic Spectra," chap. 21, McGraw-Hill Book Company, New York, 1934.

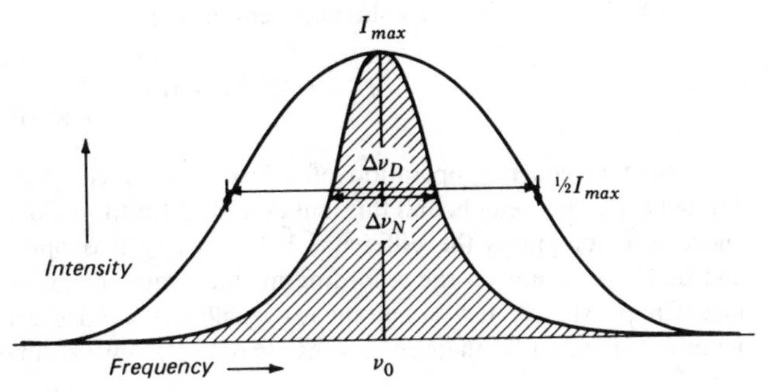

FIGURE 30V
Comparison of the *natural width* of a spectrum line with its *doppler width.*

travels in this time Δt, called the *coherent length*, is given by

$$L = c\,\Delta t = \frac{c}{\Delta \nu} \qquad (30i)$$

Hence, *the width of a spectrum line is a measure of coherence length*, and the coherence length is inversely proportional to the bandwidth of a spectrum line.

A more accurate equation for coherence length takes into account the line's actual broadening effects and is given approximately by*

$$L = \frac{c\sqrt{2}\ln 2}{\pi\,\Delta\nu} = 0.32\frac{c}{\Delta\nu} \quad \textit{for low-pressure discharges} \qquad (30j)$$

$$L = \frac{c\ln 2}{2\pi\,\Delta\nu} = 0.11\frac{c}{\Delta\nu} \quad \textit{for high-pressure discharges} \qquad (30k)$$

The spectrum lines from thermal sources have a coherence length of a few millimeters up to several tens of centimeters. A laser, on the other hand, may have a coherence length of several kilometers. One of the most coherent of nonlaser lines is the *orange line of krypton*, at $\lambda = 6058$ Å (see Fig. 30R).

EXAMPLE The Doppler width $\Delta\lambda$ of the orange line of krypton, Kr^{86}, at $\lambda =$ 6058 Å, is 0.00550 Å. Calculate (*a*) the line frequency ν, (*b*) the bandwidth $\Delta\nu$ in hertz, and (*c*) the coherence length in centimeters.

SOLUTION (*a*) The line frequency is given by $c = \nu\lambda$ as

$$\nu = \frac{3.0 \times 10^{10}\ \text{cm/s}}{6.058 \times 10^{-5}\ \text{cm}} = 4.95 \times 10^{14}\ \text{Hz}$$

(*b*) Using the well-known relationship $\Delta\nu/\nu = \Delta\lambda/\lambda$, we find

$$\Delta\nu = \nu\frac{\Delta\lambda}{\lambda} = 4.95 \times 10^{14}\frac{0.0055\ \text{Å}}{6058\ \text{Å}} = 4.50 \times 10^{8}\ \text{Hz}$$

* See Collier, Burckhardt, and Lin, "Optical Holography," p. 445, Academic Press, Inc., New York, 1971.

(*c*) By Eq. (30j,) the coherence length is

$$L = 0.32 \frac{c}{\Delta\nu} = 0.32 \frac{3 \times 10^{10}}{4.5 \times 10^8} = 21.3 \text{ cm}$$

Single-frequency operation of a laser, described above, offers an almost unlimited coherence length, and this makes it ideal for the art of *holography*. For making successful holograms the difference between any two optical paths from the light source to any point on the recording medium must be less than the coherent length (see Chap. 31). A few simultaneous oscillating modes can reduce the coherence length tremendously, thereby limiting its use to a few centimeters.

30.9 FREQUENCY DOUBLING

With the development of lasers in 1960, scientists obtained for the first time light beams intense enough to produce *light-wave harmonics*. Such phenomena had long been known in electronics and sound, where sum and difference frequencies play an important role in electronic circuitry, music, and hearing.*

In 1961 four physicists at the University of Michigan focused a beam from a ruby laser emitting 3-kW pulses of red light of wavelength 6943 Å onto a quartz crystal, thereby producing an observable number of photons of half the wavelength, or 3471.5 Å (see Fig. 30C). This new wavelength, which lies in the ultraviolet region of the spectrum, is exactly double the frequency of the laser's red light. The possibility that this was fluorescent light could be ruled out since it was emitted in a highly directional beam parallel to the incident light.†

Many related developments followed this preliminary discovery, and soon much higher efficiencies were obtained, converting laser light to harmonic frequencies. In other experiments two different wavelengths were made to interact with matter and produce sum and difference frequencies in the ultraviolet and infrared, respectively.

The classical explanation of these phenomena involves ionization of the loosely bound valence electrons, which in many crystals are shared by other atoms in the bonding of the structure. An atom giving up an electron to its neighbor leaves it with a net positive charge, and the neighbor with an extra electron has a net negative charge. As light waves pass through, these ions respond to the associated electric and magnetic fields by being set into vibration with the source frequency. When the incident light intensity is extremely high, as it is in a laser beam, the induced atomic vibrations are nonlinear in their response, just as they are with loud sounds, and higher harmonics are generated. The second harmonic is far more intense than higher modes.

From the point of view of quantum theory, when two photons interact with matter, both energy and momentum are conserved in producing a single photon.

* See Harvey E. White, "Modern College Physics," 3rd ed., p. 371, D. Van Nostrand Co., Inc., Princeton, N.J., 1956.

† P. A. Franken, A. E. Hill, C. W. Peters, and G. Weinreich, *Phys. Rev. Lett.*, **7**:118 (1961); J. A. Giordmaine, *Sci. Am.*, **210**:38 (April 1964).

30.10 OTHER LASERS

Hundreds of different kinds of lasers using many different materials have been made, emitting radiation over a wide range of wavelengths from the ultraviolet at one end of the spectrum to microwaves at the other. Many gas elements are known to lase, and the same is true of many diatomic and triatomic molecules and many metals.

One type of chemical laser derives its energy from the dissociation by light of trifluoroiodomethane (CF_3I). As this complex molecule dissociates, the carbon-iodine bond is broken and an excited iodine atom is released. On returning to the ground state, the iodine atom gives off a photon with a wavelength of 13,150 Å.

Another type of laser uses semiconductors in the form of pn junctions. Such lasers are very small, require only low voltages, and are easily modulated. The most commonly used material is gallium arsenide (GaAs) impregnated with zinc.

If a laser is fully pumped before oscillation begins, the first pulse will be consiserably higher in power than it would be under conditions of continuous operation. A short-duration pulse emitted from such a pulsed source can be amplified by passing the beam through subsequent lasers, called *amplifiers*. For example, a ruby-laser oscillator may be followed by a sequence of ruby-laser amplifiers, each consisting of a ruby rod with ends cut at Brewster's angle and unsilvered. Such a sequence can amplify a single pulse as short as a small fraction of a microsecond up to an energy of many joules.

30.11 LASER SAFETY

Laser light varies in intensity from a fraction of a milliwatt for an inexpensive He–Ne laser to many kilowatts for a CO_2 laser. Laser injuries have been few, and their dangers highly debatable. However, the greatest danger is the inadvertent direction of an undiverged laser beam directly into the eye.

The weak beam from a $\frac{1}{2}$-mW continuous He–Ne laser is probably of little danger, since eyelids can close upon sudden exposure. More intense beams, and especially pulsed beams, can cause serious injury, due primarily to the ability of the eye to focus the parallel beam onto a small area of the retina.

Good safety practice in the presence of high-powered lasers involves the use of filtering glasses and shields and awareness that a laser beam incident upon a specular reflecting surface can redirect the beam undiminished in intensity.

30.12 THE SPECKLE EFFECT

Anyone observing a diverged laser beam against a diffuse surface will notice a granular appearance. If one squints or moves back, the speckles become larger. No matter where one's eyes are focused, the speckles will appear sharp. Moving sideways causes the speckles to move too.

Curiously enough, the speckles do not exist in the reflected pattern but are created in the eye itself. Laser light reflected from a diffuse surface will enter the eye, producing bright speckles where random fluctuations cause constructive interference on the retina. Such interference maxima can be related to a local convergence, real or virtual, of laser light in the vicinity of the observed area, in the plane in which the eye is focusing. By moving the head sideways, the speckles will move in the same direction for the far-sighted person, just as an object observed on the far side of an open window will. Conversely a near-sighted person will see the speckles move in opposition. Correct vision will produce no apparent parallax.

30.13 LASER APPLICATIONS

Since the advent of the laser, many uses have arisen for it. Modulated laser beams have been used for communication. Lasers have been used by the medical profession in surgery, where retinal tissue is cauterized to weld detached retinas. They have been

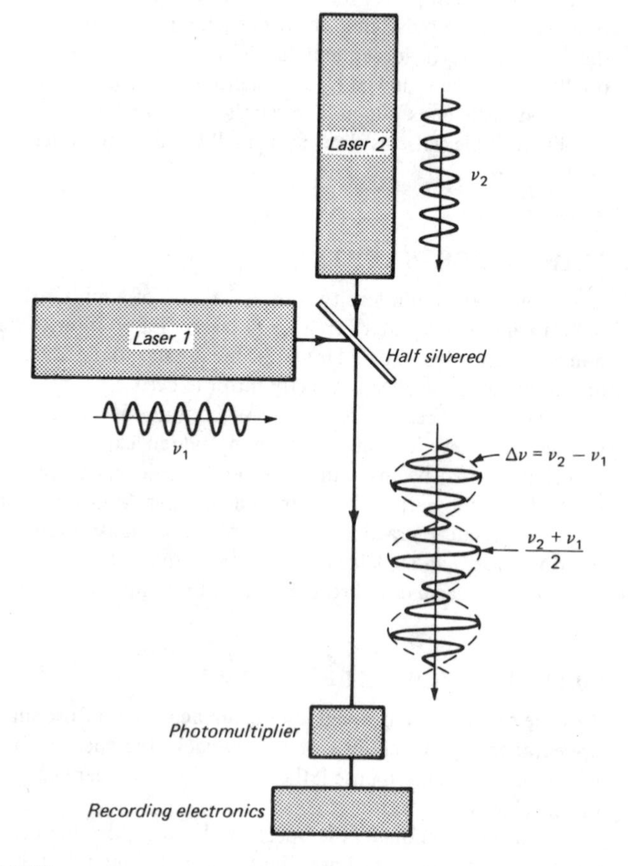

FIGURE 30W
A variation of the Michelson-Morley experiment performed with two lasers of slightly different frequency.

used by surveyors and engineers for critical alignment, as well as for ranging in metrology and determining the distance to the moon. Attenuation and scattering of laser beams have been used in atmospheric research. High-power lasers have been used to cut through diamonds and steel plates and to initiate thermonuclear reactions. One of the most significant uses of the laser has been in the production and research with holograms, the subject of the next chapter.

A variation of the Michelson-Morley experiment has been performed as a sensitive test for an ether drift.* The beams of two infrared lasers of slightly different frequency were combined by means of a beam splitter, and the resultant beat frequency was detected by means of a photomultiplier and recording electronic circuits (see Fig. 30W). The beat frequency, as with sound waves, is equal to the difference between those of the two laser beams, $\Delta\nu = \nu_1 - \nu_2$.

The exact frequency in which each laser operated was governed by the length of each resonant cavity and the speed of light inside. If both lasers, operated at about 3×10^{14} Hz, were rotated through 90°, the ether drift should affect the speed of light in the cavities and therefore the frequency difference between them. A relative change of $\Delta\nu = 3$ MHz is expected from the ether-drift hypothesis, due to the earth's orbital velocity. No change in the beat frequency was detected.

Lasers have been used, like radar, to determine distances large and small. During the Apollo moon flight 11, on July 20, 1969, Armstrong and Aldrin set up a previously constructed array of triple prisms, to reflect light from the earth back to its

* T. S. Jaseja, A. Javan, J. Murray, and C. H. Townes, Test of Special Relativity or of the Isotropy of Space by Use of Infrared Masers, *Phys. Rev.*, **133**:A1221 (1964).

Table 30B SOME COMMON LASER TYPES

Spectral type	Type	Medium	Wavelength, nm	Radiation
Ultraviolet	He–Cd	Gas	325.0	CW
	N_2	Gas	337.1	pulsed
	Kr	Gas	350.7, 356.4	CW
	Ar	Gas	351.1, 363.8	CW, pulsed
Visible	He–Cd	Gas	441.6, 537.8	CW
	Ar	Gas	457.9, 514.5	CW, pulsed
	Kr	Gas	461.9, 676.4	CW, pulsed
	Xe	Gas	460.3, 627.1	CW
	Ar–Kr	Gas	467.5, 676.4	CW
	He–Ne	Gas	632.8	CW
	Ruby $Cr^{3+}AlO_3$	Solid	694.3	pulsed
Infrared	Kr	Gas	0.753, 0.799	CW
	GaAlAs	Solid(diode)	0.850	CW
	GaAs	Solid (diode)	0.904	CW
	Nd	Solid (glass)	1.060	pulsed
	Nd	Solid (YAG)	1.060	CW, pulsed
	He–Ne	Gas	1.15, 3.39	CW
	CO_2	Gas	10.6	CW, pulsed
	H_2O	Gas	118.0	CW, pulsed
	HCN	Gas	337.0	CW, pulsed

source.* A square array of 100 of these prisms, each 4 cm in diameter, was arranged and placed approximately 20 m from the spacecraft at the landing site, the Sea of Tranquility.† A return beam of light was first picked up on the earth by a group of scientists at the Lick Observatory, University of California at Santa Cruz, Aug. 1, 1969. With a ruby laser in the 120-in. telescope, a pulsed beam of light 4 m in diameter was aimed at the moon. The return light pulses arrived approximately 2.58 s later and were accurately timed to within 0.1 μs. The accuracy in time measurements determined the reflector distance to within 6 m.

Shortly thereafter, another group, at the McDonald Observatory in Texas, picked up a return beam from the moon reflector and were able to measure the time to within 2 ns. This determines the distance to within 30 cm.

It should be pointed out that due to the relative motion of the moon and the laser transmitter, the center of the return beam will be displaced several miles (velocity aberration). Due to diffraction by each triple prism 4 cm in diameter, the light spreads to almost 15 km by the time it reaches the earth. For this reason the return beam can be picked up at the transmitter.

Much valuable information concerning the moon and the earth can be determined from the changing distance between these two astronomical bodies, and we can look forward to announcements of new findings in the future.

PROBLEMS

30.1 Using a full sheet of millimeter graph paper, draw an energy level diagram like the upper half of Fig. 30H so that it is as large as possible. Use the range 130,000 to 170,000 cm^{-1}. Use the energy levels listed below, which are given in wave numbers, and label the levels as given here. Take differences between levels to find which ones are involved in the lines at wavelengths (*a*) 6328 Å, (*b*) 11,523 Å, and (*c*) 11,177 Å.

Ans. (*a*) $\Delta\sigma$ = 15,803 cm^{-1}, $2p^55s$, $_1P^1$ jumps to $2p^53p$, *8*(2), (*b*) $\Delta\sigma$ = 8678.2 cm^{-1}, $2p^54s$, $^3P_2$ jumps to $2p^53p$, *2*(3), (*c*) $\Delta\sigma$ = 8946.9 cm^{-1}, $^1P_1$ jumps to $2p^53p$, *8*(2); see Fig. 30H

He	$1s^2$	$^1S_0 = 0$	Ne	$2p^53p$	*6*(0) = 150,918 *7*(1) = 150,773 *8*(2) = 150,856 *9*(1) = 151,039 *10*(0) = 152,971
He	$1s2s$	$^3S_1$ = 159,843 $^1S_0$ = 166,265			
Ne	$2p^6$	$^1S_0 = 0$			
Ne	$2p^53s$	$^3P_2$ = 134,042 $^3P_1$ = 134,460 $^3P_0$ = 134,820 $^1P_1$ = 135,889	Ne	$2p^54s$	$^3P_2$ = 158,605 $^3P_1$ = 158,797 $^3P_0$ = 159,381 $^1P_1$ = 159,534
Ne	$2p^53p$	*1*(1) = 148,258 *2*(3) = 149,658 *3*(2) = 149,825 *4*(1) = 150,122 *5*(2) = 150,316	Ne	$2p^55s$	$^3P_2$ = 165,829 $^3P_1$ = 165,913 $^3P_0$ = 166,607 $^1P_1$ = 166,659

* See Sec. 2.2 and Fig. 2C(*e*).
† J. E. Foller and E. J. Wampler, The Lunar Reflector, *Sci. Am.*, March 1970, p. 38.

30.2 From the energy level values given in Prob. 1, what is (*a*) the smallest energy mismatch of the helium metastable levels with the levels of neon? (*b*) What percentage of mismatch are these values?

30.3 From the energy level values given in Prob. 1, specify the three transitions not labeled in Fig. 30H and calculate their frequencies in wave numbers and their wavelengths in angstroms.

30.4 The beam from a ruby laser emitting red light of wavelength 6943 Å is used with a beam splitter to produce two coherent beams. Both beams are reflected from plane mirrors and brought together on a thin photographic emulsion. If the angle α between these two interfering beams is 10° and the plate normal bisects this angle, find the fringe separation of the interference fringes on the plate. *Ans.* 0.00398 mm

30.5 The following transitions give rise to strong lines in the neon spectrum. From the energy level values given in Prob. 1 find their wavelengths in angstroms. (*a*) $2p^5 3p$, *9*(1) to $2p^5 3s$, $^3P_2$, (*b*) $2p^5 3p$, *4*(1) to $2p^5 3s$, $^1P_1$, (*c*) $2p^5 3p$, *2*(3) to $2p^5 3s$, $^3P_2$, (*d*) $2p^5 3p$, *3*(2) to $2p^5 3s$, $^1P_1$.

30.6 Starting with the neon energy level values given in Prob. 1, the following strong lines start at levels arising from the electron configuration $2p^5 3p$ and end on the configuration $2p^5 3s$. Find their wavelengths in angstroms. (*a*) *6*(0) to $^3P_1$, (*b*) *4*(1) to $^3P_2$, (*c*) *3*(2) to $^3P_1$, (*d*) *1*(1) to $^3P_2$.

30.7 The following wavelengths are strong lines in the neon spectrum: 6144.7, 6335.0, 6403.6, and 7034.3 Å. They all end on the lowest level of the electron configuration $2p^5 3s$. (*a*) Find their frequencies in wave numbers, and using the energy level values in Prob. 1, identify the initial energy levels.

Ans. (*a*) 16,274, 15,785, 15,616, and 14,216 cm^{-1}, (*b*) *5*(2), *3*(2), *2*(3), and *1*(1)

30.8 A He–Ne laser exactly 25.0 cm long is vibrating in the TEM_{00} mode. What is (*a*) the number of loops in the standing wave pattern if $\lambda = 6328.0$ Å, and (*b*) the frequency difference between modes?

30.9 The doppler width of the red cadmium line $\lambda = 6438$ Å, produced in a low-pressure discharge, is 0.00030 Å. Calculate (*a*) the frequency of the light, (*b*) the line width in hertz, and (*c*) the coherence length.

30.10 The sodium line at $\lambda = 5890$ Å, produced in a low-pressure discharge, has a doppler width of 0.0194 Å. Calculate (*a*) the frequency of the light, (*b*) the line width in hertz, and (*c*) the coherence length in centimeters.

Ans. (*a*) 5.0934×10^{14} Hz, (*b*) 1.678×10^9 Hz, (*c*) 5.72 cm

30.11 Find the excitation energy for helium atoms raised to the $1s2s$, $^1S_0$ state (*a*) in volts and (*b*) in wave numbers. What energy is radiated by the emission of $\lambda = 6328$ Å (*c*) in volts and (*d*) in wave numbers? (*e*) What is the theoretical efficiency?

30.12 Find the excitation energy for the nitrogen molecule in the CO_2 laser shown in Fig. 30O (see Fig. 30N) (*a*) in volts and (*b*) in wave numbers. What energy is radiated when the laser emits $\lambda = 10.6\,\mu\text{m}$ (*c*) in volts and (*d*) in wave numbers? (*e*) What is the theoretical efficiency of this laser?

31
HOLOGRAPHY

The term *holography* comes from the Greek meaning *whole writing.* It is a two-step process by which (1) an object illuminated by coherent light is made to produce interference fringes in a photosensitive medium, such as a photographic emulsion, and (2) reillumination of the developed interference pattern by light of the same wavelength produces a three-dimensional image of the original object. The viewed images seen by this process have the appearance of the original object, including the differences in perspective one obtains with a change of the viewer's observing position—a full three-dimensional image.

The principles of holography were first put forward by Dennis Gabor, of the Imperial College of Science and Technology, University of London. Gabor's invention consisted of a method for improving the resolution of images obtained with an electron microscope, and his announcement of the concepts was published in 1948.* Little was made of his work at that time, and it was not until the development of the laser in 1960 that his basic ideas became more than a laboratory curiosity. He was awarded the Nobel prize in physics in 1971 for his *three-dimensional lensless method of photography* (*holography*).

* Dennis Gabor, *Nature,* **161**:777 (1948).

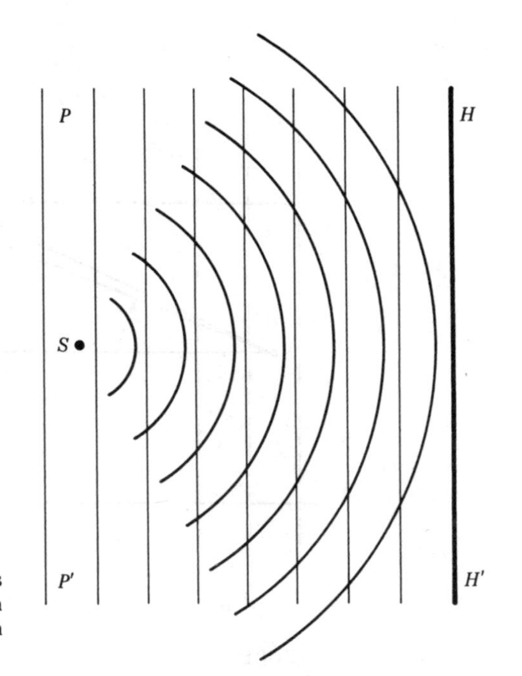

FIGURE 31A
The interference of coherent waves scattered from a point source, with plane waves, will give rise to a hologram in the form of a Gabor zone plate.

31.1 THE BASIC PRINCIPLES OF HOLOGRAPHY

In the preliminary stages Gabor's technique was to cause a beam of coherent light to be scattered from an object and then allowed to overlap an unobstructed coherent beam. The two sets of waves coming together on a photographic plate, placed in front of the object, would produce interference fringes.

Consider the interference pattern caused by coherent monochromatic plane waves incident from the left onto a point scatterer (see Fig. 31A). At the plane of the photographic plate HH' to the right, bright and dark concentric circles will be formed due to constructive and destructive interference between the *scattered light* and the direct *reference beam*. Upon development, the plate is found to contain light and dark partially absorbing fringes, as predicted.

This pattern, called a *Gabor zone plate*, is similar to a *Fresenl zone plate* treated in Chap. 18, except that the light and dark fringes shade continuously into each other (see Fig. 18I). The ring pattern is a great deal like the circular fringe pattern produced by the Michelson interferometer [see Fig. 13P(*a*) and (*b*)].

Since the reference beam is assumed to be in constant phase across the surface of the hologram plane, the interference fringes at any point P will be separated by an amount Δr, corresponding to a difference of path length of one wavelength of light λ, as measured from S* (see Fig. 31B):

● $$\lambda = \Delta r \sin \theta \qquad (31a)$$

* For finding the radius of the rings we use the geometry of Fig. 31B. The path difference $d = R - D$, and $d = n\lambda = r^2/(2R - d)$. See Newton's rings [Eq. (14k)].

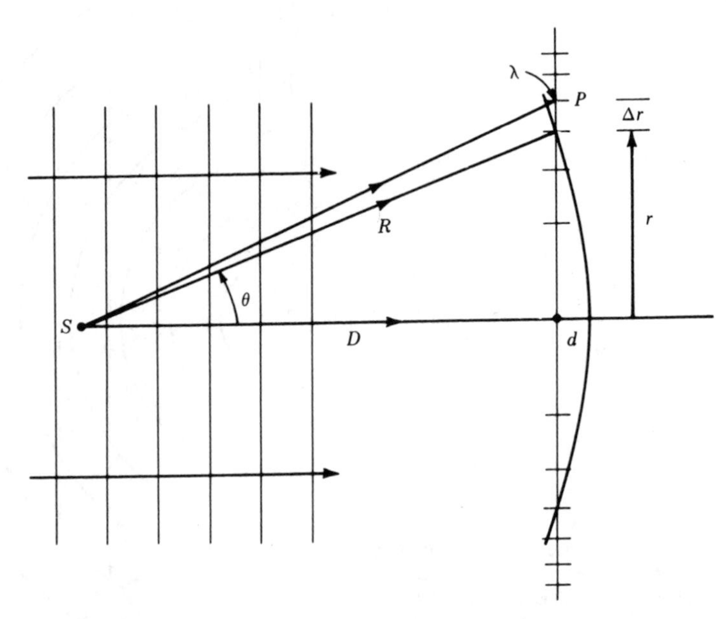

FIGURE 31B
The geometry for the fringe spacing Δr in a Gabor zone plate. P marks points of constructive interference which develop as black fringes on a hologram.

This plate is then illuminated by plane coherent waves, as it was in the making, but in the absence of the scatterer. The light formed by interference between the light and dark bands will now produce a first-order interference maximum at the angle θ as given by Eq. (31a) [see Fig. (31C)]. This light will therefore appear to diverge from S. Since all points from the holograms will produce diffracted light propagated in line with S, a virtual image is created and can be viewed from the right of the hologram.

Suppose now that two scattering centers were originally present on the left. Each now will create a Gabor zone plate. Moreover, the modulation intensity of each zone plate will be proportional to the scattered light intensity provided the photographic response is linear. The reconstruction will therefore produce a virtual image of both scattering centers, each with its proportionate intensity.

The argument can now be extended to a distributed scattering source corresponding to a continuum of scattering centers. The hologram will now consist of a continuum of superposed zone plates (see Fig. 31D). Upon reconstruction, the distributed *virtual image* should appear exactly like the original object as viewed from the right of the hologram.

Although the basic principles of Gabor's *on-axis hologram* are straightforward enough, the application of these principles suffered from several technical difficulties,

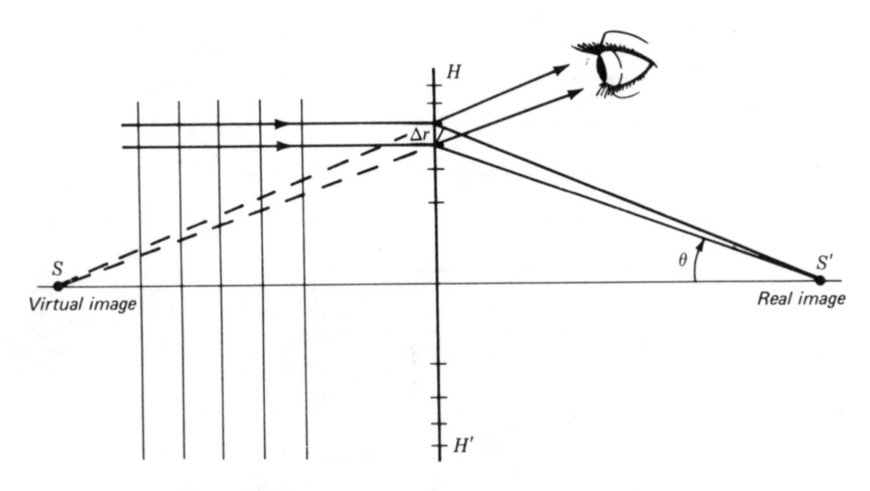

FIGURE 31C
Point images, both real and virtual, formed by plane coherent light falling on a Gabor zone-plate hologram. The virtual image can be seen at S by the eye, and the real image can be formed on a screen at S'.

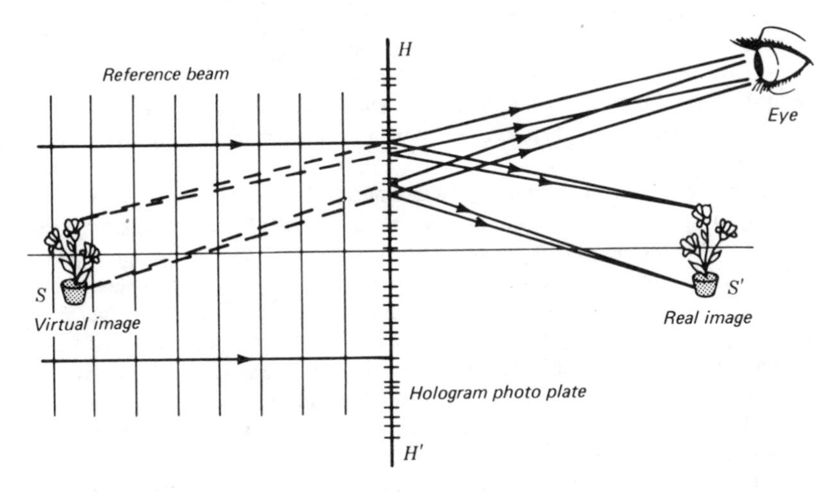

FIGURE 31D
An object at S and a reference beam form a complicated array of Gabor zone plates on HH', which upon development is illuminated by the same reference beam. The eye now observes a virtual image at S and a real image at S'. A screen or photographic plate at S' will now register this real image.

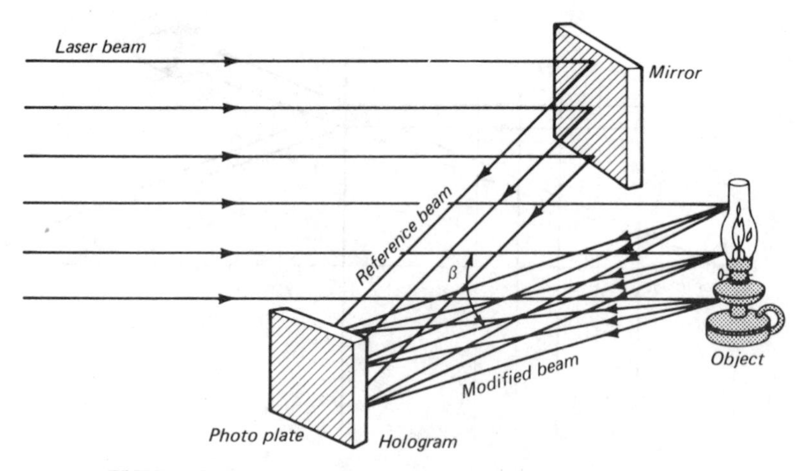

FIGURE 31E
Monochromatic, coherent laser light is reflected, unchanged, onto a photographic plate. Part of the beam is modulated by its reflection from an object to the same plate. When developed, the plate reveals interference fringes called a hologram.

the most significant being the lack of a sufficiently coherent light source. With the advent of the laser, the outlook for holography changed dramatically.

However, a second difficulty appeared in the form of a real image caused by light diffracted in the opposite direction. This image was generally observed in front of the first image, and therefore it was in the way when viewing the virtual image (see Fig. 31D).

The next major breakthrough was made in 1962 by Leith and Upatnieks, who developed the idea of the off-axis hologram.* This can be seen as a simple extension of the Gabor hologram, using an off-axis section of the holographic plate. This variation was made possible by the increased *coherence length* of the laser beam.

This simple variation not only separated the *real* from the *virtual image* line of sight but allowed for separate handling of the reference and scattered beam. The object could now be illuminated from any side or several sides. Moreover, it is not necessary that the reference beam be normally incident plane waves, provided that it is produced by the equivalent of a point source and that the reconstructing beam readily reproduces it.

One method of producing such a hologram is shown in Fig. 31E, where an incident laser beam is split into two beams, one of which changes direction as it strikes a *plane mirror* and the other is scattered by the *object*. At the photographic plate, the two beams interfere in a very irregular pattern, as shown in Fig. 31D. The angle β between the *scattered light* and the *reference beam* will determine the density of the

* G. N. Leith and J. Upatnieks, *J. Opt. Soc. Am.*, **52:**1123 (1962).

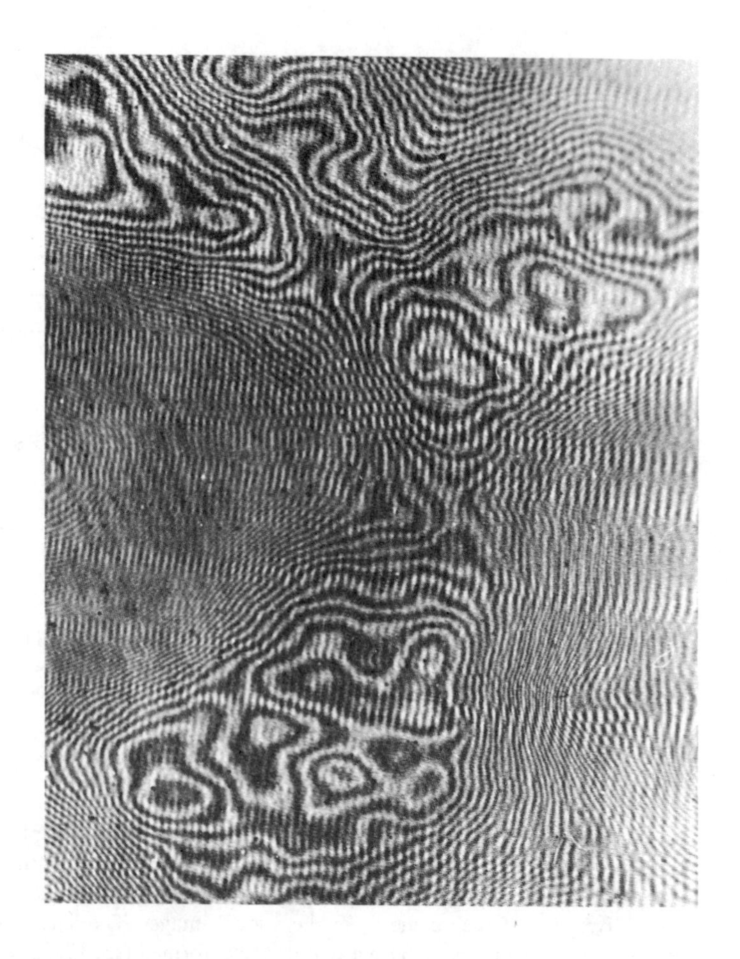

FIGURE 31F
An enlarged section of a plane hologram made with λ = 6328 Å from a He–Ne gas laser. (*Conductron Corporation.*)

fringes, or *spatial frequency*. If the angle is small, the spatial frequency will be low (fringes far apart), but visual interference of the real image will be severe. Moreover, a mottled background can be seen, called *intermodulation noise*, due to fringes produced by the interference of light from various parts of the object.

By using larger angles, these effects can be eliminated, but the resulting high spatial density will require high-resolution film, and particular care must be taken to avoid relative motion of the optical components during exposure (see Fig. 31F).

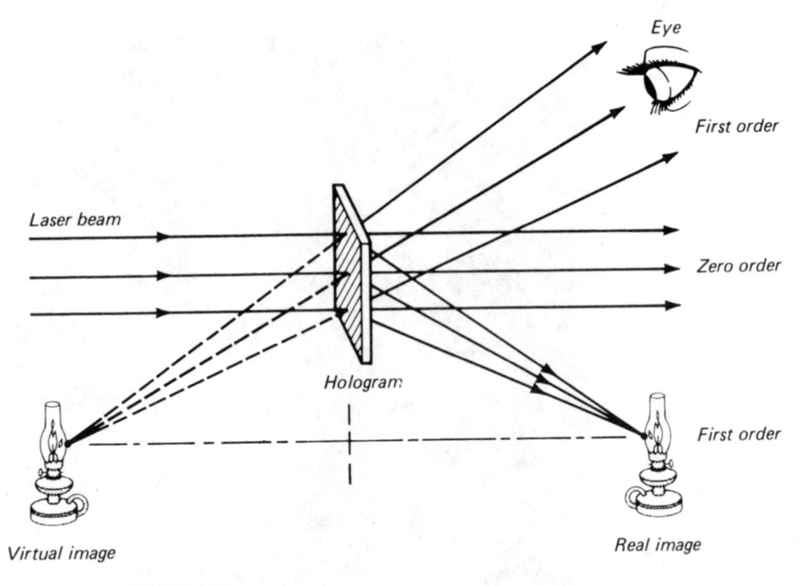

FIGURE 31G
The monochromatic, coherent laser beam is incident on a hologram, where it is modulated to produce two diffracted waves, the first order on each side. The remainder of the direct beam forms the unchanged zero order.

31.2 VIEWING A HOLOGRAM

To see the reconstructed object when a hologram is made, the photographic plate containing the interference fringes is placed in the monochromatic beam from the same laser used making the picture and with the same alignment. The diffracted waves diverge as if they came from the virtual image. The lens of the eye focuses these waves on the retina, where a real image is formed (see Fig. 31G).

The original waves producing the interference fringes and the waves reconstructing the image will be identical in all optical respects. The image is not only three dimensional but in perspective as well, and will change as the viewer moves his head. As the observer moves his eyes to different positions, the rays of light entering each pupil come through small but different sections of the fringe pattern on the hologram, and he sees the object in different perspective. If he finds an object hidden behind another, he can move his head and look around the nearby obstacle, thereby seeing the hidden object.

If the reconstruction beam does not reproduce the original reference beam geometrically, the image will be distorted. Illumination by light of wavelength different from the original will cause both a change in size and displacement of the image. Illumination by a spectral distribution will produce color fringing. The normal shrinkage of a photographic emulsion during development is sufficient to cause minor distortion similar to that caused by increasing the wavelength of the reference beam.

If the hologram is broken into many small pieces, each piece will be a hologram of the complete object scene. However, the perspective will be limited accordingly, and there may be a loss in resolution.

A hologram made in the above fashion might be thought of as a negative. Every hologram, however, is a positive print. If any hologram is copied by contact printing, thereby reversing black for white and white for black, it will produce the same images and not a reversal. This is similar to a Fresnel zone plate, where complementary zones produce identical bright spots as foci. For complementary-zone plates see Fig. 18I.

If the emulsion of a hologram is bleached by normal photographic processes after it has been fixed, the darkened silver grains are replaced by transparent media of a different refractive index. Under these conditions the film will appear uniformly transparent. This changes an *absorption* hologram into a *phase* hologram, increasing its clarity.

The *real image* from a hologram can be formed on a screen, and a photographic plate located there can be developed into a real picture. This same image can be observed by locating the eye beyond the real image, where it can intercept the waves diverging from their points of intersection in the three-dimensional image. The eye must be located far enough back, at least to the distance of most distinct vision, for the object to be seen sharply.

The undistorted real image has some visual characteristics foreign to the trained senses. As shown in Fig. 31G, the image of the lamp is illuminated on the front surface, and the real image displays that side even though it is spatially behind the other surface and should obscure it. A hologram made using an opaque object produces a *pseudoscopic image* which displays contradictory visual cues, which must be seen to be appreciated. As a result, the real image is of limited use.

31.3 THE THICK, OR VOLUME, HOLOGRAM

The holograms discussed above have been assumed to have negligible thickness and are referred to as *plane holograms*. If the recording medium is thick with respect to the spatial frequency, the interference fringes act as a series of ribbons, somewhat similar to a venetian blind. The reconstructing beam will generally pass through several sets of such fringes. This third dimension has the effect of adding an additional constraint on the diffraction pattern produced in a way similar to Bragg scattering of X rays from crystals.

In the Bragg-scattering experiments, used so much in X-ray studies, the regularly spaced atoms in the crystal act like partially reflecting planes, scattering the waves in definite preferred directions (See Fig. 31H). In these preferred directions the waves reflected from adjacent planes differ from each other by exactly one wavelength and, being in phase with each other, produce constructive interference. The Bragg-scattering relationship for these directions is given by

• $$\lambda = 2d \sin \theta \qquad \text{(31b)}$$

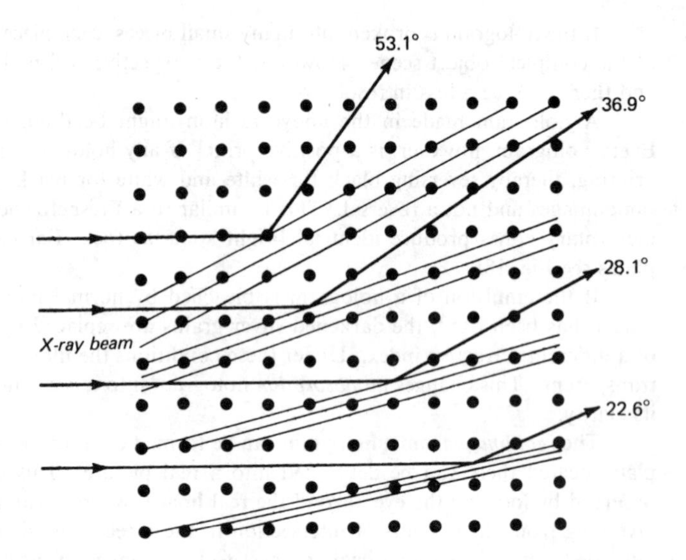

FIGURE 31H
Diagram of the reflection of X rays from the various atomic planes in a cubic crystal lattice.

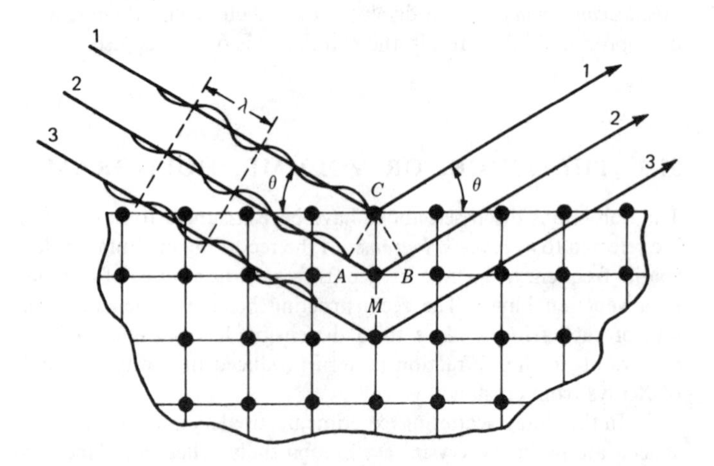

FIGURE 31I
Geometry illustrating the Bragg rule of reflection for X rays from the surface layers of a cubic crystal.

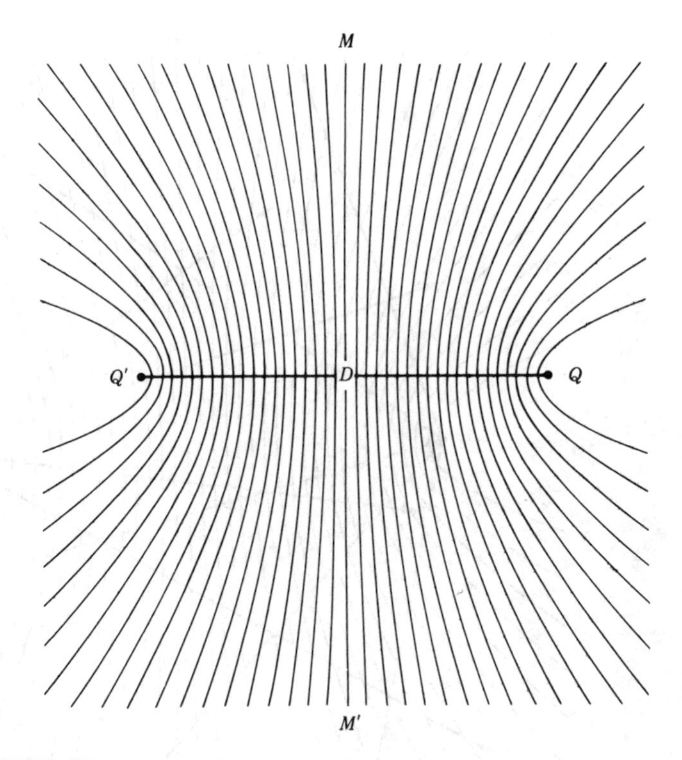

FIGURE 31J
Two point sources Q and Q' emitting monochromatic coherent waves constructively interfere along hyperboloidal surfaces.

where d is the distance between reflecting planes, λ is the wavelength of the waves, and θ is the reflection angle shown in Fig. 31I. This principle of Bragg reflection forms the basis of a particularly simple geometrical model* that can be used to account for most of the features of the thick hologram.

First consider two coherent point sources of light waves Q and Q', of wavelength λ, separated by a distance D as shown in Fig. 31J. Every point on the midplane MM', bisecting the line connecting the sources, will be equidistant from the sources and will therefore be a point of constructive interference. Other surfaces of constructive interference can be found, each of which corresponds to a difference in path length from the two sources of an integral number of wavelengths. These surfaces can be shown to be hyperboloids, which are separated by $\lambda/2$ as measured along the line connecting the sources.

* The simple geometrical model developed here for thick holograms is to be attributed to T. H. Jeong. The hyperboloids drawn in Figs. 31J, 31K, and 31L were generated by computer. See T. H. Jeong, Geometrical Model for Holography, *Amer., Jour. Phys.*, **43**: 714 (1975).

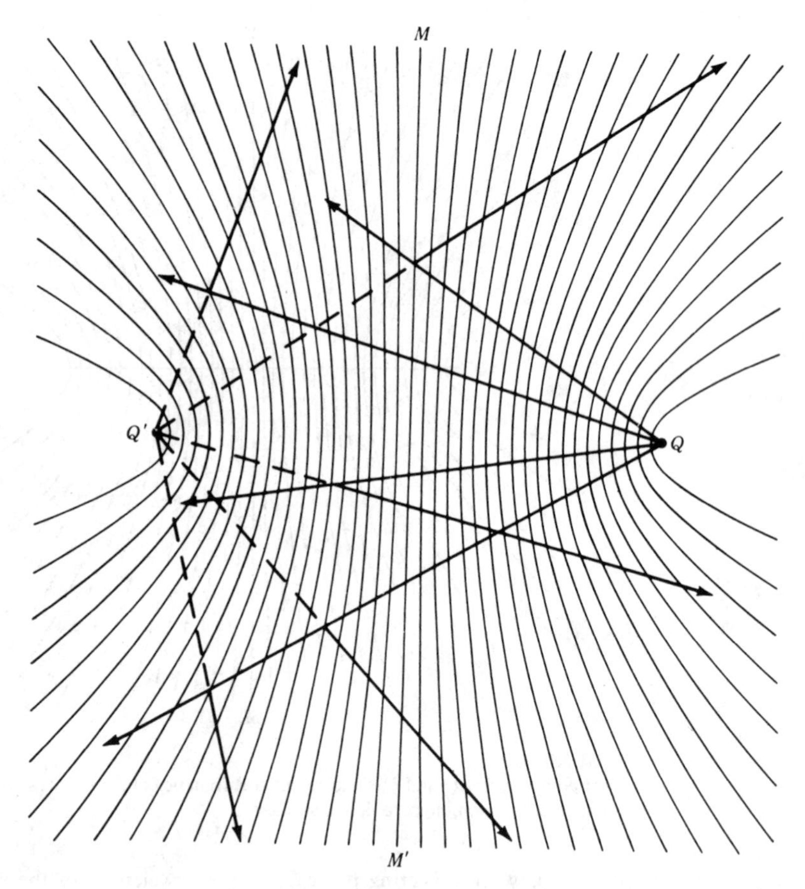

FIGURE 31K
Any ray from source Q can be reflected by any of the hyperboloidal mirrors and in such a direction that all rays appear to come from Q'.

Imagine now that each of these surfaces in the developed emulsion is a partially reflecting surface and that point Q acts as a source of coherent illumination. The midplane acts as a plane mirror, creating a virtual image at Q' (Fig. 31K); see Fig. 3E. Moreover, reflection from any portion of any of the hyperboloidal surfaces will obey the law of reflection and emerge as if they diverged from Q'. The reflected pattern from any volume occupied by the fringe surfaces will then produce a virtual image at Q'.

Consider now that Q in Fig. 31L is a primary source, e.g., a laser. Point Q' is a secondary coherent source, a scattering center exposed to the primary laser beam. A thick photographic emulsion HH' is now exposed to the interfering light at an

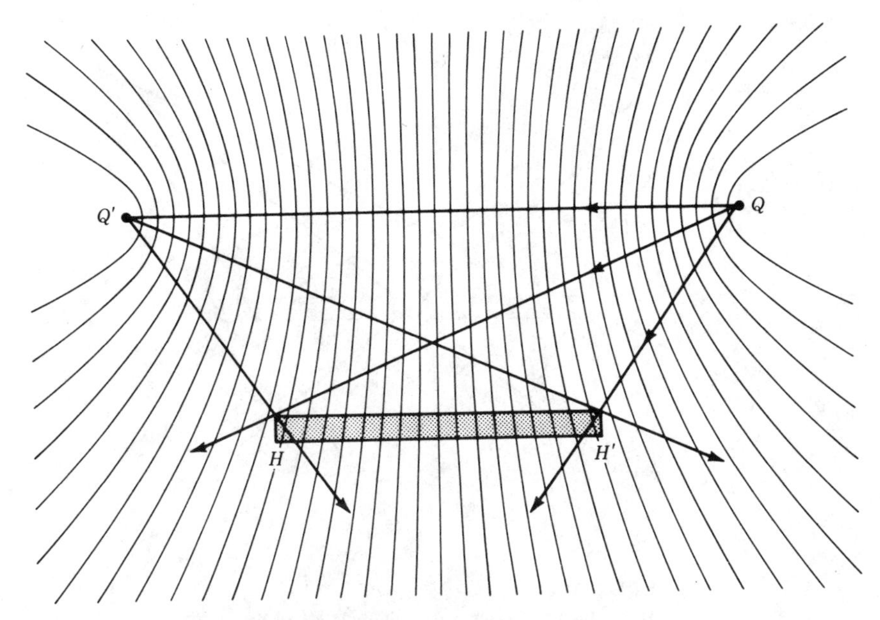

FIGURE 31L
A thick-hologram model which assumes that the interference fringe pattern between two monochromatic coherent point sources forms inside the volume of the recording medium a set of partially reflecting, absorbing, and transmitting hyperboloidal surfaces.

off-axis position. When the film is developed, it will contain darkened bands representing the portions of the hyperbolic surfaces of constructive interference. The developed image consists of grains of silver. Actually, fringes may consist of any material, or simply a change of refractive index, as in a *bleached emulsion,* is sufficient. When this hologram is illuminated from point Q and viewed on the far side, a virtual image will appear at Q' (see Fig. 31M).

As with the plane hologram, the argument can now be extended to account for the formation of a hologram capable of producing the virtual image of a *distributed object* (see Fig. 31N). Such a hologram would be thought of as a superposition of sets of hyperboloidal mirrors. When the hologram is viewed, each set reflects light from the reference beam and forms an image of a point on the object.

31.4 MULTIPLEX HOLOGRAMS

One of the remarkable features of the thick hologram is its ability to produce multiple scenes from the same photographic emulsion. If the distance between the fringes is smaller than the emulsion thickness, each ray of the reconstruction light originating from the direction of the reference beam will pass through several partially reflecting

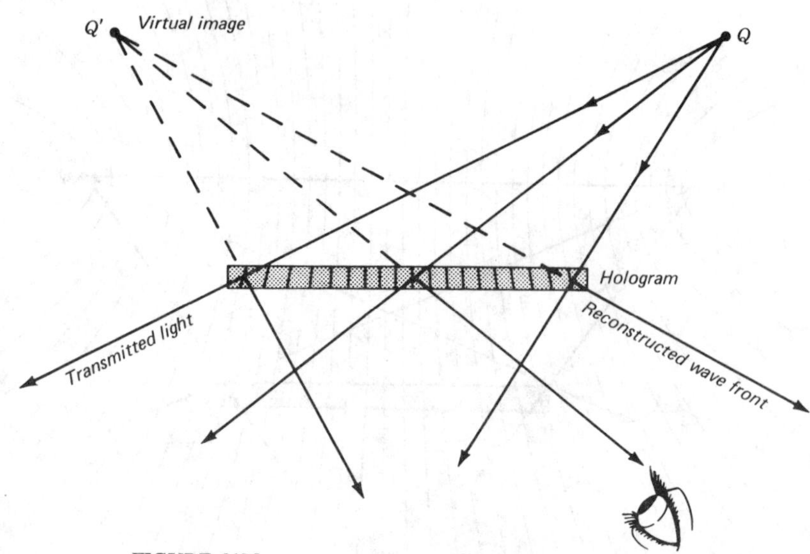

FIGURE 31M
The virtual image Q' is created by illumination of the thick hologram by a point source Q.

planes (see Fig. 31O). The reflected rays from each of these planes must be an integral number of wavelengths apart. If the reillumination beam forms an angle significantly different from the reference beam, the light reflected from the adjacent planes will no longer be in phase and the virtual image will no longer be visible.

It is therefore possible to produce many holograms in the same photosensitive medium, each with the reference beam at a different angle. When viewed later, each of these images can be separately viewed simply by varying the angle of the reference beam. This technique has been used to store hundreds of images in a single crystal of *lithium niobate*. The process is capable of storing an entire book in an appropriate medium by slightly changing the direction of the reference beam with each exposure. When viewing the finished hologram, one can "turn the page" by merely moving the reconstructing beam.

Alternatively, a multiplex hologram can be produced by appropriately moving the reference beam angle with time, thereby producing holographic motion pictures.

31.5 WHITE-LIGHT-REFLECTION HOLOGRAMS

One of the possible arrangements for producing white-light holograms is to place the photosensitive film between the reference beam and the object (see Fig. 31P). Such a hologram is produced simply by illuminating the object through the photosen-

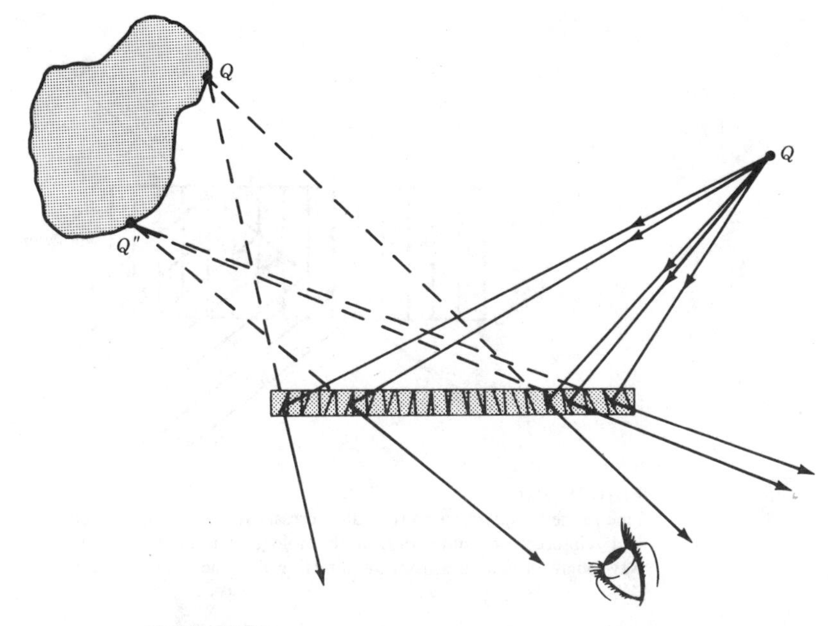

FIGURE 31N
A three-dimensional object is seen as the superposition from many sets of surfaces in the thick hologram by the interference of the reference beam with the light from points on the object.

sitive medium, thus avoiding beam splitters, mirrors, etc. In practice, the reference intensity is so high relative to the scattered intensity that the technique is limited to shiny objects located close to the recording medium. Better reflection holograms can be made by separating the object and reference beams.

Since the reference and object beams are oppositely directed, the *spatial frequency* is extremely high. A large number of reflecting planes are thereby produced, separated by about a half wavelength of light. As a result, the reconstructing light must be of the same wavelength or the reflections from adjacent planes will not be in phase for constructive interference. Alternatively, if the hologram is viewed in white light (sunlight is an excellent source), the appropriate wavelength will be selected to produce the reflected image. Ordinary photographic emulsions are of limited use as they tend to shrink during development.

The technique is especially useful in that a laser is not required for viewing. Moreover, if the hologram is produced by illumination by lasers which produce the *three additive primary colors* (*red*, *green*, and *blue*), the resulting hologram will be seen in full color when viewed in white light.

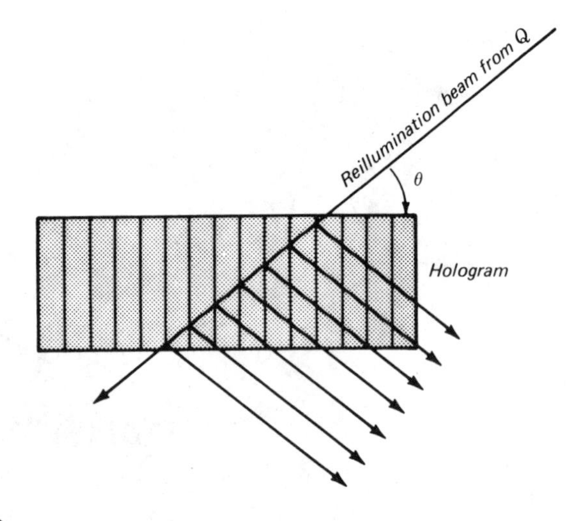

FIGURE 31O
Due to the Bragg reflection rule, all successively reflected waves will be in phase and reinforce each other only if the hologram is illuminated with the same wavelength of light and from the direction θ of the original reference beam Q.

31.6 OTHER HOLOGRAMS

A wide variety of holograms can be produced to achieve special effects. These include using lenses and mirrors and using other holographic images as objects.

One of the most impressive holographic images is formed by a 360° circular film. The technique was developed by T. H. Jeong using a photographic emulsion mounted on a cylindrical surface surrounding the object (see Fig. 31Q). The simplest, but not necessarily the best, method of illumination is to direct a diverging beam from above, illuminating the entire emulsion and object. Upon reillumination, the virtual image will be observed in the center of the cylinder, and can be viewed from all sides. If a high-intensity beam from a pulsed laser is used, there is no problem of using a jiggle-free table mounting.

At this point in the development of the art of photography, a brief comparison of lens photography and picture images with lensless photography and diffraction fringes should be mentioned. Both techniques have their advantages and disadvantages depending on the purposes for which they are used. The amount of information stored in an emulsion depends solely upon the smallness of the grain of the finished product. In the limit this appears to be determined by the size of the atoms and molecules of the storage medium itself. See Fig. 31R.

It would appear, for example, that the side-by-side storage of microscopic pictures can be equaled by the storage of superimposed sets of interference fringes in a thick hologram. On the other hand, the fine detail of the three-dimensional images

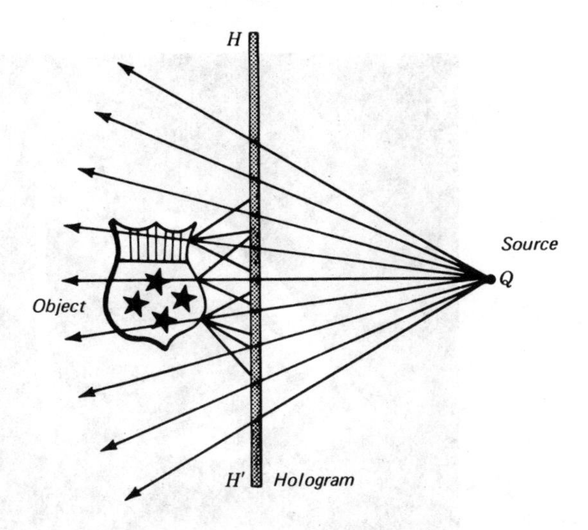

FIGURE 31P
Reflection hologram made from a single source and a transparent emulsion.

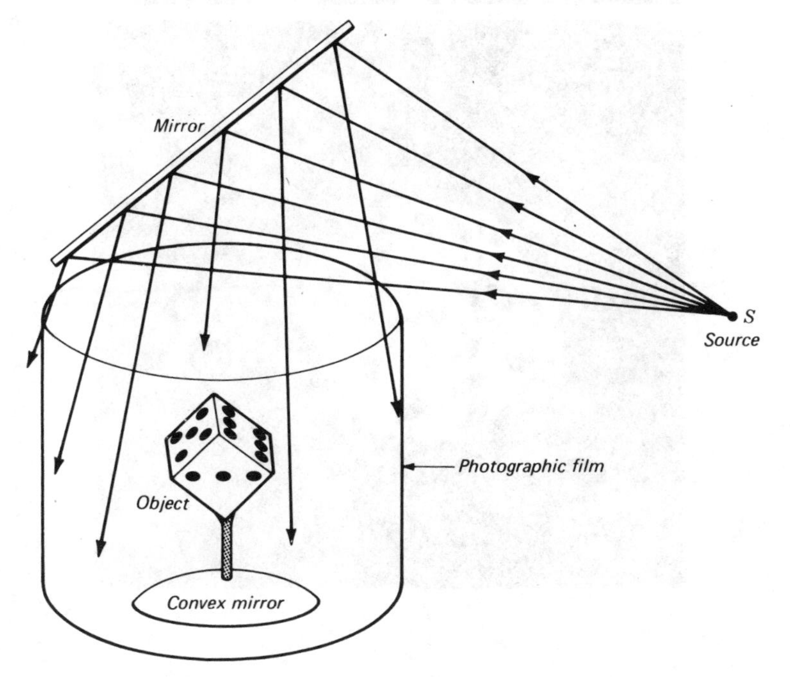

FIGURE 31Q
A 360° circular hologram can be made that can be viewed from all sides.

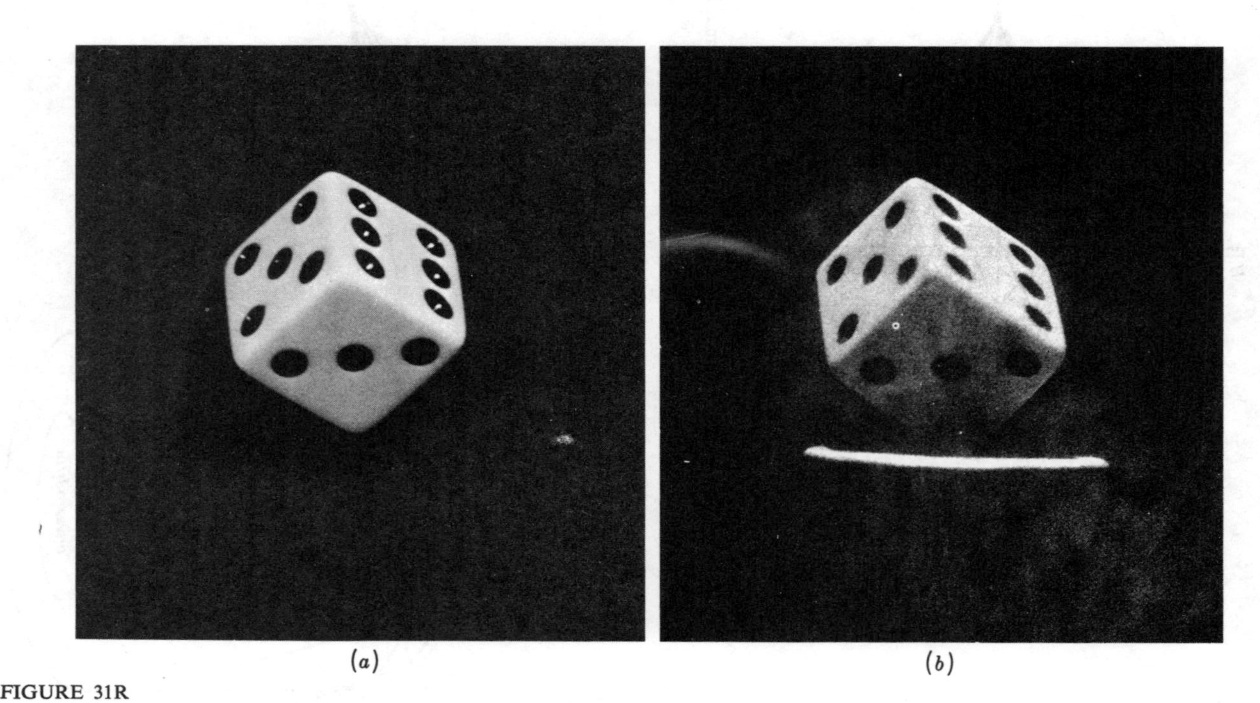

(*a*) (*b*)

FIGURE 31R
(*a*) A direct camera photograph of a 16-mm die made with an Exacta camera on 35-mm Plus X film. (*Courtesy of A. D. White.*) (*b*) Photograph of the same die seen in a 360° cylindrical hologram made with an arrangement like that shown in Fig. 31Q. (*Metrologic Instruments Inc., Bellmawr, N.J.*)

observed in full color and formed by a high-quality lens or concave mirror is to be compared with the three-dimensional images that can be stored in a hologram and used for later viewing.

31.7 STUDENT LABORATORY HOLOGRAPHY

Holography is such an intriguing subject that many students in the science laboratory wish to make and observe their own holograms. Briefly described here is an inexpensive experimental arrangement that requires a minimum of space and equipment. Since the interference maxima in a hologram are about one-half wavelength apart, very fine grain emulsions should be used and considerable care must be taken to avoid jiggling the optical components during exposure.

To reduce the vibration hazard, all components, including the laser, must be mounted on a vibration-free block or heavy plate. For this purpose a steel plate 70 to 90 cm square and 1 to 2 cm thick should be drilled and tapped with a mosaic pattern of holes for mounting the components rigidly. When all is in readiness for photography, this plate should be taken to a darkroom and placed on an inflated automobile inner tube. A valve stem mounted on the outer edge of the tube provides for easy inflation and adjustment.

A relatively popular arrangement is to construct a sandbox, fill it with dry sand, and mount it on several inner tubes. Optical components are each mounted on one end of a solid wood or plastic rod, about 4 cm in diameter and 30 cm long, pointed at the lower end. Pushed into the sand like a garden stake, this mounting is free of vibrations.

A diagram showing all components and their functions is given in Fig. 31S. M_1, M_2, and M_3 are front-silvered mirrors; MO is a microscope objective for spreading the beam. A pinhole placed at the focal point of the microscope objective will allow the undeviated laser beam to pass but will block out stray light originating in the laser or from diffraction by dust or the preceding optical components. The size of the pinhole should be about 25μm for a 10$\times$ objective and about 1 μm for a 60$\times$ objective.

Although a more uniform hologram is produced by such a *spatial filter*, it is not essential and may not be worth the effort involved in aligning the pinhole. B is a beam splitter, which is best if it reflects at least 75 percent of the light. The angle α should be 15 to 25°.

One major problem arises in the relative weakness of the modified light reflected from the object. Since the object scatters light in all directions, only a small part of it reaches the photographic plate. The maximum fringe contrast on the hologram is theoretically attained when the total light from each beam is approximately equal (see Sec. 13.4). However, in practice, the scattered beam should be 3 to 10 times weaker than the reference beam to reduce fogging of the plate due to *intermodulation noise*.

Care should be taken to approximately equalize the two path lengths in case the coherence length of the laser beam is reduced by multiple modes of oscillation. Susceptibility to vibrations should be checked before using the mounting table by

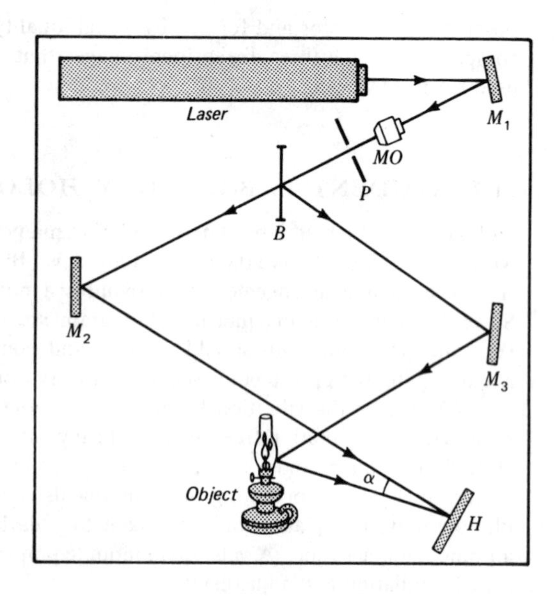

FIGURE 31S
Apparatus layout and components essential to making holograms. Components are rigidly mounted on a steel plate about 90 cm square, or on wooden stakes in the sand of a sandbox, resting on an inflated inner tube to reduce vibrations.

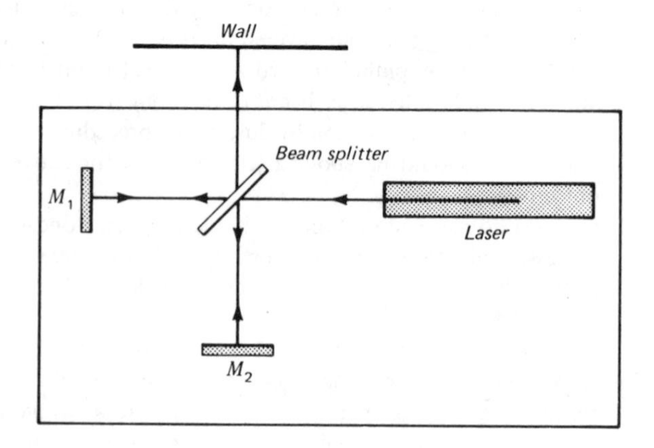

FIGURE 31T
Michelson interferometer arrangement for locating sources of vibration affecting the table set up for making holograms in the college physics laboratory.

arranging the various components to form a Michelson interferometer and projecting the fringes on a nearby wall (see Fig. 31T). A shift of one-half fringe during the time of exposure is enough to prevent any image of fringes at all, and a smaller shift is sufficient to reduce the image quality significantly. Such a test may indicate that components are creeping, that they are affected by air drafts, or that the system is jiggled by elevators, machinery, or people walking in the hall nearby. Appropriate countermeasures can then be taken. High-resolution film must be used, and several trial photographs are necessary before satisfactory holograms are obtained.

REFERENCES

CAMATINI, E.: "Optical and Acoustical Holography," Plenum Press, New York, 1972.

COLLIER, ROBERT J., CHRISTOPH B. BURCKHARDT, and LAWRENCE H. LIN: "Optical Holography," Academic Press, Inc., New York, 1971.

FRANCON, M.: "Holographie," Springer-Verlag, Berlin, 1972.

GOODMAN, J. W.: "Introduction to Fourier Optics," McGraw-Hill Book Company, New York, 1968.

HILDEBRAND, B. P., and B. B. BRENDEN: "Applications of Holography," Plenum Press, New York, 1971.

PROBLEMS

31.1 Coherent plane waves and the waves scattered from a point source fall together on a photographic plate as shown in Fig. 31A. If the wavelength of the light is 6563 Å and the perpendicular distance from the point source to the emulsion is 5.0 cm, find (*a*) the radius of the tenth bright fringe from the center of the developed pattern. (*b*) What is the distance between the tenth and eleventh bright fringes? Assume that the waves at the center of the pattern are in phase and on the developed film are black.
Ans. (*a*) 0.83016 mm, (*b*) 0.07433 mm

31.2 The beam from a ruby laser emitting red light of wavelength 6943 Å is used with a beam splitter to produce two coherent beams. Both are reflected from plane mirrors and brought together on the same photographic plate. If the angle α between these two interfering beams is 10° and the plate normal bisects this angle, find the fringe separation of the interference fringes on the plate.

31.3 Two point sources of coherent light Q and Q' are located 25.0 cm apart, as shown in Fig. 31J(*a*). (*a*) Find the fringe spacing along the center line QQ' if the wavelength of the light is 5461 Å. (*b*) How many fringes are there per millimeter?

31.4 In one part of a thick hologram a number of ribbonlike fringes are found parallel to each other and 3.750×10^{-4} mm apart. At what angle with respect to these ribbons will light be reflected in the first order if its wavelength is 6563 Å? *Ans.* 61.053°

32
MAGNETO-OPTICS AND ELECTRO-OPTICS

We have already seen in Chap. 20 and Secs. 23.9, 26.9, and 28.9 that the electromagnetic theory is capable of explaining the main features of the propagation of light through free space and through matter. In further support of the electromagnetic character of light, there is a group of optical experiments which demonstrates the interaction between light and matter when the latter is subjected to a strong external magnetic or electric field. In this group of experiments those which depend for their action on an applied magnetic field are classed under *magneto-optics* and those which depend for their action on an electric field are classed under *electro-optics*. In this chapter the following known optical effects will be treated briefly under these headings:

Magneto-optics	*Electro-optics*
Zeeman effect	Stark effect
Inverse Zeeman effect	Inverse Stark effect
Voigt effect	Electric double refraction
Cotton-Mouton effect	Kerr electro-optic effect
Faraday effect	
Kerr magneto-optic effect	

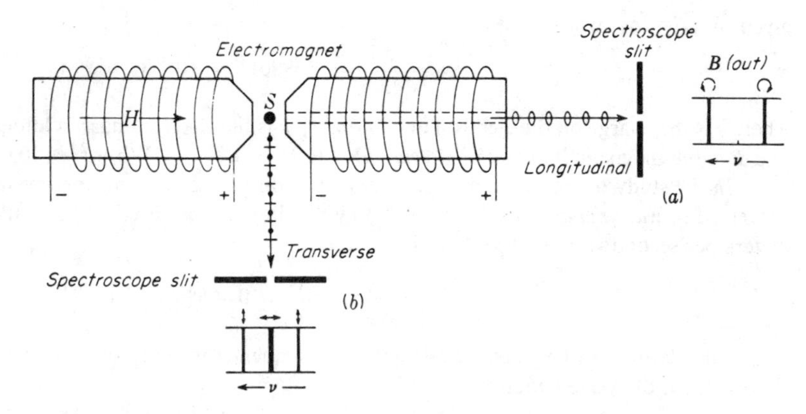

FIGURE 32A
Experimental arrangement for observing the Zeeman effect.

The four electro-optic effects in the order listed here are respectively the analogues of the first four magneto-optic effects.

32.1 ZEEMAN* EFFECT

In the year 1896, Zeeman discovered that when a sodium flame is placed between the poles of a powerful electromagnet, the two yellow lines are considerably broadened. Shortly afterward, Lorentz presented a simple theory for these observations, based upon the electron theory of matter, and predicted that each spectrum line when produced in such a field should be split into two components when viewed parallel to the field [Fig. 32A(*a*)] and into three components when viewed perpendicular to the field [Fig. 32A(*b*)]. He further predicted that in the longitudinal direction (*a*) these lines should be circularly polarized and in the transverse direction (*b*), plane-polarized. With improved experimental conditions these predictions were later verified by Zeeman, Preston, and others in the case of some spectral lines.

The Lorentz theory assumes that the electrons in matter are responsible for the origin of light waves and that they are charged particles whose motions are modified by an external magnetic field. In the special case of an electron moving in a circular orbit, the plane of which is normal to the field direction B, the electron should be speeded up or slowed down by an amount proportional to the magnetic induction B. A classical treatment of this problem shows that if ν_0 represents the orbital frequency of the electron in a field-free space, the frequency in the presence of a field will be

* P. Zeeman (1865–1943). Dutch physicist and Nobel prize winner (1902). He is most famous for his work on the splitting up of spectral lines in a magnetic field. His chief contributions are summarized in his celebrated book "Researches in Magneto-optics," Macmillan & Co., Ltd., London, 1913.

given by $\nu_0 \pm \Delta\nu$, where

$$\Delta\nu = \frac{eB}{4\pi m} = 1.399611 \times 10^{10}B \qquad s^{-1} \tag{32a}$$

where e is the charge on the electron in coulombs, m is the electron mass in kilograms, and B is the magnetic induction in *teslas*. One tesla $= 1\ \text{T} = 1\ \text{Wb/m}^2 = 10{,}000\ \text{G}$.

In the study of spectrum lines this frequency difference $\Delta\nu$ is most conveniently expressed in *wave numbers* (see Sec. 14.14) by dividing by the speed of light in centimeters per second; $c = 2.997925 \times 10^{10}$ cm/s:

$$\Delta\sigma = \frac{\Delta\nu}{c} = 0.46686B \qquad \text{cm}^{-1} \tag{32b}$$

A useful relationship between wavelength and frequency in hertz or wave numbers follows from the wave equation $c = \nu\lambda$:

$$\frac{\Delta\lambda}{\lambda} = \frac{\Delta\nu}{\nu} = \frac{\Delta\sigma}{\sigma}$$

where $\Delta\lambda$ is small compared with λ, $\Delta\nu$ is small compared with ν, and $\Delta\sigma$ is small compared with σ.

In the classical theory of the Zeeman effect we are concerned with an aggregation of atoms in which the electrons are revolving in circular or elliptical orbits oriented at random in space. It will now be shown, however, that this situation is equivalent to having one-third of the electrons vibrating in straight lines along the direction of the magnetic field and two-thirds of them revolving in circular orbits in the plane perpendicular to the field. Of the latter ones, half are revolving in one sense and half in the opposite sense. The radius of their orbits is $1/\sqrt{2}$ times the amplitude of the linear vibrations. To prove these statements, let us select any one of the electrons and resolve its elliptical motion into three mutually perpendicular linear motions as shown in Fig. 32B(*a*). For simplicity we shall assume that the electron is bound by an elastic force obeying the law

$$F = -kr \tag{32c}$$

where r is the displacement from the equilibrium position. Under this condition the three components are simple harmonic motions, but for any one electron they are not equal in amplitude or in the same phase.

If a magnetic field is now applied in the z direction, the component parallel to z will be uninfluenced, for it is equivalent to a current directed along the lines of force. The x and y vibrations will each be modified, however, since an electron which is moving across a magnetic field experiences a force

$$F_B = Bev \tag{32d}$$

perpendicular to the field and also perpendicular to its motion. The effect of this force is to change the x and y components into rosette motions such as that shown in Fig. 32B(*b*) for the y component. These can be described to better advantage in terms of circular components, y^+ and y^- for the y motion, and x^+ and x^- for the x motion [diagram (*c*) of the figure]. In the presence of the field both plus circular components have a higher frequency than the minus ones, so we may combine the x^+ and y^+ motions to get a resultant positive circular motion, as in diagram (*d*), and

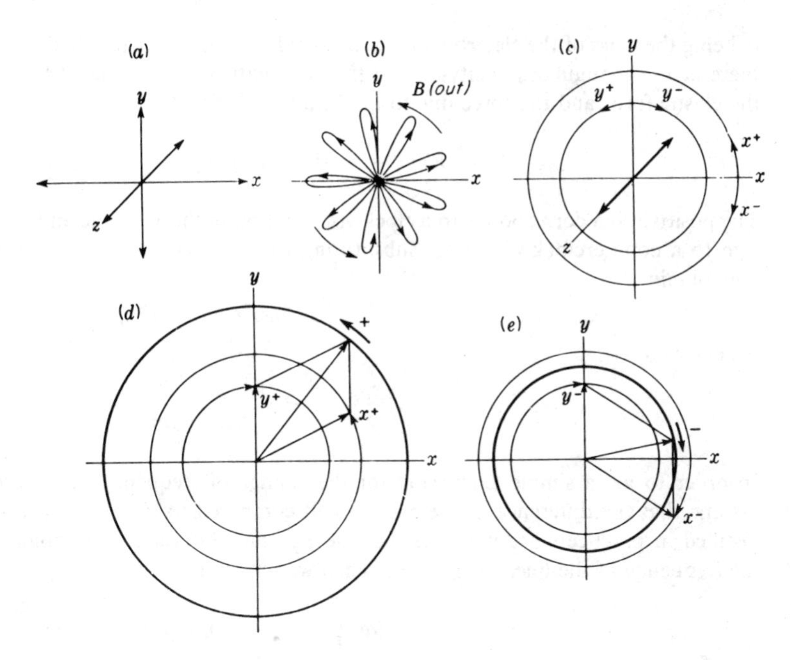

FIGURE 32B
Resolution of an orbit for the explanation of the classical Zeeman effect.

x^- and y^- to get a negative one, as in (*e*). Thus the original elliptical orbit when subjected to a magnetic field is equivalent to a linear motion of unchanged frequency along the field, plus two circular motions, one of higher and one of lower frequency, in the plane at right angles to the field.

Only the circular components will emit light along the field direction, and these give circularly polarized light of two different frequencies. The intensity of these two components must be equal when the whole aggregation of atoms is considered, because as the field goes to zero, the light is unpolarized. When we observe the light at right angles to the field, we are viewing the circular components edge-on, so these yield two different frequencies of plane-polarized light in which the vibrations are perpendicular to the field direction. Each of them has only half the intensity of the above-mentioned circularly polarized beams. In addition, the linear z motions emit light in the transverse direction. This light has the original frequency ν_0, vibrates parallel to the field, and has an intensity equal to the sum of the other two. The mean amplitude of the z components for all atoms is therefore $\sqrt{2}$ times as great as that of the x or y components.

Now let us calculate the change of frequency to be expected for the circular components. In the absence of the field the centripetal force on the electron in its circular orbit is furnished by the elastic force, so that by Eq. (32c) we have

$$F = -kr = -m\omega_0^2 r \qquad (32e)$$

m being the mass of the electron and ω_0 its angular velocity. After the field is applied, there is a new angular velocity ω, and the new centripetal force must be the sum of the elastic force and the force due to the field (Eq. 31d). Thus

$$F' = -m\omega^2 r = F \pm F_B = -kr \pm Bev$$

The positive sign corresponds to a clockwise rotation in the xy plane and the negative sign to a counterclockwise one. Substituting for $-kr$ its value from Eq. (32e), we then obtain

$$-m\omega^2 r = -m\omega_0{}^2 r \pm Bev$$

or, since $v/r = \omega$,

$$\omega^2 - \omega_0{}^2 = \pm \frac{Bev}{mr} = \mp \frac{Be\omega}{m} \tag{32f}$$

In order to get a simple expression for the change of frequency, it is necessary to assume that the difference in the ω's is small compared to either ω. This is always justified in practice since it means that the Zeeman shifts are small compared with the frequency of the lines themselves. Then we may put

$$(\omega + \omega_0)(\omega - \omega_0) \approx 2\omega(\omega - \omega_0)$$

and, from Eq. (32f),

$$\omega - \omega_0 = \pm \frac{Be\omega}{m2\omega} = \pm \frac{Be}{2m}$$

Since $\nu = \omega/2\pi$, the change in frequency becomes

$$\Delta\nu = \pm \frac{Be}{4\pi m} \tag{32g}$$

in agreement with Eq. (32a).

In this derivation it has been tacitly assumed that the radius of the circular motion remains unchanged during the application of the magnetic field. The speeding up or slowing down of the electron in its orbit occurs only while the field is changing and is due to the changing number of lines of force threading the orbit. By Faraday's law of induction this change produces an emf just as it would in a circular loop of wire. The resulting increase or decrease of velocity might be expected to change the radius, but the fact is that there is a corresponding alteration in the centripetal force which is just sufficient to maintain the radius constant. The additional force is that represented by Eq. (32d), which has the same origin as the perpendicular force on a wire carrying a current in a magnetic field.

Let us now summarize what should be the observed effect of a magnetic field on a spectrum line. The result will depend on the direction, with respect to that of the magnetic field, in which the source is viewed. When the source is viewed in the direction of the field, along the z axis, we have what is called the longitudinal Zeeman effect. From this direction only the frequencies $\nu_0 + \Delta\nu$ and $\nu_0 - \Delta\nu$ should appear,

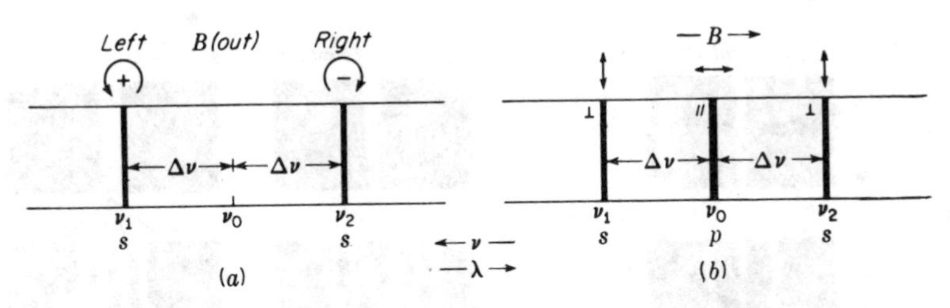

FIGURE 32C
Zeeman patterns for a normal triplet, showing the polarization of the light.

and this light should be right- or left-handed circularly polarized* [Fig. 32C(*a*)]. Since light is a transverse wave motion, the z vibrations will not emit light of frequency ν_0 in the z direction

Viewed perpendicular to the field, the z motions should be observed to give plane-polarized light with the electric vector parallel to the field (p components), and the circular motions, seen edge-on, should give plane-polarized light with the electric vector perpendicular to the field (s components). A spectrum line viewed normal to B should therefore reveal three plane-polarized components [Fig. 32C(*b*)] —a center unshifted line, and two other lines symmetrically located as shown. This is called a *normal triplet* and is observed for some spectrum lines, though by no means the majority of them.

Since the direction of rotation of the circularly polarized light depends on whether one assumes positive or negative charges as the emitters of light, it is possible to distinguish between these alternatives by using a quarter-wave plate and nicol. Figure 32C(*a*), where the positive rotation has the higher frequency, was drawn according to our assumption of negative electrons as the emitters.

In Zeeman's early investigations he was not able to split any spectrum lines into doublets or triplets, but he did observe that they were broadened and that the outside edges were polarized, as predicted by Lorentz. The polarization corresponded to emission by negative particles. He was later able to photograph the two outer components of lines arising from the elements zinc, copper, cadmium, and tin, by cutting out the p components with a nicol prism. Preston, using greater dispersion and resolving power, was able to show not only that certain lines were split up into triplets when viewed perpendicularly to the field but that others were split into as many as four, five, or even a much larger number of components. Such patterns of lines, shown in Fig. 32D, are called anomalous Zeeman patterns, and the phenomenon is called the *anomalous Zeeman effect*. The normal triplet separation $2\Delta\nu$ as

* Using the right-hand rule with the thumb pointing in the direction of the field, the fingers point in the direction of the + rotations which have the higher frequency designated by ν_1. The opposite direction gives the − rotations with the lower frequency ν_2. Looking against the light, clockwise rotations give rise to right-handed polarized light and counterclockwise rotations give rise to left-handed circularly polarized light. This latter is in agreement with the definitions used in treating optically active substances.

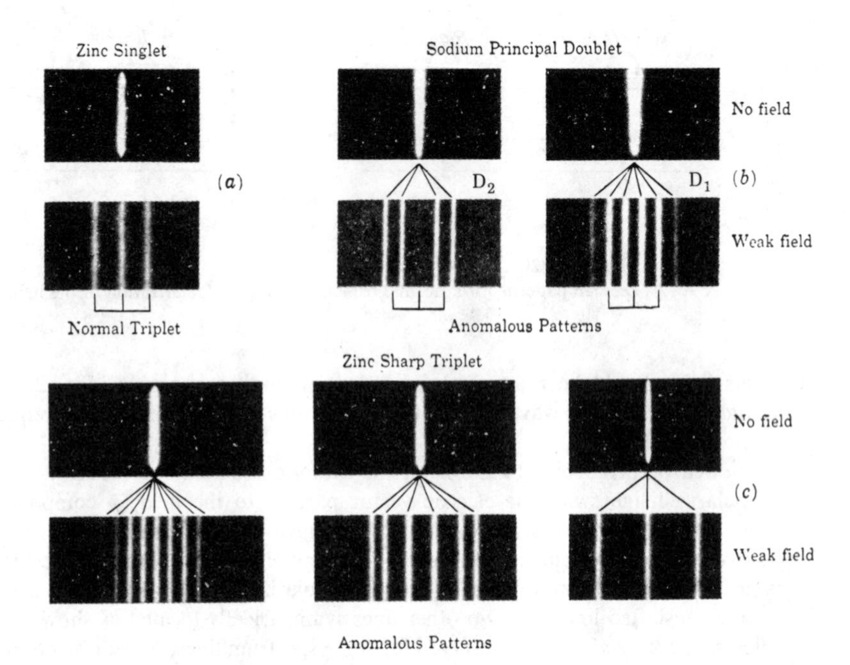

FIGURE 32D
The normal and anomalous Zeeman effects.

given by the classical theory is shown by the bracket below each pattern. From Eq. (32*a*) it is seen that each of the outer component lines should shift out by an amount proportional to the field strength, thus keeping the pattern symmetrical. In very strong magnetic fields, however, asymmetries are observed in many Zeeman patterns. This phenomenon is known as the quadratic Zeeman effect, although it may also be the beginning of a transition called the *Paschen-Back effect* according to which all anomalous patterns become normal triplets in the limit of very strong fields.

Only the normal triplet can be explained by the classical theory. The more complex patterns are now understood and are in complete agreement with the quantum theory of atomic structure and radiation.* Each line of an anomalous pattern, when viewed perpendicular to the magnetic field, is found to be plane-polarized. Usually the centerlines of a pattern are p components with their vibrations parallel to the field B, and those symmetrically placed on either side are s components with vibrations perpendicular to the field. In the longitudinal effect only frequencies corresponding to the s components are observed, and these are circularly polarized.

The quantum theory has developed to such an extent that one can now predict

* For a treatment of the anomalous Zeeman effect, see H. E. White, "Introduction to Atomic Spectra," chaps. 10, 13, and 15, McGraw-Hill Book Company, New York, 1934.

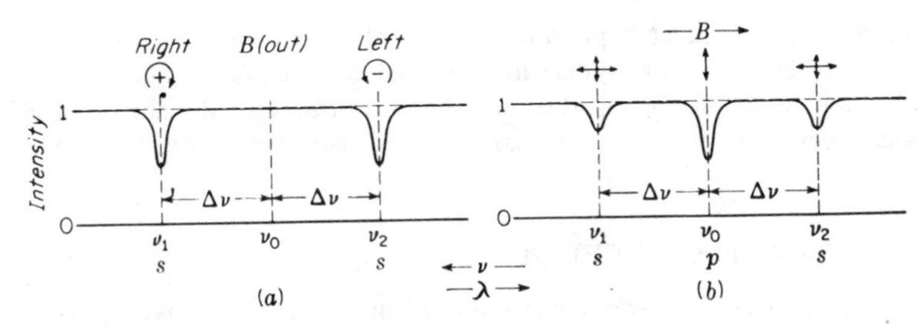

FIGURE 32E
Intensity curves for the inverse Zeeman effect. A normal triplet in absorption.

with the greatest certainty the complete Zeeman pattern for any identified spectrum line in a field of any strength. Conversely, the study of these patterns has become a potent tool in the analysis of complex spectra.

32.2 INVERSE ZEEMAN EFFECT

The Zeeman effect obtained in absorption is called the *inverse Zeeman effect*. The phenomenon is observed by sending white light through an absorbing vapor when the latter is subjected to a uniform magnetic field. In considering the longitudinal effect, analogous to Fig. 32C(*a*), unpolarized light of any particular frequency may be imagined as consisting of right and left circularly polarized components with all possible phase relations. If now ν_0 represents a natural resonance frequency of the vapor in the absence of a field, the plus circular components (see footnote, page 683) of frequency ν_1 will be strongly absorbed in the presence of a field. The corresponding minus circular components of frequency ν_1 pass on through with little decrease in intensity, since to be absorbed these must have the frequency ν_2. Hence at frequency ν_1, looking against the field direction as in Fig. 32C(*a*), right circularly polarized light is transmitted, and for a thick absorbing layer this is one-half as intense as the background of continuous light [Fig. 32E(*a*)]. A similar argument can be given for ν_2.

The Zeeman components of any spectrum line obtained in absorption along the field direction are therefore not completely absorbed, and the light that does get through is found to be circularly polarized in directions opposite to those of the corresponding components obtained in emission. This is verified by experiment even in anomalous patterns of many components.

Viewed perpendicular to the field [Fig. 32E(*b*)], the *p* and *s* components are polarized at right angles to the corresponding components in emission. For ν_0, the parallel components of all incident light vibrations are absorbed and the perpendicular components are transmitted. For ν_1, the parallel components are all transmitted. The perpendicular components, moving across the field, are absorbed by only half the oscillators (the ones having positive rotation, frequency ν_1), giving an absorption line only half as intense as that at ν_0. The result is partially polarized light with a maximum intensity for vibrations parallel to the field *B*. The same is true for the component ν_2. The absorption of the parallel component for ν_0 is analogous to the selective absorption in crystals like tourmaline (Sec. 24.6), where one component vibration is completely absorbed and the other transmitted. The frequencies of the lines observed in the inverse Zeeman effect are also given by Eqs (32*a*) and (32*b*).

32.3 FARADAY EFFECT

In 1845 Michael Faraday discovered that when a block of glass is subjected to a strong magnetic field, it becomes optically active. When plane-polarized light is sent through glass in a direction parallel to the applied magnetic field, the plane of vibration is rotated. Since Faraday's early discovery the phenomenon has been observed in many solids, liquids, and gases. The amount of rotation observed for any given substance is found by experiment to be proportional to the field strength B and to the distance the light travels through the medium. This rotation can be expressed by the relation

• $$\theta = VBl \qquad (32h)$$

where B is the magnetic induction in teslas, l is the thickness in meters, θ is the angle of rotation in minutes of arc, and V a constant to be associated with each substance. This constant, called the *Verdet constant*, is defined as the rotation per unit path per unit field strength. In gases the density must also be specified. A few values of the Verdet constant are given in Table 32A.

The Faraday effect is so closely associated with the direct and inverse Zeeman effects, presented in the two preceding sections, that its explanation follows directly from the principles given there. Because the phenomenon is best observed in vapors at wavelengths near an absorption line, the explanation given here will be confined to substances in the gaseous state. Consider the passage of light through a vapor like sodium where in the absence of a field there are certain resonance frequencies ν_0 at each of which absorption takes place. When the magnetic field is introduced, there will be for each ν_0, according to the classical theory of the Zeeman effect, two resonance frequencies, one ν_1 for left circularly polarized light and the other ν_2 for

Table 32A VALUES OF THE VERDET CONSTANT IN MINUTES OF ARC PER TESLA PER METER FOR λ5893

Substance	t, °C	V
Water	20	1.31×10^4
Glass (phosphate crown)	18	1.61×10^4
Glass (light flint)	18	3.17×10^4
Carbon disulfide (CS_2)	20	4.23×10^4
Phosphorus, P	33	13.26×10^4
Quartz (perpendicular to axis)	20	1.66×10^4
Acetone	15	1.109×10^4
Salt (NaCl)	16	3.585×10^4
Ethyl alcohol	25	1.112×10^4

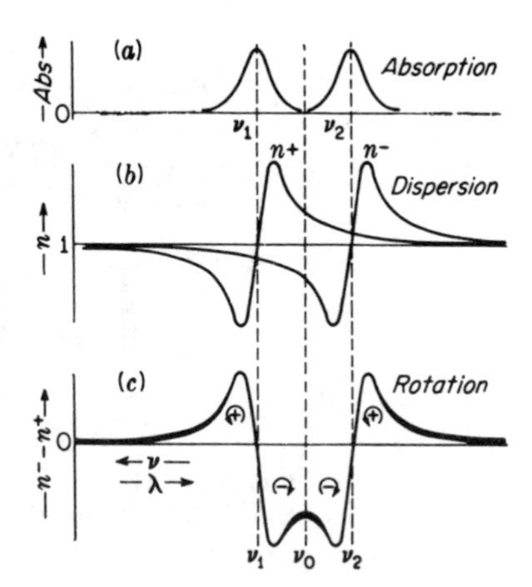

FIGURE 32F
Absorption and dispersion curves used in explaining the Faraday effect. These curves refer to the magnetic splitting of a single absorption line.

right circularly polarized light traveling along the field. For each of these directions of rotation an absorption curve and a dispersion curve [Fig. 23H(*b*)] may be drawn as shown in Fig. 32F(*a*) and (*b*).

Referring to Fig. 32F(*b*), it is observed that outside the region ν_1 to ν_2 the value of n^- is greater than n^+. Therefore positive rotations travel faster than negative, and the plane of the incident polarized light is rotated in the positive direction (see Sec. 28.3). The difference between the two dispersion curves, as given in Fig. 32F(*c*), shows that for frequencies between ν_1 and ν_2 the rotation is in the negative direction.

If plane-polarized light is reflected back and forth through the same magnetically activated vapor, the plane of vibration is found to rotate farther with each traversal. This is not the case for naturally optically active substances like quartz, where upon one reflection the light emerges vibrating in the same plane in which it entered. It should be noted that when the field direction is reversed, the direction of rotation of the plane of the incident light vibrations is also reversed. Therefore the sense of the rotation is defined in terms of the direction of the field, positive rotation being that of a right-handed screw advancing in the direction of the field, or that of the *positive* current in the coil which produces the field.

The rotation in the Faraday effect is given by Eq. (32h), which shows that the angle of rotation is proportional to the field strength. This follows from Eq. (32a) for the Zeeman effect. As the two dispersion curves separate with an increasing field, the differences in index (bottom curve) increase to a first approximation by an amount which is proportional to $\Delta\nu$ and hence to B. This is most accurately true at frequencies far from ν_1 or ν_2, where the dispersion curves over a short frequency interval may be considered as straight lines.

One of the most interesting methods developed for observing the Faraday effect is that shown in Fig. 32G. Without the right- and left-handed quartz prisms or the vapor, no light would be transmitted by the analyzer when crossed by the polarizer as shown. With the insertion of the double quartz prism the light vibrations are rotated by different amounts according to the portion of the prisms (in the plane of the figure) through which they have passed. Hence varying amounts of light will

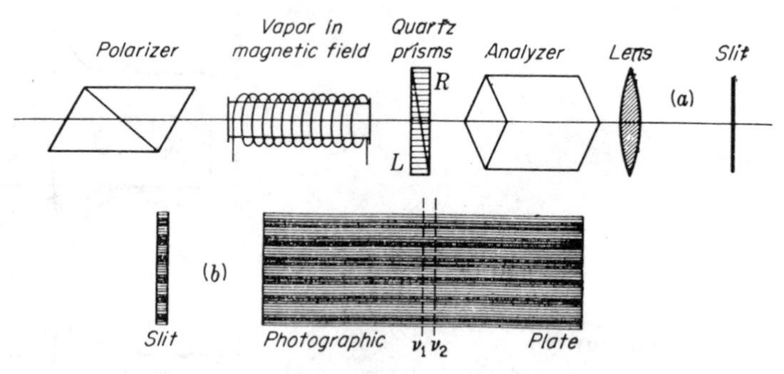

FIGURE 32G
Experimental arrangement for observing the Faraday effect.

now get through the various portions of the analyzer. When this light is focused on the slit of a spectroscope, alternate dark and light bands are formed as shown in Fig. 32G(*b*). If white light is used as a source in front of the polarizer, the spectrum as observed in the spectroscope will be crossed by a number of approximately horizontal dark and light bands. If now the vapor is introduced into the light path, absorption lines will be observed at all resonance frequencies ν_0. When the magnetic field is turned on, rotation takes place within the vapor according to Fig. 32F(*c*), thus shifting the bright bands accordingly. Close to the absorption lines the rotation is large, causing greater shifts of the bands. Since this rotation changes continuously with λ, the bands are observed to curve up or down, taking the same general form as shown in the theoretical curve of Fig. 32F(*c*). Figure 32H(*a*) is a photograph of these bands for the D lines of sodium taken under high dispersion and resolving power. They show not only the rapid increase in the positive rotation on each side of the absorption frequencies but the opposite rotation between the two. It should be noted that both these sodium lines give anomalous Zeeman patterns [Fig. 32D(*b*)]. The longitudinal effect for $\lambda5896$, D_1, however, is a doublet leading to the same kind of curves as those described above for a normal triplet. Theoretical curves for the D_2 line are left as an exercise for the student.

32.4 VOIGT EFFECT, OR MAGNETIC DOUBLE REFRACTION

In 1902 Voigt discovered that when a strong magnetic field is applied to a vapor through which light is passing perpendicular to the field, double refraction takes place* This phenomenon, now known as the *Voigt effect* or *magnetic double refraction*, is related to the transverse Zeeman effect in precisely the same way that the Faraday effect is related to the longitudinal Zeeman effect. In view of this relation the phenomenon is readily explained from absorption and dispersion curves in much the same

* W. Voigt, "Magneto- und Elektro-optik," B. G. Teubner, Leipzig, 1908.

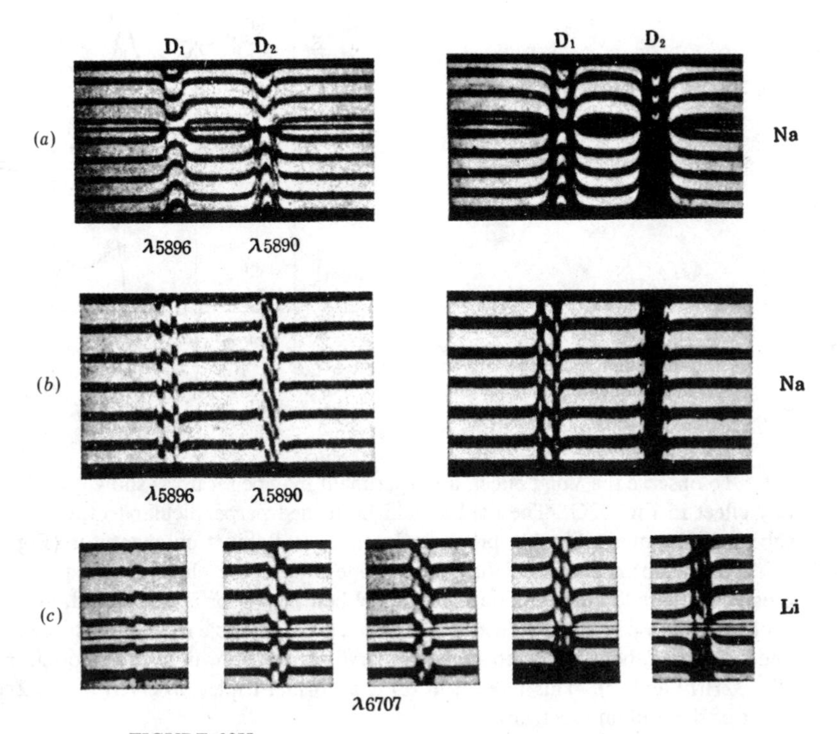

FIGURE 32H
(*a*) The Faraday effect near the sodium resonance lines D_1 and D_2, (*b*) the Voigt effect of the sodium lines, (*c*) the Voigt effect near the lithium line λ6707. (*Courtesy of Hansen.*)

way as the Faraday effect in the preceding section. Consider a vapor having a resonance frequency ν_0 which in the presence of an external field breaks up into a normal Zeeman triplet [see Fig. 32C(*b*)]. When white light is sent through this vapor, those light vibrations which have a frequency ν_0 will be in resonance with electrons of the vapor which have ν_0 as their frequency and thus be absorbed. This is represented by the central absorption and dispersion curve in Fig. 32I(*a*) and (*b*). Other light vibrations, perpendicular to the field, are in resonance with ν_1 and ν_2. These are represented by the $\perp$ absorption and dispersion curves. With unpolarized light incident on the vapor the variations in n near ν_1 and ν_2 are half as great as at ν_0, just as the absorption coefficients at ν_1 and ν_2 are only half as great as at ν_0.

The dispersion curves of Fig. 32I(*b*) show that if plane-polarized light of any frequency ν is incident on the vapor it will be broken up into two components, one perpendicular and one parallel to B. Since these components have different refractive indices (therefore different velocities), one component gets ahead of the other in phase and elliptically polarized light emerges. The relative magnitude of this phase difference varies with wavelength, as is shown by the difference curve in Fig. 32I(*c*).

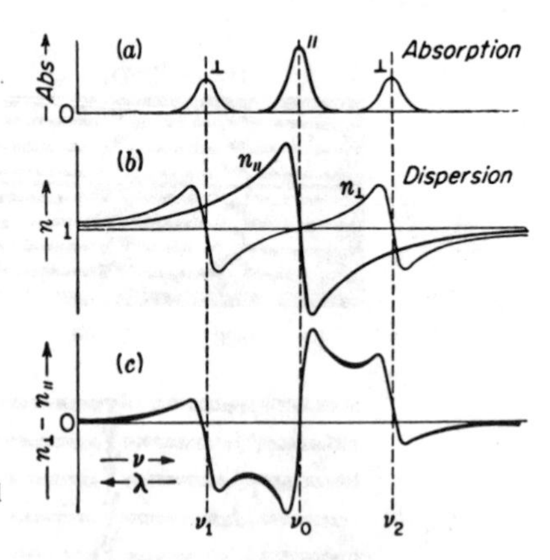

FIGURE 32I
Absorption and dispersion curves used in explaining the Voigt effect.

To observe the Voigt effect, an experiment may be set up as shown for the Faraday effect in Fig. 32G. The field should be turned perpendicular to the absorption tube and the quartz double prism replaced by a Babinet compensator (Fig. 27F). Without the absorption tube the spectroscope slit and the photographic plate will be crossed by parallel light and dark bands. When the vapor is introduced, absorption is observed at ν_0. When the field is turned on, strong double refraction close to ν_0, ν_1, and ν_2 causes these bands to curve up or down as shown in the photographs of Fig. 32H(*b*) and (*c*). This pattern in (*c*) is a normal triplet observed in the Zeeman effect of the lithium spectrum.*

The Voigt effect for anomalous Zeeman patterns like those in Fig. 32H(*b*) has been studied by Zeeman, Geest, Voigt, Landenberg, Hansen,† and others. These results are readily predicted by drawing dispersion curves similar to those shown in Fig. 32I. In any Zeeman pattern the *s* components form one continuous dispersion curve and the *p* components another. Their difference represents a plot of the double refraction as a function of the frequency. Its magnitude is proportional to the square of the field strength *B*.

32.5 COTTON-MOUTON EFFECT

This effect, which was discovered in 1907 by Cotton and Mouton, has to do with the double refraction of light in a liquid when placed in a transverse magnetic field. In pure liquids like nitrobenzene very strong double refraction is observed, the effect being some thousand times as great as the Voigt effect treated in the last section. This double refraction is attributed to a lining up of the magnetically and optically anisotropic molecules in the applied field direction. This lining up would result

* The lithium line $\lambda 6707$ is in reality a doublet, each component of which in a weak magnetic field gives an anomalous Zeeman pattern. In the strong field used to observe the Voigt effect the two have coalesced (the Paschen-Back effect) to form a normal triplet for which the above discussion has been given.

† H. M. Hansen, *Ann. Phys.*, **43**:205 (1914).

whether the magnetic dipole moments of the molecules were permanent or induced by the field. Such an effect should be by theory, and is found to be by experiment, proportional to the square of the field strength. The effect is dependent upon temperature, decreasing rapidly with a rise in temperature. The Cotton-Mouton effect is the magnetic analogue of the electro-optic Kerr effect to be discussed in Sec. 32.10 and is not related to the Zeeman effect.

32.6 KERR MAGNETO-OPTIC EFFECT

In 1888 Kerr* made the discovery that when plane-polarized light is reflected at normal incidence from the polished pole of an electromagnet, it becomes elliptically polarized to a slight degree, with the major axis of the ellipse rotated with respect to the incident vibrations. At other angles of incidence the effect is observable if one avoids the ordinary effect of elliptical polarization obtained by reflection of plane-polarized light from metals at $\phi \neq 0$ by having the electric vector of the incident light either parallel or perpendicular to the plane of incidence. Under these conditions, and without the field, the reflected beam can be extinguished by a nicol prism. Upon turning on the magnetic field the light instantly appears and cannot be extinguished by a rotation of the nicol. The introduction of a quarter-wave plate suitably oriented will now enable the light to be again extinguished, showing the reflected light to be elliptically polarized. The magnetic field has thus given rise to a vibration component called the *Kerr component* perpendicular to the incident light vibration. This is the Kerr magneto-optic effect and should be distinguished from the Kerr electro-optic effect considered in Sec. 32.10.

32.7 STARK EFFECT

In the few years following Zeeman's discovery of the splitting up of spectral lines in a magnetic field, many attempts were made to observe an analogous effect due to an external electric field. In 1913 Stark observed that when the hydrogen spectrum is excited in a strong electric field of 100 kV/cm, each line split into a symmetrical pattern. A photograph of the effect in the first line of the Balmer series of hydrogen is shown in Fig. 32J. When viewed perpendicular to the electric field, some of the components of each line pattern are observed to be plane-polarized with the electric vector parallel to the field (p components) and the others plane-polarized with the electric vector normal to the field (s components). This is the transverse Stark effect. When viewed parallel to the field, only the s components appear, but as ordinary unpolarized light. This is the longitudinal Stark effect.

The theory of the Stark effect has been developed only in terms of the quantum theory and will not be given here.†

* John Kerr (1824–1907), pronounced "car," Scottish physicist, inspired to investigate electricity and magnetism by his association with William Thomson (Lord Kelvin).

† For a more extended treatment of the Stark effect and other references to the subject see H. E. White, "Introduction to Atomic Spectra," p. 101, McGraw-Hill Book Company, New York, 1934.

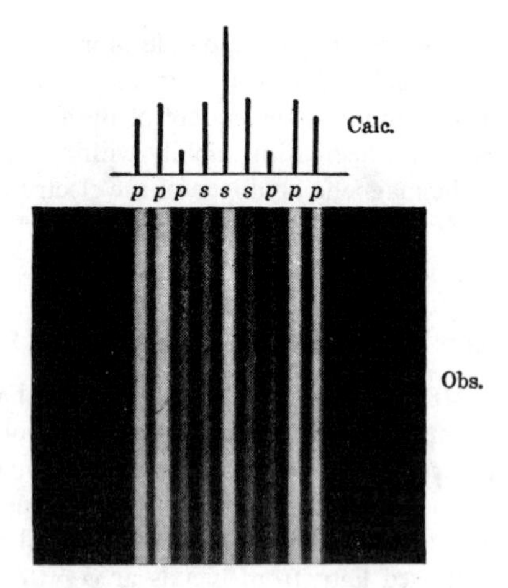

FIGURE 32J
Photograph of the Stark effect of H_α, $\lambda6563$, in hydrogen. (*Courtesy of Wierl.*)

The method used for producing strong electric fields of 100kV/cm or more, in which the light source will operate, is based upon the characteristics of the ordinary discharge of electric currents through gases at low pressures. In a discharge of the type shown in Fig. 21D, the major part of the potential drop from one electrode to the other occurs across a relatively dark region near the cathode. This region of a specially designed discharge tube, when focused on the slit of a spectrograph, may be made to give photographs of the type shown in Fig. 32K. Since the Stark effect is proportional to the field F, the pattern of $\lambda3819$, for example, may be taken to represent the field strength which is small at the top and increases toward the bottom, nearer the cathode.

The widest Stark patterns are observed in the hydrogen and helium spectra. In the case of all other spectra one seldom observes anything but a slight shift of the line, usually toward longer wavelengths. This effect is called the *quadratic Stark effect*, to distinguish it from the linear effect observed in hydrogen and helium. In the former case the shifts are proportional to the square of the electric field strength, while in the latter the splittings depend on the first power of this field. Characteristic of the Stark effect, as shown in Fig. 32K for the helium spectrum, is the appearance of new spectrum lines (marked with crosses) where the field strength is high.

32.8 INVERSE STARK EFFECT

The Stark effect with the lines appearing in absorption is called the inverse Stark effect. The phenomenon has been investigated by Grotrian and Ramsauer, using a long tube containing potassium vapor at low pressure and two long parallel metal plates only 1.5 mm apart. With a potential of 14 kV on the plates, the absorption lines $\lambda4044$, $\lambda4047$, and $\lambda3447$ were found to be shifted from the field-free position

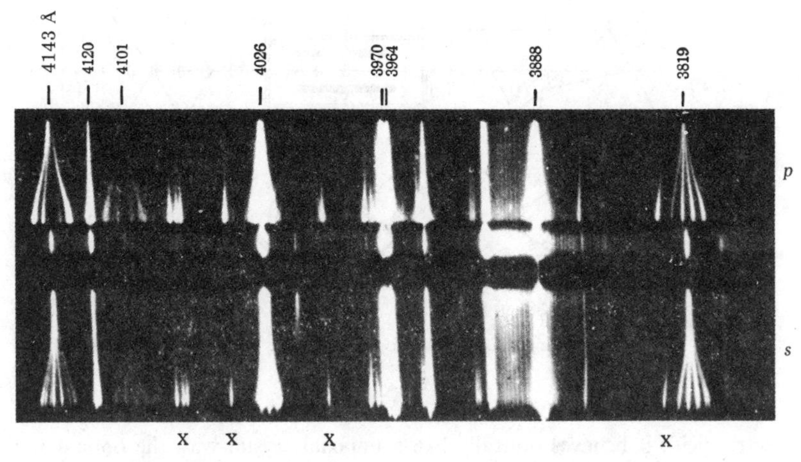

FIGURE 32K
The Stark effect in helium. (*Courtesy of Foster.*)

toward longer wavelength. This shift, although only a few hundredths of an angstrom unit in magnitude, was found to be proportional to the square of the field strength. This is therefore a case of a quadratic Stark effect.

32.9 ELECTRIC DOUBLE REFRACTION

Electric double refraction is related to the transverse Stark effect and is analogous to magnetic double refraction, or the Voigt effect, discussed in Sec. 32.4. In 1924 Ladenberg observed the absorption of the sodium resonance lines when produced with and without a strong transverse electric field applied to the vapor. Although the shift of the lines predicted by the quadratic Stark effect was too small to observe even with very high resolving power, double refraction was observed at frequencies close to the absorption lines. This double refraction is attributed to the very small difference in the frequency of the absorption line for light polarized parallel and perpendicular to the electric lines of force. The explanation is therefore analogous to that given for magnetic fields in Sec. 32.4 (see Fig. 32I).

32.10 KERR ELECTRO-OPTIC EFFECT

In 1875 Kerr discovered that when a plate of glass is subjected to a strong electric field, it becomes doubly refracting. That this effect is not due to the strains that such a field sets up in the glass is shown by the fact that the phenomenon also appears in many liquids and may even be observed in gases. When a liquid is placed in an

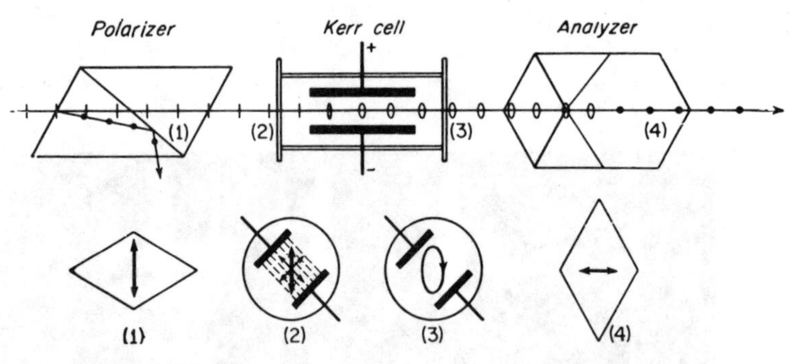

FIGURE 32L
Arrangement for the electro-optic shutter, operating by the Kerr effect.

electric field, it behaves optically like a uniaxial crystal with the optic axis parallel to the field direction, and when viewed from the perpendicular direction, it gives rise to all the phenomena of interference considered in Chap. 27.

It is convenient experimentally to observe the effect by passing light between two parallel oppositely charged plates inserted in a glass cell containing the liquid. Such a device, known as a *Kerr cell*, is shown at the center in Fig. 32L. Such a cell inserted between crossed polarizer and analyzer constitutes a very useful optical device known as the *electro-optic shutter*.* One of these uses was described in Sec. 19.5. When the electric field is off, no light is transmitted by the analyzer. When the electric field is on, the liquid becomes doubly refracting and the light is restored. With the cell oriented at 45°, the incident plane vibrations from the polarizer are broken up into two equal components parallel and perpendicular to the field, as shown at the bottom in Fig. 32L. These travel with different speeds, and hence a phase difference is introduced and the light emerges as elliptically polarized light. The horizontal component of the vibrations is transmitted by the analyzer.

The change in phase of the two vibrations in a Kerr cell is found to be proportional to the path length, i.e., the length of the electrodes l and to the square of the field strength F. The magnitude of the effect is determined by the Kerr constant K, defined by the relation

$$\Delta = K\frac{lE^2\lambda}{d^2} \tag{32i}$$

Since the phase difference δ between the two components is given by $2\pi/\lambda$ times the path difference Δ, we have

$$\delta = K\frac{2\pi lE^2}{d^2} \tag{32j}$$

where δ is in radians, l and d are in meters, E is in volts, and K is in meters per volts squared, and λ is the wavelength in the medium.

* For the theory and technique of the Kerr cell see F. G. Dunnington, *Phys. Rev.*, **38**:1506 (1931) and E. F. Kingsbury, *Rev. Sci. Instrum.*, **1**:22 (1930).

One of the substances most suitable for use in a Kerr cell is nitrobenzene, because of its relatively large Kerr constant. This is shown by the values given for a few liquids in Table 32B.

It should be pointed out that electric double refraction for gases discussed in the last section and the Kerr electro-optic effect are not the same phenomenon. In a gas the effect is due to changes inside the atom (Stark effect). In the Kerr effect it is usually due to natural or induced anistropy of the molecule and a lining up of such molecules in the field. This alignment causes the medium as a whole to be optically anistropic. As in the Cotton-Mouton effect (Sec. 32.5), the Kerr effect is dependent on temperature. In fact, the Kerr electro-optic effect is the exact electric analogue of that magnetic effect.

32.11 POCKELS ELECTRO-OPTIC EFFECT

A variety of uniaxial crystals have been found in which the induced birefringence varies linearly with the applied electric field. This effect was named after F. Pockels,* who studied the effect in 1893. Recent research has developed a variety of electro-optic crystals, such as ammonia dihydrogen phosphate ($NH_4H_2PO_4$) (ADP) and potassium dihydrogen phosphate (KH_2PO_4) (KDP), which produce sizable Pockels birefringence at relatively low voltages (see Fig. 32M).

A Pockels cell, which can be used as a fast *light modulator* or *shutter*, usually involves a crystal mounted with its optic axis and applied field parallel to the beam direction (see Fig. 32N). By placing the cell between crossed polarizers, the transmission can be modulated at frequencies well above 10^{10} Hz, and as a shutter with a response time shorter than 1 ns. Since the beam traverses the electrodes, these are frequently made of transparent metallic oxides, such as CdO, SnO, or InO, or thin metallic rings or grids.

Pockels cells, like Kerr cells, are used for a wide range of electro-optic systems, including their use as a *Q switch* to produce ultrashort laser pulses (see Sec. 30.6). These systems have been proposed as wideband laser-beam communication systems to be used, in addition to terrestrial applications, in interplanetary space.

* See R. Goldstein, Pockels Cell Primer, *Laser Focus Mag.*, (1968); R. S. Ploss, A Review of Electro-optics Materials, Methods and Uses, *Opt. Spectra*, (1969); and D. F. Nelson, Modulation of Laser Light, *Sci. Am.*, (1968).

Table 32B VALUES OF THE KERR CONSTANT FOR $\lambda = 5893$ Å

Substance	K
Benzene	0.67×10^{-14}
Carbon disulfide	3.56×10^{-14}
Water	5.10×10^{-14}
Nitrotoluene	1.37×10^{-12}
Nitrobenzene	2.44×10^{-12}

FIGURE 32M
A laboratory-grown crystal of ammonia dihydrogen phosphate ($NH_4H_2PO_4$), or ADP, for use in Pockels cells.

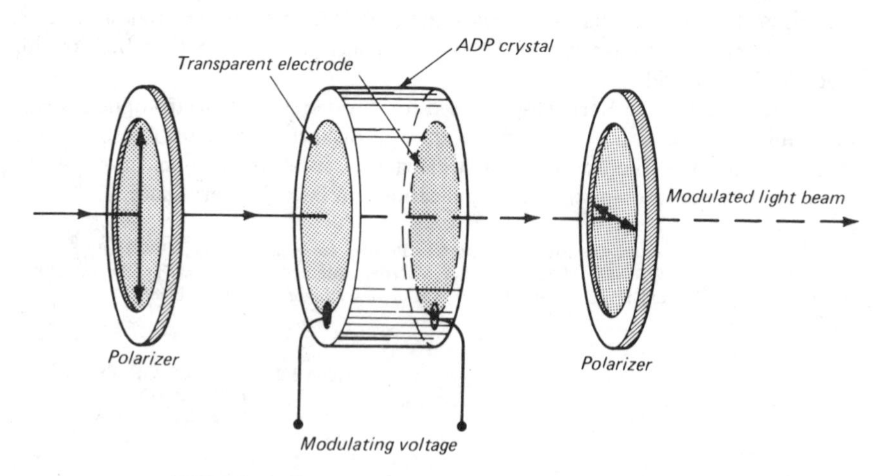

FIGURE 32N
Components of a Pockels cell for high-frequency modulation of a light beam.

PROBLEMS

32.1 Determine the Zeeman splitting $\Delta\sigma$ of a single line in the zinc spectrum, where the wavelength is 4700 Å. Express this splitting in angstroms and assume the field strength to be 2.520 T. *Ans.* $\Delta\lambda = 0.260$ Å

32.2 The photograph of the normal Zeeman effect shown in Fig. 31D(*a*) was enlarged 20 times from the original negative. The plate factor of the spectrograph used was 2.30 Å/mm at the wavelength of the line 4700 Å. What was the value of the magnetic induction?

32.3 For the first line of the Paschen series of hydrogen, the wavelength is 18,746 Å. Calculate the Zeeman shifts for a normal Zeeman triplet if the magnetic field is 1.650 T.

32.4 A diffraction grating has 50,000 lines ruled on its surface. What strength of magnetic field would have to be applied to a light source for the grating to be able to resolve a normal Zeeman triplet (*a*) in the violet at 4500 Å and (*b*) in the red at 6500 Å? Assume the first-order spectrum is used. *Ans.* (*a*) 0.952 T, (*b*) 0.659 T

32.5 In the Faraday effect a magnetic field of 0.64 T is applied to a piece of light flint glass 10.50 cm long. Find the angle of rotation in degrees.

32.6 The Faraday effect is performed with a liquid in a glass tube 20.0 cm long. If the applied magnetic induction is 0.820 T and the measured rotation of plane polarized light is 65.46°, what is the value of the Verdet constant?

32.7 The Faraday effect is performed with a piece of phosphate crown glass 5.0 cm thick. This glass is placed between Polaroids with their principal sections at 45° with each other. (*a*) What magnetic field strength applied to the glass will rotate the plane of polarization 45° in order that the light transmitted gets through with maximum intensity? (*b*) If ordinary light is sent through the system in the reverse direction, what will be the intensity of the transmitted light? (*c*) Is this a one-way optical system? (*d*) Make a diagram.

32.8 Very pure nitrobenzene is used in a Kerr cell, with a power source of 20 kV applied to its plates. If the plates of the cell are 2.5 cm long and 0.75 cm apart, find (*a*) the phase difference between the components emerging from the cell. If unpolarized light is incident on the polarizer, what is (*b*) the amplitude of the plane-polarized light incident on the cell, (*c*) the amplitude of the light emerging from the analyzer, and (*d*) the intensity of the emerging light?

Ans. (*a*) 156.44°, (*b*) $0.7071A_0$, (*c*) $0.6920A_0$, (*d*) $0.4790I_0$

32.9 What voltage applied to a Kerr cell will produce polarized light emerging from the cell? The electrodes are 3.0 cm long and 5.0 mm apart, and the cell is filled with nitrotoluene.

32.10 A Kerr cell using very pure nitrobenzene has plates that are 2.80 cm long, separated by a distance that is 0.60 cm. (*a*) What voltage should be applied to the plates to produce a maximum in the transmitted intensity? (*b*) At this field strength, what fraction of the incident unpolarized light will get through the system? Neglect losses by reflection and absorption.

33
THE DUAL NATURE OF LIGHT

In this concluding chapter we shall give a brief account of the way in which the more recently discovered corpuscular properties of light have been reconciled with the wave theory. It will not be possible to recount in any systematic way the steps which have led to our present view of the nature of light or to discuss its broad implications. This subject forms an important part of a whole field of study, that of atomic or modern physics.* Furthermore, the discussion of one part of this field presents difficulties in view of the essentially mathematical character of the quantum theory, which was first developed as a formalized set of equations and only later expressed in terms of visualizable physical concepts.

It is with the hope of at least partially satisfying the reader's curiosity about the dual nature of light, waves or particles, that the following discussion, brief and incomplete as it is, has been included.

* See, for example, H. E. White, "Introduction to Atomic and Nuclear Physics," D. Van Nostrand, Litton Educational Publishing Co., New York, 1964; H. Semat, "Introduction to Atomic and Nuclear Physics," 5th ed., Holt, Rinehart and Winston, Inc., New York, 1972; F. K. Richtmyer, E. H. Kennard, and J. N. Cooper, "Introduction to Modern Physics," 6th ed., McGraw-Hill Book Company, New York, 1969; Max Born, "Atomic Physics," 5th ed., Hafner Publishing Company, New York, 1951; and L. I. Schiff, "Quantum Mechanics," 3d ed., McGraw-Hill Book Company, New York, 1968.

33.1 SHORTCOMINGS OF THE WAVE THEORY

As long as one is dealing with questions of the interaction of light with light, such as occurs in interference and diffraction, the electromagnetic theory, and in fact any wave theory, gives a complete account of the facts. When one attempts to treat the interactions of light with matter, however, as in the emission and absorption of light, in the photoelectric effect, and in dispersion, serious difficulties at once present themselves. In many of these it is not merely a matter of slight deviations between experiment and theory, detectable only by quantitative measurements; on the contrary, the theory predicts results that are radically different from those observed. Historically the first case of this kind was encountered in the attempt to explain the distribution of energy in the spectrum of a blackbody (Sec. 21.9). Here the electromagnetic theory was used in conjunction with the classical theory of equipartition of energy, which had been so successful in explaining the specific heats of gases. The predicted curve was nearly correct at long wavelengths, but its course toward shorter wavelengths, instead of passing through a maximum and falling to zero (Fig. 21F), continued to increase indefinitely. It was only by assuming that the oscillators in the radiating source could not exist in states having all possible energies and amplitudes but only in certain definite ones for which the energy was a whole multiple of some particular value (quantum) that Planck in 1900 was able to derive the exact radiation formula [Eq. (21e)].

Other shortcomings of the older theory soon became evident. In the photoelectric effect, the measured energies of the photoelectrons ejected from metal surfaces by light were in marked disagreement with the predictions of electromagnetic theory (see the following section). The amount of energy in the waves falling on a single atom in the case of weak illumination was very much smaller than that observed in the photoelectron, and this led Einstein in 1905 to postulate the existence of photons. In the explanation of the line series observed in the atomic spectrum of hydrogen (Sec. 21.10), Bohr in 1913 had to assume that the electron revolved in a stable orbit without radiating, whereas a charge with strong centripetal acceleration should, according to the electromagnetic theory, rapidly lose energy as radiation (Sec. 20.8). This would cause the frequency to change rapidly and would make it impossible to explain the existence of sharp spectrum lines. The explanation of X rays according to the electromagnetic theory as very short pulses of radiation, caused by the sudden deceleration of electrons as they strike the target, was inconsistent with the observed continuous X-ray spectrum. As was shown by Duane and Hunt in 1917, this spectrum exhibits a sharp cutoff on the short-wavelength side, whereas the Fourier analysis of a pulse yields a continuous spectrum falling off smoothly (Sec. 12.6). The discovery in 1922 of the Compton effect, which is a shift toward lower frequency of scattered monochromatic X rays, was a striking demonstration of the inadequacy of the wave theory, since to explain it one had to postulate that photons collided with electrons in atoms and rebounded like elastic billiard balls (see below).

These constitute a few of the simpler phenomena in which the wave theory failed completely. In many of the more complex interactions between matter and radiation the theory, although giving the rough features correctly, ran into insuperable difficulties when attempts were made to give a complete quantitative account of the

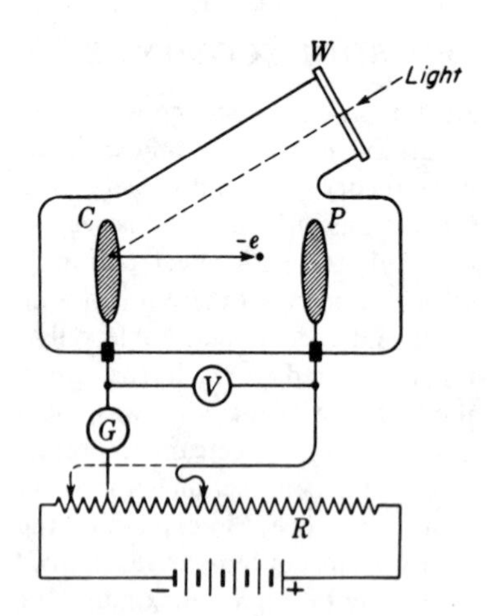

FIGURE 33A
Experimental arrangement for studying the photoelectric effect.

facts. One of the earliest phenomena in this class was the anomalous Zeeman effect (Sec. 32.1), and one of the more recent ones the Raman effect (Sec. 22.11). Others could be cited, but the list has now grown so long that it is no longer a question of introducing refinements in the wave theory to obtain agreement. Quantum theory, of which the wave theory is now recognized as an integral part, must be used in dealing with such effects.

33.2 EVIDENCE FOR LIGHT QUANTA

In reaching conclusions about the nature of a phenomenon like light we must rely upon observation of the effects it produces. An individual wave or a particle of light cannot be seen and photographed as can large-scale waves and particles of matter. We can conclude with certainty, however, that light has a wave character from the study of interference and diffraction patterns, of its velocity, of the doppler effect, etc. Evidence just as convincing exists that light consists of small packets of energy which are highly localized, and any one of which can communicate all its energy to a single atom or molecule. We have seen in Chap. 29 that these particles of energy $h\nu$ are known as light quanta or *photons*. It will be worthwhile to consider briefly three pieces of experimental evidence of this type, those selected being ones that will be useful in our further discussion of the subject.

In the *photoelectric effect* (Fig. 33A) light enters through the quartz window W and strikes the cathode C, which is a clean metal plate. A current of negative charge is observed by the galvanometer G to flow from C through the evacuated tube to the plate P, which is at some positive potential with respect to C. This shows that electrons of charge $-e$ are being ejected from the surface of the metal cathode. Their velocities and energies as they leave the surface can be studied by varying the voltage V applied to the plate. It is found that the energy is independent of the intensity of the light and

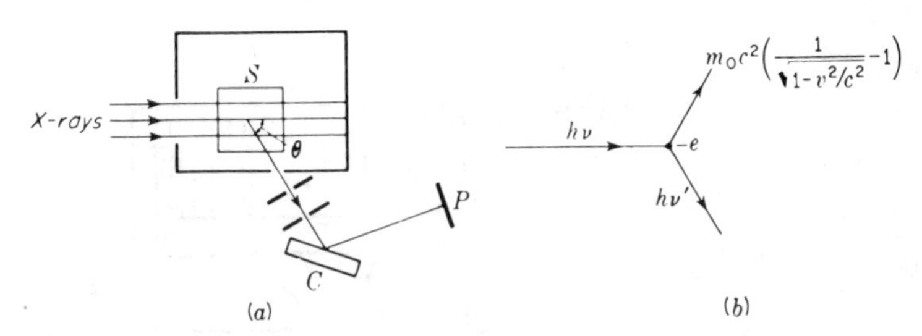

FIGURE 33B
The Compton effect: (*a*) method of observation; (*b*) energies of the incident photon, scattered photon, and recoil electron.

is determined by the frequency of the light according to Einstein's photoelectric equation

$$E = h\nu - k \qquad \textit{Energy of photoelectrons} \qquad (33a)$$

Here again h is the universal constant 6.6262×10^{-27} J/s, known as *Planck's constant*, ν is the frequency c/λ, and k is a constant characteristic of the kind of metal in the cathode. For most metals k is large enough to require that light of fairly high frequency (ultraviolet light) be used to produce the photoelectrons. The quantum character of light appears in this experiment through the fact that each electron has evidently taken on the same amount of energy $h\nu$ and emerges with the difference between this and the amount k required to get it through the surface. (This interpretation of k is verified in other ways, notably in thermionic emission.) Furthermore a very faint beam of light still causes some photoelectrons to be emitted instantaneously, and they have the full energy. Under these circumstances it is apparent that there are very few photons in the beam, each of energy $h\nu$. On the wave theory the small amount of electromagnetic energy would be spread over the whole surface, and the amount available to any one electron would be insufficient to produce the effect.

The *Compton effect* is observed in X rays that are scattered at an angle θ from a scatterer S consisting of some light element like carbon [see Fig. 33B(*a*)]. A narrow beam is defined by two lead slits and caused to fall on a crystal C. This diffracts the X rays to the photographic plate P, and by suitably turning the crystal about an axis perpendicular to the plane of the figure a spectrum can be photographed. For each monochromatic line present in the original X rays, the spectrum of the scattered rays shows a line shifted to longer wavelengths, the shift increasing with the scattering angle θ according to the equation

$$\Delta\lambda = \frac{c}{\nu'} - \frac{c}{\nu} = \frac{h}{m_0 c}(1 - \cos\theta) \qquad \textit{Compton shift} \qquad (33b)$$

where m_0 is the mass of an electron at rest and h/m_0c is called the *Compton wavelength*. This equation may be easily derived by applying the laws of conservation of energy and momentum to the collision of a photon and an electron [Fig. 33B(*b*)]. The electron in question is one that is knocked out of an atom in the scatterer, and its kinetic

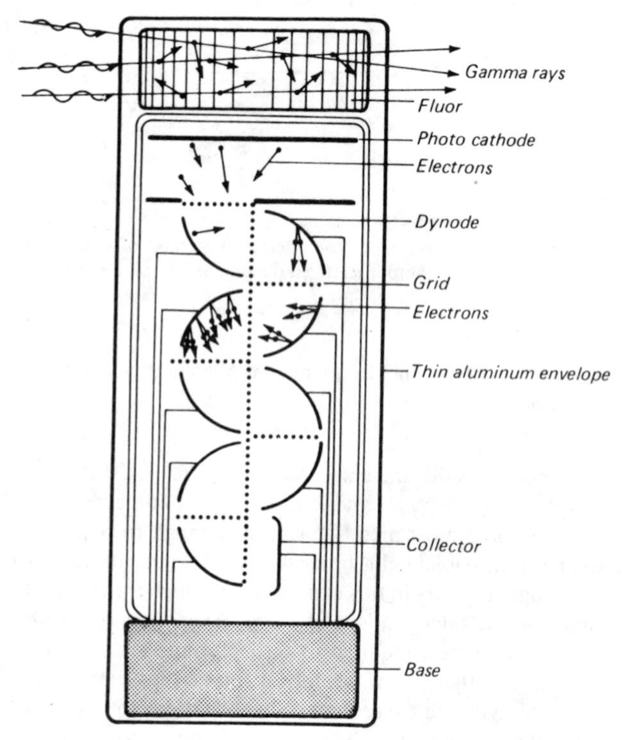

FIGURE 33C
Scintillation detector for gamma rays using a fluorescent block and a photomultiplier tube.

energy must be represented by the relativity formula given in the figure. Similarly its momentum, and also that of the photon, must be expressed in terms of relativistic equations, as will be explained below in Sec. 33.3. But the picture here presented of an elastic collision of particles is clearly foreign to any wave model of light. It has even been possible to detect the scattered photon and the recoil electron simultaneously in the two directions predicted by theory, using various detectors, such as a Wilson cloud chamber or photographic emulsion.

As a third example of the corpuscular behavior of light we may mention the *scintillation counter*, which has been a valuable instrument for measuring hard X rays and γ rays. The principle of the instrument is similar to that of the scintillation method which was employed to count α particles in the early studies of radioactivity. As is shown in Fig. 33C, photons in a beam of γ rays enter a fluorescent crystal at the top and produce visible light photons in the blue or violet part of the spectrum. The fluorescent materials frequently used are crystals of sodium iodide (NaI) and cesium iodide (CsI). The tiny flashes of light occur in the interior of the crystal as the result of the passage of each γ-ray photon. These photons strike a photocathode of a photomultiplier tube and are greatly amplified by eight or more dynodes. The resulting electric pulses activate some counting device. In this device the effects of individual photons are observed in a manner just as direct as that used for atomic particles of matter, and it leaves no doubt as to the corpuscular behavior of light when observed under these conditions.

33.3 ENERGY, MOMENTUM, AND VELOCITY OF PHOTONS

In all experiments which reveal the existence of photons, and notably in the photoelectric effect, their energy is found to be determined only by the frequency ν. The latter quantity must of course be measured independently by observing interference, a typical wave property. We have seen that the constant of proportionality between energy and frequency is Planck's constant h, so we have as an experimental result that

● $$E = h\nu \qquad \textit{Energy of a photon} \tag{33c}$$

To obtain an expression for the momentum, we make use of Einstein's equation for the equivalence of mass and energy, according to which

$$E = mc^2 \tag{33d}$$

This equation has been experimentally verified many times for matter in studies of nuclear disintegration, and it has been shown to hold in the conversion of radiation into matter that occurs in the creation of electron-position pairs by γ rays. Combining Eqs (33c) and (33d), one finds that

$$h\nu = h\frac{c}{\lambda} = mc^2$$

and therefore, since the momentum p is the product of mass and velocity,

● $$p = mc = \frac{h\nu}{c} = \frac{h}{\lambda} \qquad \textit{Momentum of a photon*} \tag{33e}$$

This result is firmly established by the experimental evidence that in order to obtain Eq. (33b) for the Compton effect the momenta of the photons must be taken as $h\nu/c$.

It is assumed in Eq. (33e) that photons always travel with the speed c, and in fact it is true without exception that

● $$\text{Velocity of a photon} = c \tag{30f}$$

In this respect photons differ from particles of matter, which can have any velocity less than c. At first sight, Eq. (33f) seems to be in contradiction to the observed fact that the measured velocity of light in matter is less than c. But this is the velocity of a group of waves (Sec. 19.8) and not that of the individual photons. As was explained in the chapter on dispersion, light waves traversing matter are retarded by the

* Einstein's general theory of relativity postulates an increase in momentum and mass of a photon as it passes through a strong gravitational field like that close to the sun. See F. R. Tangherlini, Snell's Law and the Gravitational Deflection of Light, *Am. J. Phys.*, **36**:1001 (1968); see also R. A. Houstoun, *J. Opt. Soc. Am.*, **55**:1186 (1965).

alteration of their phase through interference with the scattered waves. In the case of photons we may, at least in dilute matter like a gas, picture the photons as traveling with the velocity c in the empty space between molecules, but as having their average rate of progress retarded by the finite time consumed during the process of absorption and reemission by the molecules they encounter. In any experiment where the photon could be expected to be slowed down, for example, in an encounter with an electron in the Compton effect, it is found that the energy and frequency are decreased, not the velocity. The only slowing-down that a photon can suffer is its complete annihilation, as happens in the photoelectric effect.

33.4 DEVELOPMENT OF QUANTUM MECHANICS

The apparently irreconcilable contradiction between the corpuscular and wave pictures of light has been clarified on the basis of a new system of mechanics initiated by Heisenberg and Schrödinger in 1926 (see Chap. 29). This *quantum mechanics* is essential for the treatment of all atomic processes. It also holds for ordinary large-scale processes, although in this case the deviations from newtonian mechanics are negligible. In quantum mechanics the behavior of the electrons in an atom, for example, is calculated by the use of wave theory, and the solutions of wave equations yield the allowed energy states. Any material particle has associated with it a group of waves, and in the case of a free particle their wavelength is inversely proportional to the momentum p of the particle. This is the celebrated de Broglie relation, treated in Chap. 29, which represents an extension of Eq. (33e) to matter:

$$\lambda = \frac{h}{mv} = \frac{h}{p} \qquad \textit{Wavelength of a free particle} \qquad (33g)$$

This equation was experimentally verified by Davisson and Germer in the United States and by G. P. Thomson in England. They showed that a beam of electrons can be made to exhibit diffraction and that the pattern corresponded to that produced with X rays by the regular arrangement of atoms in a crystal lattice. Diffraction of a beam of atoms or molecules was subsequently demonstrated by Stern. The analogous behavior of electrons and light is most beautifully demonstrated in the electron microscope (Sec. 15.10). The existence both for matter and for electromagnetic radiation of the two types of behavior, as waves and as particles, was the most significant fact that was interpreted by quantum mechanics.

The physical significance of the waves that pertain to a given material particle is that the square of their amplitude at any point in space represents the probability of finding the particle at that point. The theory therefore yields the statistical distribution of the particles, and as we shall see, it denies the possibility of going any further than this. Similarly for light the wave theory gives us the statistical or average distribution of photons as the square of the amplitude of the electromagnetic wave. If we postpone for the moment the question of which model, wave or corpuscle, is the real one, and look at the achievements of quantum-mechanical theory, we find an extensive array of these, which prove beyond question the soundness of the basic assumptions of the theory. Not only are the many complex features of atomic and molecular spectra accounted for in detail, but also any process involving the extranuclear electrons and their interaction with electromagnetic radiation. Only when attempts are made to apply it to regions as small as atomic nuclei, or in general smaller than the classical radius of the electron e^2/m_0c^2, are there indications that the theory breaks down.

33.5 PRINCIPLE OF INDETERMINACY

The possibility of characterizing light as discrete packages of energy called photons would seem to rest upon our ability to determine for a given photon both the position and the momentum that it possesses at a given instant. These are usually thought of as measurable properties of a material particle. It was shown by Heisenberg, however, that for particles of atomic magnitude it is in principle impossible to determine both position and momentum simultaneously with perfect accuracy. If an experiment is designed to measure one of them exactly, the other will become completely uncertain, and vice versa. An experiment can measure both, but only within certain limits of accuracy. These limits are specified by the principle of indeterminacy (often called the *uncertainty principle*), according to which

$$\Delta p_y \Delta y \gtrsim \frac{h}{2\pi} \qquad (33h)$$

Here Δy and Δp_y represent the variations of the value of the coordinate and of the corresponding component of momentum of a particle which must be expected if we try to measure both at once, i.e., the uncertainties in these quantities. The symbol $\gtrsim$ means "is of the order of, or greater than." The reason for this semiquantitative way of stating the law will become clear through the example to be given in the following section.

The principle of indeterminacy is applicable to photons, as well as to all material particles from electrons up to the sizable bodies dealt with in ordinary mechanics. For the latter, the very small magnitude of h renders the Δp_y and Δy entirely negligible compared to the ordinary experimental errors encountered in measuring the large p_y and its corresponding y. When p_y is very small, however, as for an electron or a photon, the uncertainty may become an appreciable fraction of the momentum itself, or else the uncertainty in the position must be relatively large.

33.6 DIFFRACTION BY A SLIT

Suppose that we undertake to find the position of a photon by passing it through a narrow slit. This will specify its coordinate y in the plane of the screen to within an uncertainty Δy equal to the slit width (Fig. 33D). In doing so the momentum in the y direction, initially zero in this experiment, is rendered uncertain by an amount Δp_y given by the relation (33h), as we shall now show.

Passage of the light through the slit causes a diffraction pattern to be produced on the screen. We shall assume the screen to be far enough away relative to the width

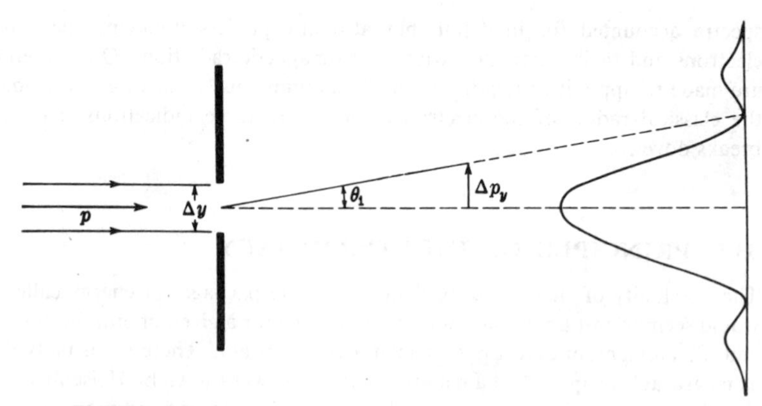

FIGURE 33D
The uncertainty principle applied to the momentum of a photon when it is diffracted by a single slit.

of the slit for Fraunhofer diffraction to be obtained. Nearly all the photons will be found within the angle θ_1, corresponding to the first zero of the pattern. We have seen in Eq. (15f) that this angle is given by

$$\sin \theta_1 = \frac{\lambda}{\Delta y} \tag{33i}$$

The corresponding uncertainty in the momentum is

$$\Delta p_y = p \sin \theta_1 = \frac{p\lambda}{\Delta y} \tag{33j}$$

Introducing the value for the momentum p given by the de Broglie relation, Eq. (33e), we find

$$\Delta p_y = \frac{h}{\lambda} \frac{\lambda}{\Delta y} = \frac{h}{\Delta y} \tag{33k}$$

This gives $\Delta p_y \, \Delta y = h$, but it will be seen that since the probability of the photon striking the center of the pattern is greatest, the uncertainty in p_y is not so large as is indicated by Eq. (33k) and hence that our result is consistent with the principle of indeterminacy

$$\Delta p_y \, \Delta y \gtrsim \frac{h}{2\pi} \tag{33h}$$

No doubt this derivation will raise some important questions in the mind of the reader. How does the photon acquire this sideways momentum? How is it possible that the width of the slit should affect a photon which passes through at one place in the slit? The answers to these questions will be postponed until we have given further consideration to the consequences of the principle of indeterminacy.

33.7 COMPLEMENTARITY

We owe to Bohr the interpretation of Heisenberg's principle in such a way as to make clear the fundamental limitations on the accuracy of measurement and their bearing on our views as to the nature of light and matter. According to the principle of complementarity stated by Bohr in 1928, the wave and corpuscular descriptions are merely complementary ways of regarding the same phenomenon. That is, to obtain the complete picture we need both these properties, but because of the principle of indeterminacy it is impossible to design an experiment which will show both of them in all detail at the same time. Any experiment will reveal the details of either the wave or the corpuscular character, according to the purpose for which the experiment is designed.

It further appears that, if one attempts to push the accuracy of measurement to the point where the experiment might be expected to reveal both aspects, there is an unavoidable interaction between the measuring apparatus and the thing measured which frustrates the attempt. This happens even in a hypothetical experiment which we imagine to be performed by an experimenter endowed with infinite skill and resource. It is therefore not a question of the usual disturbances caused by large-scale measuring instruments; these can be calculated and allowed for. The uncertainties we are concerned with here are by their very nature impossible to evaluate without spoiling the experiment in some other way. If this were not so, we would be able to overstep the boundaries prescribed by the complementarity principle. To see how these interactions occur, and that they occur just to the degree required by the principle of indeterminacy, we now describe two celebrated experiments which for technical reasons have never been performed exactly as they are here outlined, but for which the results can be confidently predicted on the basis of other actual experiments not quite as simple.

33.8 DOUBLE SLIT

The interference fringes in Young's experiment (Sec. 13.3) constitute one of the simplest manifestations of the wave character of light. Yet it should be possible to reveal the presence of photons by a suitable modification of the experiment. Such a modification would be the replacement of the observing screen by a photoelectric surface so subdivided that the individual photoelectrons from different parts of the surface could be counted. If this were done, the largest concentration of photons would be found to occur at the maxima of the interference pattern, and none at all at the minima. It is impossible to conceive of the interference between different photons going through the two slits as being responsible for such a pattern. It is even more difficult to understand how a single photon could be constrained to seek the maxima and avoid the minima, since it presumably passed through only one of the slits. The presence of the other slit should be immaterial, whereas actually it makes possible the interference pattern and its position determines the dimensions of that pattern. Nevertheless, according to quantum mechanics, the latter interpretation is correct. The fringes could be produced by single photons, going one by one through the slits. We know

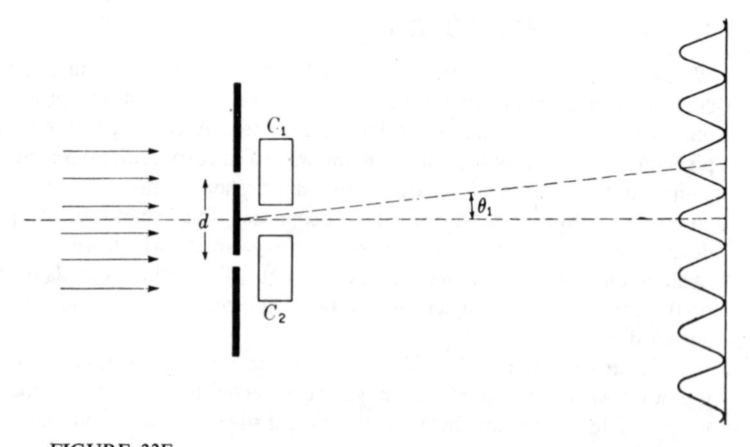

FIGURE 33E
Young's double-slit experiment as modified to demonstrate both wave and corpuscular properties of light.

that reducing the intensity of the light does not destroy the interference. The pattern is therefore a characteristic of each photon, and represents the probability of its arriving at various points on the screen. This probability must, however, be calculated by wave theory and is measured by the square of the amplitude. The experiment is one designed to show the properties of waves.

Now let us attempt to refine this experiment with the purpose of finding out through which slit any given photon passed. This might be done by putting two scintillation counters C_1 and C_2 in front of or behind the slits, as shown in Fig. 33E. With light of sufficiently high frequency, these would register each photon as it goes through one slit or the other. But in so doing we have spoiled the interference pattern because of the deflections suffered by the photons in producing the scintillations. For the fringes to be clearly visible it is necessary that these deflections be less than one-quarter of a fringe width, according to the criterion mentioned in Sec. 16.7. Thus

$$\frac{\Delta p_y}{p} < \frac{\theta_1}{4} = \frac{\lambda}{4d} \tag{33l}$$

where θ_1 is the angular separation of adjacent fringes and d the slit separation. Since the counters tell us through which slit the photon passes, they specify the y coordinate to within a distance of at most $d/2$. Hence we may write for the uncertainty in this coordinate

$$\Delta y = \frac{d}{2} \tag{33m}$$

Combination of Eqs. (33l) and 33m) then yields

$$\Delta p_y \, \Delta y < \frac{p\lambda}{4d}\frac{d}{2} = \frac{p\lambda}{8} \tag{33n}$$

Upon insertion of the de Broglie value for λ, the requirement that the interference pattern shall not be spoiled becomes

$$\Delta p_y \, \Delta y < \frac{h}{8} \tag{33o}$$

This violates the principle of indeterminacy according to which $\Delta p_y \, \Delta y \gtrsim h/2\pi$. Hence we see that it is impossible to localize individual photons and at the same time measure their wavelength. This would mean that we had simultaneously determined both position and momentum. It is possible to measure only one of these with precision, according to whether the experiment is designed for photons or waves.

33.9 DETERMINATION OF POSITION WITH A MICROSCOPE

Another idealized experiment, first discussed by Heisenberg, is that usually referred to as the *γ-ray microscope*. If it is desired to find the position of a particle as accurately as possible, the particle must be illuminated by light of the smallest possible wavelength, since the resolving power is given, according to Eq. (15l), by

$$s = \frac{\lambda}{2n \sin i} \tag{33p}$$

We can imagine, in principle at least, a microscope using γ rays which is capable of yielding an extremely small uncertainty $\Delta x \approx s$ in the position of the particle. If then the particle is at rest, its momentum p_x is exactly zero and this simultaneous knowledge of both position and momentum would appear to violate the principle of indeterminacy. One factor has been neglected, however; namely the recoil of the particle when it is hit by a photon of high energy and momentum, which is demonstrated in the Compton effect. This recoil will introduce a relatively large uncertainty in the momentum, just as the principle would predict.

To find the magnitude of the uncertainty, note that in Fig. 33F the x component of the momentum of the scattered photon can lie anywhere between $+h/\lambda \sin i$ and $-h/\lambda \sin i$, since it could have entered any part of the objective lens. The x component of the momentum of the recoiling particle is rendered uncertain by the same amount, since momentum is conserved in the collision and the momentum of the incident photons can be exactly calculated from the wavelength. Hence for the particle

$$\Delta p_x \approx \frac{2h}{\lambda} \sin i \tag{33q}$$

Multiplying by the Δx from Eq. (33p), we find

● $$\Delta p_x \, \Delta x \approx h \tag{33r}$$

as required. This is an example of the application of the principle of indeterminacy to a material particle. Complementarity is well illustrated in the experiment by the

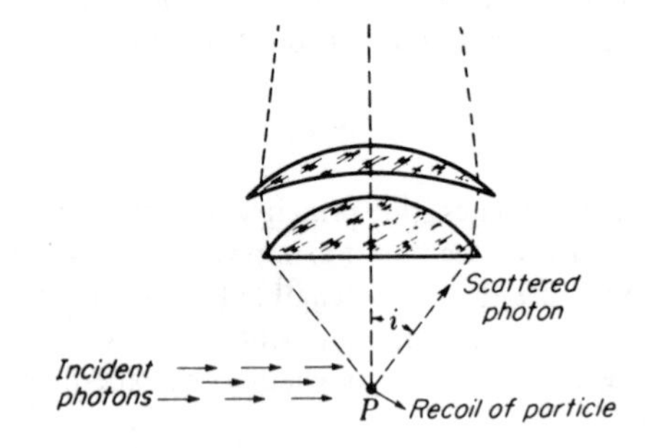

FIGURE 33F
Measurement of position with a microscope.

fact that when one uses a very short wavelength, x is found with good precision but Δp_x is made large, while the use of a longer wavelength will allow p_x to be better known at a sacrifice of the accuracy Δx in the measurement of position.

33.10 USE OF A SHUTTER

It is also instructive to consider the result of trying to localize a photon by passing light through a very rapidly acting shutter, such as that utilizing the Kerr electro-optic effect (Sec. 32.10). In Fig. 33G(*a*) let S represent schematically such a shutter, which has been opened only long enough to allow a train of N waves of uniform amplitude to pass through. The experiment could be performed with light so faint that only one photon would pass through in this time. This photon lies somewhere in the *wave packet* (Sec. 11.11) of N waves, and its probability of being found anywhere in the packet is measured by the square of the amplitude. This is uniform along the length

$$\Delta x = N\lambda_0 = N\frac{c}{\nu_0} \tag{33s}$$

The Fourier integral analysis of a finite train of N waves of equal amplitude yields a certain distribution of frequencies, and when the intensities in various frequencies are plotted, as in Fig. 33G(*b*), the resulting curve is to a very close approximation the same as that for the Fraunhofer diffraction pattern due to a single slit. The half width of the central maximum is just ν_0/N. Now such a spread of frequency corresponds, according to Eq. (33e), to an uncertainty in the momentum of the photon amounting to

$$\Delta p_x = \frac{h}{\Delta\lambda} = \frac{h\,\Delta\nu}{c} = \frac{h(\nu_0/N)}{c} \tag{33t}$$

Localization of the photon to within a distance Δx has therefore rendered its momentum indeterminate, and, as expected, the product of the two indeterminacies given by Eqs. (33s) and (33t) is again

$$\Delta p_x\,\Delta x \approx h \tag{33u}$$

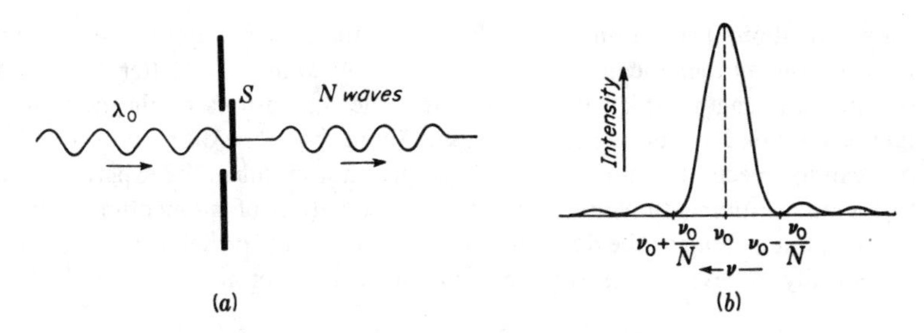

FIGURE 33G
(*a*) Idealized shutter experiment. (*b*) Result of Fourier analysis of a train of N waves.

It is to be noted that the wave packet is not the photon, nor can we speak of the photon itself as having any dimensions. The packet is merely a description of the probability of finding the photon in any given position. When the length of a wave train is measured with the Michelson interferometer (Sec. 13.12), one is not finding the length of a photon but merely that of the region in which the photon must somewhere lie.

33.11 INTERPRETATION OF THE DUAL CHARACTER OF LIGHT

Granted the truth of these principles of indeterminacy and complementarity, what can be said about the nature of light? First, it is important to realize that light (as well as the elementary particles of matter: electrons, protons, etc.) is essentially more primitive and subtle than those mechanical phenomena which we can observe on a large scale. All our information about it must be obtained indirectly. The possibility is therefore opened that it might not be feasible to describe light in the terms which we are accustomed to use for everyday things. All our experience since childhood would indicate that it should be possible to say, "Light is like a flight of bullets from a machine gun," or "Light is like a train of water waves." But such a specific statement cannot be made about light, and the complementarity principle indicates that we shall never be able to make it. We can say, "In this experiment light behaves as though it were composed of photons," and, "In that experiment it behaves as though it were waves." Since complementarity rules out any experiment in which one can measure both properties at once, the conclusion is inevitable that the concepts of photons and waves are equally justified and that each is applicable in its own sphere.

The point of view adopted in quantum mechanics with regard to such a dilemma as is presented by the double-slit experiment is simply that a classical description of the motion of a single photon has meaning only within the limits set by the principle of indeterminacy. When the interference pattern is being observed, there is no significance to the statement that the photon went through one slit or the other, i.e., to a statement about its position. When the scintillations are being counted, we can specify the position but then the momentum has lost its meaning. The latter quantity depends on the wavelength, which in turn requires the dimensions of the now nonexistent interference pattern for its determination. Similarly, in the diffraction by a single slit, one cannot specify the momentum of a single photon unless the experiment is altered to include a momentum measurement. Conservation of momentum could then be verified, but as long as the diffraction pattern exists, this principle can only be applied statistically to describe the average behavior of the photons.

33.12 REALMS OF APPLICABILITY OF WAVES AND PHOTONS

The emphasis which has in this book been placed upon the wave properties of light has a certain justification as long as one does not extend the meaning of light to include the very short-wavelength region of X rays and γ rays. The relative prominence of the wave and corpuscular properties changes steadily in favor of the latter as one proceeds through the electromagnetic spectrum in the direction of increasing frequency. Thus radio waves behave in all important respects like classical electromagnetic radiation. This is related to the fact that the photons are extremely small in their energy $h\nu$ and therefore usually very numerous. Similarly visible light of ordinary intensities contains so many photons that their average behavior is accurately given by the wave theory provided that the interactions with the individual atoms of matter do not involve the quantized energy states of the latter. This accounts for the fact that the corpuscular properties of light remained undiscovered for so many years.

The connecting link between the wave and quantum aspects of light (or of matter) is furnished by Planck's constant h. As has been emphasized by Bohr, h is the product of two variables, one characteristic of a wave, the other of a particle. Thus if we designate by T the period, or reciprocal of the frequency ν, the quantum relation can be put in the symmetrical form

$$h = ET = p\lambda \qquad \text{(33v)}$$

Now E and p are attributes of particles, while T and λ are attributes of waves. If, for example, the magnitudes of the former are large, the latter must be correspondingly small. Hence X rays and γ rays behave in most respects like photons, and their wave character is even difficult to demonstrate. The region of frequencies where particle-like properties begin to become prominent is of course determined by the magnitude of h, and its actual value, 6.6262×10^{-34} J s, is so small that very high frequencies are required before the wave character begins to be lost. Visible light lies well below this region, and its wave properties may therefore be said to be the most important. If h were much smaller than it is, the quantum theory would never have been required and classical electromagnetic theory would have sufficed to explain all experiments. It is a curious coincidence that the actual size of h, which of course is still unexplained, is such that the nature of light seems to run the whole gamut, from obvious waves at one end to obvious photons at the other, in the observed range of the electromagnetic spectrum.

PROBLEMS

33.1 Using Eqs. (29c) and (29d), calculate (*a*) the velocity and (*b*) the radius of the Bohr circular orbit, $n = 4$. (*c*) Find the de Broglie wavelength for the electron in this orbit. (*d*) How many of these wavelengths are included in the orbit curcumference?

Ans. (*a*) 5.469×10^5 m/s, (*b*) 8.4668×10^{-10} m, (*c*) 1.32997×10^{-9} m, (*d*) 4.00

33.2 Using the de Broglie relationship, find the wavelength associated with (*a*) an electron moving with one-half the speed of light, (*b*) an oxygen molecule with its mean thermal velocity of 480 m/s, and (*c*) a rifle bullet of mass 5 g moving with a velocity 550 m/s.

33.3 Find the number of photons per cubic centimeter in a monochromatic beam of radiation of intensity 3×10^{-5} W/cm^2. Take the wavelength to be (*a*) 0.20 Å and (*b*) 5000 Å.

33.4 For light of wavelength 5000 Å, calculate the magnitude of the four quantities appearing in Bohr's complementarity relation, Eq. (33v).

Ans. $E = 3.9730 \times 10^{-19}$ J, $T = 1.6678 \times 10^{-15}$ s, $p = 1.3252 \times 10^{-27}$ kg m/s, $\lambda = 5.0 \times 10^{-7}$ m

33.5 X rays of wavelength 0.4650 Å are scattered from a carbon block at an angle of 75° to the direction of the incident beam. Calculate the change of wavelength due to the Compton effect.

33.6 The radiation flux from a distant star amounts to 2.50×10^{-17} W/m^2. Assuming the effective wavelength of starlight to be 5500 Å, find how many photons per second enter the pupil of the eye under these circumstances if the pupil diameter is 6.0 mm.

33.7 When a 500-V electron is passed through a pinhole 0.0180 mm in diameter, (*a*) what uncertainty in the angle of emergence is introduced? (*b*) Make a similar calculation for a 250-g baseball thrown with a velocity of 25 m/s through a hole 16.0 cm in diameter. The relation $Ve = \frac{1}{2}mv^2$ can be used to find the electron velocity in meters per second in terms of the voltage V in volts, e in coulombs, and m in kilograms.

Ans. (*a*) 1.257 seconds of arc, (*b*) 2.197×10^{-29} second of arc

33.8 A microscope of numerical aperture 1.4 is focused on a particle of mass 0.0050 mg. If the illuminating light has a wavelength of 4800 Å, what are the values of Δp_x and Δx predicted by Heisenberg's principle of indeterminacy?

Appendices

APPENDIX I

THE PHYSICAL CONSTANTS*

Quantity	Symbol or abbreviation	Value	Error, ppm
Speed of light	c	2.9979250×10^{8} m/s	0.33
Electron charge	e	$1.6021917 \times 10^{-10}$ C	4.4
Electron mass	m	$9.1095585 \times 10^{-31}$ kg	6.0
Proton mass	M_p	$1.6726141 \times 10^{-27}$ kg	6.6
Neutron mass	m_n	$1.6749201 \times 10^{-27}$ kg	6.6
Planck's constant	h	$6.6261965 \times 10^{-34}$ J s	7.6
Unit angular momentum	$\frac{h}{2\pi}$	$1.0545915 \times 10^{-34}$ J s	7.6
Electronic ratio	$\frac{e}{m}$	1.7588028×10^{11} C/kg	3.1
Bohr radius	r_1	$5.2917716 \times 10^{-11}$ m	1.5
Proton-electron mass ratio	$\frac{m_p}{m_e}$	1836.1091	6.2
Atomic mass unit	amu	$1.6605311 \times 10^{-27}$ kg	6.6
Energy of electron mass	V_e	0.5110041 MeV	4.6
Mass energy of an electron	m_0c^2	$8.1872652 \times 10^{-14}$ J	6.5
Compton wavelength	$\frac{h}{m_0c}$	$2.4263096 \times 10^{-12}$ m	3.1
Gas constant	R	8.3143435 J/mol K	42
Stefan-Boltzmann constant	k	5.669620 J/s	170
Avogadro's number	N	6.0221694×10^{26}/kg mol	6.6
Gravitational constant	G	6.673231×10^{-11} m^3/kg s^2	460

* From *Rev. Mod. Phys.*, **41**:476 (1969).

APPENDIX II

ELECTRON SUBSHELLS

The table shows the order in which the electron subshells are filled, in building up the elements of the periodic table; atomic weights are given with respect to carbon 12 as 12 even.

$n + 1$	Sub-shells	1	2	3	4	5	6	7	8	9	10	11	12	13	14
1	1s	1.0080 H 1	4.003 He 2												
2	2s	6.939 Li 3	9.012 Be 4												
3	2p	10.81 B 5	12.000 B 6	14.007 N 7	15.999 O 8	18.998 F 9	20.183 Ne 10								
	3s	22.990 Na 11	24.312 Mg 12												
4	3p	26.98 Al 13	28.08 Si 14	30.974 P 15	32.064 S 16	35.453 Cl 17	39.948 Ar 18								
	4s	39.102 K 19	40.08 Ca 20												
	3d	44.96 Sc 21	47.90 Ti 22	50.94 V 23	51.996 Cr 24	54.94 Mn 25	55.85 Fe 26	58.93 Co 27	58.71 Ni 28	63.54 Cu 29	65.37 Zn 30				
5	4p	69.72 Ga 31	72.59 Ge 32	74.92 As 33	78.96 Se 34	79.91 Br 35	83.80 Kr 36								
	5s	85.47 Rb 37	87.62 Sr 38												

(*Continued overleaf*)

$n + 1$	Sub-shells	1	2	3	4	5	6	7	8	9	10	11	12	13	14
6	4*d*	88.90 Y 39	91.22 Zr 40	92.91 Nb 41	95.94 Mo 42	(99) Tc 43	101.1 Ru 44	102.91 Rh 45	106.4 Pd 46	107.870 Ag 47	112.40 Cd 48				
	5*p*	114.82 In 49	118.69 Sn 50	121.75 Sb 51	127.60 Te 52	126.90 I 53	131.3 Xe 54								
	6*s*	132.90 Cs 55	137.34 Ba 56												
7	4*f*	138.91 La 57	140.12 Ce 58	140.91 Pr 59	144.24 Nd 60	(145) Pm 61	150.35 Sm 62	151.96 Eu 63	157.25 Gd 64	158.92 Tb 65	162.50 Dy 66	164.93 Ho 67	167.2 Er 68	168.93 Tm 69	173.04 Yb 70
	5*d*	174.97 Lu 71	178.5 Hf 72	180.95 Ta 73	183.92 W 74	186.2 Re 75	190.2 Os 76	192.2 Ir 77	195.09 Pt 78	196.97 Au 79	200.59 Hg 80				
	6*p*	204.37 Tl 81	207.19 Pb 82	208.98 Bi 83	210 Po 84	(210) At 85	222 Rn 86								
	7*s*	(223) Fr 87	226.05 Ra 88												
8	5*f*	227 Ac 89	232.04 Th 90	231 Pa 91	238.03 U 92	(237) Np 93	(242) Pu 94	(243) Am 95	(245) Cm 96	(245) Bk 97	(248) Cf 98	(253) Es 99	(254) Fm 100	(256) Md 101	(254) No 102
	6*d*	Lw 103	(258) Rf 104	(260) Ha 105	106	107	108	109	110	111	112				

APPENDIX III

REFRACTIVE INDICES AND DISPERSIONS FOR OPTICAL GLASSES

Glass	n_C 6563 Å	n_D 5892 Å	n_F 4861 Å	n_G 4340 Å	ICT type	ν
Barium flint	1.58848	1.59144	1.59825	1.60367	591/605	60.5
Borosilicate crown, 1	1.49776	1.50000	1.50529	1.50937	500/664	66.4
2	1.51462	1.51700	1.52264	1.52708	517/645	64.5
3	1.50883	1.51124	1.51690	1.52136	511/634	63.4
Dense flint, 2	1.61216	1.61700	1.62901	1.63923	617/366	36.6
4	1.64357	1.64900	1.66270	1.67456	649/338	33.8
Extra dense flint	1.71303	1.72000	1.73780	1.75324	720/291	29.1
Fused quartz	1.45640	1.45845	1.46318	1.46690	458/676	67.6
Light barium crown	1.53828	1.54100	1.54735	1.55249	541/599	59.9
Light flint, 2	1.57100	1.57500	1.58500	1.59400	575/411	41.1
1	1.57208	1.57600	1.58606	1.59441	576/412	41.2
Spectacle crown	1.52042	1.52300	1.52933	1.53435	523/587	58.7
Strontium titanate ($SrTiO_3$)	2.37287	2.41208	1.49242	2.57373	412/345	3.45
Telescopic flint	1.52762	1.53050	1.53790	1.54379	531/516	51.6
Very dense flint	1.87900	1.89000	1.91900	1.95400	890/223	22.3

APPENDIX IV

REFRACTIVE INDICES AND DISPERSIONS OF OPTICAL CRYSTALS

Crystal	Ray	n_C 6563 Å	n_D 5892 Å	n_F 4861 Å	$n_{G'}$ 4340 Å	ICT type	ν
Quartz (SiO_2)	*O*	1.54190	1.54425	1.54968	1.55396	544/700	70.0
	E	1.55093	1.55336	1.55898	1.56340	553/687	68.7
Calcite ($CaCO_3$)	*O*	1.65438	1.65836	1.66785	1.67552	658/489	48.9
	E	1.48461	1.48641	1.49076	1.49428	486/791	79.1
Rutile (TiO_2)	*O*	2.57100	2.61310	2.73460	2.85870	613/375	3.75
	E	2.85600	2.90890	3.06310	3.22320	909/439	4.39

APPENDIX V
THE MOST INTENSE FRAUNHOFER LINES

To change wavelengths in Angstroms (Å) to nanometers (nm), move decimal one place to the left.

Designation	Origin	Wavelength, Å	Designation	Origin	Wavelength, Å
A	O_2	7594–7621*	b_4	Mg	5167.343
B	O_2	6867–6884*	c	Fe	4957.609
C	H	6562.816	F	H	4861.327
α	O_2	6276–6287*	d	Fe	4668.140
D_1	Na	5895.923	e	Fe	4383.547
D_2	Na	5889.953	G′	H	4340.465
D_3	He	5875.618	G	Fe	4307.906
E_2	Fe	5269.541	G	Ca	4307.741
b_1	Mg	5183.618	g	Ca	4226.728
b_2	Mg	5172.699	h	H	4101.735
b_3	Fe	5168.901	H	Ca^+	3968.468
b_4	Fe	5167.491	K	Ca^+	3933.666

APPENDIX VI
ABBREVIATED NUMBER SYSTEM

At a meeting held on October 14, 1960, by the International Union for Pure and Applied Physics, the following symbolism was adopted for general use.

10^3	kilo	k	10^{-3}	milli	m
10^6	mega	M	10^{-6}	micro	μ
10^9	giga	G	10^{-9}	nano	n
10^{12}	tera	T	10^{-12}	pico	p

The *angstrom* (Å) as a unit of wavelength of light is still used by many spectroscopists, but the unit *nanometer* (nm) is becoming more common.

$$1\ \text{Å} = 10^{-10}\ \text{m}$$
$$1\ \text{nm} = 10^{-9}\ \text{m}$$
$$1\ \text{nm} = 10\ \text{Å}$$

APPENDIX VII
SIGNIFICANT FIGURES

In solving the problems at the ends of the chapters in this text it is recommended that an electronic calculator be used. A number of calculators for carrying out the operations of addition, subtraction, multiplication, and division are available at a relatively low cost. Somewhat more expensive calculators, for students of science, are available today which will provide sines, cosines, and tangents of angles, squares and square roots, arcsines, cosines, and tangents, reciprocals, logarithms, exponentials, and any number of memory banks. Every student would do well to procure one of these calculators. The time saved in solving problems and the accuracy obtained is well worthwhile.

Before solving problems the student should also have a good understanding of significant figures. The table of numbers will serve to illustrate.

A Three significant figures	B Four significant figures	C Five significant figures
374	5279	24,794
21.5	63.08	6.9428
6.05	0.1062	0.37625
0.00328	0.04503	0.053177
546,000	692,700	46,009

1 All digits other than terminal zeros to the left of the decimal point are significant. The measurement 42.65 kg contains four significant figures; 42,650 also contains four significant figures.

2 The first significant figure of a number is the first digit that is not zero. The measurement 0.0132 g contains three significant figures.

3 Zeros to the right of a decimal point and to the right of a nonzero digit are significant. The reading 46.270 km contains five significant figures.

In powers of 10 notation a simple rule may be applied to specify significant figures. This assumes that at least one nonzero digit appears in front of the decimal point.

4 All zeros that appear in the base number are significant. The reading 2.40×10^5 m contains three significant figures. If this figure is desired to four significant figures it may be assumed that zeros follow the last zero shown.

Suppose the numbers like those in column *C* are to be expressed to three significant figures only. If the first figure begins with 1, 2, or 3, the number should be reduced to four figures, and if the first figure begins with 4, 5, 6, 7, 8, or 9, it should be reduced to three. The numbers in column *C* would, therefore, be written 24790, 6.94, 0.3763, 0.0532, and 46,000, respectively.

When illustrating basic scientific principles by experiment or by mathematical problems, some measurements may be specified by small whole numbers and others to several significant figures. Suppose for example that a small car is said to travel a distance of 9 m in 4.15 s and we wish to calculate the average speed. If the answer is to be expressed to three significant figures, it is common practice to assume both quantities are known to at least three figures:

$$v = \frac{9 \text{ m}}{4.15 \text{ s}} = 2.16867 \text{ m/s}$$

What has been done to obtain this answer is to assume that the numerator has the value 9.00 m, and upon dividing by 4.15 s, obtain 2.16867 m/s with a calculator, which to three-significant-figure accuracy is 2.169 m/s.

When a calculator is used in working problems, it is quite proper to carry each figure out as far as it is specified. When the final answer is obtained, it is common practice to express it to at least one more figure than the significant figures in the number containing the least number of significant figures.

Most slide rules are capable of handling the multiplication and division of numbers only to three significant figures. Most of the problems in optics require greater accuracy than this, and for them a slide rule is not adequate.

INDEX

INDEX

GLASGOW
UNIVERSITY
LIBRARY